WIRELESS COMMUNICATIONS

Wirelesss technology is a truly revolutionary paradigm shift, enabling multimedia communications between people and devices from any location. It also underpins exciting applications such as sensor networks, smart homes, telemedicine, and automated highways. This book provides a comprehensive introduction to the underlying theory, design techniques, and analytical tools of wireless communications, focusing primarily on the core principles of wireless system design.

The book begins with an overview of wireless systems and standards. The characteristics of the wireless channel are then described, including their fundamental capacity limits. Various modulation, coding, and signal processing schemes are then discussed in detail, including state-of-the-art adaptive modulation, multicarrier, spread-spectrum, and multiple-antenna techniques. The concluding chapters deal with multiuser communications, cellular system design, and ad hoc wireless network design.

Design insights and trade-offs are emphasized throughout the book. It contains many worked examples, more than 200 figures, almost 300 homework exercises, and more than 700 references. *Wireless Communications* is an ideal textbook for students as well as a valuable reference for engineers in the wireless industry.

Andrea Goldsmith received her Ph.D. from the University of California, Berkeley, and is an Associate Professor of Electrical Engineering at Stanford University. Prior to this she was an Assistant Professor at the California Institute of Technology, and she has also held positions in industry at Maxim Technologies and AT&T Bell Laboratories. She is a Fellow of the IEEE, has received numerous other awards and honors, and is the author of more than 150 technical papers in the field of wireless communications.

Wireless Communications

ANDREA GOLDSMITH
Stanford University

CAMBRIDGE
UNIVERSITY PRESS

CAMBRIDGE UNIVERSITY PRESS
Cambridge, New York, Melbourne, Madrid, Cape Town, Singapore, São Paulo

Cambridge University Press
40 West 20th Street, New York, NY 10011-4211, USA

www.cambridge.org
Information on this title: www.cambridge.org/9780521837163

First published 2005

Printed in the United States of America

A catalog record for this publication is available from the British Library.

Library of Congress Cataloging in Publication data
Goldsmith, Andrea
Wireless communications / Andrea Goldsmith.
p.   cm.
Includes bibliographical references and index.
ISBN-13: 978-0-521-83716-3
1. Wireless communication systems.   I. Title.

TK5103.2.G65   2005
621.382 – dc22   2005047075

ISBN-13   978-0-521-83716-3 hardback
ISBN-10   0-521-83716-2 hardback

To Arturo, Daniel, and Nicole

The possession of knowledge does not kill the sense of wonder and mystery.

—*Anaïs Nin*

Brief Table of Contents

Contents

Preface

Wireless communications is a broad and dynamic field that has spurred tremendous excitement and technological advances over the last few decades. The goal of this book is to provide readers with a comprehensive understanding of the fundamental principles underlying wireless communications. These principles include the characteristics and performance limits of wireless systems, the techniques and mathematical tools needed to analyze them, and the insights and trade-offs associated with their design. Current and envisioned wireless systems are used to motivate and exemplify these fundamental principles. The book can be used as a senior- or graduate-level textbook and as a reference for engineers, academic and industrial researchers, and students working in the wireless field.

ORGANIZATION OF THE BOOK

Chapter 1 begins with an overview of wireless communications, including its history, a vision for the future, and an overview of current systems and standards. Wireless channel characteristics, which drive many of the challenges in wireless system design, are described in Chapters 2 and 3. In particular, Chapter 2 covers path loss and shadowing in wireless channels, which vary over relatively large distances. Chapter 3 characterizes the flat and frequency-selective properties of multipath fading, which change over much smaller distances – on the order of the signal wavelength. Fundamental capacity limits of wireless channels along with the capacity-achieving transmission strategies are treated in Chapter 4. Although these techniques have unconstrained complexity and delay, they provide insight and motivation for many of the practical schemes discussed in later chapters. In Chapters 5 and 6 the focus shifts to digital modulation techniques and their performance in wireless channels. These chapters indicate that fading can significantly degrade performance. Thus, fading mitigation techniques are required for high-performance wireless systems.

The next several chapters cover the primary mitigation techniques for flat and frequency-selective fading. Specifically, Chapter 7 covers the underlying principles of diversity techniques, including a new mathematical tool that greatly simplifies performance analysis. These techniques can remove most of the detrimental effects of flat fading. Chapter 8 provides comprehensive coverage of coding techniques, including mature methods for block, convolutional, and trellis coding as well as recent developments in concatenated, turbo, and LDPC codes. This chapter illustrates that, though coding techniques for noisy channels have near-optimal performance, many open issues remain in the design and performance analysis of codes for wireless systems. Chapter 9 treats adaptive modulation in flat fading, which enables robust and spectrally efficient communication by leveraging the time-varying nature of the wireless channel. This chapter also ties the techniques and performance of adaptive modulation to the fundamental capacity limits of flat fading channels. Multiple-antenna techniques and space-time communication systems are covered in Chapter 10: the additional spatial dimension enables high data rates and robustness to fading. Equalization, which exploits signal processing in the receiver to compensate for frequency-selective fading, is covered in Chapter 11. Multicarrier modulation, described in Chapter 12, is simpler and more flexible than equalization for frequency-selective fading mitigation. Single-user and multiuser spread-spectrum techniques are described in Chapter 13. These techniques not only mitigate frequency-selective fading, they also allow multiple users to share the same wireless spectrum.

The last three chapters of the book focus on multiuser systems and networks. Chapter 14 treats multiple and random access techniques for sharing the wireless channel among many users with continuous or bursty data. Power control is also covered in this chapter as a mechanism to reduce interference between users while ensuring that all users meet their performance targets. The chapter closes by discussing the fundamental capacity limits of multiuser channels as well as the transmission and channel sharing techniques that achieve these limits. Chapter 15 covers the design, optimization, and performance analysis of cellular systems, along with advanced topics related to power control and fundamental limits in these systems. The last chapter, Chapter 16, discusses the fundamental principles and open research challenges associated with wireless ad hoc networks.

REQUIRED BACKGROUND

The only prerequisite knowledge for the book is a basic understanding of probability, random processes, and Fourier techniques for system and signal analysis. Background in digital communications is helpful but not required, as the underlying principles from this field are covered in the text. Three appendices summarize key background material used in different chapters of the text. Specifically, Appendix A discusses the equivalent lowpass representation of bandpass signals and systems, which simplifies bandpass system analysis. Appendix B provides a summary of the main concepts in probability and random processes that are used throughout the book. Appendix C provides definitions, results, and properties related to matrices, which are widely used in Chapters 10 and 12. The last appendix, Appendix D, summarizes the main characteristics of current wireless systems and standards.

BOOK FEATURES

The tremendous research activity in the wireless field – coupled with the complexity of wireless system design – make it impossible to provide comprehensive details on all topics discussed in the book. Thus, each chapter contains a broad list of references that build and expand on what is covered in the text. The book also contains nearly a hundred worked examples to illustrate and highlight key principles and trade-offs. In addition, the book includes about 300 homework exercises. These exercises, which fall into several broad categories, are designed to enhance and reinforce the material in the main text. Some exercises are targeted to exemplify or provide more depth to key concepts, as well as to derive or illustrate properties of wireless systems using these concepts. Exercises are also used to prove results stated but not derived in the text. Another category of exercises obtains numerical results that give insight into operating parameters and performance of wireless systems in typical environments. Exercises also introduce new concepts or system designs that are not discussed in the text. A solutions manual is available that covers all the exercises.

USING THIS BOOK IN COURSES

The book is designed to provide much flexibility as a textbook, depending on the desired length of the course, student background, and course focus. The core of the book is in Chapters 1 through 6. Thereafter, each chapter covers a different stand-alone topic that can be omitted or may be covered in other courses. Necessary prerequisites for a course using this text are an undergraduate course in signals and systems (both analog and digital) and one in probability theory and random processes. It is also helpful if students have a prerequisite or corequisite course in digital communications, in which case the material in Chapter 5 (along with overlapping material in other chapters) can be covered quickly as a review.

The book breaks down naturally into three segments: core material in Chapters 1–6, single-user wireless system design in Chapters 7–13, and multiuser wireless networks in Chapters 14–16. Most of the material in the book can be covered in two to three quarters or two semesters. A three-quarter sequence would follow the natural segmentation of the chapters, perhaps with an in-depth research project at the end. For a course sequence of two semesters or quarters, the first course could focus on Chapters 1–10 (single-user systems with flat fading) and the second course could focus on Chapters 11–16 (frequency-selective fading techniques, multiuser systems, and wireless networks). A one-quarter or semester course could focus on single-user wireless systems based on the core material in Chapters 1–6 and selected topics from Chapters 7–13. In this case a second optional quarter or semester could be offered covering multiuser systems and wireless networks (part of Chapter 13 and Chapters 14–16). I use this breakdown in a two-quarter sequence at Stanford, where the second quarter is offered every other year and includes additional reading material from the literature as well as an in-depth research project. Alternatively, a one-quarter or semester course could cover both single and multiuser systems based on Chapters 1–6 and Chapters 13–16, with some additional topics from Chapters 7–12 as time permits.

A companion Web site (http://www.cambridge.org/9780521837163) provides supplemental material for the book, including lecture slides, additional exercises, and errata.

ACKNOWLEDGMENTS

It takes a village to complete a book, and I am deeply indebted to many people for their help during the multiple phases of this project. I first want to thank the ten generations of students at Caltech and Stanford who suffered through the annual revisions of my wireless course notes: their suggestions, insights, and experiences were extremely valuable in honing the topics, coverage, and tone of the book. John Proakis and several anonymous reviewers provided valuable and in-depth comments and suggestions on early book drafts, identifying omissions and weaknesses, which greatly strengthened the final manuscript. My current graduate students Rajiv Agrawal, Shuguang Cui, Yifan Liang, Xiangheng Liu, Chris Ng, and

Taesang Yoo meticulously proofread many chapter drafts, providing new perspectives and insights, rederiving formulas, checking for typos, and catching my errors and omissions. My former graduate students Tim Holliday, Syed Jafar, Nihar Jindal, Neelesh Mehta, Stavros Toumpis, and Sriram Vishwanath carefully scrutinized one or more chapters and provided valuable input. In addition, all of my current and former students (those already mentioned as well as Mohamed-Slim Alouini, Soon-Ghee Chua, Lifang Li, and Kevin Yu) contributed to the content of the book through their research results, especially in Chapters 4, 7, 9, 10, 14, and 16. The solutions manual was developed by Rajiv Agrawal, Grace Gao, and Ankit Kumar. I am also indebted to many colleagues who took time from their busy schedules, sometimes on very short notice, to read and critique specific chapters. They were extremely gracious, generous, and honest with their comments and criticisms. Their deep and valuable insights not only greatly improved the book but also taught me a lot about wireless. For these efforts I am extremely grateful to Jeff Andrews, Tony Ephremides, Mike Fitz, Dennis Goeckel, Larry Greenstein, Ralf Koetter, P. R. Kumar, Muriel Médard, Larry Milstein, Sergio Servetto, Sergio Verdú, and Roy Yates. Don Cox was always available to share his infinite engineering wisdom and to enlighten me about many of the subtleties and assumptions associated with wireless systems. I am also grateful to my many collaborators over the years, as well as to my co-workers at Maxim Technologies and AT&T Bell Laboratories, who have enriched my knowledge of wireless communications and related fields.

I am indebted to the colleagues, students, and leadership at Stanford who created the dynamic, stimulating, and exciting research and teaching environment in which this book evolved. I am also grateful for funding support from ONR and NSF throughout the development of the book. Much gratitude is also due to my administrative assistants Joice DeBolt and Pat Oshiro for taking care of all matters big and small in support of my research and teaching, and for making sure I had enough food and caffeine to get through each day. I would also like to thank copy editor Matt Darnell for his skill and attention to detail throughout the production process. My editor Phil Meyler has followed this book from its inception ten years ago until today. His encouragement and enthusiasm about the book never waned, and he has accommodated all of my changes and delays with grace and good humor. I cannot imagine a better editor with whom to embark on such a difficult, taxing, and rewarding undertaking.

I would like to thank two people in particular for their early and ongoing support in this project and all my professional endeavors. Larry Greenstein ignited my initial interest in wireless through his deep insight and research experience. He has served as a great source of knowledge, mentoring, and friendship. Pravin Varaiya was deeply influential as a Ph.D. advisor and role model due to his breadth and depth of knowledge along with his amazing rigor, insight, and passion for excellence. He has been a constant source of encouragement, inspiration, and friendship.

My friends and family have provided much love, support, and encouragement for which I am deeply grateful. I thank them for not abandoning me despite my long absences during the final stages of finishing the manuscript, and also for providing an incredible support network without which the book could not have been completed. I am especially grateful to Remy, Penny, and Lili for their love and support, and to my mother Adrienne for her love and for instilling in me her creativity and penchant for writing. My father Werner has profoundly influenced this book and my entire career both directly and indirectly. He was the senior Professor Goldsmith, a prolific researcher, author, and pioneer in many areas of mechanical and biological engineering. His suggestion to pursue engineering launched my career, for which he was my biggest cheerleader. His pride, love, and encouragement have been a constant source of support. I was fortunate to help him complete his final paper, and I have tried in this book to mimic his rigor, attention to detail, and obsession with typos that I experienced during that collaboration.

Finally, no words are sufficient to express my gratitude and love for my husband Arturo and my children Daniel and Nicole. Arturo has provided infinite support for this book and every other aspect of my career, for which he has made many sacrifices. His pride, love, encouragement, and devotion have sustained me through the ups and downs of academic and family life. He is the best husband, father, and friend I could have dreamed of, and he enriches my life in every way. Daniel and Nicole are the sunshine in my universe – each day is brighter because of their love and sweetness. I am incredibly lucky to share my life with these three special people. This book is dedicated to them.

Abbreviations

3GPP	Third Generation Partnership Project
ACK	acknowledgment (packet)
ACL	Asynchronous Connection-Less
AFD	average fade duration
AFRD	average fade region duration
AGC	automatic gain control
AMPS	Advance Mobile Phone Service
AOA	angle of arrival
AODV	ad hoc on-demand distance vector
APP	a posteriori probability
ARQ	automatic repeat request (protocol)
ASE	area spectral efficiency
AWGN	additive white Gaussian noise
BC	broadcast channel
BCH	Bose–Chadhuri–Hocquenghem
BER	bit error rate
BICM	bit-interleaved coded modulation
BLAST	Bell Labs Layered Space Time
BPSK	binary phase-shift keying
BS	base station
CCK	complementary code keying
CD	code division
cdf	cumulative distribution function
CDI	channel distribution information
CDMA	code-division multiple access
CDPD	cellular digital packet data
CLT	central limit theorem
COVQ	channel-optimized vector quantizer
CPFSK	continuous-phase FSK
CSI	channel side information
CSIR	CSI at the receiver
CSIT	CSI at the transmitter

CSMA	carrier-sense multiple access
CTS	clear to send (packet)
DARPA	Defense Advanced Research Projects Agency
D-BLAST	diagonal BLAST
DCA	dynamic channel assignment
DCS	Digital Cellular System
DECT	Digital Enhanced Cordless Telecommunications
DFE	decision-feedback equalization
DFT	discrete Fourier transform
D-MPSK	differential M-ary PSK
DPC	dirty paper coding
DPSK	differential binary PSK
D-QPSK	differential quadrature PSK
DS	direct sequence
DSDV	destination sequenced distance vector
DSL	digital subscriber line
DSR	dynamic source routing
DSSS	direct-sequence spread spectrum
EDGE	Enhanced Data rates for GSM Evolution
EGC	equal-gain combining
ETACS	European Total Access Communication System
ETSI	European Telecommunications Standards Institute
EURO-COST	European Cooperative for Scientific and Technical Research
FAF	floor attenuation factor
FCC	Federal Communications Commission
FD	frequency division
FDD	frequency-division duplexing
FDMA	frequency-division multiple access
FFH	fast frequency hopping
FFT	fast Fourier transform
FH	frequency hopping
FHSS	frequency-hopping spread spectrum
FIR	finite impulse response
FSK	frequency-shift keying
FSMC	finite-state Markov channel
GEO	geosynchronous orbit
GFSK	Gaussian frequency-shift keying
GMSK	Gaussian minimum-shift keying
GPRS	General Packet Radio Service
GRT	general ray tracing
GSM	Global Systems for Mobile Communications
GTD	geometrical theory of diffraction
HDD	hard decision decoding
HDR	high data rate

HDSL   high–bit-rate digital subscriber line
HIPERLAN   high-performance radio local area network
HSCSD   High Speed Circuit Switched Data
HSDPA   High Speed Data Packet Access

ICI   intercarrier interference
IDFT   inverse DFT
IEEE   Institute of Electrical and Electronics Engineers
IFFT   inverse FFT
i.i.d.   independent and identically distributed
IIR   infinite impulse response
IMT   International Mobile Telephone
IP   Internet protocol
ISI   intersymbol interference
ISM   Industrial, Scientific, and Medical (spectrum band)
ITU   International Telecommunications Union

JTACS   Japanese TACS

LAN   local area network
LDPC   low-density parity-check
LEO   low-earth orbit
LLR   log likelihood ratio
LMA   local mean attenuation
LMDS   local multipoint distribution service
LMS   least mean square
LOS   line of sight

MAC   multiple access channel
MAI   multiple access interference
MAN   metropolitan area network
MAP   maximum a posteriori
MC-CDMA   multicarrier CDMA
MDC   multiple description coding
MEO   medium-earth orbit
MFSK M-ary FSK
MGF   moment generating function
MIMO   multiple-input multiple-output
MISO   multiple-input single-output
ML   maximum likelihood
MLSE   maximum likelihood sequence estimation
MMDS   multichannel multipoint distribution service
MMSE   minimum mean-square error
MPAM M-ary PAM
MPSK M-ary PSK
MQAM M-ary QAM
MRC   maximal-ratio combining
MSE   mean-square error
MSK   minimum-shift keying

MTSO	mobile telephone switching office
MUD	multiuser detector
N-AMPS	narrowband AMPS
NMT	Nordic Mobile Telephone
OFDM	orthogonal frequency-division multiplexing
OFDMA	OFDM with multiple access
O-QPSK	quadrature PSK with phase offset
OSI	open systems interconnect
OSM	Office of Spectral Management
PACS	Personal Access Communications System
PAF	partition attenuation factor
PAM	pulse amplitude modulation
PAR	peak-to-average power ratio
PBX	private branch exchange
PCS	Personal Communication Systems
PDA	personal digital assistant
PDC	Personal Digital Cellular
pdf	probability density function
PER	packet error rate
PHS	Personal Handyphone System
PLL	phase-locked loop
PN	pseudorandom
PRMA	packet-reservation multiple access
PSD	power spectral density
PSK	phase-shift keying
PSTN	public switched telephone network
QAM	quadrature amplitude modulation
QoS	quality of service
QPSK	quadrature PSK
RCPC	rate-compatible punctured convolutional
RCS	radar cross-section
RLS	root least squares
rms	root mean square
RS	Reed Solomon
RTS	request to send (packet)
RTT	radio transmission technology
SBS	symbol-by-symbol
SC	selection combining
SCO	Synchronous Connection Oriented
SDD	soft decision decoding
SDMA	space-division multiple access
SE	sequence estimator
SFH	slow frequency hopping

SHO	soft handoff
SICM	symbol-interleaved coded modulation
SIMO	single-input multiple-output
SINR	signal-to-interference-plus-noise power ratio
SIR	signal-to-interference power ratio
SISO	single-input single-output
SNR	signal-to-noise ratio
SOVA	soft output Viterbi algorithm
SSC	switch-and-stay combining
SSMA	spread-spectrum multiple access
STBC	space-time block code
STTC	space-time trellis code
SVD	singular value decomposition
TACS	Total Access Communication System
TCP	transport control protocol
TD	time division
TDD	time-division duplexing
TDMA	time-division multiple access
TIA	Telecommunications Industry Association
UEP	unequal error protection
UMTS	Universal Mobile Telecommunications System
U-NII	Unlicensed National Information Infrastructure
US	uncorrelated scattering
UWB	ultrawideband
V-BLAST	vertical BLAST
VC	vector coding
VCC	voltage-controlled clock
VCO	voltage-controlled oscillator
VQ	vector quantizer
WAN	wide area network
W-CDMA	wideband CDMA
WLAN	wireless LAN
WPAN	wireless personal area networks
WSS	wide-sense stationary
ZF	zero-forcing
ZMCSCQ	zero-mean circularly symmetric complex Gaussian
ZMSW	zero-mean spatially white
ZRP	zone routing protocol

Notation

$\approx$	approximately equal to
$\triangleq$	defined as equal to ($a \triangleq b$: a is defined as b)
$\gg$	much greater than
$\ll$	much less than
$\cdot$	multiplication operator
$*$	convolution operator
$\circledast$	circular convolution operator
$\otimes$	Kronecker product operator
$\sqrt[n]{x}$, $x^{1/n}$	nth root of x
$\arg\max[f(x)]$	value of x that maximizes the function $f(x)$
$\arg\min[f(x)]$	value of x that minimizes the function $f(x)$
$\mathrm{Co}(\mathcal{W})$	convex hull of region $\mathcal{W}$
$\delta(x)$	the delta function
$\mathrm{erfc}(x)$	the complementary error function
$\exp[x]$	e^x
$\mathrm{Im}\{x\}$	imaginary part of x
$I_0(x)$	modified Bessel function of the 0th order
$J_0(x)$	Bessel function of the 0th order
$\mathcal{L}(x)$	Laplace transform of x
$\ln(x)$	the natural log of x
$\log_x(y)$	the log, base x, of y
$\log_x \det[\mathbf{A}]$	the log, base x, of the determinant of matrix $\mathbf{A}$
$\max_x f(x)$	maximum value of $f(x)$ maximized over all x
$\mathrm{mod}_n(x)$	x modulo n
$N(\mu, \sigma^2)$	Gaussian (normal) distribution with mean μ and variance σ^2
$\bar{P}_r$	local mean received power
$Q(x)$	Gaussian Q-function
$\mathbb{R}$	field of all real numbers
$\mathrm{Re}\{x\}$	real part of x
$\mathrm{rect}(x)$	the rectangular function ($\mathrm{rect}(x) = 1$ for $|x| \le .5$, 0 else)
$\mathrm{sinc}(x)$	the sinc function ($\sin(\pi x)/(\pi x)$)

$\mathbf{E}[\cdot]$	expectation operator
$\mathbf{E}[\cdot \mid \cdot]$	conditional expectation operator
$\bar{X}$	expected (average) value of random variable X
$X \sim p_X(x)$	the random variable X has distribution $p_X(x)$
$\text{Var}[X]$	variance of random variable X
$\text{Cov}[X, Y]$	covariance of random variables X and Y
$H(X)$	entropy of random variable X
$H(Y \mid X)$	conditional entropy of random variable Y given random variable X
$I(X; Y)$	mutual information between random variables X and Y
$\mathcal{M}_X(s)$	moment generating function for random variable X
$\phi_X(s)$	characteristic function for random variable X
$\mathcal{F}[\cdot]$	Fourier transform operator ($\mathcal{F}_x[\cdot]$ is transform w.r.t. x)
$\mathcal{F}^{-1}[\cdot]$	inverse Fourier transform operator ($\mathcal{F}_x^{-1}[\cdot]$ is inverse transform w.r.t. x)
$\text{DFT}\{\cdot\}$	discrete Fourier transform operator
$\text{IDFT}\{\cdot\}$	inverse discrete Fourier transform operator
$\langle \cdot, \cdot \rangle$	inner product operator
x^*	complex conjugate of x
$\angle x$	phase of x
$\lvert x \rvert$	absolute value (amplitude) of x
$\lvert \mathcal{X} \rvert$	size of alphabet $\mathcal{X}$
$\lfloor x \rfloor$	largest integer less than or equal to x
$\lfloor x \rfloor_S$	largest number in set S less than or equal to x
$\{x : \mathcal{C}\}$	set containing all x that satisfy condition $\mathcal{C}$
$\{x_i : i = 1, \ldots, n\}, \{x_i\}_{i=1}^n$	set containing $x_1, \ldots, x_n$
$(x_i : i = 1, \ldots, n)$	the vector $\mathbf{x} = (x_1, \ldots, x_n)$
$\lVert \mathbf{x} \rVert$	norm of vector $\mathbf{x}$
$\lVert \mathbf{A} \rVert_F$	Frobenius norm of matrix $\mathbf{A}$
$\mathbf{x}^*$	complex conjugate of vector $\mathbf{x}$
$\mathbf{x}^H$	Hermitian (conjugate transpose) of vector $\mathbf{x}$
$\mathbf{x}^T$	transpose of vector $\mathbf{x}$
$\mathbf{A}^{-1}$	inverse of matrix $\mathbf{A}$
$\mathbf{A}^H$	Hermitian (conjugate transpose) of matrix $\mathbf{A}$
$\mathbf{A}^T$	transpose of matrix $\mathbf{A}$
$\det[\mathbf{A}]$	determinant of matrix $\mathbf{A}$
$\text{Tr}[\mathbf{A}]$	trace of matrix $\mathbf{A}$
$\text{vec}(\mathbf{A})$	vector obtained by stacking columns of matrix $\mathbf{A}$
$N \times M$ matrix	a matrix with N rows and M columns
$\text{diag}[x_1, \ldots, x_N]$	the $N \times N$ diagonal matrix with diagonal elements $x_1, \ldots, x_N$
$\mathbf{I}_N$	the $N \times N$ identity matrix (N omitted when size is clear from the context)

Overview of Wireless Communications

Wireless communications is, by any measure, the fastest growing segment of the communications industry. As such, it has captured the attention of the media and the imagination of the public. Cellular systems have experienced exponential growth over the last decade and there are currently about two billion users worldwide. Indeed, cellular phones have become a critical business tool and part of everyday life in most developed countries, and they are rapidly supplanting antiquated wireline systems in many developing countries. In addition, wireless local area networks currently supplement or replace wired networks in many homes, businesses, and campuses. Many new applications – including wireless sensor networks, automated highways and factories, smart homes and appliances, and remote telemedicine – are emerging from research ideas to concrete systems. The explosive growth of wireless systems coupled with the proliferation of laptop and palmtop computers suggests a bright future for wireless networks, both as stand-alone systems and as part of the larger networking infrastructure. However, many technical challenges remain in designing robust wireless networks that deliver the performance necessary to support emerging applications. In this introductory chapter we will briefly review the history of wireless networks from the smoke signals of the pre-industrial age to the cellular, satellite, and other wireless networks of today. We then discuss the wireless vision in more detail, including the technical challenges that must still be overcome. We describe current wireless systems along with emerging systems and standards. The gap between current and emerging systems and the vision for future wireless applications indicates that much work remains to be done to make this vision a reality.

1.1 History of Wireless Communications

The first wireless networks were developed in the pre-industrial age. These systems transmitted information over line-of-sight distances (later extended by telescopes) using smoke signals, torch signaling, flashing mirrors, signal flares, or semaphore flags. An elaborate set of signal combinations was developed to convey complex messages with these rudimentary signals. Observation stations were built on hilltops and along roads to relay these messages over large distances. These early communication networks were replaced first by the telegraph network (invented by Samuel Morse in 1838) and later by the telephone. In 1895, a few decades after the telephone was invented, Marconi demonstrated the first radio transmission from the Isle of Wight to a tugboat 18 miles away, and radio communications was born. Radio technology advanced rapidly to enable transmissions over larger distances with better quality, less power, and smaller, cheaper devices, thereby enabling public and private radio communications, television, and wireless networking.

Early radio systems transmitted analog signals. Today most radio systems transmit digital signals composed of binary bits, where the bits are obtained directly from a data signal or by digitizing an analog signal. A digital radio can transmit a continuous bit stream or it can group the bits into packets. The latter type of radio is called a *packet radio* and is often characterized by bursty transmissions: the radio is idle except when it transmits a packet, although it may transmit packets continuously. The first network based on packet radio, ALOHANET, was developed at the University of Hawaii in 1971. This network enabled computer sites at seven campuses spread out over four islands to communicate with a central computer on Oahu via radio transmission. The network architecture used a star topology with the central computer at its hub. Any two computers could establish a bi-directional communications link between them by going through the central hub. ALOHANET incorporated the first set of protocols for channel access and routing in packet radio systems, and many of the underlying principles in these protocols are still in use today. The U.S. military was extremely interested in this combination of packet data and broadcast radio. Throughout the 1970s and early 1980s the Defense Advanced Research Projects Agency (DARPA) invested significant resources to develop networks using packet radios for tactical communications in the battlefield. The nodes in these ad hoc wireless networks had the ability to self-configure (or reconfigure) into a network without the aid of any established infrastructure. DARPA's investment in ad hoc networks peaked in the mid 1980s, but the resulting systems fell far short of expectations in terms of speed and performance. These networks continue to be developed for military use. Packet radio networks also found commercial application in supporting wide area wireless data services. These services, first introduced in the early 1990s, enabled wireless data access (including email, file transfer, and Web browsing) at fairly low speeds, on the order of 20 kbps. No strong market for these wide area wireless data services ever really materialized, due mainly to their low data rates, high cost, and lack of "killer applications". These services mostly disappeared in the 1990s, supplanted by the wireless data capabilities of cellular telephones and wireless local area networks (WLANs).

The introduction of wired Ethernet technology in the 1970s steered many commercial companies away from radio-based networking. Ethernet's 10-Mbps data rate far exceeded anything available using radio, and companies did not mind running cables within and between their facilities to take advantage of these high rates. In 1985 the Federal Communications Commission (FCC) enabled the commercial development of wireless LANs by authorizing the public use of the Industrial, Scientific, and Medical (ISM) frequency bands for wireless LAN products. The ISM band was attractive to wireless LAN vendors because they did not need to obtain an FCC license to operate in this band. However, the wireless LAN systems were not allowed to interfere with the primary ISM band users, which forced them to use a low power profile and an inefficient signaling scheme. Moreover, the interference from primary users within this frequency band was quite high. As a result, these initial wireless LANs had very poor performance in terms of data rates and coverage. This poor performance – coupled with concerns about security, lack of standardization, and high cost (the first wireless LAN access points listed for $1400 as compared to a few hundred dollars for a wired Ethernet card) – resulted in weak sales. Few of these systems were actually used for data networking: they were relegated to low-tech applications like inventory control. The current generation of wireless LANs, based on the family of IEEE 802.11 standards, have better performance, although the data rates are still relatively low (maximum collective data rates of tens of Mbps) and the coverage area is still small (around 100 m). Wired Ethernets today offer data rates of 1 Gbps, and the performance gap between wired and wireless LANs is likely to increase over time without additional spectrum allocation. Despite their lower data rates, wireless LANs are becoming the prefered Internet access method in many homes, offices, and campus environments owing to their convenience and freedom from wires. However, most wireless LANs support applications, such as email and Web browsing, that are not bandwidth intensive. The challenge for future wireless LANs will be to support many users simultaneously with bandwidth-intensive and delay-constrained applications such as video. Range extension is also a critical goal for future wireless LAN systems.

By far the most successful application of wireless networking has been the cellular telephone system. The roots of this system began in 1915, when wireless voice transmission between New York and San Francisco was first established. In 1946, public mobile telephone service was introduced in 25 cities across the United States. These initial systems used a central transmitter to cover an entire metropolitan area. This inefficient use of the radio spectrum – coupled with the state of radio technology at that time – severely limited the system capacity: thirty years after the introduction of mobile telephone service, the New York system could support only 543 users.

A solution to this capacity problem emerged during the 1950s and 1960s as researchers at AT&T Bell Laboratories developed the cellular concept [1]. Cellular systems exploit the fact that the power of a transmitted signal falls off with distance. Thus, two users can operate on the same frequency at spatially separate locations with minimal interference between them. This allows efficient use of cellular spectrum, so that a large number of users can be accommodated. The evolution of cellular systems from initial concept to implementation was glacial. In 1947, AT&T requested spectrum for cellular service from the FCC. The design was mostly completed by the end of the 1960s; but the first field test was not until 1978, and the FCC granted service authorization in 1982 – by which time much of the original technology was out of date. The first analog cellular system, deployed in Chicago in 1983, was already saturated by 1984, when the FCC increased the cellular spectral allocation from 40 MHz to 50 MHz. The explosive growth of the cellular industry took almost everyone by surprise. In fact, a marketing study commissioned by AT&T before the first system rollout predicted that demand for cellular phones would be limited to doctors and the very rich. AT&T basically abandoned the cellular business in the 1980s to focus on fiber optic networks, eventually returning to the business after its potential became apparent. Throughout the late 1980s – as more and more cities saturated with demand for cellular service – the development of digital cellular technology for increased capacity and better performance became essential.

The second generation of cellular systems, first deployed in the early 1990s, was based on digital communications. The shift from analog to digital was driven by its higher capacity and the improved cost, speed, and power efficiency of digital hardware. Although second-generation cellular systems initially provided mainly voice services, these systems

gradually evolved to support data services such as email, Internet access, and short messaging. Unfortunately, the great market potential for cellular phones led to a proliferation
of second-generation cellular standards: three different standards in the United States alone,
other standards in Europe and Japan, and all incompatible. The fact that different cities
have different incompatible standards makes roaming throughout the United States and the
world with only one cellular phone standard impossible. Moreover, some countries have
initiated service for third-generation systems, for which there are also multiple incompatible standards. As a result of this proliferation of standards, many cellular phones today are
multimode: they incorporate multiple digital standards to faciliate nationwide and worldwide
roaming and possibly the first-generation analog standard as well, since only this standard
provides universal coverage throughout the United States.

Satellite systems are typically characterized by the height of the satellite orbit: low-earth
orbit (LEOs at roughly 2000 km altitude), medium-earth orbit (MEOs, 9000 km), or geosynchronous orbit (GEOs, 40,000 km). The geosynchronous orbits are seen as stationary from
the earth, whereas satellites with other orbits have their coverage area change over time. The
concept of using geosynchronous satellites for communications was first suggested by the
science-fiction writer Arthur C. Clarke in 1945. However, the first deployed satellites – the
Soviet Union's *Sputnik* in 1957 and the NASA/Bell Laboratories' *Echo-1* in 1960 – were not
geosynchronous owing to the difficulty of lifting a satellite into such a high orbit. The first
GEO satellite was launched by Hughes and NASA in 1963; GEOs then dominated both commercial and government satellite systems for several decades.

Geosynchronous satellites have large coverage areas, so fewer satellites (and dollars) are
necessary to provide wide area or global coverage. However, it takes a great deal of power
to reach the satellite, and the propagation delay is typically too large for delay-constrained
applications like voice. These disadvantages caused a shift in the 1990s toward lower-orbit
satellites [2; 3]. The goal was to provide voice and data service competitive with cellular
systems. However, the satellite mobile terminals were much bigger, consumed much more
power, and cost much more than contemporary cellular phones, which limited their appeal.
The most compelling feature of these systems is their ubiquitous worldwide coverage, especially in remote areas or third-world countries with no landline or cellular system infrastructure. Unfortunately, such places do not typically have large demand or the resources
to pay for satellite service either. As cellular systems became more widespread, they took
away most revenue that LEO systems might have generated in populated areas. With no real
market left, most LEO satellite systems went out of business.

A natural area for satellite systems is broadcast entertainment. Direct broadcast satellites
operate in the 12-GHz frequency band. These systems offer hundreds of TV channels and
are major competitors to cable. Satellite-delivered digital radio has also become popular.
These systems, operating in both Europe and the United States, offer digital audio broadcasts
at near-CD quality.

1.2 Wireless Vision

The vision of wireless communications supporting information exchange between people
or devices is the communications frontier of the next few decades, and much of it already

exists in some form. This vision will allow multimedia communication from anywhere in the world using a small handheld device or laptop. Wireless networks will connect palm-top, laptop, and desktop computers anywhere within an office building or campus, as well as from the corner cafe. In the home these networks will enable a new class of intelligent electronic devices that can interact with each other and with the Internet in addition to providing connectivity between computers, phones, and security/monitoring systems. Such "smart" homes can also help the elderly and disabled with assisted living, patient monitoring, and emergency response. Wireless entertainment will permeate the home and any place that people congregate. Video teleconferencing will take place between buildings that are blocks or continents apart, and these conferences can include travelers as well – from the salesperson who missed his plane connection to the CEO off sailing in the Caribbean. Wireless video will enable remote classrooms, remote training facilities, and remote hospitals anywhere in the world. Wireless sensors have an enormous range of both commercial and military applications. Commercial applications include monitoring of fire hazards, toxic waste sites, stress and strain in buildings and bridges, carbon dioxide movement, and the spread of chemicals and gasses at a disaster site. These wireless sensors self-configure into a network to process and interpret sensor measurements and then convey this information to a centralized control location. Military applications include identification and tracking of enemy targets, detection of chemical and biological attacks, support of unmanned robotic vehicles, and counterterrorism. Finally, wireless networks enable distributed control systems with remote devices, sensors, and actuators linked together via wireless communication channels. Such systems in turn enable automated highways, mobile robots, and easily reconfigurable industrial automation.

The various applications described here are all components of the wireless vision. So then what, exactly, is wireless communications? There are many ways to segment this complex topic into different applications, systems, or coverage regions [4]. Wireless applications include voice, Internet access, Web browsing, paging and short messaging, subscriber information services, file transfer, video teleconferencing, entertainment, sensing, and distributed control. Systems include cellular telephone systems, wireless LANs, wide area wireless data systems, satellite systems, and ad hoc wireless networks. Coverage regions include in-building, campus, city, regional, and global. The question of how best to characterize wireless communications along these various segments has resulted in considerable fragmentation in the industry, as evidenced by the many different wireless products, standards, and services being offered or proposed. One reason for this fragmentation is that different wireless applications have different requirements. Voice systems have relatively low data-rate requirements (around 20 kbps) and can tolerate a fairly high probability of bit error (bit error rates, or BERs, of around 10^{-3}), but the total delay must be less than about 100 ms or else it becomes noticeable to the end user.[1] On the other hand, data systems typically require much higher data rates (1–100 Mbps) and very small BERs (a BER of 10^{-8} or less, and all bits received in error must be retransmitted) but do not have a fixed delay requirement. Real-time video systems have high data-rate requirements coupled with the same delay constraints as voice systems,

---

[1] Wired telephones have a delay constraint of $\sim$30 ms. Cellular phones relax this constraint to $\sim$100 ms, and voice over the Internet relaxes the constraint even further.

while paging and short messaging have very low data-rate requirements and no hard delay constraints. These diverse requirements for different applications make it difficult to build one wireless system that can efficiently satisfy all these requirements simultaneously. Wired networks typically satisfy the diverse requirements of different applications using a single protocol, which means that the most stringent requirements for all applications must be met simultaneously. This may be possible on some wired networks – with data rates on the order of Gbps and BERs on the order of 10^{-12} – but it is not possible on wireless networks, which have much lower data rates and higher BERs. For these reasons, at least in the near future, wireless systems will continue to be fragmented, with different protocols tailored to support the requirements of different applications.

The exponential growth of cellular telephone use and wireless Internet access has led to great optimism about wireless technology in general. Obviously not all wireless applications will flourish. While many wireless systems and companies have enjoyed spectacular success, there have also been many failures along the way, including first-generation wireless LANs, the Iridium satellite system, wide area data services such as Metricom, and fixed wireless access (wireless "cable") to the home. Indeed, it is impossible to predict what wireless failures and triumphs lie on the horizon. Moreover, there must be sufficient flexibility and creativity among both engineers and regulators to allow for accidental successes. It is clear, however, that the current and emerging wireless systems of today – coupled with the vision of applications that wireless can enable – ensure a bright future for wireless technology.

1.3 Technical Issues

Many technical challenges must be addressed to enable the wireless applications of the future. These challenges extend across all aspects of the system design. As wireless terminals add more features, these small devices must incorporate multiple modes of operation in order to support the different applications and media. Computers process voice, image, text, and video data, but breakthroughs in circuit design are required to implement the same multimode operation in a cheap, lightweight, handheld device. Consumers don't want large batteries that frequently need recharging, so transmission and signal processing at the portable terminal must consume minimal power. The signal processing required to support multimedia applications and networking functions can be power intensive. Thus, wireless infrastructure-based networks, such as wireless LANs and cellular systems, place as much of the processing burden as possible on fixed sites with large power resources. The associated bottlenecks and single points of failure are clearly undesirable for the overall system. Ad hoc wireless networks without infrastructure are highly appealing for many applications because of their flexibility and robustness. For these networks, all processing and control must be performed by the network nodes in a distributed fashion, making energy efficiency challenging to achieve. Energy is a particularly critical resource in networks where nodes cannot recharge their batteries – for example, in sensing applications. Network design to meet application requirements under such hard energy constraints remains a big technological hurdle. The finite bandwidth and random variations of wireless channels also require robust applications that degrade gracefully as network performance degrades.

Design of wireless networks differs fundamentally from wired network design owing to the nature of the wireless channel. This channel is an unpredictable and difficult communications medium. First of all, the radio spectrum is a scarce resource that must be allocated to many different applications and systems. For this reason, spectrum is controlled by regulatory bodies both regionally and globally. A regional or global system operating in a given frequency band must obey the restrictions for that band set forth by the corresponding regulatory body. Spectrum can also be very expensive: in many countries spectral licenses are often auctioned to the highest bidder. In the United States, companies spent over $9 billion for second-generation cellular licenses, and the auctions in Europe for third-generation cellular spectrum garnered around $100 billion (American). The spectrum obtained through these auctions must be used extremely efficiently to receive a reasonable return on the investment, and it must also be reused over and over in the same geographical area, thus requiring cellular system designs with high capacity and good performance. At frequencies around several gigahertz, wireless radio components with reasonable size, power consumption, and cost are available. However, the spectrum in this frequency range is extremely crowded. Thus, technological breakthroughs to enable higher-frequency systems with the same cost and performance would greatly reduce the spectrum shortage. However, path loss at these higher frequencies is larger with omnidirectional antennas, thereby limiting range.

As a signal propagates through a wireless channel, it experiences random fluctuations in time if the transmitter, receiver, or surrounding objects are moving because of changing reflections and attenuation. Hence the characteristics of the channel appear to change randomly with time, which makes it difficult to design reliable systems with guaranteed performance. Security is also more difficult to implement in wireless systems, since the airwaves are susceptible to snooping by anyone with an RF antenna. The analog cellular systems have no security, and one can easily listen in on conversations by scanning the analog cellular frequency band. All digital cellular systems implement some level of encryption. However, with enough knowledge, time, and determination, most of these encryption methods can be cracked; indeed, several have been compromised. To support applications like electronic commerce and credit-card transactions, the wireless network must be secure against such listeners.

Wireless networking is also a significant challenge. The network must be able to locate a given user wherever it is among billions of globally distributed mobile terminals. It must then route a call to that user as it moves at speeds of up to 100 km/hr. The finite resources of the network must be allocated in a fair and efficient manner relative to changing user demands and locations. Moreover, there currently exists a tremendous infrastructure of wired networks: the telephone system, the Internet, and fiber optic cables – which could be used to connect wireless systems together into a global network. However, wireless systems with mobile users will never be able to compete with wired systems in terms of data rates and reliability. Interfacing between wireless and wired networks with vastly different performance capabilities is a difficult problem.

Perhaps the most significant technical challenge in wireless network design is an overhaul of the design process itself. Wired networks are mostly designed according to a layered approach, whereby protocols associated with different layers of the system operation are designed in isolation, with baseline mechanisms to interface between layers. The layers in a wireless system include: the link or physical layer, which handles bit transmissions over the communications medium; the access layer, which handles shared access to the communications medium; the network and transport layers, which route data across the network and ensure end-to-end connectivity and data delivery; and the application layer, which dictates the end-to-end data rates and delay constraints associated with the application. While a layering methodology reduces complexity and facilitates modularity and standardization, it also leads to inefficiency and performance loss due to the lack of a global design optimization. The large capacity and good reliability of wired networks make these inefficiencies relatively benign for many wired network applications, although they do preclude good performance of delay-constrained applications such as voice and video. The situation is very different in a wireless network. Wireless links can exhibit very poor performance, and this performance, along with user connectivity and network topology, changes over time. In fact, the very notion of a wireless link is somewhat fuzzy owing to the nature of radio propagation and broadcasting. The dynamic nature and poor performance of the underlying wireless communication channel indicates that high-performance networks must be optimized for this channel and must be robust and adaptive to its variations, as well as to network dynamics. Thus, these networks require integrated and adaptive protocols at all layers, from the link layer to the application layer. This cross-layer protocol design requires interdisciplinary expertise in communications, signal processing, and network theory and design.

In the next section we give an overview of the wireless systems in operation today. It will be clear from this overview that the wireless vision remains a distant goal, with many technical challenges to overcome. These challenges will be examined in detail throughout the book.

1.4 Current Wireless Systems

This section provides a brief overview of current wireless systems in operation today. The design details of these system are constantly evolving, with new systems emerging and old ones going by the wayside. Thus, we will focus mainly on the high-level design aspects of the most common systems. More details on wireless system standards can be found in [5; 6; 7]. A summary of the main wireless system standards is given in Appendix D.

1.4.1 Cellular Telephone Systems

Cellular telephone systems are extremely popular and lucrative worldwide: these are the systems that ignited the wireless revolution. Cellular systems provide two-way voice and data communication with regional, national, or international coverage. Cellular systems were initially designed for mobile terminals inside vehicles with antennas mounted on the vehicle roof. Today these systems have evolved to support lightweight handheld mobile terminals operating inside and outside buildings at both pedestrian and vehicle speeds.

The basic premise behind cellular system design is frequency reuse, which exploits the fact that signal power falls off with distance to reuse the same frequency spectrum at spatially separated locations. Specifically, the coverage area of a cellular system is divided into nonoverlapping cells, where some set of channels is assigned to each cell. This same channel set is used in another cell some distance away, as shown in Figure 1.1, where C_i denotes the channel set used in a particular cell. Operation within a cell is controlled by a centralized

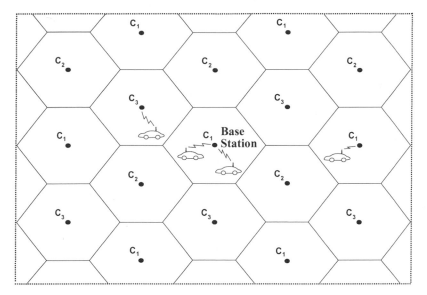

Figure 1.1: Cellular systems.

base station, as described in more detail below. The interference caused by users in different cells operating on the same channel set is called *intercell interference.* The spatial separation of cells that reuse the same channel set, the *reuse distance,* should be as small as possible so that frequencies are reused as often as possible, thereby maximizing spectral efficiency. However, as the reuse distance decreases, intercell interference increases owing to the smaller propagation distance between interfering cells. Since intercell interference must remain below a given threshold for acceptable system performance, reuse distance cannot be reduced below some minimum value. In practice it is quite difficult to determine this minimum value, since both the transmitting and interfering signals experience random power variations due to the characteristics of wireless signal propagation. In order to determine the best reuse distance and base station placement, an accurate characterization of signal propagation within the cells is needed.

Initial cellular system designs were mainly driven by the high cost of base stations, approximately $1 million each. For this reason, early cellular systems used a relatively small number of cells to cover an entire city or region. The cell base stations were placed on tall buildings or mountains and transmitted at very high power with cell coverage areas of several square miles. These large cells are called *macrocells.* Signal power radiated uniformly in all directions, so a mobile moving in a circle around the base station would have approximately constant received power unless the signal were blocked by an attenuating object. This circular contour of constant power yields a hexagonal cell shape for the system, since a hexagon is the closest shape to a circle that can cover a given area with multiple nonoverlapping cells.

Cellular systems in urban areas now mostly use smaller cells with base stations close to street level that are transmitting at much lower power. These smaller cells are called microcells or picocells, depending on their size. This evolution to smaller cells occured for two reasons: the need for higher capacity in areas with high user density and the reduced size and cost of base station electronics. A cell of any size can support roughly the same number

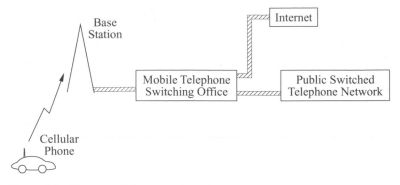

Figure 1.2: Current cellular network architecture.

of users if the system is scaled accordingly. Thus, for a given coverage area, a system with many microcells has a higher number of users per unit area than a system with just a few macrocells. In addition, less power is required at the mobile terminals in microcellular systems, since the terminals are closer to the base stations. However, the evolution to smaller cells has complicated network design. Mobiles traverse a small cell more quickly than a large cell, so handoffs must be processed more quickly. In addition, location management becomes more complicated, since there are more cells within a given area where a mobile may be located. It is also harder to develop general propagation models for small cells, since signal propagation in these cells is highly dependent on base station placement and the geometry of the surrounding reflectors. In particular, a hexagonal cell shape is generally not a good approximation to signal propagation in microcells. Microcellular systems are often designed using square or triangular cell shapes, but these shapes have a large margin of error in their approximation to microcell signal propagation [8].

All base stations in a given geographical area are connected via a high-speed communications link to a mobile telephone switching office (MTSO), as shown in Figure 1.2. The MTSO acts as a central controller for the network: allocating channels within each cell, coordinating handoffs between cells when a mobile traverses a cell boundary, and routing calls to and from mobile users. The MTSO can route voice calls through the public switched telephone network (PSTN) or provide Internet access. A new user located in a given cell requests a channel by sending a call request to the cell's base station over a separate control channel. The request is relayed to the MTSO, which accepts the call request if a channel is available in that cell. If no channels are available then the call request is rejected. A call handoff is initiated when the base station or the mobile in a given cell detects that the received signal power for that call is approaching a given minimum threshold. In this case the base station informs the MTSO that the mobile requires a handoff, and the MTSO then queries surrounding base stations to determine if one of these stations can detect that mobile's signal. If so then the MTSO coordinates a handoff between the original base station and the new base station. If no channels are available in the cell with the new base station then the handoff fails and the call is terminated. A call will also be dropped if the signal strength between a mobile and its base station falls below the minimum threshold needed for communication as a result of random signal variations.

The first generation of cellular systems used analog communications; these systems were primarily designed in the 1960s, before digital communications became prevalent. Second-generation systems moved from analog to digital because of the latter's many advantages. The components are cheaper, faster, and smaller, and they require less power. The degradation of voice quality caused by channel impairments can be mitigated with error correction coding and signal processing. Digital systems also have higher capacity than analog systems because they can use more spectrally efficient digital modulation and more efficient techniques to share the cellular spectrum. They can also take advantage of advanced compression techniques and voice activity factors. In addition, encryption techniques can be used to secure digital signals against eavesdropping. Digital systems can also offer data services in addition to voice, including short messaging, email, Internet access, and imaging capabilities (camera phones). Because of their lower cost and higher efficiency, service providers used aggressive pricing tactics to encourage user migration from analog to digital systems, and today analog systems are primarily used in areas with no digital service. However, digital systems do not always work as well as the analog ones. Users can experience poor voice quality, frequent call dropping, and spotty coverage in certain areas. System performance has certainly improved as the technology and networks mature. In some areas cellular phones provide almost the same quality as wireline service. Indeed, some people have replaced their wireline telephone service inside the home with cellular service.

Spectral sharing in communication systems, also called multiple access, is done by dividing the signaling dimensions along the time, frequency, and/or code space axes. In *frequency-division multiple access* (FDMA) the total system bandwidth is divided into orthogonal frequency channels. In *time-division multiple access* (TDMA), time is divided orthogonally and each channel occupies the entire frequency band over its assigned timeslot. TDMA is more difficult to implement than FDMA because the users must be time-synchronized. However, it is easier to accommodate multiple data rates with TDMA, since multiple timeslots can be assigned to a given user. *Code-division multiple access* (CDMA) is typically implemented using direct-sequence or frequency-hopping spread spectrum with either orthogonal or nonorthogonal codes. In direct sequence, each user modulates its data sequence by a different chip sequence that is much faster than the data sequence. In the frequency domain, the narrowband data signal is convolved with the wideband chip signal to yield a signal with a much wider bandwidth than the original data signal. In frequency hopping the carrier frequency used to modulate the narrowband data signal is varied by a chip sequence that may be faster or slower than the data sequence. This results in a modulated signal that hops over different carrier frequencies. Spread-spectrum signals are typically superimposed onto each other within the same signal bandwidth. A spread-spectrum receiver separates out each of the distinct signals by separately decoding each spreading sequence. However, for nonorthogonal codes, users within a cell interfere with each other (*intracell* interference) and codes that are reused in other cells cause intercell interference. Both the intracell and intercell interference power are reduced by the spreading gain of the code. Moreover, interference in spread-spectrum systems can be further reduced via multiuser detection or interference cancellation. More details on these different techniques for spectrum sharing and their performance analysis will be given in Chapters 13 and 14. The design trade-offs associated with spectrum sharing are very complex, and the decision of which technique is best for a given system and operating environment is never straightforward.

Efficient cellular system designs are *interference limited* – that is, the interference dominates the noise floor, since otherwise more users could be added to the system. As a result, any technique to reduce interference in cellular systems leads directly to an increase in system capacity and performance. Some methods for interference reduction in use today or proposed for future systems include cell sectorization, directional and smart antennas, multiuser detection, and dynamic resource allocation. Details of these techniques will be given in Chapter 15.

The first-generation (1G) cellular systems in the United States, called the Advance Mobile Phone Service (AMPS), used FDMA with 30-kHz FM-modulated voice channels. The FCC initially allocated 40 MHz of spectrum to this system, which was increased to 50 MHz shortly after service introduction to support more users. This total bandwidth was divided into two 25-MHz bands, one for mobile-to-base station channels and the other for base station-to-mobile channels. The FCC divided these channels into two sets that were assigned to two different service providers in each city to encourage competition. A similar system, the Total Access Communication System (TACS), emerged in Europe. AMPS was deployed worldwide in the 1980s and remains the only cellular service in some areas, including rural parts of the United States.

Many of the first-generation cellular systems in Europe were incompatible, and the Europeans quickly converged on a uniform standard for second-generation (2G) digital systems called GSM.[2] The GSM standard uses a combination of TDMA and slow frequency hopping with frequency-shift keying for the voice modulation. In contrast, the standards activities in the United States surrounding the second generation of digital cellular provoked a raging debate on spectrum-sharing techniques, resulting in several incompatible standards [9; 10; 11]. In particular, there are two standards in the 900-MHz cellular frequency band: IS-136,[3] which uses a combination of TDMA and FDMA and phase-shift keyed modulation; and IS-95, which uses direct-sequence CDMA with phase-shift keyed modulation and coding [12; 13]. The spectrum for digital cellular in the 2-GHz PCS (personal communication system) frequency band was auctioned off, so service providers could use any standard for their purchased spectrum. The end result has been three different digital cellular standards for this frequency band: IS-136, IS-95, and the European GSM standard. The digital cellular standard in Japan is similar to IS-136 but in a different frequency band, and the GSM system in Europe is at a different frequency than the GSM systems in the United States. This proliferation of incompatible standards in the United States and internationally makes it impossible to roam between systems nationwide or globally without a multimode phone and/or multiple phones (and phone numbers).

All of the second-generation digital cellular standards have been enhanced to support high-rate packet data services [14]. GSM systems provide data rates of up to 140 kbps by aggregating all timeslots together for a single user. This enhancement is called GPRS. A

---

[2] The acronym GSM originally stood for Groupe Spéciale Mobile, the name of the European charter establishing the GSM standard. As GSM systems proliferated around the world, the underlying acronym meaning was changed to Global Systems for Mobile Communications.

[3] IS-136 is the evolution of the older IS-54 standard and subsumes it.

more fundamental enhancement, Enhanced Data rates for GSM Evolution (EDGE), further increases data rates up to 384 kbps by using a high-level modulation format combined with coding. This modulation is more sensitive to fading effects, and EDGE uses adaptive techniques to mitigate that problem. Specifically, EDGE defines nine different modulation and coding combinations, each optimized to a different value of received SNR (signal-to-noise ratio). The received SNR is measured at the receiver and fed back to the transmitter, and the best modulation and coding combination for this SNR value is used. The IS-136 systems also use GPRS and EDGE enhancements to support data rates up to 384 kbps. The IS-95 systems support data rates up to 115 kbps by aggregating spreading functions [15].

The third-generation (3G) cellular systems are based on a wideband CDMA standard developed under the auspices of the International Telecommunications Union (ITU) [14]. The standard, called International Mobile Telecommunications 2000 (IMT-2000), provides different data rates depending on mobility and location, from 384 kbps for pedestrian use to 144 kbps for vehicular use to 2 Mbps for indoor office use. The 3G standard is incompatible with 2G systems, so service providers must invest in a new infrastructure before they can provide 3G service. The first 3G systems were deployed in Japan. One reason that 3G services came out first in Japan was the Japanese allocation process for 3G spectrum, which was awarded without much up-front cost. The 3G spectrum in both Europe and the United States is allocated based on auctioning, thereby requiring a huge initial investment for any company wishing to provide 3G service. European companies collectively paid over $100 billion (American) in their 3G spectrum auctions. There has been much controversy over the 3G auction process in Europe, with companies charging that the nature of the auctions caused enormous overbidding and that it will thus be difficult if not impossible to reap a profit on this spectrum. A few of the companies decided to write off their investment in 3G spectrum and not pursue system buildout. In fact, 3G systems have not grown as anticipated in Europe, and it appears that data enhancements to 2G systems may suffice to satisfy user demands at least for some time. However, the 2G spectrum in Europe is severely overcrowded, so either users will eventually migrate to 3G or regulations will change so that 3G bandwidth can be used for 2G services (this is not currently allowed in Europe). Development of 3G in the United States has lagged far behind that in Europe. The available U.S. 3G spectrum is only about half that available in Europe. Due to wrangling about which parts of the spectrum will be used, 3G spectral auctions in the United States have not yet taken place. However, U.S. regulations do allow the 1G and 2G spectrum to be used for 3G, and this flexibility has facilitated a more gradual rollout and investment than the more restrictive 3G requirements in Europe. It appears that delaying the 3G spectral auctions in the United States has allowed the FCC and U.S. service providers to learn from the mistakes and successes in Europe and Japan.

1.4.2 Cordless Phones

Cordless telephones first appeared in the late 1970s and have experienced spectacular growth ever since. Many U.S. homes today have only cordless phones, which can be a safety risk because these phones – in contrast to their wired counterparts – don't work in a power outage. Cordless phones were originally designed to provide a low-cost, low-mobility wireless connection to the PSTN, that is, a short wireless link to replace the cord connecting a telephone base unit and its handset. Since cordless phones compete with wired handsets, their voice quality must be similar. Initial cordless phones had poor voice quality and were quickly discarded by users. The first cordless systems allowed only one phone handset to connect to each base unit, and coverage was limited to a few rooms of a house or office. This is still the main premise behind cordless telephones in the United States today, although some base units now support multiple handsets and coverage has improved. In Europe and Asia, digital cordless phone systems have evolved to provide coverage over much wider areas, both in and away from home, and are similar in many ways to cellular telephone systems.

The base units of cordless phones connect to the PSTN in the exact same manner as a landline phone, and thus they impose no added complexity on the telephone network. The movement of these cordless handsets is extremely limited: a handset must remain within transmission range of its base unit. There is no coordination with other cordless phone systems, so a high density of these systems in a small area (e.g., an apartment building) can result in significant interference between systems. For this reason cordless phones today have multiple voice channels and scan between these channels to find the one with minimal interference. Many cordless phones use spread-spectrum techniques to reduce interference from other cordless phone systems and from such other systems as baby monitors and wireless LANs.

In Europe and Asia, the second-generation digital cordless phone (CT-2, for "cordless telephone, second generation") has an extended range of use beyond a single residence or office. Within a home these systems operate as conventional cordless phones. To extend the range beyond the home, base stations (also called phone-points or telepoints) are mounted in places where people congregate: shopping malls, busy streets, train stations, and airports. Cordless phones registered with the telepoint provider can place calls whenever they are in range of a telepoint. Calls cannot be received from the telepoint because the network has no routing support for mobile users, although some CT-2 handsets have built-in pagers to compensate for this deficiency. These systems also do not hand off calls if a user moves between different telepoints, so a user must remain within range of the telepoint where his call was initiated for the duration of the call. Telepoint service was introduced twice in the United Kingdom and failed both times, but these systems grew rapidly in Hong Kong and Singapore through the mid 1990s. This rapid growth deteriorated quickly after the first few years as cellular phone operators cut prices to compete with telepoint service. The main complaint about telepoint service was the incomplete radio coverage and lack of handoff. Since cellular systems avoid these problems, as long as prices were competitive there was little reason for people to use telepoint services. Most of these services have now disappeared.

Another evolution of the cordless telephone designed primarily for office buildings is the European Digital Enhanced Cordless Telecommunications (DECT) system. The main function of DECT is to provide local mobility support for users in an in-building private branch exchange (PBX). In DECT systems, base units are mounted throughout a building, and each base station is attached through a controller to the PBX of the building. Handsets communicate with the nearest base station in the building, and calls are handed off as a user walks between base stations. DECT can also ring handsets from the closest base station. The DECT standard also supports telepoint services, although this application has not received much attention (due most likely to the failure of CT-2 services). There are currently about 7 million DECT users in Europe, but the standard has not yet spread to other countries.

A more advanced cordless telephone system that emerged in Japan is the Personal Handy-phone System (PHS). The PHS system is quite similar to a cellular system, with widespread base station deployment supporting handoff and call routing between base stations. With these capabilities PHS does not suffer from the main limitations of the CT-2 system. Initially PHS systems enjoyed one of the fastest growth rates ever for a new technology. In 1997, two years after its introduction, PHS subscribers peaked at about 7 million users, but its popularity then started to decline in response to sharp price cutting by cellular providers. In 2005 there were about 4 million subscribers, attracted by the flat-rate service and relatively high speeds (128 kbps) for data. PHS operators are trying to push data rates up to 1 Mbps, which cellular providers cannot yet compete with. The main difference between a PHS system and a cellular system is that PHS cannot support call handoff at vehicle speeds. This deficiency is mainly due to the dynamic channel allocation procedure used in PHS. Dynamic channel allocation greatly increases the number of handsets that can be serviced by a single base station and their corresponding data rates, thereby lowering the system cost, but it also complicates the handoff procedure. Given the sustained popularity of PHS, it is unlikely to go the same route as CT-2 any time soon, especially if much higher data rates become available. However, it is clear from the recent history of cordless phone systems that to extend the range of these systems beyond the home requires either matching or exceeding the functionality of cellular systems or a significantly reduced cost.

1.4.3 Wireless Local Area Networks

Wireless LANs support high-speed data transmissions within a small region (e.g., a campus or small building) as users move from place to place. Wireless devices that access these LANs are typically stationary or moving at pedestrian speeds. All wireless LAN standards in the United States operate in unlicensed frequency bands. The primary unlicensed bands are the ISM bands at 900 MHz, 2.4 GHz, and 5.8 GHz and the Unlicensed National Information Infrastructure (U-NII) band at 5 GHz. In the ISM bands, unlicensed users are secondary users and so must cope with interference from primary users when such users are active. There are no primary users in the U-NII band. An FCC license is not required to operate in either the ISM or U-NII bands. However, this advantage is a double-edged sword, since other unlicensed systems operate in these bands for the same reason, which can cause a great deal of interference between systems. The interference problem is mitigated by setting a limit on the power per unit bandwidth for unlicensed systems. Wireless LANs can have either a star architecture, with wireless access points or hubs placed throughout the coverage region, or a peer-to-peer architecture, where the wireless terminals self-configure into a network.

Dozens of wireless LAN companies and products appeared in the early 1990s to capitalize on the "pent-up demand" for high-speed wireless data. These first-generation wireless LANs were based on proprietary and incompatible protocols. Most operated within the 26-MHz spectrum of the 900-MHz ISM band using direct-sequence spread spectrum, with data rates on the order of 1–2 Mbps. Both star and peer-to-peer architectures were used. The lack of standardization for these products led to high development costs, low-volume production, and small markets for each individual product. Of these original products only a handful were even mildly successful. Only one of the first-generation wireless LANs, Motorola's Altair, operated outside the 900-MHz band. This system, operating in the licensed

18-GHz band, had data rates on the order of 6 Mbps. However, performance of Altair was hampered by the high cost of components and the increased path loss at 18 GHz, and Altair was discontinued within a few years of its release.

The second-generation wireless LANs in the United States operate with 83.5 MHz of spectrum in the 2.4-GHz ISM band. A wireless LAN standard for this frequency band, the IEEE 802.11b standard, was developed to avoid some of the problems with the proprietary first-generation systems. The standard specifies direct-sequence spread spectrum with data rates of around 1.6 Mbps (raw data rates of 11 Mbps) and a range of approximately 100 m. The network architecture can be either star or peer-to-peer, although the peer-to-peer feature is rarely used. Many companies developed products based on the 802.11b standard, and after slow initial growth the popularity of 802.11b wireless LANs has expanded considerably. Many laptops come with integrated 802.11b wireless LAN cards. Companies and universities have installed 802.11b base stations throughout their locations, and many coffee houses, airports, and hotels offer wireless access, often for free, to increase their appeal.

Two additional standards in the 802.11 family were developed to provide higher data rates than 802.11b. The IEEE 802.11a wireless LAN standard operates with 300 MHz of spectrum in the 5-GHz U-NII band. The 802.11a standard is based on multicarrier modulation and provides 54-Mbps data rates at a range of about 30 m. Because 802.11a has much more bandwidth and consequently many more channels than 802.11b, it can support more users at higher data rates. There was some initial concern that 802.11a systems would be significantly more expensive than 802.11b systems, but in fact they quickly became quite competitive in price. The other standard, 802.11g, has the same design and data rates as 802.11a, but it operates in the the 2.4-GHz band with a range of about 50 m. Many wireless LAN cards and access points support all three standards to avoid incompatibilities.

In Europe, wireless LAN development revolves around the HIPERLAN (high-performance radio LAN) standards. The HIPERLAN/2 standard is similar to the IEEE 802.11a wireless LAN standard. In particular, it has a similar link layer design and also operates in a 5-GHz frequency band similar to the U-NII band. Hence it has the same maximum date rate of 54 Mbps and the same range of approximately 30 m as 802.11a. It differs from 802.11a in its access protocol and its built-in Quality-of-Service (QoS) support.

1.4.4 Wide Area Wireless Data Services

Wide area wireless data services provide wireless data to high-mobility users over a large coverage area. In these systems a given geographical region is serviced by base stations mounted on towers, rooftops, or mountains. The base stations can be connected to a backbone wired network or form a multihop ad hoc wireless network.

Initial wide area wireless data services had very low data rates, below 10 kbps, which gradually increased to 20 kbps. There were two main players providing this service: Motient and Bell South Mobile Data (formerly RAM Mobile Data). Metricom provided a similar service using architecture that consisted of a large network of small, inexpensive base stations with small coverage areas. The increased efficiency of the small coverage areas allowed for higher data rates in Metricom, 76 kbps, than in the other wide area wireless data systems. However, the high infrastructure cost for Metricom eventually forced it into bankruptcy, and the system was shut down. Some of the infrastructure was bought and is operating in a few areas as Ricochet.

The cellular digital packet data (CDPD) system is a wide area wireless data service over-layed on the analog cellular telephone network. CDPD shares the FDMA voice channels of the analog systems, since many of these channels are idle owing to the growth of digital cellular. The CDPD service provides packet data transmission at rates of 19.2 kbps and is available throughout the United States. However, since newer generations of cellular systems also provide data services and at higher data rates, CDPD is mostly being replaced by these newer services. Thus, wide area wireless data services have not been very successful, although emerging systems that offer broadband access may have more appeal.

1.4.5 Broadband Wireless Access

Broadband wireless access provides high-rate wireless communications between a fixed access point and multiple terminals. These systems were initially proposed to support interactive video service to the home, but the application emphasis then shifted to providing both high-speed data access (tens of Mbps) to the Internet and the World Wide Web as well as high-speed data networks for homes and businesses. In the United States, two frequency bands were set aside for these systems: part of the 28-GHz spectrum for local distribution systems (local multipoint distribution service, LMDS) and a band in the 2-GHz spectrum for metropolitan distribution service (multichannel multipoint distribution services, MMDS). LMDS represents a quick means for new service providers to enter the already stiff competition among wireless and wireline broadband service providers [5, Chap. 2.3]. MMDS is a television and telecommunication delivery system with transmission ranges of 30–50 km [5, Chap. 11.11]. MMDS has the capability of delivering more than a hundred digital video TV channels along with telephony and access to the Internet. MMDS will compete mainly with existing cable and satellite systems. Europe is developing a standard similar to MMDS called Hiperaccess.

WiMax is an emerging broadband wireless technology based on the IEEE 802.16 standard [16; 17]. The core 802.16 specification is a standard for broadband wireless access systems operating at radio frequencies between 2 GHz and 11 GHz for non–line-of-sight operation, and between 10 GHz and 66 GHz for line-of-sight operation. Data rates of around 40 Mbps will be available for fixed users and 15 Mbps for mobile users, with a range of several kilometers. Many manufacturers of laptops and PDAs (personal digital assistants) are planning to incorporate WiMax once it becomes available to satisfy demand for constant Internet access and email exchange from any location. WiMax will compete with wireless LANs, 3G cellular services, and possibly wireline services like cable and DSL (digital subscriber line). The ability of WiMax to challenge or supplant these systems will depend on its relative performance and cost, which remain to be seen.

1.4.6 Paging Systems

Paging systems broadcast a short paging message simultaneously from many tall base stations or satellites transmitting at very high power (hundreds of watts to kilowatts). Systems with terrestrial transmitters are typically localized to a particular geographic area, such as a city or metropolitan region, while geosynchronous satellite transmitters provide national or international coverage. In both types of systems, no location management or routing functions are needed because the paging message is broadcast over the entire coverage area. The high complexity and power of the paging transmitters allows low-complexity, low-power, pocket paging receivers capable of long usage times from small and lightweight batteries. In addition, the high transmit power allows paging signals to easily penetrate building walls. Paging service also costs less than cellular service, both for the initial device and for the monthly usage charge, although this price advantage has declined considerably in recent years as cellular prices dropped. The low cost, small and lightweight handsets, long battery life, and ability of paging devices to work almost anywhere indoors or outdoors are the main reasons for their appeal.

Early radio paging systems were analog 1-bit messages signaling a user that someone was trying to reach him or her. These systems required callback over a landline telephone to obtain the phone number of the paging party. The system evolved to allow a short digital message, including a phone number and brief text, to be sent to the pagee as well. Radio paging systems were initially extremely successful, with a peak of 50 million subscribers in the United States alone. However, their popularity began to wane with the widespread penetration and competitive cost of cellular telephone systems. Eventually the competition from cellular phones forced paging systems to provide new capabilities. Some implemented "answer-back" capability (i.e., two-way communication). This required a major change in design of the pager because now it needed to transmit signals in addition to receiving them, and the transmission distance to a satellite or base station can be very large. Paging companies also teamed up with palmtop computer makers to incorporate paging functions into these devices [18]. Despite these developments, the market for paging devices has shrunk considerably, although there is still a niche market among doctors and other professionals who must be reachable anywhere.

1.4.7 Satellite Networks

Commercial satellite systems are another major component of the wireless communications infrastructure [2; 3]. Geosynchronous systems include Inmarsat and OmniTRACS. The former is geared mainly for analog voice transmission from remote locations. For example, it is commonly used by journalists to provide live reporting from war zones. The first-generation Inmarsat-A system was designed for large (1-m parabolic dish antenna) and rather expensive terminals. Newer generations of Inmarsats use digital techniques to enable smaller, less expensive terminals, about the size of a briefcase. Qualcomm's OmniTRACS provides two-way communications as well as location positioning. The system is used primarily for alphanumeric messaging and location tracking of trucking fleets. There are several major difficulties in providing voice and data services over geosynchronous satellites. It takes a great deal of power to reach these satellites, so handsets are typically large and bulky. In addition, there is a large round-trip propagation delay; this delay is quite noticeable in two-way voice communication. Geosynchronous satellites also have fairly low data rates of less than 10 kbps. For these reasons, lower-orbit LEO satellites were thought to be a better match for voice and data communications.

LEO systems require approximately 30–80 satellites to provide global coverage, and plans for deploying such constellations were widespread in the late 1990s. One of the most ambitious of these systems, the Iridium constellation, was launched at that time. However, the cost to build, launch, and maintain these satellites is much higher than costs for terrestrial base stations. Although these LEO systems can certainly complement terrestrial systems in low-population areas and are also appealing to travelers desiring just one handset and phone number for global roaming, the growth and diminished cost of cellular prevented many ambitious plans for widespread LEO voice and data systems from materializing. Iridium was eventually forced into bankruptcy and disbanded, and most of the other systems were never launched. An exception to these failures was the Globalstar LEO system, which currently provides voice and data services over a wide coverage area at data rates under 10 kbps. Some of the Iridium satellites are still operational as well.

The most appealing use for a satellite system is the broadcast of video and audio over large geographic regions. Approximately 1 in 8 U.S. homes have direct broadcast satellite service, and satellite radio is also emerging as a popular service. Similar audio and video satellite broadcasting services are widespread in Europe. Satellites are best tailored for broadcasting, since they cover a wide area and are not compromised by an initial propagation delay. Moreover, the cost of the system can be amortized over many years and many users, making the service quite competitive in cost with that of terrestrial entertainment broadcasting systems.

1.4.8 Low-Cost, Low-Power Radios: Bluetooth and ZigBee

As radios decrease their cost and power consumption, it becomes feasible to embed them into more types of electronic devices, which can be used to create smart homes, sensor networks, and other compelling applications. Two radios have emerged to support this trend: Bluetooth and ZigBee.

Bluetooth[4] radios provide short-range connections between wireless devices along with rudimentary networking capabilities. The Bluetooth standard is based on a tiny microchip incorporating a radio transceiver that is built into digital devices. The transceiver takes the place of a connecting cable for devices such as cell phones, laptop and palmtop computers, portable printers and projectors, and network access points. Bluetooth is mainly for short-range communications – for example, from a laptop to a nearby printer or from a cell phone to a wireless headset. Its normal range of operation is 10 m (at 1-mW transmit power), and this range can be increased to 100 m by increasing the transmit power to 100 mW. The system operates in the unlicensed 2.4-GHz frequency band, so it can be used worldwide without any licensing issues. The Bluetooth standard provides one asynchronous data channel at 723.2 kbps. In this mode, also known as Asynchronous Connection-Less (ACL), there is a reverse channel with a data rate of 57.6 kbps. The specification also allows up to three synchronous channels each at a rate of 64 kbps. This mode, also known as Synchronous Connection Oriented (SCO), is mainly used for voice applications such as headsets but can also be used for data. These different modes result in an aggregate bit rate of approximately 1 Mbps. Routing of the asynchronous data is done via a packet switching protocol based on frequency hopping at 1600 hops per second. There is also a circuit switching protocol for the synchronous data.

Bluetooth uses frequency hopping for multiple access with a carrier spacing of 1 MHz. Typically, up to eighty different frequencies are used for a total bandwidth of 80 MHz. At any given time, the bandwidth available is 1 MHz, with a maximum of eight devices sharing the bandwidth. Different logical channels (different hopping sequences) can simultaneously

---

[4] The Bluetooth standard is named after Harald I Bluetooth, the king of Denmark between 940 and 985 A.D. who united Denmark and Norway. Bluetooth proposes to unite devices via radio connections, hence the inspiration for its name.

share the same 80-MHz bandwidth. Collisions will occur when devices in different piconets that are on different logical channels happen to use the same hop frequency at the same time. As the number of piconets in an area increases, the number of collisions increases and performance degrades.

The Bluetooth standard was developed jointly by 3 Com, Ericsson, Intel, IBM, Lucent, Microsoft, Motorola, Nokia, and Toshiba. The standard has now been adopted by over 1,300 manufacturers, and many consumer electronic products incorporate Bluetooth. These include wireless headsets for cell phones, wireless USB or RS232 connectors, wireless PCMCIA cards, and wireless settop boxes.

The ZigBee[5] radio specification is designed for lower cost and power consumption than Bluetooth [19]. Its specification is based on the IEEE 802.15.4 standard. The radio operates in the same ISM band as Bluetooth and is capable of connecting 255 devices per network. The specification supports data rates of up to 250 kbps at a range of up to 30 m. These data rates are slower than Bluetooth, but in exchange the radio consumes significantly less power with a larger transmission range. The goal of ZigBee is to provide radio operation for months or years without recharging, thereby targeting applications such as sensor networks and inventory tags.

1.4.9 Ultrawideband Radios

Ultrawideband (UWB) radios are extremely wideband radios with very high potential data rates [20; 21]. The concept of ultrawideband communications actually originated with Marconi's spark-gap transmitter, which occupied a very wide bandwidth. However, since only a single low-rate user could occupy the spectrum, wideband communications was abandoned in favor of more efficient communication techniques. Renewed interest in wideband communications was spurred by the FCC's decision in 2002 to allow operation of UWB devices underlayed beneath existing users in the 3.1–10.6-GHz range. The underlay in theory interferes with all systems in that frequency range, including critical safety and military systems, unlicensed systems such as 802.11 wireless LANs and Bluetooth, and cellular systems where operators paid billions of dollars for dedicated spectrum use. The FCC's ruling was quite controversial given the vested interest in interference-free spectrum of these users. To minimize the impact of UWB on primary band users, the FCC put in place severe transmit power restrictions. This requires UWB devices to be within close proximity of their intended receiver.

Ultrawideband radios come with unique advantages that have long been appreciated by the radar and communications communities. Their wideband nature provides precise ranging capabilities. Moreover, the available UWB bandwidth has the potential for extremely high data rates. Finally, power restrictions dictate that the devices be small and with low power consumption.

Initial UWB systems used ultrashort pulses with simple amplitude or position modulation. Multipath can significantly degrade performance of such systems, and proposals to mitigate the effects of multipath include equalization and multicarrier modulation. Precise

---

[5] ZigBee takes its name from the dance that honey bees use to communicate information about newly found food sources to other members of the colony.

and rapid synchronization is also a big challenge for these systems. Although many technical challenges remain, the appeal of UWB technology has sparked great interest both commercially and in the research community to address these issues.

1.5 The Wireless Spectrum

1.5.1 Methods for Spectrum Allocation

Most countries have government agencies responsible for allocating and controlling use of the radio spectrum. In the United States, spectrum is allocated by the Federal Communications Commission (FCC) for commercial use and by the Office of Spectral Management (OSM) for military use. Commercial spectral allocation is governed in Europe by the European Telecommunications Standards Institute (ETSI) and globally by the International Telecommunications Union (ITU). Governments decide how much spectrum to allocate between commercial and military use, and this decision is dynamic depending on need. Historically the FCC allocated spectral blocks for specific uses and assigned licenses to use these blocks to specific groups or companies. For example, in the 1980s the FCC allocated frequencies in the 800-MHz band for analog cellular phone service and provided spectral licenses to two operators in each geographical area based on a number of criteria. The FCC and regulatory bodies in other countries still allocate spectral blocks for specific purposes, but these blocks are now commonly assigned through spectral auctions to the highest bidder. Some argue that this market-based method is the fairest and most efficient way for governments to allocate the limited spectral resource and that it provides significant revenue to the government besides; others believe that this mechanism stifles innovation, limits competition, and hurts technology adoption. Specifically, the high cost of spectrum dictates that only large companies or conglomerates can purchase it. Moreover, the large investment required to obtain spectrum can delay the ability to invest in infrastructure for system rollout and results in high initial prices for the end user. The 3G spectral auctions in Europe, in which several companies ultimately defaulted, have provided ammunition to the opponents of spectral auctions.

In addition to spectral auctions, spectrum can be set aside in specific frequency bands that are free to use without a license according to a specific set of etiquette rules. The rules may correspond to a specific communications standard, power levels, and so forth. The purpose of these *unlicensed bands* is to encourage innovation and low-cost implementation. Many extremely successful wireless systems operate in unlicensed bands, including wireless LANs, Bluetooth, and cordless phones. A major difficulty of unlicensed bands is that they can be killed by their own success. If many unlicensed devices in the same band are used in close proximity then they interfere with each other, which can make the band unusable.

Underlay systems are another alternative for allocating spectrum. An underlay system operates as a secondary user in a frequency band with other primary users. Operation of secondary users is typically restricted so that primary users experience minimal interference. This is usually accomplished by restricting the power per hertz of the secondary users. UWB is an example of an underlay system, as are unlicensed systems in the ISM frequency bands. Such underlay systems can be extremely controversial given the complexity of characterizing how interference affects the primary users. Yet the trend toward spectrum allocation for underlays appears to be accelerating, which is mainly due to the scarcity of available spectrum for new systems and applications.

Satellite systems cover large areas spanning many countries and sometimes the globe. For wireless systems that span multiple countries, spectrum is allocated by the International Telecommunications Union Radio Communications group (ITU-R). The standards arm of this body, ITU-T, adopts telecommunication standards for global systems that must interoperate across national boundaries.

There is some movement within regulatory bodies worldwide to change the way spectrum is allocated. Indeed, the basic mechanisms for spectral allocation have not changed much since the inception of regulatory bodies in the early to mid-1900s, although spectral auctions and underlay systems are relatively new. The goal of changing spectrum allocation policy is to take advantage of the technological advances in radios to make spectrum allocation more efficient and flexible. One compelling idea is the notion of a smart or cognitive radio. This type of radio could sense its spectral environment to determine dimensions in time, space, and frequency where it would not cause interference to other users even at moderate to high transmit powers. If such radios could operate over a wide frequency band, this would open up huge amounts of new bandwidth and tremendous opportunities for new wireless systems and applications. However, many technology and policy hurdles must be overcome to allow such a radical change in spectrum allocation.

1.5.2 Spectrum Allocations for Existing Systems

Most wireless applications reside in the radio spectrum between 30 MHz and 40 GHz. These frequencies are natural for wireless systems because they are not affected by the earth's curvature, require only moderately sized antennas, and can penetrate the ionosphere. Note that the required antenna size for good reception is inversely proportional to the signal frequency, so moving systems to a higher frequency allows for more compact antennas. However, received signal power with nondirectional antennas is proportional to the inverse of frequency squared, so it is harder to cover large distances with higher-frequency signals.

As discussed in the previous section, spectrum is allocated either in licensed bands (which regulatory bodies assign to specific operators) or in unlicensed bands (which can be used by any system subject to certain operational requirements). Table 1.1 shows the licensed spectrum allocated to major commercial wireless systems in the United States today; there are similar allocations in Europe and Asia.

Note that digital TV is slated for the same bands as broadcast TV, so all broadcasters must eventually switch from analog to digital transmission. Also, the 3G broadband wireless spectrum is currently allocated to UHF television stations 60–69 but is slated to be reallocated. Both 1G analog and 2G digital cellular services occupy the same cellular band at 800 MHz, and the cellular service providers decide how much of the band to allocate between digital and analog service.

Unlicensed spectrum is allocated by the governing body within a given country. Often different countries try to match their frequency allocation for unlicensed use so that technology developed for that spectrum is compatible worldwide. Table 1.2 shows the unlicensed spectrum allocations in the United States.

Table 1.1: Licensed U.S. spectrum allocations

Service/system	Frequency span
AM radio	535–1605 kHz
FM radio	88–108 MHz
Broadcast TV (channels 2–6)	54–88 MHz
Broadcast TV (channels 7–13)	174–216 MHz
Broadcast TV (UHF)	470–806 MHz
Broadband wireless	746–764 MHz, 776–794 MHz
3G broadband wireless	1.7–1.85 MHz, 2.5–2.69 MHz
1G and 2G digital cellular phones	806–902 MHz
Personal communication systems (2G cell phones)	1.85–1.99 GHz
Wireless communications service	2.305–2.32 GHz, 2.345–2.36 GHz
Satellite digital radio	2.32–2.325 GHz
Multichannel multipoint distribution service (MMDS)	2.15–2.68 GHz
Digital broadcast satellite (satellite TV)	12.2–12.7 GHz
Local multipoint distribution service (LMDS)	27.5–29.5 GHz, 31–31.3 GHz
Fixed wireless services	38.6–40 GHz

Table 1.2: Unlicensed U.S. spectrum allocations

Band	Frequency
ISM band I (cordless phones, 1G WLANs)	902–928 MHz
ISM band II (Bluetooth, 802.11b and 802.11g WLANs)	2.4–2.4835 GHz
ISM band III (wireless PBX)	5.725–5.85 GHz
U-NII band I (indoor systems, 802.11a WLANs)	5.15–5.25 GHz
U-NII band II (short-range outdoor systems, 80211a WLANs)	5.25–5.35 GHz
U-NII band III (long-range outdoor systems, 80211a WLANs)	5.725–5.825 GHz

ISM Band I has licensed users transmitting at high power who interfere with the unlicensed users. Therefore, the requirements for unlicensed use of this band is highly restrictive and performance is somewhat poor. The U-NII bands have a total of 300 MHz of spectrum in three separate 100-MHz bands, with slightly different power restrictions on each band. Many unlicensed systems operate in these bands.

1.6 Standards

Communication systems that interact with each other require standardization. Standards are typically decided on by national or international committees; in the United States this role is played by the Telecommunications Industry Association (TIA). These committees adopt standards that are developed by other organizations. The IEEE is the major player for standards development in the United States, while ETSI plays this role in Europe. Both groups follow a lengthy process for standards development that entails input from companies and other interested parties as well as a long and detailed review process. The standards process is a large time investment, but companies participate because incorporating their ideas into the standard gives them an advantage in developing the resulting system. In general, standards do not include all the details on all aspects of the system design. This allows companies to innovate and differentiate their products from other standardized systems. The main goal of standardization is enabling systems to interoperate.

In addition to ensuring interoperability, standards also allow economies of scale and pressure prices lower. For example, wireless LANs typically operate in the unlicensed spectral bands, so they are not required to follow a specific standard. The first generation of wireless LANs were not standardized and so specialized components were needed for many systems, leading to excessively high cost that, when coupled with poor performance, led to limited adoption. This experience resulted in a strong push to standardize the next wireless LAN generation, which yielded the highly successful IEEE 802.11 family of standards.

There are, of course, disadvantages to standardization. The standards process is not perfect, as company participants often have their own agenda, which does not always coincide with the best technology or the best interests of consumers. In addition, the standards process must be completed at some point, after which it becomes more difficult to add new innovations and improvements to an existing standard. Finally, the standards process can become quite politicized. This happened with the second generation of cellular phones in the United States and ultimately led to the adoption of two different standards, a bit of an oxymoron. The resulting delays and technology split put the United States well behind Europe in the development of 2G cellular systems. Despite its flaws, standardization is clearly a necessary and often beneficial component of wireless system design and operation. However, it would benefit everyone in the wireless technology industry if some of the problems in the standardization process could be mitigated.

PROBLEMS

1-1. As storage capability increases, we can store larger and larger amounts of data on smaller and smaller storage devices. Indeed, we can envision microscopic computer chips storing terraflops of data. Suppose this data is to be transfered over some distance. Discuss the pros and cons of putting a large number of these storage devices in a truck and driving them to their destination rather than sending the data electronically.

1-2. Describe two technical advantages and disadvantages of wireless systems that use bursty data transmission rather than continuous data transmission.

1-3. Fiber optic cable typically exhibits a probability of bit error of $P_b = 10^{-12}$. A form of wireless modulation, DPSK, has $P_b = 1/2\bar{\gamma}$ in some wireless channels, where $\bar{\gamma}$ is the average SNR. Find the average SNR required to achieve the same P_b in the wireless channel as in the fiber optic cable. Because of this extremely high required SNR, wireless channels typically have P_b much larger than 10^{-12}.

1-4. Find the round-trip delay of data sent between a satellite and the earth for LEO, MEO, and GEO satellites assuming the speed of light is $3 \cdot 10^8$ m/s. If the maximum acceptable delay for a voice system is 30 ms, which of these satellite systems would be acceptable for two-way voice communication?

1-5. What applications might significantly increase the demand for wireless data?

1-6. This problem illustrates some of the economic issues facing service providers as they migrate away from voice-only systems to mixed-media systems. Suppose you are a service provider with 120 kHz of bandwidth that you must allocate between voice and data users. The voice users require 20 kHz of bandwidth and the data users require 60 kHz of bandwidth. So, for example, you could allocate all of your bandwidth to voice users, resulting in six voice channels, or you could divide the bandwidth into one data channel and three voice channels, etc. Suppose further that this is a time-division system with timeslots of duration T. All voice and data call requests come in at the beginning of a timeslot, and both types of calls last T seconds. There are six independent voice users in the system: each of these users requests a voice channel with probability .8 and pays $.20 if his call is processed. There are two independent data users in the system: each of these users requests a data channel with probability .5 and pays $1 if his call is processed. How should you allocate your bandwidth to maximize your expected revenue?

1-7. Describe three disadvantages of using a wireless LAN instead of a wired LAN. For what applications will these disadvantages be outweighed by the benefits of wireless mobility? For what applications will the disadvantages override the advantages?

1-8. Cellular systems have migrated to smaller cells in order to increase system capacity. Name at least three design issues that are complicated by this trend.

1-9. Why does minimizing the reuse distance maximize the spectral efficiency of a cellular system?

1-10. This problem demonstrates the capacity increase associated with a decrease in cell size. Consider a square city of 100 square kilometers. Suppose you design a cellular system for this city with square cells, where every cell (regardless of cell size) has 100 channels and so can support 100 active users. (In practice, the number of users that can be supported per cell is mostly independent of cell size as long as the propagation model and power scale appropriately.)

    (a) What is the total number of active users that your system can support for a cell size of 1 km^2?

    (b) What cell size would you use if your system had to support 250,000 active users?

Now we consider some financial implications based on the fact that users do not talk continuously. Assume that Friday from 5–6 P.M. is the busiest hour for cell-phone users. During this time, the average user places a single call, and this call lasts two minutes. Your system should be designed so that subscribers need tolerate no greater than a 2% blocking probability during this peak hour. (Blocking probability is computed using the Erlang B model: $P_b = (A^C/C!)/(\sum_{k=0}^{C} A^k/k!)$, where C is the number of channels and $A = U\mu H$ for U the number of users, μ the average number of call requests per unit time per user, and H the average duration of a call [5, Chap. 3.6].

    (c) How many total subscribers can be supported in the macrocell system (1-km^2 cells) and in the microcell system (with cell size from part (b))?

    (d) If a base station costs $500,000, what are the base station costs for each system?

(e) If the monthly user fee in each system is $50, what will be the monthly revenue in each case? How long will it take to recoup the infrastructure (base station) cost for each system?

1-11. How many CDPD data lines are needed to achieve the same data rate as the average rate of WiMax?

REFERENCES

[1]   V. H. McDonald, "The cellular concept," *Bell System Tech. J.,* pp. 15–49, January 1979.
[2]   F. Abrishamkar and Z. Siveski, "PCS global mobile satellites," *IEEE Commun. Mag.,* pp. 132–6, September 1996.
[3]   R. Ananasso and F. D. Priscoli, "The role of satellites in personal communication services," *IEEE J. Sel. Areas Commun.,* pp. 180–96, February 1995.
[4]   D. C. Cox, "Wireless personal communications: What is it?" *IEEE Pers. Commun. Mag.,* pp. 20–35, April 1995.
[5]   T. S. Rappaport, *Wireless Communications – Principles and Practice,* 2nd ed., Prentice-Hall, Englewood Cliffs, NJ, 2001.
[6]   W. Stallings, *Wireless Communications and Networks,* 2nd ed., Prentice-Hall, Englewood Cliffs, NJ, 2005.
[7]   K. Pahlavan and P. Krishnamurthy, *Principles of Wireless Networks: A Unified Approach,* Prentice-Hall, Englewood Cliffs, NJ, 2002.
[8]   A. J. Goldsmith and L. J. Greenstein, "A measurement-based model for predicting coverage areas of urban microcells," *IEEE J. Sel. Areas Commun.,* pp. 1013–23, September 1993.
[9]   K. S. Gilhousen, I. M. Jacobs, R. Padovani, A. J. Viterbi, L. A. Weaver, Jr., and C. E. Wheatley III, "On the capacity of a cellular CDMA system," *IEEE Trans. Veh. Tech.,* pp. 303–12, May 1991.
[10] K. Rath and J. Uddenfeldt, "Capacity of digital cellular TDMA systems," *IEEE Trans. Veh. Tech.,* pp. 323–32, May 1991.
[11] Q. Hardy, "Are claims hope or hype?" *Wall Street Journal,* p. A1, September 6, 1996.
[12] A. Mehrotra, *Cellular Radio: Analog and Digital Systems,* Artech House, Norwood, MA, 1994.
[13] J. E. Padgett, C. G. Gunther, and T. Hattori, "Overview of wireless personal communications," *IEEE Commun. Mag.,* pp. 28–41, January 1995.
[14] J. D. Vriendt, P. Lainé, C. Lerouge, and X. Xu, "Mobile network evolution: A revolution on the move," *IEEE Commun. Mag.,* pp. 104–11, April 2002.
[15] P. Bender, P. J. Black, M. S. Grob, R. Padovani, N. T. Sindhushayana, and A. J. Viterbi, "CDMA/HDR: A bandwidth efficient high speed wireless data service for nomadic users," *IEEE Commun. Mag.,* pp. 70–7, July 2000.
[16] S. J. Vaughan-Nichols, "Achieving wireless broadband with WiMax," *IEEE Computer,* pp. 10–13, June 2004.
[17] S. M. Cherry, "WiMax and Wi-Fi: Separate and Unequal," *IEEE Spectrum,* p. 16, March 2004.
[18] S. Schiesel, "Paging allies focus strategy on the Internet," *New York Times,* April 19, 1999.
[19] I. Poole, "What exactly is . . . ZigBee?" *IEEE Commun. Eng.,* pp. 44–5, August/September 2004.
[20] L. Yang and G. B. Giannakis, "Ultra-wideband communications: An idea whose time has come," *IEEE Signal Proc. Mag.,* pp. 26–54, November 2004.
[21] D. Porcino and W. Hirt, "Ultra-wideband radio technology: Potential and challenges ahead," *IEEE Commun. Mag.,* pp. 66–74, July 2003.

Path Loss and Shadowing

The wireless radio channel poses a severe challenge as a medium for reliable high-speed communication. Not only is it susceptible to noise, interference, and other channel impediments, but these impediments change over time in unpredictable ways as a result of user movement and environment dynamics. In this chapter we characterize the variation in received signal power over distance due to path loss and shadowing. Path loss is caused by dissipation of the power radiated by the transmitter as well as by effects of the propagation channel. Path-loss models generally assume that path loss is the same at a given transmit–receive distance (assuming that the path-loss model does not include shadowing effects). Shadowing is caused by obstacles between the transmitter and receiver that attenuate signal power through absorption, reflection, scattering, and diffraction. When the attenuation is strong, the signal is blocked. Received power variation due to path loss occurs over long distances (100–1000 m), whereas variation due to shadowing occurs over distances that are proportional to the length of the obstructing object (10–100 m in outdoor environments and less in indoor environments). Since variations in received power due to path loss and shadowing occur over relatively large distances, these variations are sometimes referred to as *large-scale propagation effects*. Chapter 3 will deal with received power variations due to the constructive and destructive addition of multipath signal components. These variations occur over very short distances, on the order of the signal wavelength, and so are sometimes referred to as *small-scale propagation effects*. Figure 2.1 illustrates the ratio of the received-to-transmit power in decibels[1] (dB) versus log distance for the combined effects of path loss, shadowing, and multipath.

After a brief introduction to propagation and description of our signal model, we present the simplest model for signal propagation: free-space path loss. A signal propagating between two points with no attenuation or reflection follows the free-space propagation law. We then describe ray-tracing propagation models. These models are used to approximate wave propagation according to Maxwell's equations; they are accurate when the number of multipath components is small and the physical environment is known. Ray-tracing models depend heavily on the geometry and dielectric properties of the region through which the signal propagates. We also describe empirical models with parameters based on measurements for both indoor and outdoor channels, and we present a simple generic model with few

---

[1] The decibel value of x is $10 \log_{10} x$.

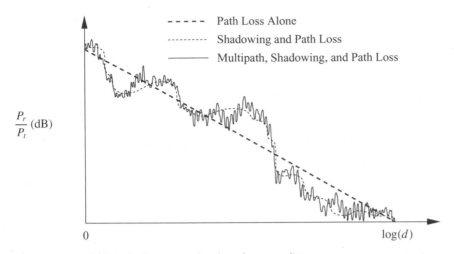

Figure 2.1: Path loss, shadowing, and multipath versus distance.

parameters that captures the primary impact of path loss in system analysis. A log-normal model for shadowing based on a large number of shadowing objects is also given. If the number of multipath components is large or if the geometry and dielectric properties of the propagation environment are unknown, then statistical multipath models must be used. These statistical multipath models will be described in Chapter 3.

Although this chapter gives a brief overview of channel models for path loss and shadowing, comprehensive coverage of channel and propagation models at different frequencies of interest merits a book in its own right, and in fact there are several excellent texts on this topic [1; 2]. Channel models for specialized systems (e.g., multiple antenna and ultrawideband systems) can be found in [3; 4].

2.1 Radio Wave Propagation

The initial understanding of radio wave propagation goes back to the pioneering work of James Clerk Maxwell, who in 1864 formulated a theory of electromagnetic propagation that predicted the existence of radio waves. In 1887, the physical existence of these waves was demonstrated by Heinrich Hertz. However, Hertz saw no practical use for radio waves, reasoning that since audio frequencies were low, where propagation was poor, radio waves could never carry voice. The work of Maxwell and Hertz initiated the field of radio communications. In 1894 Oliver Lodge used these principles to build the first wireless communication system, though its transmission distance was limited to 150 meters. By 1897 the entrepreneur Guglielmo Marconi had managed to send a radio signal from the Isle of Wight to a tugboat eighteen miles away, and in 1901 Marconi's wireless system could traverse the Atlantic ocean. These early systems used telegraph signals for communicating information. The first transmission of voice and music was made by Reginald Fessenden in 1906 using a form of amplitude modulation, which circumvented the propagation limitations at low frequencies observed by Hertz by translating signals to a higher frequency, as is done in all wireless systems today.

Electromagnetic waves propagate through environments where they are reflected, scattered, and diffracted by walls, terrain, buildings, and other objects. The ultimate details of this propagation can be obtained by solving Maxwell's equations with boundary conditions that express the physical characteristics of these obstructing objects. This requires the calculation of the radar cross-section (RCS) of large and complex structures. Since these calculations are difficult and since the necessary parameters are often not available, approximations have been developed to characterize signal propagation without resorting to Maxwell's equations.

The most common approximations use ray-tracing techniques. These techniques approximate the propagation of electromagnetic waves by representing the wavefronts as simple particles: the model determines the reflection and refraction effects on the wavefront but ignores the more complex scattering phenomenon predicted by Maxwell's coupled differential equations. The simplest ray-tracing model is the two-ray model, which accurately describes signal propagation when there is one direct path between the transmitter and receiver and one reflected path. The reflected path typically bounces off the ground, and the two-ray model is a good approximation for propagation along highways or rural roads and over water. We will analyze the two-ray model in detail, as well as more complex models with additional reflected, scattered, or diffracted components. Many propagation environments are not accurately characterized by ray-tracing models. In these cases it is common to develop analytical models based on empirical measurements, and we will discuss several of the most common of these empirical models.

Often the complexity and variability of the radio channel make it difficult to obtain an accurate deterministic channel model. For these cases, statistical models are often used. The attenuation caused by signal path obstructions such as buildings or other objects is typically characterized statistically, as described in Section 2.7. Statistical models are also used to characterize the constructive and destructive interference for a large number of multipath components, as described in Chapter 3. Statistical models are most accurate in environments with fairly regular geometries and uniform dielectric properties. Indoor environments tend to be less regular than outdoor environments, since the geometric and dielectric characteristics change dramatically depending on whether the indoor environment is an open factory, cubicled office, or metal machine shop. For these environments computer-aided modeling tools are available to predict signal propagation characteristics [5].

2.2 Transmit and Receive Signal Models

Our models are developed mainly for signals in the UHF and SHF bands, from .3–3 GHz and 3–30 GHz (respectively). This range of frequencies is quite favorable for wireless system operation because of its propagation characteristics and relatively small required antenna size. We assume the transmission distances on the earth are small enough not to be affected by the earth's curvature.

All the transmitted and received signals that we consider are real. That is because modulators are built using oscillators that generate real sinusoids (not complex exponentials). Though we model communication channels using a complex frequency response for analytical simplicity, in fact the channel simply introduces an amplitude and phase change at each frequency of the transmitted signal so that the received signal is also real. Real modulated and demodulated signals are often represented as the real part of a complex signal in order to facilitate analysis. This model gives rise to the equivalent lowpass representation of bandpass signals, which we use for our transmitted and received signals. More details on the equivalent lowpass representation of bandpass signals and systems can be found in Appendix A.

We model the transmitted signal as

$$s(t) = \text{Re}\{u(t)e^{j2\pi f_c t}\}$$
$$= \text{Re}\{u(t)\}\cos(2\pi f_c t) - \text{Im}\{u(t)\}\sin(2\pi f_c t)$$
$$= s_I(t)\cos(2\pi f_c t) - s_Q(t)\sin(2\pi f_c t), \qquad (2.1)$$

where $u(t) = s_I(t) + js_Q(t)$ is a complex baseband signal with in-phase component $s_I(t) = \text{Re}\{u(t)\}$, quadrature component $s_Q(t) = \text{Im}\{u(t)\}$, bandwidth B_u, and power P_u. The signal $u(t)$ is called the *complex envelope* or *equivalent lowpass signal* of $s(t)$. We call $u(t)$ the complex envelope of $s(t)$ because the magnitude of $u(t)$ is the magnitude of $s(t)$. The phase of $u(t)$ includes any carrier phase offset. The equivalent lowpass representation of bandpass signals with bandwidth $B \ll f_c$ allows signal manipulation via $u(t)$ irrespective of the carrier frequency. The power in the transmitted signal $s(t)$ is $P_t = P_u/2$.

The received signal will have a similar form plus an additional noise component:

$$r(t) = \text{Re}\{v(t)e^{j2\pi f_c t}\} + n(t), \qquad (2.2)$$

where $n(t)$ is the noise process introduced by the channel and the equivalent lowpass signal $v(t)$ depends on the channel through which $s(t)$ propagates. In particular, as discussed in Appendix A, if $s(t)$ is transmitted through a time-invariant channel then $v(t) = u(t) * c(t)$, where $c(t)$ is the equivalent lowpass channel impulse response for the channel. Time-varying channels will be treated in Chapter 3.

The received signal in (2.2) consists of two terms, the first term corresponding to the transmitted signal after propagation through the channel, and the second term corresponding to the noise added by the channel. The signal-to-noise power ratio (SNR) of the received signal is defined as the power of the first term divided by the power of the second term. In this chapter (and in Chapter 3) we will neglect the random noise component $n(t)$ in our analysis, since these chapters focus on signal propagation, which is not affected by noise. However, noise will play a prominent role in the capacity and performance of wireless systems studied in later chapters.

When the transmitter or receiver is moving, the received signal will have a Doppler shift of $f_D = v\cos\theta/\lambda$ associated with it, where θ is the arrival angle of the received signal relative to the direction of motion, v is the receiver velocity toward the transmitter in the direction of motion, and $\lambda = c/f_c$ is the signal wavelength ($c = 3 \cdot 10^8$ m/s is the speed of light). The geometry associated with the Doppler shift is shown in Figure 2.2. The Doppler shift results from the fact that transmitter or receiver movement over a short time interval Δt causes a slight change in distance $\Delta d = v\Delta t\cos\theta$ that the transmitted signal needs to travel to the receiver. The phase change due to this path-length difference is $\Delta\phi = 2\pi v\Delta t\cos\theta/\lambda$. The Doppler frequency is then obtained from the relationship between signal frequency and phase:

$$f_D = \frac{1}{2\pi}\frac{\Delta\phi}{\Delta t} = v\cos\frac{\theta}{\lambda}. \qquad (2.3)$$

If the receiver is moving toward the transmitter (i.e., if $-\pi/2 \le \theta \le \pi/2$), then the Doppler frequency is positive; otherwise, it is negative. We will ignore the Doppler term in the free-space and ray-tracing models of this chapter, since for typical vehicle speeds (75 km/hr) and

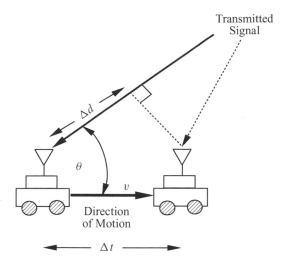

Figure 2.2: Geometry associated with Doppler shift.

frequencies (about 1 GHz) it is on the order of 100 Hz [6]. However, we will include Doppler effects in Chapter 3 on statistical fading models.

Suppose $s(t)$ of power P_t is transmitted through a given channel with corresponding received signal $r(t)$ of power P_r, where P_r is averaged over any random variations due to shadowing. We define the *linear path loss* of the channel as the ratio of transmit power to receive power:

$$P_L = \frac{P_t}{P_r}. \tag{2.4}$$

We define the *path loss* of the channel as the value of the linear path loss in decibels or, equivalently, the difference in dB between the transmitted and received signal power:

$$P_L \text{ dB} = 10 \log_{10} \frac{P_t}{P_r} \text{ dB}. \tag{2.5}$$

In general, the dB path loss is a nonnegative number; the channel does not contain active elements, and thus it can only attenuate the signal. The dB *path gain* is defined as the negative of the dB path loss: $P_G = -P_L = 10 \log_{10}(P_r/P_t)$ dB, which is generally a negative number. With shadowing, the received power is random owing to random blockage from objects, as we discuss in Section 2.7.

2.3 Free-Space Path Loss

Consider a signal transmitted through free space to a receiver located at distance d from the transmitter. Assume there are no obstructions between the transmitter and receiver and that the signal propagates along a straight line between the two. The channel model associated with this transmission is called a line-of-sight (LOS) channel, and the corresponding received signal is called the LOS signal or ray. Free-space path loss introduces a complex scale factor [1], resulting in the received signal

$$r(t) = \text{Re}\left\{ \frac{\lambda \sqrt{G_l} e^{-j2\pi d/\lambda}}{4\pi d} u(t) e^{j2\pi f_c t} \right\}, \tag{2.6}$$

where $\sqrt{G_l}$ is the product of the transmit and receive antenna field radiation patterns in the LOS direction. The phase shift $e^{-j2\pi d/\lambda}$ is due to the distance d that the wave travels.

The power in the transmitted signal $s(t)$ is P_t, so the ratio of received to transmitted power from (2.6) is

$$\frac{P_r}{P_t} = \left[\frac{\sqrt{G_l}\lambda}{4\pi d}\right]^2. \tag{2.7}$$

Thus, the received signal power falls off in inverse proportion to the square of the distance d between the transmit and receive antennas. We will see in the next section that, for other signal propagation models, the received signal power falls off more quickly relative to this distance. The received signal power is also proportional to the square of the signal wavelength, so as the carrier frequency increases the received power decreases. This dependence of received power on the signal wavelength λ is due to the effective area of the receive antenna [1]. However, directional antennas can be designed so that receive power is an increasing function of frequency for highly directional links [7]. The received power can be expressed in dBm as[2]

$$P_r \text{ dBm} = P_t \text{ dBm} + 10\log_{10}(G_l) + 20\log_{10}(\lambda) - 20\log_{10}(4\pi) - 20\log_{10}(d). \tag{2.8}$$

Free-space path loss is defined as the path loss of the free-space model:

$$P_L \text{ dB} = 10\log_{10}\frac{P_t}{P_r} = -10\log_{10}\frac{G_l\lambda^2}{(4\pi d)^2}. \tag{2.9}$$

The *free-space path gain* is thus

$$P_G = -P_L = 10\log_{10}\frac{G_l\lambda^2}{(4\pi d)^2}. \tag{2.10}$$

EXAMPLE 2.1: Consider an indoor wireless LAN with $f_c = 900$ MHz, cells of radius 100 m, and nondirectional antennas. Under the free-space path loss model, what transmit power is required at the access point in order for all terminals within the cell to receive a minimum power of $10\,\mu$W? How does this change if the system frequency is 5 GHz?

Solution: We must find a transmit power such that the terminals at the cell boundary receive the minimum required power. We obtain a formula for the required transmit power by inverting (2.7) to obtain:

$$P_t = P_r\left[\frac{4\pi d}{\sqrt{G_l}\lambda}\right]^2.$$

Substituting in $G_l = 1$ (omnidirectional antennas), $\lambda = c/f_c = .33$ m, $d = 10$ m, and $P_r = 10\,\mu$W yields $P_t = 1.45$ W $= 1.61$ dBW. (Recall that P watts equals $10\log_{10}(P)$ dbW, dB relative to 1 W). At 5 GHz only $\lambda = .06$ changes, so $P_t = 43.9$ kW $= 16.42$ dBW.

---

[2] The dBm value of x is its dB value relative to a milliwatt: $10\log_{10}(x/.001)$.

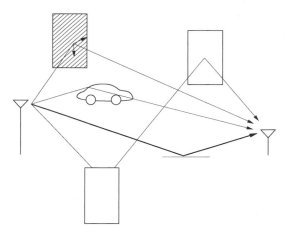

Figure 2.3: Reflected, diffracted, and scattered wave components.

2.4 Ray Tracing

In a typical urban or indoor environment, a radio signal transmitted from a fixed source will encounter multiple objects in the environment that produce reflected, diffracted, or scattered copies of the transmitted signal, as shown in Figure 2.3. These additional copies of the transmitted signal, known as *multipath signal components,* can be attenuated in power, delayed in time, and shifted in phase and/or frequency with respect to the LOS signal path at the receiver. The multipath and transmitted signal are summed together at the receiver, which often produces distortion in the received signal relative to the transmitted signal.

In ray tracing we assume a finite number of reflectors with known location and dielectric properties. The details of the multipath propagation can then be solved using Maxwell's equations with appropriate boundary conditions. However, the computational complexity of this solution makes it impractical as a general modeling tool. Ray-tracing techniques approximate the propagation of electromagnetic waves by representing the wavefronts as simple particles. Thus, the effects of reflection, diffraction, and scattering on the wavefront are approximated using simple geometric equations instead of Maxwell's more complex wave equations. The error of the ray-tracing approximation is smallest when the receiver is many wavelengths from the nearest scatterer and when all the scatterers are large relative to a wavelength and fairly smooth. Comparison of the ray-tracing method with empirical data shows it to accurately model received signal power in rural areas [8], along city streets when both the transmitter and receiver are close to the ground [8; 9; 10], and in indoor environments with appropriately adjusted diffraction coefficients [11]. Propagation effects besides received power variations, such as the delay spread of the multipath, are not always well captured with ray-tracing techniques [12].

If the transmitter, receiver, and reflectors are all immobile, then the characteristics of the multiple received signal paths are fixed. However, if the source or receiver are moving then the characteristics of the multiple paths vary with time. These time variations are deterministic when the number, location, and characteristics of the reflectors are known over time. Otherwise, statistical models must be used. Similarly, if the number of reflectors is large or if the reflector surfaces are not smooth, then we must use statistical approximations to characterize the received signal. We will discuss statistical fading models for propagation effects in Chapter 3. Hybrid models, which combine ray tracing and statistical fading, can also be found in the literature [13; 14]; however, we will not describe them here.

The most general ray-tracing model includes all attenuated, diffracted, and scattered multipath components. This model uses all of the geometrical and dielectric properties of the objects surrounding the transmitter and receiver. Computer programs based on ray tracing – such as Lucent's Wireless Systems Engineering software (WiSE), Wireless Valley's Site-Planner®, and Marconi's Planet® *EV* – are widely used for system planning in both indoor and outdoor environments. In these programs, computer graphics are combined with aerial photographs (outdoor channels) or architectural drawings (indoor channels) to obtain a three-dimensional geometric picture of the environment [5].

The following sections describe several ray-tracing models of increasing complexity. We start with a simple two-ray model that predicts signal variation resulting from a ground reflection interfering with the LOS path. This model characterizes signal propagation in isolated areas with few reflectors, such as rural roads or highways. It is not typically a good model for indoor environments. We then present a ten-ray reflection model that predicts the variation of a signal propagating along a straight street or hallway. Finally, we describe a general model that predicts signal propagation for any propagation environment. The two-ray model requires information on antenna heights only; the ten-ray model requires antenna height and street/hallway width; and the general model requires these parameters as well as detailed information about the geometry and dielectric properties of the reflectors, diffractors, and scatterers in the environment.

2.4.1 Two-Ray Model

The two-ray model is used when a single ground reflection dominates the multipath effect, as illustrated in Figure 2.4. The received signal consists of two components: the LOS component or ray, which is just the transmitted signal propagating through free space, and a reflected component or ray, which is the transmitted signal reflected off the ground.

The received LOS ray is given by the free-space propagation loss formula (2.6). The reflected ray is shown in Figure 2.4 by the segments x and x'. If we ignore the effect of surface wave attenuation[3] then, by superposition, the received signal for the two-ray model is

$$r_{2\text{-ray}}(t) = \mathrm{Re}\left\{ \frac{\lambda}{4\pi}\left[\frac{\sqrt{G_l}u(t)e^{-j2\pi l/\lambda}}{l} + \frac{R\sqrt{G_r}u(t-\tau)e^{-j2\pi(x+x')/\lambda}}{x+x'} \right]e^{j2\pi f_c t} \right\}, \quad (2.11)$$

where $\tau = (x + x' - l)/c$ is the time delay of the ground reflection relative to the LOS ray, $\sqrt{G_l} = \sqrt{G_a G_b}$ is the product of the transmit and receive antenna field radiation patterns in the LOS direction, R is the ground reflection coefficient, and $\sqrt{G_r} = \sqrt{G_c G_d}$ is the product of the transmit and receive antenna field radiation patterns corresponding to the rays of length x and x', respectively. The *delay spread* of the two-ray model equals the delay between the LOS ray and the reflected ray: $(x + x' - l)/c$.

If the transmitted signal is narrowband relative to the delay spread ($\tau \ll B_u^{-1}$) then $u(t) \approx u(t-\tau)$. With this approximation, the received power of the two-ray model for narrowband transmission is

---

[3] This is a valid approximation for antennas located more than a few wavelengths from the ground.

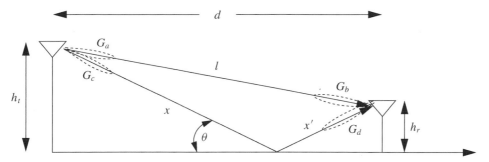

Figure 2.4: Two-ray model.

$$P_r = P_t \left[\frac{\lambda}{4\pi}\right]^2 \left|\frac{\sqrt{G_l}}{l} + \frac{R\sqrt{G_r}e^{-j\Delta\phi}}{x + x'}\right|^2, \tag{2.12}$$

where $\Delta\phi = 2\pi(x + x' - l)/\lambda$ is the phase difference between the two received signal components. Equation (2.12) has been shown [15] to agree closely with empirical data. If d denotes the horizontal separation of the antennas, h_t the transmitter height, and h_r the receiver height, then from geometry

$$x + x' - l = \sqrt{(h_t + h_r)^2 + d^2} - \sqrt{(h_t - h_r)^2 + d^2}. \tag{2.13}$$

When d is very large compared to $h_t + h_r$, we can use a Taylor series approximation in (2.13) to get

$$\Delta\phi = \frac{2\pi(x + x' - l)}{\lambda} \approx \frac{4\pi h_t h_r}{\lambda d}. \tag{2.14}$$

The ground reflection coefficient is given by

$$R = \frac{\sin\theta - Z}{\sin\theta + Z}, \tag{2.15}$$

[6; 16], where

$$Z = \begin{cases} \sqrt{\varepsilon_r - \cos^2\theta}/\varepsilon_r & \text{for vertical polarization,} \\ \sqrt{\varepsilon_r - \cos^2\theta} & \text{for horizontal polarization,} \end{cases} \tag{2.16}$$

and ε_r is the dielectric constant of the ground. For earth or road surfaces this dielectric constant is approximately that of a pure dielectric (for which ε_r is real with a value of about 15).

We see from Figure 2.4 and (2.15) that, for asymptotically large d, $x + x' \approx l \approx d$, $\theta \approx 0$, $G_l \approx G_r$, and $R \approx -1$. Substituting these approximations into (2.12) yields that, in this asymptotic limit, the received signal power is approximately

$$P_r \approx \left[\frac{\lambda\sqrt{G_l}}{4\pi d}\right]^2 \left[\frac{4\pi h_t h_r}{\lambda d}\right]^2 P_t = \left[\frac{\sqrt{G_l}h_t h_r}{d^2}\right]^2 P_t, \tag{2.17}$$

or, in dB,

$$P_r \text{ dBm} = P_t \text{ dBm} + 10\log_{10}(G_l) + 20\log_{10}(h_t h_r) - 40\log_{10}(d). \tag{2.18}$$

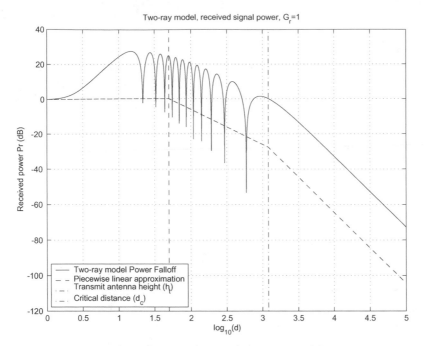

Figure 2.5: Received power versus distance for two-ray model.

Thus, in the limit of asymptotically large d, the received power falls off inversely with the fourth power of d and is independent of the wavelength λ. The received signal becomes independent of λ because directional antenna arrays have a received power that does not necessarily decrease with frequency, and combining the direct path and reflected signal effectively forms an antenna array. A plot of (2.12) as a function of distance is shown in Figure 2.5 for $f = 900$ MHz, $R = -1$, $h_t = 50$ m, $h_r = 2$ m, $G_l = 1$, $G_r = 1$, and transmit power normalized so that the plot starts at 0 dBm. This plot can be separated into three segments. For small distances $(d < h_t)$ the two rays add constructively and the path loss is slowing increasing. More precisely, it is proportional to $1/(d^2 + h_t^2)$ since, at these small distances, the distance between the transmitter and receiver is $l = \sqrt{d^2 + (h_t - h_r)^2}$; thus $1/l^2 \approx 1/(d^2 + h_t^2)$ for $h_t \gg h_r$, which is typically the case. For distances greater than h_t and up to a certain critical distance d_c, the wave experiences constructive and destructive interference of the two rays, resulting in a wave pattern with a sequence of maxima and minima. These maxima and minima are also referred to as small-scale or multipath fading, discussed in more detail in the next chapter. At the critical distance d_c the final maximum is reached, after which the signal power falls off proportionally with d^{-4}. This rapid falloff with distance is due to the fact that, for $d > d_c$, the signal components only combine destructively and so are out of phase by at least π. An approximation for d_c can be obtained by setting $\Delta\phi = \pi$ in (2.14), obtaining $d_c = 4h_t h_r / \lambda$, which is also shown in the figure. The power falloff with distance in the two-ray model can be approximated by averaging out its local maxima and minima. This results in a piecewise linear model with three segments, which is also shown in Figure 2.5 slightly offset from the actual power falloff curve for illustration purposes. In the first segment, power falloff is constant and proportional to $1/h_t^2$; for distances between

h_t and d_c, power falls off at -20 dB/decade; and at distances greater than d_c, power falls off at -40 dB/decade.

The critical distance d_c can be used for system design. For example, if propagation in a cellular system obeys the two-ray model then the critical distance would be a natural size for the cell radius, since the path loss associated with interference outside the cell would be much larger than path loss for desired signals inside the cell. However, setting the cell radius to d_c could result in very large cells, as illustrated in Figure 2.5 and in the next example. Since smaller cells are more desirable – both to increase capacity and reduce transmit power – cell radii are typically much smaller than d_c. Thus, with a two-ray propagation model, power falloff within these relatively small cells goes as distance squared. Moreover, propagation in cellular systems rarely follows a two-ray model, since cancellation by reflected rays rarely occurs in all directions.

EXAMPLE 2.2: Determine the critical distance for the two-ray model in an urban microcell ($h_t = 10$ m, $h_r = 3$ m) and an indoor microcell ($h_t = 3$ m, $h_r = 2$ m) for $f_c = 2$ GHz.

Solution: $d_c = 4h_t h_r/\lambda = 800$ m for the urban microcell and 160 m for the indoor system. A cell radius of 800 m in an urban microcell system is a bit large: urban microcells today are on the order of 100 m to maintain large capacity. However, if we did use a cell size of 800 m under these system parameters, then signal power would fall off as d^2 inside the cell, while interference from neighboring cells would fall off as d^4 and thus would be greatly reduced. Similarly, 160 m is quite large for the cell radius of an indoor system, as there would typically be many walls the signal would have to penetrate for an indoor cell radius of that size. Hence an indoor system would typically have a smaller cell radius: on the order of 10–20 m.

2.4.2 Ten-Ray Model (Dielectric Canyon)

We now examine a model for urban microcells developed by Amitay [9]. This model assumes rectilinear streets[4] with buildings along both sides of the street as well as transmitter and receiver antenna heights that are close to street level. The building-lined streets act as a dielectric canyon to the propagating signal. Theoretically, an infinite number of rays can be reflected off the building fronts to arrive at the receiver; in addition, rays may also be back-reflected from buildings behind the transmitter or receiver. However, since some of the signal energy is dissipated with each reflection, signal paths corresponding to more than three reflections can generally be ignored. When the street layout is relatively straight, back reflections are usually negligible also. Experimental data show that a model of ten reflection rays closely approximates signal propagation through the dielectric canyon [9]. The ten rays incorporate all paths with one, two, or three reflections: specifically, there is the line-of-sight (LOS) path and also the ground-reflected (GR), single-wall (SW) reflected, double-wall (DW) reflected, triple-wall (TW) reflected, wall–ground (WG) reflected, and ground–wall (GW) reflected paths. There are two of each type of wall-reflected path, one for each side of the street. An overhead view of the ten-ray model is shown in Figure 2.6.

---

[4] A rectilinear city is flat and has linear streets that intersect at $90°$ angles, as in midtown Manhattan.

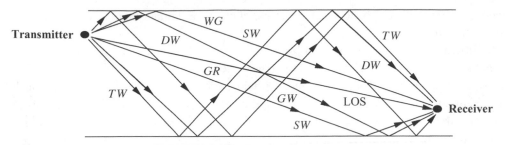

Figure 2.6: Overhead view of the ten-ray model.

For the ten-ray model, the received signal is given by

$$r_{10\text{-ray}}(t) = \text{Re}\left\{\frac{\lambda}{4\pi}\left[\frac{\sqrt{G_l}u(t)e^{-j2\pi l/\lambda}}{l} + \sum_{i=1}^{9}\frac{R_i\sqrt{G_{x_i}}u(t-\tau_i)e^{-j2\pi x_i/\lambda}}{x_i}\right]e^{j2\pi f_c t}\right\}, \quad (2.19)$$

where x_i denotes the path length of the ith reflected ray, $\tau_i = (x_i - l)/c$, and $\sqrt{G_{x_i}}$ is the product of the transmit and receive antenna gains corresponding to the ith ray. For each reflection path, the coefficient R_i is either a single reflection coefficient given by (2.15) or, if the path corresponds to multiple reflections, the product of the reflection coefficients corresponding to each reflection. The dielectric constants used in (2.19) are approximately the same as the ground dielectric, so $\varepsilon_r = 15$ is used for all the calculations of R_i. If we again assume a narrowband model such that $u(t) \approx u(t - \tau_i)$ for all i, then the received power corresponding to (2.19) is

$$P_r = P_t\left[\frac{\lambda}{4\pi}\right]^2\left|\frac{\sqrt{G_l}}{l} + \sum_{i=1}^{9}\frac{R_i\sqrt{G_{x_i}}e^{-j\Delta\phi_i}}{x_i}\right|^2, \quad (2.20)$$

where $\Delta\phi_i = 2\pi(x_i - l)/\lambda$.

Power falloff with distance in both the ten-ray model (2.20) and urban empirical data [15; 17; 18] for transmit antennas both above and below the building skyline is typically proportional to d^{-2}, even at relatively large distances. Moreover, the falloff exponent is relatively insensitive to the transmitter height. This falloff with distance squared is due to the dominance of the multipath rays, which decay as d^{-2}, over the combination of the LOS and ground-reflected rays (two-ray model), which decays as d^{-4}. Other empirical studies [19; 20; 21] have obtained power falloff with distance proportional to $d^{-\gamma}$, where γ lies anywhere between 2 and 6.

2.4.3 General Ray Tracing

General ray tracing (GRT) can be used to predict field strength and delay spread for any building configuration and antenna placement [22; 23; 24]. For this model, the building database (height, location, and dielectric properties) and the transmitter and receiver locations relative to the buildings must be specified exactly. Since this information is site specific, the GRT model is not used to obtain general theories about system performance and layout; rather, it explains the basic mechanism of urban propagation and can be used to obtain delay and

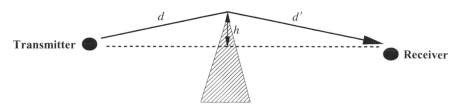

Figure 2.7: Knife-edge diffraction.

signal strength information for a particular transmitter and receiver configuration in a given environment.

The GRT method uses geometrical optics to trace the propagation of the LOS and reflected signal components as well as signal components from building diffraction and diffuse scattering. There is no limit to the number of multipath components at a given receiver location: the strength of each component is derived explicitly based on the building locations and dielectric properties. In general, the LOS and reflected paths provide the dominant components of the received signal, since diffraction and scattering losses are high. However, in regions close to scattering or diffracting surfaces – which may be blocked from the LOS and reflecting rays – these other multipath components may dominate.

The propagation model for the LOS and reflected paths was outlined in the previous section. Diffraction occurs when the transmitted signal "bends around" an object in its path to the receiver, as shown in Figure 2.7. Diffraction results from many phenomena, including the curved surface of the earth, hilly or irregular terrain, building edges, or obstructions blocking the LOS path between the transmitter and receiver [1; 5; 16]. Diffraction can be accurately characterized using the geometrical theory of diffraction (GTD) [25], but the complexity of this approach has precluded its use in wireless channel modeling. Wedge diffraction simplifies the GTD by assuming the diffracting object is a wedge rather than a more general shape. This model has been used to characterize the mechanism by which signals are diffracted around street corners, which can result in path loss exceeding 100 dB for some incident angles on the wedge [11; 24; 26; 27]. Although wedge diffraction simplifies the GTD, it still requires a numerical solution for path loss [25; 28] and thus is not commonly used. Diffraction is most commonly modeled by the *Fresnel knife-edge diffraction model* because of its simplicity. The geometry of this model is shown in Figure 2.7, where the diffracting object is assumed to be asymptotically thin, which is not generally the case for hills, rough terrain, or wedge diffractors. In particular, this model does not consider diffractor parameters such as polarization, conductivity, and surface roughness, which can lead to inaccuracies [26]. The diffracted signal of Figure 2.7 travels a distance $d + d'$, resulting in a phase shift of $\phi = 2\pi(d + d')/\lambda$. The geometry of Figure 2.7 indicates that, for h small relative to d and d', the signal must travel an additional distance relative to the LOS path of approximately

$$\Delta d \approx \frac{h^2}{2} \frac{d + d'}{dd'};$$

the corresponding phase shift relative to the LOS path is approximately

$$\Delta\phi = \frac{2\pi\Delta d}{\lambda} \approx \frac{\pi}{2}v^2, \qquad (2.21)$$

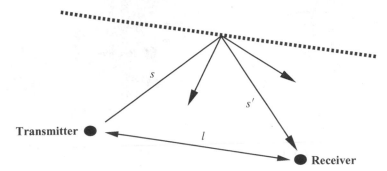

Figure 2.8: Scattering.

where

$$v = h\sqrt{\frac{2(d + d')}{\lambda dd'}} \qquad (2.22)$$

is called the *Fresnel–Kirchhoff diffraction parameter*. The path loss associated with knife-edge diffraction is generally a function of v. However, computing this diffraction path loss is fairly complex, requiring the use of Huygens's principle, Fresnel zones, and the complex Fresnel integral [1]. Moreover, the resulting diffraction loss cannot generally be found in closed form. Approximations for knife-edge diffraction path loss (in dB) relative to LOS path loss are given by Lee [16, Chap. 2] as

$$L(v) \text{ dB} = \begin{cases} 20\log_{10}[.5 - .62v] & -0.8 \le v < 0, \\ 20\log_{10}[.5e^{-.95v}] & 0 \le v < 1, \\ 20\log_{10}\left[.4 - \sqrt{.1184 - (.38 - .1v)^2}\right] & 1 \le v \le 2.4, \\ 20\log_{10}[.225/v] & v > 2.4. \end{cases} \qquad (2.23)$$

A similar approximation can be found in [29]. The knife-edge diffraction model yields the following formula for the received diffracted signal:

$$r(t) = \text{Re}\left\{L(v)\sqrt{G_d}u(t - \tau)e^{-j2\pi(d+d')/\lambda}e^{j2\pi f_c t}\right\}, \qquad (2.24)$$

where $\sqrt{G_d}$ is the antenna gain and $\tau = \Delta d/c$ is the delay associated with the defracted ray relative to the LOS path.

In addition to diffracted rays, there may also be rays that are diffracted multiple times, or rays that are both reflected and diffracted. Models exist for including all possible permutations of reflection and diffraction [30]; however, the attenuation of the corresponding signal components is generally so large that these components are negligible relative to the noise. Diffraction models can also be specialized to a given environment. For example, a model for diffraction from rooftops and buildings in cellular systems was developed by Walfisch and Bertoni in [31].

A scattered ray, shown in Figure 2.8 by the segments s and s', has a path loss proportional to the product of s and s'. This multiplicative dependence is due to the additional spreading loss that the ray experiences after scattering. The received signal due to a scattered ray is given by the bistatic radar equation [32]:

$$r(t) = \mathrm{Re}\left\{ u(t-\tau)\frac{\lambda\sqrt{G_s}\sigma e^{-j2\pi(s+s')/\lambda}}{(4\pi)^{3/2}ss'}e^{j2\pi f_c t}\right\},\tag{2.25}$$

where $\tau = (s + s' - l)/c$ is the delay associated with the scattered ray; σ (in square meters) is the radar cross-section of the scattering object, which depends on the roughness, size, and shape of the scatterer; and $\sqrt{G_s}$ is the antenna gain. The model assumes that the signal propagates from the transmitter to the scatterer based on free-space propagation and is then re-radiated by the scatterer with transmit power equal to σ times the received power at the scatterer. From (2.25), the path loss associated with scattering is

$$\begin{aligned}P_r \text{ dBm} = {}& P_t \text{ dBm} + 10\log_{10}(G_s) + 20\log_{10}(\lambda) + 10\log_{10}(\sigma) \\ & - 30\log(4\pi) - 20\log_{10}(s) - 20\log_{10}(s').\end{aligned}\tag{2.26}$$

Empirical values of $10\log_{10}\sigma$ were determined in [33] for different buildings in several cities. Results from this study indicate that $10\log_{10}\sigma$ in dBm^2 ranges from $-4.5\,\text{dBm}^2$ to $55.7\,\text{dBm}^2$, where dBm^2 denotes the dB value of the σ measurement with respect to one square meter.

The received signal is determined from the superposition of all the components due to the multiple rays. Thus, if we have a LOS ray, N_r reflected rays, N_d diffracted rays, and N_s diffusely scattered rays, the total received signal is

$$\begin{aligned}r_{\text{total}}(t) = \mathrm{Re}\Bigg\{ & \left[\frac{\lambda}{4\pi}\right]\Bigg[\frac{\sqrt{G_l}u(t)e^{j2\pi l/\lambda}}{l} + \sum_{i=1}^{N_r}\frac{R_{x_i}\sqrt{G_{x_i}}u(t-\tau_i)e^{-j2\pi x_i/\lambda}}{x_i} \\ & + \sum_{j=1}^{N_d}\frac{4\pi}{\lambda}L_j(v)\sqrt{G_{d_j}}u(t-\tau_j)e^{-j2\pi(d_j+d_j')/\lambda} \\ & + \sum_{k=1}^{N_s}\frac{\sqrt{G_{s_k}}\sigma_k u(t-\tau_k)e^{-j2\pi(s_k+s_k')/\lambda}}{\sqrt{4\pi}s_k s_k'}\Bigg]e^{j2\pi f_c t}\Bigg\},\end{aligned}\tag{2.27}$$

where τ_i, τ_j, and τ_k are (respectively) the time delays of the given reflected, diffracted, and scattered rays – normalized to the delay of the LOS ray – as defined previously. The received power P_r of $r_{\text{total}}(t)$ and the corresponding path loss P_r/P_t are then obtained from (2.27).

Any of these multipath components may have an additional attenuation factor if its propagation path is blocked by buildings or other objects. In this case, the attenuation factor of the obstructing object multiplies the component's path-loss term in (2.27). This attenuation loss will vary widely, depending on the material and depth of the object [5; 34]. Models for random loss due to attenuation are described in Section 2.7.

2.4.4 Local Mean Received Power

The path loss computed from all ray-tracing models is associated with a fixed transmitter and receiver location. In addition, ray tracing can be used to compute the *local mean received power* $\bar{P}_r$ in the vicinity of a given receiver location by adding the squared magnitude of all the received rays. This has the effect of averaging out local spatial variations due to phase changes around the given location. Local mean received power is a good indicator of link quality and is often used in cellular system functions like power control and handoff [35].

2.5 Empirical Path-Loss Models

Most mobile communication systems operate in complex propagation environments that cannot be accurately modeled by free-space path loss or ray tracing. A number of path-loss models have been developed over the years to predict path loss in typical wireless environments such as large urban macrocells, urban microcells, and, more recently, inside buildings [5, Chap. 3]. These models are mainly based on empirical measurements over a given distance in a given frequency range for a particular geographical area or building. However, applications of these models are not always restricted to environments in which the empirical measurements were made, which may compromise the accuracy of such empirically based models when applied to more general environments. Nevertheless, many wireless systems use these models as a basis for performance analysis. In our discussion we will begin with common models for urban macrocells and then describe more recent models for outdoor microcells and indoor propagation.

Analytical models characterize P_r/P_t as a function of distance, so path loss is well-defined. In contrast, empirical measurements of P_r/P_t as a function of distance include the effects of path loss, shadowing, and multipath. In order to remove multipath effects, empirical measurements for path loss typically average their received power measurements and the corresponding path loss at a given distance over several wavelengths. This average path loss is called the *local mean attenuation* (LMA) at distance d, and it generally decreases with d owing to free-space path loss and signal obstructions. The LMA in a given environment, such as a city, depends on the specific location of the transmitter and receiver corresponding to the LMA measurement. To characterize LMA more generally, measurements are typically taken throughout the environment and possibly in multiple environments with similar characteristics. Thus, the *empirical path loss* $P_L(d)$ for a given environment (a city, suburban area, or office building) is defined as the average of the LMA measurements at distance d averaged over all available measurements in the given environment. For example, empirical path loss for a generic downtown area with a rectangular street grid might be obtained by averaging LMA measurements in New York City, downtown San Francisco, and downtown Chicago. The empirical path-loss models given here are all obtained from average LMA measurements.

2.5.1 Okumura Model

One of the most common models for signal prediction in large urban macrocells is the Okumura model [36]. This model is applicable over distances of 1–100 km and frequency ranges of 150–1500 MHz. Okumura used extensive measurements of base station-to-mobile signal attenuation throughout Tokyo to develop a set of curves giving median attenuation relative to free space of signal propagation in irregular terrain. The base station heights for these measurements were 30–100 m, a range whose upper end is higher than typical base stations today. The empirical path-loss formula of Okumura at distance d parameterized by the carrier frequency f_c is given by

$$P_L(d) \text{ dB} = L(f_c, d) + A_\mu(f_c, d) - G(h_t) - G(h_r) - G_{\text{AREA}}, \qquad (2.28)$$

where $L(f_c, d)$ is free-space path loss at distance d and carrier frequency f_c, $A_\mu(f_c, d)$ is the median attenuation in addition to free-space path loss across all environments, $G(h_t)$ is the base station antenna height gain factor, $G(h_r)$ is the mobile antenna height gain factor, and G_{AREA} is the gain due to the type of environment. The values of $A_\mu(f_c, d)$ and G_{AREA} are obtained from Okumura's empirical plots [36; 5]. Okumura derived empirical formulas for $G(h_t)$ and $G(h_r)$ as follows:

$$G(h_t) = 20 \log_{10}(h_t/200), \quad 30 \text{ m} < h_t < 1000 \text{ m}; \tag{2.29}$$

$$G(h_r) = \begin{cases} 10 \log_{10}(h_r/3) & h_r \leq 3 \text{ m}, \\ 20 \log_{10}(h_r/3) & 3 \text{ m} < h_r < 10 \text{ m}. \end{cases} \tag{2.30}$$

Correction factors related to terrain are also developed in [36] that improve the model's accuracy. Okumura's model has a 10–14-dB empirical standard deviation between the path loss predicted by the model and the path loss associated with one of the measurements used to develop the model.

2.5.2 Hata Model

The Hata model [37] is an empirical formulation of the graphical path-loss data provided by Okumura and is valid over roughly the same range of frequencies, 150–1500 MHz. This empirical model simplifies calculation of path loss because it is a closed-form formula and is not based on empirical curves for the different parameters. The standard formula for empirical path loss in urban areas under the Hata model is

$$P_{L,\text{urban}}(d) \text{ dB} = 69.55 + 26.16 \log_{10}(f_c) - 13.82 \log_{10}(h_t) - a(h_r)$$
$$+ (44.9 - 6.55 \log_{10}(h_t)) \log_{10}(d). \tag{2.31}$$

The parameters in this model are the same as under the Okumura model, and $a(h_r)$ is a correction factor for the mobile antenna height based on the size of the coverage area. For small to medium-sized cities, this factor is given by

$$a(h_r) = (1.1 \log_{10}(f_c) - .7)h_r - (1.56 \log_{10}(f_c) - .8) \text{ dB},$$

[37; 5] and for larger cities at frequencies $f_c > 300$ MHz by

$$a(h_r) = 3.2(\log_{10}(11.75h_r))^2 - 4.97 \text{ dB}.$$

Corrections to the urban model are made for suburban and rural propagation, so that these models are (respectively)

$$P_{L,\text{suburban}}(d) \text{ dB} = P_{L,\text{urban}}(d) \text{ dB} - 2[\log_{10}(f_c/28)]^2 - 5.4 \tag{2.32}$$

and

$$P_{L,\text{rural}}(d) \text{ dB} = P_{L,\text{urban}}(d) \text{ dB} - 4.78[\log_{10}(f_c)]^2 + 18.33 \log_{10}(f_c) - K, \tag{2.33}$$

where K ranges from 35.94 (countryside) to 40.94 (desert). Unlike the Okumura model, the Hata model does not provide for any path-specific correction factors. The Hata model well approximates the Okumura model for distances $d > 1$ km. Hence it is a good model for first-generation cellular systems, but it does not model propagation well in current cellular systems with smaller cell sizes and higher frequencies. Indoor environments are also not captured by the Hata model.

2.5.3 COST 231 Extension to Hata Model

The Hata model was extended by the European cooperative for scientific and technical research (EURO-COST) to 2 GHz as

$$P_{L,\text{urban}}(d) \text{ dB} = 46.3 + 33.9 \log_{10}(f_c) - 13.82 \log_{10}(h_t) - a(h_r)$$
$$+ (44.9 - 6.55 \log_{10}(h_t)) \log_{10}(d) + C_M \tag{2.34}$$

[38], where $a(h_r)$ is the same correction factor as before and where C_M is 0 dB for medium-sized cities and suburbs and is 3 dB for metropolitan areas. This model is referred to as the COST 231 extension to the Hata model and is restricted to the following range of parameters: 1.5 GHz $< f_c <$ 2 GHz, 30 m $< h_t <$ 200 m, 1 m $< h_r <$ 10 m, and 1 km $< d <$ 20 km.

2.5.4 Piecewise Linear (Multislope) Model

A common empirical method for modeling path loss in outdoor microcells and indoor channels is a piecewise linear model of dB loss versus log distance. This approximation is illustrated in Figure 2.9 for dB attenuation versus log distance, where the dots represent hypothetical measurements and the piecewise linear model represents an approximation to these measurements. A piecewise linear model with N segments must specify $N-1$ breakpoints $d_1, \ldots, d_{N-1}$ as well as the slopes corresponding to each segment $s_1, \ldots, s_N$. Different methods can be used to determine the number and location of breakpoints to be used in the model. Once these are fixed, the slopes corresponding to each segment can be obtained by linear regression. The piecewise linear model has been used to characterize path loss for outdoor channels in [39] and for indoor channels in [40].

A special case of the piecewise model is the dual-slope model. The dual-slope model is characterized by a constant path-loss factor K and a path-loss exponent γ_1 above some reference distance d_0 and up to some critical distance d_c, after which power falls off with path loss exponent γ_2:

$$P_r(d) \text{ dB} = \begin{cases} P_t + K - 10\gamma_1 \log_{10}(d/d_0) & d_0 \le d \le d_c, \\ P_t + K - 10\gamma_1 \log_{10}(d_c/d_0) - 10\gamma_2 \log_{10}(d/d_c) & d > d_c. \end{cases} \tag{2.35}$$

The path-loss exponents, K, and d_c are typically obtained via a regression fit to empirical data [41; 42]. The two-ray model described in Section 2.4.1 for $d > h_t$ can be approximated by the dual-slope model, with one breakpoint at the critical distance d_c and with attenuation slope $s_1 = 20$ dB/decade and $s_2 = 40$ dB/decade.

The multiple equations in the dual-slope model can be captured by the following approximation [19; 43]:

$$P_r = \frac{P_t K}{L(d)}, \tag{2.36}$$

where

$$L(d) \triangleq \left[\frac{d}{d_0}\right]^{\gamma_1} \left(1 + \left(\frac{d}{d_c}\right)^{(\gamma_1-\gamma_2)q}\right)^{1/q}. \tag{2.37}$$

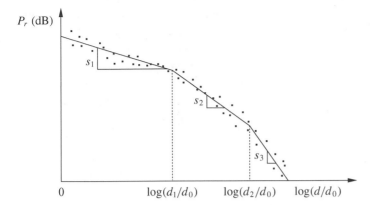

Figure 2.9: Piecewise linear model for path loss.

In this expression, q is a parameter that determines the smoothness of the path loss at the transition region close to the breakpoint distance d_c. This model can be extended to more than two regions [39].

2.5.5 Indoor Attenuation Factors

Indoor environments differ widely in the materials used for walls and floors, the layout of rooms, hallways, windows, and open areas, the location and material in obstructing objects, the size of each room, and the number of floors. All of these factors have a significant impact on path loss in an indoor environment. Thus, it is difficult to find generic models that can be accurately applied to determine empirical path loss in a specific indoor setting.

Indoor path-loss models must accurately capture the effects of attenuation across floors due to partitions as well as between floors. Measurements across a wide range of building characteristics and signal frequencies indicate that the attenuation per floor is greatest for the first floor that is passed through and decreases with each subsequent floor. Specifically, measurements in [44; 45; 46; 47] indicate that, at 900 MHz, the attenuation when transmitter and receiver are separated by a single floor ranges from 10–20 dB, while subsequent attenuation is 6–10 dB per floor for the next three floors and then a few decibels per floor for more than four floors. At higher frequencies the attenuation loss per floor is typically larger [45; 48]. The attenuation per floor is thought to decrease as the number of attenuating floors increases because of the scattering up the side of the building and reflections from adjacent buildings. Partition materials and dielectric properties vary widely and thus so do partition losses. Measurements for the partition loss at different frequencies for different partition types can be found in [5; 44; 49; 50; 51], and Table 2.1 indicates a few examples of partition losses measured at 900–1300 MHz from this data. The partition loss obtained by different researchers for the same partition type at the same frequency often varies widely, so it is difficult to make generalizations about partition loss from a specific data set.

The experimental data for floor and partition loss can be added to an analytical or empirical dB path-loss model $P_L(d)$ as

$$P_r \text{ dBm} = P_t \text{ dBm} - P_L(d) - \sum_{i=1}^{N_f} \text{FAF}_i - \sum_{i=1}^{N_p} \text{PAF}_i, \qquad (2.38)$$

Table 2.1: Typical partition losses

Partition type	Partition loss (dB)
Cloth partition	1.4
Double plasterboard wall	3.4
Foil insulation	3.9
Concrete wall	13
Aluminum siding	20.4
All metal	26

where FAF_i represents the floor attenuation factor for the ith floor traversed by the signal and PAF_i represents the partition attenuation factor associated with the ith partition traversed by the signal. The number of floors and partitions traversed by the signal are N_f and N_p, respectively.

Another important factor for indoor systems whose transmitter is located outside the building is building penetration loss. Measurements indicate that building penetration loss is a function of frequency, height, and the building materials. Building penetration loss on the ground floor typically ranges from 8 dB to 20 dB for 900 MHz to 2 GHz [1; 52; 53]. The penetration loss decreases slightly as frequency increases, and it also decreases by about 1.4 dB per floor at floors above the ground floor. This decrease in loss is typically due to reduced clutter at higher floors and the higher likelihood of an LOS path. The type and number of windows in a building also have a significant impact on penetration loss [54]. Measurements made behind windows have about 6 dB less penetration loss than measurements made behind exterior walls. Moreover, plate glass has an attenuation of around 6 dB whereas lead-lined glass has an attenuation of between 3 and 30 dB.

2.6 Simplified Path-Loss Model

The complexity of signal propagation makes it difficult to obtain a single model that characterizes path loss accurately across a range of different environments. Accurate path-loss models can be obtained from complex analytical models or empirical measurements when tight system specifications must be met or the best locations for base stations or access-point layouts must be determined. However, for general trade-off analysis of various system designs it is sometimes best to use a simple model that captures the essence of signal propagation without resorting to complicated path-loss models, which are only approximations to the real channel anyway. Thus, the following simplified model for path loss as a function of distance is commonly used for system design:

$$P_r = P_t K \left[\frac{d_0}{d} \right]^\gamma. \tag{2.39}$$

The dB attenuation is thus

$$P_r \text{ dBm} = P_t \text{ dBm} + K \text{ dB} - 10\gamma \log_{10} \left[\frac{d}{d_0} \right]. \tag{2.40}$$

In this approximation, K is a unitless constant that depends on the antenna characteristics and the average channel attenuation, d_0 is a reference distance for the antenna far field, and γ is the path-loss exponent. The values for K, d_0, and γ can be obtained to approximate either an analytical or empirical model. In particular, the free-space path-loss model, the two-ray model, the Hata model, and the COST extension to the Hata model are all of the same form as

Table 2.2: Typical path-loss exponents

Environment	γ range
Urban macrocells	3.7–6.5
Urban microcells	2.7–3.5
Office building (same floor)	1.6–3.5
Office building (multiple floors)	2–6
Store	1.8–2.2
Factory	1.6–3.3
Home	3

(2.40). Because of scattering phenomena in the antenna near field, the model (2.40) is generally valid only at transmission distances $d > d_0$, where d_0 is typically assumed to be 1–10 m indoors and 10–100 m outdoors.

When the simplified model is used to approximate empirical measurements, the value of $K < 1$ is sometimes set to the free-space path gain at distance d_0 assuming omnidirectional antennas:

$$K \text{ dB} = 20 \log_{10} \frac{\lambda}{4 \pi d_0}, \qquad (2.41)$$

and this assumption is supported by empirical data for free-space path loss at a transmission distance of 100 m [41]. Alternatively, K can be determined by measurement at d_0 or optimized (alone or together with γ) to minimize the mean-square error (MSE) between the model and the empirical measurements [41]. The value of γ depends on the propagation environment: for propagation that approximately follows a free-space or two-ray model, γ is set to 2 or 4 (respectively). The value of γ for more complex environments can be obtained via a minimum mean-square error (MMSE) fit to empirical measurements, as illustrated in Example 2.3. Alternatively, γ can be obtained from an empirically based model that takes into account frequency and antenna height [41]. Table 2.2 summarizes γ-values for different environments (data from [5; 33; 41; 44; 46; 47; 52; 55]). Path-loss exponents at higher frequencies tend to be higher [46; 51; 52; 56] whereas path-loss exponents at higher antenna heights tend to be lower [41]. Note that the wide range of empirical path-loss exponents for indoor propagation may be due to attenuation caused by floors, objects, and partitions (see Section 2.5.5).

EXAMPLE 2.3: Consider the set of empirical measurements of P_r/P_t given in Table 2.3 for an indoor system at 900 MHz. Find the path-loss exponent γ that minimizes the MSE between the simplified model (2.40) and the empirical dB power measurements, assuming that $d_0 = 1$ m and K is determined from the free-space path-gain formula at this d_0. (We minimize the MSE of the dB values rather than the linear values because this generally leads to a more accurate model.) Find the received power at 150 m for the simplified path-loss model with this path-loss exponent and a transmit power of 1 mW (0 dBm).

Solution: We first set up the MMSE error equation for the dB power measurements as

$$F(\gamma) = \sum_{i=1}^{5} [M_{\text{measured}}(d_i) - M_{\text{model}}(d_i)]^2,$$

where $M_{\text{measured}}(d_i)$ is the path-loss measurement in Table 2.3 at distance d_i and where $M_{\text{model}}(d_i) = K - 10\gamma \log_{10}(d)$ is the path loss at d_i based on (2.40). Now using the free-space path-loss formula yields $K = 20 \log_{10}(.3333/(4\pi)) = -31.54$ dB. Thus

Table 2.3: Path-loss measurements

Distance from transmitter	$M = P_r/P_t$
10 m	−70 dB
20 m	−75 dB
50 m	−90 dB
100 m	−110 dB
300 m	−125 dB

$$F(\gamma) = (-70 + 31.54 + 10\gamma)^2 + (-75 + 31.54 + 13.01\gamma)^2$$
$$+ (-90 + 31.54 + 16.99\gamma)^2 + (-110 + 31.54 + 20\gamma)^2$$
$$+ (-125 + 31.54 + 24.77\gamma)^2$$
$$= 21676.3 - 11654.9\gamma + 1571.47\gamma^2. \qquad (2.42)$$

Differentiating $F(\gamma)$ relative to γ and setting it to zero yields

$$\frac{\partial F(\gamma)}{\partial \gamma} = -11654.9 + 3142.94\gamma = 0 \implies \gamma = 3.71.$$

For the received power at 150 m under the simplified path-loss model with $K = -31.54$, $\gamma = 3.71$, and $P_t = 0$ dBm, we have $P_r = P_t + K - 10\gamma \log_{10}(d/d_0) = 0 - 31.54 - 10 \cdot 3.71 \log_{10}(150) = -112.27$ dBm. Clearly the measurements deviate from the simplified path-loss model; this variation can be attributed to shadow fading, described in Section 2.7.

2.7 Shadow Fading

A signal transmitted through a wireless channel will typically experience random variation due to blockage from objects in the signal path, giving rise to random variations of the received power at a given distance. Such variations are also caused by changes in reflecting surfaces and scattering objects. Thus, a model for the random attenuation due to these effects is also needed. The location, size, and dielectric properties of the blocking objects – as well as the changes in reflecting surfaces and scattering objects that cause the random attenuation – are generally unknown, so statistical models must be used to characterize this attenuation. The most common model for this additional attenuation is log-normal shadowing. This model has been empirically confirmed to model accurately the variation in received power in both outdoor and indoor radio propagation environments (see e.g. [41; 57]).

In the log-normal shadowing model, the ratio of transmit-to-receive power $\psi = P_t/P_r$ is assumed to be random with a log-normal distribution given by

$$p(\psi) = \frac{\xi}{\sqrt{2\pi}\sigma_{\psi_{dB}}\psi} \exp\left[-\frac{(10\log_{10}\psi - \mu_{\psi_{dB}})^2}{2\sigma_{\psi_{dB}}^2}\right], \quad \psi > 0, \qquad (2.43)$$

where $\xi = 10/\ln 10$, $\mu_{\psi_{dB}}$ is the mean of $\psi_{dB} = 10\log_{10}\psi$ in decibels, and $\sigma_{\psi_{dB}}$ is the standard deviation of ψ_{dB} (also in dB). The mean can be based on an analytical model or empirical measurements. For empirical measurements $\mu_{\psi_{dB}}$ equals the empirical path loss, since average attenuation from shadowing is already incorporated into the measurements. For analytical models, $\mu_{\psi_{dB}}$ must incorporate both the path loss (e.g., from a free-space or ray-tracing model) as well as average attenuation from blockage. Alternatively, path loss can be treated separately from shadowing, as described in the next section. A random variable with a log-normal distribution is called a *log-normal random variable*. Note that if ψ is log-normal then the received power and received SNR will also be log-normal, since these are just constant multiples of ψ. For received SNR the mean and standard deviation of this log-normal random variable are also in decibels. For log-normal received power the random variable has units of power, so its mean and standard deviation will be in dBm or dBW instead of dB. The mean of ψ (the linear average path gain) can be obtained from (2.43) as

$$\mu_\psi = \mathbf{E}[\psi] = \exp\left[\frac{\mu_{\psi_{dB}}}{\xi} + \frac{\sigma^2_{\psi_{dB}}}{2\xi^2}\right]. \tag{2.44}$$

The conversion from the linear mean (in dB) to the log mean (in dB) is derived from (2.44) as

$$10\log_{10}\mu_\psi = \mu_{\psi_{dB}} + \frac{\sigma^2_{\psi_{dB}}}{2\xi}. \tag{2.45}$$

Performance in log-normal shadowing is typically parameterized by the log mean $\mu_{\psi_{dB}}$, which is referred to as the *average dB path loss* and is given in units of dB. With a change of variables we see that the distribution of the dB value of ψ is Gaussian with mean $\mu_{\psi_{dB}}$ and standard deviation $\sigma_{\psi_{dB}}$:

$$p(\psi_{dB}) = \frac{1}{\sqrt{2\pi}\sigma_{\psi_{dB}}}\exp\left[-\frac{(\psi_{dB} - \mu_{\psi_{dB}})^2}{2\sigma^2_{\psi_{dB}}}\right]. \tag{2.46}$$

The log-normal distribution is defined by two parameters: $\mu_{\psi_{dB}}$ and $\sigma_{\psi_{dB}}$. Since $\psi = P_t/P_r$ is always greater than unity, it follows that $\mu_{\psi_{dB}}$ is always greater than or equal to zero. Note that the log-normal distribution (2.43) takes values for $0 \leq \psi \leq \infty$. Hence, for $\psi < 1$ we have $P_r > P_t$, which is physically impossible. However, this probability will be very small when $\mu_{\psi_{dB}}$ is large and positive. Thus, the log-normal model captures the underlying physical model most accurately when $\mu_{\psi_{dB}} \gg 0$.

If the mean and standard deviation for the shadowing model are based on empirical measurements, then the question arises as to whether they should be obtained by taking averages of the linear or rather the dB values of the empirical measurements. Specifically: Given empirical (linear) path-loss measurements $\{p_i\}_{i=1}^N$, should the mean path loss be determined as $\mu_\psi = (1/N)\sum_{i=1}^N p_i$ or as $\mu_{\psi_{dB}} = (1/N)\sum_{i=1}^N 10\log_{10}p_i$? A similar question arises for computing the empirical variance. In practice it is more common to determine mean path loss and variance based on averaging the dB values of the empirical measurements for several reasons. First, as we shall see, the mathematical justification for the log-normal model is based on dB measurements. In addition, the literature shows that obtaining empirical averages based on dB path-loss measurements leads to a smaller estimation error [58]. Finally, as we saw in Section 2.5.4, power falloff with distance models are often obtained by a piecewise linear approximation to empirical measurements of dB power versus the log of distance [5].

Most empirical studies for outdoor channels support a standard deviation $\sigma_{\psi_{dB}}$ ranging from 4 dB to 13 dB [6; 19; 59; 60; 61]. The mean power $\mu_{\psi_{dB}}$ depends on the path loss and building properties in the area under consideration. The mean power $\mu_{\psi_{dB}}$ varies with distance; this is due to path loss and to the fact that average attenuation from objects increases with distance owing to the potential for a larger number of attenuating objects.

The Gaussian model for the distribution of the mean received signal in dB can be justified by the following attenuation model when shadowing is dominated by the attenuation from blocking objects. The attenuation of a signal as it travels through an object of depth d is approximately equal to

$$s(d) = e^{-\alpha d}, \tag{2.47}$$

where α is an attenuation constant that depends on the object's materials and dielectric properties. If we assume that α is approximately equal for all blocking objects and that the ith blocking object has a random depth d_i, then the attenuation of a signal as it propagates through this region is

$$s(d_t) = e^{-\alpha \sum_i d_i} = e^{-\alpha d_t}, \tag{2.48}$$

where $d_t = \sum_i d_i$ is the sum of the random object depths through which the signal travels. If there are many objects between the transmitter and receiver, then by the cental limit theorem we can approximate d_t by a Gaussian random variable. Thus, $\log s(d_t) = \alpha d_t$ will have a Gaussian distribution with mean μ and standard deviation σ. The value of σ will depend on the environment.

EXAMPLE 2.4: In Example 2.3 we found that the exponent for the simplified path-loss model that best fits the measurements in Table 2.3 was $\gamma = 3.71$. Assuming the simplified path-loss model with this exponent and the same $K = -31.54$ dB, find $\sigma_{\psi_{dB}}^2$, the variance of log-normal shadowing about the mean path loss based on these empirical measurements.

Solution: The sample variance relative to the simplified path-loss model with $\gamma = 3.71$ is

$$\sigma_{\psi_{dB}}^2 = \frac{1}{5} \sum_{i=1}^{5} [M_{\text{measured}}(d_i) - M_{\text{model}}(d_i)]^2,$$

where $M_{\text{measured}}(d_i)$ is the path-loss measurement in Table 2.3 at distance d_i and $M_{\text{model}}(d_i) = K - 37.1 \log_{10}(d)$. This yields

$$\sigma_{\psi_{dB}}^2 = \tfrac{1}{5}[(-70 - 31.54 + 37.1)^2 + (-75 - 31.54 + 48.27)^2$$
$$+ (-90 - 31.54 + 63.03)^2 + (-110 - 31.54 + 74.2)^2$$
$$+ (-125 - 31.54 + 91.90)^2]$$
$$= 13.29.$$

Thus, the standard deviation of shadow fading on this path is $\sigma_{\psi_{dB}} = 3.65$ dB. Note that the bracketed term in the displayed expression equals the MMSE formula (2.42) from Example 2.3 with $\gamma = 3.71$.

Extensive measurements have been taken to characterize the empirical autocorrelation function of the shadow fading process over distance for different environments at different frequencies (see e.g. [60; 62; 63; 64; 65]). The most common analytical model for this function, first proposed by Gudmundson [60] and based on empirical measurements, assumes that the shadowing $\psi(d)$ is a first-order autoregressive process where the covariance between shadow fading at two points separated by distance δ is characterized by

$$A(\delta) = \mathbf{E}[(\psi_{\mathrm{dB}}(d) - \mu_{\psi_{\mathrm{dB}}})(\psi_{\mathrm{dB}}(d + \delta) - \mu_{\psi_{\mathrm{dB}}})] = \sigma^2_{\psi_{\mathrm{dB}}} \rho_D^{\delta/D}, \qquad (2.49)$$

where ρ_D is the normalized covariance between two points separated by a fixed distance D. This covariance must be obtained empirically, and it varies with the propagation environment and carrier frequency. Measurements indicate that for suburban macrocells (with $f_c = 900$ MHz) $\rho_D = .82$ for $D = 100$ m and that for urban microcells (with $f_c \approx 2$ GHz) $\rho_D = .3$ for $D = 10$ m [60; 64]. This model can be simplified and its empirical dependence removed by setting $\rho_D = 1/e$ for distance $D = X_c$, which yields

$$A(\delta) = \sigma^2_{\psi_{\mathrm{dB}}} e^{-\delta/X_c}. \qquad (2.50)$$

The *decorrelation distance* X_c in this model is the distance at which the signal autocovariance equals $1/e$ of its maximum value and is on the order of the size of the blocking objects or clusters of these objects. For outdoor systems, X_c typically ranges from 50 m to 100 m [63; 64]. For users moving at velocity v, the shadowing decorrelation in time τ is obtained by substituting $v\tau = \delta$ in (2.49) or (2.50). Autocorrelation relative to angular spread, which is useful for the multiple antenna systems treated in Chapter 10, has been investigated in [62; 64].

The first-order autoregressive correlation model (2.49) and its simplified form (2.50) are easy to analyze and to simulate. Specifically, one can simulate ψ_{dB} by first generating a white Gaussian noise process with power $\sigma^2_{\psi_{\mathrm{dB}}}$ and then passing it through a first-order filter with response $\rho_D^{\delta/D}$ for a covariance characterized by (2.49) or response $e^{-\delta/X_c}$ for a covariance characterized by (2.50). The filter output will produce a shadowing random process with the desired correlation properties [60; 61].

2.8 Combined Path Loss and Shadowing

Models for path loss and shadowing can be superimposed to capture power falloff versus distance along with the random attenuation about this path loss from shadowing. In this combined model, average dB path loss ($\mu_{\psi_{\mathrm{dB}}}$) is characterized by the path-loss model while shadow fading, with a mean of 0 dB, creates variations about this path loss, as illustrated by the path-loss and shadowing curve in Figure 2.1. Specifically, this curve plots the combination of the simplified path-loss model (2.39) and the log-normal shadowing random process defined by (2.46) and (2.50). For this combined model, the ratio of received to transmitted power in dB is given by

$$\frac{P_r}{P_t} \text{ dB} = 10 \log_{10} K - 10\gamma \log_{10} \frac{d}{d_0} - \psi_{\mathrm{dB}}, \qquad (2.51)$$

where ψ_{dB} is a Gauss-distributed random variable with mean zero and variance $\sigma^2_{\psi_{\mathrm{dB}}}$. In (2.51) and as shown in Figure 2.1, the path loss decreases linearly relative to $\log_{10} d$ with a slope of 10γ dB/decade, where γ is the path-loss exponent. The variations due to shadowing change more rapidly: on the order of the decorrelation distance X_c.

Examples 2.3 and 2.4 illustrated the combined model for path loss and log-normal shadowing based on the measurements in Table 2.3, where path loss obeys the simplified path-loss model with $K = -31.54$ dB and path-loss exponent $\gamma = 3.71$ and where shadowing obeys the log-normal model with mean given by the path-loss model and standard deviation $\sigma_{\psi_{dB}} = 3.65$ dB.

2.9 Outage Probability under Path Loss and Shadowing

The combined effects of path loss and shadowing have important implications for wireless system design. In wireless systems there is typically a target minimum received power level P_{min} below which performance becomes unacceptable (e.g., the voice quality in a cellular system becomes too poor to understand). However, with shadowing the received power at any given distance from the transmitter is log-normally distributed with some probability of falling below P_{min}. We define *outage probability* $P_{out}(P_{min}, d)$ under path loss and shadowing to be the probability that the received power at a given distance d, $P_r(d)$, falls below P_{min}: $P_{out}(P_{min}, d) = p(P_r(d) < P_{min})$. For the combined path-loss and shadowing model of Section 2.8 this becomes

$$p(P_r(d) \le P_{min}) = 1 - Q\left(\frac{P_{min} - (P_t + 10\log_{10} K - 10\gamma\log_{10}(d/d_0))}{\sigma_{\psi_{dB}}}\right), \qquad (2.52)$$

where the Q-function is defined as the probability that a Gaussian random variable X with mean 0 and variance 1 is greater than z:

$$Q(z) \triangleq p(X > z) = \int_z^\infty \frac{1}{\sqrt{2\pi}} e^{-y^2/2}\, dy. \qquad (2.53)$$

The conversion between the Q-function and complementary error function is

$$Q(z) = \frac{1}{2}\,\mathrm{erfc}\left(\frac{z}{\sqrt{2}}\right). \qquad (2.54)$$

We will omit the parameters of P_{out} when the context is clear or in generic references to outage probability.

EXAMPLE 2.5: Find the outage probability at 150 m for a channel based on the combined path loss and shadowing models of Examples 2.3 and 2.4, assuming a transmit power of $P_t = 10$ mW and minimum power requirement of $P_{min} = -110.5$ dBm.

Solution: We have $P_t = 10$ mW $= 10$ dBm. Hence,

$$P_{out}(-110.5\text{ dBm}, 150\text{ m})$$
$$= p(P_r(150\text{ m}) < -110.5\text{ dBm})$$
$$= 1 - Q\left(\frac{P_{min} - (P_t + 10\log_{10} K - 10\gamma\log_{10}(d/d_0))}{\sigma_{\psi_{dB}}}\right)$$
$$= 1 - Q\left(\frac{-110.5 - (10 - 31.54 - 37.1\log_{10}(150))}{3.65}\right)$$
$$= .0121.$$

An outage probability of 1% is a typical target in wireless system designs.

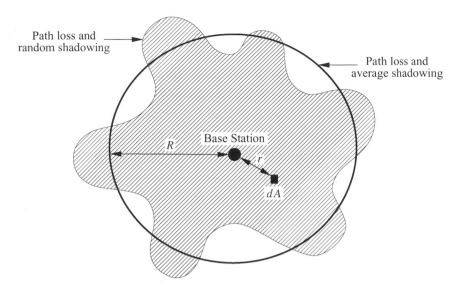

Figure 2.10: Contours of constant received power.

2.10 Cell Coverage Area

The *cell coverage area* in a cellular system is defined as the expected percentage of locations within a cell where the received power at these locations is above a given minimum. Consider a base station inside a circular cell of a given radius R. All mobiles within the cell require some minimum received SNR for acceptable performance. Assuming a given model for noise, the SNR requirement translates to a minimum received power P_{min} throughout the cell. The transmit power at the base station is designed for an *average* received power at the cell boundary of $\bar{P}_R$, averaged over the shadowing variations. However, shadowing will cause some locations within the cell to have received power below $\bar{P}_R$, and others will have received power exceeding $\bar{P}_R$. This is illustrated in Figure 2.10, where we show contours of constant received power based on a fixed transmit power at the base station for path loss and average shadowing and for path loss and random shadowing. For path loss and average shadowing, constant power contours form a circle around the base station because combined path loss and average shadowing is the same at a uniform distance from the base station. For path loss and random shadowing, the contours form an amoeba-like shape due to the random shadowing variations about the average. The constant power contours for combined path loss and random shadowing indicate the challenge that shadowing poses in cellular system design. Specifically, it is not possible for all users at the cell boundary to receive the same power level. Thus, either the base station must transmit extra power to ensure users affected by shadowing receive their minimum required power P_{min}, which causes excessive interference to neighboring cells, or some users within the cell will find their minimum received power requirement unmet. In fact, since the Gaussian distribution has infinite tails, *any* mobile in the cell has a nonzero probability of experiencing received power below its required minimum, even if the mobile is close to the base station. This makes sense intuitively because a mobile may be in a tunnel or blocked by a large building, regardless of its proximity to the base station.

We now compute cell coverage area under path loss and shadowing. The percentage of area within a cell where the received power exceeds the minimum required power P_{min} is obtained by taking an incremental area dA at radius r from the base station in the cell, as shown in Figure 2.10. Let $P_r(r)$ be the received power in dA from combined path loss and shadowing. Then the total area within the cell where the minimum power requirement is exceeded is obtained by integrating over all incremental areas where this minimum is exceeded:

$$
\begin{aligned}
C &= \mathbf{E}\left[\frac{1}{\pi R^2}\int_{\text{cell area}} 1[P_r(r) > P_{min} \text{ in } dA]\, dA\right] \\
&= \frac{1}{\pi R^2}\int_{\text{cell area}} \mathbf{E}[1[P_r(r) > P_{min} \text{ in } dA]]\, dA,
\end{aligned}
\tag{2.55}
$$

where $1[\cdot]$ denotes the indicator function. Define $P_A = p(P_r(r) > P_{min})$ in dA. Then $P_A = \mathbf{E}[1[P_r(r) > P_{min} \text{ in } dA]]$. Making this substitution in (2.55) and using polar coordinates for the integration yields

$$
C = \frac{1}{\pi R^2}\int_{\text{cell area}} P_A\, dA = \frac{1}{\pi R^2}\int_0^{2\pi}\int_0^R P_A r\, dr\, d\theta.
\tag{2.56}
$$

The *outage probability of the cell* is defined as the percentage of area within the cell that does not meet its minimum power requirement P_{min}; that is, $P_{out}^{\text{cell}} = 1 - C$.

Given the log-normal distribution for the shadowing, we have

$$
\begin{aligned}
P_A &= p(P_r(r) \geq P_{min}) = Q\left(\frac{P_{min} - (P_t + 10\log_{10} K - 10\gamma\log_{10}(r/d_0))}{\sigma_{\psi_{dB}}}\right) \\
&= 1 - P_{out}(P_{min}, r),
\end{aligned}
\tag{2.57}
$$

where P_{out} is the outage probability defined in (2.52) with $d = r$. Locations within the cell with received power below P_{min} are said to be *outage locations*.

Combining (2.56) and (2.57) yields[5]

$$
C = \frac{2}{R^2}\int_0^R rQ\left(a + b\ln\frac{r}{R}\right) dr,
\tag{2.58}
$$

where

$$
a = \frac{P_{min} - \bar{P}_r(R)}{\sigma_{\psi_{dB}}}, \qquad b = \frac{10\gamma\log_{10}(e)}{\sigma_{\psi_{dB}}},
\tag{2.59}
$$

and $\bar{P}_R = P_t + 10\log_{10} K - 10\gamma\log_{10}(R/d_0)$ is the received power at the cell boundary (distance R from the base station) due to path loss alone. This integral yields a closed-form solution for C in terms of a and b:

---

[5] Recall that (2.57) is generally valid only for $r \geq d_0$, yet to simplify the analysis we have applied the model for all r. This approximation will have little impact on coverage area, since d_0 is typically very small compared to R and the outage probability for $r < d_0$ is negligible.

$$C = Q(a) + \exp\left[\frac{2 - 2ab}{b^2}\right]Q\left(\frac{2 - ab}{b}\right). \tag{2.60}$$

If the target minimum received power equals the average power at the cell boundary, $P_{\min} = \bar{P}_r(R)$, then $a = 0$ and the coverage area simplifies to

$$C = \frac{1}{2} + \exp\left[\frac{2}{b^2}\right]Q\left(\frac{2}{b}\right). \tag{2.61}$$

Note that with this simplification C depends only on the ratio $\gamma/\sigma_{\psi_{\text{dB}}}$. Moreover, owing to the symmetry of the Gaussian distribution, under this assumption the outage probability at the cell boundary $P_{\text{out}}(\bar{P}_r(R), R) = .5$.

EXAMPLE 2.6: Find the coverage area for a cell with the combined path loss and shadowing models of Examples 2.3 and 2.4, a cell radius of 600 m, a base station transmit power of $P_t = 100$ mW $= 20$ dBm, and a minimum received power requirement of $P_{\min} = -110$ dBm and also one of $P_{\min} = -120$ dBm.

Solution: We first consider $P_{\min} = -110$ and check if $a = 0$ to see whether we should use the full formula (2.60) or the simplified formula (2.61). We have $\bar{P}_r(R) = P_t + K - 10\gamma \log_{10}(600) = 20 - 31.54 - 37.1 \log_{10}(600) = -114.6$ dBm $\neq -110$ dBm, so we use (2.60). Evaluating a and b from (2.59) yields $a = (-110 + 114.6)/3.65 = 1.26$ and $b = (37.1 \cdot .434)/3.65 = 4.41$. Substituting these into (2.60) yields

$$C = Q(1.26) + \exp\left[\frac{2 - 2(1.26 \cdot 4.41)}{4.41^2}\right]Q\left(\frac{2 - (1.26)(4.41)}{4.41}\right) = .59,$$

which would be a very low coverage value for an operational cellular system (lots of unhappy customers). Now considering the less stringent received power requirement $P_{\min} = -120$ dBm yields $a = (-120 + 114.9)/3.65 = -1.479$ and the same $b = 4.41$. Substituting these values into (2.60) yields $C = .988$, a much more acceptable value for coverage area.

EXAMPLE 2.7: Consider a cellular system designed so that $P_{\min} = \bar{P}_r(R)$. That is, the received power due to path loss and average shadowing at the cell boundary equals the minimum received power required for acceptable performance. Find the coverage area for path-loss values $\gamma = 2, 4, 6$ and $\sigma_{\psi_{\text{dB}}} = 4, 8, 12$, and explain how coverage changes as γ and $\sigma_{\psi_{\text{dB}}}$ increase.

Solution: For $P_{\min} = \bar{P}_r(R)$ we have $a = 0$, so coverage is given by the formula (2.61). The coverage area thus depends only on the value for $b = 10\gamma \log_{10}(e)/\sigma_{\psi_{\text{dB}}}$, which in turn depends only on the ratio $\gamma/\sigma_{\psi_{\text{dB}}}$. Table 2.4 contains coverage area evaluated from (2.61) for the different γ and $\sigma_{\psi_{\text{dB}}}$ values.

Not surprisingly, for fixed γ the coverage area increases as $\sigma_{\psi_{\text{dB}}}$ decreases; this is because a smaller $\sigma_{\psi_{\text{dB}}}$ means less variation about the mean path loss. Without shadowing we have 100% coverage (since $P_{\min} = \bar{P}_r(R)$) and so we expect that, as $\sigma_{\psi_{\text{dB}}}$ decreases to zero, coverage area increases to 100%. It is a bit more puzzling that for a fixed $\sigma_{\psi_{\text{dB}}}$ the coverage area increases as γ increases, since a larger γ implies that

Table 2.4: Coverage area for different γ and $\sigma_{\psi_{dB}}$

	$\sigma_{\psi_{dB}}$		
γ	4	8	12
2	.77	.67	.63
4	.85	.77	.71
6	.90	.83	.77

received signal power falls off more quickly. But recall that we have set $P_{min} = \bar{P}_r(R)$, so the faster power falloff is already taken into account (i.e., we need to transmit at much higher power with $\gamma = 6$ than with $\gamma = 2$ for this equality to hold). The reason coverage area increases with path loss exponent under this assumption is that, as γ increases, the transmit power must increase to satisfy $P_{min} = \bar{P}_r(R)$. This results in higher average power throughout the cell, yielding a higher coverage area.

PROBLEMS

2-1. Under the free-space path-loss model, find the transmit power required to obtain a received power of 1 dBm for a wireless system with isotropic antennas ($G_l = 1$) and a carrier frequency $f = 5$ GHz, assuming a distance $d = 10$ m. Repeat for $d = 100$ m.

2-2. For the two-ray model with transmitter–receiver separation $d = 100$ m, $h_t = 10$ m, and $h_r = 2$ m, find the delay spread between the two signals.

2-3. For the two-ray model, show how a Taylor series approximation applied to (2.13) results in the approximation

$$\Delta\phi = \frac{2\pi(x + x' - l)}{\lambda} \approx \frac{4\pi h_t h_r}{\lambda d}.$$

2-4. For the two-ray model, derive an approximate expression for the distance values below the critical distance d_c at which signal nulls occur.

2-5. Find the critical distance d_c under the two-ray model for a large macrocell in a suburban area with the base station mounted on a tower or building ($h_t = 20$ m), the receivers at height $h_r = 3$ m, and $f_c = 2$ GHz. Is this a good size for cell radius in a suburban macrocell? Why or why not?

2-6. Suppose that, instead of a ground reflection, a two-ray model consists of a LOS component and a signal reflected off a building to the left (or right) of the LOS path. Where must the building be located relative to the transmitter and receiver for this model to be the same as the two-ray model with a LOS component and ground reflection?

2-7. Consider a two-ray channel with impulse response $h(t) = \alpha_1\delta(t) + \alpha_2\delta(t - .022\,\mu s)$. Find the distance separating the transmitter and receiver, as well as α_1 and α_2, assuming free-space path loss on each path with a reflection coefficient of -1. Assume the transmitter and receiver are located 8 m above the ground and that the carrier frequency is 900 MHz.

2-8. Directional antennas are a powerful tool to reduce the effects of multipath as well as interference. In particular, directional antennas along the LOS path for the two-ray model can reduce the attenuation effect of ground wave cancellation, as will be illustrated in this problem. Plot the dB power ($10\log_{10} P_r$) versus log distance ($\log_{10} d$) for the two-ray model with parameters $f = 900$ MHz, $R = -1$, $h_t = 50$ m, $h_r = 2$ m, $G_l = 1$, and the following

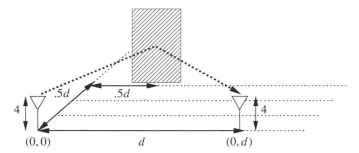

Figure 2.11: System with scattering for Problem 2-11.

values for G_r: $G_r = 1, .316, .1$, and $.01$ (i.e., $G_r = 0, -5, -10$, and -20 dB, respectively). Each of the four plots should range in distance from $d = 1$ m to $d = 100$ km. Also calculate and mark the critical distance $d_c = 4h_t h_r/\lambda$ on each plot, and normalize the plots to start at approximately 0 dB. Finally, show the piecewise linear model with flat power falloff up to distance h_t, falloff $10 \log_{10}(d^{-2})$ for $h_t < d < d_c$, and falloff $10 \log_{10}(d^{-4})$ for $d \geq d_c$. (On the power loss versus log distance plot, the piecewise linear curve becomes a set of three straight lines of slope 0, 2, and 4, respectively.) Note that at large distances it becomes increasingly difficult to have $G_r \ll G_l$ because this requires extremely precise angular directivity in the antennas.

2-9. What average power falloff with distance do you expect for the ten-ray model? Why?

2-10. For the ten-ray model, assume that the transmitter and receiver are at the same height in the middle of a street of width 20 m. The transmitter–receiver separation is 500 m. Find the delay spread for this model.

2-11. Consider a system with a transmitter, receiver, and scatterer as shown in Figure 2.11. Assume the transmitter and receiver are both at heights $h_t = h_r = 4$ m and are separated by distance d, with the scatterer at distance $.5d$ along both dimensions in a two-dimensional grid of the ground – that is, on such a grid the transmitter is located at $(0, 0)$, the receiver at $(0, d)$, and the scatterer at $(.5d, .5d)$. Assume a radar cross-section of 20 dBm^2, $G_s = 1$, and $f_c = 900$ MHz. Find the path loss of the scattered signal for $d = 1, 10, 100$, and 1000 meters. Compare with the path loss at these distances if the signal is only reflected, with reflection coefficient $R = -1$.

2-12. Under what conditions is the simplified path-loss model (2.39) the same as the free-space path-loss model (2.7)?

2-13. Consider a receiver with noise power -160 dBm within the signal bandwidth of interest. Assume a simplified path-loss model with $d_0 = 1$ m, K obtained from the free-space path-loss formula with omnidirectional antennas and $f_c = 1$ GHz, and $\gamma = 4$. For a transmit power of $P_t = 10$ mW, find the maximum distance between the transmitter and receiver such that the received signal-to-noise power ratio is 20 dB.

2-14. This problem shows how different propagation models can lead to very different SNRs (and therefore different link performance) for a given system design. Consider a linear cellular system using frequency division, as might operate along a highway or rural road (see

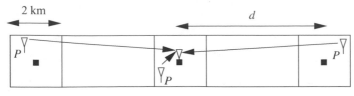

■ Base Station/Cell Center

Figure 2.12: Linear cellular system for Problem 2-14.

Figure 2.12). Each cell is allocated a certain band of frequencies, and these frequencies are reused in cells spaced a distance d away. Assume the system has square cells, 2 km per side, and that all mobiles transmit at the same power P. For the following propagation models, determine the minimum distance that the cells operating in the same frequency band must be spaced so that uplink SNR (the ratio of the minimum received signal-to-interference or S/I power from mobiles to the base station) is greater than 20 dB. You can ignore all interferers except those from the two nearest cells operating at the same frequency.

(a) Propagation for both signal and interference follow a free-space model.
(b) Propagation for both signal and interference follow the simplified path-loss model (2.39) with $d_0 = 100$ m, $K = 1$, and $\gamma = 3$.
(c) Propagation for the signal follows the simplified path-loss model with $d_0 = 100$ m, $K = 1$, and $\gamma = 2$, while propagation of the interfererence follows the same model but with $\gamma = 4$.

2-15. Find the median path loss under the Hata model assuming $f_c = 900$ MHz, $h_t = 20$ m, $h_r = 5$ m, and $d = 100$ m for a large urban city, a small urban city, a suburb, and a rural area. Explain qualitatively the path-loss differences for these four environments.

2-16. Find parameters for a piecewise linear model with three segments to approximate the two-ray model path loss (2.12) over distances between 10 and 1000 meters, assuming $h_t = 10$ m, $h_r = 2$ m, and $G_l = 1$. Plot the path loss and the piecewise linear approximation using these parameters over this distance range.

2-17. Using the indoor attentuation model, determine the required transmit power for a desired received power of -110 dBm for a signal transmitted over 100 m that goes through three floors with attenuation 15 dB, 10 dB, and 6 dB (respectively) as well as two double plasterboard walls. Assume a reference distance $d_0 = 1$, exponent $\gamma = 4$, and constant $K = 0$ dB.

2-18. Table 2.5 lists a set of empirical path loss measurements.

(a) Find the parameters of a simplified path-loss model plus log-normal shadowing that best fit this data.
(b) Find the path loss at 2 km based on this model.
(c) Find the outage probability at a distance d assuming the received power at d due to path loss alone is 10 dB above the required power for non-outage.

2-19. Consider a cellular system operating at 900 MHz where propagation follows free-space path loss with variations about this path loss due to log-normal shadowing with $\sigma = 6$ dB.

Table 2.5: Path-loss measurements for Problem 2-18

Distance from transmitter	P_r/P_t
5 m	−60 dB
25 m	−80 dB
65 m	−105 dB
110 m	−115 dB
400 m	−135 dB
1000 m	−150 dB

Suppose that for acceptable voice quality a signal-to-noise power ratio of 15 dB is required at the mobile. Assume the base station transmits at 1 W and that its antenna has a 3-dB gain. There is no antenna gain at the mobile, and the receiver noise in the bandwidth of interest is −40 dBm. Find the maximum cell size such that a mobile on the cell boundary will have acceptable voice quality 90% of the time.

2-20. In this problem we will simulate the log-normal fading process over distance based on the autocovariance model (2.50). As described in the text, the simulation first generates a white noise process and then passes it through a first-order filter with a pole at $e^{-\delta/X_c}$. Assume $X_c = 20$ m and plot the resulting log-normal fading process over a distance d ranging from 0 m to 200 m, sampling the process every meter. You should normalize your plot about 0 dB, since the mean of the log-normal shadowing is captured by path loss.

2-21. In this problem we will explore the impact of different log-normal shadowing parameters on outage probability. Consider a cellular system where the received signal power is distributed according to a log-normal distribution with mean μ dBm and standard deviation σ_ψ dBm. Assume the received signal power must be above 10 dBm for acceptable performance.

    (a) What is the outage probability when the log-normal distribution has $\mu_\psi = 15$ dBm and $\sigma_\psi = 8$ dBm?

    (b) For $\sigma_\psi = 4$ dBm, find the value of μ_ψ required for the outage probability to be less than 1% – a typical value for cellular systems.

    (c) Repeat part (b) for $\sigma_\psi = 12$ dBm.

    (d) One proposed technique for reducing outage probability is to use *macrodiversity*, where a mobile unit's signal is received by multiple base stations and then combined. This can only be done if multiple base stations are able to receive a given mobile's signal, which is typically the case for CDMA systems. Explain why this might reduce outage probability.

2-22. Derive the formula for coverage area (2.61) by applying integration by parts to (2.59).

2-23. Find the coverage area for a microcellular system where path loss follows the simplified model (with $\gamma = 3$, $d_0 = 1$, and $K = 0$ dB) and there is also log-normal shadowing with $\sigma = 4$ dB. Assume a cell radius of 100 m, a transmit power of 80 mW, and a minimum received power requirement of $P_{\min} = -100$ dBm.

2-24. Consider a cellular system where (a) path loss follows the simplified model with $\gamma = 6$ and (b) there is also log-normal shadowing with $\sigma = 8$ dB. If the received power at the cell boundary due to path loss is 20 dB higher than the minimum required received power for non-outage, find the cell coverage area.

2-25. In microcells, path-loss exponents usually range from 2 to 6 and shadowing standard deviation typically ranges from 4 to 12. Given a cellular system in which the received power due to path loss at the cell boundary equals the desired level for non-outage, find the path loss and shadowing parameters within these ranges that yield the best and worst coverage area. What is the coverage area when these parameters are in the middle of their typical ranges?

REFERENCES

[1]   D. Parsons, *The Mobile Radio Propagation Channel,* Wiley, New York, 1994.

[2]   M. Pätzold, *Mobile Fading Channels,* Wiley, New York, 2002.

[3]   *IEEE J. Sel. Areas Commun.,* Special Issue on Channel and Propagation Modeling for Wireless Systems Design, April 2002 and August 2002.

[4]   *IEEE J. Sel. Areas Commun.,* Special Issue on Ultra-Wideband Radio in Multiaccess Wireless Communications, December 2002.

[5]   T. S. Rappaport, *Wireless Communications – Principles and Practice,* 2nd ed., Prentice-Hall, Englewood Cliffs, NJ, 2001.

[6]   W. C. Jakes, Jr., *Microwave Mobile Communications,* Wiley, New York, 1974 [reprinted by IEEE Press].

[7]   A. S. Y. Poon and R. W. Brodersen, "The role of multiple-antenna systems in emerging open access environments," *EE Times Commun. Design Conf.,* October 2003.

[8]   T. Kurner, D. J. Cichon, and W. Wiesbeck, "Concepts and results for 3D digital terrain-based wave propagation models: An overview," *IEEE J. Sel. Areas Commun.,* pp. 1002–12, September 1993.

[9]   N. Amitay, "Modeling and computer simulation of wave propagation in lineal line-of-sight microcells," *IEEE Trans. Veh. Tech.,* pp. 337–42, November 1992.

[10] J. W. McKown and R. L. Hamilton, Jr., "Ray tracing as a design tool for radio networks," *IEEE Network,* pp. 27–30, November 1991.

[11] K. A. Remley, H. R. Anderson, and A. Weisshar, "Improving the accuracy of ray-tracing techniques for indoor propagation modeling," *IEEE Trans. Veh. Tech.,* pp. 2350–8, November 2000.

[12] H.-J. Li, C.-C. Chen, T.-Y. Liu, and H.-C. Lin, "Applicability of ray-tracing techniques for prediction of outdoor channel characteristics," *IEEE Trans. Veh. Tech.,* pp. 2336–49, November 2000.

[13] A. Domazetovic, L. J. Greenstein, N. Mandayan, and I. Seskar, "A new modeling approach for wireless channels with predictable path geometries," *Proc. IEEE Veh. Tech. Conf.,* September 2002.

[14] J. H. Tarng, W.-S. Liu, Y.-F. Huang, and J.-M. Huang, "A novel and efficient hybrid model of radio multipath-fading channels in indoor environments," *IEEE Trans. Ant. Prop.,* pp. 585–94, March 2003.

[15] A. J. Rustako, Jr., N. Amitay, G. J. Owens, and R. S. Roman, "Radio propagation at microwave frequencies for line-of-sight microcellular mobile and personal communications," *IEEE Trans. Veh. Tech.,* pp. 203–10, February 1991.

[16] W. C. Y. Lee, *Mobile Communications Engineering,* McGraw-Hill, New York, 1982.

[17] J.-F. Wagen, "Signal strength measurements at 881 MHz for urban microcells in downtown Tampa," *Proc. IEEE Globecom Conf.,* pp. 1313–17, December 1991.

[18] R. J. C. Bultitude and G. K. Bedal, "Propagation characteristics on microcellular urban mobile radio channels at 910 MHz," *IEEE J. Sel. Areas Commun.*, pp. 31–9, January 1989.

[19] J.-E. Berg, R. Bownds, and F. Lotse, "Path loss and fading models for microcells at 900 MHz," *Proc. IEEE Veh. Tech. Conf.*, pp. 666–71, May 1992.

[20] J. H. Whitteker, "Measurements of path loss at 910 MHz for proposed microcell urban mobile systems," *IEEE Trans. Veh. Tech.*, pp. 125–9, August 1988.

[21] H. Börjeson, C. Bergljung, and L. G. Olsson, "Outdoor microcell measurements at 1700 MHz," *Proc. IEEE Veh. Tech. Conf.*, pp. 927–31, May 1992.

[22] K. Schaubach, N. J. Davis IV, and T. S. Rappaport, "A ray tracing method for predicting path loss and delay spread in microcellular environments," *Proc. IEEE Veh. Tech. Conf.*, pp. 932–5, May 1992.

[23] F. Ikegami, S. Takeuchi, and S. Yoshida, "Theoretical prediction of mean field strength for urban mobile radio," *IEEE Trans. Ant. Prop.*, pp. 299–302, March 1991.

[24] M. C. Lawton and J. P. McGeehan, "The application of GTD and ray launching techniques to channel modeling for cordless radio systems," *Proc. IEEE Veh. Tech. Conf.*, pp. 125–30, May 1992.

[25] J. B. Keller, "Geometrical theory of diffraction," *J. Opt. Soc. Amer.*, 52, pp. 116–30, 1962.

[26] R. J. Luebbers, "Finite conductivity uniform GTD versus knife edge diffraction in prediction of propagation path loss," *IEEE Trans. Ant. Prop.*, pp. 70–6, January 1984.

[27] C. Bergljung and L. G. Olsson, "Rigorous diffraction theory applied to street microcell propagation," *Proc. IEEE Globecom Conf.*, pp. 1292–6, December 1991.

[28] R. G. Kouyoumjian and P. H. Pathak, "A uniform geometrical theory of diffraction for an edge in a perfectly conducting surface," *Proc. IEEE*, pp. 1448–61, November 1974.

[29] G. K. Chan, "Propagation and coverage prediction for cellular radio systems," *IEEE Trans. Veh. Tech.*, pp. 665–70, November 1991.

[30] K. C. Chamberlin and R. J. Luebbers, "An evaluation of Longley–Rice and GTD propagation models," *IEEE Trans. Ant. Prop.*, pp. 1093–8, November 1982.

[31] J. Walfisch and H. L. Bertoni, "A theoretical model of UHF propagation in urban environments," *IEEE Trans. Ant. Prop.*, pp. 1788–96, October 1988.

[32] M. I. Skolnik, *Introduction to Radar Systems,* 2nd ed., McGraw-Hill, New York, 1980.

[33] S. Y. Seidel, T. S. Rappaport, S. Jain, M. L. Lord, and R. Singh, "Path loss, scattering, and multipath delay statistics in four European cities for digital cellular and microcellular radiotelephone," *IEEE Trans. Veh. Tech.*, pp. 721–30, November 1991.

[34] S. T. S. Chia, "1700 MHz urban microcells and their coverage into buildings," *Proc. IEEE Ant. Prop. Conf.*, pp. 504–11, York, U.K., April 1991.

[35] D. Wong and D. C. Cox, "Estimating local mean signal power level in a Rayleigh fading environment," *IEEE Trans. Veh. Tech.*, pp. 956–9, May 1999.

[36] T. Okumura, E. Ohmori, and K. Fukuda, "Field strength and its variability in VHF and UHF land mobile service," *Rev. Elec. Commun. Lab.*, pp. 825–73, September/October 1968.

[37] M. Hata, "Empirical formula for propagation loss in land mobile radio services," *IEEE Trans. Veh. Tech.*, pp. 317–25, August 1980.

[38] European Cooperative in the Field of Science and Technical Research EURO-COST 231, "Urban transmission loss models for mobile radio in the 900 and 1800 MHz bands," rev. 2, The Hague, September 1991.

[39] E. McCune and K. Feher, "Closed-form propagation model combining one or more propagation constant segments," *Proc. IEEE Veh. Tech. Conf.*, pp. 1108–12, May 1997.

[40] D. Akerberg, "Properties of a TDMA picocellular office communication system," *Proc. IEEE Globecom Conf.*, pp. 1343–9, December 1988.

[41] V. Erceg, L. J. Greenstein, S. Y. Tjandra, S. R. Parkoff, A. Gupta, B. Kulic, A. A. Julius, and R. Bianchi, "An empirically based path loss model for wireless channels in suburban environments," *IEEE J. Sel. Areas Commun.*, pp. 1205–11, July 1999.

[42] M. Feuerstein, K. Blackard, T. Rappaport, S. Seidel, and H. Xia, "Path loss, delay spread, and outage models as functions of antenna height for microcellular system design," *IEEE Trans. Veh. Tech.*, pp. 487–98, August 1994.

[43] P. Harley, "Short distance attenuation measurements at 900 MHz and 1.8 GHz using low antenna heights for microcells," *IEEE J. Sel. Areas Commun.*, pp. 5–11, January 1989.

[44] S. Y. Seidel and T. S. Rappaport, "914 MHz path loss prediction models for indoor wireless communications in multifloored buildings," *IEEE Trans. Ant. Prop.*, pp. 207–17, February 1992.

[45] A. J. Motley and J. M. P. Keenan, "Personal communication radio coverage in buildings at 900 MHz and 1700 MHz," *Elec. Lett.*, pp. 763–4, June 1988.

[46] A. F. Toledo and A. M. D. Turkmani, "Propagation into and within buildings at 900, 1800, and 2300 MHz," *Proc. IEEE Veh. Tech. Conf.*, pp. 633–6, May 1992.

[47] F. C. Owen and C. D. Pudney, "Radio propagation for digital cordless telephones at 1700 MHz and 900 MHz," *Elec. Lett.*, pp. 52–3, September 1988.

[48] S. Y. Seidel, T. S. Rappaport, M. J. Feuerstein, K. L. Blackard, and L. Grindstaff, "The impact of surrounding buildings on propagation for wireless in-building personal communications system design," *Proc. IEEE Veh. Tech. Conf.*, pp. 814–18, May 1992.

[49] C. R. Anderson, T. S. Rappaport, K. Bae, A. Verstak, N. Tamakrishnan, W. Trantor, C. Shaffer, and L. T. Waton, "In-building wideband multipath characteristics at 2.5 and 60 GHz," *Proc. IEEE Veh. Tech. Conf.*, pp. 24–8, September 2002.

[50] L.-S. Poon and H.-S. Wang, "Propagation characteristic measurement and frequency reuse planning in an office building," *Proc. IEEE Veh. Tech. Conf.*, pp. 1807–10, June 1994.

[51] G. Durgin, T. S. Rappaport, and H. Xu, "Partition-based path loss analysis for in-home and residential areas at 5.85 GHz," *Proc. IEEE Globecom Conf.*, pp. 904–9, November 1998.

[52] A. F. Toledo, A. M. D. Turkmani, and J. D. Parsons, "Estimating coverage of radio transmission into and within buildings at 900, 1800, and 2300 MHz," *IEEE Pers. Commun. Mag.*, pp. 40–7, April 1998.

[53] R. Hoppe, G. Wölfle, and F. M. Landstorfer, "Measurement of building penetration loss and propagation models for radio transmission into buildings," *Proc. IEEE Veh. Tech. Conf.*, pp. 2298–2302, April 1999.

[54] E. H. Walker, "Penetration of radio signals into buildings in cellular radio environments," *Bell Systems Tech. J.*, pp. 2719–34, September 1983.

[55] W. C. Y. Lee, *Mobile Communication Design Fundamentals*, Sams, Indianapolis, IN, 1986.

[56] D. M. J. Devasirvathan, R. R. Murray, and D. R. Woiter, "Time delay spread measurements in a wireless local loop test bed," *Proc. IEEE Veh. Tech. Conf.*, pp. 241–5, May 1995.

[57] S. S. Ghassemzadeh, L. J. Greenstein, A. Kavcic, T. Sveinsson, and V. Tarokh, "Indoor path loss model for residential and commercial buildings," *Proc. IEEE Veh. Tech. Conf.*, pp. 3115–19, October 2003.

[58] A. J. Goldsmith, L. J. Greenstein, and G. J. Foschini, "Error statistics of real-time power measurements in cellular channels with multipath and shadowing," *IEEE Trans. Veh. Tech.*, pp. 439–46, August 1994.

[59] A. J. Goldsmith and L. J. Greenstein, "A measurement-based model for predicting coverage areas of urban microcells," *IEEE J. Sel. Areas Commun.*, pp. 1013–23, September 1993.

[60] M. Gudmundson, "Correlation model for shadow fading in mobile radio systems," *Elec. Lett.*, pp. 2145–6, November 7, 1991.

[61] G. L. Stuber, *Principles of Mobile Communications*, 2nd ed., Kluwer, Dordrecht, 2001.

[62] A. Algans, K. I. Pedersen, and P. E. Mogensen, "Experimental analysis of the joint statistical properties of azimuth spread, delay spread, and shadow fading," *IEEE J. Sel. Areas Commun.*, pp. 523–31, April 2002.

[63] M. Marsan and G. C. Hess, "Shadow variability in an urban land mobile radio environment," *Elec. Lett.*, pp. 646–8, May 1990.

[64] J. Weitzen and T. Lowe, "Measurement of angular and distance correlation properties of log-normal shadowing at 1900 MHz and its application to design of PCS systems," *IEEE Trans. Veh. Tech.*, pp. 265–73, March 2002.

[65] W. Turin, R. Jana, S. S. Ghassemzadeh, V. W. Rice, and V. Tarokh, "Autoregressive modeling of an indoor UWB channel," *Proc. IEEE Conf. UWB Syst. Technol.*, pp. 71–4, May 2002.

3

Statistical Multipath Channel Models

In this chapter we examine fading models for the constructive and destructive addition of different multipath components introduced by the channel. Although these multipath effects are captured in the ray-tracing models from Chapter 2 for deterministic channels, in practice deterministic channel models are rarely available and so we must characterize multipath channels statistically. In this chapter we model the multipath channel by a random time-varying impulse response. We will develop a statistical characterization of this channel model and describe its important properties.

If a single pulse is transmitted over a multipath channel then the received signal will appear as a pulse train, with each pulse in the train corresponding to the line-of-sight component or a distinct multipath component associated with a distinct scatterer or cluster of scatterers. The time delay spread of a multipath channel can result in significant distortion of the received signal. This delay spread equals the time delay between the arrival of the first received signal component (LOS or multipath) and the last received signal component associated with a single transmitted pulse. If the delay spread is small compared to the inverse of the signal bandwidth, then there is little time spreading in the received signal. However, if the delay spread is relatively large then there is significant time spreading of the received signal, which can lead to substantial signal distortion.

Another characteristic of the multipath channel is its time-varying nature. This time variation arises because either the transmitter or the receiver is moving and hence the location of reflectors in the transmission path, which gives rise to multipath, will change over time. Thus, if we repeatedly transmit pulses from a moving transmitter, we will observe changes in the amplitudes, delays, and number of multipath components corresponding to each pulse. However, these changes occur over a much larger time scale than the fading due to constructive and destructive addition of multipath components associated with a fixed set of scatterers. We will first use a generic time-varying channel impulse response to capture both fast and slow channel variations. We will then restrict this model to narrowband fading, where the channel bandwidth is small compared to the inverse delay spread. For this narrowband model we will assume a quasi-static environment featuring a fixed number of multipath components, each with fixed path loss and shadowing. For this quasi-static environment we then characterize the variations over short distances (small-scale variations) due to the constructive and destructive addition of multipath components. We also characterize the statistics of wideband multipath channels using two-dimensional transforms based on the underlying time-varying impulse response. Discrete-time and space-time channel models are also discussed.

3.1 Time-Varying Channel Impulse Response

Let the transmitted signal be as in Chapter 2:

$$s(t) = \text{Re}\{u(t)e^{j2\pi f_c t}\} = \text{Re}\{u(t)\}\cos(2\pi f_c t) - \text{Im}\{u(t)\}\sin(2\pi f_c t), \qquad (3.1)$$

where $u(t)$ is the equivalent lowpass signal for $s(t)$ with bandwidth B_u and where f_c is its carrier frequency. Neglecting noise, the corresponding received signal is the sum of the line-of-sight path and all resolvable multipath components:

$$r(t) = \text{Re}\left\{\sum_{n=0}^{N(t)} \alpha_n(t)u(t-\tau_n(t))e^{j(2\pi f_c(t-\tau_n(t))+\phi_{D_n})}\right\}, \qquad (3.2)$$

where $n = 0$ corresponds to the LOS path. The unknowns in this expression are: the number of resolvable multipath components $N(t)$, discussed in more detail below; and, for the LOS path and each multipath component, its path length $r_n(t)$ and corresponding delay $\tau_n(t) = r_n(t)/c$, Doppler phase shift $\phi_{D_n}(t)$, and amplitude $\alpha_n(t)$.

The nth resolvable multipath component may correspond to the multipath associated with a single reflector or with multiple reflectors clustered together that generate multipath components with similar delays, as shown in Figure 3.1. If each multipath component corresponds to just a single reflector then its corresponding amplitude $\alpha_n(t)$ is based on the path loss and shadowing associated with that multipath component, its phase change associated with delay $\tau_n(t)$ is $e^{-j2\pi f_c\tau_n(t)}$, and its Doppler shift $f_{D_n}(t) = v\cos\theta_n(t)/\lambda$ for $\theta_n(t)$ its angle of arrival relative to the direction of motion. This Doppler frequency shift leads to a Doppler phase shift of $\phi_{D_n} = \int_t 2\pi f_{D_n}(t)\,dt$. Suppose, however, that the nth multipath component results from a reflector cluster.[1] We say that two multipath components with delay τ_1 and τ_2 are *resolvable* if their delay difference significantly exceeds the inverse signal bandwidth: $|\tau_1 - \tau_2| \gg B_u^{-1}$. Multipath components that do not satisfy this resolvability criteria cannot be separated out at the receiver because $u(t-\tau_1) \approx u(t-\tau_2)$, and thus these components are *nonresolvable*. These nonresolvable components are combined into a single multipath component with delay $\tau \approx \tau_1 \approx \tau_2$ and an amplitude and phase corresponding to the sum of the different components. The amplitude of this summed signal will typically undergo fast variations due to the constructive and destructive combining of the nonresolvable multipath components. In general, wideband channels have resolvable multipath components so that each term in the summation of (3.2) corresponds to a single reflection or multiple nonresolvable components combined together, whereas narrowband channels tend to have nonresolvable multipath components contributing to each term in (3.2).

Since the parameters $\alpha_n(t)$, $\tau_n(t)$, and $\phi_{D_n}(t)$ associated with each resolvable multipath component change over time, they are characterized as random processes that we assume to be both stationary and ergodic. Thus, the received signal is also a stationary and ergodic

---

[1] Equivalently, a single "rough" reflector can create different multipath components with slightly different delays.

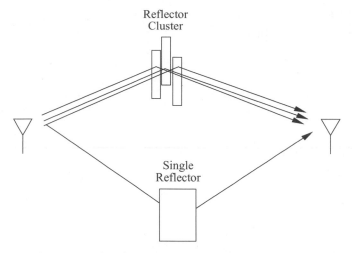

Figure 3.1: A single reflector and a reflector cluster.

random process. For wideband channels, where each term in (3.2) corresponds to a single reflector, these parameters change slowly as the propagation environment changes. For narrowband channels, where each term in (3.2) results from the sum of nonresolvable multipath components, the parameters can change quickly – on the order of a signal wavelength – owing to constructive and destructive addition of the different components.

We can simplify $r(t)$ by letting

$$\phi_n(t) = 2\pi f_c \tau_n(t) - \phi_{D_n}. \tag{3.3}$$

Then the received signal can be rewritten as

$$r(t) = \text{Re}\left\{ \left[\sum_{n=0}^{N(t)} \alpha_n(t) e^{-j\phi_n(t)} u(t - \tau_n(t)) \right] e^{j2\pi f_c t} \right\}. \tag{3.4}$$

Since $\alpha_n(t)$ is a function of path loss and shadowing while $\phi_n(t)$ depends on delay and Doppler, we typically assume that these two random processes are independent.

The received signal $r(t)$ is obtained by convolving the equivalent lowpass input signal $u(t)$ with the equivalent lowpass time-varying channel impulse response $c(\tau, t)$ of the channel and then upconverting to the carrier frequency:[2]

$$r(t) = \text{Re}\left\{ \left(\int_{-\infty}^{\infty} c(\tau, t) u(t - \tau) \, d\tau \right) e^{j2\pi f_c t} \right\}. \tag{3.5}$$

Note that $c(\tau, t)$ has two time parameters: the time t, when the impulse response is observed at the receiver; and the time $t - \tau$, when the impulse is launched into the channel relative to the observation time t. If at time t there is no physical reflector in the channel with multipath delay $\tau_n(t) = \tau$, then $c(\tau, t) = 0$. Although the definition of the time-varying channel impulse response might at first seem counterintuitive, $c(\tau, t)$ must be defined in this way to be consistent with the special case of time-invariant channels. Specifically, for time-invariant channels we have $c(\tau, t) = c(\tau, t + T)$; that is, the response at time t to an impulse at time $t - \tau$ equals the response at time $t + T$ to an impulse at time $t + T - \tau$. Setting $T = -t$,

---

[2] See Appendix A for discussion of the equivalent lowpass representation for bandpass signals and systems.

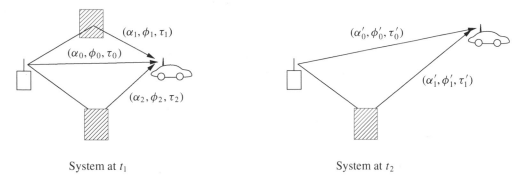

Figure 3.2: System multipath at two different measurement times.

we get that $c(\tau, t) = c(\tau, t - t) = c(\tau)$, where $c(\tau)$ is the standard time-invariant channel impulse response: the response at time τ to an impulse at time zero.[3]

We see from (3.4) and (3.5) that $c(\tau, t)$ must be given by

$$c(\tau, t) = \sum_{n=0}^{N(t)} \alpha_n(t) e^{-j\phi_n(t)} \delta(\tau - \tau_n(t)), \qquad (3.6)$$

where $c(\tau, t)$ represents the equivalent lowpass response of the channel at time t to an impulse at time $t - \tau$. Substituting (3.6) back into (3.5) yields (3.4), thereby confirming that (3.6) is the channel's equivalent lowpass time-varying impulse response:

$$
\begin{aligned}
r(t) &= \mathrm{Re}\left\{ \left[\int_{-\infty}^{\infty} c(\tau, t) u(t - \tau) \, d\tau \right] e^{j2\pi f_c t} \right\} \\
&= \mathrm{Re}\left\{ \left[\int_{-\infty}^{\infty} \sum_{n=0}^{N(t)} \alpha_n(t) e^{-j\phi_n(t)} \delta(\tau - \tau_n(t)) u(t - \tau) \, d\tau \right] e^{j2\pi f_c t} \right\} \\
&= \mathrm{Re}\left\{ \left[\sum_{n=0}^{N(t)} \alpha_n(t) e^{-j\phi_n(t)} \left(\int_{-\infty}^{\infty} \delta(\tau - \tau_n(t)) u(t - \tau) \, d\tau \right) \right] e^{j2\pi f_c t} \right\} \\
&= \mathrm{Re}\left\{ \left[\sum_{n=0}^{N(t)} \alpha_n(t) e^{-j\phi_n(t)} u(t - \tau_n(t)) \right] e^{j2\pi f_c t} \right\},
\end{aligned}
$$

where the last equality follows from the sifting property of delta functions:

$$\int \delta(\tau - \tau_n(t)) u(t - \tau) \, d\tau = \delta(t - \tau_n(t)) * u(t) = u(t - \tau_n(t)).$$

Some channel models assume a continuum of multipath delays, in which case the sum in (3.6) becomes an integral that simplifies to a time-varying complex amplitude associated with each multipath delay τ:

$$c(\tau, t) = \int \alpha(\xi, t) e^{-j\phi(\xi, t)} \delta(\tau - \xi) \, d\xi = \alpha(\tau, t) e^{-j\phi(\tau, t)}. \qquad (3.7)$$

For a concrete example of a time-varying impulse response, consider the system shown in Figure 3.2, where each multipath component corresponds to a single reflector. At time t_1

[3] By definition, $c(\tau, 0)$ is the response at time zero to an impulse at time $-\tau$, but since the channel is time invariant, this equals the response at time τ to an impulse at time zero.

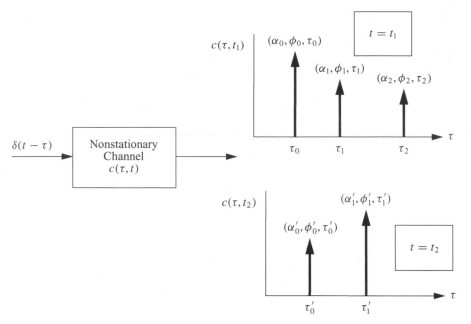

Figure 3.3: Response of nonstationary channel.

there are three multipath components associated with the received signal and with amplitude, phase, and delay triple $(\alpha_i, \phi_i, \tau_i)$, $i = 1, 2, 3$. Thus, impulses that were launched into the channel at time $t_1 - \tau_i$, $i = 1, 2, 3$, will all be received at time t_1, and impulses launched into the channel at any other time will not be received at t_1 (because there is no multipath component with the corresponding delay). The time-varying impulse response corresponding to t_1 equals

$$c(\tau, t_1) = \sum_{n=0}^{2} \alpha_n e^{-j\phi_n} \delta(\tau - \tau_n), \qquad (3.8)$$

and the channel impulse response for $t = t_1$ is shown in Figure 3.3. Figure 3.2 also shows the system at time t_2, where there are two multipath components associated with the received signal having amplitude, phase, and delay triple $(\alpha_i', \phi_i', \tau_i')$, $i = 1, 2$. Thus, impulses that were launched into the channel at time $t_2 - \tau_i'$, $i = 1, 2$, will all be received at time t_2, and impulses launched into the channel at any other time will not be received at t_2. The time-varying impulse response at t_2 equals

$$c(\tau, t_2) = \sum_{n=0}^{1} \alpha_n' e^{-j\phi_n'} \delta(\tau - \tau_n') \qquad (3.9)$$

and is also shown in Figure 3.3.

If the channel is time invariant, then the time-varying parameters in $c(\tau, t)$ become constant and $c(\tau, t) = c(\tau)$ is just a function of τ:

$$c(\tau) = \sum_{n=0}^{N} \alpha_n e^{-j\phi_n} \delta(\tau - \tau_n) \qquad (3.10)$$

for channels with discrete multipath components, and $c(\tau) = \alpha(\tau)e^{-j\phi(\tau)}$ for channels with a continuum of multipath components. For stationary channels, the response to an impulse at time t_1 is just a shifted version of its response to an impulse at time $t_2 \neq t_1$.

EXAMPLE 3.1: Consider a wireless LAN operating in a factory near a conveyor belt. The transmitter and receiver have a LOS path between them with gain α_0, phase ϕ_0, and delay τ_0. Every T_0 seconds, a metal item comes down the conveyor belt, creating an additional reflected signal path with gain α_1, phase ϕ_1, and delay τ_1. Find the time-varying impulse response $c(\tau, t)$ of this channel.

Solution: For $t \neq nT_0$ ($n = 1, 2, \ldots$), the channel impulse response simply corresponds to the LOS path. For $t = nT_0$, the channel impulse response includes both the LOS and reflected paths. Thus, $c(\tau, t)$ is given by

$$c(\tau, t) = \begin{cases} \alpha_0 e^{j\phi_0}\delta(\tau - \tau_0) & t \neq nT_0, \\ \alpha_0 e^{j\phi_0}\delta(\tau - \tau_0) + \alpha_1 e^{j\phi_1}\delta(\tau - \tau_1) & t = nT_0. \end{cases}$$

Note that, for typical carrier frequencies, the nth multipath component will have $f_c\tau_n(t) \gg 1$. For example, with $f_c = 1\,\text{GHz}$ and $\tau_n = 50\,\text{ns}$ (a typical value for an indoor system), $f_c\tau_n = 50 \gg 1$. Outdoor wireless systems have multipath delays much greater than 50 ns, so this property also holds for these systems. If $f_c\tau_n(t) \gg 1$ then a small change in the path delay $\tau_n(t)$ can lead to a large phase change in the nth multipath component with phase $\phi_n(t) = 2\pi f_c\tau_n(t) - \phi_{D_n} - \phi_0$. Rapid phase changes in each multipath component give rise to constructive and destructive addition of the multipath components constituting the received signal, which in turn causes rapid variation in the received signal strength. This phenomenon, called *fading,* will be discussed in more detail in subsequent sections.

The impact of multipath on the received signal depends on whether the spread of time delays associated with the LOS and different multipath components is large or small relative to the inverse signal bandwidth. If this channel delay spread is small then the LOS and all multipath components are typically nonresolvable, leading to the narrowband fading model described in the next section. If the delay spread is large then the LOS and all multipath components are typically resolvable into some number of discrete components, leading to the wideband fading model of Section 3.3. Observe that some of the discrete components in the wideband model consist of nonresolvable components. The delay spread is typically measured relative to the received signal component to which the demodulator is synchronized. Thus, for the time-invariant channel model of (3.10), if the demodulator synchronizes to the LOS signal component, which has the smallest delay τ_0, then the delay spread is a constant given by $T_m = \max_n[\tau_n - \tau_0]$. However, if the demodulator synchronizes to a multipath component with delay equal to the mean delay $\bar{\tau}$, then the delay spread is given by $T_m = \max_n|\tau_n - \bar{\tau}|$. In time-varying channels the multipath delays vary with time, so the delay spread T_m becomes a random variable. Moreover, some received multipath components have significantly lower power than others, so it's not clear how the delay associated with such components should be used in the characterization of delay spread. In particular, if the power of a multipath component is below the noise floor then it should not significantly contribute to the delay spread. These issues are typically dealt with by characterizing the delay spread relative to the channel power delay profile, defined in Section 3.3.1. Specifically, two common characterizations of channel delay spread – average delay spread and rms (root mean square) delay spread – are determined from the power delay profile. Other characterizations of delay spread, such as excees delay spread, the delay window, and the delay interval, are sometimes used as well [1, Chap. 5.4.1; 2, Chap. 6.7.1]. The exact characterization of delay spread is not that important for understanding the general impact of delay spread on multipath channels, as long as the characterization roughly measures the delay associated with significant multipath components. In our development below any reasonable characterization of delay spread T_m can be used, although we will typically use the rms delay spread. This is the most common characterization since, assuming the demodulator synchronizes to a signal component at the average delay spread, the rms delay spread is a good measure of the variation about this average. Channel delay spread is highly dependent on the propagation environment. In indoor channels delay spread typically ranges from 10 to 1000 nanoseconds, in suburbs it ranges from 200–2000 nanoseconds, and in urban areas it ranges from 1–30 microseconds [1, Chap. 5].

3.2 Narrowband Fading Models

Suppose the delay spread T_m of a channel is small relative to the inverse signal bandwidth B of the transmitted signal; that is, suppose $T_m \ll B^{-1}$. As discussed previously, the delay spread T_m for time-varying channels is usually characterized by the rms delay spread, but it can also be characterized in other ways. Under most delay spread characterizations, $T_m \ll B^{-1}$ implies that the delay associated with the ith multipath component $\tau_i \leq T_m$ for all i, so $u(t - \tau_i) \approx u(t)$ for all i and we can rewrite (3.4) as

$$r(t) = \text{Re}\left\{ u(t) e^{j2\pi f_c t} \left(\sum_n \alpha_n(t) e^{-j\phi_n(t)} \right) \right\}. \tag{3.11}$$

Equation (3.11) differs from the original transmitted signal by the complex scale factor in large parentheses. This scale factor is independent of the transmitted signal $s(t)$ and, in particular, of the equivalent lowpass signal $u(t)$ – as long as the narrowband assumption $T_m \ll 1/B$ is satisfied. In order to characterize the random scale factor caused by the multipath, we choose $s(t)$ to be an unmodulated carrier with random phase offset ϕ_0:

$$s(t) = \text{Re}\{e^{j(2\pi f_c t + \phi_0)}\} = \cos(2\pi f_c t + \phi_0), \tag{3.12}$$

which is narrowband for any T_m.

With this assumption the received signal becomes

$$r(t) = \text{Re}\left\{ \left[\sum_{n=0}^{N(t)} \alpha_n(t) e^{-j\phi_n(t)} \right] e^{j2\pi f_c t} \right\} = r_I(t) \cos 2\pi f_c t - r_Q(t) \sin 2\pi f_c t, \tag{3.13}$$

where the in-phase and quadrature components are given by

$$r_I(t) = \sum_{n=1}^{N(t)} \alpha_n(t) \cos \phi_n(t), \tag{3.14}$$

$$r_Q(t) = \sum_{n=1}^{N(t)} \alpha_n(t) \sin \phi_n(t) \tag{3.15}$$

and where the phase term

$$\phi_n(t) = 2\pi f_c \tau_n(t) - \phi_{D_n} - \phi_0 \tag{3.16}$$

now incorporates the phase offset ϕ_0 as well as the effects of delay and Doppler.

If $N(t)$ is large then we can invoke the central limit theorem and the fact that $\alpha_n(t)$ and $\phi_n(t)$ are independent for different components in order to approximate $r_I(t)$ and $r_Q(t)$ as jointly Gaussian random processes. The Gaussian property also holds for small N if the $\alpha_n(t)$ are Rayleigh distributed and the $\phi_n(t)$ are uniformly distributed on $[-\pi, \pi]$. This happens, again by the central limit theorem, when the nth multipath component results from a reflection cluster with a large number of nonresolvable multipath components [3].

3.2.1 Autocorrelation, Cross-Correlation, and Power Spectral Density

We now derive the autocorrelation and cross-correlation of the in-phase and quadrature received signal components $r_I(t)$ and $r_Q(t)$. Our derivations are based on some key assumptions that generally apply to propagation models without a dominant LOS component. Thus, these formulas are not typically valid when a dominant LOS component exists. We assume throughout this section that the amplitude $\alpha_n(t)$, multipath delay $\tau_n(t)$, and Doppler frequency $f_{D_n}(t)$ are changing slowly enough to be considered constant over the time intervals of interest: $\alpha_n(t) \approx \alpha_n$, $\tau_n(t) \approx \tau_n$, and $f_{D_n}(t) \approx f_{D_n}$. This will be true when each of the resolvable multipath components is associated with a single reflector. With this assumption the Doppler phase shift[4] is $\phi_{D_n}(t) = \int_t 2\pi f_{D_n} \, dt = 2\pi f_{D_n} t$ and the phase of the nth multipath component becomes $\phi_n(t) = 2\pi f_c \tau_n - 2\pi f_{D_n} t - \phi_0$.

We now make a key assumption: we assume that, for the nth multipath component, the term $2\pi f_c \tau_n$ in $\phi_n(t)$ changes rapidly relative to all other phase terms in the expression. This is a reasonable assumption because f_c is large and hence the term $2\pi f_c \tau_n$ can go through a 360° rotation for a small change in multipath delay τ_n. Under this assumption, $\phi_n(t)$ is uniformly distributed on $[-\pi, \pi]$. Thus

$$\mathbf{E}[r_I(t)] = \mathbf{E}\left[\sum_n \alpha_n \cos \phi_n(t)\right] = \sum_n \mathbf{E}[\alpha_n] \mathbf{E}[\cos \phi_n(t)] = 0, \tag{3.17}$$

where the second equality follows from the independence of α_n and ϕ_n and the last equality follows from the uniform distribution on ϕ_n. Similarly we can show that $\mathbf{E}[r_Q(t)] = 0$. Thus, the received signal also has $\mathbf{E}[r(t)] = 0$: it is a zero-mean Gaussian process. If there

---

[4] We shall assume a Doppler phase shift at $t = 0$ of zero for simplicity, because this phase offset will not affect the analysis.

is a dominant LOS component in the channel then the phase of the received signal is dominated by the phase of the LOS component, which can be determined at the receiver, so the assumption of a random uniform phase no longer holds.

Consider now the autocorrelation of the in-phase and quadrature components. Using the independence of α_n and ϕ_n, the independence of ϕ_n and ϕ_m ($n \neq m$), and the uniform distribution of ϕ_n, we get that

$$\mathbf{E}[r_I(t)r_Q(t)] = \mathbf{E}\left[\sum_n \alpha_n \cos\phi_n(t) \sum_m \alpha_m \sin\phi_m(t)\right]$$

$$= \sum_n \sum_m \mathbf{E}[\alpha_n\alpha_m]\,\mathbf{E}[\cos\phi_n(t)\sin\phi_m(t)]$$

$$= \sum_n \mathbf{E}[\alpha_n^2]\,\mathbf{E}[\cos\phi_n(t)\sin\phi_n(t)]$$

$$= 0. \tag{3.18}$$

Thus, $r_I(t)$ and $r_Q(t)$ are uncorrelated and, since they are jointly Gaussian processes, this means they are independent.

Following a derivation similar to that in (3.18), we obtain the autocorrelation of $r_I(t)$ as

$$A_{r_I}(t, t+\tau) = \mathbf{E}[r_I(t)r_I(t+\tau)] = \sum_n \mathbf{E}[\alpha_n^2]\,\mathbf{E}[\cos\phi_n(t)\cos\phi_n(t+\tau)]. \tag{3.19}$$

Now making the substitutions $\phi_n(t) = 2\pi f_c\tau_n - 2\pi f_{D_n}t - \phi_0$ and $\phi_n(t+\tau) = 2\pi f_c\tau_n - 2\pi f_{D_n}(t+\tau) - \phi_0$ we have

$$\mathbf{E}[\cos\phi_n(t)\cos\phi_n(t+\tau)]$$
$$= .5\,\mathbf{E}[\cos 2\pi f_{D_n}\tau] + .5\,\mathbf{E}[\cos(4\pi f_c\tau_n - 4\pi f_{D_n}t - 2\pi f_{D_n}\tau - 2\phi_0)]. \tag{3.20}$$

Because $2\pi f_c\tau_n$ changes rapidly relative to all other phase terms and is uniformly distributed, the second expectation term in (3.20) goes to zero and thus

$$A_{r_I}(t, t+\tau) = .5\sum_n \mathbf{E}[\alpha_n^2]\,\mathbf{E}[\cos(2\pi f_{D_n}\tau)] = .5\sum_n \mathbf{E}[\alpha_n^2]\cos\left(2\pi v\tau\cos\frac{\theta_n}{\lambda}\right), \tag{3.21}$$

since $f_{D_n} = v\cos\theta_n/\lambda$ is assumed fixed. Observe that $A_{r_I}(t, t+\tau)$ depends only on τ, $A_{r_I}(t, t+\tau) = A_{r_I}(\tau)$, and thus $r_I(t)$ is a wide-sense stationary (WSS) random process.

Using a similar derivation we can show that the quadrature component is also WSS with autocorrelation $A_{r_Q}(\tau) = A_{r_I}(\tau)$. In addition, the cross-correlation between the in-phase and quadrature components depends only on the time difference τ and is given by

$$A_{r_I,r_Q}(t, t+\tau) = A_{r_I,r_Q}(\tau) = \mathbf{E}[r_I(t)r_Q(t+\tau)]$$

$$= -.5\sum_n \mathbf{E}[\alpha_n^2]\sin\left(2\pi v\tau\cos\frac{\theta_n}{\lambda}\right) = -\mathbf{E}[r_Q(t)r_I(t+\tau)]. \tag{3.22}$$

Using these results, we can show that the received signal

$$r(t) = r_I(t)\cos(2\pi f_c t) - r_Q(t)\sin(2\pi f_c t)$$

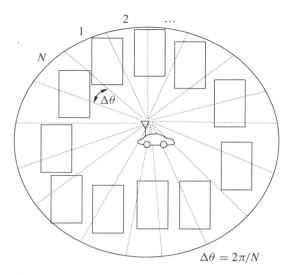

$$\Delta\theta = 2\pi/N$$

Figure 3.4: Dense scattering environment.

is also WSS with autocorrelation

$$A_r(\tau) = \mathbf{E}[r(t)r(t + \tau)] = A_{r_I}(\tau) \cos(2\pi f_c \tau) + A_{r_I, r_Q}(\tau) \sin(2\pi f_c \tau). \qquad (3.23)$$

In order to further simplify (3.21) and (3.22), we must make additional assumptions about the propagation environment. We will focus on the *uniform scattering environment* introduced by Clarke [4] and further developed by Jakes [5, Chap. 1]. In this model, the channel consists of many scatterers densely packed with respect to angle, as shown in Figure 3.4. Thus, we assume N multipath components with angle of arrival $\theta_n = n\Delta\theta$, where $\Delta\theta = 2\pi/N$. We also assume that each multipath component has the same received power and so $\mathbf{E}[\alpha_n^2] = 2P_r/N$, where P_r is the total received power. Then (3.21) becomes

$$A_{r_I}(\tau) = \frac{P_r}{N} \sum_{n=1}^{N} \cos\left(2\pi v \tau \cos \frac{n\Delta\theta}{\lambda}\right). \qquad (3.24)$$

Now making the substitution $N = 2\pi/\Delta\theta$ yields

$$A_{r_I}(\tau) = \frac{P_r}{2\pi} \sum_{n=1}^{N} \cos\left(2\pi v \tau \cos \frac{n\Delta\theta}{\lambda}\right)\Delta\theta. \qquad (3.25)$$

We now take the limit as the number of scatterers grows to infinity, which corresponds to uniform scattering from all directions. Then $N \to \infty$, $\Delta\theta \to 0$, and the summation in (3.25) becomes an integral:

$$A_{r_I}(\tau) = \frac{P_r}{2\pi} \int_0^{2\pi} \cos\left(2\pi v \tau \cos \frac{\theta}{\lambda}\right) d\theta = P_r J_0(2\pi f_D \tau), \qquad (3.26)$$

where

$$J_0(x) = \frac{1}{\pi} \int_0^{\pi} e^{-jx \cos\theta} \, d\theta$$

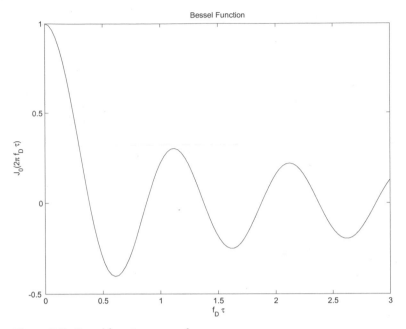

Figure 3.5: Bessel function versus $f_D\tau$.

is the Bessel function of zeroth order.[5] Similarly, for this uniform scattering environment,

$$A_{r_I,r_Q}(\tau) = \frac{P_r}{2\pi} \int \sin\left(2\pi v\tau \cos\frac{\theta}{\lambda}\right) d\theta = 0. \tag{3.27}$$

A plot of $J_0(2\pi f_D\tau)$ is shown in Figure 3.5. There are several interesting observations to make from this plot. First we see that the autocorrelation is zero for $f_D\tau \approx .4$ or, equivalently, for $v\tau \approx .4\lambda$. Thus, the signal decorrelates over a distance of approximately one half wavelength under the uniform θ_n assumption. This approximation is commonly used as a rule of thumb to determine many system parameters of interest. For example, we will see in Chapter 7 that independent fading paths obtained from multiple antennas can be exploited to remove some of the negative effects of fading. The antenna spacing must be such that each antenna receives an independent fading path; hence, based on our analysis here, an antenna spacing of $.4\lambda$ should be used. Another interesting characteristic of this plot is that the signal recorrelates after it becomes uncorrelated. Thus, we cannot assume that the signal remains independent from its initial value at $d = 0$ for separation distances greater than $.4\lambda$. Because of this recorrelation property, a Markov model is not completely accurate for Rayleigh fading. However, in many system analyses a correlation below .5 does not significantly degrade performance relative to uncorrelated fading [6, Chap. 9.6.5]. For such studies the fading process can be modeled as Markov by assuming that, once the correlation is close to zero (i.e., once the separation distance is greater than a half-wavelength), the signal remains decorrelated at all larger distances.

---

[5] Equation (3.26) can also be derived by assuming that $2\pi v\tau \cos\theta_n/\lambda$ in (3.21) and (3.22) is random with θ_n uniformly distributed, and then taking expectations with respect to θ_n. However, based on the underlying physical model, θ_n can be uniformly distributed only in a dense scattering environment. So the derivations are equivalent.

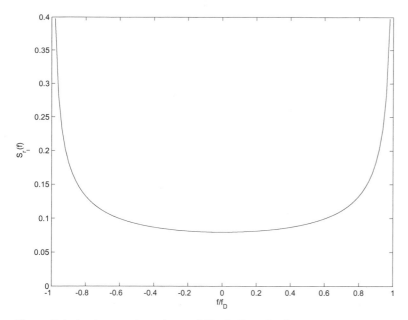

Figure 3.6: In-phase and quadrature PSD: $S_{r_I}(f) = S_{r_Q}(f)$.

The power spectral densities (PSDs) of $r_I(t)$ and $r_Q(t)$ – denoted by $S_{r_I}(f)$ and $S_{r_Q}(f)$, respectively – are obtained by taking the Fourier transform of their respective autocorrelation functions relative to the delay parameter τ. Since these autocorrelation functions are equal, so are the PSDs. Thus

$$S_{r_I}(f) = S_{r_Q}(f) = \mathcal{F}[A_{r_I}(\tau)] = \begin{cases} \dfrac{2P_r}{\pi f_D} \dfrac{1}{\sqrt{1 - (f/f_D)^2}} & |f| \leq f_D, \\ 0 & \text{else.} \end{cases} \qquad (3.28)$$

This PSD is shown in Figure 3.6.

To obtain the PSD of the received signal $r(t)$ under uniform scattering we use (3.23) with $A_{r_I,r_Q}(\tau) = 0$, (3.28), and simple properties of the Fourier transform to obtain

$$S_r(f) = \mathcal{F}[A_r(\tau)] = .25[S_{r_I}(f - f_c) + S_{r_I}(f + f_c)]$$

$$= \begin{cases} \dfrac{P_r}{2\pi f_D} \dfrac{1}{\sqrt{1 - (|f - f_c|/f_D)^2}} & |f - f_c| \leq f_D, \\ 0 & \text{else.} \end{cases} \qquad (3.29)$$

Note that this PSD integrates to P_r, the total received power.

Since the PSD models the power density associated with multipath components as a function of their Doppler frequency, it can be viewed as the probability density function (pdf) of the random frequency due to Doppler associated with multipath. We see from Figure 3.6 that the PSD $S_{r_I}(f)$ goes to infinity at $f = \pm f_D$ and, consequently, the PSD $S_r(f)$ goes to infinity at $f = \pm f_c \pm f_D$. This will not be true in practice, since the uniform scattering model is just an approximation, but for environments with dense scatterers the PSD will generally be maximized at frequencies close to the maximum Doppler frequency. The intuition

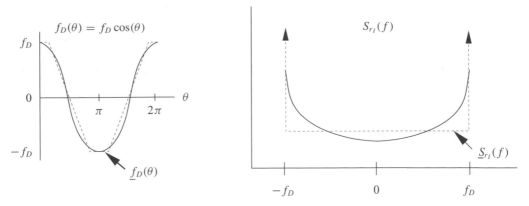

Figure 3.7: Cosine and PSD approximation by straight line segments.

for this behavior comes from the nature of the cosine function and the fact that (under our assumptions) the PSD corresponds to the pdf of the random Doppler frequency $f_D(\theta)$.

To see this, note that the uniform scattering assumption is based on many scattered paths arriving uniformly from all angles with the same average power. Thus, θ for a randomly selected path can be regarded as a uniform random variable on $[0, 2\pi]$. The pdf $p_{f_\theta}(f)$ of the random Doppler frequency $f(\theta)$ can then be obtained from the pdf of θ. By definition, $p_{f_\theta}(f)$ is proportional to the density of scatterers at Doppler frequency f. Hence, $S_{r_I}(f)$ is also proportional to this density, and we can characterize the PSD from the pdf $p_{f_\theta}(f)$. For this characterization, in Figure 3.7 we plot $f_D(\theta) = f_D \cos(\theta) = v/\lambda \cos(\theta)$ along with a (dashed) straight-line segment approximation $\underline{f_D}(\theta)$ to $f_D(\theta)$. On the right in this figure we plot the PSD $S_{r_I}(f)$ along with a dashed straight-line segment approximation to it, $\underline{S}_{r_I}(f)$, which corresponds to the Doppler approximation $\underline{f_D}(\theta)$. We see that $\cos(\theta) \approx \pm 1$ for a relatively large range of θ-values. Thus, multipath components with angles of arrival in this range of values have Doppler frequency $f_D(\theta) \approx \pm f_D$, so the power associated with all of these multipath components will add together in the PSD at $f \approx f_D$. This is shown in our approximation by the fact that the segments where $\underline{f_D}(\theta) = \pm f_D$ on the left lead to delta functions at $\pm f_D$ in the PSD approximation $\underline{S}_{r_I}(f)$ on the right. The segments where $\underline{f_D}(\theta)$ has uniform slope on the left lead to the flat part of $\underline{S}_{r_I}(f)$ on the right, since there is one multipath component contributing power at each angular increment. This explains the shape of $S_{r_I}(f)$ under uniform scattering. Formulas for the autocorrelation and PSD in nonuniform scattering – corresponding to more typical microcell and indoor environments – can be found in [5, Chap. 1; 7, Chap. 2].

The PSD is useful in constructing simulations for the fading process. A common method for simulating the envelope of a narrowband fading process is to pass two independent white Gaussian noise sources with PSD $N_0/2$ through lowpass filters with a frequency response $H(f)$ that satisfies

$$S_{r_I}(f) = S_{r_Q}(f) = \frac{N_0}{2}|H(f)|^2. \tag{3.30}$$

The filter outputs then correspond to the in-phase and quadrature components of the narrowband fading process with PSDs $S_{r_I}(f)$ and $S_{r_Q}(f)$. A similar procedure using discrete filters

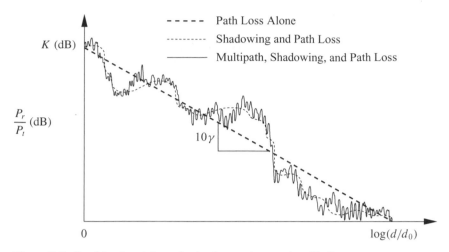

Figure 3.8: Combined path loss, shadowing, and narrowband fading.

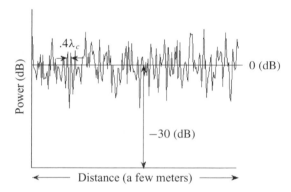

Figure 3.9: Narrowband fading.

can be used to generate discrete fading processes. Most communication simulation packages (e.g. Matlab, COSSAP) have standard modules that simulate narrowband fading based
on this method. More details on this simulation method, as well as alternative methods, can
be found in [1; 7; 8].

We have now completed our model for the three characteristics of power versus distance
exhibited in narrowband wireless channels. These characteristics are illustrated in Figure 3.8,
adding narrowband fading to the path loss and shadowing models developed in Chapter 2.
In this figure we see the decrease in signal power due to path loss decreasing as d^γ (γ is the
path-loss exponent); the more rapid variations due to shadowing, which change on the order
of the decorrelation distance X_c; and the very rapid variations due to multipath fading, which
change on the order of half the signal wavelength. If we blow up a small segment of this figure over distances where path loss and shadowing are constant we obtain Figure 3.9, which
plots dB fluctuation in received power versus linear distance $d = vt$ (not log distance). In
this figure the average received power P_r is normalized to 0 dBm. A mobile receiver trav-
eling at fixed velocity v would experience stationary and ergodic received power variations
over time as illustrated in this figure.

3.2.2 Envelope and Power Distributions

For any two Gaussian random variables X and Y, both with mean zero and equal variance σ^2, it can be shown that $Z = \sqrt{X^2 + Y^2}$ is Rayleigh distributed and that Z^2 is exponentially distributed. We have seen that, for $\phi_n(t)$ uniformly distributed, r_I and r_Q are both zero-mean Gaussian random variables. If we assume a variance of σ^2 for both in-phase and quadrature components, then the signal envelope

$$z(t) = |r(t)| = \sqrt{r_I^2(t) + r_Q^2(t)} \qquad (3.31)$$

is Rayleigh distributed with distribution

$$p_Z(z) = \frac{2z}{\bar{P}_r} \exp\left[-\frac{z^2}{\bar{P}_r}\right] = \frac{z}{\sigma^2} \exp\left[-\frac{z^2}{2\sigma^2}\right], \quad z \geq 0, \qquad (3.32)$$

where $\bar{P}_r = \sum_n \mathbf{E}[\alpha_n^2] = 2\sigma^2$ is the average received signal power of the signal – that is, the received power based on path loss and shadowing alone.

We obtain the power distribution by making the change of variables $z^2(t) = |r(t)|^2$ in (3.32) to obtain

$$p_{Z^2}(x) = \frac{1}{\bar{P}_r} e^{-x/\bar{P}_r} = \frac{1}{2\sigma^2} e^{-x/2\sigma^2}, \quad x \geq 0. \qquad (3.33)$$

Thus, the received signal power is exponentially distributed with mean $2\sigma^2$. The equivalent lowpass signal for $r(t)$ is given by $r_{LP}(t) = r_I(t) + jr_Q(t)$, which has phase $\theta = \arctan(r_Q(t)/r_I(t))$. For $r_I(t)$ and $r_Q(t)$ uncorrelated Gaussian random variables we can show that θ is uniformly distributed and independent of $|r_{LP}|$. So $r(t)$ has a Rayleigh-distributed amplitude and uniform phase, and the two are mutually independent.

EXAMPLE 3.2: Consider a channel with Rayleigh fading and average received power $\bar{P}_r = 20$ dBm. Find the probability that the received power is below 10 dBm.

Solution: We have $\bar{P}_r = 20$ dBm $= 100$ mW. We want to find the probability that $Z^2 < 10$ dBm $= 10$ mW. Thus

$$p(Z^2 < 10) = \int_0^{10} \frac{1}{100} e^{-x/100}\, dx = .095.$$

If the channel has a fixed LOS component then $r_I(t)$ and $r_Q(t)$ are not zero-mean variables. In this case the received signal equals the superposition of a complex Gaussian component and a LOS component. The signal envelope in this case can be shown to have a Rician distribution [9] given by

$$p_Z(z) = \frac{z}{\sigma^2} \exp\left[\frac{-(z^2 + s^2)}{2\sigma^2}\right] I_0\left(\frac{zs}{\sigma^2}\right), \quad z \geq 0, \qquad (3.34)$$

where $2\sigma^2 = \sum_{n,n\neq 0} \mathbf{E}[\alpha_n^2]$ is the average power in the non-LOS multipath components and $s^2 = \alpha_0^2$ is the power in the LOS component. The function I_0 is the modified Bessel function of zeroth order. The average received power in the Rician fading is given by

$$\bar{P}_r = \int_0^\infty z^2 p_Z(z)\, dz = s^2 + 2\sigma^2. \tag{3.35}$$

The Rician distribution is often described in terms of a fading parameter K, defined by

$$K = \frac{s^2}{2\sigma^2}. \tag{3.36}$$

Thus, K is the ratio of the power in the LOS component to the power in the other (non-LOS) multipath components. For $K = 0$ we have Rayleigh fading and for $K = \infty$ we have no fading (i.e., a channel with no multipath and only a LOS component). The fading parameter K is therefore a measure of the severity of the fading: a small K implies severe fading, a large K implies relatively mild fading. Making the substitutions $s^2 = KP_r/(K+1)$ and $2\sigma^2 = P_r/(K+1)$, we can write the Rician distribution in terms of K and P_r as

$$p_Z(z) = \frac{2z(K+1)}{\bar{P}_r} \exp\left[-K - \frac{(K+1)z^2}{\bar{P}_r}\right] I_0\left(2z\sqrt{\frac{K(K+1)}{\bar{P}_r}}\right), \quad z \geq 0. \tag{3.37}$$

Both the Rayleigh and Rician distributions can be obtained by using mathematics to capture the underlying physical properties of the channel models [3; 9]. However, some experimental data does not fit well into either of these distributions. Thus, a more general fading distribution was developed whose parameters can be adjusted to fit a variety of empirical measurements. This distribution is called the Nakagami fading distribution and is given by

$$p_Z(z) = \frac{2m^m z^{2m-1}}{\Gamma(m)\bar{P}_r^m} \exp\left[\frac{-mz^2}{\bar{P}_r}\right], \quad m \geq .5, \tag{3.38}$$

where $\bar{P}_r$ is the average received power and $\Gamma(\cdot)$ is the Gamma function. The Nakagami distribution is parameterized by $\bar{P}_r$ and the fading parameter m. For $m = 1$ the distribution in (3.38) reduces to Rayleigh fading. For $m = (K+1)^2/(2K+1)$ the distribution in (3.38) is approximately Rician fading with parameter K. For $m = \infty$ there is no fading: $Z = \sqrt{P_r}$ is a constant. Thus, the Nakagami distribution can model both Rayleigh and Rician distributions as well as more general ones. Note that some empirical measurements support values of the m-parameter less than unity, in which case the Nakagami fading causes more severe performance degradation than Rayleigh fading. The power distribution for Nakagami fading, obtained by a change of variables, is given by

$$p_{Z^2}(x) = \left(\frac{m}{\bar{P}_r}\right)^m \frac{x^{m-1}}{\Gamma(m)} \exp\left[\frac{-mx}{\bar{P}_r}\right]. \tag{3.39}$$

3.2.3 Level Crossing Rate and Average Fade Duration

The envelope level crossing rate L_Z is defined as the expected rate (in crossings per second) at which the signal envelope crosses the level Z in the downward direction. Obtaining L_Z requires the joint distribution $p(z, \dot{z})$ of the signal envelope $z = |r|$ and its derivative with respect to time, $\dot{z}$. We now derive L_Z based on this joint distribution.

Consider the fading process shown in Figure 3.10. The expected amount of time that the signal envelope spends in the interval $(Z, Z+dz)$ with envelope slope in the range $[\dot{z}, \dot{z}+d\dot{z}]$

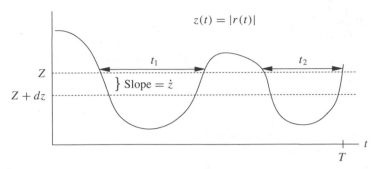

Figure 3.10: Level crossing rate and fade duration for fading process.

over time duration dt is $A = p(Z, \dot{z}) \, dz \, d\dot{z} \, dt$. The time required to cross from Z to $Z + dz$ once for a given envelope slope $\dot{z}$ is $B = dz/\dot{z}$. The ratio $A/B = \dot{z}p(Z, \dot{z}) \, d\dot{z} \, dt$ is the expected number of crossings of the envelope z within the interval $(Z, Z + dz)$ for a given envelope slope $\dot{z}$ over time duration dt. The expected number of crossings of the envelope level Z for slopes between $\dot{z}$ and $\dot{z} + d\dot{z}$ in a time interval $[0, T]$ in the downward direction is thus

$$\int_0^T \dot{z}p(Z, \dot{z}) \, d\dot{z} \, dt = \dot{z}p(Z, \dot{z}) \, d\dot{z}T. \tag{3.40}$$

Hence the expected number of crossings of the envelope level Z with negative slope over the interval $[0, T]$ is

$$N_Z = T \int_{-\infty}^0 \dot{z}p(Z, \dot{z}) \, d\dot{z}. \tag{3.41}$$

Finally, the expected number of crossings of the envelope level Z per second – that is, the level crossing rate – is

$$L_Z = \frac{N_Z}{T} = \int_{-\infty}^0 \dot{z}p(Z, \dot{z}) \, d\dot{z}. \tag{3.42}$$

Note that this is a general result that applies for any random process.

The joint pdf of z and $\dot{z}$ for Rician fading was derived in [9] and can also be found in [7, Chap. 2.1]. The level crossing rate for Rician fading is then obtained by using this pdf in (3.42), yielding

$$L_Z = \sqrt{2\pi(K + 1)} f_D \rho e^{-K - (K+1)\rho^2} I_0\left(2\rho\sqrt{K(K + 1)}\right), \tag{3.43}$$

where $\rho = Z/\sqrt{\bar{P}_r}$. It is easily shown that the rate at which the received signal power crosses a threshold value γ_0 obeys the same formula (3.43) with $\rho = \sqrt{\gamma_0/\bar{P}_r}$. For Rayleigh fading ($K = 0$) the level crossing rate simplifies to

$$L_Z = \sqrt{2\pi} f_D \rho e^{-\rho^2}, \tag{3.44}$$

where $\rho = Z/\sqrt{\bar{P}_r}$.

We define the average signal fade duration as the average time that the signal envelope stays below a given target level Z. This target level is often obtained from the signal amplitude or power level required for a given performance metric such as bit error rate. If the signal amplitude or power falls below its target then we say the system is in outage. Let t_i denote the duration of the ith fade below level Z over a time interval $[0, T]$, as illustrated in Figure 3.10. Thus t_i equals the length of time that the signal envelope stays below Z on its ith crossing. Since $z(t)$ is stationary and ergodic, for T sufficiently large we have

$$p(z(t) < Z) = \frac{1}{T} \sum_i t_i. \tag{3.45}$$

Thus, for T sufficiently large, the average fade duration is

$$\bar{t}_Z = \frac{1}{TL_Z} \sum_{i=1}^{L_Z T} t_i \approx \frac{p(z(t) < Z)}{L_Z}. \tag{3.46}$$

Using the Rayleigh distribution for $p(z(t) < Z)$ then yields

$$\bar{t}_Z = \frac{e^{\rho^2} - 1}{\rho f_D \sqrt{2\pi}} \tag{3.47}$$

with $\rho = Z/\sqrt{\bar{P}_r}$. Note that (3.47) is the average fade duration for the signal envelope (amplitude) level with Z the target amplitude and $\sqrt{\bar{P}_r}$ the average envelope level. By a change of variables it is easily shown that (3.47) also yields the average fade duration for the signal power level with $\rho = \sqrt{P_0/\bar{P}_r}$, where P_0 is the target power level and $\bar{P}_r$ is the average power level. The average fade duration (3.47) decreases with Doppler, since as a channel changes more quickly it remains below a given fade level for a shorter period of time. The average fade duration also generally increases with ρ for $\rho \gg 1$. That is because the signal is more likely to be below the target as the target level increases relative to the average. The average fade duration for Rician fading is more difficult to compute; it can be found in [7, Chap. 1.4].

The average fade duration indicates the number of bits or symbols affected by a deep fade. Specifically, consider an uncoded system with bit time T_b. Suppose the probability of bit error is high when $z < Z$. In this case, if $T_b \approx \bar{t}_Z$ then the system will likely experience single error events, where bits that are received in error have the previous and subsequent bits received correctly (since $z > Z$ for these bits). On the other hand, if $T_b \ll \bar{t}_Z$ then many subsequent bits are received with $z < Z$, so large bursts of errors are likely. Finally, if $T_b \gg \bar{t}_Z$ then, since the fading is integrated over a bit time in the demodulator, the fading gets averaged out and so can be neglected. These issues will be explored in more detail in Chapter 8, where we consider coding and interleaving.

EXAMPLE 3.3: Consider a voice system with acceptable BER when the received signal power is at or above half its average value. If the BER is below its acceptable level for more than 120 ms, users will turn off their phone. Find the range of Doppler values in a Rayleigh fading channel such that the average time duration when users have unacceptable voice quality is less than $t = 60$ ms.

Solution: The target received signal value is half the average, so $P_0 = .5\bar{P}_r$ and thus $\rho = \sqrt{.5}$. We require

$$\bar{t}_Z = \frac{e^{.5} - 1}{f_D \sqrt{\pi}} \leq t = .060$$

and thus $f_D \geq (e - 1)/(.060\sqrt{2\pi}) = 6.1 \, \text{Hz}$.

3.2.4 Finite-State Markov Channels

The complex mathematical characterization of flat fading described in the previous sections can be difficult to incorporate into wireless performance analysis. Therefore, simpler models that capture the main features of flat fading channels are needed for these analytical calculations. One such model is a finite-state Markov channel (FSMC). In this model, fading is approximated as a discrete-time Markov process with time discretized to a given interval T (typically the symbol period). Specifically, the set of all possible fading gains is modeled as a set of finite channel states. The channel varies over these states at each interval T according to a set of Markov transition probabilities. FSMCs have been used to approximate both mathematical and experimental fading models, including satellite channels [10], indoor channels [11], Rayleigh fading channels [12; 13], Rician fading channels [14], and Nakagami-m fading channels [15]. They have also been used for system design and system performance analysis [13; 16]. First-order FSMC models have been shown to be deficient for computing performance analysis, so higher-order models are generally used. The FSMC models for fading typically model amplitude variations only, although there has been some work on FSMC models for phase in fading [17] or phase-noisy channels [18].

A detailed FSMC model for Rayleigh fading was developed in [12]. In this model the time-varying SNR γ associated with the Rayleigh fading lies in the range $0 \leq \gamma \leq \infty$. The FSMC model discretizes this fading range into regions so that the jth region R_j is defined as $R_j = \{\gamma : A_j \leq \gamma < A_{j+1}\}$, where the region boundaries $\{A_j\}$ and the total number of fade regions are parameters of the model. This model assumes that γ stays within the same region over time interval T and can only transition to the same region or adjacent regions at time $T + 1$. Thus, given that the channel is in state R_j at time T, at the next time interval the channel can only transition to R_{j-1}, R_j, or R_{j+1} – a reasonable assumption when $f_D T$ is small. Under this assumption, the transition probabilities between regions are derived in [12] as

$$p_{j,j+1} = \frac{L_{j+1}T}{\pi_j}, \quad p_{j,j-1} = \frac{L_j T}{\pi_j}, \quad p_{j,j} = 1 - p_{j,j+1} - p_{j,j-1}, \tag{3.48}$$

where L_j is the level crossing rate at A_j and π_j is the steady-state distribution corresponding to the jth region: $\pi_j = p(\gamma \in R_j) = p(A_j \leq \gamma < A_{j+1})$.

3.3 Wideband Fading Models

When the signal is not narrowband we get another form of distortion due to the multipath delay spread. In this case a short transmitted pulse of duration T will result in a received signal that is of duration $T + T_m$, where T_m is the multipath delay spread. Thus, the duration of the received signal may be significantly increased. This phenomenon is illustrated in Figure 3.11. In the figure, a pulse of width T is transmitted over a multipath channel. As discussed in

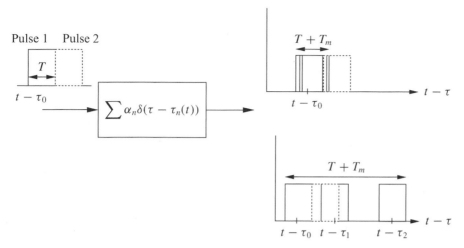

Figure 3.11: Multipath resolution.

Chapter 5, linear modulation consists of a train of pulses where each pulse carries information in its amplitude and/or phase corresponding to a data bit or symbol.[6] If the multipath
delay spread $T_m \ll T$ then the multipath components are received roughly on top of one another, as shown in the upper right of the figure. The resulting constructive and destructive
interference causes narrowband fading of the pulse, but there is little time spreading of the
pulse and therefore little interference with a subsequently transmitted pulse. On the other
hand, if the multipath delay spread $T_m \gg T$, then each of the different multipath components
can be resolved, as shown in the lower right of the figure. However, these multipath components interfere with subsequently transmitted pulses (dashed pulses in the figure). This effect
is called *intersymbol interference* (ISI).

There are several techniques to mitigate the distortion due to multipath delay spread –
including equalization, multicarrier modulation, and spread spectrum, which are discussed
in Chapters 11–13. Mitigating ISI is not necessary if $T \gg T_m$, but this can place significant constraints on data rate. Multicarrier modulation and spread spectrum actually change
the characteristics of the transmitted signal to mostly avoid intersymbol interference; however, they still experience multipath distortion due to frequency-selective fading, which is
described in Section 3.3.2.

The difference between wideband and narrowband fading models is that, as the transmit
signal bandwidth B increases so that $T_m \approx B^{-1}$, the approximation $u(t - \tau_n(t)) \approx u(t)$ is
no longer valid. Thus, the received signal is a sum of all copies of the original signal, where
each copy is delayed in time by τ_n and shifted in phase by $\phi_n(t)$. The signal copies will combine destructively when their phase terms differ significantly and will distort the direct path
signal when $u(t - \tau_n)$ differs from $u(t)$.

Although the approximation in (3.11) no longer applies when the signal bandwidth is large
relative to the inverse of the multipath delay spread, if the number of multipath components
is large and the phase of each component is uniformly distributed then the received signal

---

[6] Linear modulation typically uses nonsquare pulse shapes for bandwidth efficiency, as discussed in Section 5.4.

will still be a zero-mean complex Gaussian process with a Rayleigh-distributed envelope. However, wideband fading differs from narrowband fading in terms of the resolution of the different multipath components. Specifically, for narrowband signals, the multipath components have a time resolution that is less than the inverse of the signal bandwidth, so the multipath components characterized in equation (3.6) combine at the receiver to yield the original transmitted signal with amplitude and phase characterized by random processes. These random processes are in turn characterized by their autocorrelation (or PSD) and instantaneous distributions, as discussed in Section 3.2. However, with wideband signals, the received signal experiences distortion due to the delay spread of the different multipath components, so the received signal can no longer be characterized by just the amplitude and phase random processes. The effect of multipath on wideband signals must therefore take into account both the multipath delay spread and the time variations associated with the channel.

The starting point for characterizing wideband channels is the equivalent lowpass time-varying channel impulse response $c(\tau, t)$. Let us first assume that $c(\tau, t)$ is a continuous[7] deterministic function of τ and t. Recall that τ represents the impulse response associated with a given multipath delay and that t represents time variations. We can take the Fourier transform of $c(\tau, t)$ with respect to t as

$$S_c(\tau, \rho) = \int_{-\infty}^{\infty} c(\tau, t) e^{-j2\pi\rho t} \, dt. \tag{3.49}$$

We call $S_c(\tau, \rho)$ the *deterministic scattering function* of the equivalent lowpass channel impulse response $c(\tau, t)$. Since it is the Fourier transform of $c(\tau, t)$ with respect to the time-variation parameter t, the deterministic scattering function $S_c(\tau, \rho)$ captures the Doppler characteristics of the channel via the frequency parameter ρ.

In general, the time-varying channel impulse response $c(\tau, t)$ given by (3.6) is random instead of deterministic because of the random amplitudes, phases, and delays of the random number of multipath components. In this case we must characterize it statistically or via measurements. As long as the number of multipath components is large, we can invoke the central limit theorem to assume that $c(\tau, t)$ is a complex Gaussian process and hence that its statistical characterization is fully known from the mean, autocorrelation, and cross-correlation of its in-phase and quadrature components. As in the narrowband case, we assume that the phase of each multipath component is uniformly distributed. Thus, the in-phase and quadrature components of $c(\tau, t)$ are independent Gaussian processes with the same autocorrelation, a mean of zero, and a cross-correlation of zero. The same statistics hold for the in-phase and quadrature components if the channel contains only a small number of multipath rays – as long as each ray has a Rayleigh-distributed amplitude and uniform phase. Note that this model does not hold when the channel has a dominant LOS component.

The statistical characterization of $c(\tau, t)$ is thus determined by its *autocorrelation function,* defined as

$$A_c(\tau_1, \tau_2; t, t + \Delta t) = \mathbf{E}[c^*(\tau_1; t)c(\tau_2; t + \Delta t)]. \tag{3.50}$$

---

[7] The wideband channel characterizations in this section can also be made for discrete-time channels (discrete with respect to τ) by changing integrals to sums and Fourier transforms to discrete Fourier transforms.

Most actual channels are wide-sense stationary, so that the joint statistics of a channel measured at two different times t and $t + \Delta t$ depends only on the time difference Δt. For WSS channels, the autocorrelation of the corresponding bandpass channel $h(\tau, t) = \text{Re}\{c(\tau, t)e^{j2\pi f_c t}\}$ can be obtained [19] from $A_c(\tau_1, \tau_2; t, t + \Delta t)$ as[8] $A_h(\tau_1, \tau_2; t, t + \Delta t) = .5 \text{Re}\{A_c(\tau_1, \tau_2; t, t + \Delta t)e^{j2\pi f_c \Delta t}\}$. We will assume that our channel model is WSS, in which case the autocorrelation becomes independent of t:

$$A_c(\tau_1, \tau_2; \Delta t) = \mathbf{E}[c^*(\tau_1; t)c(\tau_2; t + \Delta t)]. \tag{3.51}$$

Moreover, in real environments the channel response associated with a given multipath component of delay τ_1 is uncorrelated with the response associated with a multipath component at a different delay $\tau_2 \neq \tau_1$, since the two components are caused by different scatterers. We say that such a channel has uncorrelated scattering (US). We denote channels that are WSS with US as WSSUS channels. The WSSUS channel model was first introduced by Bello in his landmark paper [19], where he also developed two-dimensional transform relationships associated with this autocorrelation. These relationships will be discussed in Section 3.3.4. Incorporating the US property into (3.51) yields

$$\mathbf{E}[c^*(\tau_1; t)c(\tau_2; t + \Delta t)] = A_c(\tau_1; \Delta t)\delta[\tau_1 - \tau_2] \triangleq A_c(\tau; \Delta t), \tag{3.52}$$

where $A_c(\tau; \Delta t)$ gives the average output power associated with the channel as a function of the multipath delay $\tau = \tau_1 = \tau_2$ and the difference Δt in observation time. This function assumes that, when $\tau_1 \neq \tau_2$, τ_1 and τ_2 satisfy $|\tau_1 - \tau_2| > B^{-1}$, since otherwise the receiver can't resolve the two components. In this case the two components are modeled as a single combined multipath component with delay $\tau \approx \tau_1 \approx \tau_2$.

The *scattering function* for random channels is defined as the Fourier transform of $A_c(\tau; \Delta t)$ with respect to the Δt parameter:

$$S_c(\tau, \rho) = \int_{-\infty}^{\infty} A_c(\tau, \Delta t)e^{-j2\pi\rho\Delta t} d\Delta t. \tag{3.53}$$

The scattering function characterizes the average output power associated with the channel as a function of the multipath delay τ and Doppler ρ. Note that we use the same notation $S_c(\tau, \rho)$ for the deterministic scattering and random scattering functions because the function is uniquely defined depending on whether the channel impulse response is deterministic or random. A typical scattering function is shown in Figure 3.12.

The most important characteristics of the wideband channel – including its power delay profile, coherence bandwidth, Doppler power spectrum, and coherence time – are derived from the channel autocorrelation $A_c(\tau, \Delta t)$ or the scattering function $S_c(\tau, \rho)$. These characteristics are described in subsequent sections.

---

[8] It is easily shown that the autocorrelation of the bandpass channel response $h(\tau, t)$ is $\mathbf{E}[h(\tau_1, t)h(\tau_2, t+\Delta t)] = .5 \text{Re}\{A_c(\tau_1, \tau_2; t, t + \Delta t)e^{j2\pi f_c \Delta t}\} + .5 \text{Re}\{\hat{A}_c(\tau_1, \tau_2; t, t + \Delta t)e^{j2\pi f_c(2t+\Delta t)}\}$, where $\hat{A}_c(\tau_1, \tau_2; t, t + \Delta t) = \mathbf{E}[c(\tau_1; t)c(\tau_2; t+\Delta t)]$. However, if $c(\tau, t)$ is WSS then $\hat{A}_c(\tau_1, \tau_2; t, t+\Delta t) = 0$, so $\mathbf{E}[h(\tau_1, t)h(\tau_2, t+\Delta t)] = .5 \text{Re}\{A_c(\tau_1, \tau_2; t, t + \Delta t)e^{j2\pi f_c \Delta t}\}$.

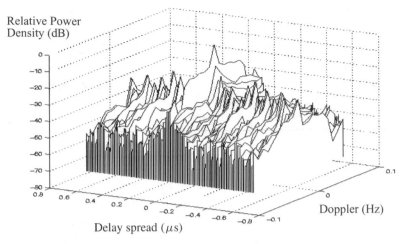

Figure 3.12: Scattering function.

3.3.1 Power Delay Profile

The *power delay profile* $A_c(\tau)$, also called the *multipath intensity profile,* is defined as the autocorrelation (3.52) with $\Delta t = 0$: $A_c(\tau) \triangleq A_c(\tau, 0)$. The power delay profile represents the average power associated with a given multipath delay, and it is easily measured empirically. The average and rms delay spread are typically defined in terms of the power delay profile $A_c(\tau)$ as

$$\mu_{T_m} = \frac{\int_0^\infty \tau A_c(\tau)\, d\tau}{\int_0^\infty A_c(\tau)\, d\tau}, \tag{3.54}$$

and

$$\sigma_{T_m} = \sqrt{\frac{\int_0^\infty (\tau - \mu_{T_m})^2 A_c(\tau)\, d\tau}{\int_0^\infty A_c(\tau)\, d\tau}}. \tag{3.55}$$

Note that $A_c(\tau) \geq 0$ for all τ, so if we define the distribution p_{T_m} of the random delay spread T_m as

$$p_{T_m}(\tau) = \frac{A_c(\tau)}{\int_0^\infty A_c(\tau)\, d\tau} \tag{3.56}$$

then μ_{T_m} and σ_{T_m} are (respectively) the mean and rms values of T_m, relative to this distribution. Defining the distribution of T_m by (3.56) – or, equivalently, defining the mean and rms delay spread by (3.54) and (3.55), respectively – weights the delay associated with a given multipath component by its relative power, so that weak multipath components contribute less to delay spread than strong ones. In particular, multipath components below the noise floor will not significantly affect these delay spread characterizations.

The time delay T where $A_c(\tau) \approx 0$ for $\tau \geq T$ can be used to roughly characterize the delay spread of the channel, and this value is often taken to be a small integer multiple of the rms delay spread. For example, we assume $A_c(\tau) \approx 0$ for $\tau > 3\sigma_{T_m}$. With this approximation, a linearly modulated signal with symbol period T_s experiences significant ISI if $T_s \ll \sigma_{T_m}$. Conversely, when $T_s \gg \sigma_{T_m}$ the system experiences negligible ISI. For calculations one can assume that $T_s \ll \sigma_{T_m}$ implies $T_s < \sigma_{T_m}/10$ and that $T_s \gg \sigma_{T_m}$ implies $T_s > 10\sigma_{T_m}$. If T_s is within an order of magnitude of σ_{T_m} then there will be some ISI, which may or may not significantly degrade performance, depending on the specifics of the system and channel. In later chapters we will study the performance degradation due to ISI in linearly modulated systems as well as ISI mitigation methods.

Although $\mu_{T_m} \approx \sigma_{T_m}$ in many channels with a large number of scatterers, the exact relationship between μ_{T_m} and σ_{T_m} depends on the shape of $A_c(\tau)$. For a channel with no LOS component and a small number of multipath components with approximately the same large delay, $\mu_{T_m} \gg \sigma_{T_m}$. In this case the large value of μ_{T_m} is a misleading metric of delay spread, since in fact all copies of the transmitted signal arrive at roughly the same time and the demodulator would synchronize to this common delay. It is typically assumed that the synchronizer locks to the multipath component at approximately the mean delay, in which case rms delay spread characterizes the time spreading of the channel.

EXAMPLE 3.4: The power delay profile is often modeled as having a one-sided exponential distribution:

$$A_c(\tau) = \frac{1}{\bar{T}_m} e^{-\tau/\bar{T}_m}, \quad \tau \geq 0.$$

Show that the average delay spread (3.54) is $\mu_{T_m} = \bar{T}_m$ and find the rms delay spread (3.55).

Solution: It is easily shown that $A_c(\tau)$ integrates to unity. The average delay spread is thus given by

$$\mu_{T_m} = \frac{1}{\bar{T}_m} \int_0^\infty \tau e^{-\tau/\bar{T}_m} \, d\tau = \bar{T}_m,$$

and the rms delay spread is

$$\sigma_{T_m} = \sqrt{\frac{1}{\bar{T}_m} \int_0^\infty \tau^2 e^{-\tau/\bar{T}_m} \, d\tau - \mu_{T_m}^2} = 2\bar{T}_m - \bar{T}_m = \bar{T}_m.$$

Thus, the average and rms delay spread are the same for exponentially distributed power delay profiles.

EXAMPLE 3.5: Consider a wideband channel with multipath intensity profile

$$A_c(\tau) = \begin{cases} e^{-\tau/.00001} & 0 \leq \tau \leq 20 \, \mu s, \\ 0 & \text{else.} \end{cases}$$

Find the mean and rms delay spreads of the channel and find the maximum symbol rate such that a linearly modulated signal transmitted through this channel does not experience ISI.

Solution: The average delay spread is

$$\mu_{T_m} = \frac{\int_0^{20 \cdot 10^{-6}} \tau e^{-\tau/.00001} \, d\tau}{\int_0^{20 \cdot 10^{-6}} e^{-\tau/.00001} \, d\tau} = 6.87 \, \mu s.$$

The rms delay spread is

$$\sigma_{T_m} = \sqrt{\frac{\int_0^{20 \cdot 10^{-6}} (\tau - \mu_{T_m})^2 e^{-\tau/.00001} \, d\tau}{\int_0^{20 \cdot 10^{-6}} e^{-\tau/.00001} \, d\tau}} = 5.25 \ \mu s.$$

We see in this example that the mean delay spread is roughly equal to its rms value. To avoid ISI we require linear modulation to have a symbol period T_s that is large relative to σ_{T_m}. Taking this to mean that $T_s > 10\sigma_{T_m}$ yields a symbol period of $T_s = 52.5 \ \mu s$ or a symbol rate of $R_s = 1/T_s = 19.04$ kilosymbols per second. This is a highly constrained symbol rate for many wireless systems. Specifically, for binary modulations where the symbol rate equals the data rate (bits per second, or bps), voice requires on the order of 32 kbps and high-speed data requires 10–100 Mbps.

3.3.2 Coherence Bandwidth

We can also characterize the time-varying multipath channel in the frequency domain by taking the Fourier transform of $c(\tau, t)$ with respect to τ. Specifically, define the random process

$$C(f; t) = \int_{-\infty}^{\infty} c(\tau; t) e^{-j2\pi f \tau} \, d\tau. \tag{3.57}$$

Because $c(\tau; t)$ is a complex zero-mean Gaussian random variable in t, the Fourier transform in (3.57) represents the sum[9] of complex zero-mean Gaussian random processes; hence $C(f; t)$ is also a zero-mean Gaussian random process that is completely characterized by its autocorrelation. Since $c(\tau; t)$ is WSS, its integral $C(f; t)$ is also. Thus, the autocorrelation of (3.57) is given by

$$A_C(f_1, f_2; \Delta t) = \mathbf{E}[C^*(f_1; t) C(f_2; t + \Delta t)]. \tag{3.58}$$

We can simplify $A_C(f_1, f_2; \Delta t)$ as follows:

$$\begin{aligned}
A_C(f_1, f_2; \Delta t) &= \mathbf{E}\left[\int_{-\infty}^{\infty} c^*(\tau_1; t) e^{j2\pi f_1 \tau_1} \, d\tau_1 \int_{-\infty}^{\infty} c(\tau_2; t + \Delta t) e^{-j2\pi f_2 \tau_2} \, d\tau_2 \right] \\
&= \int_{-\infty}^{\infty} \int_{-\infty}^{\infty} \mathbf{E}[c^*(\tau_1; t) c(\tau_2; t + \Delta t)] e^{j2\pi f_1 \tau_1} e^{-j2\pi f_2 \tau_2} \, d\tau_1 \, d\tau_2 \\
&= \int_{-\infty}^{\infty} A_c(\tau, \Delta t) e^{-j2\pi(f_2 - f_1)\tau} \, d\tau \\
&= A_C(\Delta f; \Delta t),
\end{aligned} \tag{3.59}$$

where $\Delta f = f_2 - f_1$ and the third equality follows from the WSS and US properties of $c(\tau; t)$. Thus, the autocorrelation of $C(f; t)$ in frequency depends only on the frequency difference Δf. The function $A_C(\Delta f; \Delta t)$ can be measured in practice by transmitting a pair of sinusoids through the channel that are separated in frequency by Δf and then calculating their cross-correlation at the receiver for the time separation Δt.

If we define $A_C(\Delta f) \triangleq A_C(\Delta f; 0)$ then, by (3.59),

$$A_C(\Delta f) = \int_{-\infty}^{\infty} A_c(\tau) e^{-j2\pi\Delta f \tau} \, d\tau. \tag{3.60}$$

---

[9] We can express the integral as a limit of a discrete sum.

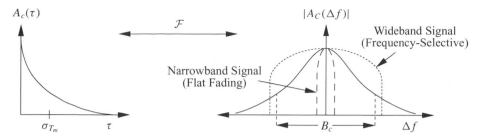

Figure 3.13: Power delay profile, rms delay spread, and coherence bandwidth.

Thus $A_C(\Delta f)$ is the Fourier transform of the power delay profile. Because $A_C(\Delta f) = \mathbf{E}[C^*(f;t)C(f + \Delta f;t)]$ is an autocorrelation, it follows that the channel response is approximately independent at frequency separations Δf where $A_C(\Delta f) \approx 0$. The frequency B_c where $A_C(\Delta f) \approx 0$ for all $\Delta f > B_c$ is called the *coherence bandwidth* of the channel. By the Fourier transform relationship between $A_c(\tau)$ and $A_C(\Delta f)$, if $A_c(\tau) \approx 0$ for $\tau > T$ then $A_C(\Delta f) \approx 0$ for $\Delta f > 1/T$. Hence, the minimum frequency separation B_c for which the channel response is roughly independent is $B_c \approx 1/T$, where T is typically taken to be the rms delay spread σ_{T_m} of $A_c(\tau)$. A more general approximation is $B_c \approx k/\sigma_{T_m}$, where k depends on the shape of $A_c(\tau)$ and the precise specification of coherence bandwidth. For example, Lee [20] has shown that $B_c \approx .02/\sigma_{T_m}$ approximates the range of frequencies over which channel correlation exceeds 0.9 whereas $B_c \approx .2/\sigma_{T_m}$ approximates the range of frequencies over which this correlation exceeds 0.5.

In general, if we are transmitting a narrowband signal with bandwidth $B \ll B_c$, then fading across the entire signal bandwidth is highly correlated; that is, the fading is roughly equal across the entire signal bandwidth. This is usually referred to as *flat fading*. On the other hand, if the signal bandwidth $B \gg B_c$, then the channel amplitude values at frequencies separated by more than the coherence bandwidth are roughly independent. Thus, the channel amplitude varies widely across the signal bandwidth. In this case the fading is called *frequency selective*. If $B \approx B_c$ then channel behavior is somewhere between flat and frequency-selective fading. Note that in linear modulation the signal bandwidth B is inversely proportional to the symbol time T_s, so flat fading corresponds to $T_s \approx 1/B \gg 1/B_c \approx \sigma_{T_m}$ – that is, the case where the channel experiences negligible ISI. Frequency-selective fading corresponds to $T_s \approx 1/B \ll 1/B_c = \sigma_{T_m}$, the case where the linearly modulated signal experiences significant ISI. Wideband signaling formats that reduce ISI, such as multicarrier modulation and spread spectrum, still experience frequency-selective fading across their entire signal bandwidth; this degrades performance, as will be discussed in Chapters 12 and 13.

We illustrate the power delay profile $A_c(\tau)$ and its Fourier transform $A_C(\Delta f)$ in Figure 3.13. This figure also shows two signals superimposed on $A_C(\Delta f)$: a narrowband signal with bandwidth much less than B_c, and a wideband signal with bandwidth much greater than B_c. We see that the autocorrelation $A_C(\Delta f)$ is flat across the bandwidth of the narrowband signal, so this signal will experience flat fading or (equivalently) negligible ISI. The autocorrelation $A_C(\Delta f)$ goes to zero within the bandwidth of the wideband signal, which means that fading will be independent across different parts of the signal bandwidth; hence fading is frequency selective, and a linearly modulated signal transmitted through this channel will experience significant ISI.

EXAMPLE 3.6: In indoor channels $\sigma_{T_m} \approx 50$ ns whereas in outdoor microcells $\sigma_{T_m} \approx 30 \ \mu s$. Find the maximum symbol rate $R_s = 1/T_s$ for these environments such that a linearly modulated signal transmitted through them experiences negligible ISI.

Solution: We assume that negligible ISI requires $T_s \gg \sigma_{T_m}$ (i.e., $T_s \geq 10\sigma_{T_m}$). This translates into a symbol rate of $R_s = 1/T_s \leq .1/\sigma_{T_m}$. For $\sigma_{T_m} \approx 50$ ns this yields $R_s \leq 2$ Mbps and for $\sigma_{T_m} \approx 30 \ \mu s$ this yields $R_s \leq 3.33$ kbps. Note that indoor systems currently support up to 50 Mbps and outdoor systems up to 2.4 Mbps. To maintain these data rates for a linearly modulated signal without severe performance degradation by ISI, some form of ISI mitigation is needed. Moreover, ISI is less severe in indoor than in outdoor systems owing to the former's lower delay spread values, which is why indoor systems tend to have higher data rates than outdoor systems.

3.3.3 Doppler Power Spectrum and Channel Coherence Time

The time variations of the channel that arise from transmitter or receiver motion cause a Doppler shift in the received signal. This Doppler effect can be characterized by taking the Fourier transform of $A_C(\Delta f; \Delta t)$ relative to Δt:

$$S_C(\Delta f; \rho) = \int_{-\infty}^{\infty} A_C(\Delta f; \Delta t) e^{-j2\pi\rho\Delta t} \, d\Delta t. \tag{3.61}$$

In order to characterize Doppler at a single frequency, we set Δf to zero and then define $S_C(\rho) \triangleq S_C(0; \rho)$. It is easily seen that

$$S_C(\rho) = \int_{-\infty}^{\infty} A_C(\Delta t) e^{-j2\pi\rho\Delta t} \, d\Delta t, \tag{3.62}$$

where $A_C(\Delta t) \triangleq A_C(\Delta f = 0; \Delta t)$. Note that $A_C(\Delta t)$ is an autocorrelation function defining how the channel impulse response decorrelates over time. In particular, $A_C(\Delta t = T) = 0$ indicates that observations of the channel impulse response at times separated by T are uncorrelated and therefore independent, since the channel is a Gaussian random process. We define the *channel coherence time* T_c to be the range of Δt values over which $A_C(\Delta t)$ is approximately nonzero. Thus, the time-varying channel decorrelates after approximately T_c seconds. The function $S_C(\rho)$ is called the *Doppler power spectrum* of the channel: as the Fourier transform of an autocorrelation it gives the PSD of the received signal as a function of Doppler ρ. The maximum ρ-value for which $|S_C(\rho)|$ is greater than zero is called the *Doppler spread* of the channel, denoted by B_D. By the Fourier transform relationship between $A_C(\Delta t)$ and $S_C(\rho)$, we have $B_D \approx 1/T_c$. If the transmitter and reflectors are all stationary and the receiver is moving with velocity v, then $B_D \leq v/\lambda = f_D$. Recall that in the narrowband fading model samples became independent at time $\Delta t = .4/f_D$, so in general $B_D \approx k/T_c$, where k depends on the shape of $S_c(\rho)$. We illustrate the Doppler power spectrum $S_C(\rho)$ and its inverse Fourier transform $A_C(\Delta_t)$ in Figure 3.14.

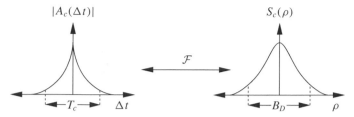

Figure 3.14: Doppler power spectrum, Doppler spread, and coherence time.

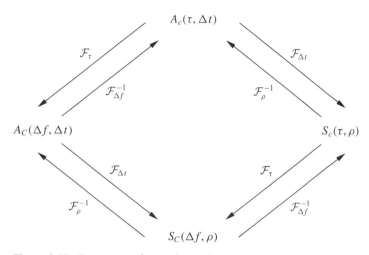

Figure 3.15: Fourier transform relationships.

EXAMPLE 3.7: For a channel with Doppler spread $B_D = 80$ Hz, find the time separation required in samples of the received signal in order for the samples to be approximately independent.

Solution: The coherence time of the channel is $T_c \approx 1/B_D = 1/80$, so samples spaced 12.5 ms apart are approximately uncorrelated. Thus, given the Gaussian properties of the underlying random process, these samples are approximately independent.

3.3.4 Transforms for Autocorrelation and Scattering Functions

From (3.61) we see that the scattering function $S_c(\tau; \rho)$ defined in (3.53) is the inverse Fourier transform of $S_C(\Delta f; \rho)$ in the Δf variable. Furthermore $S_c(\tau; \rho)$ and $A_C(\Delta f; \Delta t)$ are related by the double Fourier transform

$$S_c(\tau; \rho) = \int_{-\infty}^{\infty} \int_{-\infty}^{\infty} A_C(\Delta f; \Delta t) e^{-j2\pi \rho \Delta t} e^{j2\pi \tau \Delta f} \, d\Delta t \, d\Delta f. \qquad (3.63)$$

The relationships among the four functions $A_C(\Delta f; \Delta t)$, $A_c(\tau; \Delta t)$, $S_C(\Delta f; \rho)$, and $S_c(\tau; \rho)$ are shown in Figure 3.15.

Empirical measurements of the scattering function for a given channel are often used to approximate the channel's delay spread, coherence bandwidth, Doppler spread, and coherence

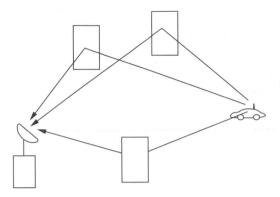

Figure 3.16: Point scatterer channel model.

time. The delay spread for a channel with empirical scattering function $S_c(\tau; \rho)$ is obtained by computing the empirical power delay profile $A_c(\tau)$ from $A_c(\tau, \Delta t) = \mathcal{F}_\rho^{-1}[S_c(\tau; \rho)]$ with $\Delta t = 0$ and then computing the mean and rms delay spread from this power delay profile. The coherence bandwidth can then be approximated as $B_c \approx 1/\sigma_{T_m}$. Similarly, the Doppler spread B_D is approximated as the range of ρ values over which $S(0; \rho)$ is roughly nonzero, with the coherence time $T_c \approx 1/B_D$.

3.4 Discrete-Time Model

Often the time-varying impulse response channel model is too complex for simple analysis. In this case a discrete-time approximation for the wideband multipath model can be used. This discrete-time model, developed by Turin in [21], is especially useful in the study of spread-spectrum systems and RAKE receivers (covered in Chapter 13). This discrete-time model is based on a physical propagation environment consisting of a composition of isolated point scatterers, as shown in Figure 3.16. In this model, the multipath components are assumed to form subpath clusters: incoming paths on a given subpath with approximate delay τ_n are combined, and incoming paths on different subpath clusters with delays τ_n and τ_m, where $|\tau_n - \tau_m| > 1/B$, can be resolved. Here B denotes the signal bandwidth.

The channel model of (3.6) is modified to include a fixed number $N + 1$ of these subpath clusters as

$$c(\tau; t) = \sum_{n=0}^{N} \alpha_n(t) e^{-j\phi_n(t)} \delta(\tau - \tau_n(t)).$$

(3.64)

The statistics of the received signal for a given t are thus given by the statistics of $\{\tau_n\}_0^N$, $\{\alpha_n\}_0^N$, and $\{\phi_n\}_0^N$. The model can be further simplified by using a discrete-time approximation as follows. For a fixed t, the time axis is divided into M equal intervals of duration T such that $MT \geq \sigma_{T_m}$, where σ_{T_m} is the rms delay spread of the channel (this is derived empirically). The subpaths are restricted to lie in one of the M time-interval bins, as shown in Figure 3.17. The multipath spread of this discrete model is MT, and the resolution between paths is T. This resolution is based on the transmitted signal bandwidth: $T \approx 1/B$. The statistics for the nth bin are that r_n ($1 \leq n \leq M$) is a binary indicator of the existence

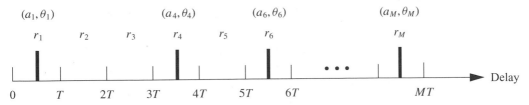

Figure 3.17: Discrete-time approximation.

of a multipath component in the nth bin: so $r_n = 1$ if there is a multipath component in the nth bin and 0 otherwise. If $r_n = 1$ then (a_n, θ_n), the amplitude and phase corresponding to this multipath component, follow an empirically determined distribution. This distribution is obtained by sample averages of (a_n, θ_n) for each n at different locations in the propagation environment. The empirical distribution of (a_n, θ_n) and (a_m, θ_m), $n \neq m$, is generally different; it may correspond to the same family of fading but with different parameters (e.g., Rician fading with different K factors) or to different fading distributions altogether (e.g., Rayleigh fading for the nth bin, Nakagami fading for the mth bin).

This completes our statistical model of the discrete-time approximation for a single snapshot. A sequence of profiles will model the signal over time as the channel impulse response changes – for example, the impulse response seen by a receiver moving at some nonzero velocity through a city. Thus, the model must include not only the first-order statistics of $(\tau_n, \alpha_n, \phi_n)$ for each profile (equivalently, each t) but also the temporal and spatial correlations (assumed to be Markov) between them. More details on the model and the empirically derived distributions for N and for $(\tau_n, \alpha_n, \phi_n)$ can be found in [21].

3.5 Space-Time Channel Models

Multiple antennas at the transmitter and/or receiver are becoming common in wireless systems because of their diversity and capacity benefits. Systems with multiple antennas require channel models that characterize both spatial (angle of arrival) and temporal characteristics of the channel. A typical model assumes the channel is composed of several scattering centers that generate the multipath [22; 23]. The location of the scattering centers relative to the receiver dictate the angle of arrival (AOA) of the corresponding multipath components. Models can be either two-dimensional or three-dimensional.

Consider a two-dimensional multipath environment in which the receiver or transmitter has an antenna array with M elements. The time-varying impulse response model (3.6) can be extended to incorporate AOA for the array as follows:

$$c(\tau, t) = \sum_{n=0}^{N(t)} \alpha_n(t) e^{-j\phi_n(t)} \mathbf{a}(\theta_n(t)) \delta(\tau - \tau_n(t)), \qquad (3.65)$$

where $\phi_n(t)$ corresponds to the phase shift at the origin of the array and $\mathbf{a}(\theta_n(t))$ is the array response vector given by

$$\mathbf{a}(\theta_n(t)) = [e^{-j\psi_{n,1}}, \ldots, e^{-j\psi_{n,M}}]^T, \qquad (3.66)$$

where $\psi_{n,i} = [x_i \cos \theta_n(t) + y_i \sin \theta_n(t)]2\pi/\lambda$ for (x_i, y_i) the antenna location relative to the origin and $\theta_n(t)$ the AOA of the multipath relative to the origin of the antenna array. Assume the AOA is stationary and identically distributed for all multipath components and denote this random AOA by θ. Let $A(\theta)$ denote the average received signal power as a function of θ. Then we define the mean and rms angular spread in terms of this power profile as

$$\mu_\theta = \frac{\int_{-\pi}^{\pi} \theta A(\theta)\, d\theta}{\int_{-\pi}^{\pi} A(\theta)\, d\theta} \tag{3.67}$$

and

$$\sigma_\theta = \sqrt{\frac{\int_{-\pi}^{\pi} (\theta - \mu_\theta)^2 A(\theta)\, d\theta}{\int_{-\pi}^{\pi} A(\theta)\, d\theta}}, \tag{3.68}$$

respectively. We say that two signals received at AOAs separated by $1/\sigma_\theta$ are roughly uncorrelated. More details on the power distribution relative to the AOA for different propagation environments (along with the corresponding correlations across antenna elements) can be found in [23].

Extending the two-dimensional models to three dimensions requires characterizing the elevation AOAs for multipath as well as the azimuth angles. Different models for such 3-D channels have been proposed in [24; 25; 26]. These ideas were used in [22] to incorporate spatiotemporal characteristics into Jakes's uniform scattering model. Several other papers on spatiotemporal modeling can be found in [27].

PROBLEMS

3-1. Consider a two-ray channel consisting of a direct ray plus a ground-reflected ray, where the transmitter is a fixed base station at height h and the receiver is mounted on a truck (also at height h). The truck starts next to the base station and moves away at velocity v. Assume that signal attenuation on each path follows a free-space path-loss model. Find the time-varying channel impulse at the receiver for transmitter–receiver separation $d = vt$ sufficiently large for the length of the reflected ray to be approximated by $r + r' \approx d + 2h^2/d$.

3-2. Find a formula for the multipath delay spread T_m for a two-ray channel model. Find a simplified formula when the transmitter–receiver separation is relatively large. Compute T_m for $h_t = 10$ m, $h_r = 4$ m, and $d = 100$ m.

3-3. Consider a time-invariant indoor wireless channel with LOS component at delay 23 ns, a multipath component at delay 48 ns, and another multipath component at delay 67 ns. Find the delay spread assuming that the demodulator synchronizes to the LOS component. Repeat assuming that the demodulator synchronizes to the first multipath component.

3-4. Show that the minimum value of $f_c \tau_n$ for a system at $f_c = 1$ GHz with a fixed transmitter and a receiver separated by more than 10 m from the transmitter is much greater than 1.

3-5. Prove, for X and Y independent zero-mean Gaussian random variables with variance σ^2, that the distribution of $Z = \sqrt{X^2 + Y^2}$ is Rayleigh distributed and that the distribution of Z^2 is exponentially distributed.

3-6. Assume a Rayleigh fading channel with average signal power $2\sigma^2 = -80\,\text{dBm}$. What is the power outage probability of this channel relative to the threshold $P_0 = -95\,\text{dBm}$? How about $P_0 = -90\,\text{dBm}$?

3-7. Suppose we have an application that requires a power outage probability of .01 for the threshold $P_0 = -80$ dBm. For Rayleigh fading, what value of the average signal power is required?

3-8. Assume a Rician fading channel with $2\sigma^2 = -80$ dBm and a target power of $P_0 = -80$ dBm. Find the outage probability assuming that the LOS component has average power $s^2 = -80$ dBm.

3-9. This problem illustrates that the tails of the Rician distribution can be quite different than its Nakagami approximation. Plot the cumulative distribution function (cdf) of the Rician distribution for $K = 1, 5, 10$ and the corresponding Nakagami distribution with $m = (K + 1)^2/(2K + 1)$. In general, does the Rician distribution or its Nakagami approximation have a larger outage probability $p(\gamma < x)$ for x large?

3-10. In order to improve the performance of cellular systems, multiple base stations can receive the signal transmitted from a given mobile unit and combine these multiple signals either by selecting the strongest one or summing the signals together, perhaps with some optimized weights. This typically increases SNR and reduces the effects of shadowing. Combining of signals received from multiple base stations is called *macrodiversity,* and here we explore the benefits of this technique. Diversity will be covered in more detail in Chapter 7.

Consider a mobile at the midpoint between two base stations in a cellular network. The received signals (in dBW) from the base stations are given by

$$P_{r,1} = W + Z_1,$$
$$P_{r,2} = W + Z_2,$$

where $Z_{1,2}$ are $N(0, \sigma^2)$ random variables. We define outage with macrodiversity to be the event that both $P_{r,1}$ and $P_{r,2}$ fall below a threshold T.

(a) Interpret the terms W, Z_1, Z_2 in $P_{r,1}$ and $P_{r,2}$.
(b) If Z_1 and Z_2 are independent, show that the outage probability is given by

$$P_{\text{out}} = [Q(\Delta/\sigma)]^2,$$

where $\Delta = W - T$ is the fade margin at the mobile's location.
(c) Now suppose that Z_1 and Z_2 are correlated in the following way:

$$Z_1 = aY_1 + bY,$$
$$Z_2 = aY_2 + bY,$$

where Y, Y_1, Y_2 are independent $N(0, \sigma^2)$ random variables and where a, b are such that $a^2 + b^2 = 1$. Show that

$$P_{\text{out}} = \int_{-\infty}^{+\infty} \frac{1}{\sqrt{2\pi}} \left[Q\left(\frac{\Delta + by\sigma}{|a|\sigma}\right)\right]^2 e^{-y^2/2}\, dy.$$

(d) Compare the outage probabilities of (b) and (c) for the special case of $a = b = 1/\sqrt{2}$, $\sigma = 8$, and $\Delta = 5$ (this will require a numerical integration).

3-11. The goal of this problem is to develop a Rayleigh fading simulator for a mobile communications channel using the method of filtering Gaussian processes that is based on the in-phase and quadrature PSDs described in Section 3.2.1. In this problem you must do the following.

(a) Develop simulation code to generate a signal with Rayleigh fading amplitude over time. Your sample rate should be at least 1000 samples per second, the average received envelope should be 1, and your simulation should be parameterized by the Doppler frequency f_D. Matlab is the easiest way to generate this simulation, but any code is fine.

(b) Write a description of your simulation that clearly explains how your code generates the fading envelope; use a block diagram and any necessary equations.

(c) Turn in your well-commented code.

(d) Provide plots of received amplitude (dB) versus time for $f_D = 1$, 10, and 100 hertz over 2 seconds.

3-12. For a Rayleigh fading channel with average power $\bar{P}_r = 30$ dB and Doppler $f_D = 10$ Hz, compute the average fade duration for target fade values of $P_0 = 0$ dB, $P_0 = 15$ dB, and $P_0 = 30$ dB.

3-13. Derive a formula for the average length of time that a Rayleigh fading process with average power $\bar{P}_r$ stays *above* a given target fade value P_0. Evaluate this average length of time for $\bar{P}_r = 20$ dB, $P_0 = 25$ dB, and $f_D = 50$ Hz.

3-14. Assume a Rayleigh fading channel with average power $\bar{P}_r = 10$ dB and Doppler $f_D = 80$ Hz. We would like to approximate the channel using a finite-state Markov model with eight states and time interval $T = 10$ ms. The regions R_j correspond to $R_1 = \{\gamma : -\infty \leq \gamma < -10$ dB$\}$, $R_2 = \{\gamma : -10$ dB $\leq \gamma < 0$ dB$\}$, $R_3 = \{\gamma : 0$ dB $\leq \gamma < 5$ dB$\}$, $R_4 = \{\gamma : 5$ dB $\leq \gamma < 10$ dB$\}$, $R_5 = \{\gamma : 10$ dB $\leq \gamma < 15$ dB$\}$, $R_6 = \{\gamma : 15$ dB $\leq \gamma < 20$ dB$\}$, $R_7 = \{\gamma : 20$ dB $\leq \gamma < 30$ dB$\}$, and $R_8 = \{\gamma : 30$ dB $\leq \gamma \leq \infty\}$. Find the transition probabilties between each region for this model.

3-15. Consider the following channel scattering function obtained by sending a 900-MHz sinusoidal input into the channel:

$$S(\tau, \rho) = \begin{cases} \alpha_1 \delta(\tau) & \rho = 70 \text{ Hz}, \\ \alpha_2 \delta(\tau - .022 \ \mu s) & \rho = 49.5 \text{ Hz}, \\ 0 & \text{else}, \end{cases}$$

where α_1 and α_2 are determined by path loss, shadowing, and multipath fading. Clearly this scattering function corresponds to a two-ray model. Assume the transmitter and receiver used to send and receive the sinusoid are located 8 m above the ground.

(a) Find the distance and velocity between the transmitter and receiver.

(b) For the distance computed in part (a), is the path loss as a function of distance proportional to d^{-2} or d^{-4}? *Hint:* Use the fact that the channel is based on a two-ray model.

(c) Does a 30-kHz voice signal transmitted over this channel experience flat or rather frequency-selective fading?

3-16. Consider a wideband channel characterized by the autocorrelation function

$$A_c(\tau, \Delta t) = \begin{cases} \text{sinc}(W\Delta t) & 0 \leq \tau \leq 10\ \mu s, \\ 0 & \text{else,} \end{cases}$$

where $W = 100$ Hz and $\text{sinc}(x) = \sin(\pi x)/(\pi x)$.

(a) Does this channel correspond to an indoor channel or an outdoor channel, and why?

(b) Sketch the scattering function of this channel.

(c) Compute the channel's average delay spread, rms delay spread, and Doppler spread.

(d) Over approximately what range of data rates will a signal transmitted via this channel exhibit frequency-selective fading?

(e) Would you expect this channel to exhibit Rayleigh or rather Rician fading statistics? Why?

(f) Assuming that the channel exhibits Rayleigh fading, what is the average length of time that the signal power is continuously below its average value?

(g) Assume a system with narrowband binary modulation sent over this channel. Your system has error correction coding that can correct two simultaneous bit errors. Assume also that you always make an error if the received signal power is below its average value and that you never make an error if this power is at or above its average value. If the channel is Rayleigh fading, then what is the maximum data rate that can be sent over this channel with error-free transmission? Make the approximation that the fade duration never exceeds twice its average value.

3-17. Let a scattering function $S_c(\tau, \rho)$ be nonzero over $0 \leq \tau \leq .1$ ms and $-.1 \leq \rho \leq .1$ Hz. Assume that the power of the scattering function is approximately uniform over the range where it is nonzero.

(a) What are the multipath spread and the Doppler spread of the channel?

(b) Suppose you input to this channel two identical sinusoids separated in time by Δt. What is the minimum value of Δf for which the channel response to the first sinusoid is approximately independent of the channel response to the second sinusoid?

(c) For two sinusoidal inputs to the channel $u_1(t) = \sin 2\pi f t$ and $u_2(t) = \sin 2\pi f(t + \Delta t)$, find the minimum value of Δt for which the channel response to $u_1(t)$ is approximately independent of the channel response to $u_2(t)$.

(d) Will this channel exhibit flat fading or frequency-selective fading for a typical voice channel with a 3-kHz bandwidth? For a cellular channel with a 30-kHz bandwidth?

REFERENCES

[1] T. S. Rappaport, *Wireless Communications – Principles and Practice*, 2nd ed., Prentice-Hall, Englewood Cliffs, NJ, 2001.

[2] D. Parsons, *The Mobile Radio Propagation Channel*, Wiley, New York, 1994.

[3] R. S. Kennedy, *Fading Dispersive Communication Channels*, Wiley, New York, 1969.

[4] R. H. Clarke, "A statistical theory of mobile radio reception," *Bell System Tech. J.*, pp. 957–1000, July/August 1968.

[5] W. C. Jakes, Jr., *Microwave Mobile Communications*, Wiley, New York, 1974.

[6] M. K. Simon and M.-S. Alouini, *Digital Communication over Fading Channels: A Unified Approach to Performance Analysis*, Wiley, New York, 2000.

[7]   G. L. Stuber, *Principles of Mobile Communications,* 2nd ed., Kluwer, Dordrecht, 2001.

[8]   M. Pätzold, *Mobile Fading Channels,* Wiley, New York, 2002.

[9]   S. O. Rice, "Mathematical analysis of random noise," *Bell System Tech. J.,* pp. 282–333, July 1944, and pp. 46–156, January 1945.

[10] F. Babich, G. Lombardi, and E. Valentinuzzi, "Variable order Markov modeling for LEO mobile satellite channels," *Elec. Lett.,* pp. 621–3, April 1999.

[11] A. M. Chen and R. R. Rao, "On tractable wireless channel models," *Proc. Internat. Sympos. Pers., Indoor, Mobile Radio Commun.,* pp. 825–30, September 1998.

[12] H. S. Wang and N. Moayeri, "Finite-state Markov channel – A useful model for radio communication channels," *IEEE Trans. Veh. Tech.,* pp. 163–71, February 1995.

[13] C. C. Tan and N. C. Beaulieu, "On first-order Markov modeling for the Rayleigh fading channel," *IEEE Trans. Commun.,* pp. 2032–40, December 2000.

[14] C. Pimentel and I. F. Blake, "Modeling burst channels using partitioned Fritchman's Markov models," *IEEE Trans. Veh. Tech.,* pp. 885–99, August 1998.

[15] Y. L. Guan and L. F. Turner, "Generalised FSMC model for radio channels with correlated fading," *IEE Proc. Commun.,* pp. 133–7, April 1999.

[16] M. Chu and W. Stark, "Effect of mobile velocity on communications in fading channels," *IEEE Trans. Veh. Tech.,* pp. 202–10, January 2000.

[17] C. Komninakis and R. D. Wesel, "Pilot-aided joint data and channel estimation in flat correlated fading," *Proc. IEEE Globecom Conf.,* pp. 2534–9, November 1999.

[18] M. Peleg, S. Shamai (Shitz), and S. Galan, "Iterative decoding for coded noncoherent MPSK communications over phase-noisy AWGN channels," *IEE Proc. Commun.,* pp. 87–95, April 2000.

[19] P. A. Bello, "Characterization of randomly time-variant linear channels," *IEEE Trans. Commun. Syst.,* pp. 360–93, December 1963.

[20] W. C. Y. Lee, *Mobile Cellular Telecommunications Systems,* McGraw-Hill, New York, 1989.

[21] G. L. Turin, "Introduction to spread spectrum antimultipath techniques and their application to urban digital radio," *Proc. IEEE,* pp. 328–53, March 1980.

[22] Y. Mohasseb and M. P. Fitz, "A 3-D spatio-temporal simulation model for wireless channels," *IEEE J. Sel. Areas Commun.,* pp. 1193–1203, August 2002.

[23] R. Ertel, P. Cardieri, K. W. Sowerby, T. Rappaport, and J. H. Reed, "Overview of spatial channel models for antenna array communication systems," *IEEE Pers. Commun. Mag.,* pp. 10–22, February 1998.

[24] T. Aulin, "A modified model for fading signal at the mobile radio channel," *IEEE Trans. Veh. Tech.,* pp. 182–202, August 1979.

[25] J. D. Parsons and M. D. Turkmani, "Characterization of mobile radio signals: Model description," *Proc. IEE,* pt. 1, pp. 549–56, December 1991.

[26] J. D. Parsons and M. D. Turkmani, "Characterization of mobile radio signals: Base station cross-correlation," *Proc. IEE,* pt. 1, pp. 557–65, December 1991.

[27] L. G. Greenstein, J. B. Andersen, H. L. Bertoni, S. Kozono, and D. G. Michelson, Eds., *IEEE J. Sel. Areas Commun.,* Special Issue on Channel and Propagation Modeling for Wireless Systems Design, August 2002.

Capacity of Wireless Channels

The growing demand for wireless communication makes it important to determine the capacity limits of the underlying channels for these systems. These capacity limits dictate the maximum data rates that can be transmitted over wireless channels with asymptotically small error probability, assuming no constraints on delay or complexity of the encoder and decoder. The mathematical theory of communication underlying channel capacity was pioneered by Claude Shannon in the late 1940s. This theory is based on the notion of mutual information between the input and output of a channel [1; 2; 3]. In particular, Shannon defined channel capacity as the channel's mutual information maximized over all possible input distributions. The significance of this mathematical construct was Shannon's coding theorem and its converse. The coding theorem proved that a code did exist that could achieve a data rate close to capacity with negligible probability of error. The converse proved that any data rate higher than capacity could not be achieved without an error probability bounded away from zero. Shannon's ideas were quite revolutionary at the time: the high data rates he predicted for telephone channels, and his notion that coding could reduce error probability without reducing data rate or causing bandwidth expansion. In time, sophisticated modulation and coding technology validated Shannon's theory and so, on telephone lines today, we achieve data rates very close to Shannon capacity with very low probability of error. These sophisticated modulation and coding strategies are treated in Chapters 5 and 8, respectively.

In this chapter we examine the capacity of a single-user wireless channel where transmitter and/or receiver have a single antenna. The capacity of single-user systems where transmitter and receiver have multiple antennas is treated in Chapter 10 and that of multiuser systems in Chapter 14. We will discuss capacity for channels that are both time invariant and time varying. We first look at the well-known formula for capacity of a time-invariant additive white Gaussian noise (AWGN) channel and then consider capacity of time-varying flat fading channels. Unlike the AWGN case, here the capacity of a flat fading channel is not given by a single formula because capacity depends on what is known about the time-varying channel at the transmitter and/or receiver. Moreover, for different channel information assumptions there are different definitions of channel capacity, depending on whether capacity characterizes the maximum rate averaged over all fading states or the maximum constant rate that can be maintained in all fading states (with or without some probability of outage).

We will first consider flat fading channel capacity where only the fading distribution is known at the transmitter and receiver. Capacity under this assumption is typically difficult to determine and is only known in a few special cases. Next we consider capacity when the channel fade level is known at the receiver only (via receiver estimation) or when the channel fade level is known at both the transmitter and the receiver (via receiver estimation and transmitter feedback). We will see that the fading channel capacity with channel fade level information at both the transmitter and receiver is achieved when the transmitter adapts its power, data rate, and coding scheme to the channel variation. The optimal power allocation in this case is a "water-filling" in time, where power and data rate are increased when channel conditions are favorable and decreased when channel conditions are not favorable.

We will also treat capacity of frequency-selective fading channels. For time-invariant frequency-selective channels the capacity is known and is achieved with an optimal power allocation that water-fills over frequency instead of time. The capacity of a time-varying frequency-selective fading channel is unknown in general. However, this channel can be approximated as a set of independent parallel flat fading channels whose capacity is the sum of capacities on each channel with power optimally allocated among the channels. The capacity of such a channel is known and the capacity-achieving power allocation water-fills over both time and frequency.

We will consider only discrete-time systems in this chapter. Most continuous-time systems can be converted to discrete-time systems via sampling, and then the same capacity results hold. However, care must be taken in choosing the appropriate sampling rate for this conversion, since time variations in the channel may increase the sampling rate required to preserve channel capacity [4].

4.1 Capacity in AWGN

Consider a discrete-time AWGN channel with channel input/output relationship $y[i] = x[i] + n[i]$, where $x[i]$ is the channel input at time i, $y[i]$ is the corresponding channel output, and $n[i]$ is a white Gaussian noise random process. Assume a channel bandwidth B and received signal power P. The received signal-to-noise ratio (SNR) – the power in $x[i]$ divided by the power in $n[i]$ – is constant and given by $\gamma = P/N_0 B$, where $N_0/2$ is the power spectral density (PSD) of the noise. The capacity of this channel is given by Shannon's well-known formula [1]:

$$C = B \log_2(1 + \gamma), \tag{4.1}$$

where the capacity units are bits per second (bps). Shannon's coding theorem proves that a code exists that achieves data rates arbitrarily close to capacity with arbitrarily small probability of bit error. The converse theorem shows that any code with rate $R > C$ has a probability of error bounded away from zero. The theorems are proved using the concept of mutual information between the channel input and output. For a discrete memoryless time-invariant channel with random input x and random output y, the channel's *mutual information* is defined as

$$I(X; Y) = \sum_{x \in \mathcal{X}, y \in \mathcal{Y}} p(x, y) \log\left(\frac{p(x, y)}{p(x)p(y)}\right), \tag{4.2}$$

where the sum is taken over all possible input and output pairs $x \in \mathcal{X}$ and $y \in \mathcal{Y}$ for $\mathcal{X}$ and $\mathcal{Y}$ the discrete input and output alphabets. The log function is typically with respect to base 2, in which case the units of mutual information are bits per second. Mutual information can also be written in terms of the *entropy* in the channel output y and conditional output $y \mid x$ as $I(X; Y) = H(Y) - H(Y \mid X)$, where $H(Y) = -\sum_{y \in \mathcal{Y}} p(y) \log p(y)$ and $H(Y \mid X) = -\sum_{x \in \mathcal{X}, y \in \mathcal{Y}} p(x, y) \log p(y \mid x)$. Shannon proved that channel capacity equals the mutual information of the channel maximized over all possible input distributions:

$$C = \max_{p(x)} I(X; Y) = \max_{p(x)} \sum_{x, y} p(x, y) \log \left(\frac{p(x, y)}{p(x)p(y)} \right). \tag{4.3}$$

For the AWGN channel, the sum in (4.3) becomes an integral over continuous alphabets and the maximizing input distribution is Gaussian, which results in the channel capacity given by (4.1). For channels with memory, mutual information and channel capacity are defined relative to input and output sequences x^n and y^n. More details on channel capacity, mutual information, the coding theorem, and its converse can be found in [2; 5; 6].

The proofs of the coding theorem and its converse place no constraints on the complexity or delay of the communication system. Therefore, Shannon capacity is generally used as an upper bound on the data rates that can be achieved under real system constraints. At the time that Shannon developed his theory of information, data rates over standard telephone lines were on the order of 100 bps. Thus, it was believed that Shannon capacity, which predicted speeds of roughly 30 kbps over the same telephone lines, was not a useful bound for real systems. However, breakthroughs in hardware, modulation, and coding techniques have brought commercial modems of today very close to the speeds predicted by Shannon in the 1940s. In fact, modems can exceed this 30-kbps limit on some telephone channels, but that is because transmission lines today are of better quality than in Shannon's day and thus have a higher received power than that used in his initial calculation. On AWGN radio channels, turbo codes have come within a fraction of a decibel of the Shannon capacity limit [7].

Wireless channels typically exhibit flat or frequency-selective fading. In the next two sections we consider capacity of flat fading and frequency-selective fading channels under different assumptions regarding what is known about the channel.

EXAMPLE 4.1: Consider a wireless channel where power falloff with distance follows the formula $P_r(d) = P_t(d_0/d)^3$ for $d_0 = 10$ m. Assume the channel has bandwidth $B = 30$ kHz and AWGN with noise PSD $N_0/2$, where $N_0 = 10^{-9}$ W/Hz. For a transmit power of 1 W, find the capacity of this channel for a transmit–receive distance of 100 m and 1 km.

Solution: The received SNR is $\gamma = P_r(d)/N_0 B = .1^3/(10^{-9} \cdot 30 \cdot 10^3) = 33 = 15$ dB for $d = 100$ m and $\gamma = .01^3/(10^{-9} \cdot 30 \cdot 10^3) = .033 = -15$ dB for $d = 1000$ m. The corresponding capacities are $C = B \log_2(1 + \gamma) = 30000 \log_2(1 + 33) = 152.6$ kbps for $d = 100$ m and $C = 30000 \log_2(1 + .033) = 1.4$ kbps for $d = 1000$ m. Note the significant decrease in capacity at greater distances due to the path-loss exponent of 3, which greatly reduces received power as distance increases.

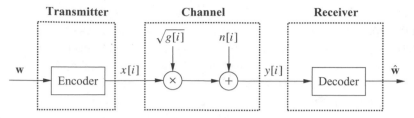

Figure 4.1: Flat fading channel and system model.

4.2 Capacity of Flat Fading Channels

4.2.1 Channel and System Model

We assume a discrete-time channel with stationary and ergodic time-varying gain $\sqrt{g[i]}$, $0 \le g[i]$, and AWGN $n[i]$, as shown in Figure 4.1. The channel power gain $g[i]$ follows a given distribution $p(g)$; for example, with Rayleigh fading $p(g)$ is exponential. We assume that $g[i]$ is independent of the channel input. The channel gain $g[i]$ can change at each time i, either as an independent and identically distributed (i.i.d.) process or with some correlation over time. In a *block fading channel*, $g[i]$ is constant over some blocklength T, after which time $g[i]$ changes to a new independent value based on the distribution $p(g)$. Let $\bar{P}$ denote the average transmit signal power, $N_0/2$ the noise PSD of $n[i]$, and B the received signal bandwidth. The instantaneous received SNR is then $\gamma[i] = \bar{P}g[i]/N_0B$, $0 \le \gamma[i] < \infty$, and its expected value over all time is $\bar{\gamma} = \bar{P}\bar{g}/N_0B$. Since $\bar{P}/N_0B$ is a constant, the distribution of $g[i]$ determines the distribution of $\gamma[i]$ and vice versa.

The system model is also shown in Figure 4.1, where an input message **w** is sent from the transmitter to the receiver, which reconstructs an estimate $\hat{\mathbf{w}}$ of the transmitted message **w** from the received signal. The message is encoded into the codeword **x**, which is transmitted over the time-varying channel as $x[i]$ at time i. The channel gain $g[i]$, also called the *channel side information* (CSI), changes during the transmission of the codeword.

The capacity of this channel depends on what is known about $g[i]$ at the transmitter and receiver. We will consider three different scenarios regarding this knowledge as follows.

1. *Channel distribution information (CDI):* The distribution of $g[i]$ is known to the transmitter and receiver.
2. *Receiver CSI:* The value of $g[i]$ is known to the receiver at time i, and both the transmitter and receiver know the distribution of $g[i]$.
3. *Transmitter and receiver CSI:* The value of $g[i]$ is known to the transmitter and receiver at time i, and both the transmitter and receiver know the distribution of $g[i]$.

Transmitter and receiver CSI allow the transmitter to adapt both its power and rate to the channel gain at time i, leading to the highest capacity of the three scenarios. Note that since the instantaneous SNR $\gamma[i]$ is just $g[i]$ multiplied by the constant $\bar{P}/N_0B$, known CSI or CDI about $g[i]$ yields the same information about $\gamma[i]$. Capacity for time-varying channels under assumptions other than these three are discussed in [8; 9].

4.2.2 Channel Distribution Information Known

We first consider the case where the channel gain distribution $p(g)$ or, equivalently, the distribution of SNR $p(\gamma)$ is known to the transmitter and receiver. For i.i.d. fading the capacity is given by (4.3), but solving for the capacity-achieving input distribution (i.e., the distribution achieving the maximum in that equation) can be quite complicated depending on the nature of the fading distribution. Moreover, fading correlation introduces channel memory, in which case the capacity-achieving input distribution is found by optimizing over input blocks, and this makes finding the solution even more difficult. For these reasons, finding the capacity-achieving input distribution and corresponding capacity of fading channels under CDI remains an open problem for almost all channel distributions.

The capacity-achieving input distribution and corresponding fading channel capacity under CDI are known for two specific models of interest: i.i.d. Rayleigh fading channels and finite-state Markov channels. In i.i.d. Rayleigh fading, the channel power gain is exponentially distributed and changes independently with each channel use. The optimal input distribution for this channel was shown in [10] to be discrete with a finite number of mass points, one of which is located at zero. This optimal distribution and its corresponding capacity must be found numerically. The lack of closed-form solutions for capacity or the optimal input distribution is somewhat surprising given that the fading follows the most common fading distribution and has no correlation structure. For flat fading channels that are not necessarily Rayleigh or i.i.d., upper and lower bounds on capacity have been determined in [11], and these bounds are tight at high SNRs.

Approximating Rayleigh fading channels via FSMCs was discussed in Section 3.2.4. This model approximates the fading correlation as a Markov process. Although the Markov nature of the fading dictates that the fading at a given time depends only on fading at the previous time sample, it turns out that the receiver must decode all past channel outputs jointly with the current output for optimal (i.e. capacity-achieving) decoding. This significantly complicates capacity analysis. The capacity of FSMCs has been derived for i.i.d. inputs in [12; 13] and for general inputs in [14]. Capacity of the FSMC depends on the limiting distribution of the channel conditioned on all past inputs and outputs, which can be computed recursively. As with the i.i.d. Rayleigh fading channel, the final result and complexity of the capacity analysis are high for this relatively simple fading model. This shows the difficulty of obtaining the capacity and related design insights on channels when only CDI is available.

4.2.3 Channel Side Information at Receiver

We now consider the case where the CSI $g[i]$ is known to the receiver at time i. Equivalently, $\gamma[i]$ is known to the receiver at time i. We also assume that both the transmitter and receiver know the distribution of $g[i]$. In this case there are two channel capacity definitions that are relevant to system design: Shannon capacity, also called *ergodic capacity,* and *capacity with outage.* As for the AWGN channel, Shannon capacity defines the maximum data rate that can be sent over the channel with asymptotically small error probability. Note that for Shannon capacity the rate transmitted over the channel is constant: the transmitter cannot adapt its transmission strategy relative to the CSI. Thus, poor channel states typically reduce Shannon capacity because the transmission strategy must incorporate the effect of these poor states. An alternate capacity definition for fading channels with receiver CSI is capacity with outage. This is defined as the maximum rate that can be transmitted over a channel with an outage probability corresponding to the probability that the transmission cannot be decoded with negligible error probability. The basic premise of capacity with outage is that a high data rate can be sent over the channel and decoded correctly except when the channel is in a slow deep fade. By allowing the system to lose some data in the event of such deep fades, a higher data rate can be maintained than if all data must be received correctly regardless of the fading state, as is the case for Shannon capacity. The probability of outage characterizes the probability of data loss or, equivalently, of deep fading.

SHANNON (ERGODIC) CAPACITY

Shannon capacity of a fading channel with receiver CSI for an average power constraint $\bar{P}$ can be obtained from results in [15] as

$$C = \int_0^\infty B \log_2(1 + \gamma) p(\gamma)\, d\gamma. \qquad (4.4)$$

Note that this formula is a probabilistic average: the capacity C is equal to Shannon capacity for an AWGN channel with SNR γ, given by $B \log_2(1 + \gamma)$ and averaged over the distribution of γ. That is why Shannon capacity is also called ergodic capacity. However, care must be taken in interpreting (4.4) as an average. In particular, it is incorrect to interpret (4.4) to mean that this average capacity is achieved by maintaining a capacity $B \log_2(1 + \gamma)$ when the instantaneous SNR is γ, for only the receiver knows the instantaneous SNR $\gamma[i]$ and so the data rate transmitted over the channel is constant, regardless of γ. Note also that the capacity-achieving code must be sufficiently long that a received codeword is affected by all possible fading states. This can result in significant delay.

By Jensen's inequality,

$$\mathbf{E}[B \log_2(1 + \gamma)] = \int B \log_2(1 + \gamma) p(\gamma)\, d\gamma$$

$$\leq B \log_2(1 + \mathbf{E}[\gamma]) = B \log_2(1 + \bar{\gamma}), \qquad (4.5)$$

where $\bar{\gamma}$ is the average SNR on the channel. Thus we see that the Shannon capacity of a fading channel with receiver CSI only is less than the Shannon capacity of an AWGN channel with the same average SNR. In other words, fading reduces Shannon capacity when only the receiver has CSI. Moreover, without transmitter CSI, the code design must incorporate the channel correlation statistics, and the complexity of the maximum likelihood decoder will be proportional to the channel decorrelation time. In addition, if the receiver CSI is not perfect, capacity can be significantly decreased [16].

EXAMPLE 4.2: Consider a flat fading channel with i.i.d. channel gain $\sqrt{g[i]}$, which can take on three possible values: $\sqrt{g_1} = .05$ with probability $p_1 = .1$, $\sqrt{g_2} = .5$ with probability $p_2 = .5$, and $\sqrt{g_3} = 1$ with probability $p_3 = .4$. The transmit power is 10 mW, the noise power spectral density $N_0/2$ has $N_0 = 10^{-9}$ W/Hz, and the channel bandwidth is 30 kHz. Assume the receiver has knowledge of the instantaneous value of $g[i]$ but the transmitter does not. Find the Shannon capacity of this channel and compare with the capacity of an AWGN channel with the same average SNR.

Solution: The channel has three possible received SNRs: $\gamma_1 = P_t g_1 / N_0 B = .01 \cdot (.05^2)/(30000 \cdot 10^{-9}) = .8333 = -.79$ dB, $\gamma_2 = P_t g_2 / N_0 B = .01 \cdot (.5^2)/(30000 \cdot 10^{-9}) = 83.333 = 19.2$ dB, and $\gamma_3 = P_t g_3 / N_0 B = .01/(30000 \cdot 10^{-9}) = 333.33 = 25$ dB. The probabilities associated with each of these SNR values are $p(\gamma_1) = .1$, $p(\gamma_2) = .5$, and $p(\gamma_3) = .4$. Thus, the Shannon capacity is given by

$$C = \sum_i B \log_2(1 + \gamma_i) p(\gamma_i)$$

$$= 30000(.1 \log_2(1.8333) + .5 \log_2(84.333) + .4 \log_2(334.33))$$

$$= 199.26 \text{ kbps.}$$

The average SNR for this channel is $\bar{\gamma} = .1(.8333) + .5(83.33) + .4(333.33) = 175.08 = 22.43$ dB. The capacity of an AWGN channel with this SNR is $C = B \log_2(1 + 175.08) = 223.8$ kbps. Note that this rate is about 25 kbps larger than that of the flat fading channel with receiver CSI and the same average SNR.

CAPACITY WITH OUTAGE

Capacity with outage applies to slowly varying channels, where the instantaneous SNR γ is constant over a large number of transmissions (a transmission burst) and then changes to a new value based on the fading distribution. With this model, if the channel has received SNR γ during a burst then data can be sent over the channel at rate $B \log_2(1 + \gamma)$ with negligible probability of error.[1] Since the transmitter does not know the SNR value γ, it must fix a transmission rate independent of the instantaneous received SNR.

Capacity with outage allows bits sent over a given transmission burst to be decoded at the end of the burst with some probability that these bits will be decoded incorrectly. Specifically, the transmitter fixes a minimum received SNR γ_{min} and encodes for a data rate $C = B \log_2(1 + \gamma_{min})$. The data is correctly received if the instantaneous received SNR is greater than or equal to γ_{min} [17; 18]. If the received SNR is below γ_{min} then the bits received over that transmission burst cannot be decoded correctly with probability approaching 1, and the receiver declares an outage. The probability of outage is thus $P_{out} = p(\gamma < \gamma_{min})$. The average rate correctly received over many transmission bursts is $C_{out} = (1 - P_{out})B \log_2(1 + \gamma_{min})$ since data is only correctly received on $1 - P_{out}$ transmissions. The value of γ_{min} is a design parameter based on the acceptable outage probability. Capacity with outage is typically characterized by a plot of capacity versus outage, as shown in Figure 4.2. In this figure we plot the normalized capacity $C/B = \log_2(1 + \gamma_{min})$ as a function of outage probability $P_{out} = p(\gamma < \gamma_{min})$ for a Rayleigh fading channel (γ exponentially distributed) with $\bar{\gamma} = 20$ dB. We see that capacity approaches zero for small outage probability, due to the requirement that bits transmitted under severe fading must be decoded correctly, and increases dramatically as outage probability increases. Note, however, that these high capacity values for large outage probabilities have higher probability of incorrect data reception. The average rate correctly received can be maximized by finding the γ_{min} (or, equivalently, the P_{out}) that maximizes C_{out}.

EXAMPLE 4.3: Assume the same channel as in the previous example, with a bandwidth of 30 kHz and three possible received SNRs: $\gamma_1 = .8333$ with $p(\gamma_1) = .1$, $\gamma_2 = 83.33$ with $p(\gamma_2) = .5$, and $\gamma_3 = 333.33$ with $p(\gamma_3) = .4$. Find the capacity versus outage for this channel, and find the average rate correctly received for outage probabilities $P_{out} < .1$, $P_{out} = .1$, and $P_{out} = .6$.

---

[1] The assumption of constant fading over a large number of transmissions is needed because codes that achieve capacity require very large blocklengths.

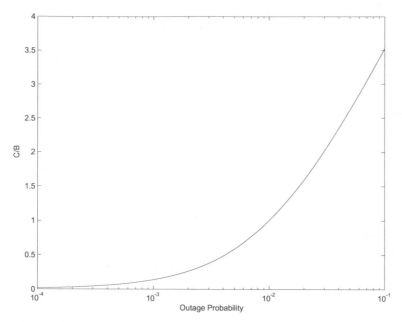

Figure 4.2: Normalized capacity (C/B) versus outage probability.

Solution: For time-varying channels with discrete SNR values, the capacity versus outage is a staircase function. Specifically, for $P_{out} < .1$ we must decode correctly in all channel states. The minimum received SNR for P_{out} in this range of values is that of the weakest channel: $\gamma_{min} = \gamma_1$, and the corresponding capacity is $C = B\log_2(1 + \gamma_{min}) = 30000\log_2(1.833) = 26.23$ kbps. For $.1 \leq P_{out} < .6$ we can decode incorrectly when the channel is in the weakest state only. Then $\gamma_{min} = \gamma_2$ and the corresponding capacity is $C = B\log_2(1 + \gamma_{min}) = 30000\log_2(84.33) = 191.94$ kbps. For $.6 \leq P_{out} < 1$ we can decode incorrectly if the channel has received SNR γ_1 or γ_2. Then $\gamma_{min} = \gamma_3$ and the corresponding capacity is $C = B\log_2(1 + \gamma_{min}) = 30000\log_2(334.33) = 251.55$ kbps. Thus, capacity versus outage has $C = 26.23$ kbps for $P_{out} < .1$, $C = 191.94$ kbps for $.1 \leq P_{out} < .6$, and $C = 251.55$ kbps for $.6 \leq P_{out} < 1$.

For $P_{out} < .1$, data transmitted at rates close to capacity $C = 26.23$ kbps are always correctly received because the channel can always support this data rate. For $P_{out} = .1$ we transmit at rates close to $C = 191.94$ kbps, but we can correctly decode these data only when the SNR is γ_2 or γ_3, so the rate correctly received is $(1 - .1)191940 = 172.75$ kbps. For $P_{out} = .6$ we transmit at rates close to $C = 251.55$ kbps but we can correctly decode these data only when the SNR is γ_3, so the rate correctly received is $(1 - .6)251550 = 125.78$ kbps. It is likely that a good engineering design for this channel would send data at a rate close to 191.94 kbps, since it would be received incorrectly at most 10% of this time and the data rate would be almost an order of magnitude higher than sending at a rate commensurate with the worst-case channel capacity. However, 10% retransmission probability is too high for some applications, in which case the system would be designed for the 26.23 kbps data rate with no retransmissions. Design issues regarding acceptable retransmission probability are discussed in Chapter 14.

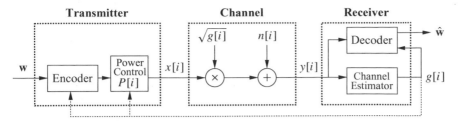

Figure 4.3: System model with transmitter and receiver CSI.

4.2.4 Channel Side Information at Transmitter and Receiver

If both the transmitter and receiver have CSI then the transmitter can adapt its transmission strategy relative to this CSI, as shown in Figure 4.3. In this case there is no notion of capacity versus outage where the transmitter sends bits that cannot be decoded, since the transmitter knows the channel and thus will not send bits unless they can be decoded correctly. In this section we will derive Shannon capacity assuming optimal power and rate adaptation relative to the CSI; we also introduce alternate capacity definitions and their power and rate adaptation strategies.

SHANNON CAPACITY

We now consider the Shannon capacity when the channel power gain $g[i]$ is known to both the transmitter and receiver at time i. The Shannon capacity of a time-varying channel with side information about the channel state at both the transmitter and receiver was originally considered by Wolfowitz for the following model. Let $s[i]$ be a stationary and ergodic stochastic process representing the channel state, which takes values on a finite set S of discrete memoryless channels. Let C_s denote the capacity of a particular channel $s \in S$ and let $p(s)$ denote the probability, or fraction of time, that the channel is in state s. The capacity of this time-varying channel is then given by Theorem 4.6.1 of [19]:

$$C = \sum_{s \in S} C_s p(s). \tag{4.6}$$

We now apply this formula to the system model in Figure 4.3. We know that the capacity of an AWGN channel with average received SNR γ is $C_\gamma = B \log_2(1 + \gamma)$. Let $p(\gamma) = p(\gamma[i] = \gamma)$ denote the distribution of the received SNR. By (4.6), the capacity of the fading channel with transmitter and receiver side information is thus[2]

$$C = \int_0^\infty C_\gamma p(\gamma)\, d\gamma = \int_0^\infty B \log_2(1 + \gamma) p(\gamma)\, d\gamma. \tag{4.7}$$

We see that, without power adaptation, (4.4) and (4.7) are the same, so transmitter side information does not increase capacity unless power is also adapted.

Now let us allow the transmit power $P(\gamma)$ to vary with γ subject to an average power constraint $\bar{P}$:

---

[2] Wolfowitz's result was for γ ranging over a finite set, but it can be extended to infinite sets [20].

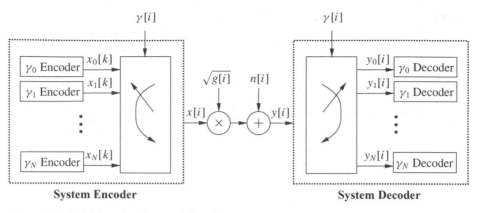

Figure 4.4: Multiplexed coding and decoding.

$$\int_0^\infty P(\gamma)p(\gamma)\,d\gamma \le \bar{P}. \tag{4.8}$$

With this additional constraint, we cannot apply (4.7) directly to obtain the capacity. However, we expect that the capacity with this average power constraint will be the average capacity given by (4.7) with the power optimally distributed over time. This motivates our definition of the fading channel capacity with average power constraint (4.8) as

$$C = \max_{P(\gamma):\,\int P(\gamma)p(\gamma)\,d\gamma = \bar{P}} \int_0^\infty B \log_2\left(1 + \frac{P(\gamma)\gamma}{\bar{P}}\right)p(\gamma)\,d\gamma. \tag{4.9}$$

It is proved in [20] that the capacity given in (4.9) can be achieved and that any rate larger than this capacity has probability of error bounded away from zero. The main idea behind the proof is a "time diversity" system with multiplexed input and demultiplexed output, as shown in Figure 4.4. Specifically, we first quantize the range of fading values to a finite set $\{\gamma_j : 1 \le j \le N\}$. For each γ_j, we design an encoder–decoder pair for an AWGN channel with SNR γ_j. The input x_j for encoder γ_j has average power $P(\gamma_j)$ and data rate $R_j = C_j$, where C_j is the capacity of a time-invariant AWGN channel with received SNR $P(\gamma_j)\gamma_j/\bar{P}$. These encoder–decoder pairs correspond to a set of input and output ports associated with each γ_j. When $\gamma[i] \approx \gamma_j$, the corresponding pair of ports are connected through the channel. The codewords associated with each γ_j are thus multiplexed together for transmission and then demultiplexed at the channel output. This effectively reduces the time-varying channel to a set of time-invariant channels in parallel, where the jth channel operates only when $\gamma[i] \approx \gamma_j$. The average rate on the channel is just the sum of rates associated with each of the γ_j channels weighted by $p(\gamma_j)$, the percentage of time that the channel SNR equals γ_j. This yields the average capacity formula (4.9).

To find the optimal power allocation $P(\gamma)$, we form the Lagrangian

$$J(P(\gamma)) = \int_0^\infty B \log_2\left(1 + \frac{\gamma P(\gamma)}{\bar{P}}\right)p(\gamma)\,d\gamma - \lambda \int_0^\infty P(\gamma)p(\gamma)\,d\gamma. \tag{4.10}$$

Next we differentiate the Lagrangian and set the derivative equal to zero:

$$\frac{\partial J(P(\gamma))}{\partial P(\gamma)} = \left[\left(\frac{B/\ln(2)}{1 + \gamma P(\gamma)/\bar{P}}\right)\frac{\gamma}{\bar{P}} - \lambda\right]p(\gamma) = 0. \tag{4.11}$$

Solving for $P(\gamma)$ with the constraint that $P(\gamma) > 0$ yields the optimal power adaptation that maximizes (4.9) as

$$\frac{P(\gamma)}{\bar{P}} = \begin{cases} 1/\gamma_0 - 1/\gamma & \gamma \geq \gamma_0, \\ 0 & \gamma < \gamma_0, \end{cases} \tag{4.12}$$

for some "cutoff" value γ_0. If $\gamma[i]$ is below this cutoff then no data is transmitted over the ith time interval, so the channel is used at time i only if $\gamma_0 \leq \gamma[i] < \infty$. Substituting (4.12) into (4.9) then yields the capacity formula:

$$C = \int_{\gamma_0}^{\infty} B \log_2\left(\frac{\gamma}{\gamma_0}\right)p(\gamma)\,d\gamma. \tag{4.13}$$

The multiplexing nature of the capacity-achieving coding strategy indicates that (4.13) is achieved with a time-varying data rate, where the rate corresponding to the instantaneous SNR γ is $B \log_2(\gamma/\gamma_0)$. Since γ_0 is constant, this means that as the instantaneous SNR increases, the data rate sent over the channel for that instantaneous SNR also increases. Note that this multiplexing strategy is not the only way to achieve capacity (4.13): it can also be achieved by adapting the transmit power and sending at a fixed rate [21]. We will see in Section 4.2.6 that for Rayleigh fading this capacity can exceed that of an AWGN channel with the same average SNR – in contrast to the case of receiver CSI only, where fading always decreases capacity.

Note that the optimal power allocation policy (4.12) depends on the fading distribution $p(\gamma)$ only through the cutoff value γ_0. This cutoff value is found from the power constraint. Specifically, rearranging the power constraint (4.8) and replacing the inequality with equality (since using the maximum available power will always be optimal) yields the power constraint

$$\int_0^{\infty} \frac{P(\gamma)}{\bar{P}}p(\gamma)\,d\gamma = 1. \tag{4.14}$$

If we now substitute the optimal power adaptation (4.12) into this expression then the cutoff value γ_0 must satisfy

$$\int_{\gamma_0}^{\infty}\left(\frac{1}{\gamma_0} - \frac{1}{\gamma}\right)p(\gamma)\,d\gamma = 1. \tag{4.15}$$

Observe that this expression depends only on the distribution $p(\gamma)$. The value for γ_0 must be found numerically [22] because no closed-form solutions exist for typical continuous distributions $p(\gamma)$.

Since γ is time varying, the maximizing power adaptation policy of (4.12) is a water-filling formula in time, as illustrated in Figure 4.5. This curve shows how much power is allocated to the channel for instantaneous SNR $\gamma(t) = \gamma$. The water-filling terminology refers to the fact that the line $1/\gamma$ sketches out the bottom of a bowl, and power is poured into the bowl to a constant water level of $1/\gamma_0$. The amount of power allocated for a given γ equals $1/\gamma_0 - 1/\gamma$, the amount of water between the bottom of the bowl ($1/\gamma$) and the constant water line ($1/\gamma_0$). The intuition behind water-filling is to take advantage of good

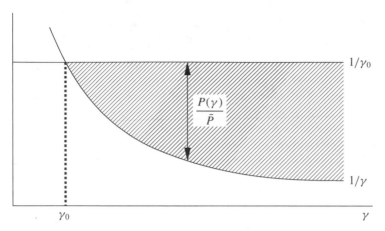

Figure 4.5: Optimal power allocation: water-filling.

channel conditions: when channel conditions are good (γ large), more power and a higher data rate are sent over the channel. As channel quality degrades (γ small), less power and rate are sent over the channel. If the instantaneous SNR falls below the cutoff value, the channel is not used. Adaptive modulation and coding techniques that follow this principle were developed in [23; 24] and are discussed in Chapter 9.

Note that the multiplexing argument sketching how capacity (4.9) is achieved applies to any power adaptation policy. That is, for any power adaptation policy $P(\gamma)$ with average power $\bar{P}$, the capacity

$$C = \int_0^\infty B \log_2\left(1 + \frac{P(\gamma)\gamma}{\bar{P}}\right) p(\gamma)\, d\gamma \tag{4.16}$$

can be achieved with arbitrarily small error probability. Of course this capacity cannot exceed (4.9), where power adaptation is optimized to maximize capacity. However, there are scenarios where a suboptimal power adaptation policy might have desirable properties that outweigh capacity maximization. In the next two sections we discuss two such suboptimal policies, which result in constant data rate systems, in contrast to the variable-rate transmission policy that achieves the capacity in (4.9).

EXAMPLE 4.4: Assume the same channel as in the previous example, with a bandwidth of 30 kHz and three possible received SNRs: $\gamma_1 = .8333$ with $p(\gamma_1) = .1$, $\gamma_2 = 83.33$ with $p(\gamma_2) = .5$, and $\gamma_3 = 333.33$ with $p(\gamma_3) = .4$. Find the ergodic capacity of this channel assuming that both transmitter and receiver have instantaneous CSI.

Solution: We know the optimal power allocation is water-filling, and we need to find the cutoff value γ_0 that satisfies the discrete version of (4.15) given by

$$\sum_{\gamma_i \geq \gamma_0}\left(\frac{1}{\gamma_0} - \frac{1}{\gamma_i}\right) p(\gamma_i) = 1. \tag{4.17}$$

We first assume that all channel states are used to obtain γ_0 (i.e., we assume $\gamma_0 \leq \min_i \gamma_i$) and see if the resulting cutoff value is below that of the weakest channel. If not then we have an inconsistency, and must redo the calculation assuming at least one of the channel states is not used. Applying (4.17) to our channel model yields

$$\sum_{i=1}^{3} \frac{p(\gamma_i)}{\gamma_0} - \sum_{i=1}^{3} \frac{p(\gamma_i)}{\gamma_i} = 1$$

$$\implies \frac{1}{\gamma_0} = 1 + \sum_{i=1}^{3} \frac{p(\gamma_i)}{\gamma_i} = 1 + \left(\frac{.1}{.8333} + \frac{.5}{83.33} + \frac{.4}{333.33} \right) = 1.13.$$

Solving for γ_0 yields $\gamma_0 = 1/1.13 = .89 > .8333 = \gamma_1$. Since this value of γ_0 is greater than the SNR in the weakest channel, this result is inconsistent because the channel should only be used for SNRs above the cutoff value. Therefore, we now redo the calculation assuming that the weakest state is not used. Then (4.17) becomes

$$\sum_{i=2}^{3} \frac{p(\gamma_i)}{\gamma_0} - \sum_{i=2}^{3} \frac{p(\gamma_i)}{\gamma_i} = 1$$

$$\implies \frac{.9}{\gamma_0} = 1 + \sum_{i=2}^{3} \frac{p(\gamma_i)}{\gamma_i} = 1 + \left(\frac{.5}{83.33} + \frac{.4}{333.33} \right) = 1.0072.$$

Solving for γ_0 yields $\gamma_0 = .89$. Hence, by assuming that the weakest channel with SNR γ_1 is not used, we obtain a consistent value for γ_0 with $\gamma_1 < \gamma_0 \leq \gamma_2$. The capacity of the channel then becomes

$$C = \sum_{i=2}^{3} B \log_2 \left(\frac{\gamma_i}{\gamma_0} \right) p(\gamma_i)$$

$$= 30000 \left(.5 \log_2 \frac{83.33}{.89} + .4 \log_2 \frac{333.33}{.89} \right) = 200.82 \text{ kbps.}$$

Comparing with the results of Example 4.3 we see that this rate is only slightly higher than for the case of receiver CSI only, and it is still significantly below that of an AWGN channel with the same average SNR. This is because the average SNR for the channel in this example is relatively high: for low-SNR channels, capacity with flat fading can exceed that of the AWGN channel with the same average SNR if we take advantage of the rare times when the fading channel is in a very good state.

ZERO-OUTAGE CAPACITY AND CHANNEL INVERSION

We now consider a suboptimal transmitter adaptation scheme where the transmitter uses the CSI to maintain a constant received power; that is, it inverts the channel fading. The channel then appears to the encoder and decoder as a time-invariant AWGN channel. This power adaptation, called *channel inversion*, is given by $P(\gamma)/\bar{P} = \sigma/\gamma$, where σ equals the constant received SNR that can be maintained with the transmit power constraint (4.8). The constant σ thus satisfies $\int (\sigma/\gamma) p(\gamma) \, d\gamma = 1$, so $\sigma = 1/\mathrm{E}[1/\gamma]$.

Fading channel capacity with channel inversion is just the capacity of an AWGN channel with SNR σ:

$$C = B \log_2[1 + \sigma] = B \log_2 \left[1 + \frac{1}{\mathrm{E}[1/\gamma]} \right]. \tag{4.18}$$

The capacity-achieving transmission strategy for this capacity uses a fixed-rate encoder and decoder designed for an AWGN channel with SNR σ. This has the advantage of maintaining a fixed data rate over the channel regardless of channel conditions. For this reason the channel capacity given in (4.18) is called *zero-outage capacity*, since the data rate is fixed under all channel conditions and there is no channel outage. Note that there exist practical coding techniques that achieve near-capacity data rates on AWGN channels, so the zero-outage capacity can be approximately achieved in practice.

Zero-outage capacity can exhibit a large data-rate reduction relative to Shannon capacity in extreme fading environments. In Rayleigh fading, for example, $E[1/\gamma]$ is infinite and thus the zero-outage capacity given by (4.18) is zero. Channel inversion is common in spread-spectrum systems with near–far interference imbalances [25]. It is also the simplest scheme to implement because the encoder and decoder are designed for an AWGN channel, independent of the fading statistics.

EXAMPLE 4.5: Assume the same channel as in the previous example, with a bandwidth of 30 kHz and three possible received SNRs: $\gamma_1 = .8333$ with $p(\gamma_1) = .1$, $\gamma_2 = 83.33$ with $p(\gamma_2) = .5$, and $\gamma_3 = 333.33$ with $p(\gamma_3) = .4$. Assuming transmitter and receiver CSI, find the zero-outage capacity of this channel.

Solution: The zero-outage capacity is $C = B \log_2[1 + \sigma]$, where $\sigma = 1/E[1/\gamma]$. Since

$$E\left[\frac{1}{\gamma}\right] = \frac{.1}{.8333} + \frac{.5}{83.33} + \frac{.4}{333.33} = .1272,$$

we have $C = 30000 \log_2[1 + 1/.1272] = 94.43$ kbps. Note that this is less than half of the Shannon capacity with optimal water-filling adaptation.

OUTAGE CAPACITY AND TRUNCATED CHANNEL INVERSION

The reason that zero-outage capacity may be significantly smaller than Shannon capacity on a fading channel is the requirement of maintaining a constant data rate in all fading states. By suspending transmission in particularly bad fading states (outage channel states), we can maintain a higher constant data rate in the other states and thereby significantly increase capacity. The *outage capacity* is defined as the maximum data rate that can be maintained in all non-outage channel states multiplied by the probability of non-outage. Outage capacity is achieved with a *truncated channel inversion* policy for power adaptation that compensates for fading only above a certain cutoff fade depth γ_0:

$$\frac{P(\gamma)}{\bar{P}} = \begin{cases} \sigma/\gamma & \gamma \geq \gamma_0, \\ 0 & \gamma < \gamma_0, \end{cases} \tag{4.19}$$

where γ_0 is based on the outage probability: $P_{\text{out}} = p(\gamma < \gamma_0)$. Since the channel is only used when $\gamma \geq \gamma_0$, the power constraint (4.8) yields $\sigma = 1/E_{\gamma_0}[1/\gamma]$, where

$$E_{\gamma_0}\left[\frac{1}{\gamma}\right] \triangleq \int_{\gamma_0}^{\infty} \frac{1}{\gamma} p(\gamma) \, d\gamma. \tag{4.20}$$

The outage capacity associated with a given outage probability P_{out} and corresponding cutoff γ_0 is given by

$$C(P_{out}) = B \log_2\left(1 + \frac{1}{\mathbf{E}_{\gamma_0}[1/\gamma]}\right) p(\gamma \geq \gamma_0). \tag{4.21}$$

We can also obtain the *maximum outage capacity* by maximizing outage capacity over all possible γ_0:

$$C = \max_{\gamma_0} B \log_2\left(1 + \frac{1}{\mathbf{E}_{\gamma_0}[1/\gamma]}\right) p(\gamma \geq \gamma_0). \tag{4.22}$$

This maximum outage capacity will still be less than Shannon capacity (4.13) because truncated channel inversion is a suboptimal transmission strategy. However, the transmit and receive strategies associated with inversion or truncated inversion may be easier to implement or have lower complexity than the water-filling schemes associated with Shannon capacity.

EXAMPLE 4.6: Assume the same channel as in the previous example, with a bandwidth of 30 kHz and three possible received SNRs: $\gamma_1 = .8333$ with $p(\gamma_1) = .1$, $\gamma_2 = 83.33$ with $p(\gamma_2) = .5$, and $\gamma_3 = 333.33$ with $p(\gamma_3) = .4$. Find the outage capacity of this channel and associated outage probabilities for cutoff values $\gamma_0 = .84$ and $\gamma_0 = 83.4$. Which of these cutoff values yields a larger outage capacity?

Solution: For $\gamma_0 = .84$ we use the channel when the SNR is γ_2 or γ_3, so $\mathbf{E}_{\gamma_0}[1/\gamma] = \sum_{i=2}^{3} p(\gamma_i)/\gamma_i = .5/83.33 + .4/333.33 = .0072$. The outage capacity is $C = B \log_2(1+1/\mathbf{E}_{\gamma_0}[1/\gamma]) p(\gamma \geq \gamma_0) = 30000 \log_2(1+138.88) \cdot .9 = 192.457$ kbps. For $\gamma_0 = 83.34$ we use the channel when the SNR is γ_3 only, so $\mathbf{E}_{\gamma_0}[1/\gamma] = p(\gamma_3)/\gamma_3 = .4/333.33 = .0012$. The capacity is $C = B \log_2(1 + 1/\mathbf{E}_{\gamma_0}[1/\gamma]) p(\gamma \geq \gamma_0) = 30000 \log_2(1 + 833.33) \cdot .4 = 116.45$ kbps. The outage capacity is larger when the channel is used for SNRs γ_2 and γ_3. Even though the SNR γ_3 is significantly larger than γ_2, the fact that this larger SNR occurs only 40% of the time makes it inefficient to only use the channel in this best state.

4.2.5 Capacity with Receiver Diversity

Receiver diversity is a well-known technique for improving the performance of wireless communications in fading channels. The main advantage of receiver diversity is that it mitigates the fluctuations due to fading so that the channel appears more like an AWGN channel. More details on receiver diversity and its performance will be given in Chapter 7. Since receiver diversity mitigates the impact of fading, an interesting question is whether it also increases the capacity of a fading channel. The capacity calculation under diversity combining requires first that the distribution of the received SNR $p(\gamma)$ under the given diversity combining technique be obtained. Once this distribution is known, it can be substituted into any of the capacity formulas already given to obtain the capacity under diversity combining. The specific capacity formula used depends on the assumptions about channel side information; for example, in the case of perfect transmitter and receiver CSI the formula (4.13) would be used. For different CSI assumptions, capacity (under both maximal ratio and selection combining diversity) was computed in [22]. It was found that, as expected, the capacity with perfect transmitter and receiver CSI is greater than with receiver CSI only, which in turn is greater

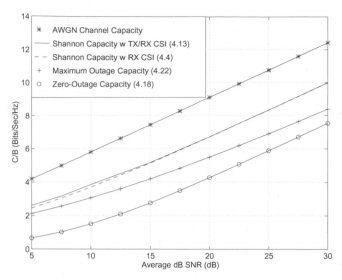

Figure 4.6: Capacity in log-normal fading.

than with channel inversion. The performance gap of these different formulas decreases as the number of antenna branches increases. This trend is expected because a large number of antenna branches makes the channel look like an AWGN channel, for which all of the different capacity formulas have roughly the same performance.

Recently there has been much research activity on systems with multiple antennas at both the transmitter and the receiver. The excitement in this area stems from the breakthrough results in [26; 27; 28] indicating that the capacity of a fading channel with multiple inputs and outputs (a MIMO channel) is M times larger than the channel capacity without multiple antennas, where $M = \min(M_t, M_r)$ for M_t the number of transmit antennas and M_r the number of receive antennas. We will discuss capacity of multiple antenna systems in Chapter 10.

4.2.6 Capacity Comparisons

In this section we compare capacity with transmitter and receiver CSI for different power allocation policies along with the capacity under receiver CSI only. Figures 4.6, 4.7, and 4.8 show plots of the different capacities (4.4), (4.13), (4.18), and (4.22) as a function of average received SNR for log-normal fading ($\sigma = 8$ dB standard deviation), Rayleigh fading, and Nakagami fading (with Nakagami parameter $m = 2$). Nakagami fading with $m = 2$ is roughly equivalent to Rayleigh fading with two-antenna receiver diversity. The capacity in AWGN for the same average power is also shown for comparison. Note that the capacity in log-normal fading is plotted relative to average dB SNR (μ_{dB}), not average SNR in dB ($10 \log_{10} \mu$): the relation between these values, as given by (2.45), is $10 \log_{10} \mu = \mu_{dB} + \sigma_{dB}^2 \ln(10)/20$.

Several observations in this comparison are worth noting. First, the figures show that the capacity of the AWGN channel is larger than that of the fading channel for all cases. However, at low SNRs the AWGN and fading channel with transmitter and receiver CSI have almost the same capacity. In fact, at low SNRs (below 0 dB), capacity of the fading channel with transmitter and receiver CSI is larger than the corresponding AWGN channel capacity.

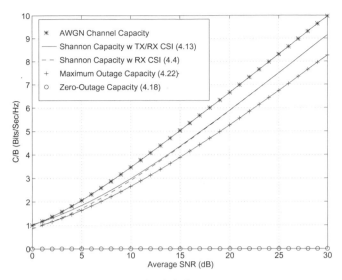

Figure 4.7: Capacity in Rayleigh fading.

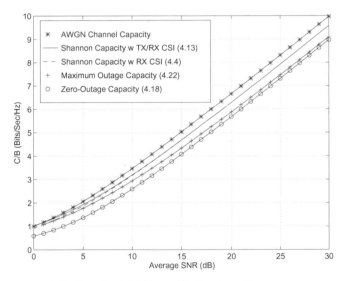

Figure 4.8: Capacity in Nakagami fading ($m = 2$).

That is because the AWGN channel always has the same low SNR, thereby limiting its capacity. A fading channel with this same low average SNR will occasionally have a high SNR, since the distribution has infinite range. Thus, if high power and rate are transmitted over the channel during these very infrequent large SNR values, the capacity will be greater than on the AWGN channel with the same low average SNR.

The severity of the fading is indicated by the Nakagami parameter m, where $m = 1$ for Rayleigh fading and $m = \infty$ for an AWGN channel without fading. Thus, comparing Figures 4.7 and 4.8 we see that, as the severity of the fading decreases (Rayleigh to Nakagami with $m = 2$), the capacity difference between the various adaptive policies also decreases, and their respective capacities approach that of the AWGN channel.

The difference between the capacity curves under transmitter and receiver CSI (4.13) and receiver CSI only (4.4) are negligible in all cases. Recalling that capacity under receiver CSI only (4.4) and under transmitter and receiver CSI without power adaptation (4.7) are the same, we conclude that, if the transmission rate is adapted relative to the channel, then adapting the power as well yields a negligible capacity gain. It also indicates that transmitter adaptation yields a negligible capacity gain relative to using only receiver side information. In severe fading conditions (Rayleigh and log-normal fading), maximum outage capacity exhibits a 1–5-dB rate penalty and zero-outage capacity yields a large capacity loss relative to Shannon capacity. However, under mild fading conditions (Nakagami with $m = 2$) the Shannon, maximum outage, and zero-outage capacities are within 3 dB of each other and within 4 dB of the AWGN channel capacity. These differences will further decrease as the fading diminishes ($m \rightarrow \infty$ for Nakagami fading).

We can view these results as a trade-off between capacity and complexity. The adaptive policy with transmitter and receiver side information requires more complexity in the transmitter (and typically also requires a feedback path between the receiver and transmitter to obtain the side information). However, the decoder in the receiver is relatively simple. The nonadaptive policy has a relatively simple transmission scheme, but its code design must use the channel correlation statistics (often unknown) and the decoder complexity is proportional to the channel decorrelation time. The channel inversion and truncated inversion policies use codes designed for AWGN channels and thus are the least complex to implement, but in severe fading conditions they exhibit large capacity losses relative to the other techniques.

In general, Shannon capacity analysis does not show how to design adaptive or nonadaptive techniques for real systems. Achievable rates for adaptive trellis-coded MQAM have been investigated in [24], where a simple four-state trellis code combined with adaptive six-constellation MQAM modulation was shown to achieve rates within 7 dB of the Shannon capacity (4.9) in Figures 4.6 and 4.7. More complex codes further close the gap to the Shannon limit of fading channels with transmitter adaptation.

4.3 Capacity of Frequency-Selective Fading Channels

In this section we examine the Shannon capacity of frequency-selective fading channels. We first consider the capacity of a time-invariant frequency-selective fading channel. This capacity analysis is like that of a flat fading channel but with the time axis replaced by the frequency axis. Then we discuss the capacity of time-varying frequency-selective fading channels.

4.3.1 Time-Invariant Channels

Consider a time-invariant channel with frequency response $H(f)$, as shown in Figure 4.9. Assume a total transmit power constraint P. When the channel is time invariant it is typically assumed that $H(f)$ is known to both the transmitter and receiver. (The capacity of time-invariant channels under different assumptions about channel knowledge are discussed in [19; 21].)

Let us first assume that $H(f)$ is *block fading*, so that frequency is divided into subchannels of bandwidth B with $H(f) = H_j$ constant over each subchannel, as shown in Figure 4.10.

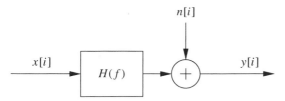

Figure 4.9: Time-invariant frequency-selective fading channel.

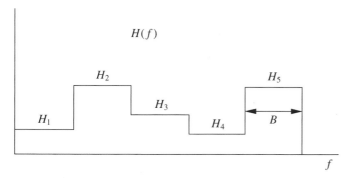

Figure 4.10: Block frequency-selective fading.

The frequency-selective fading channel thus consists of a set of AWGN channels in parallel with SNR $|H_j|^2 P_j / N_0 B$ on the jth channel, where P_j is the power allocated to the jth channel in this parallel set subject to the power constraint $\sum_j P_j \leq P$.

The capacity of this parallel set of channels is the sum of rates on each channel with power optimally allocated over all channels [5; 6]:

$$C = \sum_{\max P_j : \sum_j P_j \leq P} B \log_2\left(1 + \frac{|H_j|^2 P_j}{N_0 B}\right). \tag{4.23}$$

Note that this is similar to the capacity and optimal power allocation for a flat fading channel, with power and rate changing over frequency in a deterministic way rather than over time in a probabilistic way. The optimal power allocation is found via the same Lagrangian technique used in the flat-fading case, which leads to the water-filling power allocation

$$\frac{P_j}{P} = \begin{cases} 1/\gamma_0 - 1/\gamma_j & \gamma_j \geq \gamma_0, \\ 0 & \gamma_j < \gamma_0, \end{cases} \tag{4.24}$$

for some cutoff value γ_0, where $\gamma_j = |H_j|^2 P / N_0 B$ is the SNR associated with the jth channel assuming it is allocated the entire power budget. This optimal power allocation is illustrated in Figure 4.11. The cutoff value is obtained by substituting the power adaptation formula into the power constraint, so γ_0 must satisfy

$$\sum_j \left(\frac{1}{\gamma_0} - \frac{1}{\gamma_j}\right) = 1. \tag{4.25}$$

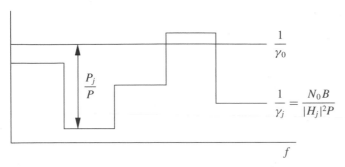

Figure 4.11: Water-filling in frequency-selective block fading.

The capacity then becomes

$$C = \sum_{j:\gamma_j \geq \gamma_0} B \log_2 \left(\frac{\gamma_j}{\gamma_0} \right). \qquad (4.26)$$

This capacity is achieved by transmitting at different rates and powers over each subchannel. Multicarrier modulation uses the same technique in adaptive loading, as discussed in more detail in Section 12.3.

When $H(f)$ is continuous, the capacity under power constraint P is similar to the case of the block fading channel, with some mathematical intricacies needed to show that the channel capacity is given by

$$C = \max_{P(f): \int P(f)\,df \leq P} \int \log_2 \left(1 + \frac{|H(f)|^2 P(f)}{N_0} \right) df. \qquad (4.27)$$

The expression inside the integral can be thought of as the incremental capacity associated with a given frequency f over the bandwidth df with power allocation $P(f)$ and channel gain $|H(f)|^2$. This result is formally proven using a Karhunen–Loeve expansion of the channel $h(t)$ to create an equivalent set of parallel independent channels [5, Chap. 8.5]. An alternate proof [29] decomposes the channel into a parallel set using the discrete Fourier transform (DFT); the same premise is used in the discrete implementation of multicarrier modulation described in Section 12.4.

The power allocation over frequency, $P(f)$, that maximizes (4.27) is found via the Lagrangian technique. The resulting optimal power allocation is water-filling over frequency:

$$\frac{P(f)}{P} = \begin{cases} 1/\gamma_0 - 1/\gamma(f) & \gamma(f) \geq \gamma_0, \\ 0 & \gamma(f) < \gamma_0, \end{cases} \qquad (4.28)$$

where $\gamma(f) = |H(f)|^2 P/N_0$. This results in channel capacity

$$C = \int_{f:\gamma(f) \geq \gamma_0} \log_2 \left(\frac{\gamma(f)}{\gamma_0} \right) df. \qquad (4.29)$$

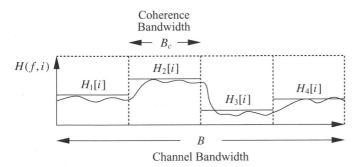

Figure 4.12: Channel division in frequency-selective fading.

EXAMPLE 4.7: Consider a time-invariant frequency-selective block fading channel that has three subchannels of bandwidth $B = 1$ MHz. The frequency responses associated with each subchannel are $H_1 = 1$, $H_2 = 2$, and $H_3 = 3$, respectively. The transmit power constraint is $P = 10$ mW and the noise PSD $N_0/2$ has $N_0 = 10^{-9}$ W/Hz. Find the Shannon capacity of this channel and the optimal power allocation that achieves this capacity.

Solution: We first find $\gamma_j = |H_j|^2 P/N_0 B$ for each subchannel, yielding $\gamma_1 = 10$, $\gamma_2 = 40$, and $\gamma_3 = 90$. The cutoff γ_0 must satisfy (4.25). Assuming that all subchannels are allocated power, this yields

$$\frac{3}{\gamma_0} = 1 + \sum_j \frac{1}{\gamma_j} = 1.14 \implies \gamma_0 = 2.64 < \gamma_j \ \forall j.$$

Since the cutoff γ_0 is less than γ_j for all j, our assumption that all subchannels are allocated power is consistent, so this is the correct cutoff value. The corresponding capacity is $C = \sum_{j=1}^{3} B \log_2(\gamma_j/\gamma_0) = 1000000(\log_2(10/2.64) + \log_2(40/2.64) + \log_2(90/2.64)) = 10.93$ Mbps.

4.3.2 Time-Varying Channels

The time-varying frequency-selective fading channel is similar to the model shown in Figure 4.9 except that $H(f) = H(f, i)$; that is, the channel varies over both frequency and time. It is difficult to determine the capacity of time-varying frequency-selective fading channels – even when the instantaneous channel $H(f, i)$ is known perfectly at the transmitter and receiver – because of the effects of self-interference (ISI). In the case of transmitter and receiver side information, the optimal adaptation scheme must consider (a) the effect of the channel on the past sequence of transmitted bits and (b) how the ISI resulting from these bits will affect future transmissions [30]. The capacity of time-varying frequency-selective fading channels is in general unknown, but there do exist upper and lower bounds as well as limiting formulas [30; 31].

We can approximate channel capacity in time-varying frequency-selective fading by taking the channel bandwidth B of interest and then dividing it up into subchannels the size of the channel coherence bandwidth B_c, as shown in Figure 4.12. We then assume that each of the resulting subchannels is independent, time varying, and flat fading with $H(f, i) = H_j[i]$ on the jth subchannel.

Under this assumption, we obtain the capacity for each of these flat fading subchannels based on the average power $\bar{P}_j$ that we allocate to each subchannel, subject to a total power constraint $\bar{P}$. Since the channels are independent, the total channel capacity is just equal to the sum of capacities on the individual narrowband flat fading channels – subject to the total average power constraint and averaged over both time and frequency:

$$C = \max_{\{\bar{P}_j\}: \sum_j \bar{P}_j \leq \bar{P}} \sum_j C_j(\bar{P}_j), \tag{4.30}$$

where $C_j(\bar{P}_j)$ is the capacity of the flat fading subchannel with average power $\bar{P}_j$ and bandwidth B_c given by (4.13), (4.4), (4.18), or (4.22) for Shannon capacity under different side information and power allocation policies. We can also define $C_j(\bar{P}_j)$ as a capacity versus outage if only the receiver has side information.

We will focus on Shannon capacity assuming perfect transmitter and receiver channel CSI, since this upperbounds capacity under any other side information assumptions or suboptimal power allocation strategies. We know that if we fix the average power per subchannel then the optimal power adaptation follows a water-filling formula. We expect that the optimal average power to be allocated to each subchannel will also follow a water-filling formula, where more average power is allocated to better subchannels. Thus we expect that the optimal power allocation is a two-dimensional water-filling in both time and frequency. We now obtain this optimal two-dimensional water-filling and the corresponding Shannon capacity.

Define $\gamma_j[i] = |H_j[i]|^2 \bar{P}/N_0 B$ to be the instantaneous SNR on the jth subchannel at time i assuming the total power $\bar{P}$ is allocated to that time and frequency. We allow the power $P_j(\gamma_j)$ to vary with $\gamma_j[i]$. The Shannon capacity with perfect transmitter and receiver CSI is given by optimizing power adaptation relative to both time (represented by $\gamma_j[i] = \gamma_j$) and frequency (represented by the subchannel index j):

$$C = \max_{P_j(\gamma_j): \sum_j \int_0^\infty P_j(\gamma_j)p(\gamma_j)\, d\gamma_j \leq \bar{P}} \sum_j \int_0^\infty B_c \log_2\left(1 + \frac{P_j(\gamma_j)\gamma_j}{\bar{P}}\right) p(\gamma_j)\, d\gamma_j. \tag{4.31}$$

To find the optimal power allocation $P_j(\gamma_j)$, we form the Lagrangian

$$J(P_j(\gamma_j))$$
$$= \sum_j \int_0^\infty B_c \log_2\left(1 + \frac{P_j(\gamma_j)\gamma_j}{\bar{P}}\right) p(\gamma_j)\, d\gamma_j - \lambda \sum_j \int_0^\infty P_j(\gamma_j)p(\gamma_j)\, d\gamma_j. \tag{4.32}$$

Note that (4.32) is similar to the Lagrangian (4.10) for the flat fading channel except that the dimension of frequency has been added by summing over the subchannels. Differentiating the Lagrangian and setting this derivative equal to zero eliminates all terms except the given subchannel and associated SNR:

$$\frac{\partial J(P_j(\gamma_j))}{\partial P_j(\gamma_j)} = \left[\left(\frac{B_c/\ln(2)}{1 + \gamma_j P_j(\gamma_j)/\bar{P}}\right)\frac{\gamma_j}{\bar{P}} - \lambda\right] p(\gamma_j) = 0. \tag{4.33}$$

Solving for $P_j(\gamma_j)$ yields the same water-filling as the flat-fading case:

$$\frac{P_j(\gamma_j)}{\bar{P}} = \begin{cases} 1/\gamma_0 - 1/\gamma_j & \gamma_j \geq \gamma_0, \\ 0 & \gamma_j < \gamma_0, \end{cases} \tag{4.34}$$

where the cutoff value γ_0 is obtained from the total power constraint over both time and frequency:

$$\sum_j \int_0^\infty P_j(\gamma_j) p(\gamma_j) \, d\gamma_j = \bar{P}. \tag{4.35}$$

Thus, the optimal power allocation (4.34) is a two-dimensional water-filling with a common cutoff value γ_0. Dividing the constraint (4.35) by $\bar{P}$ and substituting into the optimal power allocation (4.34), we get that γ_0 must satisfy

$$\sum_j \int_{\gamma_0}^\infty \left(\frac{1}{\gamma_0} - \frac{1}{\gamma_j} \right) p(\gamma_j) \, d\gamma_j = 1. \tag{4.36}$$

It is interesting to note that, in the two-dimensional water-filling, the cutoff value for all subchannels is the same. This implies that even if the fading distribution or average fade power on the subchannels is different, all subchannels suspend transmission when the instantaneous SNR falls below the common cutoff value γ_0. Substituting the optimal power allocation (4.35) into the capacity expression (4.31) yields

$$C = \sum_j \int_{\gamma_0}^\infty B_c \log_2 \left(\frac{\gamma_j}{\gamma_0} \right) p(\gamma_j) \, d\gamma_j. \tag{4.37}$$

PROBLEMS

4-1. Capacity in AWGN is given by $C = B \log_2(1 + P/N_0 B)$, where P is the received signal power, B is the signal bandwidth, and $N_0/2$ is the noise PSD. Find capacity in the limit of infinite bandwidth $B \to \infty$ as a function of P.

4-2. Consider an AWGN channel with bandwidth 50 MHz, received signal power 10 mW, and noise PSD $N_0/2$ where $N_0 = 2 \cdot 10^{-9}$ W/Hz. How much does capacity increase by doubling the received power? How much does capacity increase by doubling the channel bandwidth?

4-3. Consider two users simultaneously transmitting to a single receiver in an AWGN channel. This is a typical scenario in a cellular system with multiple users sending signals to a base station. Assume the users have equal received power of 10 mW and total noise at the receiver in the bandwidth of interest of 0.1 mW. The channel bandwidth for each user is 20 MHz.

    (a) Suppose that the receiver decodes user 1's signal first. In this decoding, user 2's signal acts as noise (assume it has the same statistics as AWGN). What is the capacity of user 1's channel with this additional interference noise?

(b) Suppose that, after decoding user 1's signal, the decoder re-encodes it and subtracts it out of the received signal. Now, in the decoding of user 2's signal, there is no interference from user 1's signal. What then is the Shannon capacity of user 2's channel?

Note: We will see in Chapter 14 that the decoding strategy of successively subtracting out decoded signals is optimal for achieving Shannon capacity of a multiuser channel with independent transmitters sending to one receiver.

4-4. Consider a flat fading channel of bandwidth 20 MHz and where, for a fixed transmit power $\bar{P}$, the received SNR is one of six values: $\gamma_1 = 20$ dB, $\gamma_2 = 15$ dB, $\gamma_3 = 10$ dB, $\gamma_4 = 5$ dB, $\gamma_5 = 0$ dB, and $\gamma_6 = -5$ dB. The probabilities associated with each state are $p_1 = p_6 = .1$, $p_2 = p_4 = .15$, and $p_3 = p_5 = .25$. Assume that only the receiver has CSI.

    (a) Find the Shannon capacity of this channel.

    (b) Plot the capacity versus outage for $0 \leq P_{\text{out}} < 1$ and find the maximum average rate that can be correctly received (maximum C_{out}).

4-5. Consider a flat fading channel in which, for a fixed transmit power $\bar{P}$, the received SNR is one of four values: $\gamma_1 = 30$ dB, $\gamma_2 = 20$ dB, $\gamma_3 = 10$ dB, and $\gamma_4 = 0$ dB. The probabilities associated with each state are $p_1 = .2$, $p_2 = .3$, $p_3 = .3$, and $p_4 = .2$. Assume that both transmitter and receiver have CSI.

    (a) Find the optimal power adaptation policy $P[i]/\bar{P}$ for this channel and its corresponding Shannon capacity per unit hertz (C/B).

    (b) Find the channel inversion power adaptation policy for this channel and associated zero-outage capacity per unit bandwidth.

    (c) Find the truncated channel inversion power adaptation policy for this channel and associated outage capacity per unit bandwidth for three different outage probabilities: $P_{\text{out}} = .1$, $P_{\text{out}} = .25$, and P_{out} (and the associated cutoff γ_0) equal to the value that achieves maximum outage capacity.

4-6. Consider a cellular system where the power falloff with distance follows the formula $P_r(d) = P_t(d_0/d)^\alpha$, where $d_0 = 100$ m and α is a random variable. The distribution for α is $p(\alpha = 2) = .4$, $p(\alpha = 2.5) = .3$, $p(\alpha = 3) = .2$, and $p(\alpha = 4) = .1$. Assume a receiver at a distance $d = 1000$ m from the transmitter, with an average transmit power constraint of $P_t = 100$ mW and a receiver noise power of .1 mW. Assume that both transmitter and receiver have CSI.

    (a) Compute the distribution of the received SNR.

    (b) Derive the optimal power adaptation policy for this channel and its corresponding Shannon capacity per unit hertz (C/B).

    (c) Determine the zero-outage capacity per unit bandwidth of this channel.

    (d) Determine the maximum outage capacity per unit bandwidth of this channel.

4-7. Assume a Rayleigh fading channel, where the transmitter and receiver have CSI and the distribution of the fading SNR $p(\gamma)$ is exponential with mean $\bar{\gamma} = 10$ dB. Assume a channel bandwidth of 10 MHz.

    (a) Find the cutoff value γ_0 and the corresponding power adaptation that achieves Shannon capacity on this channel.

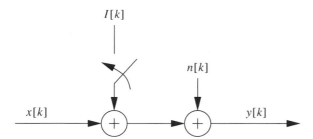

Figure 4.13: Interference channel for Problem 4-8.

(b) Compute the Shannon capacity of this channel.

(c) Compare your answer in part (b) with the channel capacity in AWGN with the same average SNR.

(d) Compare your answer in part (b) with the Shannon capacity when only the receiver knows $\gamma[i]$.

(e) Compare your answer in part (b) with the zero-outage capacity and outage capacity when the outage probability is .05.

(f) Repeat parts (b), (c), and (d) – that is, obtain the Shannon capacity with perfect transmitter and receiver side information, in AWGN for the same average power, and with just receiver side information – for the same fading distribution but with mean $\bar{\gamma} = -5$ dB. Describe the circumstances under which a fading channel has higher capacity than an AWGN channel with the same average SNR and explain why this behavior occurs.

4-8. This problem illustrates the capacity gains that can be obtained from interference estimation and also how a malicious jammer can wreak havoc on link performance. Consider the interference channel depicted in Figure 4.13. The channel has a combination of AWGN $n[k]$ and interference $I[k]$. We model $I[k]$ as AWGN. The interferer is on (i.e., the switch is down) with probability .25 and off (i.e., switch up) with probability .75. The average transmit power is 10 mW, the noise PSD has $N_0 = 10^{-8}$ W/Hz, the channel bandwidth B is 10 kHz (receiver noise power is $N_0 B$), and the interference power (when on) is 9 mW.

(a) What is the Shannon capacity of the channel if neither transmitter nor receiver know when the interferer is on?

(b) What is the capacity of the channel if both transmitter and receiver know when the interferer is on?

(c) Suppose now that the interferer is a malicious jammer with perfect knowledge of $x[k]$ (so the interferer is no longer modeled as AWGN). Assume that neither transmitter nor receiver has knowledge of the jammer behavior. Assume also that the jammer is always on and has an average transmit power of 10 mW. What strategy should the jammer use to minimize the SNR of the received signal?

4-9. Consider the malicious interferer of Problem 4-8. Suppose that the transmitter knows the interference signal perfectly. Consider two possible transmit strategies under this scenario: the transmitter can ignore the interference and use all its power for sending its signal, or it can use some of its power to cancel out the interferer (i.e., transmit the negative of the interference signal). In the first approach the interferer will degrade capacity by increasing the noise, and in the second strategy the interferer also degrades capacity because the transmitter sacrifices some power to cancel out the interference. Which strategy results in higher capacity? *Note:* There is a third strategy, in which the encoder actually exploits the structure of the interference in its encoding. This strategy is called *dirty paper coding* and is used to achieve Shannon capacity on broadcast channels with multiple antennas, as described in Chapter 14.

4-10. Show using Lagrangian techniques that the optimal power allocation to maximize the capacity of a time-invariant block fading channel is given by the water-filling formula in (4.24).

4-11. Consider a time-invariant block fading channel with frequency response

$$H(f) = \begin{cases} 1 & f_c - 20\ \text{MHz} \le f < f_c - 10\ \text{MHz}, \\ .5 & f_c - 10\ \text{MHz} \le f < f_c, \\ 2 & f_c \le f < f_c + 10\ \text{MHz}, \\ .25 & f_c + 10\ \text{MHz} \le f < f_c + 20\ \text{MHz}, \\ 0 & \text{else}, \end{cases}$$

for $f > 0$ and $H(-f) = H(f)$. For a transmit power of 10 mW and a noise PSD of .001 μW per Hertz, find the optimal power allocation and corresponding Shannon capacity of this channel.

4-12. Show that the optimal power allocation to maximize the capacity of a time-invariant frequency-selective fading channel is given by the water-filling formula in (4.28).

4-13. Consider a frequency-selective fading channel with total bandwidth 12 MHz and coherence bandwidth $B_c = 4$ MHz. Divide the total bandwidth into three subchannels of bandwidth B_c, and assume that each subchannel is a Rayleigh flat fading channel with independent fading on each subchannel. Assume the subchannels have average gains $\mathbf{E}[|H_1[i]|^2] = 1$, $\mathbf{E}[|H_2[i]|^2] = .5$, and $\mathbf{E}[|H_3[i]|^2] = .125$. Assume a total transmit power of 30 mW and a receiver noise PSD with $N_0 = .001\ \mu$W/Hz.

  (a) Find the optimal two-dimensional water-filling power adaptation for this channel and the corresponding Shannon capacity, assuming both transmitter and receiver know the instantaneous value of $H_j[i]$, $j = 1, 2, 3$.
  (b) Compare the capacity derived in part (a) with that obtained by allocating an equal average power of 10 mW to each subchannel and then water-filling on each subchannel relative to this power allocation.

REFERENCES

[1] C. E. Shannon, "A mathematical theory of communication," *Bell System Tech. J.,* pp. 379–423, 623–56, 1948.
[2] C. E. Shannon, "Communications in the presence of noise." *Proc. IRE,* pp. 10–21, 1949.
[3] C. E. Shannon and W. Weaver, *The Mathematical Theory of Communication,* University of Illinois Press, Urbana, 1949.
[4] M. Medard, "The effect upon channel capacity in wireless communications of perfect and imperfect knowledge of the channel," *IEEE Trans. Inform. Theory,* pp. 933–46, May 2000.
[5] R. G. Gallager, *Information Theory and Reliable Communication,* Wiley, New York, 1968.
[6] T. Cover and J. Thomas, *Elements of Information Theory,* Wiley, New York, 1991.

[7]   C. Heegard and S. B. Wicker, *Turbo Coding,* Kluwer, Boston, 1999.

[8]   I. Csiszár and J. Kórner, *Information Theory: Coding Theorems for Discrete Memoryless Channels,* Academic Press, New York, 1981.

[9]   I. Csiszár and P. Narayan, "The capacity of the arbitrarily varying channel," *IEEE Trans. Inform. Theory,* pp. 18–26, January 1991.

[10] I. C. Abou-Faycal, M. D. Trott, and S. Shamai, "The capacity of discrete-time memoryless Rayleigh fading channels," *IEEE Trans. Inform. Theory,* pp. 1290–1301, May 2001.

[11] A. Lapidoth and S. M. Moser, "Capacity bounds via duality with applications to multiple-antenna systems on flat-fading channels," *IEEE Trans. Inform. Theory,* pp. 2426–67, October 2003.

[12] A. J. Goldsmith and P. P. Varaiya, "Capacity, mutual information, and coding for finite-state Markov channels," *IEEE Trans. Inform. Theory,* pp. 868–86, May 1996.

[13] M. Mushkin and I. Bar-David, "Capacity and coding for the Gilbert–Elliot channel," *IEEE Trans. Inform. Theory,* pp. 1277–90, November 1989.

[14] T. Holliday, A. Goldsmith, and P. Glynn, "Capacity of finite state Markov channels with general inputs," *Proc. IEEE Internat. Sympos. Inform. Theory,* p. 289, July 2003.

[15] R. J. McEliece and W. E. Stark, "Channels with block interference," *IEEE Trans. Inform. Theory,* pp. 44–53, January 1984.

[16] A. Lapidoth and S. Shamai, "Fading channels: How perfect need 'perfect side information' be?" *IEEE Trans. Inform. Theory,* pp. 1118–34, November 1997.

[17] G. J. Foschini, D. Chizhik, M. Gans, C. Papadias, and R. A. Valenzuela, "Analysis and performance of some basic space-time architectures," *IEEE J. Sel. Areas Commun.,* pp. 303–20, April 2003.

[18] W. L. Root and P. P. Varaiya, "Capacity of classes of Gaussian channels," *SIAM J. Appl. Math.,* pp. 1350–93, November 1968.

[19] J. Wolfowitz, *Coding Theorems of Information Theory,* 2nd ed., Springer-Verlag, New York, 1964.

[20] A. J. Goldsmith and P. P. Varaiya, "Capacity of fading channels with channel side information," *IEEE Trans. Inform. Theory,* pp. 1986–92, November 1997.

[21] G. Caire and S. Shamai, "On the capacity of some channels with channel state information," *IEEE Trans. Inform. Theory,* pp. 2007–19, September 1999.

[22] M.-S. Alouini and A. J. Goldsmith, "Capacity of Rayleigh fading channels under different adaptive transmission and diversity combining techniques," *IEEE Trans. Veh. Tech.,* pp. 1165–81, July 1999.

[23] S.-G. Chua and A. J. Goldsmith, "Variable-rate variable-power MQAM for fading channels," *IEEE Trans. Commun.,* pp. 1218–30, October 1997.

[24] S.-G. Chua and A. J. Goldsmith, "Adaptive coded modulation for fading channels," *IEEE Trans. Commun.,* pp. 595–602, May 1998.

[25] K. S. Gilhousen, I. M. Jacobs, R. Padovani, A. J. Viterbi, L. A. Weaver, Jr., and C. E. Wheatley III, "On the capacity of a cellular CDMA system," *IEEE Trans. Veh. Tech.,* pp. 303–12, May 1991.

[26] G. J. Foschini, "Layered space-time architecture for wireless communication in fading environments when using multi-element antennas," *Bell System Tech. J.,* pp. 41–59, Autumn 1996.

[27] E. Teletar, "Capacity of multi-antenna Gaussian channels," AT&T Bell Labs Internal Tech. Memo, June 1995.

[28] G. J. Foschini and M. Gans, "On limits of wireless communications in a fading environment when using multiple antennas," *Wireless Pers. Commun.,* pp. 311–35, March 1998.

[29] W. Hirt and J. L. Massey, "Capacity of the discrete-time Gaussian channel with intersymbol interference," *IEEE Trans. Inform. Theory,* pp. 380–8, May 1988.

[30] A. Goldsmith and M. Medard, "Capacity of time-varying channels with channel side information," *IEEE Trans. Inform. Theory* (to appear).

[31] S. Diggavi, "Analysis of multicarrier transmission in time-varying channels," *Proc. IEEE Internat. Conf. Commun.,* pp. 1191–5, June 1997.

5

Digital Modulation and Detection

The advances over the last several decades in hardware and digital signal processing have made digital transceivers much cheaper, faster, and more power efficient than analog transceivers. More importantly, digital modulation offers a number of other advantages over analog modulation, including higher spectral efficiency, powerful error correction techniques, resistance to channel impairments, more efficient multiple access strategies, and better security and privacy. Specifically, high-level digital modulation techniques such as MQAM allow much more efficient use of spectrum than is possible with analog modulation. Advances in coding and coded modulation applied to digital signaling make the signal much less susceptible to noise and fading, and equalization or multicarrier techniques can be used to mitigate intersymbol interference (ISI). Spread-spectrum techniques applied to digital modulation can simultaneously remove or combine multipath, resist interference, and detect multiple users. Finally, digital modulation is much easier to encrypt, resulting in a higher level of security and privacy for digital systems. For all these reasons, systems currently being built or proposed for wireless applications are all digital systems.

Digital modulation and detection consist of transferring information in the form of bits over a communication channel. The bits are binary digits taking on the values of either 1 or 0. These information bits are derived from the information source, which may be a digital source or an analog source that has been passed through an A/D converter. Both digital and A/D-converted analog sources may be compressed to obtain the information bit sequence. Digital modulation consists of mapping the information bits into an analog signal for transmission over the channel. Detection consists of estimating the original bit sequence based on the signal received over the channel. The main considerations in choosing a particular digital modulation technique are

- high data rate,
- high spectral efficiency (minimum bandwidth occupancy),
- high power efficiency (minimum required transmit power),
- robustness to channel impairments (minimum probability of bit error), and
- low power/cost implementation.

Often these are conflicting requirements, and the choice of modulation is based on finding the technique that achieves the best trade-off between them.

There are two main categories of digital modulation: amplitude/phase modulation and frequency modulation. Since frequency modulation typically has a constant signal envelope and is generated using nonlinear techniques, this modulation is also called *constant envelope modulation* or *nonlinear modulation,* and amplitude/phase modulation is also called *linear modulation.* Linear modulation generally has better spectral properties than nonlinear modulation, since nonlinear processing leads to spectral broadening. However, amplitude and phase modulation embeds the information bits into the amplitude or phase of the transmitted signal, which is more susceptible to variations from fading and interference. In addition, amplitude and phase modulation techniques typically require linear amplifiers, which are more expensive and less power efficient than the nonlinear amplifiers that can be used with nonlinear modulation. Thus, the trade-off between linear versus nonlinear modulation is one of better spectral efficiency for the former technique and better power efficiency and resistance to channel impairments for the latter. Once the modulation technique is determined, the constellation size must be chosen. Modulations with large constellations have higher data rates for a given signal bandwidth, but they are more susceptible to noise, fading, and hardware imperfections. Finally, some demodulators require a coherent phase reference with respect to the transmitted signal. Obtaining this coherent reference may be difficult or significantly increase receiver complexity. Thus, modulation techniques that do not require a coherent phase reference in the receiver are desirable.

We begin this chapter with a general discussion of signal space concepts. These concepts greatly simplify the design and analysis of modulation and demodulation techniques by mapping infinite-dimensional signals to a finite-dimensional vector space. The general principles of signal space analysis will then be applied to the analysis of amplitude and phase modulation techniques, including pulse amplitude modulation (PAM), phase-shift keying (PSK), and quadrature amplitude modulation (QAM). We will also discuss constellation shaping and quadrature offset techniques for these modulations, as well as differential encoding to avoid the need for a coherent phase reference. We then describe frequency modulation techniques and their properties, including frequency-shift keying (FSK), minimum-shift keying (MSK), and continuous-phase FSK (CPFSK). Both coherent and noncoherent detection of these techniques will be discussed. Pulse-shaping techniques to improve the spectral properties of the modulated signals will also be covered, along with issues associated with carrier phase recovery and symbol synchronization.

5.1 Signal Space Analysis

Digital modulation encodes a bit stream of finite length into one of several possible transmitted signals. Intuitively, the receiver minimizes the probability of detection error by decoding the received signal as the signal in the set of possible transmitted signals that is "closest" to the one received. Determining the distance between the transmitted and received signals requires a metric for the distance between signals. By representing signals as projections onto a set of basis functions, we obtain a one-to-one correspondence between the set of transmitted signals and their vector representations. Thus, we can analyze signals in finite-dimensional vector space instead of infinite-dimensional function space, using classical notions of distance for vector spaces. In this section we show how digitally modulated signals can be

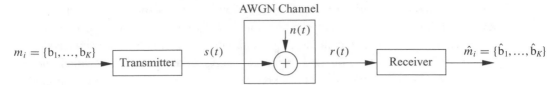

Figure 5.1: Communication system model.

represented as vectors in an appropriately defined vector space and how optimal demodulation methods can be obtained from this vector-space representation. This general analysis will then be applied to specific modulation techniques in later sections.

5.1.1 Signal and System Model

Consider the communication system model shown in Figure 5.1. Every T seconds, the system sends $K = \log_2 M$ bits of information through the channel for a data rate of $R = K/T$ bits per second (bps). There are $M = 2^K$ possible sequences of K bits, and we say that each bit sequence of length K comprises a message $m_i = \{b_1, \ldots, b_K\} \in \mathcal{M}$, where $\mathcal{M} = \{m_1, \ldots, m_M\}$ is the set of all such messages. The messages have probability p_i of being selected for transmission, where $\sum_{i=1}^{M} p_i = 1$.

Suppose message m_i is to be transmitted over the channel during the time interval $[0, T)$. Because the channel is analog, the message must be embedded into an analog signal for channel transmission. Hence, each message $m_i \in \mathcal{M}$ is mapped to a unique analog signal $s_i(t) \in \mathcal{S} = \{s_1(t), \ldots, s_M(t)\}$, where $s_i(t)$ is defined on the time interval $[0, T)$ and has energy

$$E_{s_i} = \int_0^T s_i^2(t)\, dt, \quad i = 1, \ldots, M. \tag{5.1}$$

Since each message represents a bit sequence it follows that each signal $s_i(t) \in \mathcal{S}$ also represents a bit sequence, and detection of the transmitted signal $s_i(t)$ at the receiver is equivalent to detection of the transmitted bit sequence. When messages are sent sequentially, the transmitted signal becomes a sequence of the corresponding analog signals over each time interval $[kT, (k+1)T)$: $s(t) = \sum_k s_i(t - kT)$, where $s_i(t)$ is a baseband or passband analog signal corresponding to the message m_i designated for the transmission interval $[kT, (k+1)T)$. This is illustrated in Figure 5.2, where we show the transmitted signal $s(t) = s_1(t) + s_2(t - T) + s_1(t - 2T) + s_1(t - 3T)$ corresponding to the string of messages m_1, m_2, m_1, m_1 with message m_i mapped to signal $s_i(t)$.

In the model of Figure 5.1, the transmitted signal is sent through an AWGN channel, where a white Gaussian noise process $n(t)$ of power spectral density $N_0/2$ is added to form the received signal $r(t) = s(t) + n(t)$. Given $r(t)$, the receiver must determine the best estimate of which $s_i(t) \in \mathcal{S}$ was transmitted during each transmission interval $[kT, (k+1)T)$. This best estimate for $s_i(t)$ is mapped to a best estimate of the message $m_i(t) \in \mathcal{M}$ and the receiver then outputs this best estimate $\hat{m} = \{\hat{b}_1, \ldots, \hat{b}_K\} \in \mathcal{M}$ of the transmitted bit sequence.

The goal of the receiver design in estimating the transmitted message is to minimize the probability of message error,

$$P_e = \sum_{i=1}^{M} p(\hat{m} \neq m_i \mid m_i \text{ sent}) p(m_i \text{ sent}), \tag{5.2}$$

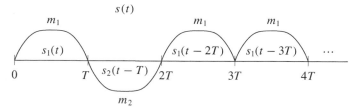

Figure 5.2: Transmitted signal for a sequence of messages.

over each time interval $[kT, (k+1)T)$. By representing the signals $\{s_i(t) : i = 1, \ldots, M\}$ geometrically, we can solve for the optimal receiver design in AWGN based on a minimum distance criterion. Note that, as we saw in previous chapters, wireless channels typically have a time-varying impulse response in addition to AWGN. We will consider the effect of an arbitrary channel impulse response on digital modulation performance in Chapter 6; methods to combat this performance degradation are discussed in Chapters 11–13.

5.1.2 Geometric Representation of Signals

The basic premise behind a geometrical representation of signals is the notion of a basis set. Specifically, using a Gram–Schmidt orthogonalization procedure [1; 2], it can be shown that any set of M real signals $\mathcal{S} = \{s_1(t), \ldots, s_M(t)\}$ defined on $[0, T)$ with finite energy can be represented as a linear combination of $N \leq M$ real orthonormal basis functions $\{\phi_1(t), \ldots, \phi_N(t)\}$. We say that these basis functions *span* the set $\mathcal{S}$. Thus, we can write each $s_i(t) \in \mathcal{S}$ in terms of its *basis function representation* as

$$s_i(t) = \sum_{j=1}^{N} s_{ij}\phi_j(t), \quad 0 \leq t < T, \tag{5.3}$$

where

$$s_{ij} = \int_0^T s_i(t)\phi_j(t)\,dt \tag{5.4}$$

is a real coefficient representing the projection of $s_i(t)$ onto the basis function $\phi_j(t)$ and

$$\int_0^T \phi_i(t)\phi_j(t)\,dt = \begin{cases} 1 & i = j, \\ 0 & i \neq j. \end{cases} \tag{5.5}$$

If the signals $\{s_i(t)\}$ are linearly independent then $N = M$, otherwise $N < M$. Moreover, the minimum number N of orthogonal basis functions needed to represent any signal $s_i(t)$ of duration T and bandwidth B is roughly $2BT$ [3, Chap. 5.3]. The signal $s_i(t)$ thus occupies $2BT$ orthogonal dimensions.

For linear passband modulation techniques, the basis set consists of the sine and cosine functions:

$$\phi_1(t) = \sqrt{\frac{2}{T}} \cos(2\pi f_c t) \tag{5.6}$$

and

$$\phi_2(t) = \sqrt{\frac{2}{T}} \sin(2\pi f_c t). \tag{5.7}$$

The $\sqrt{2/T}$ factor is needed for normalization so that $\int_0^T \phi_i^2(t)\,dt = 1, i = 1, 2$. In fact, with these basis functions we get only an approximation to (5.5), since

$$\int_0^T \phi_1^2(t)\,dt = \frac{2}{T}\int_0^T .5[1 + \cos(4\pi f_c t)]\,dt = 1 + \frac{\sin(4\pi f_c T)}{4\pi f_c T}. \tag{5.8}$$

The numerator in the second term of (5.8) is bounded by 1 and for $f_c T \gg 1$ the denominator of this term is very large. Hence this second term can be neglected. Similarly,

$$\int_0^T \phi_1(t)\phi_2(t)\,dt = \frac{2}{T}\int_0^T .5\sin(4\pi f_c t)\,dt = \frac{-\cos(4\pi f_c T)}{4\pi f_c T} \approx 0, \tag{5.9}$$

where the approximation is taken as an equality for $f_c T \gg 1$.

With the basis set $\phi_1(t) = \sqrt{2/T}\cos(2\pi f_c t)$ and $\phi_2(t) = \sqrt{2/T}\sin(2\pi f_c t)$, the basis function representation (5.3) corresponds to the equivalent lowpass representation of $s_i(t)$ in terms of its in-phase and quadrature components:

$$s_i(t) = s_{i1}\sqrt{\frac{2}{T}}\cos(2\pi f_c t) + s_{i2}\sqrt{\frac{2}{T}}\sin(2\pi f_c t). \tag{5.10}$$

Note that the carrier basis functions may have an initial phase offset ϕ_0. The basis set may also include a baseband pulse-shaping filter $g(t)$ to improve the spectral characteristics of the transmitted signal:

$$s_i(t) = s_{i1}g(t)\cos(2\pi f_c t) + s_{i2}g(t)\sin(2\pi f_c t). \tag{5.11}$$

In this case the pulse shape $g(t)$ must maintain the orthonormal properties (5.5) of basis functions; that is, we must have

$$\int_0^T g^2(t)\cos^2(2\pi f_c t)\,dt = 1 \tag{5.12}$$

and

$$\int_0^T g^2(t)\cos(2\pi f_c t)\sin(2\pi f_c t)\,dt = 0, \tag{5.13}$$

where the equalities may be approximations for $f_c T \gg 1$ as in (5.8) and (5.9). If the bandwidth of $g(t)$ satisfies $B \ll f_c$ then $g^2(t)$ is roughly constant over $T_c = 1/f_c$, so (5.13) is approximately true because the sine and cosine functions are orthogonal over one period T_c. The simplest pulse shape that satisfies (5.12) and (5.13) is the rectangular pulse shape $g(t) = \sqrt{2/T}, 0 \le t < T$.

EXAMPLE 5.1: Binary phase-shift keying (BPSK) modulation transmits the signal $s_1(t) = \alpha\cos(2\pi f_c t), 0 \le t \le T$, to send a 1-bit and the signal $s_2(t) = -\alpha\cos(2\pi f_c t)$, $0 \le t \le T$, to send a 0-bit. Find the set of orthonormal basis functions and coefficients $\{s_{ij}\}$ for this modulation.

Solution: There is only one basis function for $s_1(t)$ and $s_2(t)$:

$$\phi(t) = \sqrt{2/T}\cos(2\pi f_c t),$$

where the $\sqrt{2/T}$ is needed for normalization. The coefficients are then given by $s_1 = \alpha\sqrt{T/2}$ and $s_2 = -\alpha\sqrt{T/2}$.

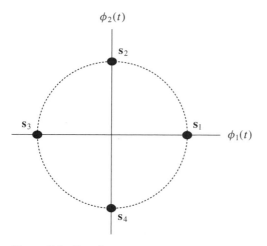

Figure 5.3: Signal space representation.

Let $\mathbf{s}_i = (s_{i1}, \ldots, s_{iN}) \in \mathbb{R}^N$ be the vector of coefficients $\{s_{ij}\}$ in the basis representation of $s_i(t)$. We call $\mathbf{s}_i$ the *signal constellation point* corresponding to the signal $s_i(t)$. The *signal constellation* consists of all constellation points $\{\mathbf{s}_1, \ldots, \mathbf{s}_M\}$. Given the basis functions $\{\phi_1(t), \ldots, \phi_N(t)\}$, there is a one-to-one correspondence between the transmitted signal $s_i(t)$ and its constellation point $\mathbf{s}_i$. Specifically, $s_i(t)$ can be obtained from $\mathbf{s}_i$ by (5.3) and $\mathbf{s}_i$ can be obtained from $s_i(t)$ by (5.4). Thus, it is equivalent to characterize the transmitted signal by $s_i(t)$ or $\mathbf{s}_i$. The representation of $s_i(t)$ in terms of its constellation point $\mathbf{s}_i \in \mathbb{R}^N$ is called its *signal space representation,* and the vector space containing the constellation is called the *signal space.* A two-dimensional signal space is illustrated in Figure 5.3, where we show $\mathbf{s}_i \in \mathbb{R}^2$ with the ith axis of $\mathbb{R}^2$ corresponding to the basis function $\phi_i(t)$, $i = 1, 2$. With this signal space representation we can analyze the infinite-dimensional functions $s_i(t)$ as vectors $\mathbf{s}_i$ in finite-dimensional vector space $\mathbb{R}^2$. This greatly simplifies the analysis of system performance as well as the derivation of optimal receiver design. Signal space representations for common modulation techniques like MPSK and MQAM are two-dimensional (corresponding to the in-phase and quadrature basis functions) and will be given later in the chapter.

In order to analyze signals via a signal space representation, we require a few definitions for vector characterization in the vector space $\mathbb{R}^N$. The length of a vector in $\mathbb{R}^N$ is defined as

$$\|\mathbf{s}_i\| \triangleq \sqrt{\sum_{j=1}^{N} s_{ij}^2}. \tag{5.14}$$

The distance between two signal constellation points $\mathbf{s}_i$ and $\mathbf{s}_k$ is thus

$$\|\mathbf{s}_i - \mathbf{s}_k\| = \sqrt{\sum_{j=1}^{N} (s_{ij} - s_{kj})^2} = \sqrt{\int_0^T (s_i(t) - s_k(t))^2 \, dt}, \tag{5.15}$$

where the second equality is obtained by writing $s_i(t)$ and $s_k(t)$ in their basis representation (5.3) and using the orthonormal properties of the basis functions. Finally, the inner product $\langle s_i(t), s_k(t) \rangle$ between two real signals $s_i(t)$ and $s_k(t)$ on the interval $[0, T]$ is

$$\langle s_i(t), s_k(t) \rangle = \int_0^T s_i(t) s_k(t) \, dt. \tag{5.16}$$

Similarly, the inner product $\langle \mathbf{s}_i, \mathbf{s}_k \rangle$ between two constellation points is

$$\langle \mathbf{s}_i, \mathbf{s}_k \rangle = \mathbf{s}_i \mathbf{s}_k^T = \int_0^T s_i(t) s_k(t) \, dt = \langle s_i(t), s_k(t) \rangle, \tag{5.17}$$

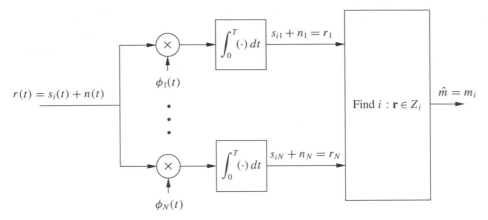

Figure 5.4: Receiver structure for signal detection in AWGN.

where the equality between the vector inner product and the corresponding signal inner product follows from the basis representation of the signals (5.3) and the orthonormal property of the basis functions (5.5). We say that two signals are *orthogonal* if their inner product is zero. Thus, by (5.5), the basis functions are orthogonal functions.

5.1.3 Receiver Structure and Sufficient Statistics

Given the channel output $r(t) = s_i(t) + n(t)$, $0 \leq t < T$, we now investigate the receiver structure to determine which constellation point $\mathbf{s}_i$ (or, equivalently, which message m_i) was sent over the time interval $[0, T)$. A similar procedure is done for each time interval $[kT, (k+1)T)$. We would like to convert the received signal $r(t)$ over each time interval into a vector, as this would allow us to work in finite-dimensional vector space when estimating the transmitted signal. However, this conversion should not be allowed to compromise the estimation accuracy. We now study a receiver that converts the received signal to a vector without compromising performance. Consider the receiver structure shown in Figure 5.4, where

$$s_{ij} = \int_0^T s_i(t)\phi_j(t)\, dt \tag{5.18}$$

and

$$n_j = \int_0^T n(t)\phi_j(t)\, dt. \tag{5.19}$$

We can rewrite $r(t)$ as

$$\sum_{j=1}^N (s_{ij} + n_j)\phi_j(t) + n_r(t) = \sum_{j=1}^N r_j\phi_j(t) + n_r(t); \tag{5.20}$$

here $r_j = s_{ij} + n_j$ and $n_r(t) = n(t) - \sum_{j=1}^N n_j\phi_j(t)$ denotes the "remainder" noise, which is the component of the noise orthogonal to the signal space. If we can show that the optimal detection of the transmitted signal constellation point $\mathbf{s}_i$ given received signal $r(t)$ does not make use of the remainder noise $n_r(t)$, then the receiver can make its estimate $\hat{m}$ of the transmitted message m_i as a function of $\mathbf{r} = (r_1, \ldots, r_N)$ alone. In other words, $\mathbf{r} = (r_1, \ldots, r_N)$ is a *sufficient statistic* for $r(t)$ in the optimal detection of the transmitted messages.

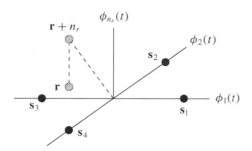

Figure 5.5: Projection of received signal onto received vector **r**.

It is intuitively clear that the remainder noise $n_r(t)$ should not help in detecting the transmitted signal $s_i(t)$, since its projection onto the signal space is zero. This is illustrated in Figure 5.5, where we assume the transmitted signal lies in a space spanned by the basis set $(\phi_1(t), \phi_2(t))$ while the remainder noise lies in a space spanned by the basis function $\phi_{n_r}(t)$, which is orthogonal to $\phi_1(t)$ and $\phi_2(t)$. Specifically, the remainder noise in Figure 5.5 is represented by n_r, where $n_r(t) = n_r \phi_{n_r}(t)$. The received signal is represented by $\mathbf{r} + n_r$, where $\mathbf{r} = (r_1, r_2)$ with $r(t) - n_r(t) = r_1 \phi_1(t) + r_2 \phi_2(t)$. From the figure it appears that projecting $\mathbf{r} + n_r$ onto $\mathbf{r}$ will not compromise the detection of which constellation $\mathbf{s}_i$ was transmitted, since n_r lies in a space orthogonal to the space in which $\mathbf{s}_i$ lies. We now proceed to show mathematically why this intuition is correct.

Let us first examine the distribution of $\mathbf{r}$. Since $n(t)$ is a Gaussian random process, if we condition on the transmitted signal $s_i(t)$ then the channel output $r(t) = s_i(t) + n(t)$ is also a Gaussian random process and $\mathbf{r} = (r_1, \ldots, r_N)$ is a Gaussian random vector. Recall that $r_j = s_{ij} + n_j$. Thus, conditioned on a transmitted constellation $\mathbf{s}_i$, we have that

$$\mu_{r_j | \mathbf{s}_i} = \mathbf{E}[r_j \mid \mathbf{s}_i] = \mathbf{E}[s_{ij} + n_j \mid s_{ij}] = s_{ij} \tag{5.21}$$

(since $n(t)$ has zero mean) and

$$\sigma_{r_j | \mathbf{s}_i} = \mathbf{E}[r_j - \mu_{r_j | \mathbf{s}_i}]^2 = \mathbf{E}[s_{ij} + n_j - s_{ij} \mid s_{ij}]^2 = \mathbf{E}[n_j^2]. \tag{5.22}$$

Moreover,

$$\begin{aligned}
\mathrm{Cov}[r_j r_k \mid \mathbf{s}_i] &= \mathbf{E}[(r_j - \mu_{r_j})(r_k - \mu_{r_k}) \mid \mathbf{s}_i] \\
&= \mathbf{E}[n_j n_k] \\
&= \mathbf{E}\left[\int_0^T n(t) \phi_j(t) \, dt \int_0^T n(\tau) \phi_k(\tau) \, d\tau \right] \\
&= \int_0^T \int_0^T \mathbf{E}[n(t) n(\tau)] \phi_j(t) \phi_k(\tau) \, dt \, d\tau \\
&= \int_0^T \int_0^T \frac{N_0}{2} \delta(t - \tau) \phi_j(t) \phi_k(\tau) \, dt \, d\tau \\
&= \frac{N_0}{2} \int_0^T \phi_j(t) \phi_k(t) \, dt \\
&= \begin{cases} N_0/2 & j = k, \\ 0 & j \neq k, \end{cases}
\end{aligned} \tag{5.23}$$

where the last equality follows from the orthonormality of the basis functions. Thus, conditioned on the transmitted constellation $\mathbf{s}_i$, the r_j are uncorrelated and, since they are Gaussian, they are also independent. Furthermore, $\mathbf{E}[n_j^2] = N_0/2$.

We have shown that, conditioned on the transmitted constellation s_i, r_j is a Gauss-distributed random variable that is independent of r_k ($k \neq j$) with mean s_{ij} and variance $N_0/2$. Thus, the conditional distribution of $\mathbf{r}$ is given by

$$p(\mathbf{r} \mid s_i \text{ sent}) = \prod_{j=1}^{N} p(r_j \mid m_i) = \frac{1}{(\pi N_0)^{N/2}} \exp\left[-\frac{1}{N_0} \sum_{j=1}^{N} (r_j - s_{ij})^2\right]. \qquad (5.24)$$

It is also straightforward to show that $\mathbf{E}[r_j n_r(t) \mid s_i] = 0$ for any t, $0 \leq t < T$. Thus, since r_j conditioned on s_i and $n_r(t)$ are Gaussian and uncorrelated, they are independent. Also, since the transmitted signal is independent of the noise, s_{ij} is independent of the process $n_r(t)$.

We now discuss the receiver design criterion and show that it is not affected by discarding $n_r(t)$. The goal of the receiver design is to minimize the probability of error in detecting the transmitted message m_i given received signal $r(t)$. In order to minimize $P_e = p(\hat{m} \neq m_i \mid r(t)) = 1 - p(\hat{m} = m_i \mid r(t))$, we maximize $p(\hat{m} = m_i \mid r(t))$. Therefore, the receiver output $\hat{m}$ given received signal $r(t)$ should correspond to the message m_i that maximizes $p(m_i$ sent $\mid r(t))$. Since there is a one-to-one mapping between messages and signal constellation points, this is equivalent to maximizing $p(s_i$ sent $\mid r(t))$. Recalling that $r(t)$ is completely described by $\mathbf{r} = (r_1, \ldots, r_N)$ and $n_r(t)$, we have

$$
\begin{aligned}
p(s_i \text{ sent} \mid r(t)) &= p((s_{i1}, \ldots, s_{iN}) \text{ sent} \mid (r_1, \ldots, r_N), n_r(t)) \\
&= \frac{p((s_{i1}, \ldots, s_{iN}) \text{ sent}, (r_1, \ldots, r_N), n_r(t))}{p((r_1, \ldots, r_N), n_r(t))} \\
&= \frac{p((s_{i1}, \ldots, s_{iN}) \text{ sent}, (r_1, \ldots, r_N)) p(n_r(t))}{p(r_1, \ldots, r_N) p(n_r(t))} \\
&= p((s_{i1}, \ldots, s_{iN}) \text{ sent} \mid (r_1, \ldots, r_N)), \qquad (5.25)
\end{aligned}
$$

where the third equality follows because $n_r(t)$ is independent of both $(r_1, \ldots, r_N)$ and $(s_{i1}, \ldots, s_{iN})$. This analysis shows that $\mathbf{r} = (r_1, \ldots, r_N)$ is a sufficient statistic for $r(t)$ in detecting m_i – in the sense that the probability of error is minimized by using only this sufficient statistic to estimate the transmitted signal and discarding the remainder noise. Since $\mathbf{r}$ is a sufficient statistic for the received signal $r(t)$, we call $\mathbf{r}$ the *received vector* associated with $r(t)$.

5.1.4 Decision Regions and the Maximum Likelihood Decision Criterion

We saw in the previous section that the optimal receiver minimizes error probability by selecting the detector output $\hat{m}$ that maximizes $1 - P_e = p(\hat{m}$ sent $\mid \mathbf{r})$. In other words, given a received vector $\mathbf{r}$, the optimal receiver selects $\hat{m} = m_i$ corresponding to the constellation s_i that satisfies $p(s_i$ sent $\mid \mathbf{r}) \geq p(s_j$ sent $\mid \mathbf{r})$ for all $j \neq i$. Let us define a set of *decision regions* $\{Z_1, \ldots, Z_M\}$ that are subsets of the signal space $\mathbb{R}^N$ by

$$Z_i = \{\mathbf{r} : p(s_i \text{ sent} \mid \mathbf{r}) > p(s_j \text{ sent} \mid \mathbf{r}) \, \forall j \neq i\}. \qquad (5.26)$$

Clearly these regions do not overlap. Moreover, they partition the signal space if there is no $\mathbf{r} \in \mathbb{R}^N$ for which $p(s_i$ sent $\mid \mathbf{r}) = p(s_j$ sent $\mid \mathbf{r})$. If such points exist then the signal space is partitioned into decision regions by arbitrarily assigning such points to decision region Z_i or Z_j. Once the signal space has been partitioned by decision regions, then for a

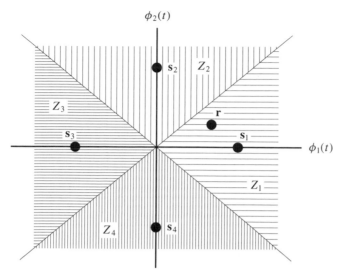

Figure 5.6: Decision regions.

received vector $\mathbf{r} \in Z_i$ the optimal receiver outputs the message estimate $\hat{m} = m_i$. Thus, the receiver processing consists of computing the received vector $\mathbf{r}$ from $r(t)$, finding which decision region Z_i contains $\mathbf{r}$, and outputting the corresponding message m_i. This process is illustrated in Figure 5.6, where we show a two-dimensional signal space with four decision regions $Z_1, \ldots, Z_4$ corresponding to four constellations $\mathbf{s}_1, \ldots, \mathbf{s}_4$. The received vector $\mathbf{r}$ lies in region Z_1, so the receiver will output the message m_1 as the best message estimate given received vector $\mathbf{r}$.

We now examine the decision regions in more detail. We will abbreviate $p(\mathbf{s}_i \text{ sent} \mid \mathbf{r}$ received) as $p(\mathbf{s}_i \mid \mathbf{r})$ and $p(\mathbf{s}_i \text{ sent})$ as $p(\mathbf{s}_i)$. By Bayes' rule,

$$p(\mathbf{s}_i \mid \mathbf{r}) = \frac{p(\mathbf{r} \mid \mathbf{s}_i)p(\mathbf{s}_i)}{p(\mathbf{r})}. \tag{5.27}$$

To minimize error probability, the receiver output $\hat{m} = m_i$ corresponds to the constellation $\mathbf{s}_i$ that maximizes $p(\mathbf{s}_i \mid \mathbf{r})$; that is, $\mathbf{s}_i$ must satisfy

$$\arg\max_{\mathbf{s}_i} \frac{p(\mathbf{r} \mid \mathbf{s}_i)p(\mathbf{s}_i)}{p(\mathbf{r})} = \arg\max_{\mathbf{s}_i} p(\mathbf{r} \mid \mathbf{s}_i)p(\mathbf{s}_i), \quad i = 1, \ldots, M, \tag{5.28}$$

where the second equality follows from the fact that $p(\mathbf{r})$ is not a function of $\mathbf{s}_i$. Assuming equally likely messages ($p(\mathbf{s}_i) = 1/M$), the receiver output $\hat{m} = m_i$ corresponds to the constellation $\mathbf{s}_i$ that satisfies

$$\arg\max_{\mathbf{s}_i} p(\mathbf{r} \mid \mathbf{s}_i), \quad i = 1, \ldots, M. \tag{5.29}$$

Let us define the likelihood function associated with our receiver as

$$L(\mathbf{s}_i) = p(\mathbf{r} \mid \mathbf{s}_i). \tag{5.30}$$

Given a received vector $\mathbf{r}$, a *maximum likelihood receiver* outputs $\hat{m} = m_i$ corresponding to the constellation $\mathbf{s}_i$ that maximizes $L(\mathbf{s}_i)$. Since the log function is increasing in its argument, maximizing $L(\mathbf{s}_i)$ is equivalent to maximizing its log. Moreover, the constant factor

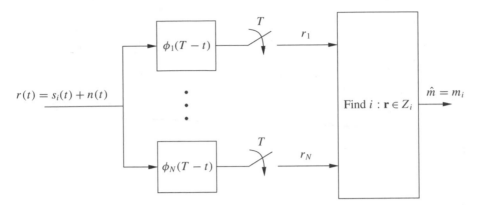

Figure 5.7: Matched filter receiver structure.

$(\pi N_0)^{-N/2}$ in (5.24) does not affect the maximization of $L(\mathbf{s}_i)$ relative to $\mathbf{s}_i$. Thus, maximizing $L(\mathbf{s}_i)$ is equivalent to maximizing the *log likelihood function,* defined as $l(\mathbf{s}_i) = \ln[(\pi N_0)^{N/2}L(\mathbf{s}_i)]$. Using (5.24) for $L(\mathbf{s}_i) = p(\mathbf{r} \mid \mathbf{s}_i)$ then yields

$$l(\mathbf{s}_i) = -\frac{1}{N_0}\sum_{j=1}^{N}(r_j - s_{ij}^2) = -\frac{1}{N_0}\|\mathbf{r} - \mathbf{s}_i\|^2. \tag{5.31}$$

Thus, the log likelihood function $l(\mathbf{s}_i)$ depends only on the distance between the received vector $\mathbf{r}$ and the constellation point $\mathbf{s}_i$. Moreover, from (5.31), $l(\mathbf{s}_i)$ is maximized by the constellation point $\mathbf{s}_i$ that is closest to the received vector $\mathbf{r}$.

The maximum likelihood receiver is implemented using the structure shown in Figure 5.4. First $\mathbf{r}$ is computed from $r(t)$, and then the signal constellation closest to $\mathbf{r}$ is determined as the constellation point $\mathbf{s}_i$ satisfying

$$\arg\min_{\mathbf{s}_i}\sum_{j=1}^{N}(r_j - s_{ij})^2 = \arg\min_{\mathbf{s}_i}\|\mathbf{r} - \mathbf{s}_i\|^2. \tag{5.32}$$

This $\mathbf{s}_i$ is determined from the decision region Z_i that contains $\mathbf{r}$, where Z_i is defined by

$$Z_i = \{\mathbf{r} : \|\mathbf{r} - \mathbf{s}_i\| < \|\mathbf{r} - \mathbf{s}_j\| \; \forall j = 1,\ldots,M, \; j \neq i\}, \quad i = 1,\ldots,M. \tag{5.33}$$

Finally, the estimated constellation $\mathbf{s}_i$ is mapped to the estimated message $\hat{m}$, which is output from the receiver. This result is intuitively satisfying, since the receiver decides that the transmitted constellation point is the one closest to the received vector. This maximum likelihood receiver structure is simple to implement because the decision criterion depends only on vector distances. This structure also minimizes the probability of message error at the receiver output when the transmitted messages are equally likely. However, if the messages and corresponding signal constellatations are not equally likely then the maximum likelihood receiver does not minimize error probability; in order to minimize error probability, the decision regions Z_i must be modified to take into account the message probabilities, as indicated in (5.27).

An alternate receiver structure is shown in Figure 5.7. This structure makes use of a bank of filters matched to each of the different basis functions. We call a filter with impulse response $\psi(t) = \phi(T-t), 0 \leq t \leq T$, the *matched filter* to the signal $\phi(t)$, so Figure 5.7 is also called a *matched filter receiver*. It can be shown that if a given input signal is passed through a filter matched to that signal then the output SNR is maximized. One can also show that the sampled matched filter outputs $(r_1, \ldots, r_n)$ in Figure 5.7 are the same as the $(r_1, \ldots, r_n)$ in Figure 5.4, so the receivers depicted in these two figures are equivalent.

EXAMPLE 5.2: For BPSK modulation, find decision regions Z_1 and Z_2 corresponding to constellations $s_1 = A$ and $s_2 = -A$ for $A > 0$.

Solution: The signal space is one-dimensional, so $\mathbf{r} = r \in \mathbb{R}$. By (5.33) the decision region $Z_1 \subset \mathbb{R}$ is defined by

$$Z_1 = \{r : \|r - A\| < \|r - (-A)\|\} = \{r : r > 0\}.$$

Thus, Z_1 contains all positive numbers on the real line. Similarly

$$Z_2 = \{r : \|r - (-A)\| < \|r - A\|\} = \{r : r < 0\}.$$

So Z_2 contains all negative numbers on the real line. For $r = 0$ the distance is the same to $s_1 = A$ and $s_2 = -A$, so we arbitrarily assign $r = 0$ to Z_2.

5.1.5 Error Probability and the Union Bound

We now analyze the error probability associated with the maximum likelihood receiver structure. For equally likely messages $p(m_i \text{ sent}) = 1/M$, we have

$$P_e = \sum_{i=1}^{M} p(\mathbf{r} \notin Z_i \mid m_i \text{ sent}) p(m_i \text{ sent})$$

$$= \frac{1}{M} \sum_{i=1}^{M} p(\mathbf{r} \notin Z_i \mid m_i \text{ sent})$$

$$= 1 - \frac{1}{M} \sum_{i=1}^{M} p(\mathbf{r} \in Z_i \mid m_i \text{ sent})$$

$$= 1 - \frac{1}{M} \sum_{i=1}^{M} \int_{Z_i} p(\mathbf{r} \mid m_i) \, d\mathbf{r}$$

$$= 1 - \frac{1}{M} \sum_{i=1}^{M} \int_{Z_i} p(\mathbf{r} = \mathbf{s}_i + \mathbf{n} \mid \mathbf{s}_i) \, d\mathbf{n}$$

$$= 1 - \frac{1}{M} \sum_{i=1}^{M} \int_{Z_i - \mathbf{s}_i} p(\mathbf{n}) \, d\mathbf{n}. \tag{5.34}$$

The integrals in (5.34) are over the N-dimensional subset $Z_i \subset \mathbb{R}^N$. We illustrate this error probability calculation in Figure 5.8, where the constellation points $\mathbf{s}_1, \ldots, \mathbf{s}_8$ are equally spaced around a circle with minimum separation $d_{\min}$. The probability of correct reception assuming the first symbol is sent, $p(\mathbf{r} \in Z_1 \mid m_1 \text{ sent})$, corresponds to the probability $p(\mathbf{r} = \mathbf{s}_1 + \mathbf{n} \in Z_1 \mid \mathbf{s}_1)$ that, when noise is added to the transmitted constellation $\mathbf{s}_1$, the resulting vector $\mathbf{r} = \mathbf{s}_1 + \mathbf{n}$ remains in the Z_1 region shown by the shaded area.

Figure 5.8 also indicates that the error probability is invariant to an angle rotation or axis shift of the signal constellation. The right side of the figure indicates a phase rotation of θ and

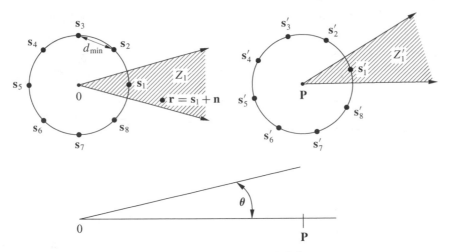

Figure 5.8: Error probability integral and its rotational/shift invariance.

axis shift of $\mathbf{P}$ relative to the constellation on the left side. Thus, $\mathbf{s}'_i = \mathbf{s}_i e^{j\theta} + \mathbf{P}$. The rotational invariance follows because the noise vector $\mathbf{n} = (n_1, \ldots, n_N)$ has components that are i.i.d. Gaussian random variables with zero mean; hence the polar representation $\mathbf{n} = |\mathbf{n}|e^{j\theta}$ has θ uniformly distributed, so the noise statistics are invariant to a phase rotation. The shift invariance follows from the fact that, if the constellation is shifted by some value $\mathbf{P} \in \mathbb{R}^N$, then the decision regions defined by (5.33) are also shifted by $\mathbf{P}$. Let $(\mathbf{s}_i, Z_i)$ denote a constellation point and corresponding decision region before the shift and $(\mathbf{s}'_i, Z'_i)$ the corresponding constellation point and decision region after the shift. It is then straightforward to show that $p(\mathbf{r} = \mathbf{s}_i + \mathbf{n} \in Z_i \mid \mathbf{s}_i) = p(\mathbf{r}' = \mathbf{s}'_i + \mathbf{n} \in Z'_i \mid \mathbf{s}'_i)$. Thus, the error probability after an axis shift of the constellation points will remain unchanged.

Although (5.34) gives an exact solution to the probability of error, we cannot solve for this error probability in closed form. Therefore, we now investigate the union bound on error probability, which yields a closed-form expression that is a function of the distance between signal constellation points. Let A_{ik} denote the event that $\|\mathbf{r} - \mathbf{s}_k\| < \|\mathbf{r} - \mathbf{s}_i\|$ given that the constellation point $\mathbf{s}_i$ was sent. If the event A_{ik} occurs, then the constellation will be decoded in error because the transmitted constellation $\mathbf{s}_i$ is not the closest constellation point to the received vector $\mathbf{r}$. However, event A_{ik} does not necessarily imply that $\mathbf{s}_k$ will be decoded instead of $\mathbf{s}_i$, since there may be another constellation point $\mathbf{s}_l$ with $\|\mathbf{r} - \mathbf{s}_l\| < \|\mathbf{r} - \mathbf{s}_k\| < \|\mathbf{r} - \mathbf{s}_i\|$. The constellation is decoded correctly if $\|\mathbf{r} - \mathbf{s}_i\| < \|\mathbf{r} - \mathbf{s}_k\|$ for all $k \neq i$. Thus

$$P_e(m_i \text{ sent}) = p\left(\bigcup_{\substack{k=1 \\ k \neq i}}^{M} A_{ik}\right) \leq \sum_{\substack{k=1 \\ k \neq i}}^{M} p(A_{ik}), \qquad (5.35)$$

where the inequality follows from the union bound on probability, defined below.

Let us now consider $p(A_{ik})$ more closely. We have

$$p(A_{ik}) = p(\|\mathbf{s}_k - \mathbf{r}\| < \|\mathbf{s}_i - \mathbf{r}\| \mid \mathbf{s}_i \text{ sent})$$
$$= p(\|\mathbf{s}_k - (\mathbf{s}_i + \mathbf{n})\| < \|\mathbf{s}_i - (\mathbf{s}_i + \mathbf{n})\|)$$
$$= p(\|\mathbf{n} + \mathbf{s}_i - \mathbf{s}_k\| < \|\mathbf{n}\|); \qquad (5.36)$$

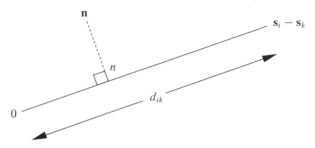

Figure 5.9: Noise projection.

that is, the probability of error equals the probability that the noise **n** is closer to the vector $\mathbf{s}_i - \mathbf{s}_k$ than to the origin. Recall that the noise has a mean of zero, so it is generally close to the origin. This probability does not depend on the entire noise component **n**: it only depends on the projection of **n** onto the line connecting the origin and the point $\mathbf{s}_i - \mathbf{s}_k$, as shown in Figure 5.9. Given the properties of **n**, the projection of **n** onto this one-dimensional line is a one-dimensional Gaussian random variable n with mean zero and variance $N_0/2$. The event A_{ik} occurs if n is closer to $\mathbf{s}_i - \mathbf{s}_k$ than to zero – that is, if $n > d_{ik}/2$, where $d_{ik} = \|\mathbf{s}_i - \mathbf{s}_k\|$ equals the distance between constellation points $\mathbf{s}_i$ and $\mathbf{s}_k$. Thus,

$$p(A_{ik}) = p\left(n > \frac{d_{ik}}{2}\right) = \int_{d_{ik}/2}^{\infty} \frac{1}{\sqrt{\pi N_0}} \exp\left[\frac{-v^2}{N_0}\right] dv = Q\left(\frac{d_{ik}}{\sqrt{2N_0}}\right). \tag{5.37}$$

Substituting (5.37) into (5.35) yields

$$P_e(m_i \text{ sent}) \le \sum_{\substack{k=1 \\ k \ne i}}^{M} Q\left(\frac{d_{ik}}{\sqrt{2N_0}}\right), \tag{5.38}$$

where the Q-function, $Q(z)$, is defined as the probability that a Gaussian random variable X with mean 0 and variance 1 is greater than z:

$$Q(z) = p(X > z) = \int_z^{\infty} \frac{1}{\sqrt{2\pi}} e^{-x^2/2} \, dx. \tag{5.39}$$

Summing (5.38) over all possible messages yields the *union bound*

$$P_e = \sum_{i=1}^{M} p(m_i) P_e(m_i \text{ sent}) \le \frac{1}{M} \sum_{i=1}^{M} \sum_{\substack{k=1 \\ k \ne i}}^{M} Q\left(\frac{d_{ik}}{\sqrt{2N_0}}\right). \tag{5.40}$$

Note that the Q-function cannot be solved for in closed form. It is related to the complementary error function as

$$Q(z) = \frac{1}{2} \operatorname{erfc}\left(\frac{z}{\sqrt{2}}\right). \tag{5.41}$$

We can also place an upper bound on $Q(z)$ with the closed-form expression

$$Q(z) \le \frac{1}{z\sqrt{2\pi}} e^{-z^2/2}, \tag{5.42}$$

and this bound is tight for $z \gg 0$.

Defining the *minimum distance* of the constellation as $d_{\min} = \min_{i,k} d_{ik}$, we can simplify (5.40) with the looser bound

$$P_e \le (M-1) Q\left(\frac{d_{\min}}{\sqrt{2N_0}}\right). \tag{5.43}$$

Using (5.42) for the Q-function yields a closed-form bound

$$P_e \le \frac{M-1}{d_{\min}\sqrt{\pi/N_0}} \exp\left[\frac{-d_{\min}^2}{4N_0}\right]. \tag{5.44}$$

Finally, P_e is sometimes approximated as the probability of error associated with constellations at the minimum distance $d_{\min}$ multiplied by the number $M_{d_{\min}}$ of neighbors at this distance:

$$P_e \approx M_{d_{\min}} Q\left(\frac{d_{\min}}{\sqrt{2N_0}}\right). \tag{5.45}$$

This approximation is called the *nearest neighbor approximation* to P_e. When different constellation points have a different number of nearest neighbors or different minimum distances, the bound can be averaged over the bound associated with each constellation point. Note that the nearest neighbor approximation will always be less than the loose bound (5.43) since $M \ge M_{d_{\min}}$. It will also be slightly less than the union bound (5.40), since the nearest neighbor approximation does not include the error associated with constellations farther apart than the minimum distance. However, the nearest neighbor approximation is quite close to the exact probability of symbol error at high SNRs, since for x and y large with $x > y$, $Q(x) \ll Q(y)$ owing to the exponential falloff of the Gaussian distribution in (5.39). This indicates that the probability of mistaking a constellation point for another point that is not one of its nearest neighbors is negligible at high SNRs. A rigorous derivation for (5.45) is made in [4] and also referenced in [5]. Moreover, [4] indicates that (5.45) captures the performance degradation due to imperfect receiver conditions such as slow carrier drift with an appropriate adjustment of the constants. The appeal of the nearest neighbor approximation is that it depends only on the minimum distance in the signal constellation and the number of nearest neighbors for points in the constellation.

EXAMPLE 5.3: Consider a signal constellation in $\mathbb{R}^2$ defined by $s_1 = (A,0)$, $s_2 = (0,A)$, $s_3 = (-A,0)$, and $s_4 = (0,-A)$. Assume $A/\sqrt{N_0} = 4$. Find the minimum distance and the union bound (5.40), looser bound (5.43), closed-form bound (5.44), and nearest neighbor approximation (5.45) on P_e for this constellation set.

Solution: The constellation is as depicted in Figure 5.3 with the radius of the circle equal to A. By symmetry, we need only consider the error probability associated with one of the constellation points, since it will be the same for the others. We focus on the error associated with transmitting constellation point s_1. The minimum distance to this constellation point is easily computed as $d_{\min} = d_{12} = d_{23} = d_{34} = d_{14} = \sqrt{A^2 + A^2} = \sqrt{2A^2}$. The distance to the other constellation points are $d_{13} = d_{24} = 2A$. By symmetry, $P_e(m_i \text{ sent}) = P_e(m_j \text{ sent})$ for $j \ne i$, so the union bound simplifies to

$$P_e \leq \sum_{j=2}^{4} Q\left(\frac{d_{1j}}{\sqrt{2N_0}}\right)$$

$$= 2Q\left(\frac{A}{\sqrt{N_0}}\right) + Q\left(\frac{\sqrt{2}A}{\sqrt{N_0}}\right) = 2Q(4) + Q(\sqrt{32}) = 3.1679 \cdot 10^{-5}.$$

The looser bound yields

$$P_e \leq 3Q(4) = 9.5014 \cdot 10^{-5},$$

which is roughly a factor of 3 looser than the union bound. The closed-form bound yields

$$P_e \leq \frac{3}{\sqrt{2\pi A^2/N_0}} \exp\left[\frac{-.5A^2}{N_0}\right] = 1.004 \cdot 10^{-4},$$

which differs from the union bound by about an order of magnitude. Finally, the nearest neighbor approximation yields

$$P_e \approx 2Q(4) = 3.1671 \cdot 10^{-5},$$

which (as expected) is approximately equal to the union bound.

Note that, for binary modulation (where $M = 2$), there is only one way to make an error and $d_{\min}$ is the distance between the two signal constellation points, so the bound (5.43) is exact:

$$P_b = Q\left(\frac{d_{\min}}{\sqrt{2N_0}}\right). \tag{5.46}$$

The square of the minimum distance $d_{\min}$ in (5.44) and (5.46) is typically proportional to the SNR of the received signal, as discussed in Chapter 6. Thus, error probability is reduced by increasing the received signal power.

Recall that P_e is the probability of a symbol (message) error: $P_e = p(\hat{m} \neq m_i \mid m_i \text{ sent})$, where m_i corresponds to a message with $\log_2 M$ bits. However, system designers are typically more interested in the bit error probability (also called the bit error rate, BER) than in the symbol error probability, because bit errors drive the performance of higher-layer networking protocols and end-to-end performance. Thus, we would like to design the mapping of the M possible bit sequences to messages m_i ($i = 1, \ldots, M$) so that a decoding error associated with an adjacent decision region, which is the most likely way to make an error, corresponds to only one bit error. With such a mapping – and with the assumption that mistaking a signal constellation point for a point other than one of its nearest neighbors has a very low probability – we can make the approximation

$$P_b \approx \frac{P_e}{\log_2 M}. \tag{5.47}$$

The most common form of mapping in which mistaking a constellation point for one of its nearest neighbors results in a single bit error is called Gray coding. Mapping by Gray coding is discussed in more detail in Section 5.3. Signal space concepts are applicable to any modulation where bits are encoded as one of several possible analog signals, including the amplitude, phase, and frequency modulations discussed in what follows.

5.2 Passband Modulation Principles

The basic principle of passband digital modulation is to encode an information bit stream into a carrier signal, which is then transmitted over a communications channel. Demodulation is the process of extracting this information bit stream from the received signal. Corruption of the transmitted signal by the channel can lead to bit errors in the demodulation process. The goal of modulation is to send bits at a high data rate while minimizing the probability of data corruption.

In general, modulated carrier signals encode information in the amplitude $\alpha(t)$, frequency $f(t)$, or phase $\theta(t)$ of a carrier signal. Thus, the modulated signal can be represented as

$$s(t) = \alpha(t)\cos[2\pi(f_c + f(t))t + \theta(t) + \phi_0] = \alpha(t)\cos(2\pi f_c t + \phi(t) + \phi_0), \quad (5.48)$$

where $\phi(t) = 2\pi f(t)t + \theta(t)$ and ϕ_0 is the phase offset of the carrier. This representation combines frequency and phase modulation into angle modulation.

We can rewrite the right-hand side of (5.48) in terms of its in-phase and quadrature components as

$$s(t) = \alpha(t)\cos(\phi(t) + \phi_0)\cos(2\pi f_c t) - \alpha(t)\sin(\phi(t) + \phi_0)\sin(2\pi f_c t)$$
$$= s_I(t)\cos(2\pi f_c t) - s_Q(t)\sin(2\pi f_c t), \quad (5.49)$$

where $s_I(t) = \alpha(t)(\cos\phi(t) + \phi_0)$ is the in-phase component of $s(t)$ and where $s_Q(t) = \alpha(t)(\sin\phi(t) + \phi_0)$ is its quadrature component. We can also write $s(t)$ in terms of its equivalent lowpass representation as

$$s(t) = \text{Re}\{u(t)e^{j2\pi f_c t}\}, \quad (5.50)$$

where $u(t) = s_I(t) + js_Q(t)$. This representation, described in more detail in Appendix A, is useful because receivers typically process the in-phase and quadrature signal components separately.

5.3 Amplitude and Phase Modulation

In amplitude and phase modulation, the information bit stream is encoded in the amplitude and/or phase of the transmitted signal. Specifically: over a time interval of T_s, $K = \log_2 M$ bits are encoded into the amplitude and/or phase of the transmitted signal $s(t)$, $0 \leq t < T_s$. The transmitted signal over this period, $s(t) = s_I(t)\cos(2\pi f_c t) - s_Q(t)\sin(2\pi f_c t)$, can be written in terms of its signal space representation as $s(t) = s_{i1}\phi_1(t) + s_{i2}\phi_2(t)$ with basis functions $\phi_1(t) = g(t)\cos(2\pi f_c t + \phi_0)$ and $\phi_2(t) = -g(t)\sin(2\pi f_c t + \phi_0)$, where $g(t)$ is a shaping pulse. For $\phi_0 = 0$, to send the ith message over the time interval $[kT, (k+1)T)$ we set $s_I(t) = s_{i1}g(t)$ and $s_Q(t) = s_{i2}g(t)$. These in-phase and quadrature signal components are baseband signals with spectral characteristics determined by the pulse shape $g(t)$. In particular, their bandwidth B equals the bandwidth of $g(t)$, and the transmitted signal $s(t)$ is a passband signal with center frequency f_c and passband bandwidth $2B$. In practice we take $B = K_g/T_s$, where K_g depends on the pulse shape: for rectangular pulses $K_g = .5$ and for

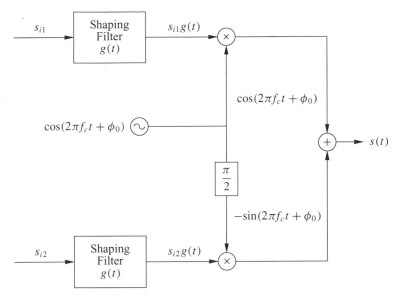

Figure 5.10: Amplitude/phase modulator.

raised cosine pulses $.5 \leq K_g \leq 1$, as discussed in Section 5.5. Thus, for rectangular pulses the bandwidth of $g(t)$ is $.5/T_s$ and the bandwidth of $s(t)$ is $1/T_s$. The signal constellation for amplitude and phase modulation is defined based on the constellation points $\{(s_{i1}, s_{i2}) \in \mathbb{R}^2,$ $i = 1, \ldots, M\}$. The equivalent lowpass representation of $s(t)$ is

$$s(t) = \text{Re}\{x(t)e^{j\phi_0}e^{j2\pi f_c t}\}, \tag{5.51}$$

where $x(t) = (s_{i1} + js_{i2})g(t)$. The constellation point $\mathbf{s}_i = (s_{i1}, s_{i2})$ is called the *symbol* associated with the $\log_2 M$ bits, and T_s is called the *symbol time*. The bit rate for this modulation is K bits per symbol or $R = \log_2 M/T_s$ bits per second.

There are three main types of amplitude/phase modulation:

- pulse amplitude modulation (MPAM) – information encoded in amplitude only;
- phase-shift keying (MPSK) – information encoded in phase only;
- quadrature amplitude modulation (MQAM) – information encoded in both amplitude and phase.

The number of bits per symbol $K = \log_2 M$, the signal constellation $\{\mathbf{s}_i, i = 1, \ldots, M\}$, and the choice of pulse shape $g(t)$ determine the digital modulation design. The pulse shape $g(t)$ is chosen to improve spectral efficiency and combat ISI, as discussed in Section 5.5.

Amplitude and phase modulation over a given symbol period can be generated using the modulator structure shown in Figure 5.10. Note that the basis functions in this figure have an arbitrary phase ϕ_0 associated with the transmit oscillator. Demodulation over each symbol period is performed using the demodulation structure of Figure 5.11, which

In-Phase Branch

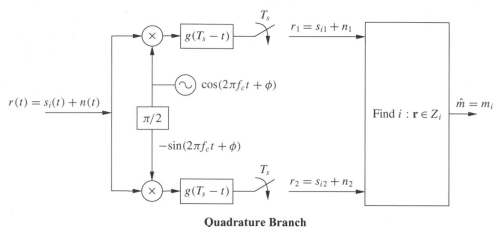

Quadrature Branch

Figure 5.11: Amplitude/phase demodulator (coherent: $\phi = \phi_0$).

is equivalent to the structure of Figure 5.7 for $\phi_1(t) = g(t)\cos(2\pi f_c t + \phi)$ and $\phi_2(t) = -g(t)\sin(2\pi f_c t + \phi)$. Typically the receiver includes some additional circuitry for *carrier phase recovery* that matches the carrier phase ϕ at the receiver to the carrier phase ϕ_0 at the transmitter;[1] this is known as *coherent detection*. If $\phi - \phi_0 = \Delta\phi \neq 0$ then the in-phase branch will have an unwanted term associated with the quadrature branch and vice versa; that is, $r_1 = s_{i1}\cos(\Delta\phi) + s_{i2}\sin(\Delta\phi) + n_1$ and $r_2 = -s_{i1}\sin(\Delta\phi) + s_{i2}\cos(\Delta\phi) + n_2$, which can result in significant performance degradation. The receiver structure also assumes that the sampling function every T_s seconds is synchronized to the start of the symbol period, which is called *synchronization* or *timing recovery*. Receiver synchronization and carrier phase recovery are complex receiver operations that can be highly challenging in wireless environments. These operations are discussed in more detail in Section 5.6. We will assume perfect carrier recovery in our discussion of MPAM, MPSK, and MQAM and therefore set $\phi = \phi_0 = 0$ for their analysis.

5.3.1 Pulse Amplitude Modulation (MPAM)

We will start by looking at the simplest form of linear modulation, one-dimensional MPAM, which has no quadrature component ($s_{i2} = 0$). For MPAM, all of the information is encoded into the signal amplitude A_i. The transmitted signal over one symbol time is given by

$$s_i(t) = \text{Re}\{A_i g(t) e^{j2\pi f_c t}\} = A_i g(t)\cos(2\pi f_c t), \quad 0 \le t \le T_s \gg 1/f_c, \qquad (5.52)$$

where $A_i = (2i - 1 - M)d$, $i = 1, 2, \dots, M$. The signal constellation is thus $\{A_i, i = 1, \dots, M\}$, which is parameterized by the distance d. This distance is typically a function of the signal energy. The pulse shape $g(t)$ must satisfy (5.12) and (5.13). The minimum distance between constellation points is $d_{\min} = \min_{i,j}|A_i - A_j| = 2d$. The amplitude of the transmitted signal takes on M different values, which implies that each pulse conveys $\log_2 M = K$ bits per symbol time T_s.

---

[1] In fact, an additional phase term of $-2\pi f_c \tau$ will result from a propagation delay of τ in the channel. Thus, coherent detection requires the receiver phase $\phi = \phi_0 - 2\pi f_c \tau$, as discussed in more detail in Section 5.6.

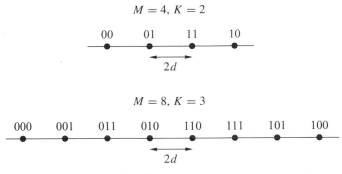

Figure 5.12: Gray encoding for MPAM.

Over each symbol period, the MPAM signal associated with the ith constellation has energy

$$E_{s_i} = \int_0^{T_s} s_i^2(t)\, dt = \int_0^{T_s} A_i^2 g^2(t) \cos^2(2\pi f_c t)\, dt = A_i^2, \qquad (5.53)$$

since the pulse shape must satisfy (5.12).[2] Note that the energy is not the same for each signal $s_i(t)$, $i = 1, \ldots, M$. Assuming equally likely symbols, the average energy is

$$\bar{E}_s = \frac{1}{M} \sum_{i=1}^M A_i^2. \qquad (5.54)$$

The constellation mapping is usually done by Gray encoding, where the messages associated with signal amplitudes that are adjacent to each other differ by one bit value, as illustrated in Figure 5.12. With this encoding method, if noise causes the demodulation process to mistake one symbol for an adjacent one (the most likely type of error), the result is only a single bit error in the sequence of K bits. Gray codes can also be designed for MPSK and square MQAM constellations but not for rectangular MQAM.

EXAMPLE 5.4: For $g(t) = \sqrt{2/T_s}$ $(0 \le t < T_s)$ a rectangular pulse shape, find the average energy of 4-PAM modulation.

Solution: For 4-PAM the A_i values are $A_i = \{-3d, -d, d, 3d\}$, so the average energy is

$$\bar{E}_s = \frac{d^2}{4}(9 + 1 + 1 + 9) = 5d^2.$$

The decision regions Z_i, $i = 1, \ldots, M$, associated with pulse amplitude $A_i = (2i - 1 - M)d$ for $M = 4$ and $M = 8$ are shown in Figure 5.13. Mathematically, for any M these decision regions are defined by

$$Z_i = \begin{cases} (-\infty, A_i + d) & i = 1, \\ [A_i - d, A_i + d) & 2 \le i \le M - 1, \\ [A_i - d, \infty) & i = M. \end{cases}$$

---

[2] Recall from (5.8) that (5.12) and hence (5.53) are not exact equalities but rather very good approximations for $f_c T_s \gg 1$.

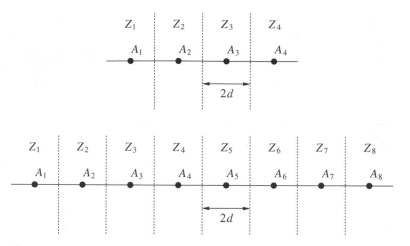

Figure 5.13: Decision regions for MPAM.

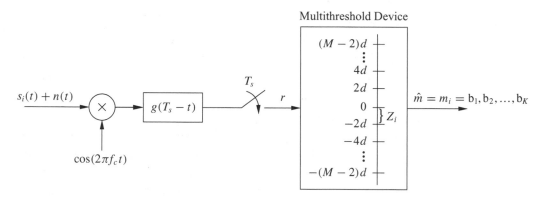

Figure 5.14: Coherent demodulator for MPAM.

From (5.52) we see that MPAM has only a single basis function $\phi_1(t) = g(t)\cos(2\pi f_c t)$. Thus, the coherent demodulator of Figure 5.11 for MPAM reduces to the demodulator shown in Figure 5.14, where the multithreshold device maps r to a decision region Z_i and outputs the corresponding bit sequence $\hat{m} = m_i = \{b_1, \ldots, b_K\}$.

5.3.2 Phase-Shift Keying (MPSK)

For MPSK, all of the information is encoded in the phase of the transmitted signal. Thus, the transmitted signal over one symbol time T_s is given by

$$
\begin{aligned}
s_i(t) &= \text{Re}\{Ag(t)e^{j2\pi(i-1)/M}e^{j2\pi f_c t}\} \\
&= Ag(t)\cos\left[2\pi f_c t + \frac{2\pi(i-1)}{M}\right] \\
&= Ag(t)\cos\left[\frac{2\pi(i-1)}{M}\right]\cos 2\pi f_c t - Ag(t)\sin\left[\frac{2\pi(i-1)}{M}\right]\sin 2\pi f_c t \quad (5.55)
\end{aligned}
$$

for $0 \le t \le T_s$. Therefore, the constellation points or symbols (s_{i1}, s_{i2}) are given by $s_{i1} = A\cos[2\pi(i-1)/M]$ and $s_{i2} = A\sin[2\pi(i-1)/M]$ for $i = 1, \ldots, M$. The pulse shape $g(t)$

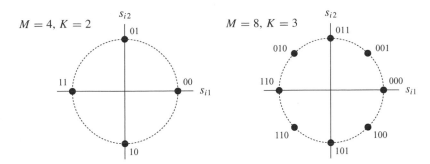

Figure 5.15: Gray encoding for MPSK.

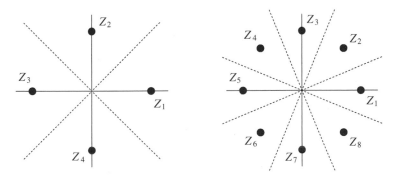

Figure 5.16: Decision regions for MPSK.

satisfies (5.12) and (5.13), and the $\theta_i = 2\pi(i-1)/M$ $(i = 1, 2, \ldots, M = 2^K)$ are the different phases in the signal constellation points that convey the information bits. The minimum distance between constellation points is $d_{\min} = 2A\sin(\pi/M)$, where A is typically a function of the signal energy. Note that 2-PSK is often referred to as binary PSK or BPSK, while 4-PSK is often called quadrature phase shift keying (QPSK) and is the same as MQAM with $M = 4$ (defined in Section 5.3.3).

All possible transmitted signals $s_i(t)$ have equal energy:

$$E_{s_i} = \int_0^{T_s} s_i^2(t)\, dt = A^2. \tag{5.56}$$

Observe that for $g(t) = \sqrt{2/T_s}$ over a symbol time (i.e., a rectangular pulse) this signal has constant envelope, unlike the other amplitude modulation techniques MPAM and MQAM. However, rectangular pulses are spectrally inefficient, and more efficient pulse shapes give MPSK a nonconstant signal envelope. As for MPAM, constellation mapping is usually done by Gray encoding, where the messages associated with signal phases that are adjacent to each other differ by one bit value; see Figure 5.15. With this encoding method, mistaking a symbol for an adjacent one causes only a single bit error.

The decision regions Z_i $(i = 1, \ldots, M)$ associated with MPSK for $M = 4$ and $M = 8$ are shown in Figure 5.16. If we represent $\mathbf{r} = r_1 + jr_2 = re^{j\theta} \in \mathbb{R}^2$ in polar coordinates, then these decision regions for any M are defined by

$$Z_i = \{re^{j\theta} : 2\pi(i-1.5)/M \leq \theta < 2\pi(i-.5)/M\}. \tag{5.57}$$

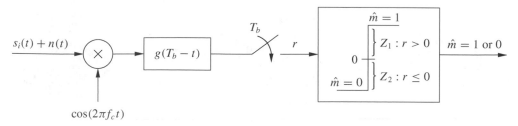

Figure 5.17: Coherent demodulator for BPSK.

From (5.55) we see that MPSK has both in-phase and quadrature components, and thus the coherent demodulator is as shown in Figure 5.11. For the special case of BPSK, the decision regions as given in Example 5.2 simplify to $Z_1 = (r : r > 0)$ and $Z_2 = (r : r \leq 0)$. Moreover, BPSK has only a single basis function $\phi_1(t) = g(t)\cos(2\pi f_c t)$ and, since there is only a single bit transmitted per symbol time T_s, the bit time $T_b = T_s$. Thus, the coherent demodulator of Figure 5.11 for BPSK reduces to the demodulator shown in Figure 5.17, where the threshold device maps r to the positive or negative half of the real line and then outputs the corresponding bit value. We have assumed in this figure that the message corresponding to a bit value of 1, $m_1 = 1$, is mapped to constellation point $s_1 = A$ and that the message corresponding to a bit value of 0, $m_2 = 0$, is mapped to the constellation point $s_2 = -A$.

5.3.3 Quadrature Amplitude Modulation (MQAM)

For MQAM, the information bits are encoded in both the amplitude and phase of the transmitted signal. Thus, whereas both MPAM and MPSK have one degree of freedom in which to encode the information bits (amplitude or phase), MQAM has two degrees of freedom. As a result, MQAM is more spectrally efficient than MPAM and MPSK in that it can encode the most number of bits per symbol for a given average energy.

The transmitted signal is given by

$$
\begin{aligned}
s_i(t) &= \mathrm{Re}\{A_i e^{j\theta_i} g(t) e^{j2\pi f_c t}\} \\
&= A_i \cos(\theta_i) g(t) \cos(2\pi f_c t) - A_i \sin(\theta_i) g(t) \sin(2\pi f_c t), \quad 0 \leq t \leq T_s, \quad (5.58)
\end{aligned}
$$

where the pulse shape $g(t)$ satisfies (5.12) and (5.13). The energy in $s_i(t)$ is

$$
E_{s_i} = \int_0^{T_s} s_i^2(t) = A_i^2, \tag{5.59}
$$

the same as for MPAM. The distance between any pair of symbols in the signal constellation is

$$
d_{ij} = \|\mathbf{s}_i - \mathbf{s}_j\| = \sqrt{(s_{i1} - s_{j1})^2 + (s_{i2} - s_{j2})^2}. \tag{5.60}
$$

For square signal constellations, where s_{i1} and s_{i2} take values on $(2i - 1 - L)d$ with $i = 1, 2, \ldots, L$, the minimum distance between signal points reduces to $d_{\min} = 2d$, the same as for MPAM. In fact, MQAM with square constellations of size L^2 is equivalent to MPAM modulation with constellations of size L on each of the in-phase and quadrature signal components. Common square constellations are 4-QAM and 16-QAM, which are shown

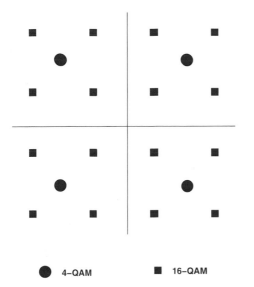

● 4–QAM   ■ 16–QAM

Figure 5.18: 4-QAM and 16-QAM constellations.

in Figure 5.18. These square constellations have $M = L^2 = 2^{2l}$ constellation points, which are used to send $2l$ bits/symbol or l bits per dimension, where $l = .5 \log_2 M$. It can be shown that the average power of a square signal constellation with l bits per dimension, P_l, is proportional to $4^l/3$, and it follows that the average power for one more bit per dimension $P_{l+1} \approx 4P_l$. Thus, for square constellations it takes approximately 6 dB more power to send an additional 1 bit/dimension or 2 bits/symbol while maintaining the same minimum distance between constellation points.

Good constellation mappings can be hard to find for QAM signals, especially for irregular constellation shapes. In particular, it is hard to find a Gray code mapping where all adjacent symbols differ by a single bit. The decision regions Z_i ($i = 1, \dots, M$) associated with MQAM for $M = 16$ are shown in Figure 5.19. From (5.58) we see that MQAM has both in-phase and quadrature components, and thus the coherent demodulator is as shown in Figure 5.11.

5.3.4 Differential Modulation

The information in MPSK and MQAM signals is carried in the signal phase. These modulation techniques therefore require coherent demodulation; that is, the phase of the transmitted signal carrier ϕ_0 must be matched to the phase of the receiver carrier ϕ. Techniques for phase recovery typically require more complexity and cost in the receiver, and they are also susceptible to phase drift of the carrier. Moreover, obtaining a coherent phase reference in a rapidly fading channel can be difficult. Issues associated with carrier phase recovery are discussed in more detail in Section 5.6. The difficulties as well as the cost and complexity associated with carrier phase recovery motivate the use of differential modulation techniques, which do not require a coherent phase reference at the receiver.

Differential modulation falls in the more general class of modulation with memory, where the symbol transmitted over time $[kT_s, (k+1)T_s)$ depends on the bits associated with the current message to be transmitted *and* on the bits transmitted over prior symbol times. The basic principle of differential modulation is to use the previous symbol as a phase reference for the current symbol, thus avoiding the need for a coherent phase reference at the receiver. Specifically, the information bits are encoded as the differential phase between the current symbol and the previous symbol. For example, in differential BPSK (referred to as DPSK), if the symbol over time $[(k-1)T_s, kT_s)$ has phase $\theta(k-1) = e^{j\theta_i}$ for $\theta_i = 0, \pi$, then to encode a 0-bit over $[kT_s, (k+1)T_s)$ the symbol would have phase $\theta(k) = e^{j\theta_i}$ and to encode a 1-bit the symbol would have phase $\theta(k) = e^{j(\theta_i+\pi)}$. In other words: a 0-bit is encoded by no change in phase, whereas a 1-bit is encoded as a phase change of π. Similarly, in 4-PSK modulation

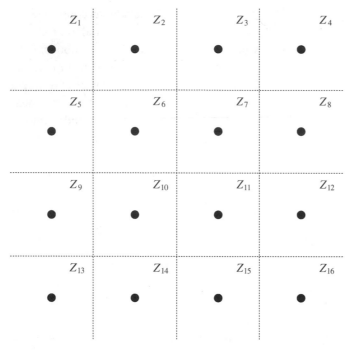

Figure 5.19: Decision regions for MQAM with $M = 16$.

with differential encoding, the symbol phase over symbol interval $[kT_s, (k + 1)T_s)$ depends on the current information bits over this time interval and the symbol phase over the previous symbol interval. The phase transitions for DQPSK modulation are summarized in Table 5.1.

Specifically, suppose the symbol over time $[(k - 1)T_s, kT_s)$ has phase $\theta(k - 1) = e^{j\theta_i}$. Then, over symbol time $[kT_s, (k + 1)T_s)$, if the information bits are 00 then the corresponding symbol would have phase $\theta(k) = e^{j\theta_i}$; that is, to encode the bits 00, the symbol from symbol interval $[(k - 1)T_s, kT_s)$ is repeated over the next interval $[kT_s, (k + 1)T_s)$. If the two information bits to be sent at time interval $[kT_s, (k + 1)T_s)$ are 01, then the corresponding symbol has phase $\theta(k) = e^{j(\theta_i + \pi/2)}$. For information bits 10 the symbol phase is $\theta(k) = e^{j(\theta_i - \pi/2)}$, and for information bits 11 the symbol phase is $\theta(n) = e^{j(\theta_k + \pi)}$. We see that the symbol phase over symbol interval $[kT_s, (k + 1)T_s)$ depends on the current information bits over this time interval and on the symbol phase θ_i over the previous symbol interval. Note that this mapping of bit sequences to phase transitions ensures that the most likely detection error – that of mistaking a received symbol for one of its nearest neighbors – results in a single bit error. For example, if the bit sequence 00 is encoded in the kth symbol then the kth symbol has the same phase as the $(k - 1)$th symbol. Assume this phase is θ_i. The most likely detection error of the kth symbol is to decode it as one of its nearest neighbor symbols, which have phase $\theta_i \pm \pi/2$. But decoding the received symbol with phase $\theta_i \pm \pi/2$ would result in a decoded information sequence of either 01 or 10 – that is, it would differ by a single bit from the original sequence 00. More generally, we can use Gray encoding for the phase transitions in differential MPSK for any M, so that a message of all 0-bits results in no phase change, a message with a single 1-bit and the rest 0-bits results in the minimum phase change

Table 5.1: Mapping for DQPSK with Gray encoding

Bit sequence	Phase transition
00	0
01	$\pi/2$
10	$-\pi/2$
11	π

of $2\pi/M$, a message with two 1-bits and the rest 0-bits results in a phase change of $4\pi/M$, and so forth. Differential encoding is most common for MPSK signals, since the differential mapping is relatively simple. Differential encoding can also be done for MQAM with a more complex differential mapping. Differential encoding of MPSK is denoted by DMPSK, and for BPSK and QPSK by DPSK and DQPSK, respectively.

EXAMPLE 5.5: Find the sequence of symbols transmitted using DPSK for the bit sequence 101110 starting at the kth symbol time, assuming the transmitted symbol at the $(k-1)$th symbol time was $s(k-1) = Ae^{j\pi}$.

Solution: The first bit, a 1, results in a phase transition of π, so $s(k) = A$. The next bit, a 0, results in no transition, so $s(k+1) = A$. The next bit, a 1, results in another transition of π, so $s(k+1) = Ae^{j\pi}$, and so on. The full symbol sequence corresponding to 101110 is $A, A, Ae^{j\pi}, A, Ae^{j\pi}, Ae^{j\pi}$.

The demodulator for differential modulation is shown in Figure 5.20. Assume the transmitted constellation at time k is $s(k) = Ae^{j(\theta(k)+\phi_0)}$. Then the received vector associated with the sampler outputs is

$$\mathbf{r}(k) = r_1(k) + jr_2(k) = Ae^{j(\theta(k)+\phi_0)} + n(k), \tag{5.61}$$

where $n(k)$ is complex white Gaussian noise. The received vector at the previous time sample $k-1$ is thus

$$\mathbf{r}(k-1) = r_1(k-1) + jr_2(k-1) = Ae^{j(\theta(k-1)+\phi_0)} + n(k-1). \tag{5.62}$$

The phase difference between $\mathbf{r}(k)$ and $\mathbf{r}(k-1)$ determines which symbol was transmitted. Consider

$$\mathbf{r}(k)\mathbf{r}^*(k-1) = A^2 e^{j(\theta(k)-\theta(k-1))} + Ae^{j(\theta(k)+\phi_0)}n^*(k-1)$$
$$+ Ae^{-j(\theta(k-1)+\phi_0)}n(k) + n(k)n^*(k-1). \tag{5.63}$$

In the absence of noise ($n(k) = n(k-1) = 0$) only the first term in (5.63) is nonzero, and this term yields the desired phase difference. The phase comparator in Figure 5.20 extracts this phase difference and outputs the corresponding symbol.

Differential modulation is less sensitive to a random drift in the carrier phase. However, if the channel has a nonzero Doppler frequency then the signal phase can decorrelate between symbol times, making the previous symbol a noisy phase reference. This decorrelation gives rise to an irreducible error floor for differential modulation over wireless channels with Doppler, as we shall discuss in Chapter 6.

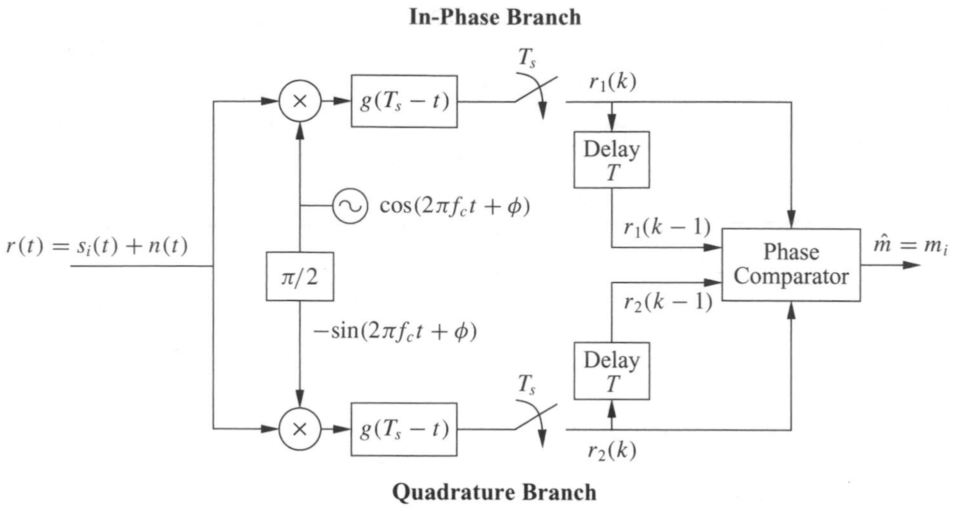

In-Phase Branch

Quadrature Branch

Figure 5.20: Differential PSK demodulator.

5.3.5 Constellation Shaping

Rectangular and hexagonal constellations have a better power efficiency than the square or circular constellations associated with MQAM and MPSK, respectively. These irregular constellations can save up to 1.3 dB of power at the expense of increased complexity in the constellation map [6]. The optimal constellation shape is a sphere in N-dimensional space, which must be mapped to a sequence of constellations in two-dimensional space in order to be generated by the modulator shown in Figure 5.10. The general conclusion in [6] is that, for uncoded modulation, the increased complexity of spherical constellations is not worth their energy gains, since coding can provide much better performance at less complexity cost. However, if a complex channel code is already being used and little further improvement can be obtained by a more complex code, constellation shaping may obtain around 1 dB of additional gain. An in-depth discussion of constellation shaping (and of constellations that allow a noninteger number of bits per symbol) can be found in [6].

5.3.6 Quadrature Offset

A linearly modulated signal with symbol $\mathbf{s}_i = (s_{i1}, s_{i2})$ will lie in one of the four quadrants of the signal space. At each symbol time kT_s the transition to a new symbol value in a different quadrant can cause a phase transition of up to 180°, which may cause the signal amplitude to transition through the zero point; these abrupt phase transitions and large amplitude variations can be distorted by nonlinear amplifiers and filters. The abrupt transitions are avoided by offsetting the quadrature branch pulse $g(t)$ by half a symbol period, as shown in Figure 5.21. This *quadrature offset* makes the signal less sensitive to distortion during symbol transitions.

Phase modulation with quadrature offset is usually abbreviated as OMPSK, where the O indicates the offset. For example, QPSK modulation with quadrature offset is referred to as OQPSK. Offset QPSK has the same spectral properties as QPSK for linear amplification, but it has higher spectral efficiency under nonlinear amplification because the maximum phase transition of the signal is 90°, corresponding to the maximum phase transition in either the

In-Phase Branch

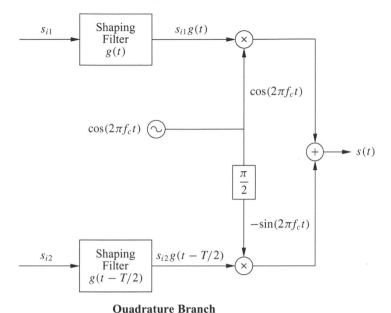

Quadrature Branch

Figure 5.21: Modulator with quadrature offset.

in-phase or quadrature branch but not both simultaneously. Another technique to mitigate the amplitude fluctuations of a 180° phase shift used in the IS-136 standard for digital cellular is $\pi/4$-QPSK [7; 8]. This technique allows for a maximum phase transition of 135° degrees, versus 90° for offset QPSK and 180° for QPSK. Thus, $\pi/4$-QPSK has worse spectral properties than OQPSK under nonlinear amplification. However, $\pi/4$-QPSK can be differentially encoded to eliminate the need for a coherent phase reference, which is a significant advantage. Using differential encoding with $\pi/4$-QPSK is called $\pi/4$-DQPSK. The $\pi/4$-DQPSK modulation works as follows: the information bits are first differentially encoded as in DQPSK, which yields one of the four QPSK constellation points. Then, every other symbol transmission is shifted in phase by $\pi/4$. This periodic phase shift has a similar effect as the time offset in OQPSK: it reduces the amplitude fluctuations at symbol transitions, which makes the signal more robust against noise and fading.

5.4 Frequency Modulation

Frequency modulation encodes information bits into the frequency of the transmitted signal. Specifically: at each symbol time, $K = \log_2 M$ bits are encoded into the frequency of the transmitted signal $s(t)$, $0 \le t < T_s$, resulting in a transmitted signal $s_i(t) = A\cos(2\pi f_i t + \phi_i)$, where i is the index of the ith message corresponding to the $\log_2 M$ bits and ϕ_i is the phase associated with the ith carrier. The signal space representation is $s_i(t) = \sum_j s_{ij}\phi_j(t)$, where $s_{ij} = A\delta(i - j)$ and $\phi_j(t) = \cos(2\pi f_j t + \phi_j)$, so the basis functions correspond to carriers at different frequencies and only one such basis function is transmitted in each symbol period. The basis functions are orthogonal for a minimum carrier frequency separation of $\Delta f = \min_{ij}|f_j - f_i| = .5/T_s$ for $\phi_i = \phi_j$ and of $\Delta f = 1/T_s$ for $\phi_i \ne \phi_j$.

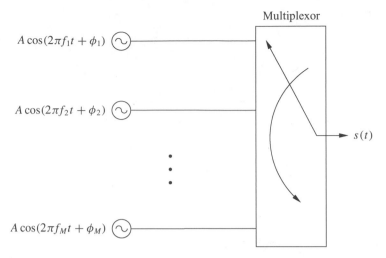

Figure 5.22: Frequency modulator.

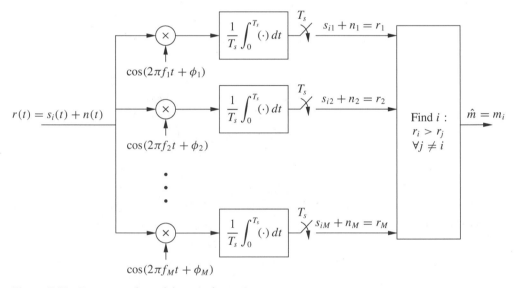

Figure 5.23: Frequency demodulator (coherent).

Because frequency modulation encodes information in the signal frequency, the transmitted signal $s(t)$ has a constant envelope A. Since the signal is constant envelope, nonlinear amplifiers can be used with high power efficiency and hence the modulated signal is less sensitive to amplitude distortion introduced by the channel or the hardware. The price exacted for this robustness is a lower spectral efficiency: because the modulation technique is nonlinear, it tends to have a higher bandwidth occupancy than the amplitude and phase modulation techniques described in Section 5.3.

In its simplest form, frequency modulation over a given symbol period can be generated using the modulator structure shown in Figure 5.22. Demodulation over each symbol period is performed using the demodulation structure of Figure 5.23. Note that the demodulator of Figure 5.23 requires the jth carrier signal to be matched in phase to the jth carrier signal at the transmitter; this is similar to the coherent phase reference requirement in amplitude and

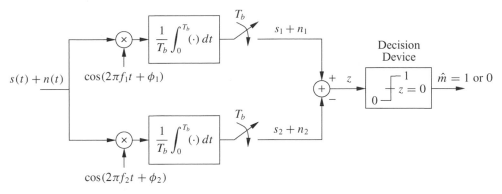

Figure 5.24: Demodulator for binary FSK.

phase modulation. An alternate receiver structure that does not require this coherent phase reference will be discussed in Section 5.4.3. Another issue in frequency modulation is that the different carriers shown in Figure 5.22 have different phases, $\phi_i \neq \phi_j$ for $i \neq j$, so at each symbol time T_s there will be a phase discontinuity in the transmitted signal. Such discontinuities can significantly increase signal bandwidth. Thus, in practice an alternate modulator is used that generates a frequency-modulated signal with continuous phase, as will be discussed in Section 5.4.2.

5.4.1 Frequency-Shift Keying (FSK) and Minimum-Shift Keying (MSK)

In MFSK the modulated signal is given by

$$s_i(t) = A \cos[2\pi f_c t + 2\pi \alpha_i \Delta f_c t + \phi_i], \quad 0 \leq t < T_s, \tag{5.64}$$

where $\alpha_i = (2i - 1 - M)$ for $i = 1, 2, \ldots, M = 2^K$. The minimum frequency separation between FSK carriers is thus $2\Delta f_c$. MFSK consists of M basis functions $\phi_i(t) = \sqrt{2/T_s} \cos[2\pi f_c t + 2\pi \alpha_i \Delta f_c t + \phi_i]$, $i = 1, \ldots, M$, where the $\sqrt{2/T_s}$ is a normalization factor to ensure that $\int_0^{T_s} \phi_i^2(t) = 1$. Over any given symbol time, only one basis function is transmitted through the channel.

A simple way to generate the MFSK signal is as shown in Figure 5.22, where M oscillators are operating at the different frequencies $f_i = f_c + \alpha_i \Delta f_c$ and the modulator switches between these different oscillators each symbol time T_s. However, this implementation entails a discontinuous phase transition at the switching times due to phase offsets between the oscillators, and this discontinuous phase leads to undesirable spectral broadening. (An FSK modulator that maintains continuous phase is discussed in the next section.) Coherent detection of MFSK uses the standard structure of Figure 5.23. For binary signaling the structure can be simplified to that shown in Figure 5.24, where the decision device outputs a 1-bit if its input is greater than zero and a 0-bit if its input is less than zero.

MSK is a special case of binary FSK where $\phi_1 = \phi_2$ and the frequency separation is $2\Delta f_c = .5/T_s$. Note that this is the minimum frequency separation that ensures $\langle s_i(t), s_j(t) \rangle = 0$ over a symbol time for $i \neq j$. Since signal orthogonality is required for demodulation, it follows that $2\Delta f_c = .5/T_s$ is the minimum possible frequency separation in FSK and so MSK is the minimum bandwidth FSK modulation.

5.4.2 Continuous-Phase FSK (CPFSK)

A better way to generate MFSK – one that eliminates the phase discontinuity – is to frequency modulate a single carrier with a modulating waveform, as in analog FM. In this case the modulated signal will be given by

$$s_i(t) = A \cos\left[2\pi f_c t + 2\pi\beta \int_{-\infty}^{t} u(\tau)\, d\tau\right] = A \cos[2\pi f_c t + \theta(t)], \qquad (5.65)$$

where $u(t) = \sum_k a_k g(t - kT_s)$ is an MPAM signal modulated with the information bit stream, as described in Section 5.3.1. Clearly the phase $\theta(t)$ is continuous with this implementation. This form of MFSK is therefore called continuous-phase FSK, or CPFSK.

By Carson's rule [9], for β small the transmission bandwidth of $s(t)$ is approximately

$$B_s \approx 2M\Delta f_c + 2B_g, \qquad (5.66)$$

where B_g is the bandwidth of the pulse shape $g(t)$ used in the MPAM modulating signal $u(t)$. By comparison, the bandwidth of a linearly modulated waveform with pulse shape $g(t)$ is roughly $B_s \approx 2B_g$. Thus, the spectral occupancy of a CPFSK-modulated signal is larger than that of a linearly modulated signal by $M\Delta f_c \geq .5M/T_s$. The spectral efficiency penalty of CPFSK relative to linear modulation increases with data rate, in particular with the number of bits per symbol $K = \log_2 M$ and with the symbol rate $R_s = 1/T_s$.

Coherent detection of CPFSK can be done symbol-by-symbol or over a sequence of symbols. The sequence estimator is the optimal detector, since a given symbol depends on previously transmitted symbols and so it is optimal to detect (or estimate) all symbols simultaneously. However, sequence estimation can be impractical owing to the memory and computational requirements associated with making decisions based on sequences of symbols. Details on both symbol-by-symbol and sequence detectors for coherent demodulation of CPFSK can be found in [10, Chap. 5.3].

5.4.3 Noncoherent Detection of FSK

The receiver requirement for a coherent phase reference associated with each FSK carrier can be difficult and expensive to meet. The need for a coherent phase reference can be eliminated if the receiver first detects the energy of the signal at each frequency and, if the ith branch has the highest energy of all branches, then outputs message m_i. The modified receiver is shown in Figure 5.25.

Suppose the transmitted signal corresponds to frequency f_i:

$$s(t) = A \cos(2\pi f_i t + \phi_i)$$
$$= A \cos(\phi_i) \cos(2\pi f_i t) - A \sin(\phi_i) \sin(2\pi f_i t), \quad 0 \leq t < T_s. \qquad (5.67)$$

Let the phase ϕ_i represent the phase offset between the transmitter and receiver oscillators at frequency f_i. A coherent receiver with carrier signal $\cos(2\pi f_i t)$ detects only the first term $A \cos(\phi_i) \cos(2\pi f_i t)$ associated with the received signal, which can be close to zero for a phase offset $\phi_i \approx \pm\pi/2$. To get around this problem, in Figure 5.25 the receiver splits the received signal into M branches corresponding to each frequency f_j, $j = 1, \ldots, M$. For

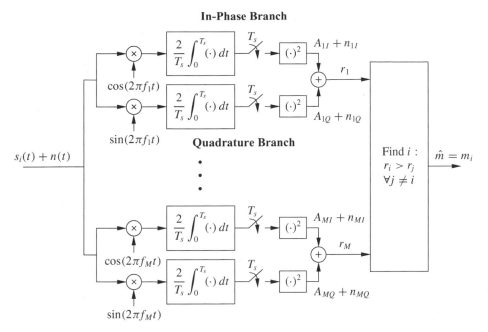

Figure 5.25: Noncoherent FSK demodulator.

each such carrier frequency f_j, the received signal is multiplied by a noncoherent in-phase and quadrature carrier at that frequency, integrated over a symbol time, sampled, and then squared. For the jth branch the squarer output associated with the in-phase component is denoted as $A_{jI} + n_{jI}$ and the corresponding output associated with the quadrature component is denoted as $A_{jQ} + n_{jQ}$, where n_{jI} and n_{jQ} are due to the noise $n(t)$ at the receiver input. Then, if $i = j$, we have $A_{jI} = A^2 \cos^2(\phi_i)$ and $A_{jQ} = A^2 \sin^2(\phi_i)$; if $i \neq j$ then $A_{jI} = A_{jQ} = 0$. In the absence of noise, the input to the decision device of the ith branch will be $A^2 \cos^2(\phi_i) + A^2 \sin^2(\phi_i) = A^2$, independent of ϕ_i, and all other branches will have an input of zero. Thus, over each symbol period, the decision device outputs the bit sequence corresponding to frequency f_j if the jth branch has the largest input to the decision device. A similar structure – in which each branch consists of a filter matched to the carrier frequency followed by an envelope detector and sampler – can also be used [1, Chap. 6.8]. Note that the noncoherent receiver of Figure 5.25 still requires accurate synchronization for sampling. Synchronization issues are discussed in Section 5.6.

5.5 Pulse Shaping

For amplitude and phase modulation, the bandwidth of the baseband and passband modulated signal is a function of the bandwidth of the pulse shape $g(t)$. If $g(t)$ is a rectangular pulse of width T_s, then the envelope of the signal is constant. However, a rectangular pulse has high spectral sidelobes, which can cause adjacent channel interference. Pulse shaping is a method for reducing sidelobe energy relative to a rectangular pulse; however, the shaping must be done in such a way that intersymbol interference between pulses in the received signal is not introduced. Note that – prior to sampling the received signal – the transmitted pulse $g(t)$ is convolved with the channel impulse response $c(t)$ and the matched filter $g^*(-t)$; hence, in order to eliminate ISI prior to sampling, we must ensure that the effective received pulse $p(t) = g(t) * c(t) * g^*(-t)$ has no ISI. Since the channel model is AWGN, we assume $c(t) = \delta(t)$ so $p(t) = g(t) * g^*(-t)$ (in Chapter 11 we will analyze ISI for more general channel impulse responses $c(t)$). To avoid ISI between samples of the received pulses, the effective pulse shape $p(t)$ must satisfy the *Nyquist criterion,* which requires the pulse to equal zero at the ideal sampling point associated with past or future symbols:

$$p(kT_s) = \begin{cases} p_0 = p(0) & k = 0, \\ 0 & k \neq 0. \end{cases}$$

In the frequency domain this translates to

$$\sum_{l=-\infty}^{\infty} P\left(f + \frac{l}{T_s}\right) = p_0 T_s. \tag{5.68}$$

The following pulse shapes all satisfy the Nyquist criterion.

1. *Rectangular pulses:* $g(t) = \sqrt{2/T_s}$ ($0 \leq t \leq T_s$), which yields the triangular effective pulse shape

$$p(t) = \begin{cases} 2 + 2t/T_s & -T_s \leq t < 0, \\ 2 - 2t/T_s & 0 \leq t < T_s, \\ 0 & \text{else.} \end{cases}$$

   This pulse shape leads to constant envelope signals in MPSK but has poor spectral properties as a result of its high sidelobes.

2. *Cosine pulses:* $p(t) = \sin \pi t/T_s, 0 \leq t \leq T_s$. Cosine pulses are mostly used in OMPSK modulation, where the quadrature branch of the modulation is shifted in time by $T_s/2$. This leads to a constant amplitude modulation with sidelobe energy that is 10 dB lower than that of rectangular pulses.

3. *Raised cosine pulses:* These pulses are designed in the frequency domain according to the desired spectral properties. Thus, the pulse $p(t)$ is first specified relative to its Fourier transform:

$$P(f) = \begin{cases} T_s & 0 \leq |f| \leq \dfrac{1-\beta}{2T_s}, \\ \dfrac{T_s}{2}\left[1 - \sin \dfrac{\pi T_s}{\beta}\left(f - \dfrac{1}{2T_s}\right)\right] & \dfrac{1-\beta}{2T_s} \leq |f| \leq \dfrac{1+\beta}{2T_s}; \end{cases}$$

   here β is defined as the rolloff factor, which determines the rate of spectral rolloff (see Figure 5.26). Setting $\beta = 0$ yields a rectangular pulse. The pulse $p(t)$ in the time domain corresponding to $P(f)$ is

$$p(t) = \frac{\sin \pi t/T_s}{\pi t/T_s} \frac{\cos \beta \pi t/T_s}{1 - 4\beta^2 t^2/T_s^2}.$$

   The frequency- and time-domain properties of the raised cosine pulse are shown in Figures 5.26 and 5.27, respectively. The tails of this pulse in the time domain decay as $1/t^3$ (faster

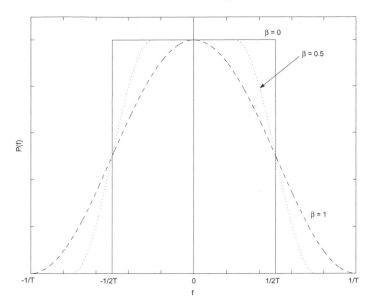

Figure 5.26: Frequency-domain (spectral) properties of the raised cosine pulse ($T = T_s$).

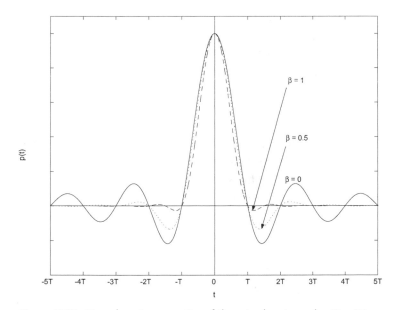

Figure 5.27: Time-domain properties of the raised cosine pulse ($T = T_s$).

than for the previous pulse shapes), so a mistiming error in sampling leads to a series of inter-symbol interference components that converge. A variation of the raised cosine pulse is the root cosine pulse, derived by taking the square root of the frequency response for the raised cosine pulse. The root cosine pulse has better spectral properties than the raised cosine pulse but decays less rapidly in the time domain, which makes performance degradation due to synchronization errors more severe. Specifically, a mistiming error in sampling leads to a series of ISI components that may diverge.

Pulse shaping is also used with CPFSK to improve spectral efficiency, specifically in the MPAM signal that is frequency modulated to form the FSK signal. The most common pulse shape used in CPFSK is the Gaussian pulse shape, defined as

$$g(t) = \frac{\sqrt{\pi}}{\alpha} e^{-\pi^2 t^2 / \alpha^2}, \qquad (5.69)$$

where α is a parameter that dictates spectral efficiency. The spectrum of $g(t)$, which dictates the spectrum of the CPFSK signal, is given by

$$G(f) = e^{-\alpha^2 f^2}. \qquad (5.70)$$

The parameter α is related to the 3-dB bandwidth of $g(t)$, B_g, by

$$\alpha = \frac{.5887}{B_g}. \qquad (5.71)$$

Clearly, increasing α results in a higher spectral efficiency.

When the Gaussian pulse shape is applied to MSK modulation, it is abbreviated as GMSK. In general, GMSK signals have a high power efficiency (since they have a constant amplitude) and a high spectal efficiency (since the Gaussian pulse shape has good spectral properties for large α). For these reasons, GMSK is used in the GSM standard for digital cellular systems. Although this is a good choice for voice modulation, it is not the best choice for data. The Gaussian pulse shape does not satisfy the Nyquist criterion and so the pulse shape introduces ISI, which increases as α increases. Thus, improving spectral efficiency by increasing α leads to a higher ISI level, thereby creating an irreducible error floor from this self-interference. Since the required BER for voice is a relatively high $P_b \approx 10^{-3}$, the ISI can be fairly high and still maintain this target BER. In fact, it is generally used as a rule of thumb that $B_g T_s = .5$ is a tolerable amount of ISI for voice transmission with GMSK. However, a much lower BER is required for data, which will put more stringent constraints on the maximum α and corresponding minimum B_g, thereby decreasing the spectral efficiency of GMSK for data transmission. Techniques such as equalization can be used to mitigate the ISI in this case so that a tolerable BER is possible without significantly compromising spectral efficiency. However, it is more common to use linear modulation for spectrally efficient data transmission. Indeed, the data enhancements to GSM use linear modulation.

5.6 Symbol Synchronization and Carrier Phase Recovery

One of the most challenging tasks of a digital demodulator is to acquire accurate symbol timing and carrier phase information. Timing information, obtained via synchronization, is needed to delineate the received signal associated with a given symbol. In particular, timing information is used to drive the sampling devices associated with the demodulators for amplitude, phase, and frequency demodulation shown in Figures 5.11 and 5.23. Carrier phase information is needed in all coherent demodulators for both amplitude/phase and frequency modulation, as discussed in Sections 5.3 and 5.4.

This section gives a brief overview of standard techniques for synchronization and carrier phase recovery in AWGN channels. In this context the estimation of symbol timing and carrier phase falls under the broader category of signal parameter estimation in noise. Estimation theory provides the theoretical framework for studying this problem and for developing the maximum likelihood estimator of the carrier phase and symbol timing. However, most wireless channels suffer from time-varying multipath in addition to AWGN. Synchronization and carrier phase recovery is particularly challenging in such channels because multipath and time variations can make it extremely difficult to estimate signal parameters prior to demodulation. Moreover, there is little theory addressing good methods for estimation of carrier phase and symbol timing when these parameters are corrupted by time-varying multipath in addition to noise. In most performance analysis of wireless communication systems it is assumed that the receiver synchronizes to the multipath component with delay equal to the average delay spread;[3] then the channel is treated as AWGN for recovery of timing information and carrier phase. In practice, however, the receiver will sychronize to either the strongest multipath component or the first multipath component that exceeds a given power threshold. The other multipath components will then compromise the receiver's ability to acquire timing and carrier phase, especially in wideband systems like UWB. Multicarrier and spread-spectrum systems have additional considerations related to synchronization and carrier recovery, which will be discussed in Chapters 12 and 13.

The importance of synchronization and carrier phase estimation cannot be overstated: without them, wireless systems could not function. Moreover, as data rates increase and channels become more complex by adding additional degrees of freedom (e.g., multiple antennas), the tasks of receiver synchronizaton and phase recovery become even more complex and challenging. Techniques for synchronization and carrier recovery have been developed and analyzed extensively for many years, and they are continually evolving to meet the challenges associated with higher data rates, new system requirements, and more challenging channel characteristics. We give only a brief introduction to synchronizaton and carrier phase recovery techniques in this section. Comprehensive coverage of this topic and performance analysis of these techniques can be found in [11; 12]; more condensed treatments can be found in [10, Chap. 6; 13].

5.6.1 Receiver Structure with Phase and Timing Recovery

The carrier phase and timing recovery circuitry for the amplitude and phase demodulator is shown in Figure 5.28. For BPSK only the in-phase branch of this demodulator is needed. For the coherent frequency demodulator of Figure 5.23, a carrier phase recovery circuit is needed for *each* of the distinct M carriers; the resulting circuit complexity motivates the need for noncoherent demodulators as described in Section 5.4.3. We see in Figure 5.28 that the carrier phase and timing recovery circuits operate directly on the received signal prior to demodulation.

Assuming an AWGN channel, the received signal $r(t)$ is a delayed version of the transmitted signal $s(t)$ plus AWGN $n(t)$: $r(t) = s(t-\tau) + n(t)$, where τ is the random propagation delay. Using the equivalent lowpass form we have $s(t) = \text{Re}\{x(t)e^{j\phi_0}e^{j2\pi f_c t}\}$ and thus

$$r(t) = \text{Re}\{x(t-\tau)e^{j\phi}e^{j2\pi f_c t}\} + n(t), \tag{5.72}$$

---

[3] That is why delay spread is typically characterized by its rms value about its mean, as discussed in more detail in Chapter 2.

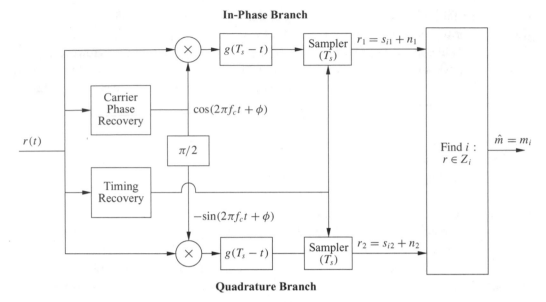

Figure 5.28: Receiver structure with carrier and timing recovery.

where $\phi = \phi_0 - 2\pi f_c \tau$ results from the transmit carrier phase and the propagation delay. Estimation of τ is needed for symbol timing, and estimation of ϕ is needed for carrier phase recovery. Let us express these two unknown parameters as a vector $\psi = (\phi, \tau)$. Then we can express the received signal in terms of ψ as

$$r(t) = s(t; \psi) + n(t). \tag{5.73}$$

Parameter estimation must take place over some finite time interval $T_0 \geq T_s$. We call T_0 the *observation interval*. In practice, however, parameter estimation is initially done over this interval and is thereafter performed continually by updating the initial estimate using tracking loops. Our development here focuses on the initial parameter estimation over T_0. Discussion of parameter tracking can be found in [11; 12].

There are two common estimation methods for signal parameters in noise: the maximum likelihood (ML) criterion discussed in Section 5.1.4 in the context of receiver design, and the maximum a posteriori (MAP) criterion. The ML criterion chooses the estimate $\hat{\psi}$ that maximizes $p(r(t) \mid \psi)$ over the observation interval T_0, whereas the MAP criterion assumes some probability distribution $p(\psi)$ on ψ and then chooses the estimate $\hat{\psi}$ that maximizes

$$p(\psi \mid r(t)) = \frac{p(r(t) \mid \psi)p(\psi)}{p(r(t))}$$

over T_0. We assume that there is no prior knowledge of $\hat{\psi}$, so that $p(\psi)$ becomes uniform and hence the MAP and ML criteria are equivalent.

To characterize the distribution $p(r(t) \mid \psi)$, $0 \leq t < T_0$, let us expand $r(t)$ over the observation interval along a set of orthonormal basis functions $\{\phi_k(t)\}$ as

$$r(t) = \sum_{k=1}^{K} r_k \phi_k(t), \quad 0 \le t < T_0.$$

Because $n(t)$ is white with zero mean and power spectral density $N_0/2$, the distribution of the vector $\mathbf{r} = (r_1, \ldots, r_K)$ conditioned on the unknown parameter $\boldsymbol{\psi}$ is given by

$$p(\mathbf{r} \mid \boldsymbol{\psi}) = \left(\frac{1}{\sqrt{\pi N_0}} \right)^K \exp\left[-\sum_{k=1}^{K} \frac{(r_k - s_k(\boldsymbol{\psi}))^2}{N_0} \right], \qquad (5.74)$$

where (by the basis expansion)

$$r_k = \int_{T_0} r(t) \phi_k(t) \, dt$$

and

$$s_k(\boldsymbol{\psi}) = \int_{T_0} s(t; \boldsymbol{\psi}) \phi_k(t) \, dt.$$

From these basis expansions we can show that

$$\sum_{k=1}^{K} [r_k - s_k(\boldsymbol{\psi})]^2 = \int_{T_0} [r(t) - s(t; \boldsymbol{\psi})]^2 \, dt. \qquad (5.75)$$

Using this in (5.74) yields that maximizing $p(\mathbf{r} \mid \boldsymbol{\psi})$ is equivalent to maximizing the *likelihood function*

$$\Lambda(\boldsymbol{\psi}) = \exp\left[-\frac{1}{N_0} \int_{T_0} [r(t) - s(t; \boldsymbol{\psi})]^2 \right] dt. \qquad (5.76)$$

Maximizing the likelihood function (5.76) results in the joint ML estimate of the carrier phase and symbol timing. Maximum likelihood estimation of the carrier phase and symbol timing can also be done separately, and in subsequent sections we will discuss this separate estimation in more detail. Techniques for joint estimation are more complex; details of such techniques can be found in [10, Chap. 6.4; 11, Chaps. 8–9].

5.6.2 Maximum Likelihood Phase Estimation

In this section we derive the maximum likelihood phase estimate assuming the timing is known. The likelihood function (5.76) with timing known reduces to

$$\Lambda(\phi) = \exp\left[-\frac{1}{N_0} \int_{T_0} [r(t) - s(t; \phi)]^2 \, dt \right]$$

$$= \exp\left[-\frac{1}{N_0} \int_{T_0} x^2(t) \, dt + \frac{2}{N_0} \int_{T_0} r(t) s(t; \phi) \, dt - \frac{1}{N_0} \int_{T_0} s^2(t; \phi) \, dt \right]. \qquad (5.77)$$

We estimate the carrier phase as the value $\hat{\phi}$ that maximizes this function. Note that the first term in (5.77) is independent of ϕ. Moreover, we assume that the third integral, which measures the energy in $s(t; \phi)$ over the observation interval, is relatively constant in ϕ. Given these assumptions, we see that the $\hat{\phi}$ that maximizes (5.77) also maximizes

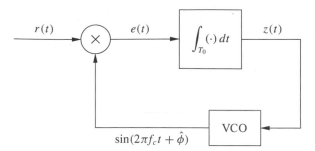

Figure 5.29: Phase-locked loop for carrier phase recovery (unmodulated carrier).

$$\Lambda'(\phi) = \int_{T_0} r(t)s(t;\phi)\,dt. \tag{5.78}$$

We can solve directly for the maximizing $\hat{\phi}$ in the simple case where the received signal is just an unmodulated carrier plus noise: $r(t) = A\cos(2\pi f_c t + \phi) + n(t)$. Then $\hat{\phi}$ must maximize

$$\Lambda'(\phi) = \int_{T_0} r(t)\cos(2\pi f_c t + \phi)\,dt. \tag{5.79}$$

Differentiating $\Lambda'(\phi)$ relative to ϕ and then setting it to zero yields that $\hat{\phi}$ satisfies

$$\int_{T_0} r(t)\sin(2\pi f_c t + \hat{\phi})\,dt = 0. \tag{5.80}$$

Solving (5.80) for $\hat{\phi}$ yields

$$\hat{\phi} = -\tan^{-1}\left[\frac{\int_{T_0} r(t)\sin(2\pi f_c t)\,dt}{\int_{T_0} r(t)\cos(2\pi f_c t)\,dt}\right]. \tag{5.81}$$

We could build a circuit to compute (5.81) from the received signal $r(t)$; in practice, however, carrier phase recovery is accomplished by using a phase lock loop to satisfy (5.80), as shown in Figure 5.29. In this figure, the integrator input in the absence of noise is given by $e(t) = r(t)\sin(2\pi f_c t + \hat{\phi})$ and the integrator output is

$$z(t) = \int_{T_0} r(t)\sin(2\pi f_c t + \hat{\phi})\,dt,$$

which is precisely the left-hand side of (5.80). Thus, if $z(t) = 0$ then the estimate $\hat{\phi}$ is the maximum likelihood estimate for ϕ. If $z(t) \neq 0$ then the voltage-controlled oscillator (VCO) adjusts its phase estimate $\hat{\phi}$ up or down depending on the polarity of $z(t)$: for $z(t) > 0$ it decreases $\hat{\phi}$ to reduce $z(t)$, and for $z(t) < 0$ it increases $\hat{\phi}$ to increase $z(t)$. In practice the integrator in Figure 5.29 is replaced with a *loop filter* whose output $.5A\sin(\hat{\phi} - \phi) \approx .5A(\hat{\phi} - \phi)$ is a function of the low-frequency component of its input $e(t) = A\cos(2\pi f_c t + \phi)\sin(2\pi f_c t + \hat{\phi}) = .5A\sin(\hat{\phi} - \phi) + .5A\sin(2\pi f_c t + \phi + \hat{\phi})$. This discussion of phase-locked loop (PLL) operation assumes that $\hat{\phi} \approx \phi$ because otherwise the polarity of $z(t)$ may not indicate the correct phase adjustment; that is, we would not necessarily have $\sin(\hat{\phi} - \phi) \approx \hat{\phi} - \phi$. The PLL typically exhibits some transient behavior in its initial estimation of the carrier phase. The advantage of a PLL is that it continually adjusts its estimate $\hat{\phi}$ to maintain $z(t) \approx 0$, which corrects for slow phase variations due to oscillator drift at the transmitter or changes in the propagation delay. In fact, the PLL is an example of a feedback control loop. More details on the PLL and its performance can be found in [10; 11].

The PLL derivation is for an unmodulated carrier, yet amplitude and phase modulation embed the message bits into the amplitude and phase of the carrier. For such signals there are two common carrier phase recovery approaches to deal with the effect of the data sequence on the received signal: the data sequence is either (a) assumed known or (b) treated as random with the phase estimate averaged over the data statistics. The first scenario is referred to as *decision-directed* parameter estimation, and this scenario typically results from sending a known training sequence. The second scenario is referred to as *non–decision-directed* parameter estimation. With this technique the likelihood function (5.77) is maximized by averaging over the statistics of the data. One decision-directed technique uses data decisions to remove the modulation of the received signal: the resulting unmodulated carrier is then passed through a PLL. This basic structure is called a *decision-feedback PLL* because data decisions are fed back into the PLL for processing. The structure of a non–decision-directed carrier phase recovery loop depends on the underlying distribution of the data. For large constellations, most distributions lead to highly nonlinear functions of the parameter to be estimated. In this case the symbol distribution is often assumed to be Gaussian along each signal dimension, which greatly simplifies the recovery loop structure. An alternate non–decision-directed structure takes the Mth power of the signal ($M = 2$ for PAM and M for MPSK modulation), passes it through a bandpass filter at frequency Mf_c, and then uses a PLL. The nonlinear operation removes the effect of the amplitude or phase modulation so that the PLL can operate on an unmodulated carrier at frequency Mf_c. Many other structures for both decision-directed and non–decision-directed carrier recovery can be used, with different trade-offs in performance and complexity. A more comprehensive discussion of design and performance of carrier phase recovery can be found in [10, Chaps. 6.2.4–6.2.5; 11].

5.6.3 Maximum Likelihood Timing Estimation

In this section we derive the maximum likelihood estimate of delay τ assuming the carrier phase is known. Since we assume that the phase ϕ is known, the timing recovery will not affect the carrier phase recovery loop and associated downconversion shown in Figure 5.28. Thus, it suffices to consider timing estimation for the in-phase or quadrature equivalent lowpass signals of $r(t)$ and $s(t; \tau)$. We denote the in-phase and quadrature components for $r(t)$ as $r_I(t)$ and $r_Q(t)$ and for $s(t; \tau)$ as $s_I(t; \tau)$ and $s_Q(t; \tau)$. We focus on the in-phase branch since the timing recovered from this branch can be used for the quadrature branch. The equivalent lowpass in-phase signal is given by

$$s_I(t; \tau) = \sum_k s_I(k) g(t - kT_s - \tau), \tag{5.82}$$

where $g(t)$ is the pulse shape and $s_I(k)$ denotes the amplitude associated with the in-phase component of the message transmitted over the kth symbol period. The in-phase equivalent lowpass received signal is $r_I(t) = s_I(t; \tau) + n_I(t)$. As in the case of phase synchronization, there are two categories of timing estimators: those for which the information symbols

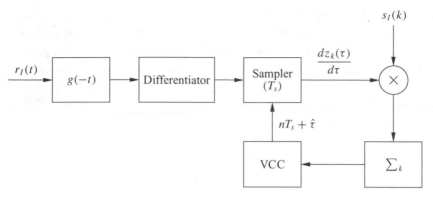

Figure 5.30: Decision-directed timing estimation.

output from the demodulator are assumed known (decision-directed estimators), and those for which this sequence is not assumed known (non–decision-directed estimators).

The likelihood function (5.76) with known phase ϕ has a form similar to (5.77), the case of known delay:

$$\Lambda(\tau) = \exp\left[-\frac{1}{N_0}\int_{T_0}[r_I(t) - s_I(t; \tau)]^2\, dt\right]$$

$$= \exp\left[-\frac{1}{N_0}\int_{T_0} r_I^2(t)\, dt + \frac{2}{N_0}\int_{T_0} r_I(t)s_I(t; \tau)\, dt - \frac{1}{N_0}\int_{T_0} s_I^2(t; \tau)\, dt\right]. \quad (5.83)$$

Since the first and third terms in (5.83) do not change significantly with τ, the delay estimate $\hat{\tau}$ that maximizes (5.83) also maximizes

$$\Lambda'(\tau) = \int_{T_0} r_I(t)s_I(t; \tau)\, dt$$

$$= \sum_k s_I(k)\int_{T_0} r(t)g(t - kT_s - \tau)\, dt = \sum_k s_I(k)z_k(\tau), \quad (5.84)$$

where

$$z_k(\tau) = \int_{T_0} r(t)g(t - kT_s - \tau)\, dt. \quad (5.85)$$

Differentiating (5.84) relative to τ and then setting it to zero yields that the timing estimate $\hat{\tau}$ must satisfy

$$\sum_k s_I(k)\frac{d}{d\tau}z_k(\tau) = 0. \quad (5.86)$$

For decision-directed estimation, (5.86) gives rise to the estimator shown in Figure 5.30. The input to the voltage-controlled clock (VCC) is (5.86). If this input is zero, then the timing estimate $\hat{\tau} = \tau$. If not, the clock (i.e., the timing estimate $\hat{\tau}$) is adjusted to drive the VCC input to zero. This timing estimation loop is also an example of a feedback control loop.

One structure for non–decision-directed timing estimation is the *early–late gate synchronizer* shown in Figure 5.31. This structure exploits two properties of the autocorrelation of $g(t)$, $R_g(\tau) = \int_0^{T_s} g(t)g(t-\tau)\, dt$ – namely, its symmetry ($R_g(\tau) = R_g(-\tau)$) and the fact that

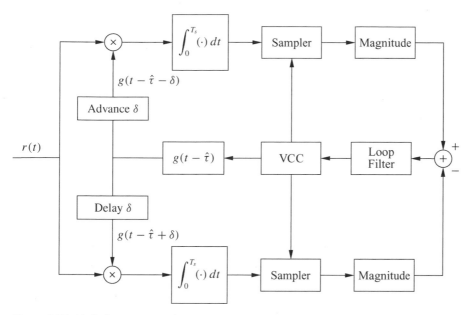

Figure 5.31: Early–late gate synchronizer.

its maximum value is at $\tau = 0$. The input to the sampler in the upper branch of Figure 5.31 is proportional to the autocorrelation $R_g(\hat{\tau} - \tau + \delta) = \int_0^{T_s} g(t - \tau)g(t - \hat{\tau} - \delta)\,dt$, and the input to the sampler in the lower branch is proportional to the autocorrelation $R_g(\hat{\tau} - \tau - \delta) = \int_0^{T_s} g(t - \tau)g(t - \hat{\tau} + \delta)\,dt$. If $\hat{\tau} = \tau$ then, since $R_g(\delta) = R_g(-\delta)$, the input to the loop filter will be zero and the voltage-controlled clock will maintain its correct timing estimate. If $\hat{\tau} > \tau$ then $R_g(\hat{\tau} - \tau + \delta) < R_g(\hat{\tau} - \tau - \delta)$, and this negative input to the VCC will cause it to decrease its estimate of $\hat{\tau}$. Conversely, if $\hat{\tau} < \tau$ then $R_g(\hat{\tau} - \tau + \delta) > R_g(\hat{\tau} - \tau - \delta)$, and this positive input to the VCC will cause it to increase its estimate of $\hat{\tau}$.

More details on these and other structures for decision-directed and non–decision-directed timing estimation – as well as their performance trade-offs – can be found in [10, Chaps. 6.2.4–6.2.5; 11].

PROBLEMS

5-1. Using properties of orthonormal basis functions, show that if $s_i(t)$ and $s_j(t)$ have constellation points $\mathbf{s}_i$ and $\mathbf{s}_j$ (respectively) then

$$\|\mathbf{s}_i - \mathbf{s}_j\|^2 = \int_0^T (s_i(t) - s_j(t))^2\,dt.$$

5-2. Find an alternate set of orthonormal basis functions for the space spanned by $\cos(2\pi t/T)$ and $\sin(2\pi t/T)$.

5-3. Consider a set of M orthogonal signal waveforms $s_m(t)$, for $1 \le m \le M$ and $0 \le t \le T$, where each waveform has the same energy $\mathcal{E}$. Define a new set of M waveforms as

$$s'_m(t) = s_m(t) - \frac{1}{M}\sum_{i=1}^{M} s_i(t), \quad 1 \le m \le M,\ 0 \le t \le T.$$

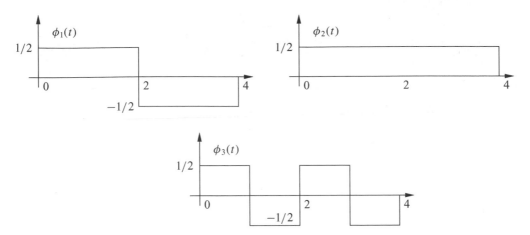

Figure 5.32: Signal waveforms for Problem 5-4.

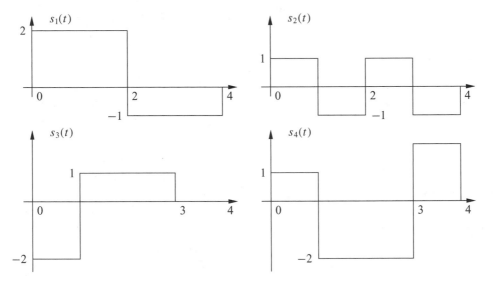

Figure 5.33: Signal waveforms for Problem 5-5.

Show that the M signal waveforms $\{s_m'(t)\}$ have equal energy, given by

$$\mathcal{E}' = (M-1)\mathcal{E}/M.$$

What is the inner product between any two waveforms?

5-4. Consider the three signal waveforms $\{\phi_1(t), \phi_2(t), \phi_3(t)\}$ shown in Figure 5.32.

    (a) Show that these waveforms are orthonormal.

    (b) Express the waveform $x(t)$ as a linear combination of $\{\phi_i(t)\}$ and find the coefficients, where $x(t)$ is given as

$$x(t) = \begin{cases} 2 & 0 \le t < 2, \\ 4 & 2 \le t \le 4. \end{cases}$$

5-5. Consider the four signal waveforms as shown in Figure 5.33.

    (a) Determine the dimensionality of the waveforms and a set of basis functions.

    (b) Use the basis functions to represent the four waveforms by vectors.

    (c) Determine the minimum distance between all the vector pairs.

5-6. Derive a mathematical expression for decision regions Z_i that minimize error probability assuming that messages are not equally likely – that is, assuming $p(m_i) = p_i$ ($i = 1, \ldots, M$), where p_i is not necessarily equal to $1/M$. Solve for these regions in the case of QPSK modulation with $s_1 = (A_c, 0)$, $s_2 = (0, A_c)$, $s_3 = (-A_c, 0)$, and $s_4 = (0, A_c)$, assuming $p(s_1) = p(s_3) = .2$ and $p(s_1) = p(s_3) = .3$.

5-7. Show that the remainder noise term $n_r(t)$ is independent of the correlator outputs r_i for all i. In other words, show that $\mathbf{E}[n_r(t)r_i] = 0$ for all i. Thus, since r_j (conditioned on $\mathbf{s}_i$) and $n_r(t)$ are Gaussian and uncorrelated, they are independent.

5-8. Show that output SNR is maximized when a given input signal is passed through a filter that is matched to that signal.

5-9. Find the matched filters $g(T - t)$, $0 \le t \le T$, and find $\int_0^T g(t)g(T - t)\,dt$ for the following waveforms.

(a) Rectangular pulse: $g(t) = \sqrt{2/T}$.
(b) Sinc pulse: $g(t) = \mathrm{sinc}(t)$.
(c) Gaussian pulse: $g(t) = (\sqrt{\pi}/\alpha)e^{-\pi^2 t^2/\alpha^2}$.

5-10. Show that the ML receiver of Figure 5.4 is equivalent to the matched filter receiver of Figure 5.7.

5-11. Compute the three bounds (5.40), (5.43), (5.44) as well as the approximation (5.45) for an asymmetric signal constellation $s_1 = (A_c, 0)$, $s_2 = (0, 2A_c)$, $s_3 = (-2A_c, 0)$, and $s_4 = (0, -A_c)$, assuming that $A_c/\sqrt{N_0} = 4$.

5-12. Find the input to each branch of the decision device in Figure 5.11 if the transmit carrier phase ϕ_0 differs from the receiver carrier phase ϕ by $\Delta\phi$.

5-13. Consider a 4-PSK constellation with $d_{\min} = \sqrt{2}$. What is the additional energy required to send one extra bit (8-PSK) while keeping the same minimum distance (and thus with the same bit error probability)?

5-14. Show that the average power of a square signal constellation with l bits per dimension, P_l, is proportional to $4^l/3$ and that the average power for one more bit per dimension, keeping the same minimum distance, is $P_{l+1} \approx 4P_l$. Find P_l for $l = 2$ and compute the average energy of MPSK and MPAM constellations with the same number of bits per symbol.

5-15. For MPSK with differential modulation, let $\Delta\phi$ denote the phase drift of the channel over a symbol time T_s. In the absence of noise, how large must $\Delta\phi$ be in order for a detection error to occur?

5-16. Find the Gray encoding of bit sequences to phase transitions in differential 8-PSK. Then find the sequence of symbols transmitted using differential 8-PSK modulation with this Gray encoding for the bit sequence 101110100101110 starting at the kth symbol time, assuming the transmitted symbol at the $(k - 1)$th symbol time is $\mathbf{s}(k - 1) = Ae^{j\pi/4}$.

5-17. Consider the octal signal point constellation shown in Figure 5.34.

(a) The nearest neighbor signal points in the 8-QAM signal constellation are separated by a distance of A. Determine the radii a and b of the inner and outer circles.
(b) The adjacent signal points in the 8-PSK are separated by a distance of A. Determine the radius r of the circle.

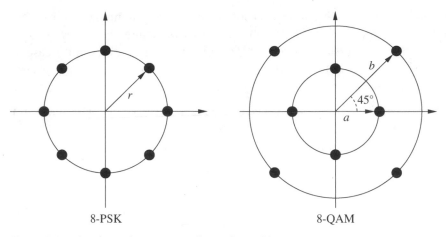

8-PSK   8-QAM

Figure 5.34: Octal signal point constellation for Problem 5-17.

(c) Determine the average transmitter powers for the two signal constellations and compare the two powers. What is the relative power advantage of one constellation over
the other? (Assume that all signal points are equally probable.)

(d) Is it possible to assign three data bits to each point of the signal constellation such
that nearest neighbor (adjacent) points differ in only one bit position?

(e) Determine the symbol rate if the desired bit rate is 90 Mbps.

5-18. The $\pi/4$-QPSK modulation may be considered as two QPSK systems offset by $\pi/4$
radians.

(a) Sketch the signal space diagram for a $\pi/4$-QPSK signal.

(b) Using Gray encoding, label the signal points with the corresponding data bits.

(c) Determine the sequence of symbols transmitted via $\pi/4$-QPSK for the bit sequence
0100100111100101.

(d) Repeat part (c) for $\pi/4$-DQPSK, assuming that the last symbol transmitted on the in-
phase branch had a phase of π and that the last symbol transmitted on the quadrature
branch had a phase of $-3\pi/4$.

5-19. Show that the minimum frequency separation for FSK such that the $\cos(2\pi f_j t)$ and
$\cos(2\pi f_i t)$ are orthogonal is $\Delta f = \min_{ij}|f_j - f_i| = .5/T_s$.

5-20. Show that the Nyquist criterion for zero ISI pulses given by $p(kT_s) = p_0\delta(k)$ is equivalent to the frequency domain condition (5.68).

5-21. Show that the Gaussian pulse shape does not satisfy the Nyquist criterion.

REFERENCES

[1]   S. Haykin, *Communication Systems,* Wiley, New York, 2002.

[2]   J. Proakis and M. Salehi, *Communication Systems Engineering,* Prentice-Hall, Englewood Cliffs,
NJ, 2002.

[3]   J. M. Wozencraft and I. M. Jacobs, *Principles of Communication Engineering,* Wiley, New York,
1965.

[4]   M. Fitz, "Further results in the unified analysis of digital communication systems," *IEEE Trans. Commun.,* pp. 521–32, March 1992.

[5]   R. Ziemer, "An overview of modulation and coding for wireless communications," *Proc. IEEE Veh. Tech. Conf.,* pp. 26–30, April 1996.

[6]   G. D. Forney, Jr., and L.-F. Wei, "Multidimensional constellations – Part I: Introduction, figures of merit, and generalized cross constellations," *IEEE J. Sel. Areas Commun.,* pp. 877–92, August 1989.

[7]   T. S. Rappaport, *Wireless Communications – Principles and Practice,* 2nd ed., Prentice-Hall, Englewood Cliffs, NJ, 2001.

[8]   G. L. Stuber, *Principles of Mobile Communications,* 2nd ed., Kluwer, Dordrecht, 2001.

[9]   S. Haykin, *An Introduction to Analog and Digital Communications,* Wiley, New York, 1989.

[10]   J. G. Proakis, *Digital Communications,* 4th ed., McGraw-Hill, New York, 2001.

[11]   U. Mengali and A. N. D'Andrea, *Synchronization Techniques for Digital Receivers,* Plenum, New York, 1997.

[12]   H. Meyr, M. Moeneclaey, and S. A. Fechtel, *Digital Communication Receivers,* vol. 2, *Synchronization, Channel Estimation, and Signal Processing,* Wiley, New York, 1997.

[13]   L. E. Franks, "Carrier and bit synchronization in data communication – A tutorial review," *IEEE Trans. Commun.,* pp. 1107–21, August 1980.

Performance of Digital Modulation over Wireless Channels

We now consider the performance of the digital modulation techniques discussed in the previous chapter when used over AWGN channels and channels with flat fading. There are two performance criteria of interest: the probability of error, defined relative to either symbol or bit errors; and the outage probability, defined as the probability that the instantaneous signal-to-noise ratio falls below a given threshold. Flat fading can cause a dramatic increase in either the average bit error probability or the signal outage probability. Wireless channels may also exhibit frequency-selective fading and Doppler shift. Frequency-selective fading gives rise to intersymbol interference (ISI), which causes an irreducible error floor in the received signal. Doppler causes spectral broadening, which leads to adjacent channel interference (small at typical user velocities) and also to an irreducible error floor in signals with differential phase encoding (e.g. DPSK), since the phase reference of the previous symbol partially decorrelates over a symbol time. This chapter describes the impact on digital modulation performance of noise, flat fading, frequency-selective fading, and Doppler.

6.1 AWGN Channels

In this section we define the signal-to-noise power ratio (SNR) and its relation to energy per bit (E_b) and energy per symbol (E_s). We then examine the error probability on AWGN channels for different modulation techniques as parameterized by these energy metrics. Our analysis uses the signal space concepts of Section 5.1.

6.1.1 Signal-to-Noise Power Ratio and Bit/Symbol Energy

In an AWGN channel the modulated signal $s(t) = \text{Re}\{u(t)e^{j2\pi f_c t}\}$ has noise $n(t)$ added to it prior to reception. The noise $n(t)$ is a white Gaussian random process with mean zero and power spectral density (PSD) $N_0/2$. The received signal is thus $r(t) = s(t) + n(t)$.

We define the received SNR as the ratio of the received signal power P_r to the power of the noise within the bandwidth of the transmitted signal $s(t)$. The received power P_r is determined by the transmitted power and the path loss, shadowing, and multipath fading, as described in Chapters 2 and 3. The noise power is determined by the bandwidth of the transmitted signal and the spectral properties of $n(t)$. Specifically, if the bandwidth of the complex envelope $u(t)$ of $s(t)$ is B then the bandwidth of the transmitted signal $s(t)$ is $2B$.

Since the noise $n(t)$ has uniform PSD $N_0/2$, the total noise power within the bandwidth $2B$ is $N = N_0/2 \cdot 2B = N_0 B$. Hence the received SNR is given by

$$\text{SNR} = \frac{P_r}{N_0 B}.$$

In systems with interference, we often use the received signal-to-interference-plus-noise power ratio (SINR) in place of SNR for calculating error probability. This is a reasonable approximation if the interference statistics approximate those of Gaussian noise. The received SINR is given by

$$\text{SINR} = \frac{P_r}{N_0 B + P_I},$$

where P_I is the average power of the interference.

The SNR is often expressed in terms of the signal energy per bit E_b (or per symbol, E_s) as

$$\text{SNR} = \frac{P_r}{N_0 B} = \frac{E_s}{N_0 B T_s} = \frac{E_b}{N_0 B T_b}, \tag{6.1}$$

where T_s is the symbol time and T_b is the bit time (for binary modulation $T_s = T_b$ and $E_s = E_b$). For pulse shaping with $T_s = 1/B$ (e.g., raised cosine pulses with $\beta = 1$), we have SNR $= E_s/N_0$ for multilevel signaling and SNR $= E_b/N_0$ for binary signaling. For general pulses, $T_s = k/B$ for some constant k, in which case $k \cdot$ SNR $= E_s/N_0$.

The quantities $\gamma_s = E_s/N_0$ and $\gamma_b = E_b/N_0$ are sometimes called the SNR per symbol and the SNR per bit, respectively. For performance specification, we are interested in the bit error probability P_b as a function of γ_b. However, with M-ary signaling (e.g., MPAM and MPSK) the bit error probability depends on both the symbol error probability and the mapping of bits to symbols. Thus, we typically compute the symbol error probability P_s as a function of γ_s based on the signal space concepts of Section 5.1 and then obtain P_b as a function of γ_b using an exact or approximate conversion. The approximate conversion typically assumes that the symbol energy is divided equally among all bits and that Gray encoding is used, so that (at reasonable SNRs) one symbol error corresponds to exactly one bit error. These assumptions for M-ary signaling lead to the approximations

$$\gamma_b \approx \frac{\gamma_s}{\log_2 M} \tag{6.2}$$

and

$$P_b \approx \frac{P_s}{\log_2 M}. \tag{6.3}$$

6.1.2 Error Probability for BPSK and QPSK

We first consider BPSK modulation with coherent detection and perfect recovery of the carrier frequency and phase. With binary modulation each symbol corresponds to one bit, so the symbol and bit error rates are the same. The transmitted signal is $s_1(t) = Ag(t) \cos(2\pi f_c t)$ to send a 0-bit and $s_2(t) = -Ag(t) \cos(2\pi f_c t)$ to send a 1-bit for $A > 0$. From (5.46) we have that the probability of error is

$$P_b = Q\left(\frac{d_{\min}}{\sqrt{2N_0}}\right). \tag{6.4}$$

From Section 5.3.2, $d_{\min} = \|s_1 - s_0\| = \|A - (-A)\| = 2A$. Let us now relate A to the energy per bit. We have

$$E_b = \int_0^{T_b} s_1^2(t)\, dt = \int_0^{T_b} s_2^2(t)\, dt = \int_0^{T_b} A^2 g^2(t) \cos^2(2\pi f_c t)\, dt = A^2 \qquad (6.5)$$

by (5.56). Thus, the signal constellation for BPSK in terms of energy per bit is given by $s_0 = \sqrt{E_b}$ and $s_1 = -\sqrt{E_b}$. This yields the minimum distance $d_{\min} = 2A = 2\sqrt{E_b}$. Substituting this into (6.4) yields

$$P_b = Q\left(\frac{2\sqrt{E_b}}{\sqrt{2N_0}}\right) = Q\left(\sqrt{\frac{2E_b}{N_0}}\right) = Q(\sqrt{2\gamma_b}). \qquad (6.6)$$

QPSK modulation consists of BPSK modulation on both the in-phase and quadrature components of the signal. With perfect phase and carrier recovery, the received signal components corresponding to each of these branches are orthogonal. Therefore, the bit error probability on each branch is the same as for BPSK: $P_b = Q(\sqrt{2\gamma_b})$. The symbol error probability equals the probability that either branch has a bit error:

$$P_s = 1 - \left[1 - Q(\sqrt{2\gamma_b})\right]^2. \qquad (6.7)$$

Since the symbol energy is split between the in-phase and quadrature branches, we have $\gamma_s = 2\gamma_b$. Substituting this into (6.7) yields P_s is terms of γ_s as

$$P_s = 1 - \left[1 - Q(\sqrt{\gamma_s})\right]^2. \qquad (6.8)$$

From Section 5.1.5, the union bound (5.40) on P_s for QPSK is

$$P_s \leq 2Q\left(A/\sqrt{N_0}\right) + Q\left(\sqrt{2}A/\sqrt{N_0}\right). \qquad (6.9)$$

Writing this in terms of $\gamma_s = 2\gamma_b = A^2/N_0$ yields

$$P_s \leq 2Q(\sqrt{\gamma_s}) + Q(\sqrt{2\gamma_s}) \leq 3Q(\sqrt{\gamma_s}). \qquad (6.10)$$

The closed-form bound (5.44) becomes

$$P_s \leq \frac{3}{\sqrt{2\pi\gamma_s}} \exp[-.5\gamma_s]. \qquad (6.11)$$

Using the fact that the minimum distance between constellation points is $d_{\min} = \sqrt{2A^2}$ in (5.45), we obtain the nearest neighbor approximation

$$P_s \approx 2Q\left(\sqrt{A^2/N_0}\right) = 2Q(\sqrt{\gamma_s}). \qquad (6.12)$$

Note that with Gray encoding we can approximate P_b from P_s by $P_b \approx P_s/2$, since QPSK has two bits per symbol.

EXAMPLE 6.1: Find the bit error probability P_b and symbol error probability P_s of QPSK assuming $\gamma_b = 7$ dB. Compare the exact P_b with the approximation $P_b \approx P_s/2$ based on the assumption of Gray coding. Finally, compute P_s based on the nearest neighbor bound using $\gamma_s = 2\gamma_b$ and then compare with the exact P_s.

Solution: We have $\gamma_b = 10^{7/10} = 5.012$, so

$$P_b = Q(\sqrt{2\gamma_b}) = Q(\sqrt{10.024}) = 7.726 \cdot 10^{-4}.$$

The exact symbol error probability P_s is

$$P_s = 1 - [1 - Q(\sqrt{2\gamma_b})]^2 = 1 - [1 - Q(\sqrt{10.02})]^2 = 1.545 \cdot 10^{-3}.$$

The bit error probability approximation assuming Gray coding yields $P_b \approx P_s/2 = 7.723 \cdot 10^{-4}$, which is quite close to the exact P_b. The nearest neighbor approximation to P_s yields

$$P_s \approx 2Q(\sqrt{\gamma_s}) = 2Q(\sqrt{10.024}) = 1.545 \cdot 10^{-3},$$

which matches well with the exact P_s.

6.1.3 Error Probability for MPSK

The signal constellation for MPSK has $s_{i1} = A\cos[2\pi(i-1)/M]$ and $s_{i2} = A\sin[2\pi(i-1)/M]$ for $A > 0$ and $i = 1, \dots, M$. The symbol energy is $E_s = A^2$, so $\gamma_s = A^2/N_0$. From (5.57) it follows that, for the received vector $\mathbf{r} = re^{j\theta}$ represented in polar coordinates, an error occurs if the ith signal constellation point is transmitted and $\theta \notin (2\pi(i-1-.5)/M, 2\pi(i-1+.5)/M)$. The joint distribution of r and θ can be obtained through a bivariate transformation of the noise n_1 and n_2 on the in-phase and quadrature branches [1, Chap. 5.2.7], which yields

$$p(r, \theta) = \frac{r}{\pi N_0} \exp\left[-\frac{1}{N_0}\left(r^2 - 2\sqrt{E_s}\,r\cos(\theta) + E_s\right)\right]. \tag{6.13}$$

Since the error probability depends only on the distribution of θ, we can integrate out the dependence on r to obtain

$$p(\theta) = \int_0^\infty p(r, \theta)\,dr = \frac{1}{2\pi}e^{-\gamma_s \sin^2(\theta)}\int_0^\infty z\exp\left[-\frac{(z - \sqrt{2\gamma_s}\cos(\theta))^2}{2}\right]dz. \tag{6.14}$$

By symmetry, the probability of error is the same for each constellation point. Thus, we can derive P_s from the probability of error assuming the constellation point $\mathbf{s}_1 = (A, 0)$ is transmitted, which is

$$P_s = 1 - \int_{-\pi/M}^{\pi/M} p(\theta)\,d\theta$$

$$= 1 - \int_{-\pi/M}^{\pi/M} \frac{1}{2\pi}e^{-\gamma_s \sin^2(\theta)}\int_0^\infty z\exp\left[-\frac{(z - \sqrt{2\gamma_s}\cos(\theta))^2}{2}\right]dz. \tag{6.15}$$

A closed-form solution to this integral does not exist for $M > 4$ and so the exact value of P_s must be computed numerically.

Each point in the MPSK constellation has two nearest neighbors at distance $d_{\min} = 2A \sin(\pi/M)$. Thus, the nearest neighbor approximation (5.45) to P_s is given by

$$P_s \approx 2Q\left(\sqrt{2}A \sin(\pi/M)/\sqrt{N_0}\right) = 2Q\left(\sqrt{2\gamma_s} \sin(\pi/M)\right). \tag{6.16}$$

This nearest neighbor approximation can differ significantly from the exact value of P_s. However, it is much simpler to compute than the numerical integration of (6.15) that is required to obtain the exact P_s. This formula can also be obtained by approximating $p(\theta)$ as

$$p(\theta) \approx \sqrt{\gamma_s/\pi} \cos(\theta) e^{-\gamma_s \sin^2(\theta)}. \tag{6.17}$$

Using this in the first line of (6.15) yields (6.16).

EXAMPLE 6.2: Compare the probability of bit error for 8-PSK and 16-PSK assuming $\gamma_b = 15$ dB and using the P_s approximation given in (6.16) along with the approximations (6.3) and (6.2).

Solution: From (6.2) we have that, for 8-PSK, $\gamma_s = (\log_2 8) \cdot 10^{15/10} = 94.87$. Substituting this into (6.16) yields

$$P_s \approx 2Q\left(\sqrt{189.74} \sin(\pi/8)\right) = 1.355 \cdot 10^{-7}. \tag{6.18a}$$

Now, using (6.3), we get $P_b = P_s/3 = 4.52 \cdot 10^{-8}$. For 16-PSK we have $\gamma_s = (\log_2 16) \cdot 10^{15/10} = 126.49$. Substituting this into (6.16) yields

$$P_s \approx 2Q\left(\sqrt{252.98} \sin(\pi/16)\right) = 1.916 \cdot 10^{-3}, \tag{6.18b}$$

and by using (6.3) we get $P_b = P_s/4 = 4.79 \cdot 10^{-4}$. Note that P_b is much larger for 16-PSK than for 8-PSK given the same γ_b. This result is expected because 16-PSK packs more bits per symbol into a given constellation and so, for a fixed energy per bit, the minimum distance between constellation points will be smaller.

The error probability derivation for MPSK assumes that the carrier phase is perfectly known to the receiver. Under phase estimation error, the distribution of $p(\theta)$ used to obtain P_s must incorporate the distribution of the phase rotation associated with carrier phase offset. This distribution is typically a function of the carrier phase estimation technique and the SNR. The impact of phase estimation error on coherent modulation is studied in [1, Apx. C; 2, Chap. 4.3.2; 3; 4]. These works indicate that, as expected, significant phase offset leads to an irreducible bit error probability. Moreover, nonbinary signaling is more sensitive than BPSK to phase offset because of the resulting cross-coupling between in-phase and quadrature signal components. The impact of phase estimation error can be especially severe in fast fading, where the channel phase changes rapidly owing to constructive and destructive multipath interference. Even with differential modulation, phase changes over and between symbol times can produce irreducible errors [5]. Timing errors can also degrade performance; analysis of timing errors in MPSK performance can be found in [2, Chap. 4.3.3; 6].

6.1.4 Error Probability for MPAM and MQAM

The constellation for MPAM is $A_i = (2i - 1 - M)d$, $i = 1, 2, \ldots, M$. Each of the $M - 2$ inner constellation points of this constellation have two nearest neighbors at distance $2d$.

The probability of making an error when sending one of these inner constellation points is just the probability that the noise exceeds d in either direction: $P_s(\mathbf{s}_i) = p(|\mathbf{n}| > d), i = 2, \ldots, M-1$. For the outer constellation points there is only one nearest neighbor, so an error occurs if the noise exceeds d in one direction only: $P_s(\mathbf{s}_i) = p(\mathbf{n} > d) = .5p(|\mathbf{n}| > d), i = 1, M$. The probability of error is thus

$$P_s = \frac{1}{M} \sum_{i=1}^{M} P_s(\mathbf{s}_i)$$

$$= \frac{M-2}{M} 2Q\left(\sqrt{\frac{2d^2}{N_0}}\right) + \frac{2}{M} Q\left(\sqrt{\frac{2d^2}{N_0}}\right) = \frac{2(M-1)}{M} Q\left(\sqrt{\frac{2d^2}{N_0}}\right). \quad (6.19)$$

From (5.54), the average energy per symbol for MPAM is

$$\bar{E}_s = \frac{1}{M} \sum_{i=1}^{M} A_i^2 = \frac{1}{M} \sum_{i=1}^{M} (2i - 1 - M)^2 d^2 = \frac{1}{3}(M^2 - 1)d^2. \quad (6.20)$$

Thus we can write P_s in terms of the average energy $\bar{E}_s$ as

$$P_s = \frac{2(M-1)}{M} Q\left(\sqrt{\frac{6\bar{\gamma}_s}{M^2 - 1}}\right). \quad (6.21)$$

Consider now MQAM modulation with a square signal constellation of size $M = L^2$. This system can be viewed as two MPAM systems with signal constellations of size L transmitted over the in-phase and quadrature signal components, each with half the energy of the original MQAM system. The constellation points in the in-phase and quadrature branches take values $A_i = (2i - 1 - L)d, i = 1, 2, \ldots, L$. The symbol error probability for each branch of the MQAM system is thus given by (6.21) with M replaced by $L = \sqrt{M}$ and $\bar{\gamma}_s$ equal to the average energy per symbol in the MQAM constellation:

$$P_{s, \text{branch}} = \frac{2(\sqrt{M} - 1)}{\sqrt{M}} Q\left(\sqrt{\frac{3\bar{\gamma}_s}{M - 1}}\right). \quad (6.22)$$

Note that $\bar{\gamma}_s$ is multiplied by a factor of 3 in (6.22) instead of the factor of 6 in (6.21), since the MQAM constellation splits its total average energy $\bar{\gamma}_s$ between its in-phase and quadrature branches. The probability of symbol error for the MQAM system is then

$$P_s = 1 - \left(1 - \frac{2(\sqrt{M} - 1)}{\sqrt{M}} Q\left(\sqrt{\frac{3\bar{\gamma}_s}{M - 1}}\right)\right)^2. \quad (6.23)$$

The nearest neighbor approximation to probability of symbol error depends on whether the constellation point is an inner or outer point. Inner points have four nearest neighbors, while outer points have either two or three nearest neighbors; in both cases the distance between nearest neighbors is $2d$. If we take a conservative approach and set the number of nearest neighbors to be four, we obtain the nearest neighbor approximation

$$P_s \approx 4Q\left(\sqrt{\frac{3\bar{\gamma}_s}{M - 1}}\right). \quad (6.24)$$

For nonrectangular constellations, it is relatively straightforward to show that the probability of symbol error is upper bounded as

$$P_s \le 1 - \left[1 - 2Q\left(\sqrt{\frac{3\bar{\gamma}_s}{M-1}}\right)\right]^2 \le 4Q\left(\sqrt{\frac{3\bar{\gamma}_s}{M-1}}\right), \tag{6.25}$$

which is the same as (6.24) for square constellations. The nearest neighbor approximation for nonrectangular constellations is

$$P_s \approx M_{d_{\min}} Q\left(\frac{d_{\min}}{\sqrt{2N_0}}\right), \tag{6.26}$$

where $M_{d_{\min}}$ is the largest number of nearest neighbors for any constellation point in the constellation and $d_{\min}$ is the minimum distance in the constellation.

EXAMPLE 6.3: For 16-QAM with $\gamma_b = 15$ dB ($\gamma_s = \log_2 M \cdot \gamma_b$), compare the exact probability of symbol error (6.23) with (a) the nearest neighbor approximation (6.24) and (b) the symbol error probability for 16-PSK with the same γ_b (which was obtained in Example 6.2).

Solution: The average symbol energy $\gamma_s = 4 \cdot 10^{1.5} = 126.49$. The exact P_s is then given by

$$P_s = 1 - \left(1 - \frac{2(4-1)}{4}Q\left(\sqrt{\frac{3 \cdot 126.49}{15}}\right)\right)^2 = 7.37 \cdot 10^{-7}.$$

The nearest neighbor approximation is given by

$$P_s \approx 4Q\left(\sqrt{\frac{3 \cdot 126.49}{15}}\right) = 9.82 \cdot 10^{-7},$$

which is slightly larger than the exact value owing to the conservative approximation that every constellation point has four nearest neighbors. The symbol error probability for 16-PSK from Example 6.2 is $P_s \approx 1.916 \cdot 10^{-3}$, which is roughly four orders of magnitude larger than the exact P_s for 16-QAM. The larger P_s for MPSK versus MQAM with the same M and same γ_b is due to the fact that MQAM uses both amplitude and phase to encode data whereas MPSK uses just the phase. Thus, for the same energy per symbol or bit, MQAM makes more efficient use of energy and therefore has better performance.

The MQAM demodulator requires both amplitude and phase estimates of the channel so that the decision regions used in detection to estimate the transmitted symbol are not skewed in amplitude or phase. The analysis of performance degradation due to phase estimation error is similar to the case of MPSK discussed previously. The channel amplitude is used to scale the decision regions so that they correspond to the transmitted symbol: this scaling is called automatic gain control (AGC). If the channel gain is estimated in error then the AGC improperly scales the received signal, which can lead to incorrect demodulation even in the absence of noise. The channel gain is typically obtained using pilot symbols to estimate the channel gain at the receiver. However, pilot symbols do not lead to perfect channel estimates, and the estimation error can lead to bit errors. More details on the impact of amplitude and phase estimation errors on the performance of MQAM modulation can be found in [7, Chap. 10.3; 8].

6.1.5 Error Probability for FSK and CPFSK

Let us first consider the error probability of binary FSK with the coherent demodulator of Figure 5.24. Since demodulation is coherent, we can neglect any phase offset in the carrier signals. The transmitted signal is defined by

$$s_i(t) = A\sqrt{2T_b}\cos(2\pi f_i t), \quad i = 1, 2. \tag{6.27}$$

Hence $E_b = A^2$ and $\gamma_b = A^2/N_0$. The input to the decision device is

$$z = s_1 + n_1 - s_2 - n_2. \tag{6.28}$$

The device outputs a 1-bit if $z > 0$ or a 0-bit if $z \leq 0$. Let us assume that $s_1(t)$ is transmitted; then

$$z \mid 1 = A + n_1 - n_2. \tag{6.29}$$

An error occurs if $z = A + n_1 - n_2 \leq 0$. On the other hand, if $s_2(t)$ is transmitted then

$$z \mid 0 = n_1 - A - n_2, \tag{6.30}$$

and an error occurs if $z = n_1 - A - n_2 > 0$. For n_1 and n_2 independent white Gaussian random variables with mean zero and variance $N_0/2$, their difference is a white Gaussian random variable with mean zero and variance equal to the sum of variances $N_0/2 + N_0/2 = N_0$. Then, for equally likely bit transmissions,

$$P_b = .5p(A + n_1 - n_2 \leq 0) + .5p(n_1 - A - n_2 > 0) = Q(A/\sqrt{N_0}) = Q(\sqrt{\gamma_b}). \tag{6.31}$$

The derivation of P_s for coherent MFSK with $M > 2$ is more complex and does not lead to a closed-form solution [2, eq. (4.92)]. The probability of symbol error for noncoherent MFSK is derived in [9, Chap. 8.1] as

$$P_s = \sum_{m=1}^{M} (-1)^{m+1} \binom{M-1}{m} \frac{1}{m+1} \exp\left[\frac{-m\gamma_s}{m+1}\right]. \tag{6.32}$$

The error probability of CPFSK depends on whether the detector is coherent or non-coherent and also on whether it uses symbol-by-symbol detection or sequence estimation. Analysis of error probability for CPFSK is complex because the memory in the modulation requires error probability analysis over multiple symbols. The formulas for error probability can also become quite complicated. Detailed derivations of error probability for these different CPFSK structures can be found in [1; Chap. 5.3]. As with linear modulations, FSK performance degrades under frequency and timing errors. A detailed analysis of the impact of such errors on FSK performance can be found in [2, Chap. 5.2; 10; 11].

Table 6.1: Approximate symbol and bit error probabilities for coherent modulations

Modulation	$P_s(\gamma_s)$	$P_b(\gamma_b)$
BFSK		$P_b = Q(\sqrt{\overline{\gamma_b}})$
BPSK		$P_b = Q(\sqrt{2\gamma_b})$
QPSK, 4-QAM	$P_s \approx 2Q(\sqrt{\overline{\gamma_s}})$	$P_b \approx Q(\sqrt{2\gamma_b})$
MPAM	$P_s = \dfrac{2(M-1)}{M} Q\left(\sqrt{\dfrac{6\overline{\gamma_s}}{M^2-1}}\right)$	$P_b \approx \dfrac{2(M-1)}{M\log_2 M} Q\left(\sqrt{\dfrac{6\overline{\gamma_b}\log_2 M}{M^2-1}}\right)$
MPSK	$P_s \approx 2Q\left(\sqrt{2\gamma_2}\sin\left(\dfrac{\pi}{M}\right)\right)$	$P_b \approx \dfrac{2}{\log_2 M} Q\left(\sqrt{2\gamma_b\log_2 M}\sin\left(\dfrac{\pi}{M}\right)\right)$
Rectangular MQAM	$P_s \approx 4Q\left(\sqrt{\dfrac{3\overline{\gamma_s}}{M-1}}\right)$	$P_b \approx \dfrac{4}{\log_2 M} Q\left(\sqrt{\dfrac{3\overline{\gamma_b}\log_2 M}{M-1}}\right)$
Nonrectangular MQAM	$P_s \approx 4Q\left(\sqrt{\dfrac{3\overline{\gamma_s}}{M-1}}\right)$	$P_b \approx \dfrac{4}{\log_2 M} Q\left(\sqrt{\dfrac{3\overline{\gamma_b}\log_2 M}{M-1}}\right)$

6.1.6 Error Probability Approximation for Coherent Modulations

Many of the approximations or exact values for P_s derived so far for coherent modulation are in the following form:

$$P_s(\gamma_s) \approx \alpha_M Q(\sqrt{\beta_M \gamma_s}), \tag{6.33}$$

where α_M and β_M depend on the type of approximation and the modulation type. In particular, the nearest neighbor approximation has this form, where α_M is the number of nearest neighbors to a constellation at the minimum distance and β_M is a constant that relates minimum distance to average symbol energy. In Table 6.1 we summarize the specific values of α_M and β_M for common P_s expressions for PSK, QAM, and FSK modulations based on the derivations in prior sections.

Performance specifications are generally more concerned with the bit error probability P_b as a function of the bit energy γ_b. To convert from P_s to P_b and from γ_s to γ_b we use the approximations (6.3) and (6.2), which assume Gray encoding and high SNR. Using these approximations in (6.33) yields a simple formula for P_b as a function of γ_b:

$$P_b(\gamma_b) = \hat{\alpha}_M Q(\sqrt{\hat{\beta}_M \gamma_b}), \tag{6.34}$$

where $\hat{\alpha}_M = \alpha_M/\log_2 M$ and $\hat{\beta}_M = (\log_2 M)\beta_M$ for α_M and β_M in (6.33). This conversion is used in what follows to obtain P_b versus γ_b from the general form of P_s versus γ_s in (6.33).

6.1.7 Error Probability for Differential Modulation

The probability of error for differential modulation is based on the phase difference associated with the phase comparator input of Figure 5.20. Specifically, the phase comparator extracts the phase of

$$\mathbf{r}(k)\mathbf{r}^*(k-1) = A^2 e^{j(\theta(k)-\theta(k-1))} + Ae^{j(\theta(k)+\phi_0)}n^*(k-1)$$

$$+ Ae^{-j(\theta(k-1)+\phi_0)}n(k) + n(k)n^*(k-1) \tag{6.35}$$

in order to determine the transmitted symbol. By symmetry we can assume a given phase difference when computing the error probability. Assuming then a phase difference of zero, $\theta(k) - \theta(k - 1) = 0$, yields

$$
\mathbf{r}(k)\mathbf{r}^*(k - 1) = A^2 + Ae^{j(\theta(k)+\phi_0)}n^*(k - 1)
$$
$$
+ Ae^{-j(\theta(k-1)+\phi_0)}n(k) + n(k)n^*(k - 1). \qquad (6.36)
$$

Next we define new random variables

$$
\tilde{n}(k) = n(k)e^{-j(\theta(k-1)+\phi_0)} \quad \text{and} \quad \tilde{n}(k - 1) = n(k - 1)e^{-j(\theta(k)+\phi_0)},
$$

which have the same statistics as $n(k)$ and $n(k - 1)$. Then

$$
\mathbf{r}(k)\mathbf{r}^*(k - 1) = A^2 + A(\tilde{n}^*(k - 1) + \tilde{n}(k)) + \tilde{n}(k)\tilde{n}^*(k - 1). \qquad (6.37)
$$

There are three terms in (6.37): the first term, with the desired phase difference of zero; and the second and third terms, which contribute noise. At reasonable SNRs the third noise term is much smaller than the second, so we neglect it. Dividing the remaining terms by A yields

$$
\tilde{z} = A + \text{Re}\{\tilde{n}^*(k - 1) + \tilde{n}(k)\} + j\,\text{Im}\{\tilde{n}^*(k - 1) + \tilde{n}(k)\}. \qquad (6.38)
$$

Let us define $x = \text{Re}\{\tilde{z}\}$ and $y = \text{Im}\{\tilde{z}\}$. The phase of $\tilde{z}$ is then given by

$$
\theta_{\tilde{z}} = \tan^{-1} y/x. \qquad (6.39)
$$

Given that the phase difference was zero, an error occurs if $|\theta_{\tilde{z}}| \geq \pi/M$. Determining $p(|\theta_{\tilde{z}}| \geq \pi/M)$ is identical to the case of coherent PSK except that, by (6.38), we have two noise terms instead of one and so the noise power is twice that of the coherent case. This will lead to a performance of differential modulation that is roughly 3 dB worse than that of coherent modulation.

In DPSK modulation we need only consider the in-phase branch of Figure 5.20 when making a decision, so we set $x = \text{Re}\{\tilde{z}\}$ in our analysis. In particular, assuming a zero is transmitted, if $x = A + \text{Re}\{\tilde{n}^*(k - 1) + \tilde{n}(k)\} < 0$ then a decision error is made. This probability can be obtained by finding the characteristic or moment generating function for x, taking the inverse Laplace transform to get the distribution of x, and then integrating over the decision region $x < 0$. This technique is quite general and can be applied to a wide variety of different modulation and detection types in both AWGN and fading [9, Chap. 1.1]: we will use it later to compute the average probability of symbol error for linear modulations in fading both with and without diversity. In DPSK the characteristic function for x is obtained using the general quadratic form of complex Gaussian random variables [1, Apx. B; 12, Apx. B], and the resulting bit error probability is given by

$$
P_b = \tfrac{1}{2}e^{-\gamma_b}. \qquad (6.40)
$$

For DQPSK the characteristic function for $\tilde{z}$ is obtained in [1, Apx. C], which yields the bit error probability

$$
P_b \approx \int_b^\infty x \exp\left[\frac{-(a^2 + x^2)}{2}\right]I_0(ax)\,dx - \frac{1}{2}\exp\left[\frac{-(a^2 + b^2)}{2}\right]I_0(ab), \qquad (6.41)
$$

where $a \approx .765\sqrt{\gamma_b}$ and $b \approx 1.85\sqrt{\gamma_b}$ and where $I_0(x)$ is the modified Bessel function of the first kind and zeroth order.

6.2 Alternate Q-Function Representation

In (6.33) we saw that P_s for many coherent modulation techniques in AWGN is approximated in terms of the Gaussian Q-function. Recall that $Q(z)$ is defined as the probability that a Gaussian random variable X with mean 0 and variance 1 exceeds the value z:

$$Q(z) = p(X \geq z) = \int_z^\infty \frac{1}{\sqrt{2\pi}} e^{-x^2/2} \, dx. \qquad (6.42)$$

The Q-function is not that easy to work with since the argument z is in the lower limit of the integrand, the integrand has infinite range, and the exponential function in the integral doesn't lead to a closed-form solution.

In 1991 an alternate representation of the Q-function was obtained by Craig [13]. The alternate form is given by

$$Q(z) = \frac{1}{\pi} \int_0^{\pi/2} \exp\left[\frac{-z^2}{2\sin^2\phi}\right] d\phi, \quad z > 0. \qquad (6.43)$$

This representation can also be deduced from the work of Weinstein [14] or Pawula et al. [5]. In this alternate form, the integrand is over a finite range that is independent of the function argument z, and the integral is Gaussian with respect to z. These features will prove important in using the alternate representation to derive average error probability in fading.

Craig's motivation for deriving the alternate representation was to simplify the probability of error calculation for AWGN channels. In particular, we can write the probability of bit error for BPSK using the alternate form as

$$P_b = Q(\sqrt{2\gamma_b}) = \frac{1}{\pi} \int_0^{\pi/2} \exp\left[\frac{-\gamma_b}{\sin^2\phi}\right] d\phi. \qquad (6.44)$$

Similarly, the alternate representation can be used to obtain a simple *exact* formula for the P_s of MPSK in AWGN as

$$P_s = \frac{1}{\pi} \int_0^{(M-1)\pi/M} \exp\left[\frac{-g\gamma_s}{\sin^2\phi}\right] d\phi \qquad (6.45)$$

(see [13]), where $g = \sin^2(\pi/M)$. Note that this formula does not correspond to the general form $\alpha_M Q(\sqrt{\beta_M \gamma_s})$: the general form is an approximation, whereas (6.45) is exact. Note also that (6.45) is obtained via a finite-range integral of simple trigonometric functions that is easily computed using a numerical computer package or calculator.

6.3 Fading

In AWGN the probability of symbol error depends on the received SNR or, equivalently, on γ_s. In a fading environment the received signal power varies randomly over distance or time as a result of shadowing and/or multipath fading. Thus, in fading, γ_s is a random variable with distribution $p_{\gamma_s}(\gamma)$ and so $P_s(\gamma_s)$ is also random. The performance metric when γ_s is random depends on the rate of change of the fading. There are three different performance criteria that can be used to characterize the random variable P_s:

- the outage probability, P_{out}, defined as the probability that γ_s falls below a given value corresponding to the maximum allowable P_s;
- the average error probability, $\bar{P}_s$, averaged over the distribution of γ_s;
- combined average error probability and outage, defined as the average error probability that can be achieved some percentage of time or some percentage of spatial locations.

The average probability of symbol error applies when the fading coherence time is on the order of a symbol time ($T_s \approx T_c$), so that the signal fade level is roughly constant over a symbol period. Since many error correction coding techniques can recover from a few bit errors and since end-to-end performance is typically not seriously degraded by a few simultaneous bit errors (since the erroneous bits can be dropped or retransmitted), the average error probability is a reasonably good figure of merit for the channel quality under these conditions.

However, if the signal fading is changing slowly ($T_s \ll T_c$) then a deep fade will affect many simultaneous symbols. Hence fading may lead to large error bursts, which cannot be corrected for with coding of reasonable complexity. Therefore, these error bursts can seriously degrade end-to-end performance. In this case acceptable performance cannot be guaranteed over all time – or, equivalently, throughout a cell – without drastically increasing transmit power. Under these circumstances, an outage probability is specified so that the channel is deemed unusable for some fraction of time or space. Outage and average error probability are often combined when the channel is modeled as a combination of fast and slow fading (e.g., log-normal shadowing with fast Rayleigh fading).

Note that if $T_c \ll T_s$ then the fading will be averaged out by the matched filter in the demodulator. Thus, for very fast fading, performance is the same as in AWGN.

6.3.1 Outage Probability

The outage probability relative to γ_0 is defined as

$$P_{\text{out}} = p(\gamma_s < \gamma_0) = \int_0^{\gamma_0} p_{\gamma_s}(\gamma)\, d\gamma, \tag{6.46}$$

where γ_0 typically specifies the minimum SNR required for acceptable performance. For example, if we consider digitized voice, $P_b = 10^{-3}$ is an acceptable error rate because it generally can't be detected by the human ear. Thus, for a BPSK signal in Rayleigh fading, $\gamma_b < 7$ dB would be declared an outage; hence we set $\gamma_0 = 7$ dB.

In Rayleigh fading the outage probability becomes

$$P_{\text{out}} = \int_0^{\gamma_0} \frac{1}{\bar{\gamma}_s} e^{-\gamma_s/\bar{\gamma}_s}\, d\gamma_s = 1 - e^{-\gamma_0/\bar{\gamma}_s}. \tag{6.47}$$

Inverting this formula shows that, for a given outage probability, the required average SNR $\bar{\gamma}_s$ is

$$\bar{\gamma}_s = \frac{\gamma_0}{-\ln(1 - P_{\text{out}})}. \tag{6.48}$$

In decibels this means that $10 \log \gamma_s$ must exceed the target $10 \log \gamma_0$ by

$$F_d = -10 \log[-\ln(1 - P_{\text{out}})]$$

in order to maintain acceptable performance more than $100 \cdot (1 - P_{\text{out}})$ percent of the time. The quantity F_d is typically called the *dB fade margin*.

EXAMPLE 6.4: Determine the required $\bar{\gamma}_b$ for BPSK modulation in slow Rayleigh fading such that, for 95% of the time (or in 95% of the locations), $P_b(\gamma_b) < 10^{-4}$.

Solution: For BPSK modulation in AWGN the target BER is obtained at $\gamma_b = 8.5$ dB. That is, for $P_b(\gamma_b) = Q(\sqrt{2\gamma_b})$ we have $P_b(10^{.85}) = 10^{-4}$. Thus, $\gamma_0 = 8.5$ dB. We want $P_{\text{out}} = p(\gamma_b < \gamma_0) = .05$, so

$$\bar{\gamma}_b = \frac{\gamma_0}{-\ln(1 - P_{\text{out}})} = \frac{10^{.85}}{-\ln(1 - .05)} = 21.4 \text{ dB}. \tag{6.49}$$

6.3.2 Average Probability of Error

The average probability of error is used as a performance metric when $T_s \approx T_c$. We can therefore assume that γ_s is roughly constant over a symbol time. Then the average probability of error is computed by integrating the error probability in AWGN over the fading distribution:

$$\bar{P}_s = \int_0^\infty P_s(\gamma) p_{\gamma_s}(\gamma) \, d\gamma, \tag{6.50}$$

where $P_s(\gamma)$ is the probability of symbol error in AWGN with SNR γ, which can be approximated by the expressions in Table 6.1. For a given distribution of the fading amplitude r (e.g., Rayleigh, Rician, log-normal), we compute $p_{\gamma_s}(\gamma)$ by making the change of variable

$$p_{\gamma_s}(\gamma) \, d\gamma = p(r) \, dr. \tag{6.51}$$

For example, in Rayleigh fading the received signal amplitude r has the Rayleigh distribution

$$p(r) = \frac{r}{\sigma^2} e^{-r^2/2\sigma^2}, \quad r \geq 0, \tag{6.52}$$

and the signal power is exponentially distributed with mean $2\sigma^2$. The SNR per symbol for a given amplitude r is

$$\gamma = \frac{r^2 T_s}{2\sigma_n^2}, \tag{6.53}$$

where $\sigma_n^2 = N_0/2$ is the PSD of the noise in the in-phase and quadrature branches. Differentiating both sides of this expression yields

$$d\gamma = \frac{r T_s}{\sigma_n^2} \, dr. \tag{6.54}$$

Substituting (6.53) and (6.54) into (6.52) and then (6.51) yields

$$p_{\gamma_s}(\gamma) = \frac{\sigma_n^2}{\sigma^2 T_s} e^{-\gamma \sigma_n^2/\sigma^2 T_s}. \tag{6.55}$$

Since the average SNR per symbol $\bar{\gamma}_s$ is just $\sigma^2 T_s / \sigma_n^2$, we can rewrite (6.55) as

$$p_{\gamma_s}(\gamma) = \frac{1}{\bar{\gamma}_s} e^{-\gamma/\bar{\gamma}_s}, \tag{6.56}$$

which is the exponential distribution. For binary signaling this reduces to

$$p_{\gamma_b}(\gamma) = \frac{1}{\bar{\gamma}_b} e^{-\gamma/\bar{\gamma}_b}. \tag{6.57}$$

Integrating (6.6) over the distribution (6.57) yields the following average probability of error for BPSK in Rayleigh fading:

$$\bar{P}_b = \frac{1}{2}\left[1 - \sqrt{\frac{\bar{\gamma}_b}{1 + \bar{\gamma}_b}}\right] \approx \frac{1}{4\bar{\gamma}_b} \quad \text{(BPSK)}, \tag{6.58}$$

where the approximation holds for large $\bar{\gamma}_b$. A similar integration of (6.31) over (6.57) yields the average probability of error for binary FSK in Rayleigh fading as

$$\bar{P}_b = \frac{1}{2}\left[1 - \sqrt{\frac{\bar{\gamma}_b}{2 + \bar{\gamma}_b}}\right] \approx \frac{1}{4\bar{\gamma}_b} \quad \text{(binary FSK)}. \tag{6.59}$$

Thus, the performance of BPSK and binary FSK converge at high SNRs. For noncoherent modulation, if we assume the channel phase is relatively constant over a symbol time then we obtain the probability of error by again integrating the error probability in AWGN over the fading distribution. For DPSK this yields

$$\bar{P}_b = \frac{1}{2(1 + \bar{\gamma}_b)} \approx \frac{1}{2\bar{\gamma}_b} \quad \text{(DPSK)}, \tag{6.60}$$

where again the approximation holds for large $\bar{\gamma}_b$. Note that, in the limit of large $\bar{\gamma}_b$, there is an approximate 3-dB power penalty in using DPSK instead of BPSK. This was also observed in AWGN and represents the power penalty of differential detection. In practice the power penalty is somewhat smaller, since DPSK can correct for slow phase changes introduced in the channel or receiver, which are not taken into account in these error calculations.

If we use the general approximation $P_s \approx \alpha_M Q(\sqrt{\beta_M \gamma_s})$ then the average probability of symbol error in Rayleigh fading can be approximated as

$$\bar{P}_s \approx \int_0^\infty \alpha_M Q(\sqrt{\beta_M \gamma}) \cdot \frac{1}{\bar{\gamma}_s} e^{-\gamma/\bar{\gamma}_s} \, d\gamma_s = \frac{\alpha_M}{2}\left[1 - \sqrt{\frac{.5\beta_M \bar{\gamma}_s}{1 + .5\beta_M \bar{\gamma}_s}}\right] \approx \frac{\alpha_M}{2\beta_M \bar{\gamma}_s}, \tag{6.61}$$

where the last approximation is in the limit of high SNR.

It is interesting to compare the bit error probabilities of the different modulation schemes in AWGN and in fading. For binary PSK, FSK, and DPSK, the bit error probability in AWGN decreases exponentially with increasing γ_b. However, in fading the bit error probability for all the modulation types decreases just linearly with increasing $\bar{\gamma}_b$. Similar behavior occurs for nonbinary modulation. Thus, the power necessary to maintain a given P_b, particularly for small values, is much higher in fading channels than in AWGN channels. For example,

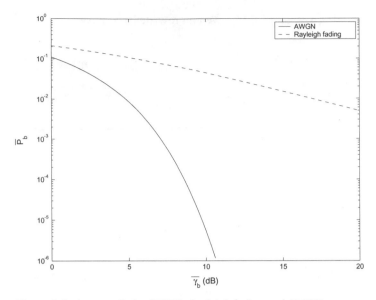

Figure 6.1: Average P_b for BPSK in Rayleigh fading and AWGN.

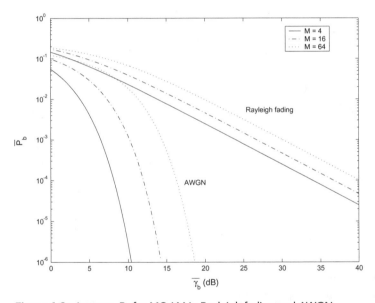

Figure 6.2: Average P_b for MQAM in Rayleigh fading and AWGN.

in Figure 6.1 we plot the error probability of BPSK in AWGN and in flat Rayleigh fading. We see that it requires approximately 8-dB SNR to maintain a 10^{-3} bit error rate in AWGN, whereas it takes approximately 24-dB SNR to maintain the same error rate in fading. A similar plot for the error probabilities of MQAM, based on the approximations (6.24) and (6.61), is shown in Figure 6.2. From these figures it is clear that minimizing transmit power requires some technique to remove the effects of fading. We will discuss some of these techniques – including diversity combining, spread spectrum, and RAKE receivers – in later chapters.

Rayleigh fading is one of the worst-case fading scenarios. In Figure 6.3 we show the average bit error probability of BPSK in Nakagami fading for different values of the Nakagami-m

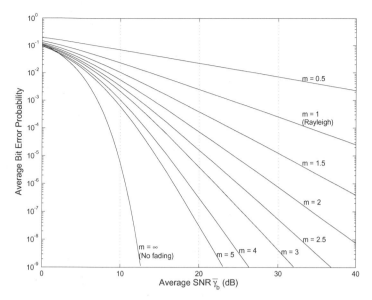

Figure 6.3: Average P_b for BPSK in Nakagami fading.

parameter. We see that, as m increases, the fading decreases and the average bit error probability converges to that of an AWGN channel.

6.3.3 Moment Generating Function Approach to Average Error Probability

The *moment generating function* (MGF) is a useful tool for performance analysis of modulation in fading both with and without diversity. In this section we discuss how it can be used to simplify performance analysis of average probability of symbol error in fading. In the next chapter we will see that it also greatly simplifies analysis in fading channels with diversity.

The MGF for a nonnegative random variable γ with distribution $p_\gamma(\gamma)$, $\gamma \geq 0$, is defined as

$$\mathcal{M}_\gamma(s) = \int_0^\infty p_\gamma(\gamma) e^{s\gamma} \, d\gamma. \tag{6.62}$$

Note that this function is just the Laplace transform of the distribution $p_\gamma(\gamma)$ with the argument reversed in sign: $\mathcal{L}[p_\gamma(\gamma)] = \mathcal{M}_\gamma(-s)$. Thus, the MGF for most fading distributions of interest can be computed either in closed-form using classical Laplace transforms or through numerical integration. In particular, the MGF for common multipath fading distributions are as follows [9, Chap. 5.1].

■ Rayleigh:

$$\mathcal{M}_{\gamma_s}(s) = (1 - s\bar{\gamma}_s)^{-1}. \tag{6.63}$$

■ Rician with factor K:

$$\mathcal{M}_{\gamma_s}(s) = \frac{1+K}{1+K-s\bar{\gamma}_s} \exp\left[\frac{Ks\bar{\gamma}_s}{1+K-s\bar{\gamma}_s}\right]. \tag{6.64}$$

■ Nakagami-m:

$$\mathcal{M}_{\gamma_s}(s) = \left(1 - \frac{s\bar{\gamma}_s}{m}\right)^{-m}. \tag{6.65}$$

As indicated by its name, the moments $\mathbf{E}[\gamma^n]$ of γ can be obtained from $\mathcal{M}_\gamma(s)$ as

$$\mathbf{E}[\gamma^n] = \frac{\partial^n}{\partial s^n}[\mathcal{M}_{\gamma_s}(s)]|_{s=0}. \tag{6.66}$$

The basic premise of the MGF approach for computing average error probability in fading is to express the probability of error P_s in AWGN for the modulation of interest either as an exponential function of γ_s,

$$P_s = c_1 \exp[-c_2 \gamma_s] \tag{6.67}$$

for constants c_1 and c_2, or as a finite-range integral of such an exponential function:

$$P_s = \int_A^B c_1 \exp[-c_2(x)\gamma_s]\,dx, \tag{6.68}$$

where the constant $c_2(x)$ may depend on the integrand but the SNR γ_s does not (and is not in the limits of integration, either). These forms allow the average probability of error to be expressed in terms of the MGF for the fading distribution. Specifically, if $P_s = c_1 \exp[-c_2 \gamma_s]$, then

$$\bar{P}_s = \int_0^\infty c_1 \exp[-c_2 \gamma] p_{\gamma_s}(\gamma)\,d\gamma = c_1 \mathcal{M}_{\gamma_s}(-c_2). \tag{6.69}$$

Since DPSK is in this form with $c_1 = 1/2$ and $c_2 = 1$, we see that the average probability of bit error for DPSK in any type of fading is

$$\bar{P}_b = \tfrac{1}{2}\mathcal{M}_{\gamma_s}(-1), \tag{6.70}$$

where $\mathcal{M}_{\gamma_s}(s)$ is the MGF of the fading distribution. For example, using $\mathcal{M}_{\gamma_s}(s)$ for Rayleigh fading given by (6.63) with $s = -1$ yields $\bar{P}_b = [2(1 + \bar{\gamma}_b)]^{-1}$, which is the same as we obtained in (6.60). If P_s is in the integral form of (6.68) then

$$\begin{aligned}
\bar{P}_s &= \int_0^\infty \int_A^B c_1 \exp[-c_2(x)\gamma]\,dx\, p_{\gamma_s}(\gamma)\,d\gamma \\
&= c_1 \int_A^B \left[\int_0^\infty \exp[-c_2(x)\gamma] p_{\gamma_s}(\gamma)\,d\gamma \right] dx \\
&= c_1 \int_A^B \mathcal{M}_{\gamma_s}(-c_2(x))\,dx. \tag{6.71}
\end{aligned}$$

In this latter case, the average probability of symbol error is a single finite-range integral of the MGF of the fading distribution, which typically can be found in closed form or easily evaluated numerically.

Let us now apply the MGF approach to specific modulations and fading distributions. In (6.33) we gave a general expression for P_s of coherent modulation in AWGN in terms of the Gaussian Q-function. We now make a slight change of notation in (6.33), setting $\alpha = \alpha_M$ and $g = .5\beta_M$ to obtain

$$P_s(\gamma_s) = \alpha Q(\sqrt{2g\gamma_s}), \tag{6.72}$$

where α and g are constants that depend on the modulation. The notation change is to obtain the error probability as an exact MGF, as we now show.

Using the alternate Q-function representation (6.43), we get that

$$P_s = \frac{\alpha}{\pi} \int_0^{\pi/2} \exp\left[\frac{-g\gamma}{\sin^2\phi}\right] d\phi, \tag{6.73}$$

which is in the desired form (6.68). Thus, the average error probability in fading for modulations with $P_s = \alpha Q(\sqrt{2g\gamma_s})$ in AWGN is given by

$$
\begin{aligned}
\bar{P}_s &= \frac{\alpha}{\pi} \int_0^\infty \int_0^{\pi/2} \exp\left[\frac{-g\gamma}{\sin^2\phi}\right] d\phi \; p_{\gamma_s}(\gamma) \, d\gamma \\
&= \frac{\alpha}{\pi} \int_0^{\pi/2} \left[\int_0^\infty \exp\left[\frac{-g\gamma}{\sin^2\phi}\right] p_{\gamma_s}(\gamma) \, d\gamma\right] d\phi \\
&= \frac{\alpha}{\pi} \int_0^{\pi/2} \mathcal{M}_{\gamma_s}\left(\frac{-g}{\sin^2\phi}\right) d\phi, \tag{6.74}
\end{aligned}
$$

where $\mathcal{M}_{\gamma_s}(s)$ is the MGF associated with the distribution $p_{\gamma_s}(\gamma)$ as defined by (6.62). Recall that Table 6.1 approximates the error probability in AWGN for many modulations of interest as $P_s \approx \alpha Q(\sqrt{2g\gamma_s})$, so (6.74) gives an approximation for the average error probability of these modulations in fading. Moreover, the exact average probability of symbol error for coherent MPSK can be obtained in a form similar to (6.74) by noting that Craig's formula for P_s of MPSK in AWGN given by (6.45) is in the desired form (6.68). Thus, the exact average probability of error for MPSK becomes

$$
\begin{aligned}
\bar{P}_s &= \int_0^\infty \frac{1}{\pi} \int_0^{(M-1)\pi/M} \exp\left[\frac{-g\gamma_s}{\sin^2\phi}\right] d\phi \; p_{\gamma_s}(\gamma) \, d\gamma \\
&= \frac{1}{\pi} \int_0^{(M-1)\pi/M} \left[\int_0^\infty \exp\left[\frac{-g\gamma_s}{\sin^2\phi}\right] p_{\gamma_s}(\gamma) \, d\gamma\right] d\phi \\
&= \frac{1}{\pi} \int_0^{(M-1)\pi/M} \mathcal{M}_{\gamma_s}\left(-\frac{g}{\sin^2\phi}\right) d\phi, \tag{6.75}
\end{aligned}
$$

where $g = \sin^2(\pi/M)$ depends on the size of the MPSK constellation. The MGF $\mathcal{M}_{\gamma_s}(s)$ for Rayleigh, Rician, and Nakagami-m distributions were given by (6.63), (6.64), and (6.65), respectively. Substituting $s = -g/\sin^2\phi$ in these expressions yields the following equations.

■ Rayleigh:

$$\mathcal{M}_{\gamma_s}\left(-\frac{g}{\sin^2\phi}\right) = \left(1 + \frac{g\bar{\gamma}_s}{\sin^2\phi}\right)^{-1}. \tag{6.76}$$

■ Rician with factor K:

$$\mathcal{M}_{\gamma_s}\left(-\frac{g}{\sin^2\phi}\right) = \frac{(1+K)\sin^2\phi}{(1+K)\sin^2\phi + g\bar{\gamma}_s} \exp\left[-\frac{Kg\bar{\gamma}_s}{(1+K)\sin^2\phi + g\bar{\gamma}_s}\right]. \tag{6.77}$$

■ Nakagami-m:

$$\mathcal{M}_{\gamma_s}\left(-\frac{g}{\sin^2\phi}\right) = \left(1 + \frac{g\bar{\gamma}_s}{m\sin^2\phi}\right)^{-m}. \tag{6.78}$$

All of these functions are simple trigonometrics and are therefore easy to integrate over the finite range in (6.74) or (6.75).

EXAMPLE 6.5: Use the MGF technique to find an expression for the average probability of error for BPSK modulation in Nakagami fading.

Solution: We use the fact that BPSK for an AWGN channel has $P_b = Q(\sqrt{2\gamma_b})$, so $\alpha = 1$ and $g = 1$ in (6.72). The moment generating function for Nakagami-m fading is given by (6.78), and substituting this into (6.74) with $\alpha = g = 1$ yields

$$\bar{P}_b = \frac{1}{\pi} \int_0^{\pi/2} \left(1 + \frac{\bar{\gamma}_b}{m \sin^2 \phi} \right)^{-m} d\phi.$$

From (6.23) we see that the exact probability of symbol error for MQAM in AWGN contains both the Q-function and its square. Fortunately, an alternate form of $Q^2(z)$ allows us to apply the same techniques used here for MPSK to MQAM modulation. Specifically, an alternate representation of $Q^2(z)$ is derived in [15] as

$$Q^2(z) = \frac{1}{\pi} \int_0^{\pi/4} \exp\left[\frac{-z^2}{2 \sin^2 \phi} \right] d\phi. \tag{6.79}$$

Note that this is identical to the alternate representation for $Q(z)$ given in (6.43) except that the upper limit of the integral is $\pi/4$ instead of $\pi/2$. Thus we can write (6.23) in terms of the alternate representations for $Q(z)$ and $Q^2(z)$ as

$$P_s(\gamma_s) = \frac{4}{\pi} \left(1 - \frac{1}{\sqrt{M}} \right) \int_0^{\pi/2} \exp\left(-\frac{g\gamma_s}{\sin^2 \phi} \right) d\phi$$
$$- \frac{4}{\pi} \left(1 - \frac{1}{\sqrt{M}} \right)^2 \int_0^{\pi/4} \exp\left[-\frac{g\gamma_s}{\sin^2 \phi} \right] d\phi, \tag{6.80}$$

where $g = 1.5/(M - 1)$ is a function of the MQAM constellation size. Then the average probability of symbol error in fading becomes

$$\bar{P}_s = \int_0^\infty P_s(\gamma) p_{\gamma_s}(\gamma) \, d\gamma$$
$$= \frac{4}{\pi} \left(1 - \frac{1}{\sqrt{M}} \right) \int_0^{\pi/2} \int_0^\infty \exp\left[-\frac{g\gamma}{\sin^2 \phi} \right] p_{\gamma_s}(\gamma) \, d\gamma \, d\phi$$
$$- \frac{4}{\pi} \left(1 - \frac{1}{\sqrt{M}} \right)^2 \int_0^{\pi/4} \int_0^\infty \exp\left[-\frac{g\gamma}{\sin^2 \phi} \right] p_{\gamma_s}(\gamma) \, d\gamma \, d\phi$$
$$= \frac{4}{\pi} \left(1 - \frac{1}{\sqrt{M}} \right) \int_0^{\pi/2} \mathcal{M}_{\gamma_s}\left(-\frac{g}{\sin^2 \phi} \right) d\phi$$
$$- \frac{4}{\pi} \left(1 - \frac{1}{\sqrt{M}} \right)^2 \int_0^{\pi/4} \mathcal{M}_{\gamma_s}\left(-\frac{g}{\sin^2 \phi} \right) d\phi. \tag{6.81}$$

Thus, the exact average probability of symbol error is obtained via two finite-range integrals of the MGF over the fading distribution, which typically can be found in closed form or easily evaluated numerically.

The MGF approach can also be applied to noncoherent and differential modulations. For example, consider noncoherent MFSK, with P_s in AWGN given by (6.32), which is a finite sum of the desired form (6.67). Thus, in fading, the average symbol error probability of noncoherent MFSK is given by

$$
\begin{aligned}
\bar{P}_s &= \int_0^\infty \sum_{m=1}^M (-1)^{m+1} \binom{M-1}{m} \frac{1}{m+1} \exp\left[\frac{-m\gamma}{m+1}\right] p_{\gamma_s}(\gamma)\, d\gamma \\
&= \sum_{m=1}^M (-1)^{m+1} \binom{M-1}{m} \frac{1}{m+1} \left[\int_0^\infty \exp\left[\frac{-m\gamma}{m+1}\right] p_{\gamma_s}(\gamma)\, d\gamma\right] \\
&= \sum_{m=1}^M (-1)^{m+1} \binom{M-1}{m} \frac{1}{m+1} \mathcal{M}_{\gamma_s}\left(-\frac{m}{m+1}\right).
\end{aligned}
\tag{6.82}
$$

Finally, for differential MPSK it can be shown [16] that the average probability of symbol error is given by

$$
P_s = \frac{\sqrt{g}}{2\pi} \int_{-\pi/2}^{\pi/2} \frac{\exp\left[-\gamma_s\left(1 - \sqrt{1-g}\cos\theta\right)\right]}{1 - \sqrt{1-g}\cos\theta}\, d\theta
\tag{6.83}
$$

for $g = \sin^2(\pi/M)$, which is in the desired form (6.68). Thus we can express the average probability of symbol error in terms of the MGF of the fading distribution as

$$
\bar{P}_s = \frac{\sqrt{g}}{2\pi} \int_{-\pi/2}^{\pi/2} \frac{\mathcal{M}_{\gamma_s}\left(-\left(1 - \sqrt{1-g}\cos\theta\right)\right)}{1 - \sqrt{1-g}\cos\theta}\, d\theta.
\tag{6.84}
$$

A more extensive discussion of the MGF technique for finding average probability of symbol error for different modulations and fading distributions can be found in [9, Chap. 8.2].

6.3.4 Combined Outage and Average Error Probability

When the fading environment is a superposition of both fast and slow fading (e.g., log-normal shadowing and Rayleigh fading), a common performance metric is combined outage and average error probability, where outage occurs when the slow fading falls below some target value and the average performance in non-outage is obtained by averaging over the fast fading. We use the following notation.

- $\bar{\bar{\gamma}}_s$ denotes the average SNR per symbol for a fixed path loss with averaging over fast fading and shadowing.
- $\bar{\gamma}_s$ denotes the (random) SNR per symbol for a fixed path loss and random shadowing but averaged over fast fading. Its average value, averaged over the shadowing, is $\bar{\bar{\gamma}}_s$.
- γ_s denotes the random SNR due to fixed path loss, shadowing, and multipath. Its average value, averaged over multipath only, is $\bar{\gamma}_s$. Its average value, averaged over both multipath and shadowing, is $\bar{\bar{\gamma}}_s$.

With this notation we can specify an average error probability $\bar{P}_s$ with some probability $1 - P_{\text{out}}$. An outage is declared when the received SNR per symbol due to shadowing and path loss alone, $\bar{\gamma}_s$, falls below a given target value $\bar{\gamma}_{s_0}$. When not in outage ($\bar{\gamma}_s \geq \bar{\gamma}_{s_0}$), the average probability of error is obtained by averaging over the distribution of the fast fading conditioned on the mean SNR:

$$\bar{P}_s = \int_0^\infty P_s(\gamma_s) p(\gamma_s \mid \bar{\gamma}_s) \, d\gamma_s. \tag{6.85}$$

The criterion used to determine the outage target $\bar{\gamma}_{s_0}$ is typically based on a given maximum acceptable average probability of error $\bar{P}_{s_0}$. The target $\bar{\gamma}_{s_0}$ must then satisfy

$$\bar{P}_{s_0} = \int_0^\infty P_s(\gamma_s) p(\gamma_s \mid \bar{\gamma}_{s_0}) \, d\gamma_s. \tag{6.86}$$

It is clear that, whenever $\bar{\gamma}_s > \bar{\gamma}_{s_0}$, the average error probability $\bar{P}_s$ will be below the target maximum value $\bar{P}_{s_0}$.

EXAMPLE 6.6: Consider BPSK modulation in a channel with both log-normal shadowing ($\sigma_{\psi_{\text{dB}}} = 8$ dB) and Rayleigh fading. The desired maximum average error probability is $\bar{P}_{b_0} = 10^{-4}$, which requires $\bar{\gamma}_{b_0} = 34$ dB. Determine the value of $\bar{\bar{\gamma}}_b$ that will ensure $\bar{P}_b \leq 10^{-4}$ with probability $1 - P_{\text{out}} = .95$.

Solution: We must find $\bar{\bar{\gamma}}_b$, the average of γ_b in both the fast and slow fading, such that $p(\bar{\gamma}_b > \bar{\gamma}_{b_0}) = 1 - P_{\text{out}}$. For log-normal shadowing we compute this as

$$p(\bar{\gamma}_b > 34) = p\left(\frac{\bar{\gamma}_b - \bar{\bar{\gamma}}_b}{\sigma_{\psi_{\text{dB}}}} \geq \frac{34 - \bar{\bar{\gamma}}_b}{\sigma_{\psi_{\text{dB}}}}\right) = Q\left(\frac{34 - \bar{\bar{\gamma}}_b}{\sigma_{\psi_{\text{dB}}}}\right) = 1 - P_{\text{out}}, \tag{6.87}$$

since, assuming units in dB, $(\bar{\gamma}_b - \bar{\bar{\gamma}}_b)/\sigma_{\psi_{\text{dB}}}$ is a Gauss-distributed random variable with mean 0 and standard deviation 1. Thus, the value of $\bar{\bar{\gamma}}_b$ is obtained by substituting the values of P_{out} and $\sigma_{\psi_{\text{dB}}}$ in (6.87) and using a table of Q-functions or an inversion program, which yields $(34 - \bar{\bar{\gamma}}_b)/8 = -1.6$ or $\bar{\bar{\gamma}}_b = 46.8$ dB.

6.4 Doppler Spread

One consequence of Doppler spread is an irreducible error floor for modulation techniques using differential detection. This is due to the fact that in differential modulation the signal phase associated with one symbol is used as a phase reference for the next symbol. If the channel phase decorrelates over a symbol, then the phase reference becomes extremely noisy, leading to a high symbol error rate that is independent of received signal power. The phase correlation between symbols and consequent degradation in performance are functions of the Doppler frequency $f_D = v/\lambda$ and the symbol time T_s.

The first analysis of the irreducible error floor due to Doppler was done by Bello and Nelin in [17]. In that work, analytical expressions for the irreducible error floor of noncoherent FSK and DPSK due to Doppler are determined for a Gaussian Doppler power spectrum. However, these expressions are not in closed form, so they must be evaluated numerically. Closed-form expressions for the bit error probability of DPSK in fast Rician fading – where

Table 6.2: Correlation coefficients for different Doppler power spectra models

Type	Doppler power spectrum $S_C(f)$	$\rho_C = A_C(T)/A_C(0)$
Rectangular	$P_0/2B_D, \ \lvert f \rvert < B_D$	$\text{sinc}(2B_D T)$
Gaussian	$(P_0/\sqrt{\pi}B_D)e^{-f^2/B_D^2}$	$e^{-(\pi B_D T)^2}$
Uniform scattering	$P_0/\pi\sqrt{B_D^2 - f^2}, \ \lvert f \rvert < B_D$	$J_0(2\pi B_D T)$
1st-order Butterworth	$P_0 B_D/\pi(f^2 + B_D^2)$	$e^{-2\pi B_D T}$

the channel decorrelates over a bit time – can be obtained using the MGF technique, with the MGF obtained based on the general quadratic form of complex Gaussian random variables [1, Apx. B; 12, Apx. B]. A different approach utilizing alternate forms of the Marcum Q-function can also be used [9, Chap. 8.2.5]. The resulting average bit error probability for DPSK is

$$\bar{P}_b = \frac{1}{2}\left(\frac{1 + K + \bar{\gamma}_b(1 - \rho_C)}{1 + K + \bar{\gamma}_b}\right)\exp\left[-\frac{K\bar{\gamma}_b}{1 + K + \bar{\gamma}_b}\right], \tag{6.88}$$

where ρ_C is the channel correlation coefficient after a bit time T_b, K is the fading parameter of the Rician distribution, and $\bar{\gamma}_b$ is the average SNR per bit. For Rayleigh fading ($K = 0$) this simplifies to

$$\bar{P}_b = \frac{1}{2}\left(\frac{1 + \bar{\gamma}_b(1 - \rho_C)}{1 + \bar{\gamma}_b}\right). \tag{6.89}$$

Letting $\bar{\gamma}_b \to \infty$ in (6.88) yields the irreducible error floor:

$$\bar{P}_{\text{floor}} = \frac{(1 - \rho_C)e^{-K}}{2} \quad \text{(DPSK)}. \tag{6.90}$$

A similar approach is used in [18] to bound the bit error probability of DQPSK in fast Rician fading as

$$\bar{P}_b \leq \frac{1}{2}\left(1 - \sqrt{\frac{\left(\rho_C\bar{\gamma}_s/\sqrt{2}\right)^2}{(\bar{\gamma}_s + 1)^2 - \left(\rho_C\bar{\gamma}_s/\sqrt{2}\right)^2}}\right)\exp\left[-\frac{(2 - \sqrt{2})K\bar{\gamma}_s/2}{(\bar{\gamma}_s + 1) - \left(\rho_C\bar{\gamma}_s/\sqrt{2}\right)}\right], \tag{6.91}$$

where K is as before, ρ_C is the channel correlation coefficient after a symbol time T_s, and $\bar{\gamma}_s$ is the average SNR per symbol. Letting $\bar{\gamma}_s \to \infty$ yields the irreducible error floor:

$$\bar{P}_{\text{floor}} = \frac{1}{2}\left(1 - \sqrt{\frac{\left(\rho_C/\sqrt{2}\right)^2}{1 - \left(\rho_C/\sqrt{2}\right)^2}}\right)\exp\left[-\frac{(2 - \sqrt{2})(K/2)}{1 - \rho_C/\sqrt{2}}\right] \quad \text{(DQPSK)}. \tag{6.92}$$

As discussed in Section 3.2.1, the channel correlation $A_C(\tau)$ over time τ equals the inverse Fourier transform of the Doppler power spectrum $S_C(f)$ as a function of Doppler frequency f. The correlation coefficient is thus $\rho_C = A_C(T)/A_C(0)$ evaluated at $T = T_s$ for DQPSK or at $T = T_b$ for DPSK. Table 6.2, from [19], gives the value of ρ_C for several different Doppler power spectra models, where B_D is the Doppler spread of the channel. Assuming the uniform scattering model ($\rho_C = J_0(2\pi f_D T_b)$) and Rayleigh fading ($K = 0$) in (6.90) yields an irreducible error for DPSK of

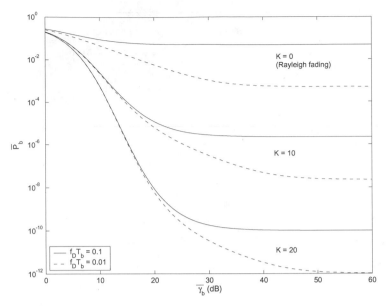

Figure 6.4: Average P_b for DPSK in fast Rician fading with uniform scattering.

$$\bar{P}_{\text{floor}} = \frac{1 - J_0(2\pi f_D T_b)}{2} \approx .5(\pi f_D T_b)^2, \tag{6.93}$$

where $B_D = f_D = v/\lambda$ is the maximum Doppler in the channel. Note that in this expression the error floor decreases with data rate $R = 1/T_b$. This is true in general for irreducible error floors of differential modulation due to Doppler, since the channel has less time to decorrelate between transmitted symbols. This phenomenon is one of the few instances in digital communications where performance improves as data rate increases.

A plot of (6.88), the error probability of DPSK in fast Rician fading, for uniform scattering ($\rho_C = J_0(2\pi f_D T_b)$) and different values of $f_D T_b$ is shown in Figure 6.4. We see from this figure that the error floor starts to dominate at $\bar{\gamma}_b = 15$ dB in Rayleigh fading ($K = 0$), and as K increases the value of $\bar{\gamma}_b$ where the error floor dominates also increases. We also see that increasing the data rate $R_b = 1/T_b$ by an order of magnitude decreases the error floor by roughly two orders of magnitude.

EXAMPLE 6.7: Assume a Rayleigh fading channel with uniform scattering and a maximum Doppler of $f_D = 80$ Hz. For what approximate range of data rates will the irreducible error floor of DPSK be below 10^{-4}?

Solution: We have $\bar{P}_{\text{floor}} \approx .5(\pi f_D T_b)^2 < 10^{-4}$. Solving for T_b with $f_D = 80$ Hz, we get

$$T_b < \frac{\sqrt{2 \cdot 10^{-4}}}{\pi \cdot 80} = 5.63 \cdot 10^{-5},$$

which yields $R > 17.77$ kbps.

Deriving analytical expressions for the irreducible error floor becomes intractable with more complex modulations, in which case simulations are often used. In particular, simulations of the irreducible error floor for $\pi/4$-DQPSK with square-root raised cosine filtering have been conducted (since this modulation is used in the IS-136 TDMA standard) in [20; 21]. These simulation results indicate error floors between 10^{-3} and 10^{-4}. As expected, in these simulations the error floor increases with vehicle speed, since at higher vehicle speeds the channel typically decorrelates more over a given symbol time.

6.5 Intersymbol Interference

Frequency-selective fading gives rise to intersymbol interference, where the received symbol over a given symbol period experiences interference from other symbols that have been delayed by multipath. Since increasing signal power also increases the power of the ISI, this interference gives rise to an irreducible error floor that is independent of signal power. The irreducible error floor is difficult to analyze because it depends on the modulation format and the ISI characteristics, which in turn depend on the characteristics of the channel and the sequence of transmitted symbols.

The first extensive analysis of the degradation in symbol error probability due to ISI was done by Bello and Nelin [22]. In that work, analytical expressions for the irreducible error floor of coherent FSK and noncoherent DPSK are determined assuming a Gaussian delay profile for the channel. To simplify the analysis, only ISI associated with adjacent symbols was taken into account. Even with this simplification, the expressions are complex and must be approximated for evaluation. The irreducible error floor can also be evaluated analytically based on the worst-case sequence of transmitted symbols, or it can be averaged over all possible symbol sequences [23, Chap. 8.2]. These expressions are also complex to evaluate owing to their dependence on the channel and symbol sequence characteristics. An approximation to symbol error probability with ISI can be obtained by treating the ISI as uncorrelated white Gaussian noise [24]. Then the SNR becomes

$$\hat{\gamma}_s = \frac{P_r}{N_0 B + I}, \tag{6.94}$$

where P_r is the received power associated with the line-of-sight signal component, and I is the received power associated with the ISI. In a static channel the resulting probability of symbol error will be $P_s(\hat{\gamma}_s)$, where P_s is the probability of symbol error in AWGN. If both the LOS signal component and the ISI experience flat fading, then $\hat{\gamma}_s$ will be a random variable with distribution $p(\hat{\gamma}_s)$, and the average symbol error probability is then $\bar{P}_s = \int P_s(\hat{\gamma}_s) p(\hat{\gamma}_s) \, d\gamma_s$. Note that $\hat{\gamma}_s$ is the ratio of two random variables – the LOS received power P_r and the ISI received power I – and thus the resulting distribution $p(\hat{\gamma}_s)$ may be hard to obtain and typically is not in closed form.

Irreducible error floors due to ISI are often obtained by simulation, which can easily incorporate different channel models, modulation formats, and symbol sequence characteristics [20; 21; 24; 25; 26]. The most extensive simulations for determining irreducible error floor due to ISI were done by Chuang in [25]. In this work BPSK, DPSK, QPSK, OQPSK and MSK modulations were simulated for different pulse shapes and for channels with different power delay profiles, including a Gaussian, exponential, equal-amplitude two-ray, and empirical power delay profile. The results of [25] indicate that the irreducible error floor is

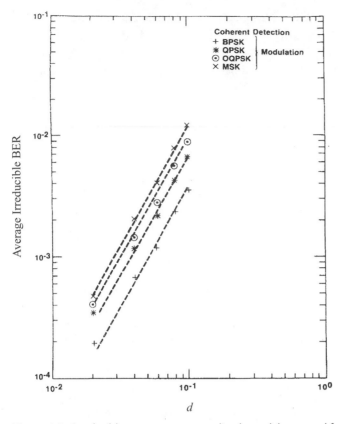

Figure 6.5: Irreducible error versus normalized rms delay spread for Gaussian power delay profile. (Reprinted by permission from [25, Fig. 9], © 1987 IEEE.)

more sensitive to the rms delay spread of the channel than to the shape of its power delay profile. Moreover, pulse shaping can significantly impact the error floor: for the raised cosine pulses discussed in Section 5.5, increasing β from 0 to 1 can reduce the error floor by over an order of magnitude.

An example of Chuang's simulation results is shown in Figure 6.5. This figure plots the irreducible bit error rate as a function of normalized rms delay spread $d = \sigma_{T_m}/T_s$ for BPSK, QPSK, OQPSK, and MSK modulation assuming a static channel with a Gaussian power delay profile. We see from the figure that for all modulations we can approximately bound the irreducible error floor as $P_{\text{floor}} \leq d^2$ for $.02 \leq d \leq .1$. Other simulation results [24] support this bound as well. This bound imposes severe constraints on data rate even when symbol error probabilities on the order of 10^{-2} are acceptable. For example, the rms delay spread in a typical urban environment is approximately $\sigma_{T_m} = 2.5\ \mu$s. To keep $\sigma_{T_m} < .1T_s$ requires that the data rate not exceed 40 kbaud, which generally isn't enough for high-speed data applications. In rural environments, where multipath is not attenuated to the same degree as in cities, $\sigma_{T_m} \approx 25\ \mu$s, which reduces the maximum data rate to 4 kbaud.

EXAMPLE 6.8: Using the approximation $\bar{P}_{\text{floor}} \leq (\sigma_{T_m}/T_s)^2$, find the maximum data rate that can be transmitted through a channel with delay spread $\sigma_{T_m} = 3\ \mu$s, using either BPSK or QPSK modulation, such that the probability of bit error P_b is less than 10^{-3}.

Solution: For BPSK, we set $\bar{P}_{\text{floor}} = (\sigma_{T_m}/T_b)^2$ and so require $T_b \geq \sigma_{T_m}/\sqrt{\bar{P}_{\text{floor}}} = 94.87\ \mu s$, which leads to a data rate of $R = 1/T_b = 10.54$ kbps. For QPSK, the same calculation yields $T_s \geq \sigma_{T_m}/\sqrt{\bar{P}_{\text{floor}}} = 94.87\ \mu s$. Since there are two bits per symbol, this leads to a data rate of $R = 2/T_s = 21.01$ kbps. This indicates that, for a given data rate, QPSK is more robust to ISI than BPSK because its symbol time is slower. The result holds also when using the more accurate error floors associated with Figure 6.5 rather than the bound in this example.

PROBLEMS

6-1. Consider a system in which data is transferred at a rate of 100 bits per second over the channel.

(a) Find the symbol duration if we use a sinc pulse for signaling and the channel bandwidth is 10 kHz.

(b) Suppose the received SNR is 10 dB. Find the SNR per symbol and the SNR per bit if 4-QAM is used.

(c) Find the SNR per symbol and the SNR per bit for 16-QAM, and compare with these metrics for 4-QAM.

6-2. Consider BPSK modulation where the a priori probability of 0 and 1 is not the same. Specifically, $p(s_n = 0) = 0.3$ and $p(s_n = 1) = 0.7$.

(a) Find the probability of bit error P_b in AWGN assuming we encode a 1 as $s_1(t) = A\cos(2\pi f_c t)$ and a 0 as $s_2(t) = -A\cos(2\pi f_c t)$ for $A > 0$, assuming the receiver structure is as shown in Figure 5.17.

(b) Suppose you can change the threshold value in the receiver of Figure 5.17. Find the threshold value that yields equal error probability regardless of which bit is transmitted – that is, the threshold value that yields $p(\hat{m} = 0 \mid m = 1)p(m = 1) = p(\hat{m} = 1 \mid m = 0)p(m = 0)$.

(c) Now suppose we change the modulation so that $s_1(t) = A\cos(2\pi f_c t)$ and $s_2(t) = -B\cos(2\pi f_c t)$. Find $A > 0$ and $B > 0$ so that the receiver of Figure 5.17 with threshold at zero has $p(\hat{m} = 0 \mid m = 1)p(m = 1) = p(\hat{m} = 1 \mid m = 0)p(m = 0)$.

(d) Compute and compare the expression for P_b in parts (a), (b), and (c) assuming $E_b/N_0 = 10$ dB and $N_0 = .1$. For which system is P_b minimized?

6-3. Consider a BPSK receiver whose demodulator has a phase offset of ϕ relative to the transmitted signal, so for a transmitted signal $s(t) = \pm g(t)\cos(2\pi f_c t)$ the carrier in the demodulator of Figure 5.17 is $\cos(2\pi f_c t + \phi)$. Determine the threshold level in the threshold device of Figure 5.17 that minimizes probability of bit error, and find this minimum error probability.

6-4. Assume a BPSK demodulator in which the receiver noise is added after the integrator, as shown in Figure 6.6. The decision device outputs a 1 if its input x has $\text{Re}\{x\} \geq 0$, and a 0 otherwise. Suppose the tone jammer $n(t) = 1.1e^{j\theta}$, where $p(\theta = n\pi/3) = 1/6$ for

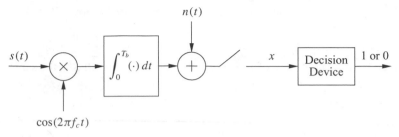

Figure 6.6: BPSK demodulator for Problem 6-4.

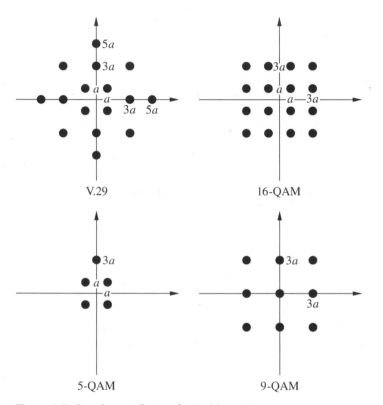

Figure 6.7: Signal constellations for Problem 6-5.

$n = 0, 1, 2, 3, 4, 5$. What is the probability of making a decision error in the decision device (i.e., outputting the wrong demodulated bit), assuming $A_c = \sqrt{2/T_b} = 1$ and that information bits corresponding to a 1 ($s(t) = A_c \cos(2\pi f_c t)$) or a 0 ($s(t) = -A_c \cos(2\pi f_c t)$) are equally likely.

6-5. Find an approximation to P_s for the signal constellations shown in Figure 6.7.

6-6. Plot the exact symbol error probability and the approximation from Table 6.1 of 16-QAM with $0 \le \gamma_s \le 30$ dB. Does the error in the approximation increase or decrease with γ_s? Why?

6-7. Plot the symbol error probability P_s for QPSK using the approximation in Table 6.1 and Craig's exact result for $0 \le \gamma_s \le 30$ dB. Does the error in the approximation increase or decrease with γ_s? Why?

6-8. In this problem we derive an algebraic proof of the alternate representation of the Q-function (6.43) from its original representation (6.42). We will work with the complementary error function (erfc) for simplicity and make the conversion at the end. The $\text{erfc}(x)$ function is traditionally defined by

$$\text{erfc}(x) = \frac{2}{\sqrt{\pi}} \int_x^\infty e^{-t^2} \, dt. \tag{6.95}$$

The alternate representation of this, corresponding to the alternate representation of the Q-function (6.43), is

$$\text{erfc}(x) = \frac{2}{\pi} \int_0^{\pi/2} e^{-x^2/\sin^2\theta} \, d\theta. \tag{6.96}$$

(a) Consider the integral

$$I_x(a) \triangleq \int_0^\infty \frac{e^{-at^2}}{x^2 + t^2} \, dt, \tag{6.97}$$

and show that $I_x(a)$ satisfies the following differential equation:

$$x^2 I_x(a) - \frac{\partial I_x(a)}{\partial a} = \frac{1}{2} \sqrt{\frac{\pi}{a}}. \tag{6.98}$$

(b) Solve the differential equation (6.98) and deduce that

$$I_x(a) \triangleq \int_0^\infty \frac{e^{-at^2}}{x^2 + t^2} \, dt = \frac{\pi}{2x} e^{ax^2} \text{erfc}\left(x\sqrt{a}\right). \tag{6.99}$$

Hint: $I_x(a)$ is a function in two variables x and a. However, since all our manipulations deal with a only, you can assume x to be a constant while solving the differential equation.

(c) Setting $a = 1$ in (6.99) and making a suitable change of variables in the left-hand side of (6.99), derive the alternate representation of the erfc function:

$$\text{erfc}(x) = \frac{2}{\pi} \int_0^{\pi/2} e^{-x^2/\sin^2\theta} \, d\theta.$$

(d) Convert this alternate representation of the erfc function to the alternate representation of the Q-function.

6-9. Consider a communication system that uses BPSK signaling, with average signal power of 100 W and noise power at the receiver of 4 W. Based on its error probability, can this system be used for transmission of data? Can it be used for voice? Now consider the presence of fading with an average SNR $\bar{\gamma}_b = 20$ dB. How do your answers to the previous questions change?

6-10. Consider a cellular system at 900 MHz with a transmission rate of 64 kbps and multipath fading. Explain which performance metric – average probability of error or outage probability – is more appropriate (and why) for user speeds of 1 mph, 10 mph, and 100 mph.

6-11. Derive the expression for the moment generating function for SNR in Rayleigh fading.

6-12. This problem illustrates why satellite systems that must compensate for shadow fading are going bankrupt. Consider an LEO satellite system orbiting 500 km above the earth. Assume that the signal follows a free-space path-loss model with no multipath fading or shadowing. The transmitted signal has a carrier frequency of 900 MHz and a bandwidth of 10 kHz. The handheld receivers have noise power spectral density of 10^{-16} mW/Hz (total noise power is $N_0 B$). Assume nondirectional antennas (0-dB gain) at both the transmitter and receiver. Suppose the satellite must support users in a circular cell on the earth of radius 100 km at a BER of 10^{-6}.

(a) For DPSK modulation, find the transmit power needed for all users in the cell to meet the 10^{-6} BER target.

(b) Repeat part (a) assuming that the channel also experiences log-normal shadowing with $\sigma_{\psi_{dB}} = 8$ dB and that users in a cell must have $P_b = 10^{-6}$ (for each bit) with probability 0.9.

6-13. In this problem we explore the power penalty involved in going from BPSK to the higher-level signal modulation of 16-PSK.

(a) Find the minimum distance between constellation points in 16-PSK modulation as a function of signal energy E_s.

(b) Find α_M and β_M such that the symbol error probability of 16-PSK in AWGN is approximately

$$P_s \approx \alpha_M Q\left(\sqrt{\beta_M \gamma_s}\right).$$

(c) Using your expression in part (b), find an approximation for the average symbol error probability of 16-PSK in Rayleigh fading in terms of $\bar{\gamma}_s$.

(d) Convert the expressions for average symbol error probability of 16-PSK in Rayleigh fading to an expression for average bit error probability, assuming Gray coding.

(e) Find the approximate value of $\bar{\gamma}_b$ required to obtain a BER of 10^{-3} in Rayleigh fading for BPSK and 16-PSK. What is the power penalty in going to the higher-level signal constellation at this BER?

6-14. Find a closed-form expression for the average probability of error for DPSK modulation in Nakagami-m fading. Evalute for $m = 4$ and $\bar{\gamma}_b = 10$ dB.

6-15. The Nakagami distribution is parameterized by m, which ranges from $m = .5$ to $m = \infty$. The m-parameter measures the ratio of LOS signal power to multipath power, so $m = 1$ corresponds to Rayleigh fading, $m = \infty$ corresponds to an AWGN channel with no fading, and $m = .5$ corresponds to fading that results in performance that is worse than with a Rayleigh distribution. In this problem we explore the impact of the parameter m on the performance of BPSK modulation in Nakagami fading.

Plot the average bit error $\bar{P}_b$ of BPSK modulation in Nakagami fading with average SNR ranging from 0 dB to 20 dB for m parameters $m = 1$ (Rayleigh), $m = 2$, and $m = 4$. (The moment generating function technique of Section 6.3.3 should be used to obtain the average error probability.) At an average SNR of 10 dB, what is the difference in average BER?

6-16. Assume a cellular system with log-normal shadowing plus Rayleigh fading. The signal modulation is DPSK. The service provider has determined that it can deal with an outage probability of .01 – that is, 1 in 100 customers can be unhappy at any given time. In non-outage, the voice BER requirement is $\bar{P}_b = 10^{-3}$. Assume a noise PSD $N_0/2$ with $N_0 = 10^{-16}$ mW/Hz, a signal bandwidth of 30 kHz, a carrier frequency of 900 MHz, free-space path-loss propagation with nondirectional antennas, and a shadowing standard deviation of $\sigma_{\psi_{dB}} = 6$ dB. Find the maximum cell size that can achieve this performance if the transmit power at the mobiles is limited to 100 mW.

6-17. Consider a cellular system with circular cells of radius 100 meters. Assume that propagation follows the simplified path-loss model with $K = 1$, $d_0 = 1$ m, and $\gamma = 3$. Assume the signal experiences (in addition to path loss) log-normal shadowing with $\sigma_{\psi_{dB}} = 4$ as well as Rayleigh fading. The transmit power at the base station is $P_t = 100$ mW, the system bandwidth is $B = 30$ kHz, and the noise PSD $N_0/2$ has $N_0 = 10^{-14}$ W/Hz. Assuming BPSK modulation, we want to find the cell coverage area (percentage of locations in the cell) where users have average P_b of less than 10^{-3}.

(a) Find the received power due to path loss at the cell boundary.
(b) Find the minimum average received power (due to path loss and shadowing) such that, with Rayleigh fading about this average, a BPSK modulated signal with this average received power at a given cell location has $\bar{P}_b < 10^{-4}$.
(c) Given the propagation model for this system (simplified path loss, shadowing, and Rayleigh fading), find the percentage of locations in the cell where $\bar{P}_b < 10^{-4}$ under BPSK modulation.

6-18. In this problem we derive the probability of bit error for DPSK in fast Rayleigh fading. By symmetry, the probability of error is the same for transmitting a 0-bit or a 1-bit. Let us assume that over time kT_b a 0-bit is transmitted, so the transmitted symbol at time kT_b is the same as at time $k - 1$: $\mathbf{s}(k) = \mathbf{s}(k - 1)$. In fast fading, the corresponding received symbols are $\mathbf{r}(k - 1) = g_{k-1}\mathbf{s}(k - 1) + n(k - 1)$ and $\mathbf{r}(k) = g_k\mathbf{s}(k - 1) + n(k)$, where g_{k-1} and g_k are the fading channel gains associated with transmissions over times $(k - 1)T_b$ and kT_b.

(a) Show that the decision variable that is input to the phase comparator of Figure 5.20 in order to extract the phase difference is $\mathbf{r}(k)\mathbf{r}^*(k - 1) = g_k g_{k-1}^* + g_k \mathbf{s}(k - 1)^* n_{k-1} + g_{k-1}^* s_{k-1}^* n_k + n_k n_{k-1}^*$.

Assuming a reasonable SNR, the last term $n_k n_{k-1}^*$ of this expression can be neglected. So neglecting this term and then defining $\tilde{n}_k = s_{k-1}^* n_k$ and $\tilde{n}_{k-1} = s_{k-1}^* n_{k-1}$, we get a new random variable $\tilde{z} = g_k g_{k-1}^* + g_k \tilde{n}_{k-1}^* + g_{k-1}^* \tilde{n}_k$. Given that a 0-bit was transmitted over time kT_b, an error is made if $x = \text{Re}\{\tilde{z}\} < 0$, so we must determine the distribution of x. The characteristic function for x is the two-sided Laplace transform of the distribution of x:

$$\Phi_X(s) = \int_{-\infty}^{\infty} p_X(s)e^{-sx}\, dx = \mathbf{E}[e^{-sx}].$$

This function will have a left-plane pole p_1 and a right-plane pole p_2, so it can be written as

$$\Phi_X(s) = \frac{p_1 p_2}{(s - p_1)(s - p_2)}.$$

The left-plane pole p_1 corresponds to the distribution $p_X(x)$ for $x \geq 0$, and the right-plane pole corresponds to the distribution $p_X(x)$ for $x < 0$.

(b) Show through partial fraction expansion that $\Phi_X(s)$ can be written as

$$\Phi_X(s) = \frac{p_1 p_2}{p_1 - p_2}\frac{1}{s - p_1} + \frac{p_1 p_2}{p_2 - p_1}\frac{1}{s - p_2}.$$

An error is made if $x = \text{Re}\{\tilde{z}\} < 0$, so here we need only consider the distribution $p_X(x)$ for $x < 0$ corresponding to the second term of $\Phi_X(s)$.

(c) Show that the inverse Laplace transform of the second term of $\Phi_X(s)$ from part (b) is

$$p_X(x) = \frac{p_1 p_2}{p_2 - p_1} e^{p_2 x}, \quad x < 0.$$

(d) Use part (c) to show that $P_b = -p_1/(p_2 - p_1)$.

In $x = \text{Re}\{\tilde{z}\} = \text{Re}\{g_k g_{k-1}^* + g_k \tilde{n}_{k-1}^* + g_{k-1}^* \tilde{n}_k\}$, the channel gains g_k, g_{k-1} and noises $\tilde{n}_k, \tilde{n}_{k-1}$ are complex Gaussian random variables. Thus, the poles p_1, p_2 in $p_X(x)$ are derived using the general quadratic form of complex Gaussian random variables [1, Apx. B; 12, Apx. B] as

$$p_1 = \frac{-1}{N_0(\bar{\gamma}_b[1 + \rho_c] + 1)} \quad \text{and} \quad p_2 = \frac{1}{N_0(\bar{\gamma}_b[1 - \rho_c] + 1)}$$

for ρ_C the correlation coefficient of the channel over the bit time T_b.

(e) Find a general expression for P_b in fast Rayleigh fading using these values of p_1 and p_2 in the P_b expression from part (d).

(f) Show that this reduces to the average probability of error $\bar{P}_b = 1/2(1 + \bar{\gamma}_b)$ for a slowly fading channel that does not decorrelate over a bit time.

6-19. Plot the bit error probability for DPSK in fast Rayleigh fading for $\bar{\gamma}_b$ ranging from 0 dB to 60 dB and $\rho_C = J_0(2\pi B_D T)$ with $B_D T = .01, .001,$ and $.0001$. For each value of $B_D T$, at approximately what value of $\bar{\gamma}_b$ does the error floor dominate the error probability?

6-20. Find the irreducible error floor due to Doppler for DQPSK modulation with a data rate of 40 kbps, assuming a Gaussian Doppler power spectrum with $B_D = 80$ Hz and Rician fading with $K = 2$.

6-21. Consider a wireless channel with an average delay spread of 100 ns and a Doppler spread of 80 Hz. Given the error floors due to Doppler and ISI – and assuming DQPSK modulation in Rayleigh fading and uniform scattering – approximately what range of data rates can be transmitted over this channel with a BER of less than 10^{-4}?

6-22. Using the error floors of Figure 6.5, find the maximum data rate that can be transmitted through a channel with delay spread $\sigma_{T_m} = 3 \mu s$ (using BPSK, QPSK, or MSK modulation) such that the probability of bit error P_b is less than 10^{-3}.

REFERENCES

[1]   J. G. Proakis, *Digital Communications,* 4th ed., McGraw-Hill, New York, 2001.

[2]   M. K. Simon, S. M. Hinedi, and W. C. Lindsey, *Digital Communication Techniques: Signal Design and Detection,* Prentice-Hall, Englewood Cliffs, NJ, 1995.

[3]   S. Rhodes, "Effect of noisy phase reference on coherent detection of offset-QPSK signals," *IEEE Trans. Commun.,* pp. 1046–55, August 1974.

[4]   N. R. Sollenberger and J. C.-I. Chuang, "Low-overhead symbol timing and carrier recovery for portable TDMA radio systems," *IEEE Trans. Commun.,* pp. 1886–92, October 1990.

[5]   R. Pawula, S. Rice, and J. Roberts, "Distribution of the phase angle between two vectors perturbed by Gaussian noise," *IEEE Trans. Commun.,* pp. 1828–41, August 1982.

[6]   W. Cowley and L. Sabel, "The performance of two symbol timing recovery algorithms for PSK demodulators," *IEEE Trans. Commun.,* pp. 2345–55, June 1994.

[7]   W. T. Webb and L. Hanzo, *Modern Quadrature Amplitude Modulation,* IEEE/Pentech Press, London, 1994.

[8]   X. Tang, M.-S. Alouini, and A. Goldsmith, "Effect of channel estimation error on M-QAM BER performance in Rayleigh fading," *IEEE Trans. Commun.*, pp. 1856–64, December 1999.

[9]   M. K. Simon and M.-S. Alouini, *Digital Communication over Fading Channels: A Unified Approach to Performance Analysis*, Wiley, New York, 2000.

[10] S. Hinedi, M. Simon, and D. Raphaeli, "The performance of noncoherent orthogonal M-FSK in the presence of timing and frequency errors," *IEEE Trans. Commun.*, pp. 922–33, February–April 1995.

[11] E. Grayver and B. Daneshrad, "A low-power all-digital FSK receiver for deep space applications," *IEEE Trans. Commun.*, pp. 911–21, May 2001.

[12] M. Schwartz, W. R. Bennett, and S. Stein, *Communication Systems and Techniques*, McGraw-Hill, New York, 1966 [reprinted 1995 by Wiley/IEEE Press].

[13] J. Craig, "New, simple and exact result for calculating the probability of error for two-dimensional signal constellations," *Proc. Military Commun. Conf.*, pp. 25.5.1–25.5.5, November 1991.

[14] F. S. Weinstein, "Simplified relationships for the probability distribution of the phase of a sine wave in narrow-band normal noise," *IEEE Trans. Inform. Theory*, pp. 658–61, September 1974.

[15] M. K. Simon and D. Divsalar, "Some new twists to problems involving the Gaussian probability integral," *IEEE Trans. Commun.*, pp. 200–10, February 1998.

[16] R. F. Pawula, "A new formula for MDPSK symbol error probability," *IEEE Commun. Lett.*, pp. 271–2, October 1998.

[17] P. A. Bello and B. D. Nelin, "The influence of fading spectrum on the bit error probabilities of incoherent and differentially coherent matched filter receivers," *IEEE Trans. Commun. Syst.*, pp. 160–8, June 1962.

[18] P. Y. Kam, "Tight bounds on the bit-error probabilities of 2DPSK and 4DPSK in nonselective Rician fading," *IEEE Trans. Commun.*, pp. 860–2, July 1998.

[19] P. Y. Kam, "Bit error probabilities of MDPSK over the nonselective Rayleigh fading channel with diversity reception," *IEEE Trans. Commun.*, pp. 220–4, February 1991.

[20] V. Fung, R. S. Rappaport, and B. Thoma, "Bit error simulation for $\pi/4$ DQPSK mobile radio communication using two-ray and measurement based impulse response models," *IEEE J. Sel. Areas Commun.*, pp. 393–405, April 1993.

[21] S. Chennakeshu and G. J. Saulnier, "Differential detection of $\pi/4$-shifted-DQPSK for digital cellular radio," *IEEE Trans. Veh. Tech.*, pp. 46–57, February 1993.

[22] P. A. Bello and B. D. Nelin, "The effects of frequency selective fading on the binary error probabilities of incoherent and differentially coherent matched filter receivers," *IEEE Trans. Commun. Syst.*, pp. 170–86, June 1963.

[23] M. B. Pursley, *Introduction to Digital Communications*, Prentice-Hall, Englewood Cliffs, NJ, 2005.

[24] S. Gurunathan and K. Feher, "Multipath simulation models for mobile radio channels," *Proc. IEEE Veh. Tech. Conf.*, pp. 131–4, May 1992.

[25] J. C.-I. Chuang, "The effects of time delay spread on portable radio communications channels with digital modulation," *IEEE J. Sel. Areas Commun.*, pp. 879–89, June 1987.

[26] C. Liu and K. Feher, "Bit error rate performance of $\pi/4$ DQPSK in a frequency selective fast Rayleigh fading channel," *IEEE Trans. Veh. Tech.*, pp. 558–68, August 1991.

Diversity

In Chapter 6 we saw that both Rayleigh fading and log-normal shadowing exact a large power penalty on the performance of modulation over wireless channels. One of the best techniques to mitigate the effects of fading is diversity combining of independently fading signal paths. Diversity combining exploits the fact that independent signal paths have a low probability of experiencing deep fades simultaneously. Thus, the idea behind diversity is to send the same data over independent fading paths. These independent paths are combined in such a way that the fading of the resultant signal is reduced. For example, consider a system with two antennas at either the transmitter or receiver that experience independent fading. If the antennas are spaced sufficiently far apart, it is unlikely that they both experience deep fades at the same time. By selecting the antenna with the strongest signal, a technique known as *selection combining,* we obtain a much better signal than if we had just one antenna. This chapter focuses on common methods used at the transmitter and receiver to achieve diversity. Other diversity techniques that have potential benefits beyond these schemes in terms of performance or complexity are discussed in [1, Chap. 9.10].

Diversity techniques that mitigate the effect of multipath fading are called *microdiversity,* and that is the focus of this chapter. Diversity to mitigate the effects of shadowing from buildings and objects is called *macrodiversity.* Macrodiversity is generally implemented by combining signals received by several base stations or access points, which requires coordination among these different stations or points. Such coordination is implemented as part of the networking protocols in infrastructure-based wireless networks. We will therefore defer discussion of macrodiversity until Chapter 15, where we examine the design of such networks.

7.1 Realization of Independent Fading Paths

There are many ways of achieving independent fading paths in a wireless system. One method is to use multiple transmit or receive antennas, also called an antenna array, where the elements of the array are separated in distance. This type of diversity is referred to as *space diversity.* Note that with receiver space diversity, independent fading paths are realized without an increase in transmit signal power or bandwidth. Moreover, coherent combining of the diversity signals increases the signal-to-noise power ratio at the receiver over the SNR that would be obtained with just a single receive antenna. This SNR increase, called array gain, can also be obtained with transmitter space diversity by appropriately weighting the antenna transmit powers relative to the channel gains. In addition to array gain, space diversity also provides diversity gain, defined as the change in slope of the error probability resulting from the diversity combining. We will describe both the array gain and diversity gain for specific diversity-combining techniques in subsequent sections.

The maximum diversity gain for either transmitter or receiver space diversity typically requires that the separation between antennas be such that the fading amplitudes correspond-ing to each antenna are approximately independent. For example, from equation (3.26) we know that, in a uniform scattering environment with omnidirectional transmit and receive antennas, the minimum antenna separation required for independent fading on each antenna is approximately one half-wavelength (.38λ, to be exact). If the transmit or receive anten-nas are directional (which is common at the base station if the system has cell sectorization), then the multipath is confined to a small angle relative to the LOS ray, which means that a larger antenna separation is required to obtain independent fading samples [2].

A second method of achieving diversity is by using either two transmit antennas or two re-ceive antennas with different polarization (e.g., vertically and horizontally polarized waves). The two transmitted waves follow the same path. However, since the multiple random reflec-tions distribute the power nearly equally relative to both polarizations, the average receive power corresponding to either polarized antenna is approximately the same. Since the scatter-ing angle relative to each polarization is random, it is highly improbable that signals received on the two differently polarized antennas would be simultaneously in a deep fade. There are two disadvantages of polarization diversity. First, you can have at most two diversity branches, corresponding to the two types of polarization. The second disadvantage is that polarization diversity loses effectively half the power (3 dB) because the transmit or receive power is divided between the two differently polarized antennas.

Directional antennas provide angle (or directional) diversity by restricting the receive an-tenna beamwidth to a given angle. In the extreme, if the angle is very small then at most one of the multipath rays will fall within the receive beamwidth, so there is no multipath fading from multiple rays. However, this diversity technique requires either a sufficient number of directional antennas to span all possible directions of arrival or a single antenna whose di-rectivity can be steered to the arrival angle of one of the multipath components (preferably the strongest one). Note also that with this technique the SNR may decrease owing to the loss of multipath components that fall outside the receive antenna beamwidth – unless the directional gain of the antenna is sufficiently large to compensate for this lost power. *Smart antennas* are antenna arrays with adjustable phase at each antenna element: such arrays form directional antennas that can be steered to the incoming angle of the strongest multipath com-ponent [3].

Frequency diversity is achieved by transmitting the same narrowband signal at differ-ent carrier frequencies, where the carriers are separated by the coherence bandwidth of the channel. This technique requires additional transmit power to send the signal over multiple frequency bands. Spread-spectrum techniques, discussed in Chapter 13, are sometimes de-scribed as providing frequency diversity in that the channel gain varies across the bandwidth of the transmitted signal. However, this is not equivalent to sending the same information signal over independently fading paths. As discussed in Section 13.2.4, spread spectrum with RAKE reception does provide independently fading paths of the information signal and thus is a form of path diversity. Time diversity is achieved by transmitting the same signal at different times, where the time difference is greater than the channel coherence time (the inverse of the channel Doppler spread). Time diversity does not require increased transmit power but it does lower the data rate, since data is repeated in the diversity time slots rather than sending new data in those time slots. Time diversity can also be achieved through coding and interleaving, as will be discussed in Chapter 8. Clearly time diversity cannot be used for stationary applications, since the channel coherence time is infinite and thus fading is highly correlated over time.

In this chapter we focus on space diversity as a reference when describing diversity systems and the different combining techniques, although the techniques can be applied to any type of diversity. Thus, the combining techniques will be defined as operations on an antenna array. Receiver and transmitter diversity are treated separately, since their respective system models and combining techniques have important differences.

7.2 Receiver Diversity

7.2.1 System Model

In receiver diversity, the independent fading paths associated with multiple receive antennas are combined to obtain a signal that is then passed through a standard demodulator. The combining can be done in several ways, which vary in complexity and overall performance. Most combining techniques are linear: the output of the combiner is just a weighted sum of the different fading paths or *branches,* as shown in Figure 7.1 for M-branch diversity. Specifically, when all but one of the complex α_i are zero, only one path is passed to the combiner output. If more than one of the α_i are nonzero then the combiner adds together multiple paths, though each path may be weighted by a different value. Combining more than one branch signal requires *co-phasing,* where the phase θ_i of the ith branch is removed through multiplication by $\alpha_i = a_i e^{-j\theta_i}$ for some real-valued a_i. This phase removal requires coherent detection of each branch to determine its phase θ_i. Without co-phasing, the branch signals would not add up coherently in the combiner, so the resulting output could still exhibit significant fading due to constructive and destructive addition of the signals in all the branches.

The multiplication by α_i can be performed either before detection (predetection) or after detection (postdetection) with essentially no difference in performance. Combining is typically performed postdetection, since the branch signal power and/or phase is required to determine the appropriate α_i value. Postdetection combining of multiple branches requires a dedicated receiver for each branch to determine the branch phase, which increases the hardware complexity and power consumption, particularly for a large number of branches.

The main purpose of diversity is to coherently combine the independent fading paths so that the effects of fading are mitigated. The signal output from the combiner equals the original transmitted signal $s(t)$ multiplied by a random complex amplitude term $\alpha_\Sigma = \sum_i a_i r_i$. This complex amplitude term results in a random SNR γ_Σ at the combiner output, where the distribution of γ_Σ is a function of the number of diversity paths, the fading distribution on each path, and the combining technique, as we shall describe in more detail.

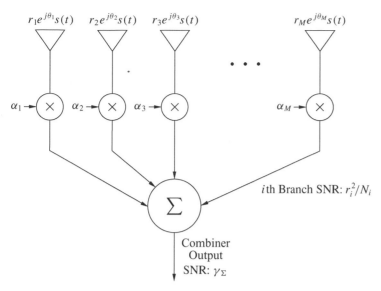

Figure 7.1: Linear combiner.

The array gain in receiver space diversity results from coherent combining of multiple receive signals. Even in the absence of fading, this can lead to an increase in average received SNR. For example, suppose there is no fading so that $r_i = \sqrt{E_s}$ for E_s the energy per symbol of the transmitted signal. Assume identical noise PSD (power spectral density) $N_0/2$ on each branch and pulse shaping such that $BT_s = 1$. Then each branch has the same SNR $\gamma_i = E_s/N_0$. Let us set $a_i = r_i/\sqrt{N_0}$ (we will see later that these weights are optimal for maximal-ratio combining in fading). Then the received SNR is

$$\gamma_\Sigma = \frac{\left(\sum_{i=1}^{M} a_i r_i\right)^2}{N_0 \sum_{i=1}^{M} a_i^2} = \frac{\left(\sum_{i=1}^{M} \dfrac{E_s}{\sqrt{N_0}}\right)^2}{N_0 \sum_{i=1}^{M} \dfrac{E_s}{N_0}} = \frac{ME_s}{N_0}. \tag{7.1}$$

Thus, in the absence of fading, with appropriate weighting there is an M-fold increase in SNR due to the coherent combining of the M signals received from the different antennas. This SNR increase in the absence of fading is referred to as the *array gain*. More precisely, array gain A_g is defined as the increase in the average combined SNR $\bar{\gamma}_\Sigma$ over the average branch SNR $\bar{\gamma}$:

$$A_g = \frac{\bar{\gamma}_\Sigma}{\bar{\gamma}}.$$

Array gain occurs for all diversity combining techniques but is most pronounced in maximal-ratio combining. The array gain allows a system with multiple transmit or receive antennas in a fading channel to achieve better performance than a system without diversity in an AWGN channel with the same average SNR. We will see this effect in performance curves for both maximal-ratio and equal-gain combining with a large number of antennas.

With fading, the combining of multiple independent fading paths leads to a more favorable distribution for γ_Σ than would be the case with just a single path. In particular, the performance of a diversity system (whether it uses space diversity or another form of diversity) in terms of $\bar{P}_s$ and P_{out} is as defined in Sections 6.3.1 and 6.3.2:

$$\bar{P}_s = \int_0^\infty P_s(\gamma) p_{\gamma_\Sigma}(\gamma) \, d\gamma, \tag{7.2}$$

where $P_s(\gamma)$ is the probability of symbol error for demodulation of $s(t)$ in AWGN with SNR γ; and

$$P_{\text{out}} = p(\gamma_\Sigma \leq \gamma_0) = \int_0^{\gamma_0} p_{\gamma_\Sigma}(\gamma) \, d\gamma \tag{7.3}$$

for some target SNR value γ_0. The more favorable distribution for γ_Σ leads to a decrease in $\bar{P}_s$ and P_{out} as a result of diversity combining, and the resulting performance advantage is called the *diversity gain*. In particular, the average probability of error for some diversity systems can be expressed in the form $\bar{P}_s = c\bar{\gamma}^{-M}$, where c is a constant that depends on the specific modulation and coding, $\bar{\gamma}$ is the average received SNR per branch, and M is the *diversity order* of the system. The diversity order indicates how the *slope* of the average probability of error as a function of average SNR changes with diversity. (Figures 7.3 and 7.6 below show these slope changes as a function of M for different combining techniques.) Recall from (6.61) that a general approximation for average error probability in Rayleigh fading with no diversity is $\bar{P}_s \approx \alpha_M/(2\beta_M\bar{\gamma})$. This expression has a diversity order of 1, which is consistent with a single receive antenna. The maximum diversity order of a system with M antennas is M, and when the diversity order equals M the system is said to achieve *full diversity order*.

In the following sections we will describe the different combining techniques and their performance in more detail. These techniques entail various trade-offs between performance and complexity.

7.2.2 Selection Combining

In selection combining (SC), the combiner outputs the signal on the branch with the highest SNR r_i^2/N_i. This is equivalent to choosing the branch with the highest $r_i^2 + N_i$ if the noise power $N_i = N$ is the same on all branches.[1] Because only one branch is used at a time, SC often requires just one receiver that is switched into the active antenna branch. However, a dedicated receiver on each antenna branch may be needed for systems that transmit continuously in order to monitor SNR on each branch simultaneously. With SC, the path output from the combiner has an SNR equal to the maximum SNR of all the branches. Moreover, since only one branch output is used, co-phasing of multiple branches is not required; hence this technique can be used with either coherent or differential modulation.

For M-branch diversity, the cumulative distribution function (cdf) of γ_Σ is given by

$$P_{\gamma_\Sigma}(\gamma) = p(\gamma_\Sigma < \gamma) = p(\max[\gamma_1, \gamma_2, \ldots, \gamma_M] < \gamma) = \prod_{i=1}^M p(\gamma_i < \gamma). \tag{7.4}$$

---

[1] In practice $r_i^2 + N_i$ is easier to measure than SNR, since the former involves finding only the total power in the received signal.

We obtain the distribution of γ_Σ by differentiating $P_{\gamma_\Sigma}(\gamma)$ relative to γ, and we obtain the outage probability by evaluating $P_{\gamma_\Sigma}(\gamma)$ at $\gamma = \gamma_0$. Assume that we have M branches with uncorrelated Rayleigh fading amplitudes r_i. The instantaneous SNR on the ith branch is therefore given by $\gamma_i = r_i^2/N$. Defining the average SNR on the ith branch as $\bar{\gamma}_i = \mathbf{E}[\gamma_i]$, the SNR distribution will be exponential:

$$p(\gamma_i) = \frac{1}{\bar{\gamma}_i}e^{-\gamma_i/\bar{\gamma}_i}. \tag{7.5}$$

From (6.47), the outage probability for a target γ_0 on the ith branch in Rayleigh fading is

$$P_{\text{out}}(\gamma_0) = 1 - e^{-\gamma_0/\bar{\gamma}_i}. \tag{7.6}$$

The outage probability of the selection combiner for the target γ_0 is then

$$P_{\text{out}}(\gamma_0) = \prod_{i=1}^{M} p(\gamma_i < \gamma_0) = \prod_{i=1}^{M} [1 - e^{-\gamma_0/\bar{\gamma}_i}]. \tag{7.7}$$

If the average SNR for all of the branches are the same ($\bar{\gamma}_i = \bar{\gamma}$ for all i), then this reduces to

$$P_{\text{out}}(\gamma_0) = p(\gamma_\Sigma < \gamma_0) = [1 - e^{-\gamma_0/\bar{\gamma}}]^M. \tag{7.8}$$

Differentiating (7.8) relative to γ_0 yields the distribution for γ_Σ:

$$p_{\gamma_\Sigma}(\gamma) = \frac{M}{\bar{\gamma}}[1 - e^{-\gamma/\bar{\gamma}}]^{M-1}e^{-\gamma/\bar{\gamma}}. \tag{7.9}$$

From (7.9) we see that the average SNR of the combiner output in independent and identically distributed Rayleigh fading is

$$\begin{aligned}
\bar{\gamma}_\Sigma &= \int_0^\infty \gamma p_{\gamma_\Sigma}(\gamma)\,d\gamma \\
&= \int_0^\infty \frac{\gamma M}{\bar{\gamma}}[1 - e^{-\gamma/\bar{\gamma}}]^{M-1}e^{-\gamma/\bar{\gamma}}\,d\gamma \\
&= \bar{\gamma}\sum_{i=1}^{M}\frac{1}{i}.
\end{aligned} \tag{7.10}$$

Thus, the average SNR gain and corresponding array gain increase with M, but not linearly. The biggest gain is obtained by going from no diversity to two-branch diversity. Increasing the number of diversity branches from two to three will give much less gain than going from one to two, and in general increasing M yields diminishing returns in terms of the array gain. This trend is also illustrated in Figure 7.2, which shows P_{out} versus $\bar{\gamma}/\gamma_0$ for different M in i.i.d. Rayleigh fading. We see that there is dramatic improvement even with two-branch selection combining: going from $M = 1$ to $M = 2$ at 1% outage probability yields an approximate 12-dB reduction in required SNR, and at .01% outage probability there is an approximate 20-dB reduction in required SNR. However, at .01% outage, going from two-branch to three-branch diversity results in an additional reduction of about 7 dB,

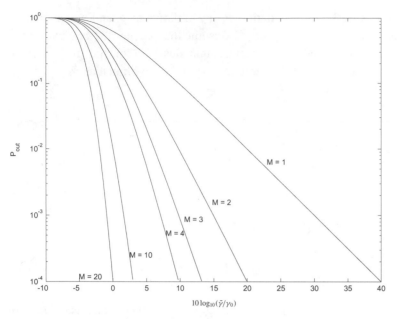

Figure 7.2: Outage probability of selection combining in Rayleigh fading.

and going from three-branch to four-branch results in an additional reduction of about 4 dB. Clearly the power savings is most substantial when going from no diversity to two-branch diversity, with diminishing returns as the number of branches is increased. It should be noted also that, even with Rayleigh fading on all branches, the distribution of the combiner output SNR is no longer exponential.

EXAMPLE 7.1: Find the outage probability of BPSK modulation at $P_b = 10^{-3}$ for a Rayleigh fading channel with SC diversity for $M = 1$ (no diversity), $M = 2$, and $M = 3$. Assume equal branch SNRs of $\bar{\gamma} = 15$ dB.

Solution: A BPSK modulated signal with $\gamma_b = 7$ dB has $P_b = 10^{-3}$. Thus, we have $\gamma_0 = 7$ dB. Substituting $\gamma_0 = 10^{.7}$ and $\bar{\gamma} = 10^{1.5}$ into (7.8) yields $P_{\text{out}} = .1466$ for $M = 1$, $P_{\text{out}} = .0215$ for $M = 2$, and $P_{\text{out}} = .0031$ for $M = 3$. We see that each additional branch reduces outage probability by almost an order of magnitude.

The average probability of symbol error is obtained from (7.2) with $P_s(\gamma)$ the probability of symbol error in AWGN for the signal modulation and $p_{\gamma_\Sigma}(\gamma)$ the distribution of the combiner SNR. For most fading distributions and coherent modulations, this result cannot be obtained in closed form and must be evaluated numerically or by approximation. In Figure 7.3 we plot $\bar{P}_b$ versus $\bar{\gamma}_b$ in i.i.d. Rayleigh fading, obtained by a numerical evaluation of $\int Q(\sqrt{2\gamma})p_{\gamma_\Sigma}(\gamma)\,d\gamma$ for $p_{\gamma_\Sigma}(\gamma)$ given by (7.9). The figure shows that the diversity system for $M \geq 8$ has a lower error probability than an AWGN channel with the same SNR, owing to the array gain of the combiner. The same will be true for the performance of maximal-ratio and equal-gain combining. Closed-form results do exist for differential modulation under i.i.d. Rayleigh fading on each branch [1, Chap. 9.7; 4, Chap. 6.1]. For example, it can be

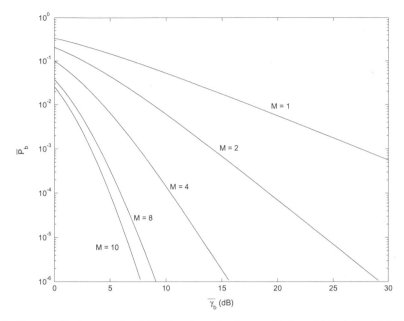

Figure 7.3: Average P_b of BPSK under selection combining with i.i.d. Rayleigh fading.

shown that – for DPSK with $p_{\gamma_\Sigma}(\gamma)$ given by (7.9) – the average probability of symbol error is given by

$$\bar{P}_b = \int_0^\infty \frac{1}{2} e^{-\gamma} p_{\gamma_\Sigma}(\gamma) \, d\gamma = \frac{M}{2} \sum_{m=0}^{M-1} (-1)^m \frac{\binom{M-1}{m}}{1+m+\bar{\gamma}}. \qquad (7.11)$$

In the foregoing derivations we assume that there is no correlation between the branch amplitudes. If the correlation is nonzero then there is a slight degradation in performance, which is almost negligible for correlations below .5. Derivation of the exact performance degradation due to branch correlation can be found in [1, Chap. 9.7; 2].

7.2.3 Threshold Combining

Selection combining for systems that transmit continuously may require a dedicated receiver on each branch to continuously monitor branch SNR. A simpler type of combining, called threshold combining, avoids the need for a dedicated receiver on each branch by scanning each of the branches in sequential order and outputting the first signal whose SNR is above a given threshold γ_T. As in SC, co-phasing is not required because only one branch output is used at a time. Hence this technique can be used with either coherent or differential modulation.

Once a branch is chosen, the combiner outputs that signal as long as the SNR on that branch remains above the desired threshold. If the SNR on the selected branch falls below the threshold, the combiner switches to another branch. There are several criteria the combiner can use for determining which branch to switch to [5]. The simplest criterion is to switch randomly to another branch. With only two-branch diversity this is equivalent to switching to the other branch when the SNR on the active branch falls below γ_T. This method is called *switch-and-stay combining* (SSC). The switching process and SNR associated with SSC is

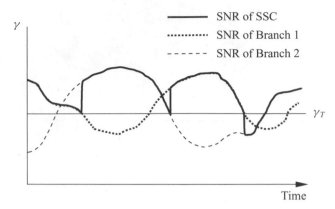

Figure 7.4: SNR of the SSC technique.

illustrated in Figure 7.4. Since the SSC does not select the branch with the highest SNR, its performance is between that of no diversity and ideal SC.

Let us denote the SNR on the ith branch by γ_i and the SNR of the combiner output by γ_Σ. The cdf of γ_Σ will depend on the threshold level γ_T and the cdf of γ_i. For two-branch diversity with i.i.d. branch statistics, the cdf of the combiner output $P_{\gamma_\Sigma}(\gamma) = p(\gamma_\Sigma \leq \gamma)$ can be expressed in terms of the cdf $P_{\gamma_i}(\gamma) = p(\gamma_i \leq \gamma)$ and the distribution $p_{\gamma_i}(\gamma)$ of the individual branch SNRs as

$$P_{\gamma_\Sigma}(\gamma) = \begin{cases} P_{\gamma_1}(\gamma_T) P_{\gamma_2}(\gamma) & \gamma < \gamma_T, \\ p(\gamma_T \leq \gamma_2 \leq \gamma) + P_{\gamma_1}(\gamma_T) P_{\gamma_2}(\gamma) & \gamma \geq \gamma_T. \end{cases} \tag{7.12}$$

For Rayleigh fading in each branch with $\bar{\gamma}_i = \bar{\gamma}$ $(i = 1, 2)$, this yields

$$P_{\gamma_\Sigma}(\gamma) = \begin{cases} 1 - e^{-\gamma_T/\bar{\gamma}} - e^{-\gamma/\bar{\gamma}} + e^{-(\gamma_T+\gamma)/\bar{\gamma}} & \gamma < \gamma_T, \\ 1 - 2e^{-\gamma/\bar{\gamma}} + e^{-(\gamma_T+\gamma)/\bar{\gamma}} & \gamma \geq \gamma_T. \end{cases} \tag{7.13}$$

The outage probability P_{out} associated with a given γ_0 is obtained by evaluating $P_{\gamma_\Sigma}(\gamma)$ at $\gamma = \gamma_0$:

$$P_{\text{out}}(\gamma_0) = P_{\gamma_\Sigma}(\gamma_0) = \begin{cases} 1 - e^{-\gamma_T/\bar{\gamma}} - e^{-\gamma_0/\bar{\gamma}} + e^{-(\gamma_T+\gamma_0)/\bar{\gamma}} & \gamma_0 < \gamma_T, \\ 1 - 2e^{-\gamma_0/\bar{\gamma}} + e^{-(\gamma_T+\gamma_0)/\bar{\gamma}} & \gamma_0 \geq \gamma_T. \end{cases} \tag{7.14}$$

The performance of SSC under other types of fading, as well as the effects of fading correlation, is studied in [1, Chap. 9.8; 6; 7]. In particular, it is shown in [1, Chap. 9.8] that, for any fading distribution, SSC with an optimized threshold of $\gamma_T = \gamma_0$ has the same outage probability as SC.

EXAMPLE 7.2: Find the outage probability of BPSK modulation at $P_b = 10^{-3}$ for two-branch SSC diversity with i.i.d. Rayleigh fading on each branch for threshold values of $\gamma_T = 5$ dB, 7 dB, and 10 dB. Assume the average branch SNR is $\bar{\gamma} = 15$ dB. Discuss how the outage probability changes with γ_T. Also compare outage probability under SSC with that of SC and no diversity from Example 7.1.

Solution: As in Example 7.1, $\gamma_0 = 7$ dB. For $\gamma_T = 5$ dB we have $\gamma_0 \geq \gamma_T$, so we use the second line of (7.14) to get

$$P_{\text{out}} = 1 - 2e^{-10^{.7}/10^{1.5}} + e^{-(10^{.5}+10^{.7})/10^{1.5}} = .0654.$$

For $\gamma_T = 7$ dB we have $\gamma_0 = \gamma_T$, so we again use the second line of (7.14) and obtain

$$P_{\text{out}} = 1 - 2e^{-10^{.7}/10^{1.5}} + e^{-(10^{.7}+10^{.7})/10^{1.5}} = .0215.$$

For $\gamma_T = 10$ dB we have $\gamma_0 < \gamma_T$, so we use the first line of (7.14) to get

$$P_{\text{out}} = 1 - e^{-10/10^{1.5}} - e^{-10^{.7}/10^{1.5}} + e^{-(10+10^{.7})/10^{1.5}} = .0397.$$

We see that the outage probability is smaller for $\gamma_T = 7$ dB than for the other two values. At $\gamma_T = 5$ dB the threshold is too low, so the active branch can be below the target γ_0 for a long time before a switch is made; this contributes to a large outage probability. At $\gamma_T = 10$ dB the threshold is too high: the active branch will often fall below this threshold value, which will cause the combiner to switch to the other antenna even though that other antenna may have a lower SNR than the active one. This example shows that the threshold γ_T minimizing P_{out} equals γ_0.

From Example 7.1, SC has $P_{\text{out}} = .0215$. Thus, $\gamma_T = 7$ dB is the optimal threshold where SSC performs the same as SC. We also see that performance with an unoptimized threshold can be much worse than SC. However, the performance of SSC under all three thresholds is better than the performance without diversity, derived as $P_{\text{out}} = .1466$ in Example 7.1.

We obtain the distribution of γ_Σ by differentiating (7.12) relative to γ. Then the average probability of error is obtained from (7.2) with $P_s(\gamma)$ the probability of symbol error in AWGN and $p_{\gamma_\Sigma}(\gamma)$ the distribution of the SSC output SNR. For most fading distributions and coherent modulations, this result cannot be obtained in closed form and so must be evaluated numerically or by approximation. However, for i.i.d. Rayleigh fading we can differentiate (7.13) to obtain

$$p_{\gamma_\Sigma}(\gamma) = \begin{cases} (1 - e^{-\gamma_T/\bar{\gamma}})(1/\bar{\gamma})e^{-\gamma/\bar{\gamma}} & \gamma < \gamma_T, \\ (2 - e^{-\gamma_T/\bar{\gamma}})(1/\bar{\gamma})e^{-\gamma/\bar{\gamma}} & \gamma \geq \gamma_T. \end{cases} \tag{7.15}$$

As with SC, for most fading distributions and coherent modulations the resulting average probability of error is not in closed form and must be evaluated numerically. However, closed form results do exist for differential modulation under i.i.d. Rayleigh fading on each branch. In particular, the average probability of symbol error for DPSK is given by

$$\bar{P}_b = \int_0^\infty \frac{1}{2} e^{-\gamma} p_{\gamma_\Sigma}(\gamma) \, d\gamma = \frac{1}{2(1+\bar{\gamma})}(1 - e^{-\gamma_T/\bar{\gamma}} + e^{-\gamma_T}e^{-\gamma_T/\bar{\gamma}}). \tag{7.16}$$

EXAMPLE 7.3: Find the average probability of error for DPSK modulation under two-branch SSC diversity with i.i.d. Rayleigh fading on each branch for threshold values of $\gamma_T = 3$ dB, 7 dB, and 10 dB. Assume the average branch SNR is $\bar{\gamma} = 15$ dB. Discuss how the average proability of error changes with γ_T. Also compare average error probability under SSC with that of SC and with no diversity.

Solution: Evaluating (7.16) with $\bar{\gamma} = 15$ dB and $\gamma_T = 3$, 7, and 10 dB yields (respectively) $\bar{P}_b = .0029$, $\bar{P}_b = .0023$, and $\bar{P}_b = .0042$. As in the previous example, there is an optimal threshold that minimizes average probability of error. Setting the threshold too high or too low degrades performance. From (7.11) we have that SC yields $\bar{P}_b = .5(1 + 10^{1.5})^{-1} - .5(2 + 10^{1.5})^{-1} = 4.56 \cdot 10^{-4}$, which is roughly an order of magnitude less than with SSC and an optimized threshold. With no diversity we have $\bar{P}_b = .5(1 + 10^{1.5})^{-1} = .0153$, which is roughly an order of magnitude worse than with two-branch SSC.

7.2.4 Maximal-Ratio Combining

In SC and SSC, the output of the combiner equals the signal on one of the branches. In maximal-ratio combining (MRC) the output is a weighted sum of all branches, so the α_i in Figure 7.1 are all nonzero. The signals are co-phased and so $\alpha_i = a_i e^{-j\theta_i}$, where θ_i is the phase of the incoming signal on the ith branch. Thus, the envelope of the combiner output will be $r = \sum_{i=1}^{M} a_i r_i$. Assuming the same noise PSD $N_0/2$ in each branch yields a total noise PSD $N_{tot}/2$ at the combiner output of $N_{tot}/2 = \sum_{i=1}^{M} a_i^2 N_0/2$. Thus, the output SNR of the combiner is

$$\gamma_\Sigma = \frac{r^2}{N_{tot}} = \frac{1}{N_0} \frac{\left(\sum_{i=1}^{M} a_i r_i\right)^2}{\sum_{i=1}^{M} a_i^2}. \tag{7.17}$$

The goal is to choose the a_i to maximize γ_Σ. Intuitively, branches with a high SNR should be weighted more than branches with a low SNR, so the weights a_i^2 should be proportional to the branch SNRs r_i^2/N_0. We find the a_i that maximize γ_Σ by taking partial derivatives of (7.17) or using the Cauchy–Schwartz inequality [2]. Solving for the optimal weights yields $a_i^2 = r_i^2/N_0$, and the resulting combiner SNR becomes $\gamma_\Sigma = \sum_{i=1}^{M} r_i^2/N_0 = \sum_{i=1}^{M} \gamma_i$. Thus, the SNR of the combiner output is the sum of SNRs on each branch. Hence, the average combiner SNR and corresponding array gain increase linearly with the number of diversity branches M, in contrast to the diminishing returns associated with the average combiner SNR in SC given by (7.10). As with SC, the distribution of the combiner output SNR does not remain exponential even when there is Rayleigh fading on all branches.

To obtain the distribution of γ_Σ, we take the product of the exponential moment generating functions or characteristic functions. Assuming i.i.d. Rayleigh fading on each branch with equal average branch SNR $\bar{\gamma}$, the distribution of γ_Σ is χ^2 with $2M$ degrees of freedom, expected value $\bar{\gamma}_\Sigma = M\bar{\gamma}$, and variance $2M\bar{\gamma}$:

$$p_{\gamma_\Sigma}(\gamma) = \frac{\gamma^{M-1} e^{-\gamma/\bar{\gamma}}}{\bar{\gamma}^M (M-1)!}, \quad \gamma \geq 0. \tag{7.18}$$

The corresponding outage probability for a given threshold γ_0 is given by

$$P_{out} = p(\gamma_\Sigma < \gamma_0) = \int_0^{\gamma_0} p_{\gamma_\Sigma}(\gamma)\, d\gamma = 1 - e^{-\gamma_0/\bar{\gamma}} \sum_{k=1}^{M} \frac{(\gamma_0/\bar{\gamma})^{k-1}}{(k-1)!}. \tag{7.19}$$

Figure 7.5 is a plot of P_{out} for maximal-ratio combining indexed by the number of diversity branches.

The average probability of symbol error is obtained from (7.2) with $P_s(\gamma)$ the probability of symbol error in AWGN for the signal modulation and $p_{\gamma_\Sigma}(\gamma)$ the distribution of γ_Σ.

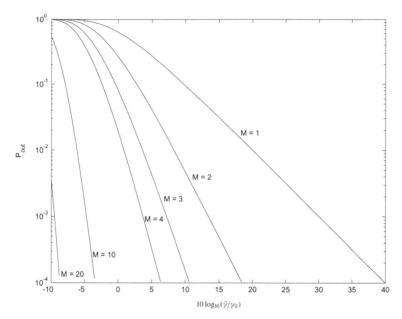

Figure 7.5: P_{out} for maximal-ratio combining with i.i.d. Rayleigh fading.

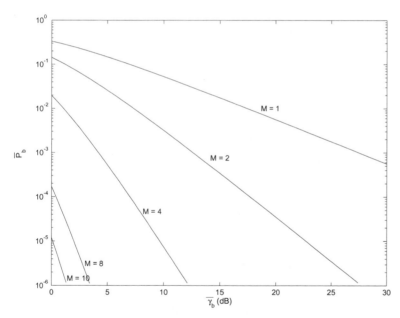

Figure 7.6: Average P_b for maximal-ratio combining with i.i.d. Rayleigh fading.

For BPSK modulation with i.i.d. Rayleigh fading, where $p_{\gamma_\Sigma}(\gamma)$ is given by (7.18), it can be shown [4, Chap. 6.3] that

$$\bar{P}_b = \int_0^\infty Q(\sqrt{2\gamma})\,p_{\gamma_\Sigma}(\gamma)\,d\gamma = \left(\frac{1-\Gamma}{2}\right)^M \sum_{m=0}^{M-1}\binom{M-1+m}{m}\left(\frac{1+\Gamma}{2}\right)^m, \quad (7.20)$$

where $\Gamma = \sqrt{\bar{\gamma}/(1+\bar{\gamma})}$. This equation is plotted in Figure 7.6. Comparing the outage probability for MRC in Figure 7.5 with that of SC in Figure 7.2 – or comparing the average probability of error for MRC in Figure 7.6 with that of SC in Figure 7.3 – indicates that MRC has significantly better performance than SC. In Section 7.4 we will use a different analysis based on moment generating functions to compute average error probability under MRC, which can be applied to any modulation type, any number of diversity branches, and any fading distribution on the different branches.

We can obtain a simple upper bound on the average probability of error by applying the Chernoff bound $Q(x) \le e^{-x^2/2}$ to the Q-function. Recall that, for static channel gains with MRC, we can approximate the probability of error as

$$P_s = \alpha_M Q\left(\sqrt{\beta_M \gamma_\Sigma}\right) \le \alpha_M e^{-\beta_M \gamma_\Sigma/2} = \alpha_M e^{-\beta_M(\gamma_1 + \cdots + \gamma_M)/2}. \tag{7.21}$$

Integrating over the χ^2 distribution for γ_Σ yields

$$\bar{P}_s \le \alpha_M \prod_{i=1}^{M} \frac{1}{1 + \beta_M \bar{\gamma}_i/2}. \tag{7.22}$$

In the limit of high SNR and assuming that the γ_i are identically distributed with $\bar{\gamma}_i = \bar{\gamma}$, we have

$$\bar{P}_s \approx \alpha_M \left(\frac{\beta_M \bar{\gamma}}{2}\right)^{-M}. \tag{7.23}$$

Thus, at high SNR, the diversity order of MRC is M (the number of antennas) and so MRC achieves full diversity order.

7.2.5 Equal-Gain Combining

Maximal-ratio combining requires knowledge of the time-varying SNR on each branch, which can be difficult to measure. A simpler technique is equal-gain combining (EGC), which co-phases the signals on each branch and then combines them with equal weighting, $\alpha_i = e^{-\theta_i}$. The SNR of the combiner output, assuming equal noise PSD $N_0/2$ in each branch, is then given by

$$\gamma_\Sigma = \frac{1}{N_0 M} \left(\sum_{i=1}^{M} r_i\right)^2. \tag{7.24}$$

The distribution and cdf of γ_Σ do not exist in closed form. For i.i.d. Rayleigh fading with two-branch diversity and average branch SNR $\bar{\gamma}$, an expression for the cdf in terms of the Q-function can be derived [4, Chap. 6.4; 8, Chap. 5.6] as

$$P_{\gamma_\Sigma}(\gamma) = 1 - e^{-2\gamma/\bar{\gamma}} - \sqrt{\pi \gamma/\bar{\gamma}}\, e^{-\gamma/\bar{\gamma}}\left(1 - 2Q\left(\sqrt{2\gamma/\bar{\gamma}}\right)\right). \tag{7.25}$$

The resulting outage probability is given by

$$P_{\text{out}}(\gamma_0) = 1 - e^{-2\gamma_R} - \sqrt{\pi \gamma_R}\, e^{-\gamma_R}\left(1 - 2Q\left(\sqrt{2\gamma_R}\right)\right), \tag{7.26}$$

where $\gamma_R = \gamma_0/\bar{\gamma}$. Differentiating (7.25) with respect to γ yields the distribution

$$p_{\gamma_\Sigma}(\gamma) = \frac{1}{\bar{\gamma}} e^{-2\gamma/\bar{\gamma}} + \sqrt{\pi}\, e^{-\gamma/\bar{\gamma}}\left(\frac{1}{\sqrt{4\gamma \bar{\gamma}}} - \frac{1}{\bar{\gamma}}\sqrt{\frac{\gamma}{\bar{\gamma}}}\right)\left(1 - 2Q\left(\sqrt{\frac{2\gamma}{\bar{\gamma}}}\right)\right). \tag{7.27}$$

Substituting this into (7.2) for BPSK yields the average probability of bit error:

$$\bar{P}_b = \int_0^\infty Q(\sqrt{2\gamma})\,p_{\gamma_\Sigma}(\gamma)\,d\gamma = .5\left(1 - \sqrt{1 - \left(\frac{1}{1+\bar{\gamma}}\right)^2}\right). \tag{7.28}$$

It is shown in [8, Chap. 5.7] that performance of EGC is quite close to that of MRC, typically exhibiting less than 1 dB of power penalty. This is the price paid for the reduced complexity of using equal gains. A more extensive performance comparison between SC, MRC, and EGC can be found in [1, Chap. 9].

EXAMPLE 7.4: Compare the average probability of bit error with BPSK under MRC and EGC two-branch diversity, assuming i.i.d. Rayleigh fading with an average SNR of 10 dB on each branch.

Solution: By (7.20), under MRC we have

$$\bar{P}_b = \left(\frac{1 - \sqrt{10/11}}{2}\right)^2 (2 + \sqrt{10/11}) = 1.60 \cdot 10^{-3}.$$

By (7.28), under EGC we have

$$\bar{P}_b = .5\left(1 - \sqrt{1 - \left(\frac{1}{11}\right)^2}\right) = 2.07 \cdot 10^{-3}.$$

Thus we see that the performance of MRC and EGC are almost the same.

7.3 Transmitter Diversity

In transmit diversity there are multiple transmit antennas, and the transmit power is divided among these antennas. Transmit diversity is desirable in systems where more space, power, and processing capability is available on the transmit side than on the receive side, as best exemplified by cellular systems. Transmit diversity design depends on whether or not the complex channel gain is known to the transmitter. When this gain is known, the system is quite similar to receiver diversity. However, without this channel knowledge, transmit diversity gain requires a combination of space and time diversity via a novel technique called the *Alamouti scheme* and its extensions. We now discuss transmit diversity under different assumptions about channel knowledge at the transmitter, assuming the channel gains are known at the receiver.

7.3.1 Channel Known at Transmitter

Consider a transmit diversity system with M transmit antennas and one receive antenna. We assume that the path gain $r_i e^{j\theta_i}$ associated with the ith antenna is known at the transmitter; this is referred to as having channel side information (CSI) at the transmitter, or CSIT. Let $s(t)$ denote the transmitted signal with total energy per symbol E_s. The signal is multiplied by a complex gain $\alpha_i = a_i e^{-j\theta_i}$ $(0 \le a_i \le 1)$ and then sent through the ith antenna. This complex multiplication performs both co-phasing and weighting relative to the channel gains. Because of the average total energy constraint E_s, the weights must satisfy $\sum_{i=1}^{M} a_i^2 = 1$. The weighted signals transmitted over all antennas are added "in the air", which leads to a received signal given by

$$r(t) = \sum_{i=1}^{M} a_i r_i s(t). \tag{7.29}$$

Let $N_0/2$ denote the noise PSD in the receiver.

Suppose we wish to set the branch weights to maximize received SNR. Using a similar analysis as in receiver MRC diversity, we see that the weights a_i that achieve the maximum SNR are given by

$$a_i = \frac{r_i}{\sqrt{\sum_{i=1}^{M} r_i^2}}, \tag{7.30}$$

and the resulting SNR is

$$\gamma_\Sigma = \frac{E_s}{N_0} \sum_{i=1}^{M} r_i^2 = \sum_{i=1}^{M} \gamma_i \tag{7.31}$$

for $\gamma_i = r_i^2 E_s/N_0$ equal to the branch SNR between the ith transmit antenna and the receive antenna. Thus we see that transmit diversity when the channel gains are known to the transmitter is very similar to receiver diversity with MRC: the received SNR is the sum of SNRs on each of the individual branches. In particular, if all antennas have the same gain $r_i = r$ then $\gamma_\Sigma = Mr^2 E_s/N_0$, so there is an array gain of M corresponding to an M-fold increase in SNR over a single antenna transmitting with full power. Using the Chernoff bound as in Section 7.2.4 we see that, for static gains,

$$P_s = \alpha_M Q\left(\sqrt{\beta_M \gamma_\Sigma}\right) \leq \alpha_M e^{-\beta_M \gamma_\Sigma/2} = \alpha_M e^{-\beta_M (\gamma_1 + \cdots + \gamma_M)/2}. \tag{7.32}$$

Integrating over the χ^2 distribution for γ_Σ yields the same bound as (7.22):

$$\bar{P}_s \leq \alpha_M \prod_{i=1}^{M} \frac{1}{1 + \beta_M \bar{\gamma}_i/2}. \tag{7.33}$$

In the limit of high SNR and assuming that the γ_i are identically distributed with $\bar{\gamma}_i = \bar{\gamma}$, we have

$$\bar{P}_s \approx \alpha_M \left(\frac{\beta_M \bar{\gamma}}{2}\right)^{-M}. \tag{7.34}$$

Thus, at high SNR, the diversity order of transmit diversity with MRC is M, so both transmit and receive MRC achieve full diversity order. The analysis for EGC and SC assuming transmitter channel knowledge is the same as under receiver diversity.

The complication of transmit diversity is to obtain the channel phase – and, for SC and MRC, the channel gain – at the transmitter. These channel values can be measured at the receiver using a pilot technique and then fed back to the transmitter. Alternatively, in cellular systems with time division, the base station can measure the channel gain and phase on transmissions from the mobile to the base and then use these measurements when transmitting back to the mobile, since under time division the forward and reverse links are reciprocal.

7.3.2 Channel Unknown at Transmitter – The Alamouti Scheme

We now consider the same model as in Section 7.3.1 but assume here that the transmitter no longer knows the channel gains $r_i e^{j\theta_i}$, so there is no CSIT. In this case it is not obvious how to obtain diversity gain. Consider, for example, a naive strategy whereby for a two-antenna system we divide the transmit energy equally between the two antennas. Thus, the transmit signal on antenna i will be $s_i(t) = \sqrt{.5}s(t)$ for $s(t)$ the transmit signal with energy per symbol E_s. Assume that each antenna has a complex Gaussian channel gain $h_i = r_i e^{j\theta_i}$ ($i = 1, 2$) with mean 0 and variance 1. The received signal is then

$$r(t) = \sqrt{.5}(h_1 + h_2)s(t). \tag{7.35}$$

Note that $h_1 + h_2$ is the sum of two complex Gaussian random variables and thus is itself a complex Gaussian, with mean equal to the sum of means (0) and variance equal to the sum of variances (2). Hence $\sqrt{.5}(h_1 + h_2)$ is a complex Gaussian random variable with mean 0 and variance 1, so the received signal has the same distribution as if we had just used one antenna with the full energy per symbol. In other words, we have obtained no performance advantage from the two antennas, since we could neither divide our energy intelligently between them nor obtain coherent combining via co-phasing.

Transmit diversity gain can be obtained even in the absence of channel information, given an appropriate scheme to exploit the antennas. A particularly simple and prevalent scheme for this diversity that combines both space and time diversity was developed by Alamouti in [9]. Alamouti's scheme is designed for a digital communication system with two-antenna transmit diversity. The scheme works over two symbol periods and it is assumed that the channel gain is constant over this time. Over the first symbol period, two different symbols s_1 and s_2 (each with energy $E_s/2$) are transmitted simultaneously from antennas 1 and 2, respectively. Over the next symbol period, symbol $-s_2^*$ is transmitted from antenna 1 and symbol s_1^* is transmitted from antenna 2, each again with symbol energy $E_s/2$.

Assume complex channel gains $h_i = r_i e^{j\theta_i}$ ($i = 1, 2$) between the ith transmit antenna and the receive antenna. The received symbol over the first symbol period is $y_1 = h_1 s_1 + h_2 s_2 + n_1$ and the received symbol over the second symbol period is $y_2 = -h_1 s_2^* + h_2 s_1^* + n_2$, where n_i ($i = 1, 2$) is the AWGN sample at the receiver associated with the ith symbol transmission. We assume that the noise sample has mean 0 and power N.

The receiver uses these sequentially received symbols to form the vector $\mathbf{y} = [y_1 \quad y_2^*]^T$ given by

$$\mathbf{y} = \begin{bmatrix} h_1 & h_2 \\ h_2^* & -h_1^* \end{bmatrix} \begin{bmatrix} s_1 \\ s_2 \end{bmatrix} + \begin{bmatrix} n_1 \\ n_2^* \end{bmatrix} = \mathbf{H}_A \mathbf{s} + \mathbf{n},$$

where $\mathbf{s} = [s_1 \quad s_2]^T$, $\mathbf{n} = [n_1 \quad n_2]^T$, and

$$\mathbf{H}_A = \begin{bmatrix} h_1 & h_2 \\ h_2^* & -h_1^* \end{bmatrix}.$$

Let us define the new vector $\mathbf{z} = \mathbf{H}_A^H \mathbf{y}$. The structure of $\mathbf{H}_A$ implies that

$$\mathbf{H}_A^H \mathbf{H}_A = (|h_1^2| + |h_2^2|)\mathbf{I}_2 \tag{7.36}$$

is diagonal and thus

$$\mathbf{z} = [z_1 \ z_2]^T = (|h_1^2| + |h_2^2|)\mathbf{I}_2 \mathbf{s} + \tilde{\mathbf{n}}, \tag{7.37}$$

where $\tilde{\mathbf{n}} = \mathbf{H}_A^H \mathbf{n}$ is a complex Gaussian noise vector with mean 0 and covariance matrix $E[\tilde{\mathbf{n}}\tilde{\mathbf{n}}^*] = (|h_1^2| + |h_2^2|)N_0 \mathbf{I}_2$. The diagonal nature of $\mathbf{z}$ effectively decouples the two symbol transmissions, so that each component of $\mathbf{z}$ corresponds to one of the transmitted symbols:

$$z_i = (|h_1^2| + |h_2^2|)s_i + \tilde{n}_i, \quad i = 1, 2. \tag{7.38}$$

The received SNR thus corresponds to the SNR for z_i given by

$$\gamma_i = \frac{(|h_1^2| + |h_2^2|)E_s}{2N_0}, \tag{7.39}$$

where the factor of 2 comes from the fact that s_i is transmitted using half the total symbol energy E_s. The received SNR is thus equal to the sum of SNRs on each branch divided by 2. By (7.38) the Alamouti scheme achieves a diversity order of 2 – the maximum possible for a two-antenna transmit system – despite the fact that channel knowledge is not available at the transmitter. However, by (7.39) it only achieves an array gain of 1, whereas MRC can achieve an array gain and a diversity gain of 2. The Alamouti scheme can be generalized for $M > 2$; this generalization falls into the category of orthogonal space-time block code design [10, Chap. 6.3.3; 11, Chap. 7.4].

7.4 Moment Generating Functions in Diversity Analysis

In this section we use the MGFs introduced in Section 6.3.3 to greatly simplify the analysis of average error probability under diversity. The use of MGFs in diversity analysis arises from the difficulty in computing the distribution $p_{\gamma_\Sigma}(\gamma)$ of the combiner SNR γ_Σ. Specifically, although the average probability of error and outage probability associated with diversity combining are given by the simple formulas (7.2) and (7.3), these formulas require integration over the distribution $p_{\gamma_\Sigma}(\gamma)$. This distribution is often not in closed form for an arbitrary number of diversity branches with different fading distributions on each branch, regardless of the combining technique that is used. In particular, the distribution for $p_{\gamma_\Sigma}(\gamma)$ is often in the form of an infinite-range integral, in which case the expressions for (7.2) and (7.3) become double integrals that can be difficult to evaluate numerically. Even when $p_{\gamma_\Sigma}(\gamma)$ is in closed form, the corresponding integrals (7.2) and (7.3) may not lead to closed-form solutions and may also be difficult to evaluate numerically. A large body of work over many decades has addressed approximations and numerical techniques for computing the integrals associated with average probability of symbol error for different modulations, fading distributions, and combining techniques (see [12] and the references therein). Expressing the average error probability in terms of the MGF for γ_Σ instead of its distribution often eliminates these integration difficulties. Specifically, when the diversity fading paths are independent but not necessarily identically distributed, the average error probability based on the MGF of γ_Σ is typically in closed form or consists of a single finite-range integral that can be easily computed numerically.

The simplest application of MGFs in diversity analysis is for coherent modulation with MRC, so this is treated first. We then discuss the use of MGFs in the analysis of average error probability under EGC and SC.

7.4.1 Diversity Analysis for MRC

The simplicity of using MGFs in the analysis of maximal-ratio combining stems from the fact that, as derived in Section 7.2.4, the combiner SNR γ_Σ is the sum of the γ_i, the branch SNRs:

$$\gamma_\Sigma = \sum_{i=1}^{M} \gamma_i. \tag{7.40}$$

As in the analysis of average error probability without diversity (Section 6.3.3), let us again assume that the probability of error in AWGN for the modulation of interest can be expressed either as an exponential function of γ_s, as in (6.67), or as a finite range integral of such a function, as in (6.68).

We first consider the case where P_s is in the form of (6.67). Then the average probability of symbol error under MRC is

$$\bar{P}_s = \int_0^\infty c_1 \exp[-c_2\gamma] p_{\gamma_\Sigma}(\gamma)\, d\gamma. \tag{7.41}$$

We assume that the branch SNRs are independent, so that their joint distribution becomes a product of the individual distributions: $p_{\gamma_1,\ldots,\gamma_M}(\gamma_1,\ldots,\gamma_M) = p_{\gamma_1}(\gamma_1)\ldots p_{\gamma_M}(\gamma_M)$. Using this factorization and substituting $\gamma = \gamma_1 + \cdots + \gamma_M$ in (7.41) yields

$$\bar{P}_s = c_1 \underbrace{\int_0^\infty \int_0^\infty \cdots \int_0^\infty}_{M\text{-fold}} \exp[-c_2(\gamma_1+\cdots+\gamma_M)] p_{\gamma_1}(\gamma_1)\ldots p_{\gamma_M}(\gamma_M)\, d\gamma_1 \ldots d\gamma_M. \tag{7.42}$$

Now using the product forms $\exp[-c_2(\gamma_1 + \cdots + \gamma_M)] = \prod_{i=1}^{M} \exp[-c_2\gamma_i]$ and $p_{\gamma_1}(\gamma_1)\ldots p_{\gamma_M}(\gamma_M) = \prod_{i=1}^{M} p_{\gamma_i}(\gamma_i)$ in (7.42) yields

$$\bar{P}_s = c_1 \underbrace{\int_0^\infty \int_0^\infty \cdots \int_0^\infty}_{M\text{-fold}} \prod_{i=1}^{M} \exp[-c_2\gamma_i] p_{\gamma_i}(\gamma_i)\, d\gamma_i. \tag{7.43}$$

Finally, switching the order of integration and multiplication in (7.43) yields our desired final form:

$$\bar{P}_s = c_1 \prod_{i=1}^{M} \int_0^\infty \exp[-c_2\gamma_i] p_{\gamma_i}(\gamma_i)\, d\gamma_i = c_1 \prod_{i=1}^{M} \mathcal{M}_{\gamma_i}(-c_2), \tag{7.44}$$

where $\mathcal{M}_{\gamma_i}(s)$ is the MGF of the fading distribution for the ith diversity branch as given by (6.63), (6.64), and (6.65) for (respectively) Rayleigh, Rician, and Nakagami fading. Thus, the average probability of symbol error is just the product of MGFs associated with the SNR on each branch.

Similarly, when P_s is in the form of (6.68), we get

$$\bar{P}_s = \int_0^\infty \int_A^B c_1 \exp[-c_2(x)\gamma]\, dx\, p_{\gamma_\Sigma}(\gamma)\, d\gamma$$

$$= \underbrace{\int_0^\infty \int_0^\infty \cdots \int_0^\infty}_{M\text{-fold}} \int_A^B c_1 \prod_{i=1}^{M} \exp[-c_2(x)\gamma_i] p_{\gamma_i}(\gamma_i)\, d\gamma_i. \tag{7.45}$$

Again, switching the order of integration and multiplication yields our desired final form:

$$\bar{P}_s = c_1 \int_A^B \prod_{i=1}^M \int_0^\infty \exp[-c_2(x)\gamma_i] p_{\gamma_i}(\gamma_i)\, d\gamma_i = c_1 \int_A^B \prod_{i=1}^M \mathcal{M}_{\gamma_i}(-c_2(x))\, dx. \quad (7.46)$$

Thus, the average probability of symbol error is just a single finite-range integral of the product of MGFs associated with the SNR on each branch. The simplicity of (7.44) and (7.46) are quite remarkable, given that these expressions apply for any number of diversity branches and any type of fading distribution on each branch (as long as the branch SNRs are independent).

We now apply these general results to specific modulations and fading distributions. Let us first consider DPSK, where $P_b(\gamma_b) = .5e^{-\gamma_b}$ in AWGN is in the form of (6.67) with $c_1 = 1/2$ and $c_2 = 1$. Then, by (7.44), the average probability of bit error in DPSK under M-fold MRC diversity is

$$\bar{P}_b = \frac{1}{2} \prod_{i=1}^M \mathcal{M}_{\gamma_i}(-1). \quad (7.47)$$

Note that this reduces to the probability of average bit error without diversity given by (6.60) for $M = 1$.

EXAMPLE 7.5: Compute the average probability of bit error for DPSK modulation under three-branch MRC, assuming i.i.d. Rayleigh fading in each branch with $\bar{\gamma}_1 = 15$ dB and $\bar{\gamma}_2 = \bar{\gamma}_3 = 5$ dB. Compare with the case of no diversity for $\bar{\gamma} = 15$ dB.

Solution: From (6.63), $\mathcal{M}_{\gamma_i}(s) = (1 - s\bar{\gamma}_i)^{-1}$. Using this MGF in (7.47) with $s = -1$ yields

$$\bar{P}_b = \frac{1}{2} \frac{1}{1 + 10^{1.5}} \left(\frac{1}{1 + 10^{.5}}\right)^2 = 8.85 \cdot 10^{-4}.$$

With no diversity we have

$$\bar{P}_b = \frac{1}{2(1 + 10^{1.5})} = 1.53 \cdot 10^{-2}.$$

This indicates that additional diversity branches can significantly reduce average BER, even when the SNR on these branches is somewhat low.

EXAMPLE 7.6: Compute the average probability of bit error for DPSK modulation under three-branch MRC, assuming: Nakagami fading in the first branch with $m = 2$ and $\bar{\gamma}_1 = 15$ dB; Rician fading in the second branch with $K = 3$ and $\bar{\gamma}_2 = 5$ dB; and Nakagami fading in the third branch with $m = 4$ and $\bar{\gamma}_3 = 5$ dB. Compare with the results of the prior example.

Solution: From (6.64) and (6.65), for Nakagami fading $\mathcal{M}_{\gamma_i}(s) = (1 - s\bar{\gamma}_i/m)^{-m}$ and for Rician fading

$$\mathcal{M}_{\gamma_s}(s) = \frac{1 + K}{1 + K - s\bar{\gamma}_s} \exp\left[\frac{Ks\bar{\gamma}_s}{1 + K - s\bar{\gamma}_s}\right].$$

Using these MGFs in (7.47) with $s = -1$ yields

$$\bar{P}_b = \frac{1}{2}\left(\frac{1}{1 + 10^{1.5}/2}\right)^2 \frac{4}{4 + 10^{.5}} \exp\left[\frac{-3 \cdot 10^{.5}}{4 + 10^{.5}}\right]\left(\frac{1}{1 + 10^{.5}/4}\right)^4 = 6.9 \cdot 10^{-5},$$

which is more than an order of magnitude lower than the average error probability under i.i.d. Rayleigh fading with the same branch SNRs derived in the previous problem. This indicates that Nakagami and Rician fading are much more benign distributions than Rayleigh, especially when multiple branches are combined under MRC. This example also illustrates the power of the MGF approach: computing average probability of error when the branch SNRs follow different distributions simply consists of multiplying together different functions in closed form; the result is then also in closed form. In contrast, computing the distribution of the sum of random variables from different families involves the convolution of their distributions, which rarely leads to a closed-form distribution.

For BPSK we see from (6.44) that P_b has the same form as (6.68) with $x = \phi$, $c_1 = 1/\pi$, $A = 0$, $B = \pi/2$, and $c_2(\phi) = 1/\sin^2 \phi$. Thus we obtain the average bit error probability for BPSK with M-fold diversity as

$$\bar{P}_b = \frac{1}{\pi} \int_0^{\pi/2} \prod_{l=1}^{M} \mathcal{M}_{\gamma_i}\left(-\frac{1}{\sin^2 \phi}\right) d\phi. \tag{7.48}$$

Similarly, if $P_s = \alpha Q(\sqrt{2g\gamma_s})$ then P_s has the same form as (6.68) with $x = \phi$, $c_1 = 1/\pi$, $A = 0$, $B = \pi/2$, and $c_2(\phi) = g/\sin^2 \phi$; here the resulting average symbol error probability with M-fold diversity is given by

$$\bar{P}_s = \frac{\alpha}{\pi} \int_0^{\pi/2} \prod_{i=1}^{M} \mathcal{M}_{\gamma_i}\left(-\frac{g}{\sin^2 \phi}\right) d\phi. \tag{7.49}$$

If the branch SNRs are i.i.d. then this expression simplifies to

$$\bar{P}_s = \frac{\alpha}{\pi} \int_0^{\pi/2} \left(\mathcal{M}_{\gamma}\left(-\frac{g}{\sin^2 \phi}\right)\right)^M d\phi, \tag{7.50}$$

where $\mathcal{M}_\gamma(s)$ is the common MGF for the branch SNRs. The probability of symbol error for MPSK in (6.45) is also in the form (6.68), leading to average symbol error probability

$$\bar{P}_s = \frac{1}{\pi} \int_0^{(M-1)\pi/M} \prod_{i=1}^{M} \mathcal{M}_{\gamma_i}\left(-\frac{g}{\sin^2 \phi}\right) d\phi, \tag{7.51}$$

where $g = \sin^2(\pi/M)$. For i.i.d. fading this simplifies to

$$\bar{P}_s = \frac{1}{\pi} \int_0^{(M-1)\pi/M} \left(\mathcal{M}_{\gamma}\left(-\frac{g}{\sin^2 \phi}\right)\right)^M d\phi. \tag{7.52}$$

EXAMPLE 7.7: Find an expression for the average symbol error probability for 8-PSK modulation for two-branch MRC combining, where each branch is Rayleigh fading with average SNR of 20 dB.

Solution: The MGF for Rayleigh fading is $\mathcal{M}_{\gamma_i}(s) = (1 - s\bar{\gamma}_i)^{-1}$. Using this MGF in (7.52) with $s = -(\sin^2 \pi/8)/\sin^2 \phi$ and $\bar{\gamma} = 100$ yields

$$\bar{P}_s = \frac{1}{\pi} \int_0^{7\pi/8} \left(\frac{1}{1 + (100 \sin^2 \pi/8)/\sin^2 \phi} \right)^2 d\phi.$$

This expression does not lead to a closed-form solution and so must be evaluated numerically, which results in $\bar{P}_s = 1.56 \cdot 10^{-3}$.

We can use similar techniques to extend the derivation of the exact error probability for MQAM in fading, given by (6.81), to include MRC diversity. Specifically, we first integrate the expression for P_s in AWGN, expressed in (6.80) using the alternate representation of Q and Q^2, over the distribution of γ_Σ. Since $\gamma_\Sigma = \sum_i \gamma_i$ and the SNRs are independent, the exponential function and distribution in the resulting expression can be written in product form. Then we use the same reordering of integration and multiplication as in the MPSK derivation. The resulting average probability of symbol error for MQAM modulation with MRC combining is given by

$$\bar{P}_s = \frac{4}{\pi} \left(1 - \frac{1}{\sqrt{M}} \right) \int_0^{\pi/2} \prod_{i=1}^M \mathcal{M}_{\gamma_i} \left(-\frac{g}{\sin^2 \phi} \right) d\phi$$

$$- \frac{4}{\pi} \left(1 - \frac{1}{\sqrt{M}} \right)^2 \int_0^{\pi/4} \prod_{i=1}^M \mathcal{M}_{\gamma_i} \left(-\frac{g}{\sin^2 \phi} \right) d\phi. \qquad (7.53)$$

More details on the use of MGFs to obtain average probability of error under M-fold MRC diversity for a broad class of modulations can be found in [1, Chap. 9.2].

7.4.2 Diversity Analysis for EGC and SC

Moment generating functions are less useful in the analysis of EGC and SC than in MRC. The reason is that, with MRC, $\gamma_\Sigma = \sum_i \gamma_i$ and so $\exp[-c_2 \gamma_\Sigma] = \prod_i \exp[-c_2 \gamma_i]$. This factorization leads directly to the simple formulas whereby probability of symbol error is based on a product of MGFs associated with each of the branch SNRs. Unfortunately, neither EGC nor SC leads to this type of factorization. However, working with the MGF of γ_Σ can sometimes lead to simpler results than working directly with its distribution. This is illustrated in [1, Chap. 9.3.3]: the exact probability of symbol error for MPSK with EGC is obtained based on the characteristic function associated with each branch SNR, where the characteristic function is just the MGF evaluated at $s = j2\pi f$ (i.e., it is the Fourier transform of the distribution). The resulting average error probability, given by [1, eq. (9.78)], is a finite-range integral over a sum of closed-form expressions and thus is easily evaluated numerically.

7.4.3 Diversity Analysis for Noncoherent and Differentially Coherent Modulation

A similar MGF approach to determining the average symbol error probability of noncoherent and differentially coherent modulations with diversity combining is presented in [1; 13]. This approach differs from that of the coherent modulation case in that it relies on an alternate form of the Marcum Q-function instead of the Gaussian Q-function, since the BER of noncoherent and differentially coherent modulations in AWGN are given in terms of the Marcum Q-function. Otherwise the approach is essentially the same as in the coherent case, and it leads to BER expressions involving a single finite-range integral that can be readily evaluated numerically. More details on this approach are found in [1] and [13].

PROBLEMS

7-1. Find the outage probability of QPSK modulation at $P_s = 10^{-3}$ for a Rayleigh fading channel with SC diversity for $M = 1$ (no diversity), $M = 2$, and $M = 3$. Assume branch SNRs of $\bar{\gamma}_1 = 10$ dB, $\bar{\gamma}_2 = 15$ dB, and $\bar{\gamma}_3 = 20$ dB.

7-2. Plot the distribution $p_{\gamma_\Sigma}(\gamma)$ given by (7.9) for the selection combiner SNR in Rayleigh fading with M-branch diversity assuming $M = 1, 2, 4, 8$, and 10. Assume that each branch has an average SNR of 10 dB. Your plot should be linear on both axes and should focus on the range of linear γ values $0 \leq \gamma \leq 60$. Discuss how the distribution changes with increasing M and why this leads to lower probability of error.

7-3. Derive the average probability of bit error for DPSK under SC with i.i.d. Rayleigh fading on each branch as given by (7.11).

7-4. Derive a general expression for the cdf of the two-branch SSC output SNR for branch statistics that are not i.i.d., and show that it reduces to (7.12) for i.i.d. branch statistics. Evaluate your expression assuming Rayleigh fading in each branch with different average SNRs $\bar{\gamma}_1$ and $\bar{\gamma}_2$.

7-5. Derive the average probability of bit error for DPSK under SSC with i.i.d. Rayleigh fading on each branch as given by (7.16).

7-6. Plot the average probability of bit error for DPSK under MRC with $M = 2, 3$, and 4, assuming i.i.d. Rayleigh fading on each branch and an average branch SNR ranging from 0 dB to 20 dB.

7-7. Show that the weights a_i maximizing γ_Σ under receiver diversity with MRC are $a_i^2 = r_i^2/N_0$ for $N_0/2$ the common noise PSD on each branch. Also show that, with these weights, $\gamma_\Sigma = \sum_i \gamma_i$.

7-8. This problem illustrates that, because of array gain, you can get performance gains from diversity combining even without fading. Consider an AWGN channel with N-branch diversity combining and $\gamma_i = 10$ dB per branch. Assume MQAM modulation with $M = 4$ and use the approximation $P_b = .2e^{-1.5\gamma/(M-1)}$ for bit error probability, where γ is the received SNR.

   (a) Find P_b for $N = 1$.
   (b) Find N so that, under MRC, $P_b < 10^{-6}$.

7-9. Derive the average probability of bit error for BPSK under MRC with i.i.d. Rayleigh fading on each branch as given by (7.20).

7-10. Derive the average probability of bit error for BPSK under EGC with i.i.d. Rayleigh fading on each branch as given by (7.28).

7-11. Compare the average probability of bit error for BPSK modulation under no diversity, two-branch SC, two-branch SSC with $\gamma_T = \gamma_0$, two-branch EGC, and two-branch MRC. Assume i.i.d. Rayleigh fading on each branch with equal branch SNRs of 10 dB and of 20 dB. How does the relative performance change as the branch SNRs increase?

7-12. Plot the average probability of bit error for BPSK under both MRC and EGC assuming two-branch diversity with i.i.d. Rayleigh fading on each branch and average branch SNR ranging from 0 dB to 20 dB. What is the maximum dB penalty of EGC as compared to MRC?

7-13. Compare the outage probability of BPSK modulation at $P_b = 10^{-3}$ under MRC versus EGC, assuming two-branch diversity with i.i.d. Rayleigh fading on each branch and average branch SNR $\bar{\gamma} = 10$ dB.

7-14. Compare the average probability of bit error for BPSK under MRC versus EGC, assuming two-branch diversity with i.i.d. Rayleigh fading on each branch and average branch SNR $\bar{\gamma} = 10$ dB.

7-15. Compute the average BER of a channel with two-branch transmit diversity under the Alamouti scheme, assuming the average branch SNR is 10 dB.

7-16. Consider a fading distribution $p(\gamma)$ where $\int_0^\infty p(\gamma)e^{-x\gamma}\,d\gamma = .01\bar{\gamma}/\sqrt{x}$. Find the average P_b for a BPSK modulated signal where (a) the receiver has two-branch diversity with MRC combining and (b) each branch has an average SNR of 10 dB and experiences independent fading with distribution $p(\gamma)$.

7-17. Consider a fading channel that is BPSK modulated and has three-branch diversity with MRC, where each branch experiences independent fading with an average received SNR of 15 dB. Compute the average BER of this channel for Rayleigh fading and for Nakagami fading with $m = 2$. *Hint:* Using the alternate Q-function representation greatly simplifies this computation, at least for Nakagami fading.

7-18. Plot the average probability of error as a function of branch SNR for a two-branch MRC system with BPSK modulation, where the first branch has Rayleigh fading and the second branch has Nakagami-m fading with $m = 2$. Assume the two branches have the same average SNR; your plots should have this average branch SNR ranging from 5 dB to 20 dB.

7-19. Plot the average probability of error as a function of branch SNR for an M-branch MRC system with 8-PSK modulation for $M = 1, 2, 4, 8$. Assume that each branch has Rayleigh fading with the same average SNR. Your plots should have this SNR ranging from 5 dB to 20 dB.

7-20. Derive the average probability of symbol error for MQAM modulation under MRC diversity given by (7.53) from the probability of error in AWGN (6.80) by utilizing the alternate representation of Q and Q^2.

7-21. Compare the average probability of symbol error for 16-PSK and 16-QAM modulation, assuming three-branch MRC diversity with Rayleigh fading on the first branch and Rician fading on the second and third branches with $K = 2$. Assume equal average branch SNRs of 10 dB.

7-22. Plot the average probability of error as a function of branch SNR for an M-branch MRC system with 16-QAM modulation for $M = 1, 2, 4, 8$. Assume that each branch has Rayleigh fading with the same average SNR. Your plots should have an SNR ranging from 5 dB to 20 dB.

REFERENCES

[1] M. K. Simon and M.-S. Alouini, *Digital Communication over Fading Channels: A Unified Approach to Performance Analysis,* Wiley, New York, 2000.

[2] W. C. Y. Lee, *Mobile Communications Engineering,* McGraw-Hill, New York, 1982.

[3] J. Winters, "Signal acquisition and tracking with adaptive arrays in the digital mobile radio system IS-54 with flat fading," *IEEE Trans. Veh. Tech.,* pp. 1740–51, November 1993.

[4] G. L. Stuber, *Principles of Mobile Communications,* 2nd ed., Kluwer, Dordrecht, 2001.

[5] M. Blanco and K. Zdunek, "Performance and optimization of switched diversity systems for the detection of signals with Rayleigh fading," *IEEE Trans. Commun.,* pp. 1887–95, December 1979.

[6] A. Abu-Dayya and N. Beaulieu, "Switched diversity on microcellular Ricean channels," *IEEE Trans. Veh. Tech.,* pp. 970–6, November 1994.

[7] A. Abu-Dayya and N. Beaulieu, "Analysis of switched diversity systems on generalized-fading channels," *IEEE Trans. Commun.,* pp. 2959–66, November 1994.

[8] M. Yacoub, *Principles of Mobile Radio Engineering,* CRC Press, Boca Raton, FL, 1993.

[9] S. Alamouti, "A simple transmit diversity technique for wireless communications," *IEEE J. Sel. Areas Commun.,* pp. 1451–8, October 1998.

[10] A. Paulraj, R. Nabar, and D. Gore, *Introduction to Space-Time Wireless Communications,* Cambridge University Press, 2003.

[11] E. G. Larsson and P. Stoica, *Space-Time Block Coding for Wireless Communications,* Cambridge University Press, 2003.

[12] M. K. Simon and M.-S. Alouini, "A unified approach to the performance analysis of digital communications over generalized fading channels," *Proc. IEEE,* pp. 1860–77, September 1998.

[13] M. K. Simon and M.-S. Alouini, "A unified approach for the probability of error for noncoherent and differentially coherent modulations over generalized fading channels," *IEEE Trans. Commun.,* pp. 1625–38, December 1998.

8

Coding for Wireless Channels

Coding allows bit errors introduced by transmission of a modulated signal through a wireless channel to be either detected or corrected by a decoder in the receiver. Coding can be considered as the embedding of signal constellation points in a higher-dimensional signaling space than is needed for communications. By going to a higher-dimensional space, the distance between points can be increased, which provides for better error correction and detection.

In this chapter we describe codes designed for additive white Gaussian noise channels and for fading channels. Codes designed for AWGN channels typically do not work well on fading channels because they cannot correct for long error bursts that occur in deep fading. Codes for fading channels are mainly based on using an AWGN channel code combined with interleaving, but the criterion for the code design changes to provide fading diversity. Other coding techniques to combat performance degradation due to fading include unequal error protection codes and joint source and channel coding.

We first provide an overview of code design in both fading and AWGN, along with basic design parameters such as minimum distance, coding gain, bandwidth expansion, and diversity order. Sections 8.2 and 8.3 provide a basic overview of block and convolutional code designs for AWGN channels. Although these designs are not directly applicable to fading channels, codes for fading channels and other codes used in wireless systems (e.g., spreading codes in CDMA) require background in these fundamental techniques. Concatenated codes and their evolution to turbo codes – as well as low-density parity-check (LDPC) codes – for AWGN channels are also described. These extremely powerful codes exhibit near-capacity performance with reasonable complexity levels. Coded modulation was invented in the late 1970s as a technique to obtain error correction through a joint design of the modulation and coding. We will discuss the basic design principles behind trellis and more general lattice coded modulation along with their performance in AWGN.

Code designs for fading channels are covered in Section 8.8. These designs combine block or convolutional codes designed for AWGN channels with interleaving and then modify the AWGN code design metric to incorporate maximum fading diversity. Diversity gains can also be obtained by combining coded modulation with symbol or bit interleaving, although bit interleaving generally provides much higher diversity gain. Thus, coding combined with interleaving provides diversity gain in the same manner as other forms of diversity, with the diversity order built into the code design. Unequal error protection is an alternative to

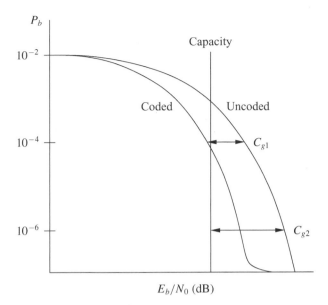

Figure 8.1: Coding gain in AWGN channels.

diversity in fading mitigation. In these codes bits are prioritized, and high-priority bits are encoded with stronger error protection against deep fades. Since bit priorities are part of the source code design, unequal error protection is a special case of joint source and channel coding, which we also describe.

Coding is a very broad and deep subject, with many excellent books devoted solely to this topic. This chapter assumes no background in coding, and thus it provides an in-depth discussion of code designs for AWGN channels before designs for wireless systems can be treated. This in-depth discussion can be omitted for a more cursory treatment of coding for wireless channels by focusing on Sections 8.1 and 8.8.

8.1 Overview of Code Design

The main reason to apply error correction coding in a wireless system is to reduce the probability of bit or block error. The bit error probability P_b for a coded system is the probability that a bit is decoded in error. The block error probability P_{bl}, also called the packet error rate, is the probability that one or more bits in a block of coded bits is decoded in error. Block error probability is useful for packet data systems where bits are encoded and transmitted in blocks. The amount of error reduction provided by a given code is typically characterized by its coding gain in AWGN and its diversity gain in fading.

Coding gain in AWGN is defined as the amount that the bit energy or signal-to-noise power ratio can be reduced under the coding technique for a given P_b or P_{bl}. We illustrate coding gain for P_b in Figure 8.1. We see in this figure that the gain C_{g1} at $P_b = 10^{-4}$ is less than the gain C_{g2} at $P_b = 10^{-6}$, and there is negligible coding gain at $P_b = 10^{-2}$. In fact, codes designed for high-SNR channels can have negative coding gain at low SNRs, since the extra redundancy of the code does not provide sufficient performance gain in P_b or P_{bl} at low SNRs to compensate for spreading the bit energy over multiple coded bits. Thus, unexpected fluctuations in channel SNR can significantly degrade code performance. The coding gain in AWGN is generally a function of the minimum Euclidean distance of the code, which equals the minimum distance in signal space between codewords or error events. Thus, codes designed for AWGN channels maximize their Euclidean distance for good performance.

Error probability with or without coding tends to fall off with SNR as a waterfall shape at low to moderate SNRs. Whereas this waterfall shape holds at all SNRs for uncoded systems and many coded systems, some codes (such as turbo codes) exhibit error floors as SNR grows. The error floor, also shown in Figure 8.1, kicks in at a threshold SNR that depends on the code design. For SNRs above this threshold, the slope of the error probability curve decreases because minimum distance error events dominate code performance in this SNR regime.

Code performance is also commonly measured against channel capacity. The capacity curve is associated with the SNR (E_b/N_0) where Shannon capacity $B \log_2(1 + \text{SNR})$ equals the data rate of the system. At rates up to capacity, the capacity-achieving code has a probability of error that goes to zero, as indicated by the straight line in Figure 8.1. The capacity curve thus indicates the best possible performance that any practical code can achieve.

For many codes, the error correction capability of a code does not come for free. This performance enhancement is paid for by increased complexity and – for block codes, convolutional codes, turbo codes, and LDPC codes – by either a decrease in data rate or an increase in signal bandwidth. Consider a code with n coded bits for every k uncoded bits. This code effectively embeds a k-dimensional subspace into a larger n-dimensional space to provide larger distances between coded symbols. However, if the data rate through the channel is fixed at R_b, then the information rate for a code that uses n coded bits for every k uncoded bits is $(k/n)R_b$; that is, coding decreases the data rate by the fraction k/n. We can keep the information rate constant and introduce coding gain by decreasing the bit time by k/n. This typically results in an expanded bandwidth of the transmittted signal by n/k. Coded modulation uses a joint design of the code and modulation to obtain coding gain without this bandwidth expansion, as discussed in more detail in Section 8.7.

Codes designed for AWGN channels do not generally work well in fading owing to bursts of errors that cannot be corrected for. However, good performance in fading can be obtained by combining AWGN channel codes with interleaving and by designing the code to optimize its inherent diversity. The interleaver spreads out bursts of errors over time, so it provides a form of time diversity. This diversity is exploited by the inherent diversity in the code. In fact, codes designed in this manner exhibit performance similar to MRC diversity, with diversity order equal to the minimum Hamming distance of the code. Hamming distance is the number of coded symbols that differ between different codewords or error events. Thus, coding and interleaving designed for fading channels maximize their Hamming distance for good performance.

8.2 Linear Block Codes

Linear block codes are conceptually simple codes that are basically an extension of single-bit parity-check codes for error detection. A single-bit parity-check code is one of the most common forms of detecting transmission errors. This code uses one extra bit in a block of n data bits to indicate whether the number of 1-bits in a block is odd or even. Thus, if a single error occurs, either the parity bit is corrupted or the number of detected 1-bits in the information bit sequence will be different from the number used to compute the parity bit; in either case the parity bit will not correspond to the number of detected 1-bits in the information bit sequence, so the single error is detected. Linear block codes extend this notion by using a larger number of parity bits to either detect more than one error or correct for one or more errors. Unfortunately, linear block codes – along with convolutional codes – trade their error detection or correction capability for either bandwidth expansion or a lower data rate, as we shall discuss in more detail. We will restrict our attention to binary codes, where both the original information and the corresponding code consist of bits taking a value of either 0 or 1.

8.2.1 Binary Linear Block Codes

A binary block code generates a block of n coded bits from k information bits. We call this an (n, k) binary block code. The coded bits are also called *codeword symbols*. The n codeword symbols can take on 2^n possible values corresponding to all possible combinations of the n binary bits. We select 2^k codewords from these 2^n possibilities to form the code, where each k bit information block is uniquely mapped to one of these 2^k codewords. The rate of the code is $R_c = k/n$ information bits per codeword symbol. If we assume that codeword symbols are transmitted across the channel at a rate of R_s symbols per second, then the information rate associated with an (n, k) block code is $R_b = R_c R_s = (k/n) R_s$ bits per second. Thus we see that block coding reduces the data rate compared to what we obtain with uncoded modulation by the code rate R_c.

A block code is called a *linear* code when the mapping of the k information bits to the n codeword symbols is a linear mapping. In order to describe this mapping and the corresponding encoding and decoding functions in more detail, we must first discuss properties of the vector space of binary n-tuples and its corresponding subspaces. The set of all binary n-tuples B_n is a vector space over the binary field, which consists of the two elements 0 and 1. All fields have two operations, addition and multiplication: for the binary field these operations correspond to binary addition (modulo 2 addition) and standard multiplication. A subset S of B_n is called a *subspace* if it satisfies the following conditions.

1. The all-zero vector is in S.
2. The set S is closed under addition; that is, if $S_i \in S$ and $S_j \in S$ then $S_i + S_j \in S$.

An (n, k) block code is linear if the 2^k length-n codewords of the code form a subspace of B_n. Thus, if $\mathbf{C}_i$ and $\mathbf{C}_j$ are two codewords in an (n, k) linear block code, then $\mathbf{C}_i + \mathbf{C}_j$ must form another codeword of the code.

EXAMPLE 8.1: The vector space B_3 consists of all binary tuples of length 3:

$$B_3 = \{[000], [001], [010], [011], [100], [101], [110], [111]\}.$$

Note that B_3 is a subspace of itself, since it contains the all-zero vector and is closed under addition. Determine which of the following subsets of B_3 form a subspace:

- $A_1 = \{[000], [001], [100], [101]\}$;
- $A_2 = \{[000], [100], [110], [111]\}$;
- $A_3 = \{[001], [100], [101]\}$.

Solution: It is easily verified that A_1 is a subspace, since it contains the all-zero vector and the sum of any two tuples in A_1 is also in A_1. A_2 is not a subspace because it is not closed under addition, as $110 + 111 = 001 \notin A_2$. A_3 is not a subspace because it is not closed under addition ($001 + 001 = 000 \notin A_3$) and it does not contain the all-zero vector.

Intuitively, the greater the distance between codewords in a given code, the less chance that errors introduced by the channel will cause a transmitted codeword to be decoded as a different codeword. We define the *Hamming distance* between two codewords $\mathbf{C}_i$ and $\mathbf{C}_j$, denoted as $d(\mathbf{C}_i, \mathbf{C}_j)$ or d_{ij}, as the number of elements in which they differ:

$$d_{ij} = \sum_{l=1}^{n} (\mathbf{C}_i(l) + \mathbf{C}_j(l)), \tag{8.1}$$

where $\mathbf{C}_m(l)$ denotes the lth bit in $\mathbf{C}_m$. For example, if $\mathbf{C}_i = [00101]$ and $\mathbf{C}_j = [10011]$ then $d_{ij} = 3$. We define the *weight* of a given codeword $\mathbf{C}_i$ as the number of 1-bits in the codeword, so $\mathbf{C}_i = [00101]$ has weight 2. The weight of a given codeword $\mathbf{C}_i$ is just its Hamming distance d_{0i} from the all-zero codeword $\mathbf{C}_0 = [00 \ldots 0]$ or, equivalently, the sum of its elements:

$$w(\mathbf{C}_i) = \sum_{l=1}^{n} \mathbf{C}_i(l). \tag{8.2}$$

Since $0 + 0 = 1 + 1 = 0$, the Hamming distance between $\mathbf{C}_i$ and $\mathbf{C}_j$ is equal to the weight of $\mathbf{C}_i + \mathbf{C}_j$. For example, with $\mathbf{C}_i = [00101]$ and $\mathbf{C}_j = [10011]$ as given before, $w(\mathbf{C}_i) = 2$, $w(\mathbf{C}_j) = 3$, and $d_{ij} = w(\mathbf{C}_i + \mathbf{C}_j) = w([10110]) = 3$. Since the Hamming distance between any two codewords equals the weight of their sum and since this sum is also a codeword, we can determine the minimum distance between all codewords in a code by just looking at the minimum distance between all nonzero codewords and the all-zero codeword $\mathbf{C}_0$. Thus, we define the minimum distance of a code as

$$d_{\min} = \min_{i, i \neq 0} d_{0i}. \tag{8.3}$$

We will see in Section 8.2.6 that the minimum distance of a linear block code is a critical parameter in determining its probability of error.

8.2.2 Generator Matrix

The generator matrix is a compact description of how codewords are generated from information bits in a linear block code. The design goal in linear block codes is to find generator matrices such that their corresponding codes are easy to encode and decode yet have powerful error correction/detection capabilities. Consider an (n, k) code with k information bits, denoted as

$$\mathbf{U}_i = [u_{i1}, \ldots, u_{ik}],$$

that are encoded into the codeword

$$\mathbf{C}_i = [c_{i1}, \ldots, c_{in}].$$

We represent the encoding operation as a set of n equations defined by

$$c_{ij} = u_{i1}g_{1j} + u_{i2}g_{2j} + \cdots + u_{ik}g_{kj}, \quad j = 1, \ldots, n, \tag{8.4}$$

where g_{ij} is binary (0 or 1) and where binary (standard) multiplication is used. We can write these n equations in matrix form as

$$\mathbf{C}_i = \mathbf{U}_i \mathbf{G}, \tag{8.5}$$

where the $k \times n$ *generator matrix* $\mathbf{G}$ for the code is defined as

$$\mathbf{G} = \begin{bmatrix} g_{11} & g_{12} & \cdots & g_{1n} \\ g_{21} & g_{22} & \cdots & g_{2n} \\ \vdots & \vdots & \vdots & \vdots \\ g_{k1} & g_{k2} & \cdots & g_{kn} \end{bmatrix}. \tag{8.6}$$

If we denote the lth row of $\mathbf{G}$ as $\mathbf{g}_l = [g_{l1}, \ldots, g_{ln}]$, then we can write any codeword $\mathbf{C}_i$ as linear combinations of these row vectors as follows:

$$\mathbf{C}_i = u_{i1}\mathbf{g}_1 + u_{i2}\mathbf{g}_2 + \cdots + u_{ik}\mathbf{g}_k. \tag{8.7}$$

Since a linear (n, k) block code is a subspace of dimension k in the larger n-dimensional space, it follows that the k row vectors $\{\mathbf{g}_l\}_{l=1}^{k}$ of $\mathbf{G}$ must be linearly independent so that they span the k-dimensional subspace associated with the 2^k codewords. Hence, $\mathbf{G}$ has rank k. Since the set of basis vectors for this subspace is not unique, the generator matrix is also not unique.

A *systematic* linear block code is described by a generator matrix of the form

$$\mathbf{G} = [\mathbf{I}_k \mid \mathbf{P}] = \begin{bmatrix} 1 & 0 & \cdots & 0 & p_{11} & p_{12} & \cdots & p_{1(n-k)} \\ 0 & 1 & \cdots & 0 & p_{21} & p_{22} & \cdots & p_{2(n-k)} \\ \vdots & \vdots & \vdots & \vdots & \vdots & \vdots & \vdots & \vdots \\ 0 & 0 & \cdots & 1 & p_{k1} & p_{k2} & \cdots & p_{k(n-k)} \end{bmatrix}, \tag{8.8}$$

where $\mathbf{I}_k$ is the $k \times k$ identity matrix and $\mathbf{P}$ is a $k \times (n - k)$ matrix that determines the redundant, or parity, bits to be used for error correction or detection. The codeword output from a systematic encoder is of the form

$$\mathbf{C}_i = \mathbf{U}_i \mathbf{G} = \mathbf{U}_i [\mathbf{I}_k \mid \mathbf{P}] = [u_{i1}, \ldots, u_{ik}, p_1, \ldots, p_{(n-k)}], \tag{8.9}$$

where the first k bits of the codeword are the original information bits and the last $(n - k)$ bits of the codeword are the parity bits obtained from the information bits as

$$p_j = u_{i1}p_{1j} + \cdots + u_{ik}p_{kj}, \quad j = 1, \ldots, n - k. \tag{8.10}$$

Note that any generator matrix for an (n, k) linear block code can be reduced by row operations and column permutations to a generator matrix in systematic form.

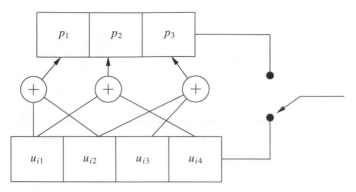

Figure 8.2: Implementation of (7, 4) binary code.

EXAMPLE 8.2: Systematic linear block codes are typically implemented with $n - k$ modulo-2 adders tied to the appropriate stages of a shift register. The resulting parity bits are appended to the end of the information bits to form the codeword. Find the corresponding implementation for generating a (7, 4) binary code with the generator matrix

$$\mathbf{G} = \begin{bmatrix} 1 & 0 & 0 & 0 & 1 & 1 & 0 \\ 0 & 1 & 0 & 0 & 1 & 0 & 1 \\ 0 & 0 & 1 & 0 & 0 & 0 & 1 \\ 0 & 0 & 0 & 1 & 0 & 1 & 0 \end{bmatrix}. \tag{8.11}$$

Solution: The matrix $\mathbf{G}$ is already in systematic form with

$$\mathbf{P} = \begin{bmatrix} 1 & 1 & 0 \\ 1 & 0 & 1 \\ 0 & 0 & 1 \\ 0 & 1 & 0 \end{bmatrix}. \tag{8.12}$$

Let P_{lj} denote the ljth element of $\mathbf{P}$. By (8.10) we see that the first parity bit in the codeword is $p_1 = u_{i1}P_{11} + u_{i2}P_{21} + u_{i3}P_{31} + u_{i4}P_{41} = u_{i1} + u_{i2}$. Similarly, the second parity bit is $p_2 = u_{i1}P_{12} + u_{i2}P_{22} + u_{i3}P_{32} + u_{i4}P_{42} = u_{i1} + u_{i4}$ and the third parity bit is $p_3 = u_{i1}P_{13} + u_{i2}P_{23} + u_{i3}P_{33} + u_{i4}P_{43} = u_{i2} + u_{i3}$. The shift register implementation to generate these parity bits is shown in Figure 8.2. The codeword output is $[u_{i1}u_{i2}u_{i3}u_{i4}p_1p_2p_3]$, where the switch is in the down position to output the systematic bits u_{ij} ($j = 1, \ldots, 4$) of the code or in the up position to output the parity bits p_j ($j = 1, 2, 3$) of the code.

8.2.3 Parity-Check Matrix and Syndrome Testing

The parity-check matrix is used to decode linear block codes with generator matrix $\mathbf{G}$. The parity-check matrix $\mathbf{H}$ corresponding to a generator matrix $\mathbf{G} = [\mathbf{I}_k \mid \mathbf{P}]$ is defined as

$$\mathbf{H} = [\mathbf{P}^T \mid \mathbf{I}_{n-k}]. \tag{8.13}$$

It is easily verified that $\mathbf{GH}^T = \mathbf{0}_{k,n-k}$, where $\mathbf{0}_{k,n-k}$ denotes an all-zero $k \times (n - k)$ matrix. Recall that a given codeword $\mathbf{C}_i$ in the code is obtained by multiplication of the information bit sequence $\mathbf{U}_i$ by the generator matrix $\mathbf{G}$: $\mathbf{C}_i = \mathbf{U}_i\mathbf{G}$. Thus,

$$\mathbf{C}_i\mathbf{H}^T = \mathbf{U}_i\mathbf{G}\mathbf{H}^T = \mathbf{0}_{n-k} \tag{8.14}$$

for any input sequence $\mathbf{U}_i$, where $\mathbf{0}_{n-k}$ denotes the all-zero row vector of length $n-k$. Thus, multiplication of any valid codeword with the parity-check matrix results in an all-zero vector. This property is used to determine whether the received vector is a valid codeword or has been corrupted, based on the notion of *syndrome testing*, which we now define.

Let $\mathbf{R}$ be the received codeword resulting from transmission of codeword $\mathbf{C}$. In the absence of channel errors, $\mathbf{R} = \mathbf{C}$. However, if the transmission is corrupted, then one or more of the codeword symbols in $\mathbf{R}$ will differ from those in $\mathbf{C}$. We therefore write the received codeword as

$$\mathbf{R} = \mathbf{C} + \mathbf{e}, \tag{8.15}$$

where $\mathbf{e} = [e_1, e_2, \ldots, e_n]$ is the *error pattern* indicating which codeword symbols were corrupted by the channel. We define the *syndrome* of $\mathbf{R}$ as

$$\mathbf{S} = \mathbf{R}\mathbf{H}^T. \tag{8.16}$$

If $\mathbf{R}$ is a valid codeword (i.e., $\mathbf{R} = \mathbf{C}_i$ for some i) then $\mathbf{S} = \mathbf{C}_i\mathbf{H}^T = \mathbf{0}_{n-k}$ by (8.14). Thus, the syndrome equals the all-zero vector if the transmitted codeword is not corrupted – or is corrupted in a manner such that the received codeword is a valid codeword in the code but is different from the transmitted codeword. If the received codeword $\mathbf{R}$ contains detectable errors, then $\mathbf{S} \neq \mathbf{0}_{n-k}$. If the received codeword contains correctable errors, the syndrome identifies the error pattern corrupting the transmitted codeword, and these errors can then be corrected. Note that the syndrome is a function only of the error pattern $\mathbf{e}$ and not the transmitted codeword $\mathbf{C}$, since

$$\mathbf{S} = \mathbf{R}\mathbf{H}^T = (\mathbf{C} + \mathbf{e})\mathbf{H}^T = \mathbf{C}\mathbf{H}^T + \mathbf{e}\mathbf{H}^T = \mathbf{0}_{n-k} + \mathbf{e}\mathbf{H}^T. \tag{8.17}$$

Because $\mathbf{S} = \mathbf{e}\mathbf{H}^T$ corresponds to $n-k$ equations in n unknowns, there are 2^k possible error patterns that can produce a given syndrome $\mathbf{S}$. However, since the probability of bit error is typically small and independent for each bit, the most likely error pattern is the one with minimal weight, corresponding to the least number of errors introduced in the channel. Thus, if an error pattern $\hat{\mathbf{e}}$ is the most likely error associated with a given syndrome $\mathbf{S}$, the transmitted codeword is typically decoded as

$$\hat{\mathbf{C}} = \mathbf{R} + \hat{\mathbf{e}} = \mathbf{C} + \mathbf{e} + \hat{\mathbf{e}}. \tag{8.18}$$

If the most likely error pattern does occur, then $\hat{\mathbf{e}} = \mathbf{e}$ and $\hat{\mathbf{C}} = \mathbf{C}$ – that is, the corrupted codeword is correctly decoded. The decoding process and associated error probability will be covered in Section 8.2.6.

Let $\mathbf{C}_w$ denote a codeword in a given (n, k) code with minimum weight (excluding the all-zero codeword). Then $\mathbf{C}_w\mathbf{H}^T = \mathbf{0}_{n-k}$ is just the sum of $d_{\min}$ columns of $\mathbf{H}^T$, since $d_{\min}$ equals the number of 1-bits (the weight) in the minimum weight codeword of the code. Since the rank of $\mathbf{H}^T$ is at most $n-k$, this implies that the minimum distance of an (n, k) block code is upper bounded by

$$d_{\min} \leq n - k + 1, \tag{8.19}$$

which is referred to as the *Singelton bound*.

8.2.4 Cyclic Codes

Cyclic codes are a subclass of linear block codes in which all codewords in a given code are cyclic shifts of one another. Specifically, if the codeword $\mathbf{C} = [c_0, c_1, \ldots, c_{n-1}]$ is a codeword in a given code, then a cyclic shift by 1, denoted as $\mathbf{C}^{(1)} = [c_{n-1}, c_0, \ldots, c_{n-2}]$, is also a codeword. More generally, any cyclic shift $C^{(i)} = [c_{n-i}, c_{n-i+1}, \ldots, c_{n-i-1}]$ is also a codeword. The cyclic nature of cyclic codes creates a nice structure that allows their encoding and decoding functions to be of much lower complexity than the matrix multiplications associated with encoding and decoding for general linear block codes. Thus, most linear block codes used in practice are cyclic codes.

Cyclic codes are generated via a *generator polynomial* instead of a generator matrix. The generator polynomial $g(X)$ for an (n, k) cyclic code has degree $n - k$ and is of the form

$$g(X) = g_0 + g_1 X + \cdots + g_{n-k} X^{n-k}, \tag{8.20}$$

where g_i is binary (0 or 1) and $g_0 = g_{n-k} = 1$. The k-bit information sequence $[u_0, \ldots, u_{k-1}]$ is also written in polynomial form as the *message polynomial*

$$u(X) = u_0 + u_1 X + \cdots + u_{k-1} X^{k-1}. \tag{8.21}$$

The codeword associated with a given k-bit information sequence is obtained from the polynomial coefficients of the generator polynomial multiplied by the message polynomial; thus, the codeword $C = [c_0, \ldots, c_{n-1}]$ is obtained from

$$c(X) = u(X)g(X) = c_0 + c_1 X + \cdots + c_{n-1} X^{n-1}. \tag{8.22}$$

A codeword described by a polynomial $c(X)$ is a valid codeword for a cyclic code with generator polynomial $g(X)$ if and only if $g(X)$ divides $c(X)$ with no remainder (no remainder polynomial terms) – that is, if and only if

$$\frac{c(X)}{g(X)} = q(X) \tag{8.23}$$

for a polynomial $q(X)$ of degree less than k.

EXAMPLE 8.3: Consider a $(7, 4)$ cyclic code with generator polynomial $g(X) = 1 + X^2 + X^3$. Determine if the codewords described by polynomials $c_1(X) = 1 + X^2 + X^5 + X^6$ and $c_2(X) = 1 + X^2 + X^3 + X^5 + X^6$ are valid codewords for this generator polynomial.

Solution: Division of binary polynomials is similar to division of standard polynomials except that, under binary addition, subtraction is the same as addition. Dividing $c_1(X) = 1 + X^2 + X^5 + X^6$ by $g(X) = 1 + X^2 + X^3$, we have:

$$
\begin{array}{r}
X^3 + 1 \\
X^3 + X^2 + 1 \overline{)\, X^6 + X^5 + X^2 + 1} \\
\underline{X^6 + X^5 + X^3} \\
X^3 + X^2 + 1 \\
\underline{X^3 + X^2 + 1} \\
0.
\end{array}
$$

Since $g(X)$ divides $c(X)$ with no remainder, it is a valid codeword. In fact, we have $c_1(X) = (1 + X^3)g(X) = u(X)g(X)$ and so the information bit sequence corresponding to $c_1(X)$ is $\mathbf{U} = [1001]$, corresponding to the coefficients of the message polynomial $u(X) = 1 + X^3$.

Dividing $c_2(X) = 1 + X^2 + X^3 + X^5 + X^6$ by $g(X) = 1 + X^2 + X^3$ yields

$$
\begin{array}{r}
X^3 + 1 \\
X^3 + X^2 + 1 \overline{)\, X^6 + X^5 + X^3 + X^2 + 1} \\
\underline{X^6 + X^5 + X^3} \\
X^2 + 1,
\end{array}
$$

where we note that there is a remainder of $X^2 + 1$ in the division. Thus, $c_2(X)$ is not a valid codeword for the code corresponding to this generator polynomial.

Recall that systematic linear block codes have the first k codeword symbols equal to the information bits and the remaining codeword symbols equal to the parity bits. A cyclic code can be put in systematic form by first multiplying the message polynomial $u(X)$ by X^{n-k}, yielding

$$X^{n-k}u(X) = u_0 X^{n-k} + u_1 X^{n-k+1} + \cdots + u_{k-1}X^{n-1}. \tag{8.24}$$

This shifts the message bits to the k rightmost digits of the codeword polynomial. If we next divide (8.24) by $g(X)$, we obtain

$$\frac{X^{n-k}u(X)}{g(X)} = q(X) + \frac{p(X)}{g(X)}, \tag{8.25}$$

where $q(X)$ is a polynomial of degree at most $k - 1$ and $p(X)$ is a remainder polynomial of degree at most $n - k - 1$. Multiplying (8.25) through by $g(X)$, we now have

$$X^{n-k}u(X) = q(X)g(X) + p(X). \tag{8.26}$$

Adding $p(X)$ to both sides yields

$$p(X) + X^{n-k}u(X) = q(X)g(X). \tag{8.27}$$

This implies that $p(X) + X^{n-k}u(X)$ is a valid codeword, since it is divisible by $g(X)$ with no remainder. The codeword is described by the n coefficients of the codeword polynomial $p(X) + X^{n-k}u(X)$. Note that we can express $p(X)$ (of degree $n - k - 1$) as

$$p(X) = p_0 + p_1 X + \cdots + p_{n-k-1}X^{n-k-1}. \tag{8.28}$$

Combining (8.24) and (8.28), we get

$$
\begin{aligned}
&p(X) + X^{n-k}u(X) \\
&\quad = p_0 + p_1 X + \cdots + p_{n-k-1}X^{n-k-1} + u_0 X^{n-k} + u_1 X^{n-k+1} + \cdots + u_{k-1}X^{n-1}.
\end{aligned} \tag{8.29}
$$

Thus, the codeword corresponding to this polynomial has the first k bits consisting of the message bits $[u_0, \ldots, u_k]$ and the last $n - k$ bits consisting of the parity bits $[p_0, \ldots, p_{n-k-1}]$, as is required for the systematic form.

Note that the systematic codeword polynomial is generated in three steps: first multiplying the message polynomial $u(X)$ by X^{n-k}; then dividing $X^{n-k}u(X)$ by $g(X)$ to obtain the remainder polynomial $p(X)$ (along with the quotient polynomial $q(X)$, which is not used); and finally adding $p(X)$ to $X^{n-k}u(X)$ to get (8.29). The polynomial multiplications are straightforward to implement, and the polynomial division is easily implemented with a feedback shift register [1, Chap. 8.1; 2, Chap. 6.7]. Thus, codeword generation for systematic cyclic codes has very low cost and low complexity.

Let us now consider how to characterize channel errors for cyclic codes. The codeword polynomial corresponding to a transmitted codeword is of the form

$$c(X) = u(X)g(X). \tag{8.30}$$

The received codeword can also be written in polynomial form as

$$r(X) = c(X) + e(X) = u(X)g(X) + e(X), \tag{8.31}$$

where $e(X)$ is the error polynomial with coefficients equal to 1 where errors occur. For example, if the transmitted codeword is $\mathbf{C} = [1011001]$ and the received codeword is $\mathbf{R} = [1111000]$, then $e(X) = X + X^{n-1}$. The *syndrome polynomial* $s(X)$ for the received codeword is defined as the remainder when $r(X)$ is divided by $g(X)$, so $s(X)$ has degree $n-k-1$. But by (8.31), the syndrome polynomial $s(X)$ is equivalent to the error polynomial $e(X)$ modulo $g(X)$. Moreover, we obtain the syndrome through a division circuit similar to the one used for generating the code. As stated previously, this division circuit is typically implemented using a feedback shift register, resulting in a low-cost implementation of low complexity.

8.2.5 Hard Decision Decoding (HDD)

The probability of error for linear block codes depends on whether the decoder uses soft decisions or hard decisions. In hard decision decoding (HDD), each coded bit is demodulated as a 0 or 1 – that is, the demodulator detects each coded bit (symbol) individually. For example, in BPSK, the received symbol is decoded as a 1 if it is closer to $\sqrt{E_b}$ and as a 0 if it is closer to $-\sqrt{E_b}$. This form of demodulation removes information that can be used by the channel decoder. In particular, for the BPSK example the distance of the received bit from $\sqrt{E_b}$ and $-\sqrt{E_b}$ can be used in the channel decoder to make better decisions about the transmitted codeword. In soft decision decoding, these distances are used in the decoding process. Soft decision decoding of linear block codes, which is more common in wireless systems than hard decision decoding, is treated in Section 8.2.7.

Hard decision decoding uses *minimum distance* decoding based on Hamming distance. In minimum distance decoding the n bits corresponding to a codeword are first demodulated to a 0 or 1, and the demodulator output is then passed to the decoder. The decoder compares this received codeword to the 2^k possible codewords that constitute the code and decides in favor of the codeword that is closest in Hamming distance to (differs in the least number of bits from) the received codeword. Mathematically, for a received codeword $\mathbf{R}$, the decoder uses the formula

$$\text{pick } \mathbf{C}_j \text{ s.t. } d(\mathbf{C}_j, \mathbf{R}) \leq d(\mathbf{C}_i, \mathbf{R}) \; \forall i \neq j. \tag{8.32}$$

If there is more than one codeword with the same minimum distance to $\mathbf{R}$, one of these is chosen at random by the decoder.

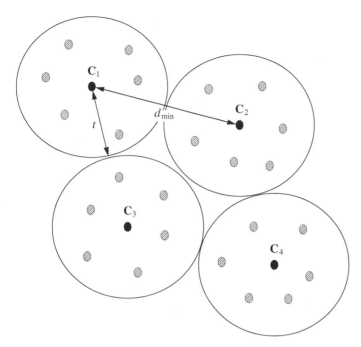

Figure 8.3: Maximum likelihood decoding in code space.

Maximum likelihood decoding picks the transmitted codeword that has the highest probability of having produced the received codeword. In other words, given the received codeword $\mathbf{R}$, the maximum likelihood decoder choses the codeword $\mathbf{C}_j$ as

$$\mathbf{C}_j = \arg \max_i p(\mathbf{R} \mid \mathbf{C}_i), \quad i = 0, \ldots, 2^k - 1. \tag{8.33}$$

Since the most probable error event in an AWGN channel is the event with the minimum number of errors needed to produce the received codeword, the minimum distance criterion (8.32) and the maximum likelihood criterion (8.33) are equivalent. Once the maximum likelihood codeword $\mathbf{C}_i$ is determined, it is decoded to the k bits that produce codeword $\mathbf{C}_i$.

Because maximum likelihood detection of codewords is based on a distance decoding metric, we can best illustrate this process in code space, as shown in Figure 8.3. The minimum Hamming distance between codewords, which are illustrated by the black dots in this figure, is $d_{\min}$. Each codeword is centered inside a sphere of radius $t = \lfloor .5d_{\min} \rfloor$, where $\lfloor x \rfloor$ denotes the largest integer less than or equal to x. The shaded dots represent received sequences where one or more bits differ from those of the transmitted codeword. The figure indicates the Hamming distance between $\mathbf{C}_1$ and $\mathbf{C}_2$.

Minimum distance decoding can be used either to detect or to correct errors. Detected errors in a data block cause either the data to be dropped or a retransmission of the data. Error correction allows the corruption in the data to be reversed. For error correction the minimum distance decoding process ensures that a received codeword lying within a Hamming distance t from the transmitted codeword will be decoded correctly. Thus, the decoder can correct up to t errors, as can be seen from Figure 8.3: since received codewords corresponding to t or fewer errors will lie within the sphere centered around the correct codeword, it will be decoded as that codeword using minimum distance decoding. We see from Figure 8.3

that the decoder can detect all error patterns of $d_{\min} - 1$ errors. In fact, a decoder for an (n, k) code can detect $2^n - 2^k$ possible error patterns. The reason is that there are $2^k - 1$ nondetectable errors, corresponding to the case where a corrupted codeword is exactly equal to a codeword in the set of possible codewords (of size 2^k) that is not equal to the transmitted codeword. Since there are $2^n - 1$ total possible error patterns, this yields $2^n - 2^k$ detectable error patterns. Note that this is not hard decision decoding because we are not correcting errors, just detecting them.

EXAMPLE 8.4: A $(5, 2)$ code has codewords $\mathbf{C}_0 = [00000]$, $\mathbf{C}_1 = [01011]$, $\mathbf{C}_2 = [10101]$, and $\mathbf{C}_3 = [11110]$. Suppose the all-zero codeword $\mathbf{C}_0$ is transmitted. Find the set of error patterns corresponding to nondetectable errors for this codeword transmission.

Solution: The nondetectable error patterns correspond to the three nonzero codewords. That is, $\mathbf{e}_1 = [01011]$, $\mathbf{e}_2 = [10101]$, and $\mathbf{e}_3 = [11110]$ are nondetectable error patterns, since adding any of these to $\mathbf{C}_0$ results in a valid codeword.

8.2.6 Probability of Error for HDD in AWGN

The probability of codeword error, P_e, is defined as the probability that a transmitted codeword is decoded in error. Under hard decision decoding a received codeword *may* be decoded in error if it contains more than t errors (it will not be decoded in error if there is not an alternative codeword closer to the received codeword than the transmitted codeword). The error probability is thus bounded above by the probability that more than t errors occur. Since the bit errors in a codeword occur independently on an AWGN channel, this probability is given by

$$P_e \leq \sum_{j=t+1}^{n} \binom{n}{j} p^j (1 - p)^{n-j}, \tag{8.34}$$

where p is the probability of error associated with transmission of the bits in the codeword. Thus, p corresponds to the error probability associated with uncoded modulation for the given energy per codeword symbol, as treated in Chapter 6 for AWGN channels. For example, if the codeword symbols are sent via coherent BPSK modulation then $p = Q(\sqrt{2E_c/N_0})$, where E_c is the energy per codeword symbol and $N_0/2$ is the noise power spectral density. Since there are k/n information bits per codeword symbol, the relationship between the energy per bit and the energy per symbol is $E_c = kE_b/n$. Thus, powerful block codes with a large number of parity bits (k/n small) reduce the channel energy per symbol and therefore increase the error probability in demodulating the codeword symbols. However, the error correction capability of these codes typically more than compensates for this reduction, especially at high SNRs. At low SNRs this may not happen, in which case the code exhibits *negative coding gain* – it has a higher error probability than uncoded modulation. The bound (8.34) holds with equality when the decoder corrects exactly t or fewer errors in a codeword and cannot correct for more than t errors in a codeword. A code with this property is called a *perfect* code.

At high SNRs, the most likely way to make a codeword error is to mistake a codeword for one of its nearest neighbors. Nearest neighbor errors yield a pair of upper and lower bounds on error probability. The lower bound is the probability of mistaking a codeword for a given nearest neighbor at distance $d_{\min}$:

$$P_e \geq \sum_{j=t+1}^{d_{\min}} \binom{d_{\min}}{j} p^j (1-p)^{d_{\min}-j}. \tag{8.35}$$

The upper bound, a union bound, assumes that all of the other $2^k - 1$ codewords are at distance $d_{\min}$ from the transmitted codeword. Thus, the union bound is just $2^k - 1$ times (8.35), the probability of mistaking a given codeword for a nearest neighbor at distance $d_{\min}$:

$$P_e \leq (2^k - 1) \sum_{j=t+1}^{d_{\min}} \binom{d_{\min}}{j} p^j (1-p)^{d_{\min}-j}. \tag{8.36}$$

When the number of codewords is large or the SNR is low, both of these bounds are quite loose.

A tighter upper bound can be obtained by applying the Chernoff bound, $P(X \geq x) \leq e^{-x^2/2}$ for X a zero-mean, unit-variance, Gaussian random variable, to compute codeword error probability. Using this bound, it can be shown [3, Chap. 5.2] that the probability of decoding the all-zero codeword as the jth codeword with weight w_j is upper bounded by

$$P(w_j) \leq [4p(1-p)]^{w_j/2}. \tag{8.37}$$

Since the probability of decoding error is upper bounded by the probability of mistaking the all-zero codeword for any of the other codewords, we obtain the upper bound

$$P_e \leq \sum_{j=1}^{2^k-1} [4p(1-p)]^{w_j/2}. \tag{8.38}$$

This bound requires the weight distribution $\{w_j\}_{j=1}^{2^k-1}$ for all codewords (other than the all-zero codeword corresponding to $j = 0$) in the code. A simpler, slightly looser upper bound is obtained from (8.38) by using $d_{\min}$ instead of the individual codeword weights. This simplification yields the bound

$$P_e \leq (2^k - 1)[4p(1-p)]^{d_{\min}/2}. \tag{8.39}$$

Note that the probability of codeword error P_e depends on p, which is a function of the Euclidean distance between modulation points associated with the transmitted codeword symbols. In fact, the best codes for AWGN channels should not be based on Hamming distance: they should be based on maximizing the Euclidean distance between the codewords after modulation. However, this requires that the channel code be designed jointly with the modulation. This is the basic concept of coded modulation, which will be discussed in Section 8.7. However, Hamming distance is a better measure of code performance in fading when codes are combined with interleaving, as discussed in Section 8.8.

The probability of bit error after decoding the received codeword depends in general on the specific code and decoder and in particular on how bits are mapped to codewords, as in the bit mapping procedure associated with nonbinary modulation. This bit error probability is often approximated as

$$P_b \approx \frac{1}{n} \sum_{j=t+1}^{n} j \binom{n}{j} p^j (1-p)^{n-j} \tag{8.40}$$

[2, Chap. 6.5]; for $t = 1$, this can be simplified [2] to $P_b \approx p - p(1-p)^{n-1}$.

EXAMPLE 8.5: Consider a $(24, 12)$ linear block code with a minimum distance $d_{\min} = 8$ (an extended Golay code, discussed in Section 8.2.8, is one such code). Find P_e based on the loose bound (8.39), assuming the codeword symbols are transmitted over the channel using BPSK modulation with $E_b/N_0 = 10$ dB. Also find P_b for this code using the approximation $P_b = P_e/k$ and compare with the bit error probability for uncoded modulation.

Solution: For $E_b/N_0 = 10$ dB we have $E_c/N_0 = \frac{12}{24}10 = 5$. Thus, $p = Q(\sqrt{10}) = 7.82 \cdot 10^{-4}$. Using this value in (8.39) with $k = 12$ and $d_{\min} = 8$ yields $P_e \leq 3.92 \cdot 10^{-7}$. Using the P_b approximation we get $P_b \approx (1/k)P_e = 3.27 \cdot 10^{-8}$. For uncoded modulation we have $P_b = Q(\sqrt{2E_b/N_0}) = Q(\sqrt{20}) = 3.87 \cdot 10^{-6}$. Thus we obtain over two orders of magnitude performance gain with this code. Note that the loose bound can be orders of magnitude away from the true error probability, so this calculation may significantly underestimate the performance gain of the code.

8.2.7 Probability of Error for SDD in AWGN

The HDD described in the previous section discards information that can reduce probability of codeword error. For example, in BPSK, the transmitted signal constellation is $\pm\sqrt{E_b}$ and the received symbol after matched filtering is decoded as a 1 if it is closer to $\sqrt{E_b}$ and as a 0 if it is closer to $-\sqrt{E_b}$. Thus, the distance of the received symbol from $\sqrt{E_b}$ and $-\sqrt{E_b}$ is not used in decoding, yet this information can be used to make better decisions about the transmitted codeword. When these distances are used in the channel decoder it is called *soft decision decoding* (SDD), since the demodulator does not make a hard decision about whether a 0 or 1 bit was transmitted but rather makes a soft decision corresponding to the distance between the received symbol and the symbol corresponding to a 0-bit or a 1-bit transmission. We now describe the basic premise of SDD for BPSK modulation; these ideas are easily extended to higher-level modulations.

Consider a codeword transmitted over a channel using BPSK. As in the case of HDD, the energy per codeword symbol is $E_c = (k/n)E_b$. If the jth codeword symbol is a 1 it will be received as $r_j = \sqrt{E_c} + n_j$ and if it is a 0 it will be received as $r_j = -\sqrt{E_c} + n_j$, where n_j is the AWGN sample of mean zero and variance $N_0/2$ associated with the receiver. In SDD, given a received codeword $\mathbf{R} = [r_1, \ldots, r_n]$, the decoder forms a *correlation metric* $C(\mathbf{R}, \mathbf{C}_i)$ for each codeword $\mathbf{C}_i$ ($i = 0, \ldots, 2^k - 1$) in the code and then the decoder chooses the codeword $\mathbf{C}_i$ with the highest correlation metric. The correlation metric is defined as

$$C(\mathbf{R}, \mathbf{C}_i) = \sum_{j=1}^{n} (2c_{ij} - 1)r_j, \tag{8.41}$$

where c_{ij} denotes the jth coded bit in the codeword $\mathbf{C}_i$. If $c_{ij} = 1$ then $2c_{ij} - 1 = 1$ and if $c_{ij} = 0$ then $2c_{ij} - 1 = -1$. Thus the received codeword symbol is weighted by the polarity associated with the corresponding symbol in the codeword for which the correlation metric is being computed. Hence $C(\mathbf{R}, \mathbf{C}_i)$ is large when most of the received symbols have a large magnitude and the same polarity as the corresponding symbols in $\mathbf{C}_i$, is smaller when most of the received symbols have a small magnitude and the same polarity as the corresponding symbols in $\mathbf{C}_i$, and is typically negative when most of the received symbols have a different polarity than the corresponding symbols in $\mathbf{C}_i$. In particular, at very high SNRs, if $\mathbf{C}_i$ is transmitted then $C(\mathbf{R}, \mathbf{C}_i) \approx n\sqrt{E_c}$ while $C(\mathbf{R}, \mathbf{C}_j) < n\sqrt{E_c}$ for $j \neq i$.

For an AWGN channel, the probability of codeword error is the same for any codeword of a linear code. Error analysis is typically easiest when assuming transmission of the all-zero codeword. Let us therefore assume that the all-zero codeword $\mathbf{C}_0$ is transmitted and the corresponding received codeword is $\mathbf{R}$. To correctly decode $\mathbf{R}$, we must have that $C(\mathbf{R}, \mathbf{C}_0) > C(\mathbf{R}, \mathbf{C}_i)$, $i = 1, \ldots, 2^k - 1$. Let w_i denote the Hamming weight of the ith codeword $\mathbf{C}_i$, which equals the number of 1-bits in $\mathbf{C}_i$. Then, conditioned on the transmitted codeword $\mathbf{C}_i$, it follows that $C(\mathbf{R}, \mathbf{C}_i)$ is Gauss distributed with mean $\sqrt{E_c}n(1 - 2w_i/n)$ and variance $nN_0/2$. Note that the correlation metrics are not independent, since they are all functions of $\mathbf{R}$. The probability $P_e(\mathbf{C}_i) = p(C(\mathbf{R}, \mathbf{C}_0) < C(\mathbf{R}, \mathbf{C}_i))$ can be shown to equal the probability that a Gauss-distributed random variable with variance $2w_i N_0$ is less than $-2w_i\sqrt{E_c}$; that is,

$$P_e(\mathbf{C}_i) = Q\left(\frac{2w_i\sqrt{E_c}}{\sqrt{2w_i N_0}}\right) = Q\left(\sqrt{2w_i \gamma_b R_c}\right). \tag{8.42}$$

Then, by the union bound, the probability of error is upper bounded by the sum of pairwise error probabilities relative to each $\mathbf{C}_i$:

$$P_e \leq \sum_{i=1}^{2^k-1} P_e(\mathbf{C}_i) = \sum_{i=1}^{2^k-1} Q\left(\sqrt{2w_i \gamma_b R_c}\right). \tag{8.43}$$

Computing (8.43) requires the weight distribution w_i $(i = 1, \ldots, 2^k - 1)$ of the code. This bound can be simplified by noting that $w_i \geq d_{\min}$, so

$$P_e \leq (2^k - 1)Q\left(\sqrt{2\gamma_b R_c d_{\min}}\right). \tag{8.44}$$

The Chernoff bound on the Q-function is $Q(\sqrt{2x}) < e^{-x}$. Applying this bound to (8.43) yields

$$P_e \leq (2^k - 1)e^{-\gamma_b R_c d_{\min}} < 2^k e^{-\gamma_b R_c d_{\min}} = e^{-\gamma_b R_c d_{\min} + k \ln 2}. \tag{8.45}$$

Comparing this bound with that of uncoded BPSK modulation,

$$P_b = Q\left(\sqrt{2\gamma_b}\right) < e^{-\gamma_b}, \tag{8.46}$$

we get a dB coding gain of approximately

$$G_c = 10 \log_{10}[(\gamma_b R_c d_{\min} - k \ln 2)/\gamma_b] = 10 \log_{10}[R_c d_{\min} - (k \ln 2)/\gamma_b]. \tag{8.47}$$

Note that the coding gain depends on the code rate, the number of information bits per codeword, the minimum distance of the code, and the channel SNR. In particular, the coding gain decreases as γ_b decreases, and it becomes negative at sufficiently low SNRs. In general the performance of SDD is about 2–3 dB better than HDD [1, Chap. 8.1].

EXAMPLE 8.6: Find the approximate coding gain of SDD over uncoded modulation for the $(24, 12)$ code with $d_{\min} = 8$ considered in Example 8.5, with $\gamma_b = 10$ dB.

Solution: Setting $\gamma_b = 10$, $R_c = 12/24$, $d_{\min} = 8$, and $k = 12$ in (8.47) yields $G_c = 5$ dB. This significant coding gain is a direct result of the large minimum distance of the code.

8.2.8 Common Linear Block Codes

We now describe some common linear block codes. More details can be found in [1; 2; 3; 4]. The most common type of block code is a Hamming code, which is parameterized by an integer $m \geq 2$. For an (n, k) Hamming code, $n = 2^m - 1$ and $k = 2^m - m - 1$, so $n - k = m$ redundant bits are introduced by the code. The minimum distance of all Hamming codes is $d_{\min} = 3$, so $t = 1$ error in the $n = 2^m - 1$ codeword symbols can be corrected. Although Hamming codes are not very powerful, they are perfect codes and thus have probability of error given exactly by the right side of (8.34).

Golay and extended Golay codes are another class of channel codes with good performance. The Golay code is a linear $(23, 12)$ code with $d_{\min} = 7$ and $t = 3$. The extended Golay code is obtained by adding a single parity bit to the Golay code, resulting in a $(24, 12)$ block code with $d_{\min} = 8$ and $t = 3$. The extra parity bit does not change the error correction capability (since t remains the same), but it greatly simplifies implementation because the information bit rate is exactly half the coded bit rate. Thus, both uncoded and coded bit streams can be generated by the same clock, using every other clock sample to generate the uncoded bits. These codes have higher $d_{\min}$ and thus better error correction capabilities than Hamming codes, but at a cost of more complex decoding and a lower code rate $R_c = k/n$. The lower code rate implies that the code either has a lower data rate or requires additional bandwidth.

Another powerful class of block codes is the Bose–Chadhuri–Hocquenghem (BCH) codes. These are cyclic codes, and at high rates they typically outperform all other block codes with the same n and k at moderate to high SNRs. This code class provides a large selection of blocklengths, code rates, and error correction capabilities. In particular, the most common BCH codes have $n = 2^m - 1$ for any integer $m \geq 3$.

The P_b for a number of BCH codes under hard decision decoding and coherent BPSK modulation is shown in Figure 8.4. The plot is based on the approximation (8.40), where for coherent BPSK we have

$$p = Q\left(\sqrt{\frac{2E_c}{N_0}}\right) = Q(\sqrt{2R_c\gamma_b}). \tag{8.48}$$

In the figure, the BCH $(127, 36)$ code actually has a negative coding gain at low SNRs. This is not uncommon for powerful channel codes owing to their reduced energy per symbol, as discussed in Section 8.2.6.

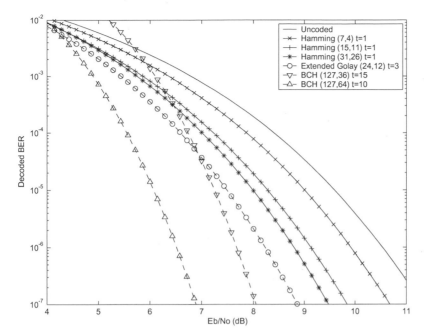

Figure 8.4: P_b for different BCH codes.

8.2.9 Nonbinary Block Codes: The Reed Solomon Code

A nonbinary block code has similar properties as the binary code: it has K information symbols mapped into codewords of length N. However, the N codeword symbols of each codeword are chosen from a nonbinary alphabet of size $q > 2$. Thus the codeword symbols can take any value in $\{0, 1, \dots, q-1\}$. Usually $q = 2^k$, so that k bits can be mapped into one symbol.

The most common nonbinary block code is the Reed Solomon (RS) code, used in a range of applications from magnetic recording to cellular digital packet data (CDPD). Reed Solomon codes have $N = q - 1 = 2^k - 1$ and $K = 1, 2, \dots, N - 1$. The value of K dictates the error correction capability of the code. Specifically, an RS code can correct up to $t = .5 \lfloor N - K \rfloor$ codeword symbol errors. In nonbinary codes the minimum distance between codewords is defined as the number of codeword symbols in which the codewords differ. Reed Solomon codes achieve a minimum distance of $d_{\min} = N - K + 1$, which is the largest possible minimum distance between codewords for any linear code with the same encoder input and output blocklengths. Reed Solomon codes are often *shortened* to meet the requirements of a given system [4, Chap. 5.10].

Because nonbinary codes – and RS codes in particular – generate symbols corresponding to 2^k bits, they are sometimes used with M-ary modulation where $M = 2^k$. In particular, with 2^k-ary modulation each codeword symbol is transmitted over the channel as one of 2^k possible constellation points. If the error probability associated with the modulation (the probability of mistaking the received constellation point for a constellation point other than the transmitted point) is P_M, then the probability of codeword error associated with the nonbinary code is upper bounded by

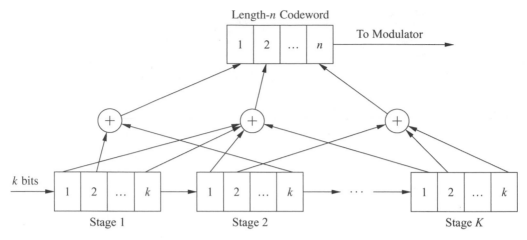

Figure 8.5: Convolutional encoder.

$$P_e \leq \sum_{j=t+1}^{N} \binom{N}{j} P_M^j (1 - P_M)^{N-j}, \tag{8.49}$$

which is similar to the form of (8.34) for the binary code. We can then approximate the probability of information symbol error as

$$P_s \approx \frac{1}{N} \sum_{j=t+1}^{N} j \binom{N}{j} P_M^j (1 - P_M)^{N-j}. \tag{8.50}$$

8.3 Convolutional Codes

A convolutional code generates coded symbols by passing the information bits through a linear finite-state shift register, as shown in Figure 8.5. The shift register consists of K stages with k bits per stage. There are n binary addition operators with inputs taken from all K stages: these operators produce a codeword of length n for each k-bit input sequence. Specifically, the binary input data is shifted into each stage of the shift register k bits at a time, and each of these shifts produces a coded sequence of length n. The rate of the code is $R_c = k/n$. The maximum span of output symbols that can be influenced by a given input bit in a convolutional code is called the *constraint length* of the code. It is clear from Figure 8.5 that a length-n codeword depends on kK input bits – in contrast to a block code, which only depends on k input bits. Thus, the constraint length of the encoder is kK bits or, equivalently, K k-bit bytes. Convolutional codes are said to have memory since the current codeword depends on more input bits (kK) than the number input to the encoder to generate it (k). Note that in general a convolutional encoder may not have the same number of bits per stage.

8.3.1 Code Characterization: Trellis Diagrams

When a length-n codeword is generated by the convolutional encoder of Figure 8.5, this codeword depends both on the k bits input to the first stage of the shift register as well as the *state* of the encoder, defined as the contents in the other $K - 1$ stages of the shift register. In order

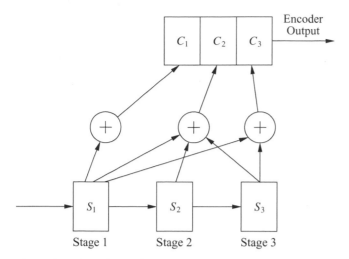

Figure 8.6: Convolutional encoder example ($n = 3$, $k = 1$, $k = 3$).

to characterize this convolutional code, we must characterize how the codeword generation depends both on the k input bits and the encoder state, which has $2^{k(K-1)}$ possible values. There are multiple ways to characterize convolutional codes, including a tree diagram, state diagram, and trellis diagram [1, Chap. 8.2]. The tree diagram represents the encoder in the form of a tree, where each branch represents a different encoder state and the corresponding encoder output. A state diagram is a graph showing the different states of the encoder and the possible state transitions and corresponding encoder outputs. A trellis diagram uses the fact that the tree representation repeats itself once the number of stages in the tree exceeds the constraint length of the code. The trellis diagram simplifies the tree representation by merging nodes in the tree corresponding to the same encoder state. In this section we focus on the trellis representation of a convolutional code, since this is the most common characterization. The details of trellis diagram representation are best described by an example.

Consider the convolutional encoder shown in Figure 8.6 with $n = 3$, $k = 1$, and $K = 3$. In this encoder, one bit at a time is shifted into Stage 1 of the three-stage shift register. At a given time t we denote the bit in Stage i of the shift register as S_i. The three stages of the shift register are used to generate a codeword of length 3, $C_1 C_2 C_3$; from the figure we see that $C_1 = S_1$, $C_2 = S_1 + S_2 + S_3$, and $C_3 = S_1 + S_3$. A bit sequence **U** shifted into the encoder generates a sequence of coded symbols, which we denote by **C**. Note that the coded symbols corresponding to C_1 are just the original information bits. As with block codes, when one of the coded symbols in a convolutional code corresponds to the original information bits we say that the code is *systematic*. We define the encoder state as $S = S_2 S_3$ (i.e., the contents of the last two stages of the encoder), and there are $2^2 = 4$ possible values for this encoder state. To characterize the encoder, we must show for each input bit and each possible encoder state what the encoder output will be, and we must also show how the new input bit changes the encoder state for the next input bit.

The trellis diagram for this code is shown in Figure 8.7. The solid lines in Figure 8.7 indicate the encoder state transition when a 0-bit is input to Stage 1 of the encoder, and the dashed lines indicate the state transition corresponding to a 1-bit input. For example, starting

$S = S_2 S_3$

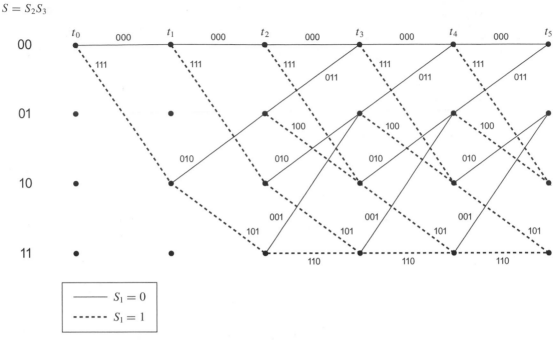

Figure 8.7: Trellis diagram.

at state $S = 00$, if a 0-bit is input to Stage 1 then, when the shift register transitions, the new state will remain as $S = 00$ (since the 0 in Stage 1 transitions to Stage 2, and the 0 in Stage 2 transitions to Stage 3, resulting in the new state $S = S_2 S_3 = 00$). On the other hand, if a 1-bit is input to Stage 1 then, when the shift register transitions, the new state will become $S = 10$ (since the 1 in Stage 1 transitions to Stage 2, and the 0 in Stage 2 transitions to Stage 3, re-sulting in the new state $S = S_2 S_3 = 10$). The encoder output corresponding to a particular encoder state S and input S_1 is written next to the transition lines in the figure. This output is the encoder output that results from the encoder addition operations on the bits S_1, S_2, and S_3 in each stage of the encoder. For example, if $S = 00$ and $S_1 = 1$ then the encoder output $C_1 C_2 C_3$ has $C_1 = S_1 = 1$, $C_2 = S_1 + S_2 + S_3 = 1$, and $C_3 = S_1 + S_3 = 1$. This output 111 is drawn next to the dashed line transitioning from state $S = 00$ to state $S = 10$ in Figure 8.7. Note that the encoder output for $S_1 = 0$ and $S = 00$ is always the all-zero codeword regard-less of the addition operations that form the codeword $C_1 C_2 C_3$, since summing together any number of 0s always yields 0. The portion of the trellis between time t_i and t_{i+1} is called the ith *branch* of the trellis. Figure 8.7 indicates that the initial state at time t_0 is the all-zero state. The trellis achieves *steady state,* defined as the point where all states can be entered from either of two preceding states, at time t_3. After this steady state is reached, the trellis repeats itself in each time interval. Note also that, in steady state, each state transitions to one of two possible new states. In general, trellis structures starting from the all-zero state at time t_0 achieve steady state at time t_K.

For general values of k and K, the trellis diagram will have 2^{K-1} states, where each state has 2^k paths entering each node and 2^k paths leaving each node. Thus, the number of paths through the trellis grows exponentially with k, K, and the length of the trellis path.

EXAMPLE 8.7: Consider the convolutional code represented by the trellis in Figure 8.7. For an initial state $S = S_2 S_3 = 01$, find the state sequence S and the encoder output C for input bit sequence $\mathbf{U} = 011$.

Solution: The first occurrence of $S = 01$ in the trellis is at time t_2. We see at t_2 that if the information bit $S_1 = 0$ then we follow the solid line in the trellis from $S = 01$ at t_2 to $S = 00$ at t_3, and the output corresponding to this path through the trellis is $C = 011$. Now at t_3, starting at $S = 00$, for the information bit $S_1 = 1$ we follow the dashed line in the trellis to $S = 10$ at t_4, and the output corresponding to this path through the trellis is $C = 111$. Finally, at t_4, starting at $S = 10$, for the information bit $S_1 = 1$ we follow the dashed line in the trellis to $S = 11$ at t_5, and the output corresponding to this path through the trellis is $C = 101$.

8.3.2 Maximum Likelihood Decoding

The convolutional code generated by the finite state shift register is basically a finite-state machine. Thus, unlike an (n, k) block code – where maximum likelihood detection entails finding the length-n codeword that is closest to the received length-n codeword – maximum likelihood detection of a convolutional code entails finding the most likely sequence of coded symbols $\mathbf{C}$ given the received sequence of coded symbols, which we denote by $\mathbf{R}$. In particular, for a received sequence $\mathbf{R}$, the decoder decides that coded symbol sequence $\mathbf{C}^*$ was transmitted if

$$p(\mathbf{R} \mid \mathbf{C}^*) \geq p(\mathbf{R} \mid \mathbf{C}) \quad \forall \mathbf{C}. \tag{8.51}$$

Since each possible sequence $\mathbf{C}$ corresponds to one path through the trellis diagram of the code, maximum likelihood decoding corresponds to finding the maximum likelihood path through the trellis diagram. For an AWGN channel, noise affects each coded symbol independently. Thus, for a convolutional code of rate $1/n$, we can express the likelihood (8.51) for a path of length L through the trellis as

$$p(\mathbf{R} \mid \mathbf{C}) = \prod_{i=0}^{L-1} p(R_i \mid C_i) = \prod_{i=0}^{L-1} \prod_{j=1}^{n} p(R_{ij} \mid C_{ij}), \tag{8.52}$$

where C_i is the portion of the code sequence $\mathbf{C}$ corresponding to the ith branch of the trellis, R_i is the portion of the received code sequence $\mathbf{R}$ corresponding to the ith branch of the trellis, C_{ij} is the jth coded symbol corresponding to C_i, and R_{ij} is the jth received coded symbol corresponding to R_i. The log likelihood function is defined as the log of $p(\mathbf{R} \mid \mathbf{C})$, given as

$$\log p(\mathbf{R} \mid \mathbf{C}) = \sum_{i=0}^{L-1} \log p(R_i \mid C_i) = \sum_{i=0}^{L-1} \sum_{j=1}^{n} \log p(R_{ij} \mid C_{ij}). \tag{8.53}$$

The expression

$$B_i = \sum_{j=1}^{n} \log p(R_{ij} \mid C_{ij}) \tag{8.54}$$

is called the *branch metric* because it indicates the component of (8.53) associated with the ith branch of the trellis. The sequence or path that maximizes the likelihood function also maximizes the log likelihood function, since the log is monotonically increasing. However, it is computationally more convenient for the decoder to use the log likelihood function because it involves a summation rather than a product. The log likelihood function associated with a given path through the trellis is also called the *path metric,* which by (8.53) is equal to the sum of branch metrics along each branch of the path. The path through the trellis with the maximum path metric corresponds to the maximum likelihood path.

The decoder can use either hard or soft decisions for the expressions $\log p(R_{ij} \mid C_{ij})$ in the log likelihood metric. For hard decision decoding, the R_{ij} is decoded as a 1 or a 0. The probability of hard decision decoding error depends on the modulation and is denoted as p. If $\mathbf{R}$ and $\mathbf{C}$ are N symbols long and differ in d places (i.e., their Hamming distance is d), then

$$p(\mathbf{R} \mid \mathbf{C}) = p^d(1 - p)^{N-d}$$

and

$$\log p(\mathbf{R} \mid \mathbf{C}) = -d \log \frac{1 - p}{p} + N \log(1 - p). \tag{8.55}$$

Since $p < .5$, (8.55) is maximized when d is minimized. So the coded sequence $\mathbf{C}$ with minimum Hamming distance to the received sequence $\mathbf{R}$ corresponds to the maximum likelihood sequence.

In soft decision decoding, the value of the received coded symbols (R_{ij}) are used directly in the decoder, rather than quantizing them to 1 or 0. For example, if the C_{ij} are sent via BPSK over an AWGN channel with a 1 mapped to $\sqrt{E_c}$ and a 0 mapped to $-\sqrt{E_c}$, then

$$R_{ij} = \sqrt{E_c}(2C_{ij} - 1) + n_{ij}, \tag{8.56}$$

where $E_c = kE_b/n$ is the energy per coded symbol and n_{ij} denotes Gaussian noise of mean zero and variance $\sigma^2 = .5N_0$. Thus,

$$p(R_{ij} \mid C_{ij}) = \frac{1}{\sqrt{2\pi}\sigma} \exp\left[-\frac{(R_{ij} - \sqrt{E_c}(2C_{ij} - 1))^2}{2\sigma^2}\right]. \tag{8.57}$$

Maximizing this likelihood function is equivalent to choosing the C_{ij} that is closest in Euclidean distance to R_{ij}. In determining which sequence $\mathbf{C}$ maximizes the log likelihood function (8.53), any terms that are common to two different sequences $\mathbf{C}_1$ and $\mathbf{C}_2$ can be neglected, since they contribute the same amount to the summation. Similarly, we can scale all terms in (8.53) without changing the maximizing sequence. Thus, by neglecting scaling factors and terms in (8.57) that are common to any C_{ij}, we can replace $\sum_{j=1}^{n} \log p(R_{ij} \mid C_{ij})$ in (8.53) with the *equivalent branch metric*

$$\mu_i = \sum_{j=1}^{n} R_{ij}(2C_{ij} - 1) \tag{8.58}$$

and obtain the same maximum likelihood output.

We now illustrate the path metric computation under both hard and soft decisions for the convolutional code of Figure 8.6 with the trellis diagram in Figure 8.7. For simplicity, we will consider only two possible paths through the trellis and compute their corresponding likelihoods for a given received sequence $\mathbf{R}$. Assume we start at time t_0 in the all-zero state.

The first path we consider is the all-zero path, corresponding to the all-zero input sequence. The second path we consider starts in state $S = 00$ at time t_0 and transitions to state $S = 10$ at time t_1, then to state $S = 01$ at time t_2, and finally to state $S = 00$ at time t_3, at which point this path merges with the all-zero path. Since the paths and therefore their branch metrics at times $t < t_0$ and $t \geq t_3$ are the same, the maximum likelihood path corresponds to the path whose sum of branch metrics over the branches in which the two paths differ is smaller. From Figure 8.7 we see that the all-zero path through the trellis generates the coded sequence $\mathbf{C}_0 = 000000000$ over the first three branches in the trellis. The second path generates the coded sequence $\mathbf{C}_1 = 111010011$ over the first three branches in the trellis.

Let us first consider hard decision decoding with error probability p. Suppose the received sequence over these three branches is $\mathbf{R} = 100110111$. Note that the Hamming distance between $\mathbf{R}$ and $\mathbf{C}_0$ is 6 while the Hamming distance between $\mathbf{R}$ and $\mathbf{C}_1$ is 4. As discussed previously, the most likely path therefore corresponds to $\mathbf{C}_1$, since it has minimum Hamming distance to $\mathbf{R}$. The path metric for the all-zero path is

$$M_0 = \sum_{i=0}^{2} \sum_{j=1}^{3} \log P(R_{ij} \mid C_{ij}) = 6 \log p + 3 \log(1 - p), \qquad (8.59)$$

while the path metric for the other path is

$$M_1 = \sum_{i=0}^{2} \sum_{j=1}^{3} \log P(R_{ij} \mid C_{ij}) = 4 \log p + 5 \log(1 - p). \qquad (8.60)$$

Assuming $p \ll 1$, which is generally the case, this yields $M_0 \approx 6 \log p$ and $M_1 \approx 4 \log p$. Since $\log p < 1$, this confirms that the second path has a larger path metric than the first.

Let us now consider soft decision decoding over time t_0 to t_3. Suppose the received sequence (before demodulation) over these three branches, for $E_c = 1$, is $\mathbf{R} = (.8, -.35, -.15, 1.35, 1.22, -.62, .87, 1.08, .91)$. The path metric for the all-zero path is

$$M_0 = \sum_{i=0}^{2} \mu_i = \sum_{i=0}^{2} \sum_{j=1}^{3} R_{ij}(2C_{ij} - 1) = \sum_{i=0}^{2} \sum_{j=1}^{3} -R_{ij} = -5.11,$$

and the path metric for the second path is

$$M_1 = \sum_{i=0}^{2} \sum_{j=1}^{3} R_{ij}(2C_{ij} - 1) = 1.91.$$

Thus, the second path has a higher path metric than the first. In order to determine if the second path is the maximum likelihood path, we must compare its path metric to that of all other paths through the trellis.

The difficulty with maximum likelihood decoding is that the complexity of computing the log likelihood function (8.53) grows exponentially with the memory of the code, and this computation must be done for every possible path through the trellis. The Viterbi algorithm, discussed in the next section, reduces the complexity of maximum likelihood decoding by taking advantage of the structure of the path metric computation.

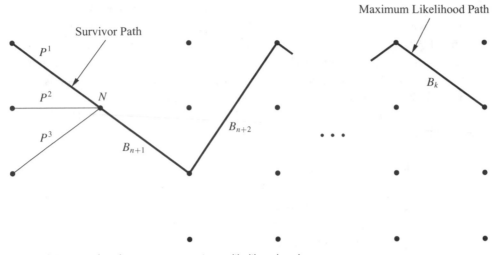

Figure 8.8: Partial path metrics on maximum likelihood path.

8.3.3 The Viterbi Algorithm

The Viterbi algorithm, introduced by Viterbi in 1967 [5], reduces the complexity of maximum likelihood decoding by systematically removing paths from consideration that cannot achieve the highest path metric. The basic premise is to look at the partial path metrics associated with all paths *entering* a given node (node N) in the trellis. Since the possible paths through the trellis *leaving* node N are the same for each *entering* path, the complete trellis path with the highest path metric that goes through node N must coincide with the path that has the highest partial path metric up to node N. This is illustrated in Figure 8.8, where path 1, path 2, and path 3 enter node N (at trellis depth n) with partial path metrics $P^l = \sum_{k=0}^{n-1} B_k^l$ ($l = 1, 2, 3$) up to this node. Assume P^1 is the largest of these partial path metrics. The complete path with the highest metric has branch metrics $\{B_k\}$ after node N. The maximum likelihood path starting from node N (i.e., the path starting from node N with the largest path metric) has partial path metric $\sum_{k=n}^{\infty} B_k$. The complete path metric for path l ($l = 1, 2, 3$) up to node N and the maximum likelihood path after node N is $P^l + \sum_{k=n}^{\infty} B_k$ ($l = 1, 2, 3$), and thus the path with the maximum partial path metric P^l up to node N (path 1 in this example) must correspond to the path with the largest path metric that goes through node N.

The Viterbi algorithm takes advantage of this structure by discarding all paths entering a given node except the path with the largest partial path metric up to that node. The path that is not discarded is called the *survivor path*. Thus, for the example of Figure 8.8, path 1 is the survivor at node N and paths 2 and 3 are discarded from further consideration. Hence, at every stage in the trellis there are 2^{K-1} surviving paths, one for each possible encoder state. A branch for a given stage of the trellis cannot be decoded until all surviving paths at a subsequent trellis stage overlap with that branch; see Figure 8.9, which shows the surviving paths at time t_{k+3}. We see in the figure that all of these surviving paths can be traced back to a *common stem* from time t_k to t_{k+1}. At this point the decoder can output the codeword symbol C_i associated with this branch of the trellis. Note that there is not a fixed decoding delay associated with how far back in the trellis a common stem occurs for a given set of surviving paths,

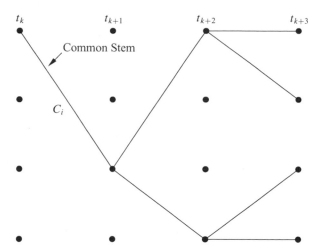

Figure 8.9: Common stem for all survivor paths in the trellis.

since this delay depends on k, K, and the specific code properties. To avoid a random decoding delay, the Viterbi algorithm is typically modified so that, at a given stage in the trellis, the most likely branch n stages back is decided upon based on the partial path metrics up to that point. Although this modification does not yield exact maximum likelihood decoding, for n sufficiently large (typically $n \geq 5K$) it is a good approximation.

The Viterbi algorithm must keep track of $2^{k(K-1)}$ surviving paths and their corresponding metrics. At each stage, in order to determine the surviving path, 2^k metrics must be computed for each node corresponding to the 2^k paths entering each node. Thus, the number of computations in decoding and the memory requirements for the algorithm increase exponentially with k and K. This implies that practical implementations of convolutional codes are restricted to relatively small values of k and K.

8.3.4 Distance Properties

As with block codes, the error correction capability of convolutional codes depends on the distance between codeword sequences. Since convolutional codes are linear, the minimum distance between all codeword sequences can be found by determining the minimum distance from any sequence or, equivalently, determining any trellis path to the all-zero sequence/trellis path. Clearly the trellis path with minimum distance to the all-zero path will diverge and remerge with the all-zero path, so that the two paths coincide except over some number of trellis branches. To find this minimum distance path we must consider all paths that diverge from the all-zero state and then remerge with this state. As an example, in Figure 8.10 we draw all paths in Figure 8.7 between times t_0 and t_5 that diverge and remerge with the all-zero state. Note that path 2 is identical to path 1 – just shifted in time – and thus is not considered as a separate path. Note also that we could look over a longer time interval, but any paths that diverge and remerge over this longer interval would traverse the same branches (shifted in time) as one of these paths plus some additional branches and would therefore have larger path metrics. In particular, we see that path 4 traverses the trellis branches 00-10-01-10-01-00, whereas path 1 traverses the branches 00-10-01-00. Since

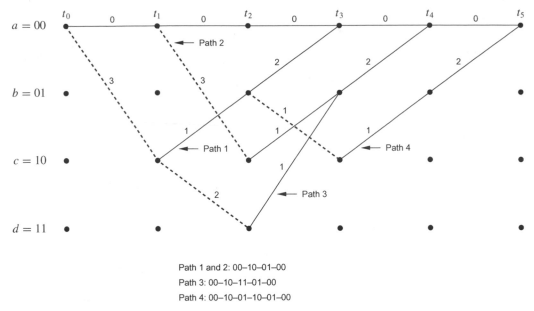

Path 1 and 2: 00–10–01–00
Path 3: 00–10–11–01–00
Path 4: 00–10–01–10–01–00

Figure 8.10: Path distances to the all-zero path.

path 4 traverses the same branches as path 1 on its first, second, and last transition – and since it has additional transitions – its path metric will be no smaller than the meteric of path 1. Thus we need not consider a longer time interval to find the minimum distance path. For each path in Figure 8.10 we label the Hamming distance of the codeword on each branch to the all-zero codeword in the corresponding branch of the all-zero path. By summing up the Hamming distances on all branches of each path, we see that path 1 has a Hamming distance of 6 and that paths 3 and 4 have Hamming distances of 8. Recalling that dashed lines indicate 1-bit inputs while solid lines indicate 0-bit inputs, we see that path 1 corresponds to an input bit sequence from t_0 to t_5 of 10000, path 3 corresponds to an input bit sequence of 11000, and path 4 corresponds to an input bit sequence of 10100. Thus, path 1 results in one bit error relative to the all-zero squence, and paths 3 and 4 result in two bit errors.

We define the *minimum free distance* d_f of a convolutional code, also called simply the free distance, to be the minimum Hamming distance of all paths through the trellis to the all-zero path, which for this example is 6. The error correction capability of the code is obtained in the same manner as for block codes, with $d_{\min}$ replaced by d_f, so that the code can correct t channels errors with $t = \lfloor .5 d_f \rfloor$.

8.3.5 State Diagrams and Transfer Functions

The transfer function of a convolutional code is used to characterize paths that diverge and remerge from the all-zero path, and it is also used to obtain probability of error bounds. The transfer function is obtained from the code's state diagram representing possible transitions from the all-zero state to the all-zero state. The state diagram for the code illustrated in Figure 8.7 is shown in Figure 8.11, with the all-zero state $a = 00$ split into a second node e to facilitate representing paths that begin and end in this state. Transitions between states due to a 0 input bit are represented by solid lines, while transitions due to a 1 input bit are represented

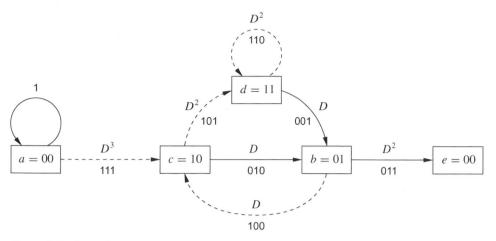

Figure 8.11: State diagram.

by dashed lines. The branches of the state diagram are labeled as either $D^0 = 1$, D^1, D^2, or D^3, where the exponent of D corresponds to the Hamming distance between the codeword (which is shown for each branch transition) and the all-zero codeword in the all-zero path. The self-loop in node a can be ignored because it does not contribute to the distance properties of the code.

The state diagram can be represented by state equations for each state. For the example of Figure 8.11, we obtain state equations corresponding to the four states

$$X_c = D^3 X_a + D X_b, \quad X_b = D X_c + D X_d, \quad X_d = D^2 X_c + D^2 X_d, \quad X_e = D^2 X_b, \quad (8.61)$$

where $X_a, \ldots, X_e$ are dummy variables characterizing the partial paths. The transfer function of the code, describing the paths from state a to state e, is defined as $T(D) = X_e/X_a$. By solving the state equations for the code, which can be done using standard techniques such as Mason's formula, we obtain a transfer function of the form

$$T(D) = \sum_{d=d_f}^{\infty} a_d D^d, \quad (8.62)$$

where a_d is the number of paths with Hamming distance d from the all-zero path. As stated before, the minimum Hamming distance to the all-zero path is d_f, and the transfer function $T(D)$ indicates that there are a_{d_f} paths with this minimum distance. For the example of Figure 8.11, we can solve the state equations given in (8.61) to get the transfer function

$$T(D) = \frac{D^6}{1 - 2D^2} = D^6 + 2D^8 + 4D^{10} + \cdots. \quad (8.63)$$

We see from the transfer function that there is one path with minimum distance $d_f = 6$ and two paths with Hamming distance 8, which is consistent with Figure 8.10. The transfer function is a convenient shorthand for enumerating the number and corresponding Hamming distance of all paths in a particular code that diverge and later remerge with the all-zero path.

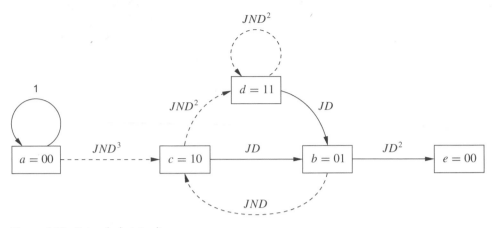

Figure 8.12: Extended state diagram.

Although the transfer function is sufficient to capture the number and Hamming distance of paths in the trellis to the all-zero path, we need a more detailed characterization to compute the bit error probability of the convolutional code. We therefore introduce two additional parameters into the transfer function, N and J, for this additional characterization. The factor N is introduced on all branch transitions associated with a 1 input bit (dashed lines in Figure 8.11). The factor J is introduced to every branch in the state diagram such that the exponent of J in the transfer function equals the number of branches in any given path from node a to node e. The extended state diagram corresponding to the trellis of Figure 8.7 is shown in Figure 8.12.

The extended state diagram can also be represented by state equations. For the example of Figure 8.12, these are given by:

$$X_c = JND^3X_a + JNDX_b, \qquad X_b = JDX_c + JDX_d,$$
$$X_d = JND^2X_c + JND^2X_d, \qquad X_e = JD^2X_b. \tag{8.64}$$

Similarly to the previous transfer function definition, the transfer function associated with this extended state is defined as $T(D, N, J) = X_e/X_a$, which for this example yields

$$T(D, N, J) = \frac{J^3ND^6}{1 - JND^2(1 + J)}$$
$$= J^3ND^6 + J^4N^2D^8 + J^5N^2D^8 + J^5N^3D^{10} + \cdots. \tag{8.65}$$

The factor J is most important when we are interested in transmitting finite-length sequences; for infinite-length sequences we typically set $J = 1$ to obtain the transfer function for the extended state:

$$T(D, N) = T(D, N, J = 1). \tag{8.66}$$

The transfer function for the extended state tells us more information about the diverging and remerging paths than just their Hamming distance; namely, the minimum distance path with Hamming distance 6 is of length 3 and results in a single bit error (exponent of N is unity), one path of Hamming distance 8 is of length 4 and results in two bit errors, and the other path of Hamming distance 8 is of length 5 and results in two bit errors, consistent with Figure 8.10. The extended transfer function is a convenient shorthand for representing the Hamming distance, length, and number of bit errors that correspond to each diverging and remerging path of a code from the all-zero path. In the next section we show that this representation is useful in characterizing the probability of error for convolutional codes.

8.3.6 Error Probability for Convolutional Codes

Since convolutional codes are linear codes, the probability of error can be obtained by first assuming that the all-zero sequence is transmitted and then determining the probability that the decoder decides in favor of a different sequence. We will consider error probability for both hard decision and soft decision decoding; soft decisions are much more common in wireless systems owing to their superior performance.

First consider soft decision decoding. We are interested in the probability that the all-zero sequence is sent but a different sequence is decoded. If the coded symbols output from the convolutional encoder are sent over an AWGN channel using coherent BPSK modulation with energy $E_c = R_c E_b$, then it can be shown [1] that, if the all-zero sequence is transmitted, the probability of mistaking this sequence with a sequence Hamming distance d away is

$$P_2(d) = Q\left(\sqrt{\frac{2E_c}{N_0}}d\right) = Q(\sqrt{2\gamma_b R_c d}). \tag{8.67}$$

We call this probability the *pairwise error probability,* since it is the error probability associated with a pairwise comparison of two paths that differ in d bits. The transfer function enumerates all paths that diverge and remerge with the all zero path, so by the union bound we can upper bound the probability of mistaking the all-zero path for another path through the trellis as

$$P_e \leq \sum_{d=d_f}^{\infty} a_d Q(\sqrt{2\gamma_b R_c d}), \tag{8.68}$$

where a_d denotes the number of paths of distance d from the all-zero path. This bound can be expressed in terms of the transfer function itself if we use the Chernoff upper bound for the Q-function, which yields

$$Q(\sqrt{2\gamma_b R_c d}) \leq e^{-\gamma_b R_c d}.$$

Using this in (8.68) we obtain the upper bound

$$P_e < T(D)|_{D=e^{-\gamma_b R_c}}. \tag{8.69}$$

This upper bound tells us the probability of mistaking one sequence for another, but it does not yield the more fundamental probability of bit error. We know that the exponent in the factor N of $T(D, N)$ indicates the number of information bit errors associated with selecting an incorrect path through the trellis. Specifically, we can express $T(D, N)$ as

$$T(D, N) = \sum_{d=d_f}^{\infty} a_d D^d N^{f(d)}, \tag{8.70}$$

where $f(d)$ denotes the number of bit errors associated with a path of distance d from the all-zero path. Then we can upper bound the bit error probability for $k = 1$ as

$$P_b \leq \sum_{d=d_f}^{\infty} a_d f(d) Q\left(\sqrt{2\gamma_b R_c d}\right) \tag{8.71}$$

[1, Chap. 8.2], which differs from (8.68) only in the weighting factor $f(d)$ corresponding to the number of bit errors in each incorrect path. If the Q-function is upper bounded using the Chernoff bound as before, we get the upper bound

$$P_b < \left. \frac{dT(D, N)}{dN} \right|_{N=1, D=e^{-\gamma_b R_c}}. \tag{8.72}$$

If $k > 1$ then we divide (8.71) or (8.72) by k to obtain P_b.

All of these bounds assume coherent BPSK transmission (or coherent QPSK, which is equivalent to two independent BPSK transmissions). For other modulations, the pairwise error probability $P_2(d)$ must be recomputed based on the probability of error associated with the given modulation.

Let us now consider hard decision decoding. The probability of selecting an incorrect path at distance d from the all-zero path, for d odd, is given by

$$P_2(d) = \sum_{k=.5(d+1)}^{d} \binom{d}{k} p^k (1-p)^{(d-k)}, \tag{8.73}$$

where p is the probability of error on the channel. This follows because the incorrect path will be selected only if the decoded path is closer to the incorrect path than to the all-zero path – that is, the decoder makes at least $.5(d+1)$ errors. If d is even then the incorrect path is selected when the decoder makes more than $.5d$ errors, and the decoder makes a choice at random if the number of errors is exactly $.5d$. We can bound the pairwise error probability as

$$P_2(d) < [4p(1-p)]^{d/2}. \tag{8.74}$$

Following the same approach as in soft decision decoding, we then obtain the error probability bound as

$$P_e < \sum_{d=d_f}^{\infty} a_d [4p(1-p)]^{d/2} < T(D)|_{D=\sqrt{4p(1-p)}}, \tag{8.75}$$

and

$$P_b < \sum_{d=d_f}^{\infty} a_d f(d) P_2(d) = \left. \frac{dT(D, N)}{dN} \right|_{N=1, D=\sqrt{4p(1-p)}}. \tag{8.76}$$

8.4 Concatenated Codes

A concatenated code uses two levels of coding: an inner code and an outer code, as shown in Figure 8.13. The inner code is typically designed to remove most of the errors introduced by the channel, and the outer code is typically a less powerful code that further reduces error

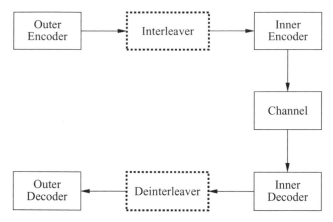

Figure 8.13: Concatenated coding.

probability when the received coded bits have a relatively low probability of error (since most errors are corrected by the inner code). Concatenated codes can be particularly effective at correcting bursts of errors, which are common in wireless channels as a result of deep fades. In addition, at low SNRs Viterbi decoding of a convolutional code tends to have errors that occur in bursts. To compensate for these error bursts, an inner convolutional code is often concatenated with an outer Reed Solomon code, since RS codes have good burst error correcting properties. In addition, concatenated codes frequently have the inner and outer codes separated by an interleaver to break up bursts of errors. Interleaver design for different coding techniques is described in Section 8.8.

Concatenated codes typically achieve very low error probability with less complexity than a single code with the same error probability performance. The decoding of concatenated codes is usually done in two stages, as indicated in the figure: first the inner code is decoded, and then the outer code is decoded separately. This is a suboptimal technique, since in fact both codes are working in tandem to reduce error probability. However, the maximum likelihood decoder for a concatenated code, which performs joint decoding, is highly complex. It was discovered in the mid-1990s that a near-optimal decoder for concatenated codes can be obtained based on iterative decoding. This is the basic premise behind turbo codes, described in the next section.

8.5 Turbo Codes

Turbo codes, introduced in 1993 in a landmark paper by Berrou, Glavieux, and Thitimajshima ([6]; see also [7]), are powerful codes that can come within a fraction of a decibel of the Shannon capacity limit on AWGN channels. Turbo codes and the more general family of codes on graphs with iterative decoding algorithms [8; 9] have been studied extensively, yet some of their characteristics are still not well understood. The main ideas behind codes on graphs were introduced by Gallager in the early sixties [10]; at the time, however, these coding techniques were thought impractical and were generally not pursued by researchers in the field. The landmark 1993 paper on turbo codes [6] provided more than enough motivation to revisit the work of Gallager and others on iterative, graph-based decoding techniques.

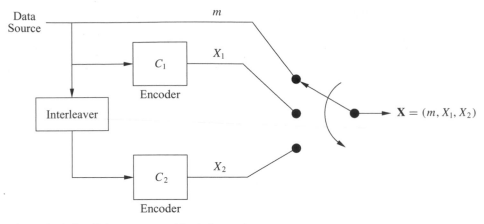

Figure 8.14: Parallel concatenated (turbo) encoder.

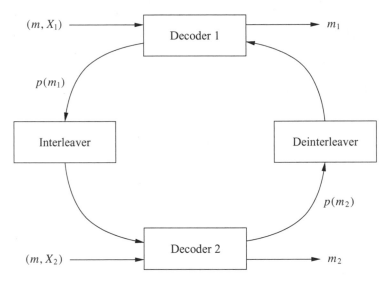

Figure 8.15: Turbo decoder.

As first described by Berrou et al., turbo codes consist of two key components: parallel concatenated encoding and iterative, "turbo" decoding [6; 11]. A typical parallel concatenated encoder is shown in Figure 8.14. It consists of two parallel convolutional encoders separated by an interleaver, with the input to the channel being the data bits m along with the parity bits X_1 and X_2 output from each of the encoders in response to input m. Since the m information bits are transmitted as part of the codeword, we consider this a systematic turbo code. The key to parallel concatenated encoding lies in the recursive nature of the encoders and the impact of the interleaver on the information stream. Interleavers also play a significant role in the reduction of error floors [11], which are commonly exhibited in turbo codes.

Iterative or "turbo" decoding exploits the component-code substructure of the turbo encoder by associating a component decoder with each of the component encoders. More specifically, each decoder performs soft input–soft output decoding, as shown in Figure 8.15 for the example encoder of Figure 8.14. In this figure decoder 1 generates a soft decision in

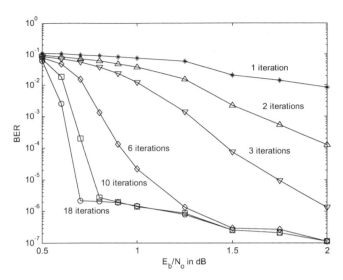

Figure 8.16: Turbo code performance (rate 1/2, $K = 5$ component codes with interleaver depth 2^{16}).

the form of a probability measure $p(m_1)$ on the transmitted information bits based on the received codeword (m, X_1). The probability measure is generated based on either a maximum a posteriori (MAP) probability or a soft output Viterbi algorithm (SOVA), which attaches a reliability indicator to the VA hard decision outputs [4, Chap. 12.5]. This probability information is passed to decoder 2, which generates its own probability measure $p(m_2)$ from its received codeword (m, X_2) and the probability measure $p(m_1)$. This reliability information is input to decoder 1, which revises its measure $p(m_1)$ based on this information and the original received codeword. Decoder 1 sends the new reliability information to decoder 2, which revises its measure using this new information. Turbo decoding proceeds in an iterative manner, with the two component decoders alternately updating their probability measures. Ideally the decoders will eventually agree on probability measures that reduce to hard decisions $m = m_1 = m_2$. However, the stopping condition for turbo decoding is not well-defined, in part because there are many cases in which the turbo decoding algorithm does not converge: the decoders cannot agree on the value of m. Several methods have been proposed for detecting convergence (if it occurs), including bit estimate variance [7] and neural net–based techniques [12].

The simulated performance of turbo codes over multiple iterations of the decoder is shown in Figure 8.16 for a code composed of two convolutional codes of rate 1/2 with constraint length $K = 5$ separated by an interleaver of depth $d = 2^{16} = 65536$. The decoder converges after approximately eighteen iterations. This curve indicates several important aspects of turbo codes. First, note their exceptional performance: bit error probability of 10^{-6} at an E_b/N_0 of less than 1 dB. In fact, the original turbo code proposed in [6] performed within .5 dB of the Shannon capacity limit at $P_b = 10^{-5}$. The intuitive explanation for the amazing performance of turbo codes is that the code complexity introduced by the encoding structure is similar to the codes that achieve Shannon capacity. The iterative procedure of the turbo decoder allows these codes to be decoded without excessive complexity. However, note that the turbo code exhibits an error floor: in Figure 8.16 this floor occurs at 10^{-6}. This floor is problematic for systems that require extremely low bit error rates. Several mechanisms have been investigated to lower the error floor, including bit interleaving and increasing the constraint length of the component codes.

An alternative to parallel concatenated coding is serial concatenated coding [13]. In this coding technique, one component code serves as an outer code, and then the output of this first encoder is interleaved and passed to a second encoder. The output of the second encoder comprises the coded bits. Iterative decoding between the inner and outer codes is used for decoding. There has been much work comparing serial and parallel concatenated code performance (see e.g. [13; 14; 15]). Whereas both codes perform very well under similar delay and complexity conditions, in some cases serial concatenated coding performs better at low bit error rates and also can exhibit a lower error floor.

8.6 Low-Density Parity-Check Codes

Low-density parity-check (LDPC) codes were originally invented by Gallager [10]. However, these codes were largely ignored until the introduction of turbo codes, which rekindled some of the same ideas. Subsequent to the landmark paper [6] on turbo codes in 1993, LDPC codes were announced by Mackay and Neal [16] and Wiberg [17]. Shortly thereafter it was recognized that these new code designs were actually reinventions of Gallager's original ideas, and subsequently much work has been devoted to finding the capacity limits, encoder and decoder designs, and practical implementation of LDPC codes for different channels.

Low-density parity-check codes are linear block codes with a particular structure for the parity check matrix $\mathbf{H}$, which was defined in Section 8.2.3. Specifically, a (d_v, d_c) regular binary LDPC has a parity-check matrix $\mathbf{H}$ with d_v 1s in each column and d_c 1s in each row, where d_v and d_c are chosen as part of the codeword design and are small relative to the codeword length. Since the fraction of nonzero entries in $\mathbf{H}$ is small, the parity-check matrix for the code has a low density – hence the name low-density parity-check codes.

Provided that the codeword length is long, LDPC codes achieve performance close to the Shannon limit and in some cases surpass the performance of parallel or serially concatenated codes [18]. The fundamental practical difference between turbo codes and LDPC codes is that turbo codes tend to have low encoding complexity (linear in blocklength) but high decoding complexity (due to their iterative nature and message passing). In contrast, LDPC codes tend to have relatively high encoding complexity but low decoding complexity. In particular, like turbo codes, LDPC decoding uses iterative techniques that are related to Pearl's belief propagation, which is commonly used by the artificial intelligence community (see [19]). However, the belief propagation corresponding to LDPC decoding may be simpler than for turbo decoding [19; 20]. In addition, this belief propagation decoding is parallelizable and can be closely approximated with decoders of very low complexity [21], although this may also be possible for turbo decoding. Finally, the decoding algorithm for LDPC codes can determine when a correct codeword has been detected, which is not necessarily the case for turbo codes. The trade-offs between turbo and LDPC codes is an active research area, and many open questions remain regarding their relative performance.

Additional work in the area of LDPC codes includes finding capacity limits for these codes [21], determining effective code designs [22] and efficient encoding and decoding algorithms [21; 23], and expanding the code designs to include nonregular [18] and nonbinary LDPC codes [24] as well as coded modulation [25].

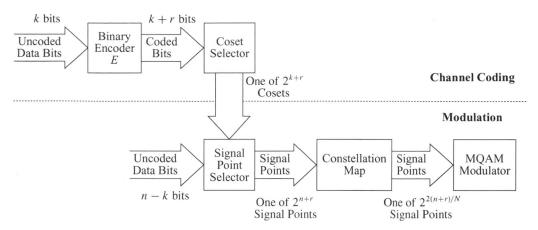

Figure 8.17: General coding scheme.

8.7 Coded Modulation

Although Shannon proved the capacity theorem for AWGN channels in the late 1940s, it wasn't until the 1990s that rates approaching the Shannon limit were attained – primarily for AWGN channels with binary modulation using turbo codes. Shannon's theorem predicted the possibility of reducing both energy and bandwidth simultaneously through coding. However, as described in Section 8.1, traditional error correction coding schemes (e.g., block, convolutional, and turbo codes) provide coding gain at the expense of increased bandwidth or reduced data rate.

The spectrally efficient coding breakthrough came when Ungerboeck [26] introduced a coded modulation technique to jointly optimize both channel coding and modulation. This joint optimization results in significant coding gains without bandwidth expansion. Ungerboeck's trellis coded modulation, which uses multilevel/phase signal modulation and simple convolutional coding with mapping by set partitioning, compares favorably to later developments in coded modulation (coset and lattice codes) as well as to more complex trellis codes [27]. We now outline the general principles of this coding technique. Comprehensive treatments of trellis, lattice, and coset codes can be found in [26; 27; 28].

The basic scheme for trellis and lattice coding – or, more generally, for any type of coset coding – is depicted in Figure 8.17. There are five elements required to generate the coded modulation:

1. a binary encoder E, block or convolutional, that operates on k uncoded data bits to produce $k + r$ coded bits;
2. a subset (coset) selector, which uses the coded bits to choose one of 2^{k+r} subsets from a partition of the N-dimensional signal constellation;
3. a point selector, which uses $n - k$ additional uncoded bits to choose one of the 2^{n-k} signal points in the selected subset;
4. a constellation map, which maps the selected point from N-dimensional space to a sequence of $N/2$ points in two-dimensional space; and
5. an MQAM modulator (or other M-ary modulator).

The first two elements constitute the channel coding, and the remaining elements are the modulation. The receiver essentially reverses the modulation and coding steps. After MQAM demodulation and an inverse $2/N$ constellation mapping, decoding is done in essentially two stages: first, the points within each subset that are closest to the received signal point are determined; then, the maximum likelihood subset sequence is calculated. When the encoder E is a convolutional encoder, this scheme is referred to as *trellis coded modulation*; for E a block encoder, it is called *lattice coded modulation*. These schemes are also referred to as trellis or lattice codes.

Steps 1–5 essentially decouple the channel coding gain from the gain associated with signal shaping in the modulation. Specifically, the code distance properties – and thus the channel coding gain – are determined by the encoder (E) properties and the subset partitioning, which are essentially decoupled from signal shaping. We will discuss the channel coding gain in more detail below. Optimal shaping of the signal constellation provides up to an additional 1.53 dB of shaping gain (for asymptotically large N), independent of the channel coding scheme.[1] However, the performance improvement from shaping gain is offset by the corresponding complexity of the constellation map, which grows exponentially with N. The size of the transmit constellation is determined by the average power constraint and does not affect the shaping or coding gain.

The channel coding gain results from a selection of sequences among all possible sequences of signal points. If we consider a sequence of N input bits as a point in N-dimensional space (the *sequence space*), then this selection is used to guarantee some minimum distance $d_{\min}$ in the sequence space between possible input sequences. Errors generally occur when a sequence is mistaken for its closest neighbor, and in AWGN channels this error probability is a decreasing function of $d_{\min}^2$. We can thus decrease the BER by increasing the separation between each point in the sequence space by a fixed amount ("stretching" the space). However, this will result in a proportional power increase, so no net coding gain is realized. The effective power gain of the channel code is therefore the minimum squared distance between selected sequence points (the sequence points obtained through coding) multiplied by the density of the selected sequence points. Specifically, if the minimum distance and density of all points in the sequence space are denoted by d_0 and Δ_0, respectively, and if the minimum distance and density of points in the sequence space selected through coding are denoted by $d_{\min}$ and Δ, respectively, then maximum likelihood sequence detection yields a channel coding gain of

$$G_c = \left(\frac{d_{\min}^2}{d_0^2}\right)\left(\frac{\Delta}{\Delta_0}\right). \tag{8.77}$$

The second term on the right side of this expression is also referred to as the *constellation expansion factor* and equals 2^{-r} (per N dimensions) for a redundancy of r bits in the encoder E [27].

Some of the nominal coding gain in (8.77) is lost owing to selected sequences having more than one nearest neighbor in the sequence space, which increases the possibility of incorrect sequence detection. This loss in coding gain is characterized by the *error coefficient,*

---

[1] A square constellation has 0 dB of shaping gain; a circular constellation, which is the geometrical figure with the least average energy for a given area, achieves the maximum shape gain for a given N [27].

which is tabulated for most common lattice and trellis coded modulations in [27]. In general, the error coefficient is larger for lattice codes than for trellis codes with comparable values of G_c.

Channel coding is done using set partitioning of lattices. A *lattice* is a discrete set of vectors in real Euclidean N-dimensional space that forms a group under ordinary vector addition, so the sum or difference of any two vectors in the lattice is also in the lattice. A *sublattice* is a subset of a lattice that is itself a lattice. The sequence space for *uncoded* MQAM modulation is just the N-cube,[2] so the minimum distance between points is no different than in the two-dimensional case. By restricting input sequences to lie on a lattice in N-dimensional space that is denser than the N-cube, we can increase $d_{\min}$ while maintaining the same density (or, equivalently, the same average power) in the transmit signal constellation; hence, there is no constellation expansion. The N-cube is a lattice, but for every $N > 1$ there are denser lattices in N-dimensional space. Finding the densest lattice in N dimensions is a well-known mathematical problem, and it has been solved for all N for which the decoder complexity is manageable.[3] Once the densest lattice is known, we can form partitioning subsets, or *cosets,* of the lattice via translation of any sublattice. The choice of the partitioning sublattice will determine the size of the partition – that is, the number of subsets that the subset selector in Figure 8.17 has to choose from. Data bits are then conveyed in two ways: through the sequence of cosets from which constellation points are selected, and through the points selected within each coset. The density of the lattice determines the distance between points within a coset, while the distance between subset sequences is essentially determined by the binary code properties of the encoder E and its redundancy r. If we let d_p denote the minimum distance between points within a coset and d_s the minimum distance between the coset sequences, then the minimum distance code is $d_{\min} = \min(d_p, d_s)$. The effective coding gain is given by

$$G_c = 2^{-2r/N} d_{\min}^2, \tag{8.78}$$

where $2^{-2r/N}$ is the constellation expansion factor (in two dimensions) from the r extra bits introduced by the binary channel encoder.

Returning to Figure 8.17, suppose we want to send $m = n + r$ bits per dimension, so that an N sequence conveys mN bits. If we use the densest lattice in N-dimensional space that lies within an N-dimensional sphere, where the radius of the sphere is just large enough to enclose 2^{mN} points, then we achieve a total coding gain that combines the coding gain (resulting from the lattice density and the encoder properties) with the shaping gain of the N-dimensional sphere over the N-dimensional rectangle. Clearly, the coding gain is decoupled from the shaping gain. An increase in signal power would allow us to use a larger N-dimensional sphere and hence transmit more uncoded bits. It is possible to generate maximum-density N-dimensional lattices for $N = 4, 8, 16$, and 24 using a simple partition of the two-dimensional rectangular lattice combined with either conventional block or convolutional coding. Details of this type of code construction, and the corresponding decoding

---

[2] The Cartesian product of two-dimensional rectangular lattices with points at odd integers.
[3] The complexity of the maximum likelihood decoder implemented with the Viterbi algorithm is roughly proportional to N.

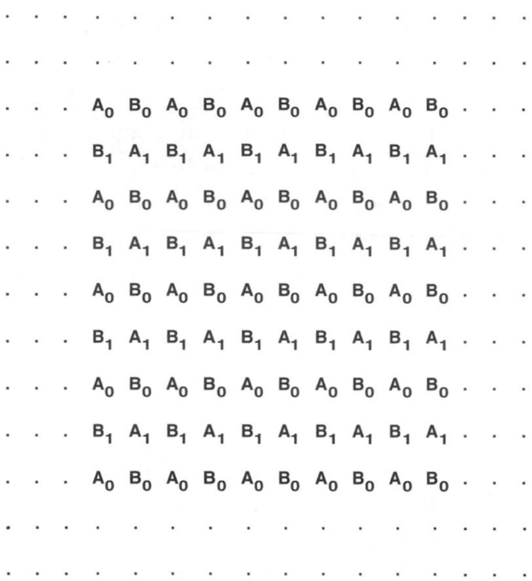

```
 ·  ·  ·  ·  ·  ·  ·  ·  ·  ·  ·  ·  ·  ·  ·
 ·  ·  ·  ·  ·  ·  ·  ·  ·  ·  ·  ·  ·  ·  ·
 ·  ·  · A₀ B₀ A₀ B₀ A₀ B₀ A₀ B₀ A₀ B₀ ·  ·  ·
 ·  ·  · B₁ A₁ B₁ A₁ B₁ A₁ B₁ A₁ B₁ A₁ ·  ·  ·
 ·  ·  · A₀ B₀ A₀ B₀ A₀ B₀ A₀ B₀ A₀ B₀ ·  ·  ·
 ·  ·  · B₁ A₁ B₁ A₁ B₁ A₁ B₁ A₁ B₁ A₁ ·  ·  ·
 ·  ·  · A₀ B₀ A₀ B₀ A₀ B₀ A₀ B₀ A₀ B₀ ·  ·  ·
 ·  ·  · B₁ A₁ B₁ A₁ B₁ A₁ B₁ A₁ B₁ A₁ ·  ·  ·
 ·  ·  · A₀ B₀ A₀ B₀ A₀ B₀ A₀ B₀ A₀ B₀ ·  ·  ·
 ·  ·  · B₁ A₁ B₁ A₁ B₁ A₁ B₁ A₁ B₁ A₁ ·  ·  ·
 ·  ·  · A₀ B₀ A₀ B₀ A₀ B₀ A₀ B₀ A₀ B₀ ·  ·  ·
 ·  ·  ·  ·  ·  ·  ·  ·  ·  ·  ·  ·  ·  ·  ·
 ·  ·  ·  ·  ·  ·  ·  ·  ·  ·  ·  ·  ·  ·  ·
```

Figure 8.18: Subset partition for an 8-dimensional lattice.

algorithms, can be found in [28] for both lattice and trellis codes. For these constructions, an effective coding gain of approximately 1.5, 3.0, 4.5, and 6.0 dB is obtained with lattice codes for $N = 4$, 8, 16, and 24, respectively. Trellis codes exhibit higher coding gains with comparable complexity.

We conclude this section with an example of coded modulation: the $N = 8$, 3-dB–gain lattice code proposed in [28]. First, the two-dimensional signal constellation is partitioned into four subsets as shown in Figure 8.18, where the subsets are represented by the points A_0, A_1, B_0, and B_1. For example, a 16-QAM constellation would have four subsets, each consisting of four constellation points. Note that the distance between points in each subset is twice the distance between points in the (uncoded) constellation. From this subset partition, we form an 8-dimensional lattice by taking all sequences of four points in which (i) all points are either A-points or B-points and (ii) within a four-point sequence, the point subscripts satisfy the parity check $i_1 + i_2 + i_3 + i_4 = 0$ (so the sequence subscripts must be codewords in the $(4, 3)$ parity-check code, which has a minimum Hamming distance of 2). Thus, three data bits and one parity check bit are used to determine the lattice subset. The square of the minimum distance resulting from this subset partition is four times that of the uncoded signal constellation, yielding a 6-dB gain. However, the extra parity check bit expands the constellation by $1/2$ bit per dimension, which by Section 5.3.3 costs an additional power factor of $4^{.5} = 2$, or 3 dB. Thus, the net coding gain is $6 - 3 = 3$ dB. The remaining data bits are used to choose a point within the selected subset and so, for a data rate of m bits per symbol, the four lattice subsets must each have 2^{m-1} points.[4] For example, with a 16-QAM constellation the four subsets each have $2^{m-1} = 4$ points, so the data rate is 3 data bits per 16-QAM symbol.

---

[4] This yields $m - 1$ bits/symbol, with the additional bit/symbol conveyed by the channel code.

Coded modulation using turbo codes has also been investigated [29; 30; 31]. This work shows that turbo trellis coded modulation can come very close to the Shannon limit for non-binary signaling.

8.8 Coding with Interleaving for Fading Channels

Block, convolutional, and coded modulation are designed for good performance in AWGN channels. In fading channels, errors associated with the demodulator tend to occur in bursts, corresponding to the times when the channel is in a deep fade. Most codes designed for AWGN channels cannot correct for the long bursts of errors exhibited in fading channels. Hence, codes designed for AWGN channels can exhibit worse performance in fading than an uncoded system.

To improve performance of coding in fading channels, coding is typically combined with *interleaving* to mitigate the effect of error bursts. The basic premise of coding and interleaving is to spread error bursts due to deep fades over many codewords so that each received codeword exhibits at most a few simultaneous symbol errors, which can be corrected for. The spreading out of burst errors is accomplished by the interleaver and the error correction is accomplished by the code. The size of the interleaver must be large enough that fading is independent across a received codeword. Slowly fading channels require large interleavers, which in turn can lead to large delays.

Coding and interleaving is a form of diversity, and performance of coding and interleaving is often characterized by the diversity order associated with the resulting probability of error. This diversity order is typically a function of the minimum Hamming distance of the code. Thus, designs for coding and interleaving on fading channels must focus on maximizing the diversity order of the code rather than on metrics like Euclidean distance, which are used as a performance criterion in AWGN channels. In the following sections we discuss coding and interleaving for block, convolutional, and coded modulation in more detail. We will assume that the receiver has knowledge of the channel fading, which greatly simplifies both the analysis and the decoder. Estimates of channel fading are commonly obtained through pilot symbol transmissions [32; 33]. Maximum likelihood detection of coded signals in fading without this channel knowledge is computationally intractable [34] and so usually requires approximations to either the ML decoding metric or the channel [34; 35; 36; 37; 38; 39]. Note that turbo codes designed for AWGN channels, described in Section 8.5, have an interleaver inherent to the code design. However, the interleaver design considerations for AWGN channels are different than for fading channels. A discussion of interleaver design and performance analysis for turbo codes in fading channels can be found in [40, Chap. 8; 41; 42]. Low-density parity-check codes can also be designed for inherent diversity against fading, with performance similar to that of turbo codes optimized for fading diversity [43].

8.8.1 Block Coding with Interleaving

Block codes are typically combined with block interleaving to spread out burst errors from fading. A block interleaver is an array with d rows and n columns, as shown in Figure 8.19. For block interleavers designed for an (n, k) block code, codewords are read into the interleaver

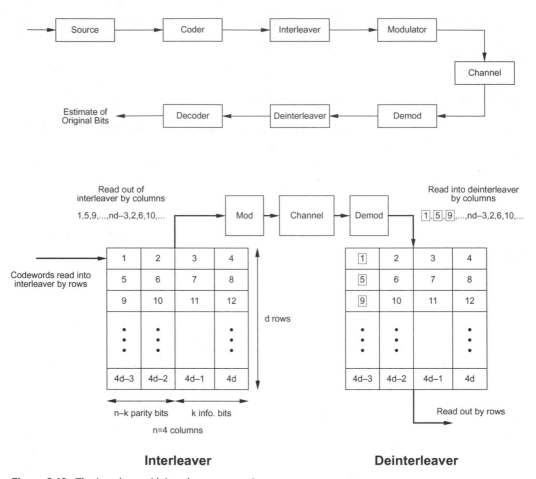

Interleaver   **Deinterleaver**

Figure 8.19: The interleaver/deinterleaver operation.

by rows so that each row contains an (n, k) codeword. The interleaver contents are read out by columns into the modulator for subsequent transmission over the channel. During transmission, codeword symbols in the same codeword are separated by $d - 1$ other symbols, so symbols in the same codeword experience approximately independent fading if their separation in time is greater than the channel coherence time – that is, if $dT_s > T_c \approx 1/B_D$, where T_s is the codeword symbol duration, T_c is the channel coherence time, and B_D is the channel Doppler spread. An interleaver is called a *deep interleaver* if the condition $dT_s > T_c$ is satisfied. The deinterleaver is an array identical to the interleaver. Bits are read into the deinterleaver from the demodulator by column so that each row of the deinterleaver contains a codeword (whose bits may have been corrupted by the channel). The deinterleaver output is read into the decoder by rows, one codeword at a time.

Figure 8.19 illustrates the ability of coding and interleaving to correct for bursts of errors. Suppose our coding scheme is an (n, k) binary block code with error correction capability $t = 2$. If this codeword is transmitted through a channel with an error burst of three symbols, then three out of four of the codeword symbols will be received in error. Since the code can only correct two or fewer errors, the codeword will be decoded in error. However, if the codeword is put through an interleaver then, as shown in Figure 8.19, the error burst of three symbols will be spread out over three separate codewords. Since a single symbol error can be easily corrected by an (n, k) code with $t = 2$, the original information bits can be decoded without error. Convolutional interleavers are similar in concept to block interleavers and are better suited to convolutional codes, as will be discussed in Section 8.8.2.

Performance analysis of coding and interleaving requires pairwise error probability analysis or the application of Chernoff or union bounds. Details of this analysis can be found in [1, Chap. 14.6]. The union bound provides a simple approximation to performance. Assume a Rayleigh fading channel with deep interleaving such that each coded symbol fades independently. Then the union bound for an (n, k) block code with soft decision decoding under noncoherent FSK modulation yields a codeword error given as

$$P_e < (2^k - 1)[4p(1 - p)]^{d_{\min}}, \tag{8.79}$$

where $d_{\min}$ is the minimum Hamming distance of the code and

$$p = \frac{1}{2 + R_c \bar{\gamma}_b}. \tag{8.80}$$

Similarly, for slowly fading channels in which a coherent phase reference can be obtained, the union bound on the codeword error probability of an (n, k) block code with soft decision decoding and BPSK modulation yields

$$P_e < 2^k \binom{2d_{\min} - 1}{d_{\min}} \left(\frac{1}{4R_c \bar{\gamma}_b} \right)^{d_{\min}}. \tag{8.81}$$

Note that both (8.79) and (8.81) are similar to the formula for error probability under MRC diversity combining given by (7.23), with $d_{\min}$ providing the diversity order. Similar formulas apply for hard decoding, with diversity order reduced by a factor of 2 relative to soft decision decoding. Thus, designs for block coding and interleaving over fading channels optimize their performance by maximizing the Hamming distance of the code.

Coding and interleaving is a suboptimal coding technique, since the correlation of the fading that affects subsequent bits contains information about the channel that could be used in a true maximum likelihood decoding scheme. By essentially throwing away this information, the inherent capacity of the channel is decreased [44]. Despite this capacity loss, coding with interleaving using codes designed for AWGN channels is a common coding technique for fading channels, since the complexity required for maximum likelihood decoding on correlated coded symbols is prohibitive.

EXAMPLE 8.8: Consider a Rayleigh fading channel with a Doppler of $B_D = 80$ Hz. The system uses a $(5, 2)$ Hamming code with interleaving to compensate for the fading. If the codeword symbols are sent through the channel at 30 kbps, find the required interleaver depth needed to obtain independent fading on each symbol. What is the longest burst of codeword symbol errors that can be corrected and the total interleaver delay for this depth?

Solution: The $(5, 2)$ Hamming code has a minimum distance of 3, so it can correct $t = \lfloor .5 \cdot 3 \rfloor = 1$ codeword symbol error. The codeword symbols are sent through the channel at a rate $R_s = 30$ kbps, so the symbol time is $T_s = 1/R_s = 3.3 \cdot 10^{-5}$. Assume a coherence time for the channel of $T_c = 1/B_D = .0125$ s. The bits in the interleaver are separated by dT_s, so we require $dT_s \geq T_c$ for independent fading on each codeword symbol. Solving for d yields $d \geq T_c/T_s = 375$. Since the interleaver spreads a burst of errors over the depth d of the interleaver, a burst of d symbol errors in the interleaved codewords will result in just one symbol error per codeword after deinterleaving, which can be corrected. The system can therefore tolerate an error burst of 375 symbols. However, all rows of the interleaver must be filled before it can read out by columns, so the total delay of the interleaver is $ndT_s = 5 \cdot 375 \cdot 3.3 \cdot 10^{-5} = 62.5$ ms. This delay can degrade quality in a voice system. We thus see that the price paid for correcting long error bursts through coding and interleaving is significant delay.

8.8.2 Convolutional Coding with Interleaving

As with block codes, convolutional codes suffer performance degradation in fading channels because the code is not designed to correct for bursts of errors. Thus, it is common to use an interleaver to spread out error bursts. In block coding the interleaver spreads errors across different codewords. Since there is no similar notion of a codeword in convolutional codes, a slightly different interleaver design is needed to mitigate the effect of burst errors. The interleaver commonly used with convolutional codes, called a *convolutional interleaver,* is designed both to spread out burst errors and to work well with the incremental nature of convolutional code generation [45; 46].

A block diagram for a convolutional interleaver is shown in Figure 8.20. The encoder output is multiplexed into buffers of increasing size, from no buffering to a buffer of size $N - 1$. The channel input is similarly multiplexed from these buffers into the channel. The reverse operation is performed at the decoder. Thus, the convolutional interleaver delays the transmission through the channel of the encoder output by progressively larger amounts, and this delay schedule is reversed at the receiver. This interleaver takes sequential outputs of the encoder and separates them by $N - 1$ other symbols in the channel transmission, thereby breaking up burst errors in the channel. Note that a convolutional encoder can also be used with a block code, but it is most commonly used with a convolutional code. The total memory associated with the convolutional interleaver is $.5N(N - 1)$ and the delay is $N(N - 1)T_s$ [2, Chap. 8.2.2], where T_s is the symbol time for transmitting the coded symbols over the channel.

The probability of error analysis for convolutional coding and interleaving is given in [1, Chap. 14.6] under assumptions similar to those used in our block fading analysis. The Chernoff bound again yields probability of error under soft decision decoding with a diversity order based on the minimum free distance of the code. Hard decision decoding reduces this diversity by a factor of 2.

EXAMPLE 8.9: Consider a channel with coherence time $T_c = 12.5$ ms and a coded bit rate of $R_s = 100$ kilosymbols per second. Find the average delay of a convolutional interleaver that achieves independent fading between subsequent coded bits.

Solution: For the convolutional interleaver, each subsequent coded bit is separated by NT_s and we require $NT_s \geq T_c$ for independent fading, where $T_s = 1/R_s$. Thus we

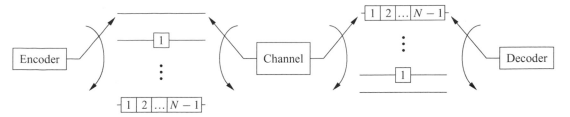

Figure 8.20: Convolutional coding and interleaving.

have $N \geq T_c/T_s = .0125/.00001 = 1250$. Note that this is the same as the required depth for a block interleaver to get independent fading on each coded bit. The total delay is $N(N-1)T_s = 15$ s. This is quite a high delay for either voice or data.

8.8.3 Coded Modulation with Symbol/Bit Interleaving

As with block and convolutional codes, coded modulation designed for an AWGN channel performs poorly in fading. This leads to the notion of coded modulation with interleaving for fading channels. However, unlike block and convolutional codes, there are two options for interleaving in coded modulation. One option is to interleave the bits and then map them to modulated symbols. This is called bit-interleaved coded modulation (BICM). Alternatively, the modulation and coding can be done jointly as in coded modulation for AWGN channels and the resulting symbols interleaved prior to transmission. This technique is called symbol-interleaved coded modulation (SICM).

Symbol-interleaved coded modulation seems at first like the best approach because it preserves joint coding and modulation, the main design premise behind coded modulation. However, the coded modulation design criterion must be changed in fading, since performance in fading depends on the code diversity as characterized by its Hamming distance rather than its Euclidean distance. Initial work on coded modulation for fading channels focused on techniques to maximize diversity in SICM. However, good design criteria were hard to obtain, and the performance of these codes was somewhat disappointing [47; 48; 49].

A major breakthrough in the design of coded modulation for fading channels was the discovery of bit-interleaved coded modulation [50; 51]. In BICM the code diversity is equal to the smallest number of distinct bits (rather than channel symbols) along any error event. This is achieved by bitwise interleaving at the encoder output prior to symbol mapping, with an appropriate soft decision bit metric as an input to the Viterbi decoder. Although this breaks the coded modulation paradigm of joint modulation and coding, it provides much better performance than SICM. Moreover, analytical tools for evaluating the performance of BICM as well as design guidelines for good performance are known [50]. BICM is now the dominant technique for coded modulation in fading channels.

8.9 Unequal Error Protection Codes

When not all bits transmitted over the channel have the same priority or bit error probability requirement, multiresolution or unequal error protection (UEP) codes can be used. This scenario arises, for example, in voice and data systems where voice is typically more tolerant to bit errors than data: data received in error must be retransmitted, so $P_b < 10^{-6}$ is

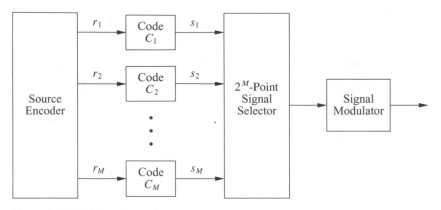

Figure 8.21: Multilevel encoder.

typically required, whereas good quality voice requires only on the order of $P_b < 10^{-3}$. This scenario also arises for certain types of compression. For example, in image compression, bits corresponding to the low-resolution reproduction of the image are required, whereas high-resolution bits simply refine the image. With multiresolution channel coding, all bits are received correctly with a high probability under benign channel conditions. However, if the channel is in a deep fade, only the high-priority bits or those requiring low P_b will be received correctly with high probability.

Practical implementation of a UEP code was first studied by Imai and Hirakawa [52]. Binary UEP codes were later considered for combined speech and channel coding [53] as well as for combined image and channel coding [54]. These implementations use traditional (block or convolutional) error correction codes, so coding gain is directly proportional to bandwidth expansion. Subsequently, two bandwidth-efficient implementations for UEP codes were proposed: time multiplexing of bandwidth-efficient coded modulation [55], and coded modulation techniques applied to both uniform and nonuniform signal constellations [56; 57]. All of these multilevel codes can be designed for either AWGN or fading channels. We now briefly summarize these UEP coding techniques; specifically, we describe the principles behind multilevel coding and multistate decoding as well as the more complex bandwidth-efficient implementations.

A block diagram of a general multilevel encoder is shown in Figure 8.21. The source encoder first divides the information sequence into M parallel bit streams of decreasing priority. The channel encoder consists of M different binary error correcting codes $C_1, \ldots, C_M$ with decreasing codeword distances. The ith-priority bit stream enters the ith encoder, which generates the coded bits s_i. If the 2^M points in the signal constellation are numbered from 0 to $2^M - 1$, then the point selector chooses the constellation point s corresponding to

$$s = \sum_{i=1}^{M} s_i \cdot 2^{i-1}. \tag{8.82}$$

For example, if $M = 3$ and the signal constellation is 8-PSK, then the chosen signal point will have phase $2\pi s/8$.

Optimal decoding of the multilevel code uses a maximum likelihood decoder, which determines the input sequence that maximizes the received sequence probability. The ML

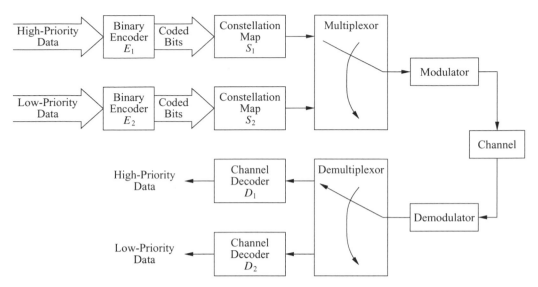

Figure 8.22: Transceiver for time-multiplexed coded modulation.

decoder must therefore jointly decode the code sequences $\{s_1\}, \ldots, \{s_m\}$. This can entail significant complexity even if the individual codes in the multilevel code have low complexity. For example, if the component codes are convolutional codes with 2^{μ_i} states, $i = 1, \ldots, M$, then the number of states in the optimal decoder is $2^{\mu_1 + \cdots + \mu_M}$. Because of the high complexity of optimal decoding, the suboptimal technique of multistage decoding, introduced in [52], is used for most implementations. Multistage decoding is accomplished by decoding the component codes sequentially. First the most powerful code, C_1, is decoded, then C_2, and so forth. Once the code sequence corresponding to encoder C_i is estimated, it is assumed to be correct for code decisions on the weaker code sequences.

The binary encoders of this multilevel code require extra code bits to achieve their coding gain, so they are not bandwidth efficient. An alternative approach proposed in [56] uses time multiplexing of the coded modulations described in Section 8.7. In this approach, different lattice or trellis coded modulations with different coding gains are used for each priority class of input data. The transmit signal constellations corresponding to each encoder may differ in size (number of signal points), but the average power of each constellation is the same. The signal points output by each of the individual encoders are then time-multiplexed together for transmission over the channel, as shown in Figure 8.22 for two bit streams of different priority. Let R_i denote the bit rate of encoder C_i in this figure for $i = 1, 2$. If T_1 equals the fraction of time that the high-priority C_1 code is transmitted and if T_2 equals the fraction of time that the C_2 code is transmitted, then the total bit rate is $(R_1 T_1 + R_2 T_2)/(T_1 + T_2)$, with the high-priority bits constituting $R_1 T_1/(R_1 T_1 + R_2 T_2)$ percent of this total.

The time-multiplexed coding method yields a higher gain if the constellation maps S_1 and S_2 of Figure 8.22 are designed jointly. This revised scheme is shown in Figure 8.23 for two encoders, where the extension to M encoders is straightforward. In trellis and lattice coded modulation, recall that bits are encoded to select the lattice subset and that uncoded bits choose the constellation point within the subset. The binary encoder properties reduce the P_b for the encoded bits only; the P_b for the uncoded bits is determined by the separation

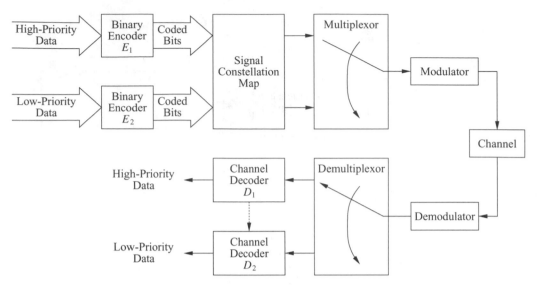

Figure 8.23: Joint optimization of signal constellation.

of the constellation signal points. We can easily modify this scheme to yield two levels of coding gain, where the high-priority bits are heavily encoded and are used to choose the subset of the partitioned constellation, while the low-priority bits are uncoded or lightly coded and are used to select the constellation signal point.

8.10 Joint Source and Channel Coding

The underlying premise of UEP codes is that the bit error probabilities of the channel code should be matched to the priority or P_b requirements associated with the bits to be transmitted. These bits are often taken from the output of a compression algorithm acting on the original data source. Hence, UEP coding can be considered as a joint design between compression (also called *source coding*) and channel coding. Although Shannon determined that the source and channel codes can be designed separately on an AWGN channel with no loss in optimality [58], this result holds only in the limit of infinite source code dimension, infinite channel code blocklength, and infinite complexity and delay. Thus, there has been much work on investigating the benefits of joint source and channel coding under more realistic system assumptions.

Work in the area of joint source and channel coding falls into several broad categories: source-optimized channel coding; channel-optimized source coding; and iterative algorithms, which combine these two code designs. In source-optimized channel coding, the source code is designed for a noiseless channel. A channel code is then designed for this source code to minimize end-to-end distortion over the given channel based on the distortion associated with corruption of the different transmitted bits. Unequal error protection channel coding – where the P_b of the different component channel codes is matched to the bit priorities associated with the source code – is an example of this technique. Source-optimized channel coding has been applied to image compression with convolution channel coding and with rate-compatible punctured convolutional (RCPC) channel codes in [54; 59; 60]. A comprehensive treatment of matching RCPC channel codes or MQAM to subband and linear predictive speech coding – in both AWGN and Rayleigh fading channels – can be found in [61]. In source-optimized modulation, the source code is designed for a noiseless channel and then the modulation is optimized to minimize end-to-end distortion. An example of this approach is given in [62], where compression by a vector quantizer (VQ) is followed by multicarrier modulation, and the modulation provides unequal error protection to the different source bits by assigning different powers to each subcarrier.

Channel-optimized source coding is another approach to joint source and channel coding. In this technique the source code is optimized based on the error probability associated with the channel code, where the channel code is designed independently of the source. Examples of work taking this approach include the channel-optimized vector quantizer (COVQ) and its scalar variation [63; 64]. Source-optimized channel coding and modulation can be combined with channel-optimized source coding via an iterative design. This approach is used for the joint design of a COVQ and multicarrier modulation in [65] and for the joint design of a COVQ and RCPC channel code in [66]. Combined trellis coded modulation and *trellis coded quantization,* a source coding strategy that borrows from the basic premise of trellis coded modulation, is investigated in [67; 68]. All of this work on joint source and channel code design indicates that significant performance advantages are possible when the source and channel codes are jointly designed. Moreover, many sophisticated channel code designs, such as turbo and LDPC codes, have not yet been combined with source codes in a joint optimization. Thus, much more work is needed in the broad area of joint source and channel coding to optimize performance for different applications.

PROBLEMS

8-1. Consider a $(3, 1)$ linear block code where each codeword consists of three data bits and one parity bit.

(a) Find all codewords in this code.
(b) Find the minimum distance of the code.

8-2. Consider a $(7, 4)$ code with generator matrix
$$\mathbf{G} = \begin{bmatrix} 0 & 1 & 0 & 1 & 1 & 0 & 0 \\ 1 & 0 & 1 & 0 & 1 & 0 & 0 \\ 0 & 1 & 1 & 0 & 0 & 1 & 0 \\ 1 & 1 & 0 & 0 & 0 & 0 & 1 \end{bmatrix}.$$

(a) Find all the codewords of the code.
(b) What is the minimum distance of the code?
(c) Find the parity check matrix of the code.
(d) Find the syndrome for the received vector $\mathbf{R} = [1101011]$.
(e) Assuming an information bit sequence of all 0s, find all minimum weight error patterns **e** that result in a valid codeword that is not the all-zero codeword.
(f) Use row and column operations to reduce **G** to systematic form and find its corresponding parity-check matrix. Sketch a shift register implementation of this systematic code.

8-3. All Hamming codes have a minimum distance of 3. What are the error correction and error detection capabilities of a Hamming code?

8-4. The $(15, 11)$ Hamming code has generator polynomial $g(X) = 1 + X + X^4$. Determine if the codewords described by polynomials $c_1(X) = 1 + X + X^3 + X^7$ and $c_2(X) = 1 + X^3 + X^5 + X^6$ are valid codewords for this generator polynomial. Also find the systematic form of this polynomial $p(X) + X^{n-k}u(X)$ that generates the codewords in systematic form.

8-5. The $(7, 4)$ cyclic Hamming code has a generator polynomial $g(X) = 1 + X^2 + X^3$.

(a) Find the generator matrix for this code in systematic form.
(b) Find the parity-check matrix for the code.
(c) Suppose the codeword $\mathbf{C} = [1011010]$ is transmitted through a channel and the corresponding received codeword is $\mathbf{C} = [1010011]$. Find the syndrome polynomial associated with this received codeword.
(d) Find all possible received codewords such that, for the transmitted codeword $\mathbf{C} = [1011010]$, the received codeword has a syndrome polynomial of zero.

8-6. The weight distribution of a Hamming code of blocklength n is given by

$$N(x) = \sum_{i=0}^{n} N_i x^i = \frac{1}{n+1}[(1+x)^n + n(1+x)^{.5(n-1)}(1-x)^{.5(n+1)}],$$

where N_i denotes the number of codewords of weight i.

(a) Use this formula to determine the weight distribution of a Hamming $(7, 4)$ code.
(b) Use the weight distribution from part (a) to find the union upper bound based on weight distribution (8.38) for a Hamming $(7, 4)$ code, assuming BPSK modulation of the coded bits with an SNR of 10 dB. Compare with the probability of error from the looser bound (8.39) for the same modulation.

8-7. Find the union upper bound on probability of codeword error for a Hamming code with $m = 7$. Assume the coded bits are transmitted over an AWGN channel using 8-PSK modulation with an SNR of 10 dB. Compute the probability of bit error for the code assuming a codeword error corresponds to one bit error, and compare with the bit error probability for uncoded modulation.

8-8. Plot P_b versus γ_b for a $(24, 12)$ linear block code with $d_{\min} = 8$ and $0 \leq E_b/N_0 \leq 20$ dB using the union bound for probability of codeword error. Assume that the coded bits are transmitted over the channel using QPSK modulation. Over what range of E_b/N_0 does the code exhibit negative coding gain?

8-9. Use (8.47) to find the approximate coding gain of a $(7, 4)$ Hamming code with SDD over uncoded modulation, assuming $\gamma_b = 15$ dB.

8-10. Plot the probability of codeword error for a $(24, 12)$ code with $d_{\min} = 8$ for $0 \leq \gamma_b \leq 10$ dB under both hard and soft decoding, using the union bound (8.36) for hard decoding and the approximation (8.44) for soft decoding. What is the difference in coding gain at high SNR for the two decoding techniques?

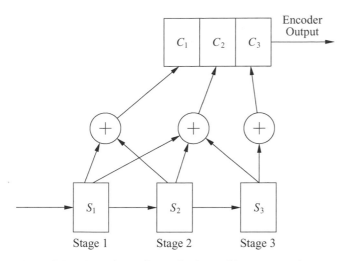

Figure 8.24: Convolutional encoder for Problems 8-14 and 8-15.

8-11. Evalute the upper and lower bounds on codeword error probability, (8.35) and (8.36) respectively, for an extended Golay code with HDD, assuming an AWGN channel with BPSK modulation and an SNR of 10 dB.

8-12. Consider a Reed Solomon code with $k = 3$ and $K = 4$, mapping to 8-PSK modulation. Find the number of errors that can be corrected with this code and its minimum distance. Also find its probability of bit error assuming the coded symbols transmitted over the channel via 8-PSK have a symbol error probability of 10^{-3}.

8-13. For the trellis of Figure 8.7, determine the state sequence and encoder output assuming an initial state $S = 00$ and information bit sequence $\mathbf{U} = [0110101101]$.

8-14. Consider the convolutional code generated by the encoder shown in Figure 8.24.

(a) Sketch the trellis diagram of the code.
(b) For a received sequence $\mathbf{R} = [001010001]$, find the path metric for the all-zero path assuming probability of symbol error $p = 10^{-3}$.
(c) Find one path at a minimum Hamming distance from the all-zero path and compute its path metric for the same $\mathbf{R}$ and p as in part (b).

8-15. This problem is based on the convolutional encoder of Figure 8.24.

(a) Draw the state diagram for this convolutional encoder.
(b) Determine its transfer function $T(D, N, J)$.
(c) Determine the minimum distance of paths through the trellis to the all-zero path.
(d) Compute the upper bound (8.72) on probability of bit error for this code assuming SDD and BPSK modulation with $\gamma_b = 10$ dB.
(e) Compute the upper bound (8.76) on probability of bit error for this code assuming HDD and BPSK modulation with $\gamma_b = 10$ dB. How much coding gain is achieved with soft versus hard decoding?

Uncoded Bits

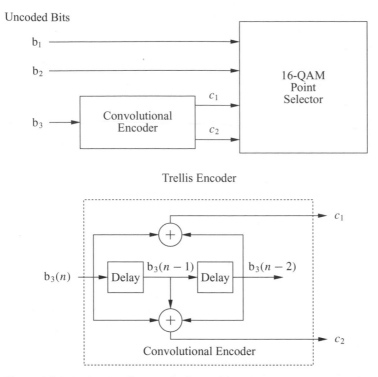

Figure 8.25: 16-QAM trellis encoder for Problem 8-16.

8-16. Suppose you have a 16-QAM signal constellation that is trellis encoded via the scheme of Figure 8.25. Assume the set partitioning for 16-QAM shown in Figure 8.18.

   (a) Assuming that parallel transitions – that is, transitions within a subset – dominate the error probability, find the coding gain of this trellis code relative to uncoded 8-PSK, given that d_0 for the 16-QAM is .632 and for the 8-PSK is .765.

   (b) Draw the trellis for the convolutional encoder and assign subsets to the trellis transitions according to the following heuristic rules of Ungerboeck: (i) if k bits are to be encoded per modulation interval then the trellis must allow for $2k$ possible transitions from each state to the next state; (ii) more than one transition may occur between pairs of states; (iii) all waveforms should occur with equal frequency and with a fair amount of regularity and symmetry; (iv) transitions originating from the same state are assigned waveforms either from the A or B subsets but not from both; (v) transitions entering into the same state are assigned waveforms either from the A or B subsets but not from both; and (vi) parallel transitions are assigned from only one subset.

   (c) What is the minimum distance error event through the trellis relative to the path generated by the all-zero bit stream?

   (d) Assuming that your answer to part (c) is the minimum distance error event for the trellis, what is $d_{\min}$ of the code?

8-17. Consider a channel with coherence time $T_c = 10$ ms and a coded bit rate of $R_s = 50$ kilosymbols per second. Find the average delay of a convolutional interleaver that achieves independent fading between subsequent coded bits. Also find the memory requirements of this system.

8-18. In a Rayleigh fading channel, determine an upper bound for the bit error probability P_b of a Golay (23, 12) code with deep interleaving ($dT_s \gg T_c$), BPSK modulation, soft decision decoding, and an average coded E_c/N_0 of 15 dB. Compare with the uncoded P_b in Rayleigh fading.

8-19. Consider a Rayleigh fading channel with BPSK modulation, average SNR of 10 dB, and a Doppler of 80 Hz. The data rate over the channel is 30 kbps. Assume that bit errors occur on this channel whenever $P_b(\gamma) \geq 10^{-2}$. Design an interleaver and associated (n, k) block code that corrects essentially all of the bit errors, where the interleaver delay is constrained to be less than 50 ms. Your design should include the dimensions of the interleaver as well as the block code type and the values of n and k.

8-20. Assume a multilevel encoder as in Figure 8.21, where the information bits have three different error protection levels ($M = 3$) and the three encoder outputs are modulated using 8-PSK modulation with an SNR of 10 dB. Assume the code C_i associated with the ith bit stream b_i is a Hamming code with parameter m_i, where $m_1 = 2$, $m_2 = 3$, and $m_3 = 4$.

    (a) Find the probability of error for each Hamming code C_i, assuming it is decoded individually using HDD.

    (b) If the symbol time of the 8-PSK modulation is $T_s = 10 \ \mu$s, what is the data rate for each of the three bit streams?

    (c) For what size code must the maximum likelihood decoder of this UEP code be designed?

8-21. Design a two-level UEP code using either Hamming or Golay codes and such that, for a channel with an SNR of 10 dB, the UEP code has $P_b = 10^{-3}$ for the low-priority bits and $P_b = 10^{-6}$ for the high priority bits.

REFERENCES

[1] J. G. Proakis, *Digital Communications,* 4th ed., McGraw-Hill, New York, 2001.
[2] B. Sklar, *Digital Communications – Fundamentals and Applications,* Prentice-Hall, Englewood Cliffs, NJ, 1988.
[3] D. G. Wilson, *Digital Modulation and Coding,* Prentice-Hall, Englewood Cliffs, NJ, 1996.
[4] S. Lin and J. D. J. Costello, *Error Control Coding,* 2nd ed., Prentice-Hall, Englewood Cliffs, NJ, 2004.
[5] A. J. Viterbi, "Error bounds for convolutional codes and asymptotically optimum decoding algorithm," *IEEE Trans. Inform. Theory,* pp. 260–9, April 1967.
[6] C. Berrou, A. Glavieux, and P. Thitimajshima, "Near Shannon limit error-correcting coding and decoding: Turbo-codes," *Proc. IEEE Internat. Conf. Commun.,* pp. 54–8, May 1993.
[7] C. Berrou and A. Glavieux, "Near optimum error correcting coding and decoding: Turbo-codes," *IEEE Trans. Commun.,* pp. 1261–71, October 1996.
[8] *IEEE Trans. Inform. Theory,* Special Issue on Codes and Graphs and Iterative Algorithms, February 2001.
[9] S. B. Wicker and S. Kim, *Codes, Graphs, and Iterative Decoding,* Kluwer, Boston, 2002.
[10] R. G. Gallager, "Low-density parity-check codes," *IRE Trans. Inform. Theory,* pp. 21–8, January 1962.

[11] C. Heegard and S. B. Wicker, *Turbo Coding*, Kluwer, Boston, 1999.

[12] M. E. Buckley and S. B. Wicker, "The design and performance of a neural network for predicting decoder error in turbo-coded ARQ protocols," *IEEE Trans. Commun.*, pp. 566–76, April 2000.

[13] S. Benedetto, D. Divsalar, G. Montorsi, and F. Pollara, "Serial concatenation of interleaved codes: Performance analysis, design and iterative decoding," *IEEE Trans. Inform. Theory*, pp. 909–26, May 1998.

[14] H. Jin and R. J. McEliece, "Coding theorems for turbo code ensembles," *IEEE Trans. Inform. Theory*, pp. 1451–61, June 2002.

[15] I. Sasan and S. Shamai, "Improved upper bounds on the ML decoding error probability of parallel and serial concatenated turbo codes via their ensemble distance spectrum," *IEEE Trans. Inform. Theory*, pp. 24–47, January 2000.

[16] D. J. C. MacKay and R. M. Neal, "Near Shannon limit performance of low density parity check codes," *Elec. Lett.*, p. 1645, August 1996.

[17] N. Wiberg, N.-A. Loeliger, and R. Kotter, "Codes and iterative decoding on general graphs," *Euro. Trans. Telecommun.*, pp. 513–25, June 1995.

[18] T. Richardson, A. Shokrollahi, and R. Urbanke, "Design of capacity-approaching irregular low-density parity-check codes," *IEEE Trans. Inform. Theory*, pp. 619–37, February 2001.

[19] R. McEliece, D. J. C. MacKay, and J.-F. Cheng, "Turbo decoding as an instance of Pearl's 'belief propagation' algorithm," *IEEE J. Sel. Areas Commun.*, pp. 140–52, February 1998.

[20] F. R. Kschischang and D. Frey, "Iterative decoding of compound codes by probability propagation in graphical models," *IEEE J. Sel. Areas Commun.*, pp. 219–30, February 1998.

[21] T. Richardson and R. Urbanke, "The capacity of low-density parity-check codes under message passing decoding," *IEEE Trans. Inform. Theory*, pp. 599–618, February 2001.

[22] S.-Y. Chung, G. D. Forney, T. Richardson, and R. Urbanke, "On the design of low-density parity-check codes within 0.0045 dB of the Shannon limit," *IEEE Commun. Lett.*, pp. 58–60, February 2001.

[23] M. Fossorier, "Iterative reliability-based decoding of low-density parity check codes," *IEEE J. Sel. Areas Commun.*, pp. 908–17, May 2001.

[24] M. C. Davey and D. MacKay, "Low density parity-check codes over GF(q)," *IEEE Commun. Lett.*, pp. 165–7, June 1998.

[25] J. Hou, P. Siegel, L. Milstein, and H. D. Pfister, "Capacity-approaching bandwidth efficient coded modulation schemes based on low-density parity-check codes," *IEEE Trans. Inform. Theory*, pp. 2141–55, September 2003.

[26] G. Ungerboeck, "Channel coding with multi-level/phase signals," *IEEE Trans. Inform. Theory*, pp. 55–67, January 1982.

[27] G. D. Forney, "Coset codes, I: Introduction and geometrical classification, and II: Binary lattices and related codes," *IEEE Trans. Inform. Theory*, pp. 1123–87, September 1988.

[28] G. D. Forney, Jr., R. G. Gallager, G. R. Lang, F. M. Longstaff, and S. U. Quereshi, "Efficient modulation for band-limited channels," *IEEE J. Sel. Areas Commun.*, pp. 632–47, September 1984.

[29] S. Benedetto, D. Divsalar, G. Montorsi, and F. Pollara, "Parallel concatenated trellis coded modulation," *Proc. IEEE Internat. Conf. Commun.*, pp. 974–8, June 1996.

[30] C. Fragouli and R. D. Wesel, "Turbo-encoder design for symbol-interleaved parallel concatenated trellis-coded modulation," *IEEE Trans. Commun.*, pp. 425–35, March 2001.

[31] P. Robertson and T. Worz, "Bandwidth-efficient turbo trellis-coded modulation using punctured component codes," *IEEE J. Sel. Areas Commun.*, pp. 206–18, February 1998.

[32] G. T. Irvine and P. J. Mclane, "Symbol-aided plus decision-directed reception for PSK TCM modulation on shadowed mobile satellite fading channels," *IEEE J. Sel. Areas Commun.*, pp. 1289–99, October 1992.

[33] D. Subasinghe-Dias and K. Feher, "A coded 16-QAM scheme for fast fading mobile radio channels," *IEEE Trans. Commun.*, pp. 1906–16, February–April 1995.

[34] P. Y. Kam and H. M. Ching, "Sequence estimation over the slow nonselective Rayleigh fading channel with diversity reception and its application to Viterbi decoding," *IEEE J. Sel. Areas Commun.*, pp. 562–70, April 1992.

[35] D. Makrakis, P. T. Mathiopoulos, and D. P. Bouras, "Optimal decoding of coded PSK and QAM signals in correlated fast fading channels and AWGN – A combined envelope, multiple differential and coherent detection approach," *IEEE Trans. Commun.*, pp. 63–75, January 1994.

[36] M. J. Gertsman and J. H. Lodge, "Symbol-by-symbol MAP demodulation of CPM and PSK signals on Rayleigh flat-fading channels," *IEEE Trans. Commun.*, pp. 788–99, July 1997.

[37] H. Kong and E. Shwedyk, "Sequence detection and channel state estimation over finite state Markov channels," *IEEE Trans. Veh. Tech.*, pp. 833–9, May 1999.

[38] G. M. Vitetta and D. P. Taylor, "Maximum-likelihood decoding of uncoded and coded PSK signal sequences transmitted over Rayleigh flat-fading channels," *IEEE Trans. Commun.*, pp. 2750–8, November 1995.

[39] L. Li and A. J. Goldsmith, "Low-complexity maximum-likelihood detection of coded signals sent over finite-state Markov channels," *IEEE Trans. Commun.*, pp. 524–31, April 2002.

[40] B. Vucetic and J. Yuan, *Turbo Codes: Principles and Applications,* Kluwer, Dordrecht, 2000.

[41] E. K. Hall and S. G. Wilson, "Design and analysis of turbo codes on Rayleigh fading channels," *IEEE J. Sel. Areas Commun.*, pp. 160–74, February 1998.

[42] C. Komninakis and R. D. Wesel, "Joint iterative channel estimation and decoding in flat correlated Rayleigh fading," *IEEE J. Sel. Areas Commun.* pp. 1706–17, September 2001.

[43] J. Hou, P. H. Siegel, and L. B. Milstein, "Performance analysis and code optimization of low-density parity-check codes on Rayleigh fading channels," *IEEE J. Sel. Areas Commun.*, pp. 924–34, May 2001.

[44] A. J. Goldsmith and P. P. Varaiya, "Capacity, mutual information, and coding for finite-state Markov channels," *IEEE Trans. Inform. Theory,* pp. 868–86, May 1996.

[45] G. D. Forney, "Burst error correcting codes for the classic bursty channel," *IEEE Trans. Commun. Tech.,* pp. 772–81, October 1971.

[46] J. L. Ramsey, "Realization of optimum interleavers," *IEEE Trans. Inform. Theory,* pp. 338–45, 1970.

[47] C.-E. W. Sundberg and N. Seshadri, "Coded modulation for fading channels – An overview," *Euro. Trans. Telecommun.,* pp. 309–24, May/June 1993.

[48] L.-F. Wei, "Coded M-DPSK with built-in time diversity for fading channels," *IEEE Trans. Inform. Theory,* pp. 1820–39, November 1993.

[49] S. H. Jamali and T. Le-Ngoc, *Coded-Modulation Techniques for Fading Channels,* Kluwer, New York, 1994.

[50] G. Caire, G. Taricco, and E. Biglieri, "Bit-interleaved coded modulation," *IEEE Trans. Inform. Theory,* pp. 927–46, May 1998.

[51] E. Zehavi, "8-PSK trellis codes for a Rayleigh channel," *IEEE Trans. Commun.,* pp. 873–84, May 1992.

[52] H. Imai and S. Hirakawa, "A new multilevel coding method using error correcting codes," *IEEE Trans. Inform. Theory,* pp. 371–7, May 1977.

[53] R. V. Cox, J. Hagenauer, N. Seshadri, and C.-E. W. Sundberg, "Variable rate sub-band speech coding and matched convolutional channel coding for mobile radio channels," *IEEE Trans. Signal Proc.,* pp. 1717–31, August 1991.

[54] J. W. Modestino and D. G. Daut, "Combined source-channel coding of images," *IEEE Trans. Commun.,* pp. 1644–59, November 1979.

[55] A. R. Calderbank and N. Seshadri, "Multilevel codes for unequal error protection," *IEEE Trans. Inform. Theory,* pp. 1234–48, July 1993.

[56] L.-F. Wei, "Coded modulation with unequal error protection," *IEEE Trans. Commun.,* pp. 1439–49, October 1993.

[57] N. Seshadri and C.-E. W. Sundberg, "Multilevel trellis coded modulations for the Rayleigh fading channel," *IEEE Trans. Commun.,* pp. 1300–10, September 1993.

[58] C. E. Shannon, "Coding theorems for a discrete source with a fidelity criterion," *IRE National Convention Record,* part 4, pp. 142–63, 1959.

[59] N. Tanabe and N. Farvardin, "Subband image coding using entropy-coded quantization over noisy channels," *IEEE J. Sel. Areas Commun.,* pp. 926–43, June 1992.

[60] H. Jafarkhani, P. Ligdas, and N. Farvardin, "Adaptive rate allocation in a joint source/channel coding framework for wireless channels," *Proc. IEEE Veh. Tech. Conf.,* pp. 492–6, April 1996.

[61] W. C. Wong, R. Steele, and C.-E. W. Sundberg, *Source-Matched Mobile Communications,* Pentech and IEEE Press, London and New York, 1995.

[62] K.-P. Ho and J. M. Kahn, "Transmission of analog signals using multicarrier modulation: A combined source-channel coding approach," *IEEE Trans. Commun.,* pp. 1432–43, November 1996.

[63] N. Farvardin and V. Vaishampayan, "On the performance and complexity of channel-optimized vector quantizers," *IEEE Trans. Inform. Theory,* pp. 155–60, January 1991.

[64] N. Farvardin and V. Vaishampayan, "Optimal quantizer design for noisy channels: An approach to combined source-channel coding," *IEEE Trans. Inform. Theory,* pp. 827–38, November 1987.

[65] K.-P. Ho and J. M. Kahn, "Combined source-channel coding using channel-optimized quantizer and multicarrier modulation," *Proc. IEEE Internat. Conf. Commun.,* pp. 1323–7, June 1996.

[66] A. J. Goldsmith and M. Effros, "Joint design of fixed-rate source codes and multiresolution channel codes," *IEEE Trans. Commun.,* pp. 1301–12, October 1998.

[67] E. Ayanoglu and R. M. Gray, "The design of joint source and channel trellis waveform coders," *IEEE Trans. Inform. Theory,* pp. 855–65, November 1987.

[68] T. R. Fischer and M. W. Marcellin, "Joint trellis coded quantization/modulation," *IEEE Trans. Commun.,* pp. 172–6, February 1991.

Adaptive Modulation and Coding

Adaptive modulation and coding enable robust and spectrally efficient transmission over time-varying channels. The basic premise is to estimate the channel at the receiver and feed this estimate back to the transmitter, so that the transmission scheme can be adapted relative to the channel characteristics. Modulation and coding techniques that do not adapt to fading conditions require a fixed link margin to maintain acceptable performance when the channel quality is poor. Thus, these systems are effectively designed for worst-case channel conditions. Since Rayleigh fading can cause a signal power loss of up to 30 dB, designing for the worst-case channel conditions can result in very inefficient utilization of the channel. Adapting to the channel fading can increase average throughput, reduce required transmit power, or reduce average probability of bit error by taking advantage of favorable channel conditions to send at higher data rates or lower power – and by reducing the data rate or increasing power as the channel degrades. In Section 4.2.4 we derived the optimal adaptive transmission scheme that achieves the Shannon capacity of a flat fading channel. In this chapter we describe more practical adaptive modulation and coding techniques to maximize average spectral efficiency while maintaining a given average or instantaneous bit error probability. The same basic premise can be applied to MIMO channels, frequency-selective fading channels with equalization, OFDM or CDMA, and cellular systems. The application of adaptive techniques to these systems will be described in subsequent chapters.

Adaptive transmission was first investigated in the late sixties and early seventies [1; 2]. Interest in these techniques was short-lived, perhaps due to hardware constraints, lack of good channel estimation techniques, and/or systems focusing on point-to-point radio links without transmitter feedback. As technology evolved these issues became less constraining, resulting in a revived interest in adaptive modulation methods for 3G wireless systems [3; 4; 5; 6; 7; 8; 9; 10; 11; 12]. As a result, many wireless systems – including both GSM and CDMA cellular systems as well as wireless LANs – use adaptive transmission techniques [13; 14; 15; 16].

There are several practical constraints that determine when adaptive modulation should be used. Adaptive modulation requires a feedback path between the transmitter and receiver, which may not be feasible for some systems. Moreover, if the channel is changing faster than it can be reliably estimated and fed back to the transmitter, adaptive techniques will perform poorly. Many wireless channels exhibit variations on different time scales; examples include multipath fading, which can change very quickly, and shadowing, which changes more slowly. Often only the slow variations can be tracked and adapted to, in which case flat fading mitigation is needed to address the effects of multipath. Hardware constraints may dictate how often the transmitter can change its rate and/or power, and this may limit the performance gains possible with adaptive modulation. Finally, adaptive modulation typically varies the rate of data transmission relative to channel conditions. We will see that average spectral efficiency of adaptive modulation under an average power constraint is maximized by setting the data rate to be small or zero in poor channel conditions. However, with this scheme the quality of fixed-rate applications (such as voice or video) with hard delay constraints may be significantly compromised. Thus, in delay-constrained applications the adaptive modulation should be optimized to minimize outage probability for a fixed data rate [17].

9.1 Adaptive Transmission System

In this section we describe the system associated with adaptive transmission. The model is the same as the model of Section 4.2.1 used to determine the capacity of flat fading channels. We assume linear modulation, where the adaptation takes place at a multiple of the symbol rate $R_s = 1/T_s$. We also assume the modulation uses ideal Nyquist data pulses ($\text{sinc}[t/T_s]$), so the signal bandwidth $B = 1/T_s$. We model the flat fading channel as a discrete-time channel in which each channel use corresponds to one symbol time T_s. The channel has stationary and ergodic time-varying gain $\sqrt{g[i]}$ that follows a given distribution $p(g)$ and AWGN $n[i]$, with power spectral density $N_0/2$. Let $\bar{P}$ denote the average transmit signal power, $B = 1/T_s$ the received signal bandwidth, and $\bar{g}$ the average channel gain. The instantaneous received signal-to-noise power ratio (SNR) is then $\gamma[i] = \bar{P}g[i]/N_0 B$, $0 \leq \gamma[i] < \infty$, and its expected value over all time is $\bar{\gamma} = \bar{P}\bar{g}/N_0 B$. Since $g[i]$ is stationary, the distribution of $\gamma[i]$ is independent of i, and we denote this distribution by $p(\gamma)$.

In adaptive transmission we estimate the power gain or the received SNR at time i and then adapt the modulation and coding parameters accordingly. The most common parameters to adapt are the data rate $R[i]$, transmit power $P[i]$, and coding parameters $C[i]$. For M-ary modulation the data rate $R[i] = \log_2 M[i]/T_s = B \log_2 M[i]$ bps. The *spectral efficiency* of the M-ary modulation is $R[i]/B = \log_2 M[i]$ bps/Hz. We denote the SNR estimate as $\hat{\gamma}[i] = \bar{P}\hat{g}[i]/N_0 B$, which is based on the power gain estimate $\hat{g}[i]$. Suppose the transmit power is adapted relative to $\hat{\gamma}[i]$. We denote this adaptive transmit power at time i by $P(\hat{\gamma}[i]) = P[i]$, and the received power at time i is then $\gamma[i]P(\hat{\gamma}[i])/\bar{P}$. Similarly, we can adapt the data rate of the modulation $R(\hat{\gamma}[i]) = R[i]$ and/or the coding parameters $C(\hat{\gamma}[i]) = C[i]$ relative to the estimate $\hat{\gamma}[i]$. When the context is clear, we will omit the time reference i relative to γ, $P(\gamma)$, $R(\gamma)$, and $C(\gamma)$.

The system model is illustrated in Figure 9.1. We assume that an estimate $\hat{g}[i]$ of the channel power gain $g[i]$ at time i is available to the receiver after an estimation time delay of i_e and that this same estimate is available to the transmitter after a combined estimation and feedback path delay of $i_d = i_e + i_f$. The availability of this channel information at the transmitter allows it to adapt its transmission scheme relative to the channel variation. The adaptive strategy may take into account the estimation error and delay in $\hat{g}[i - i_d]$ or it may treat $\hat{g}[i - i_d]$ as the true gain: this issue will be discussed in more detail in Section 9.3.7.

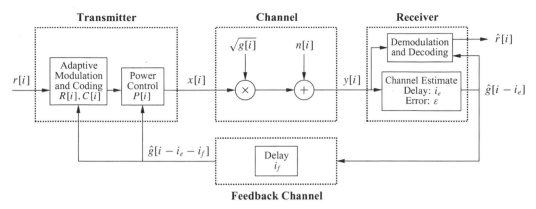

Figure 9.1: System model.

We assume that the feedback path does not introduce any errors, which is a reasonable assumption if strong error correction and detection codes are used on the feedback path and if packets associated with detected errors are retransmitted.

The rate of channel variation will dictate how often the transmitter must adapt its transmission parameters, and it will also affect the estimation error of $g[i]$. When the channel gain consists of both fast and slow fading components, the adaptive transmission may adapt to both if $g[i]$ changes sufficiently slowly, or it may adapt to just the slow fading. In particular, if $g[i]$ corresponds to shadowing and multipath fading, then at low speeds the shadowing is essentially constant, and the multipath fading is sufficiently slow that it can be estimated and fed back to the transmitter with estimation error and delay that do not significantly degrade performance. At high speeds the system can no longer effectively estimate and feed back the multipath fading in order to adapt to it. In this case, the adaptive transmission responds to the shadowing variations only, and the error probability of the modulation must be averaged over the fast fading distribution. Adaptive techniques for combined fast and slow fading are discussed in Section 9.5.

9.2 Adaptive Techniques

There are many parameters that can be varied at the transmitter relative to the channel gain γ. In this section we discuss adaptive techniques associated with variation of the most common parameters: data rate, power, coding, error probability, and combinations of these adaptive techniques.

9.2.1 Variable-Rate Techniques

In variable-rate modulation the data rate $R[\gamma]$ is varied relative to the channel gain γ. This can be done by fixing the symbol rate $R_s = 1/T_s$ of the modulation and using multiple modulation schemes or constellation sizes, or by fixing the modulation (e.g. BPSK) and changing the symbol rate. Symbol rate variation is difficult to implement in practice because a varying signal bandwidth is impractical and complicates bandwidth sharing. In contrast, changing the constellation size or modulation type with a fixed symbol rate is fairly easy, and these techniques are used in current systems. Specifically, the GSM and IS-136 EDGE system as well as 802.11a wireless LANs vary their modulation and coding relative to channel quality [15]. In general the modulation parameters that dictate the transmission rate are fixed over a block or frame of symbols, where the frame size is a parameter of the design. Frames may also include pilot symbols for channel estimation and other control information.

When a discrete set of modulation types or constellation sizes are used, each value of γ must be mapped to one of the possible modulation schemes. This is often done to maintain the bit error probability of each scheme below a given value. These ideas are illustrated in the following example as well as in subsequent sections on specific adaptive modulation techniques.

EXAMPLE 9.1: Consider an adaptive modulation system that uses QPSK and 8-PSK for a target P_b of approximately 10^{-3}. If the target P_b cannot be met with either scheme, then no data is transmitted. Find the range of γ-values associated with the three possible transmission schemes (no transmission, QPSK, and 8-PSK) as well as the average spectral efficiency of the system, assuming Rayleigh fading with $\bar{\gamma} = 20$ dB.

Solution: First note that the SNR $\gamma = \gamma_s$ for both QPSK and 8-PSK. From Section 6.1 we have $P_b \approx Q(\sqrt{\gamma})$ for QPSK and $P_b \approx .666 Q(\sqrt{2\gamma} \sin(\pi/8))$ for 8-PSK. Since $\gamma > 14.79$ dB yields $P_b < 10^{-3}$ for 8-PSK, the adaptive modulation uses 8-PSK modulation for $\gamma > 14.79$ dB. Since $\gamma > 10.35$ dB yields $P_b < 10^{-3}$ for QPSK, the adaptive modulation uses QPSK modulation for 14.79 dB $\geq \gamma > 10.35$ dB. The channel is not used for $\gamma \leq 10.35$ dB.

We determine the average rate by analyzing how often each of the different transmission schemes is used. Since 8-PSK is used when $\gamma > 14.79$ dB $= 30.1$, in Rayleigh fading with $\bar{\gamma} = 20$ dB the spectral efficiency $R[\gamma]/B = \log_2 8 = 3$ bps/Hz is transmitted a fraction of time equal to $P_8 = \int_{30.1}^{\infty} \frac{1}{100} e^{-\gamma/100} \, d\gamma = .74$. QPSK is used when $10.35 < \gamma \leq 14.79$ dB, where 10.35 dB $= 10.85$ in linear units. So $R[\gamma] = \log_2 4 = 2$ bps/Hz is transmitted a fraction of time equal to $P_4 = \int_{10.85}^{30.1} \frac{1}{100} e^{-\gamma/100} \, d\gamma = .157$. During the remaining .103 portion of time there is no data transmission. So the average spectral efficiency is $.74 \cdot 3 + .157 \cdot 2 + .103 \cdot 0 = 2.534$ bps/Hz.

Note that if $\gamma \leq 10.35$ dB then – rather than suspending transmission, which leads to an outage probability of roughly .1 – either just one signaling dimension could be used (i.e., BPSK could be transmitted) or error correction coding could be added to the QPSK to meet the P_b target. If block or convolutional codes were used then the spectral efficiency for $\gamma \leq 10.35$ dB would be less than 2 bps/Hz but larger than a spectral efficiency of zero corresponding to no transmission. These variable-coding techniques are described in Section 9.2.4.

9.2.2 Variable-Power Techniques

Adapting the transmit power alone is generally used to compensate for SNR variation due to fading. The goal is to maintain a fixed bit error probability or, equivalently, a constant received SNR. The power adaptation thus inverts the channel fading so that the channel appears as an AWGN channel to the modulator and demodulator.[1] The power adaptation for channel inversion is given by

---

[1] Channel inversion and truncated channel inversion were discussed in Section 4.2.4 in the context of fading channel capacity.

$$\frac{P(\gamma)}{\bar{P}} = \frac{\sigma}{\gamma}, \tag{9.1}$$

where σ equals the constant received SNR. The average power constraint $\bar{P}$ implies that

$$\int \frac{P(\gamma)}{\bar{P}} p(\gamma) \, d\gamma = \int \frac{\sigma}{\gamma} p(\gamma) \, d\gamma = 1. \tag{9.2}$$

Solving (9.2) for σ yields that $\sigma = 1/\mathbf{E}[1/\gamma]$, so σ is determined by $p(\gamma)$, which in turn depends on the average transmit power $\bar{P}$ through $\bar{\gamma}$. Thus, for a given average power $\bar{P}$, if the value for σ required to meet the target bit error rate is greater than $1/\mathbf{E}[1/\gamma]$ then this target cannot be met. Note that for Rayleigh fading, where γ is exponentially distributed, $\mathbf{E}[1/\gamma] = \infty$ and so no target P_b can be met using channel inversion.

The fading can also be inverted above a given cutoff γ_0, which leads to a truncated channel inversion for power adaptation. In this case the power adaptation is given by

$$\frac{P(\gamma)}{\bar{P}} = \begin{cases} \sigma/\gamma & \gamma \geq \gamma_0, \\ 0 & \gamma < \gamma_0. \end{cases} \tag{9.3}$$

The cutoff value γ_0 can be based on a desired outage probability $P_{\text{out}} = p(\gamma < \gamma_0)$ or on a desired target BER above a cutoff that is determined by the target BER and $p(\gamma)$. Since the channel is only used when $\gamma \geq \gamma_0$, given an average power $\bar{P}$ we have $\sigma = 1/\mathbf{E}_{\gamma_0}[1/\gamma]$, where

$$\mathbf{E}_{\gamma_0}\left[\frac{1}{\gamma}\right] \triangleq \int_{\gamma_0}^{\infty} \frac{1}{\gamma} p(\gamma) \, d\gamma. \tag{9.4}$$

EXAMPLE 9.2: Find the power adaptation for BPSK modulation that maintains a fixed $P_b = 10^{-3}$ in non-outage for a Rayleigh fading channel with $\bar{\gamma} = 10$ dB. Also find the resulting outage probability.

Solution: The power adaptation is truncated channel inversion, so we need only find σ and γ_0. For BPSK modulation, with a constant SNR of $\sigma = 4.77$ we get $P_b = Q(\sqrt{2\sigma}) = 10^{-3}$. Setting $\sigma = 1/\mathbf{E}_{\gamma_0}[1/\gamma]$ and solving for γ_0, which must be done numerically, yields $\gamma_0 = .7423$. So $P_{\text{out}} = p(\gamma < \gamma_0) = 1 - e^{-\gamma_0/10} = .379$. Hence there is a high outage probability, which results from requiring $P_b = 10^{-3}$ in this relatively weak channel.

9.2.3 Variable Error Probability

We can also adapt the instantaneous BER subject to an average BER constraint $\bar{P}_b$. In Section 6.3.2 we saw that in fading channels the instantaneous error probability varies as the received SNR γ varies, resulting in an average BER of $\bar{P}_b = \int P_b(\gamma) p(\gamma) \, d\gamma$. This is not considered an adaptive technique, since the transmitter does not adapt to γ. Thus, in adaptive modulation, error probability is typically adapted along with some other form of adaptation such as constellation size or modulation type. Adaptation based on varying both data rate and error probability to reduce transmit energy was first proposed by Hayes in [1], where a 4-dB power savings was obtained at a target average bit error probability of 10^{-4}.

9.2.4 Variable-Coding Techniques

In adaptive coding, different channel codes are used to provide different amounts of coding gain to the transmitted bits. For example, a stronger error correction code may be used when γ is small, with a weaker code or no coding used when γ is large. Adaptive coding can be implemented by multiplexing together codes with different error correction capabilities. However, this approach requires that the channel remain roughly constant over the block length or constraint length of the code [7]. On such slowly varying channels, adaptive coding is particularly useful when the modulation must remain fixed, as may be the case owing to complexity or peak-to-average power ratio constraints.

An alternative technique to code multiplexing is rate-compatible punctured convolutional (RCPC) codes [18]. This is a family of convolutional codes at different code rates $R_c = k/n$. The basic premise of RCPC codes is to have a single encoder and decoder whose error correction capability can be modified by not transmitting certain coded bits (*puncturing* the code). Moreover, RCPC codes have a rate compatibility constraint so that the coded bits associated with a high-rate (weaker) code are also used by all lower-rate (stronger) codes. Thus, to increase the error correction capability of the code, the coded bits of the weakest code are transmitted along with additional coded bits to achieve the desired level of error correction. The rate compatibility makes it easy to adapt the error protection of the code, since the same encoder and decoder are used for all codes in the RCPC family, with puncturing at the transmitter to achieve the desired error correction. Decoding is performed by a Viterbi algorithm operating on the trellis associated with the lowest rate code, with the puncturing incorporated into the branch metrics. Puncturing is a very effective and powerful adaptive coding technique, and it forms the basis of adaptive coding in the GSM and IS-136 EDGE protocol for data transmission [13].

Adaptive coding through either multiplexing or puncturing can be done for fixed modulation or combined with adaptive modulation as a hybrid technique. When the modulation is fixed, adaptive coding is often the only practical mechanism to address the channel variations [6; 7]. The focus of this chapter is on systems where adaptive modulation is possible, so adaptive coding on its own will not be further discussed.

9.2.5 Hybrid Techniques

Hybrid techniques can adapt multiple parameters of the transmission scheme, including rate, power, coding, and instantaneous error probability. In this case joint optimization of the different techniques is used to meet a given performance requirement. Rate adaptation is often combined with power adaptation to maximize spectral efficiency, and we apply this joint optimization to different modulations in subsequent sections. Adaptive modulation and coding has been widely investigated in the literature and is currently used in both cellular systems and wireless LANs [13; 15].

9.3 Variable-Rate Variable-Power MQAM

In the previous section we discussed general approaches to adaptive modulation and coding. In this section we describe a specific form of adaptive modulation where the rate and power of MQAM are varied to maximize spectral efficiency while meeting a given instantaneous

P_b target. We study this specific form of adaptive modulation because it provides insight into the benefits of adaptive modulation and since, moreover, the same scheme for power and rate adaptation that achieves capacity also optimizes this adaptive MQAM design. We will show that there is a constant power gap between the spectral efficiency of this adaptive MQAM technique and capacity in flat fading, and this gap can be partially closed by coded modulation using a trellis or lattice code superimposed on the adaptive modulation.

Consider a family of MQAM signal constellations with a fixed symbol time T_s, where M denotes the number of points in each signal constellation. We assume $T_s = 1/B$ based on ideal Nyquist pulse shaping. Let $\bar{P}$, N_0, $\gamma = \bar{P}g/N_0B$, and $\bar{\gamma} = \bar{P}/N_0B$ be as given in our system model. Then the average E_s/N_0 equals the average SNR:

$$\frac{\bar{E}_s}{N_0} = \frac{\bar{P}T_s}{N_0} = \bar{\gamma}. \tag{9.5}$$

The spectral efficiency for fixed M is $R/B = \log_2 M$, the number of bits per symbol. This efficiency is typically parameterized by the average transmit power $\bar{P}$ and the BER of the modulation technique.

9.3.1 Error Probability Bounds

In [19] the BER for an AWGN channel with MQAM modulation, ideal coherent phase detection, and SNR γ is bounded by

$$P_b \leq 2e^{-1.5\gamma/(M-1)}. \tag{9.6}$$

A tighter bound, good to within 1 dB for $M \geq 4$ and $0 \leq \gamma \leq 30$ dB, is

$$P_b \leq .2e^{-1.5\gamma/(M-1)}. \tag{9.7}$$

Note that these expressions are only bounds, so they don't match the error probability expressions from Table 6.1. We use these bounds because they are easy to invert: so we can obtain M as a function of the target P_b and the power adaptation policy, as we will see shortly. Adaptive modulation designs can also be based on BER expressions that are not invertible or on BER simulation results, with numerical inversion used to obtain the constellation size and SNR associated with a given BER target.

In a fading channel with nonadaptive transmission (constant transmit power and rate), the average BER is obtained by integrating the BER in AWGN over the fading distribution $p(\gamma)$. Thus, we use the average BER expression to find the maximum data rate that can achieve a given average BER for a given average SNR. Similarly, if the data rate and average BER are fixed, we can find the required average SNR to achieve this target, as illustrated in the next example.

EXAMPLE 9.3: Find the average SNR required to achieve an average BER of $\bar{P}_b = 10^{-3}$ for nonadaptive BPSK modulation in Rayleigh fading. What is the spectral efficiency of this scheme?

Solution: From Section 6.3.2, BPSK in Rayleigh fading has $\bar{P}_b \approx 1/4\bar{\gamma}$. Thus, without transmitter adaptation, for a target average BER of $\bar{P}_b = 10^{-3}$ we require

$\bar{\gamma} = 1/4\bar{P_b} = 250 = 24$ dB. The spectral efficiency is $R/B = \log_2 2 = 1$ bps/Hz. We will see that adaptive modulation provides a much higher spectral efficiency at this same SNR and target BER.

9.3.2 Adaptive Rate and Power Schemes

We now consider adapting the transmit power $P(\gamma)$ relative to γ, subject to the average power constraint $\bar{P}$ and an instantaneous BER constraint $P_b(\gamma) = P_b$. The received SNR is then $\gamma P(\gamma)/\bar{P}$, and the P_b bound for each value of γ, using the tight bound (9.7), becomes

$$P_b(\gamma) \leq .2 \exp\left[\frac{-1.5\gamma}{M-1}\frac{P(\gamma)}{\bar{P}}\right]. \tag{9.8}$$

We adjust M and $P(\gamma)$ to maintain the target P_b. Rearranging (9.8) yields the following maximum constellation size for a given P_b:

$$M(\gamma) = 1 + \frac{1.5\gamma}{-\ln(5P_b)}\frac{P(\gamma)}{\bar{P}} = 1 + K\gamma\frac{P(\gamma)}{\bar{P}}, \tag{9.9}$$

where

$$K = \frac{-1.5}{\ln(5P_b)} < 1. \tag{9.10}$$

We maximize spectral efficiency by maximizing

$$\mathbf{E}[\log_2 M(\gamma)] = \int_0^\infty \log_2\left(1 + \frac{K\gamma P(\gamma)}{\bar{P}}\right)p(\gamma)\,d\gamma \tag{9.11}$$

subject to the power constraint

$$\int_0^\infty P(\gamma)p(\gamma)\,d\gamma = \bar{P}. \tag{9.12}$$

The power adaptation policy that maximizes (9.11) has the same form as the optimal power adaptation policy (4.12) that achieves capacity:

$$\frac{P(\gamma)}{\bar{P}} = \begin{cases} 1/\gamma_0 - 1/\gamma K & \gamma \geq \gamma_0/K, \\ 0 & \gamma < \gamma_0/K, \end{cases} \tag{9.13}$$

where γ_0/K is the optimized cutoff fade depth below which the channel is not used, for K given by (9.10). If we define $\gamma_K = \gamma_0/K$ and multiply both sides of (9.13) by K, we obtain

$$\frac{KP(\gamma)}{\bar{P}} = \begin{cases} 1/\gamma_K - 1/\gamma & \gamma \geq \gamma_K, \\ 0 & \gamma < \gamma_K, \end{cases} \tag{9.14}$$

where γ_K is a cutoff fade depth below which the channel is not used. This cutoff must satisfy the power constraint (9.12):

$$\int_{\gamma_K}^\infty \left(\frac{1}{\gamma_K} - \frac{1}{\gamma}\right)p(\gamma)\,d\gamma = K. \tag{9.15}$$

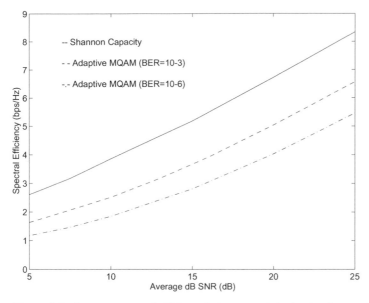

Figure 9.2: Average spectral efficiency in log-normal shadowing ($\sigma_{\psi_{dB}} = 8$ dB).

Substituting (9.13) or (9.14) into (9.9), we get that the instantaneous rate is given by

$$\log_2 M(\gamma) = \log_2\left(\frac{\gamma}{\gamma_K}\right), \tag{9.16}$$

and the corresponding average spectral efficiency (9.11) is given by

$$\frac{R}{B} = \int_{\gamma_K}^{\infty} \log_2\left(\frac{\gamma}{\gamma_K}\right) p(\gamma)\, d\gamma. \tag{9.17}$$

Comparing the power adaptation and average spectral efficiency – (4.12) and (4.13) – associated with the Shannon capacity of a fading channel with (9.13) and (9.17), the optimal power adaptation and average spectral efficiency of adaptive MQAM, we see that the power and rate adaptation are the same and lead to the same average spectral efficiency, with an effective power loss of K for adaptive MQAM as compared to the capacity-achieving scheme. Moreover, this power loss is independent of the fading distribution. Thus, if the capacity of a fading channel is R bps/Hz at SNR $\bar{\gamma}$, then uncoded adaptive MQAM requires a received SNR of $\bar{\gamma}/K$ to achieve the same rate. Equivalently, K is the maximum possible coding gain for this variable rate and power MQAM method. We discuss superimposing a trellis or lattice code on the adaptive MQAM to obtain some of this coding gain in Section 9.3.8.

We plot the average spectral efficiency (9.17) of adaptive MQAM at a target P_b of 10^{-3} and 10^{-6} for both log-normal shadowing and Rayleigh fading in Figures 9.2 and 9.3, respectively. We also plot the capacity in these figures for comparison. Note that the gap between the spectral efficiency of variable-rate variable-power MQAM and capacity is the constant K, which from (9.10) is a simple function of the BER.

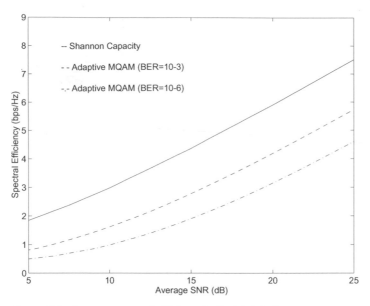

Figure 9.3: Average spectral efficiency in Rayleigh fading.

9.3.3 Channel Inversion with Fixed Rate

We can also apply channel inversion power adaptation to maintain a fixed received SNR. We then transmit a single fixed-rate MQAM modulation that achieves the target P_b. The constellation size M that meets this target P_b is obtained by substituting the channel inversion power adaptation $P(\gamma)/\bar{P} = \sigma/\gamma$ of (9.2) into (9.9) with $\sigma = 1/\mathbf{E}[1/\gamma]$. Since the resulting spectral efficiency $R/B = M$, this yields the spectral efficiency of the channel inversion power adaptation as

$$\frac{R}{B} = \log_2\left(1 + \frac{-1.5}{\ln(5P_b)\,\mathbf{E}[1/\gamma]}\right). \tag{9.18}$$

This spectral efficiency is based on the tight bound (9.7); if the resulting $M = R/B < 4$ then the loose bound (9.6) must be used, in which case $\ln(5P_b)$ is replaced by $\ln(.5P_b)$ in (9.18).

With truncated channel inversion the channel is only used when $\gamma > \gamma_0$. Thus, the spectral efficiency with truncated channel inversion is obtained by substituting $P(\gamma)/\bar{P} = \sigma/\gamma$ ($\gamma > \gamma_0$) into (9.9) and multiplying by the probability that $\gamma > \gamma_0$. The maximum value is obtained by optimizing relative to the cutoff level γ_0:

$$\frac{R}{B} = \max_{\gamma_0} \log_2\left(1 + \frac{-1.5}{\ln(5P_b)\,\mathbf{E}_{\gamma_0}[1/\gamma]}\right) p(\gamma > \gamma_0). \tag{9.19}$$

The spectral efficiency of adaptive MQAM (with the optimal water-filling and truncated channel inversion power adaptation) in a Rayleigh fading channel with a target BER of 10^{-3} is shown in Figure 9.4, along with the capacity under the same two power adaptation policies. We see, surprisingly, that truncated channel inversion with fixed-rate transmission has almost the same spectral efficiency as optimal variable-rate variable-power MQAM. This suggests

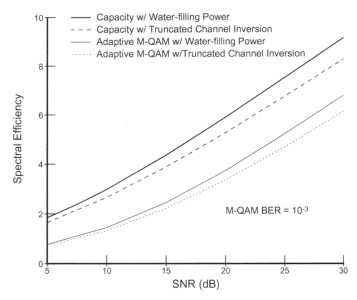

Figure 9.4: Spectral efficiency with different power adaptation policies (Rayleigh fading).

that truncated channel inversion is more desirable in practice, since it achieves almost the same spectral efficiency as variable-rate variable-power transmission but does not require that we vary the rate. However, this assumes there is no restriction on constellation size. Specifically, the spectral efficiencies (9.17), (9.18), and (9.19) assume that M can be any real number and that the power and rate can vary continuously with γ. Though MQAM modulation for noninteger values of M is possible, the complexity is quite high [20]. Moreover, it is difficult in practice to continually adapt the transmit power and constellation size to the channel fading, particularly in fast fading environments. Thus, we now consider restricting the constellation size to just a handful of values. This will clearly affect the spectral efficiency, though (as shown in the next section) not by very much.

9.3.4 Discrete-Rate Adaptation

We now assume the same model as in the previous section, but we restrict the adaptive MQAM to a limited set of constellations. Specifically, we assume a set of square constellations of size $M_0 = 0$, $M_1 = 2$, and $M_j = 2^{2(j-1)}$, $j = 2, \ldots, N-1$ for some N. We assume square constellations for $M > 2$ since they are easier to implement than rectangular ones [21]. We first analyze the impact of this restriction on the spectral efficiency of the optimal adaptation policy. We then determine the effect on the channel inversion policies.

Consider a variable-rate variable-power MQAM transmission scheme subject to these constellation restrictions. Thus, at each symbol time we transmit a symbol from a constellation in the set $\{M_j : j = 0, 1, \ldots, N-1\}$; the choice of constellation depends on the fade level γ over that symbol time. Choosing the M_0 constellation corresponds to no data transmission. For each value of γ, we must decide which constellation to transmit and what the associated transmit power should be. The rate at which the transmitter must change its constellation and power is analyzed below. Since the power adaptation is continuous while the constellation size is discrete, we call this a continuous-power discrete-rate adaptation scheme.

Table 9.1: Rate and power adaptation for five regions

Region (j)	γ range	M_j	$S_j(\gamma)/\bar{S}$
0	$0 \leq \gamma/\gamma_K^* < 2$	0	0
1	$2 \leq \gamma/\gamma_K^* < 4$	2	$1/K\gamma$
2	$4 \leq \gamma/\gamma_K^* < 16$	4	$3/K\gamma$
3	$16 \leq \gamma/\gamma_K^* < 64$	16	$15/K\gamma$
4	$64 \leq \gamma/\gamma_K^* < \infty$	64	$63/K\gamma$

We determine the constellation size associated with each γ by discretizing the range of channel fade levels. Specifically, we divide the range of γ into N *fading regions* $R_j = [\gamma_{j-1}, \gamma_j)$, $j = 0, \ldots, N - 1$, where $\gamma_{-1} = 0$ and $\gamma_{N-1} = \infty$. We transmit constellation M_j when $\gamma \in R_j$. The spectral efficiency for $\gamma \in R_j$ is thus $\log_2 M_j$ bps/Hz for $j > 0$.

The adaptive MQAM design requires that the boundaries of the R_j regions be determined. Although these boundaries can be optimized to maximize spectral efficiency, as derived in Section 9.4.2, the optimal boundaries cannot be found in closed form and require an exhaustive search to obtain. Thus, we will use a suboptimal technique to determine boundaries. These suboptimal boundaries are much easier to find than the optimal ones and have almost the same performance. Define

$$M(\gamma) = \frac{\gamma}{\gamma_K^*}, \tag{9.20}$$

where $\gamma_K^* > 0$ is a parameter that will later be optimized to maximize spectral efficiency. Note that substituting (9.13) into (9.9) yields (9.20) with $\gamma_K^* = \gamma_K$. Therefore, the appropriate choice of γ_K^* in (9.20) defines the optimal constellation size for each γ when there is no constellation restriction.

Assume now that γ_K^* is fixed and define $M_N = \infty$. To obtain the constellation size M_j ($j = 0, \ldots, N - 1$) for a given SNR γ, we first compute $M(\gamma)$ from (9.20). We then find j such that $M_j \leq M(\gamma) < M_{j+1}$ and assign constellation M_j to this γ-value. Thus, for a fixed γ, we transmit the largest constellation in our set $\{M_j : j = 0, \ldots, N - 1\}$ that is smaller than $M(\gamma)$. For example, if the fade level γ satisfies $2 \leq \gamma/\gamma_K^* < 4$ then we transmit BPSK. The region boundaries other than $\gamma_{-1} = 0$ and $\gamma_{N-1} = \infty$ are located at $\gamma_j = \gamma_K^* M_{j+1}$, $j = 0, \ldots, N - 2$. Clearly, increasing the number of discrete signal constellations N yields a better approximation to the continuous adaptation (9.9), resulting in a higher spectral efficiency.

Once the regions and associated constellations are fixed, we must find a power adaptation policy that satisfies the BER requirement and the power constraint. By (9.9) we can maintain a fixed BER for the constellation $M_j > 0$ using the power adaptation policy

$$\frac{P_j(\gamma)}{\bar{P}} = \begin{cases} (M_j - 1)(1/\gamma K) & M_j < \gamma/\gamma_K^* \leq M_{j+1}, \\ 0 & M_j = 0 \end{cases} \tag{9.21}$$

for $\gamma \in R_j$, since this power adaptation policy leads to a fixed received E_s/N_0 for the constellation M_j of

$$\frac{E_s(j)}{N_0} = \frac{\gamma P_j(\gamma)}{\bar{P}} = \frac{M_j - 1}{K}. \tag{9.22}$$

By definition of K, MQAM modulation with constellation size M_j and E_s/N_0 given by (9.22) results in the desired target P_b. In Table 9.1 we tabulate the constellation size and power adaptation as a function of γ and γ_K^* for five fading regions.

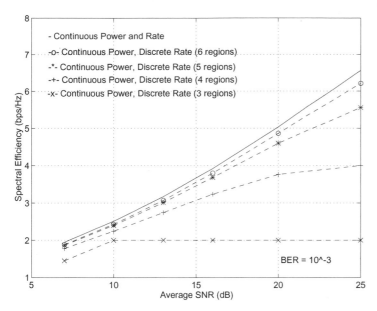

Figure 9.5: Discrete-rate efficiency in log-normal shadowing ($\sigma_{\psi_{dB}} = 8$ dB).

The spectral efficiency for this discrete-rate policy is just the sum of the data rates associated with each of the regions multiplied by the probability that γ falls in that region:

$$\frac{R}{B} = \sum_{j=1}^{N-1} \log_2(M_j) p\left(M_j \leq \frac{\gamma}{\gamma_K^*} < M_{j+1}\right). \tag{9.23}$$

Since M_j is a function of γ_K^*, we can maximize (9.23) relative to γ_K^*, subject to the power constraint

$$\sum_{j=1}^{N-1} \int_{\gamma_K^* M_j}^{\gamma_K^* M_{j+1}} \frac{P_j(\gamma)}{\bar{P}} p(\gamma) \, d\gamma = 1, \tag{9.24}$$

where $P_j(\gamma)/\bar{P}$ is defined in (9.21). There is no closed-form solution for the optimal γ_K^*: in the calculations that follow it was found using numerical search techniques.

In Figures 9.5 and 9.6 we show the maximum of (9.23) versus the number of fading regions N for log-normal shadowing and Rayleigh fading, respectively. We assume a BER of 10^{-3} for both plots. From Figure 9.5 we see that restricting our adaptive policy to just six fading regions ($M_j = 0, 2, 4, 16, 64, 256$) results in a spectral efficiency that is within 1 dB of the efficiency obtained with continuous-rate adaptation (9.17) under log-normal shadowing. A similar result holds for Rayleigh fading using five regions ($M_j = 0, 2, 4, 16, 64$).

We can simplify our discrete-rate policy even further by using a constant transmit power for each constellation M_j. Thus, each fading region is associated with one signal constellation and one transmit power. This policy is called discrete-power discrete-rate adaptive MQAM. Since the transmit power and constellation size are fixed in each region, the BER will vary with γ in each region. Thus, the region boundaries and transmit power must be set to achieve a given target average BER.

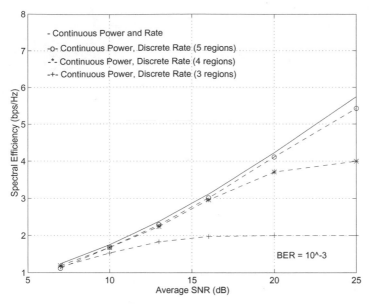

Figure 9.6: Discrete-rate efficiency in Rayleigh fading.

A restriction on allowable signal constellations will also affect the total channel inversion and truncated channel inversion policies. Specifically, suppose that, with the channel inversion policies, the constellation must be chosen from a fixed set of possible constellations $\mathcal{M} = \{M_0 = 0, \dots, M_{N-1}\}$. For total channel inversion the spectral efficiency with this restriction is thus

$$\frac{R}{B} = \log_2 \left\lfloor \left(1 + \frac{-1.5}{\ln(5P_b)\, \mathbf{E}[1/\gamma]}\right) \right\rfloor_{\mathcal{M}}, \tag{9.25}$$

where $\lfloor x \rfloor_{\mathcal{M}}$ denotes the largest number in the set $\mathcal{M}$ less than or equal to x. The spectral efficiency with this policy will be restricted to values of $\log_2 M$ ($M \in \mathcal{M}$), with discrete jumps at the $\bar{\gamma}$-values where the spectral efficiency without constellation restriction (9.18) equals $\log_2 M$. For truncated channel inversion the spectral efficiency is given by

$$\frac{R}{B} = \max_{\gamma_0} \log_2 \left\lfloor \left(1 + \frac{-1.5}{\ln(5P_b)\, \mathbf{E}_{\gamma_0}[1/\gamma]}\right) \right\rfloor_{\mathcal{M}} p(\gamma > \gamma_0). \tag{9.26}$$

In Figures 9.7 and 9.8 we show the impact of constellation restriction on adaptive MQAM for the different power adaptation policies. When the constellation is restricted we assume six fading regions, so $\mathcal{M} = \{M_0 = 0, 2, 4, \dots, 256\}$. The power associated with each fading region for the discrete-power discrete-rate policy was chosen to have an average BER equal to the instantaneous BER of the discrete-rate continuous-power adaptative policy. We see from these figures that, for variable-rate MQAM with a small set of constellations, restricting the power to a single value for each constellation degrades spectral efficiency by about 1–2 dB relative to continuous-power adaptation. For comparison, we also plot the maximum efficiency (9.17) for continuous-power and -rate adaptation. All discrete-rate policies have performance that is within 3 dB of this theoretical maximum.

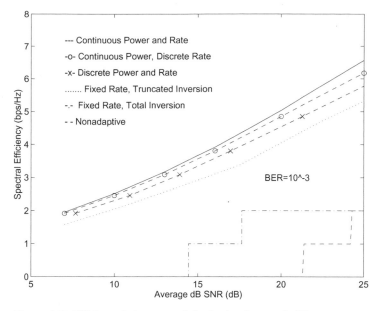

Figure 9.7: Efficiency in log-normal shadowing ($\sigma_{\psi_{dB}} = 8$ dB).

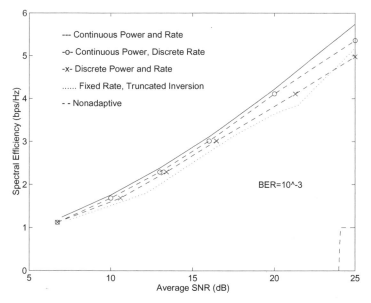

Figure 9.8: Efficiency in Rayleigh fading.

These figures also show the spectral efficiency of fixed-rate transmission with truncated channel inversion (9.26). The efficiency of this scheme is quite close to that of the discrete-power discrete-rate policy. However, to achieve this high efficiency, the optimal γ_0 is quite large, with a corresponding outage probability $P_{\text{out}} = p(\gamma \leq \gamma_0)$ ranging from .1 to .6. Thus, this policy is similar to packet radio, with bursts of high-speed data when the channel conditions are favorable. The efficiency of total channel inversion (9.25) is also shown for log-normal shadowing; this efficiency equals zero in Rayleigh fading. We also plot the spectral efficiency of nonadaptive transmission, where both the transmission rate and power are constant. As discussed in Section 9.3.1, the average BER in this case is obtained by averaging the probability of error (9.31) over the fade distribution $p(\gamma)$. The spectral efficiency is obtained by determining the value of M that yields a 10^{-3} average BER for the given value of $\bar{\gamma}$, as illustrated in Example 9.3. Nonadaptive transmission clearly suffers a large spectral efficiency loss in exchange for its simplicity. However, if the channel varies rapidly and cannot be accurately estimated, nonadaptive transmission may be the best alternative. Similar curves can be obtained for a target BER of 10^{-6}, with roughly the same spectral efficiency loss relative to a 10^{-3} BER as was exhibited in Figures 9.2 and 9.3.

9.3.5 Average Fade Region Duration

The choice of the number of regions to use in the adaptive policy will depend on how fast the channel is changing as well as on the hardware constraints, which dictate how many constellations are available to the transmitter and at what rate the transmitter can change its constellation and power. Channel estimation and feedback considerations along with hardware constraints may dictate that the constellation remains constant over tens or even hundreds of symbols. In addition, power-amplifier linearity requirements and out-of-band emission constraints may restrict the rate at which power can be adapted. An in-depth discussion of hardware implementation issues can be found in [22]. However, determining how long the SNR γ remains within a particular fading region R_j is of interest, since it determines the trade-off between the number of regions and the rate of power and constellation adaptation. We now investigate the time duration over which the SNR remains within a given fading region.

Let $\bar{\tau}_j$ denote the average time duration that γ stays within the jth fading region. Let $A_j = \gamma_K^* M_j$ for γ_K^* and M_j as previously defined. The jth fading region is then defined as $\{\gamma : A_j \leq \gamma < A_{j+1}\}$. We call $\bar{\tau}_j$ the jth average fade region duration (AFRD). This definition is similar to the average fade duration (AFD; see Section 3.2.3), except that the AFD measures the average time that γ stays below a single level, whereas we are interested in the average time that γ stays between two levels. For the worst-case region ($j = 0$), these two definitions coincide.

Determining the exact value of $\bar{\tau}_j$ requires a complex derivation based on the joint density $p(\gamma, \dot{\gamma})$, and it remains an open problem. However, a good approximation can be obtained using the finite-state Markov model described in Section 3.2.4. In this model, fading is approximated as a discrete-time Markov process with time discretized to a given interval T, typically the symbol time. It is assumed (i) that the fade value γ remains within one region over a symbol period and (ii) that from a given region the process can only transition to the same region or to adjacent regions. Note that this approximation can lead to longer deep fade durations than more accurate models [23]. The transition probabilities between regions under this assumption are given as

$$p_{j,j+1} = \frac{L_{j+1}T}{\pi_j}, \quad p_{j,j-1} = \frac{L_j T}{\pi_j}, \quad p_{j,j} = 1 - p_{j,j+1} - p_{j,j-1}, \qquad (9.27)$$

where L_j is the level-crossing rate at A_j and π_j is the steady-state distribution corresponding to the jth region: $\pi_j = p(A_j \leq \gamma < A_{j+1})$. Since the time over which the Markov process stays in a given state is geometrically distributed [24, Chap. 2.3], $\bar{\tau}_j$ is given by

Table 9.2: Average fade region duration $\bar{\tau}_j$ for $f_D = 100$ Hz

Region (j)	$\bar{\gamma} = 10$ dB	$\bar{\gamma} = 20$ dB
0	2.23 ms	0.737 ms
1	0.83 ms	0.301 ms
2	3.00 ms	1.06 ms
3	2.83 ms	2.28 ms
4	1.43 ms	3.84 ms

$$\bar{\tau}_j = \frac{T}{p_{j,j+1} + p_{j,j-1}} = \frac{\pi_j}{L_{j+1} + L_j}. \tag{9.28}$$

The value of $\bar{\tau}_j$ is thus a simple function of the level crossing rate and the fading distribution. Whereas the level crossing rate is known for Rayleigh fading [25, Chap. 1.3.4], it cannot be obtained for log-normal shadowing because the joint distribution $p(\gamma, \dot{\gamma})$ for this fading type is unknown.

In Rayleigh fading, the level crossing rate is given by (3.44) as

$$L_j = \sqrt{\frac{2\pi A_j}{\bar{\gamma}}} f_D e^{-A_j/\bar{\gamma}}, \tag{9.29}$$

where $f_D = v/\lambda$ is the Doppler frequency. Substituting (9.29) into (9.28), we easily see that $\bar{\tau}_j$ is inversely proportional to the Doppler frequency. Moreover, π_j and A_j do not depend on f_D and so, if we compute $\bar{\tau}_j$ for a given Doppler frequency f_D, then we can compute $\hat{\bar{\tau}}_j$ corresponding to another Doppler frequency $\hat{f}_D$ as

$$\hat{\bar{\tau}}_j = \frac{f_D}{\hat{f}_D} \bar{\tau}_j. \tag{9.30}$$

Table 9.2 shows the $\bar{\tau}_j$ values corresponding to five regions ($M_j = 0, 2, 4, 16, 64$) in Rayleigh fading[2] for $f_D = 100$ Hz and two average power levels: $\bar{\gamma} = 10$ dB ($\gamma_K^* = 1.22$) and $\bar{\gamma} = 20$ dB ($\gamma_K^* = 1.685$). The AFRD for other Doppler frequencies is easily obtained using the table values and (9.30). This table indicates that, even at high velocities, for rates of 100 kilosymbols/second the discrete-rate discrete-power policy will maintain the same constellation and transmit power over tens to hundreds of symbols.

EXAMPLE 9.4: Find the AFRDs for a Rayleigh fading channel with $\bar{\gamma} = 10$ dB, $M_j = 0, 2, 4, 16, 64, 64$, and $f_D = 50$ Hz.

Solution: We first note that all parameters are the same as used in the calculation of Table 9.2 except that the Doppler $\hat{f}_D = 50$ Hz is half the Doppler of $f_D = 100$ Hz used to compute the table values. Thus, from (9.30), we obtain the AFRDs with this new Doppler by multiplying each value in the table by $f_D/\hat{f}_D = 2$.

---

[2] The validity of the finite-state Markov model for Rayleigh fading channels has been investigated in [26].

In shadow fading we can obtain a coarse approximation of $\bar{\tau}_j$ based on the shadowing auto-correlation function (2.50): $A(\delta) = \sigma_{\psi_{dB}}^2 e^{-\delta/X_c}$, where $\delta = v\tau$ for v the mobile's velocity. Specifically, we can approximate the AFRD for all regions as $\bar{\tau}_j \approx .1X_c/v$, since then the correlation between fade levels separated in time by $\bar{\tau}_j$ is .9. Thus, for a small number of regions it is likely that γ will remain within the same region over this time period.

9.3.6 Exact versus Approximate Bit Error Probability

The adaptive policies described in prior sections are based on the BER upper bounds of Section 9.3.1. Since these are upper bounds, they will lead to a lower BER than the target. We would like to see how the BER achieved with these policies differs from the target BER. A more accurate value for the BER achieved with these policies can be obtained by simulation or by using a better approximation for BER than the upper bounds. From Table 6.1, the BER of MQAM with Gray coding at high SNRs is well approximated by

$$P_b \approx \frac{4}{\log_2 M} Q\left(\sqrt{\frac{3\gamma}{M-1}}\right). \tag{9.31}$$

Moreover, for the continuous-power discrete-rate policy, $\gamma = E_s/N_0$ for the jth signal constellation is

$$\frac{E_s(j)}{N_0} = \frac{M_j - 1}{K}. \tag{9.32}$$

Thus, we can obtain a more accurate analytical expression for the average BER associated with our adaptive policies by averaging over the BER (9.31) for each signal constellation as

$$\bar{P}_b = \frac{\sum_{j=1}^{N-1} 4Q(\sqrt{3/K}) \int_{\gamma_K^* M_j}^{\gamma_K^* M_{j+1}} p(\gamma)\,d\gamma}{\sum_{j=1}^{N-1} \log_2 M_j \int_{\gamma_K^* M_j}^{\gamma_K^* M_{j+1}} p(\gamma)\,d\gamma} \tag{9.33}$$

with $M_N = \infty$.

We plot the analytical expression (9.33) along with the simulated BER for the variable rate and power MQAM with a target BER of 10^{-3} in Figures 9.9 and 9.10 for log-normal shadowing and Rayleigh fading, respectively. The simulated BER is slightly better than the analytical calculation of (9.33) because this equation is based on the nearest neighbor bound for the maximum number of nearest neighbors. Both the simulated and analytical BER are smaller than the target BER of 10^{-3} for $\bar{\gamma} > 10$ dB. The BER bound of 10^{-3} breaks down at low SNRs, since (9.7) is not applicable to BPSK, so we must use the looser bound (9.6). Since the adaptive policy often uses the BPSK constellation at low SNRs, the P_b will be larger than that predicted from the tight bound (9.7). That the simulated BER is less than the target at high SNRs implies that the analytical calculations in Figures 9.5 and 9.6 are pessimistic; a slightly higher efficiency could be achieved while still maintaining the target P_b of 10^{-3}.

9.3.7 Channel Estimation Error and Delay

In this section we examine the effects of estimation error and delay, where the estimation error $\varepsilon = \hat{\gamma}/\gamma \neq 1$ and the delay $i_d = i_f + i_e \neq 0$. We first consider the estimation error. Suppose the transmitter adapts its power and rate relative to a target BER P_b based on the channel estimate $\hat{\gamma}$ instead of the true value γ. By (9.8), the BER is then bounded by

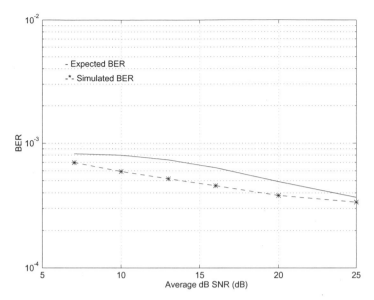

Figure 9.9: BER for log-normal shadowing (six regions).

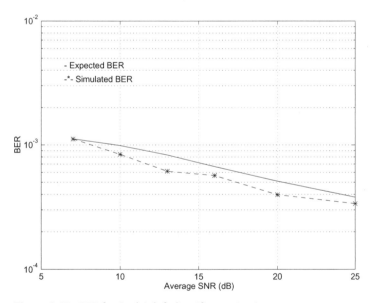

Figure 9.10: BER for Rayleigh fading (five regions).

$$P_b(\gamma, \hat{\gamma}) \le .2 \exp\left[\frac{-1.5\gamma}{M(\hat{\gamma}) - 1} \frac{P(\hat{\gamma})}{\bar{P}}\right] = .2[5P_b]^{1/\varepsilon}, \qquad (9.34)$$

where the right-hand equality is obtained by substituting the optimal rate (9.9) and power (9.13) policies. For $\varepsilon = 1$, (9.34) reduces to the target P_b. For $\varepsilon \ne 1$: $\varepsilon > 1$ yields an increase in BER above the target, and $\varepsilon < 1$ yields a decrease in BER.

The effect of estimation error on BER is given by

$$\bar{P}_b \le \int_0^\infty \int_{\gamma_K}^\infty .2[5P_b]^{\gamma/\hat{\gamma}} p(\gamma, \hat{\gamma}) \, d\hat{\gamma} \, d\gamma = \int_0^\infty .2[5P_b]^{1/\varepsilon} p(\varepsilon) \, d\varepsilon. \qquad (9.35)$$

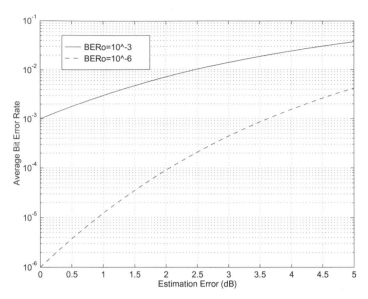

Figure 9.11: Effect of estimation error on BER.

The distribution $p(\varepsilon)$ is a function of the joint distribution $p(\gamma, \hat{\gamma})$, which in turn depends on the channel estimation technique. It has been shown [27] that, when the channel is estimated using pilot symbols, the joint distribution of the signal envelope and its estimate is bivariate Rayleigh. This joint distribution was used in [27] to obtain the probability of error for nonadaptive modulation with channel estimation errors. This analysis can be extended to adaptive modulation using a similar methodology.

If the estimation error stays within some finite range then we can bound its effect using (9.34). We plot the BER increase as a function of a constant ε in Figure 9.11. This figure shows that for a target BER of 10^{-3} the estimation error should be less than 1 dB, and for a target BER of 10^{-6} it should be less than .5 dB. These values are pessimistic, since they assume a constant value of estimation error. Even so, the estimation error can be kept within this range using the pilot-symbol assisted estimation technique described in [28] with appropriate choice of parameters. When the channel is underestimated ($\varepsilon < 1$), the BER decreases but there will also be some loss in spectral efficiency, since the mean of the channel estimate $\hat{\bar{\gamma}}$ will differ from the true mean $\bar{\gamma}$. The effect of this average power estimation error is characterized in [29].

Suppose now that the channel is estimated perfectly ($\varepsilon = 1$) but the delay i_d of the estimation and feedback path is nonzero. Thus, at time i the transmitter will use the delayed version of the channel estimate $\hat{\gamma}[i] = \gamma[i - i_d]$ to adjust its power and rate. It was shown in [30] that, conditioned on the outdated channel estimates, the received signal follows a Rician distribution, and the probability of error can then be computed by averaging over the distribution of the estimates. Moreover, [30] develops adaptive coding designs to mitigate the effect of estimation delay on the performance of adaptive modulation. Alternatively, channel prediction can be used to mitigate these effects [31].

The increase in BER from estimation delay can also be examined in the same manner as in (9.34). Given the exact channel SNR $\gamma[i]$ and its delayed value $\gamma[i - i_d]$, we have

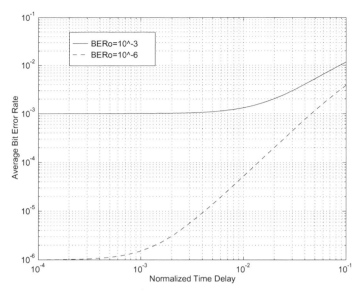

Figure 9.12: Effect of normalized delay $(i_d f_D)$ on BER.

$$P_b(\gamma[i], \gamma[i - i_d]) \leq .2\exp\left[\frac{-1.5\gamma[i]}{M(\gamma[i - i_d]) - 1}\frac{P(\gamma[i - i_d])}{\bar{P}}\right]$$

$$= .2[5P_{b_0}]^{\gamma[i]/\gamma[i - i_d]}.\tag{9.36}$$

Define $\xi[i, i_d] = \gamma[i]/\gamma[i - i_d]$. Since $\gamma[i]$ is stationary and ergodic, the distribution of $\xi[i, i_d]$ conditioned on $\gamma[i]$ depends only on i_d and the value of $\gamma = \gamma[i]$. We denote this distribution by $p_{i_d}(\xi \mid \gamma)$. The average BER is obtained by integrating over ξ and γ. Specifically, it is shown in [32] that

$$P_b[i_d] = \int_{\gamma_K}^{\infty}\left[\int_0^{\infty} .2[5P_{b_0}]^{\xi} p_{i_d}(\xi \mid \gamma)\, d\xi\right] p(\gamma)\, d\gamma,\tag{9.37}$$

where γ_K is the cutoff level of the optimal policy and $p(\gamma)$ is the fading distribution. The distribution $p_{i_d}(\xi \mid \gamma)$ will depend on the autocorrelation of the fading process. A closed-form expression for $p_{i_d}(\xi \mid \gamma)$ in Nakagami fading (of which Rayleigh fading is a special case) is derived in [32]. Using this distribution in (9.37), we obtain the average BER in Rayleigh fading as a function of the delay parameter i_d. A plot of (9.37) versus the normalized time delay $i_d f_D$ is shown in Figure 9.12. From this figure we see that the total estimation and feedback path delay must be kept to within $.001/f_D$ in order to keep the BER near its desired target.

9.3.8 Adaptive Coded Modulation

Additional coding gain can be achieved with adaptive modulation by superimposing trellis or lattice codes on the adaptive modulation. Specifically, by using the subset partitioning inherent to coded modulation, trellis (or lattice) codes designed for AWGN channels can be superimposed directly onto the adaptive modulation with the same approximate coding gain. The basic idea of adaptive coded modulation is to exploit the separability of code and constellation design that is characteristic of coded modulation, as described in Section 8.7.

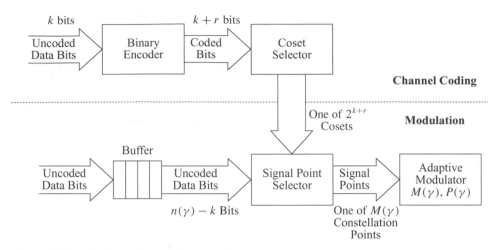

Figure 9.13: Adaptive coded modulation scheme.

Coded modulation is a natural coding scheme to use with variable-rate variable-power MQAM, since the channel coding gain is essentially independent of the modulation. We can therefore adjust the power and rate (number of levels or signal points) in the transmit constellation relative to the instantaneous SNR *without* affecting the channel coding gain, as we now describe in more detail.

The coded modulation scheme is shown in Figure 9.13. The coset code design is the same as it would be for an AWGN channel; that is, the lattice structure and conventional encoder follow the trellis or lattice coded modulation designs outlined in Section 8.7. Let G_c denote the coding gain of the coded modulation, as given by (8.78). The modulation works as follows. The signal constellation is a square lattice with an adjustable number of constellation points M. The size of the MQAM signal constellation from which the signal point is selected is determined by the transmit power, which is adjusted relative to the instantaneous SNR and the desired BER, as in the uncoded case described in Section 9.3.2.

Specifically, if the BER approximation (9.7) is adjusted for the coding gain, then for a particular SNR $= \gamma$ we have

$$P_b \approx .2e^{-1.5(\gamma G_c/M-1)}, \tag{9.38}$$

where M is the size of the transmit signal constellation. As in the uncoded case, using the tight bound (9.7) allows us to adjust the number of constellation points M and signal power relative to the instantaneous SNR in order to maintain a fixed BER:

$$M(\gamma) = 1 + \frac{1.5\gamma G_c}{-\ln(5P_b)} \frac{P(\gamma)}{\bar{P}}. \tag{9.39}$$

The number of uncoded bits required to select the coset point is

$$n(\gamma) - 2k/N = \log_2 M(\gamma) - 2(k+r)/N.$$

Since this value varies with time, these uncoded bits must be queued until needed, as shown in Figure 9.13.

The bit rate per transmission is $\log_2 M(\gamma)$, and the data rate is $\log_2 M(\gamma) - 2r/N$. Therefore, we maximize the data rate by maximizing $\mathbf{E}[\log_2 M]$ relative to the average power constraint. From this maximization, we obtain the optimal power adaptation policy for this modulation scheme:

$$\frac{P(\gamma)}{\bar{P}} = \begin{cases} 1/\gamma_0 - 1/\gamma K_c & \gamma \geq \gamma_0/K_c, \\ 0 & \gamma < \gamma_0/K_c, \end{cases} \tag{9.40}$$

where $\gamma_{K_c} = \gamma_0/K_c$ is the cutoff fade depth for $K_c = KG_c$ with K given by (9.10). This is the same as the optimal policy for the uncoded case (9.13) with K replaced by K_c. Thus, the coded modulation increases the effective transmit power by G_c relative to the uncoded variable-rate variable-power MQAM performance. The adaptive data rate is obtained by substituting (9.40) into (9.39) to get

$$M(\gamma) = \left(\frac{\gamma}{\gamma_{K_c}}\right). \tag{9.41}$$

The resulting spectral efficiency is

$$\frac{R}{B} = \int_{\gamma_{K_c}}^{\infty} \log_2\left(\frac{\gamma}{\gamma_{K_c}}\right) p(\gamma) \, d\gamma. \tag{9.42}$$

If the constellation expansion factor is not included in the coding gain G_c, then we must subtract $2r/N$ from (9.42) to get the data rate. More details on this adaptive coded modulation scheme can be found in [33], along with plots of the spectral efficiency for adaptive trellis coded modulation of varying complexity. These results indicate that adaptive trellis coded modulation can achieve within 5 dB of Shannon capacity at reasonable complexity and that the coding gains of superimposing a given trellis code onto uncoded adaptive modulation are roughly equal to the coding gains of the trellis code in an AWGN channel.

9.4 General *M*-ary Modulations

The variable rate and power techniques already described for MQAM can be applied to other M-ary modulations. For any modulation, the basic premise is the same: the transmit power and constellation size are adapted to maintain a given fixed instantaneous BER for each symbol while maximizing average data rate. In this section we will consider optimal rate and power adaptation for both continuous-rate and discrete-rate variation of general M-ary modulations.

9.4.1 Continuous-Rate Adaptation

We first consider the case where both rate and power can be adapted continuously. We want to find the optimal power $P(\gamma)$ and rate $k(\gamma) = \log_2 M(\gamma)$ adaptation for general M-ary modulation that maximizes the average data rate $\mathbf{E}[k(\gamma)]$ with average power $\bar{P}$ while meeting a given BER target. This optimization is simplified when the exact or approximate probability of bit error for the modulation can be written in the following form:

$$P_b(\gamma) \approx c_1 \exp\left[\frac{-c_2 \gamma (P(\gamma)/\bar{P})}{2^{c_3 k(\gamma)} - c_4}\right], \tag{9.43}$$

where c_1, c_2, and c_3 are positive fixed constants and c_4 is a real constant. For example, in the BER bounds for MQAM given by (9.6) and (9.7), $c_1 = 2$ or $.2$, $c_2 = 1.5$, $c_3 = 1$, and

$c_4 = 1$. The probability of bit error for most M-ary modulations can be approximated in this form with appropriate curve fitting.

The advantage of (9.43) is that, when $P_b(\gamma)$ is in this form, we can invert it to express the rate $k(\gamma)$ as a function of the power adaptation $P(\gamma)$ and the BER target P_b as follows:

$$
k(\gamma) = \log_2 M(\gamma) = \begin{cases} \dfrac{1}{c_3} \log_2 \left[c_4 - \dfrac{c_2 \gamma}{\ln(P_b/c_1)} \dfrac{P(\gamma)}{\bar{P}} \right] & P(\gamma) \geq 0, \ k(\gamma) \geq 0, \\ 0 & \text{else.} \end{cases} \tag{9.44}
$$

To find the power and rate adaptation that maximize spectral efficiency $\mathbf{E}[k(\gamma)]$, we create the Lagrangian

$$
J(P(\gamma)) = \int_0^\infty k(\gamma) p(\gamma)\, d\gamma + \lambda \left[\int_0^\infty P(\gamma) p(\gamma)\, d\gamma - \bar{P} \right]. \tag{9.45}
$$

The optimal adaptation policy maximizes this Lagrangian with nonnegative rate and power, so it satisfies

$$
\frac{\partial J}{\partial P(\gamma)} = 0, \quad P(\gamma) \geq 0, \ k(\gamma) \geq 0. \tag{9.46}
$$

Solving (9.46) for $P(\gamma)$ with (9.44) for $k(\gamma)$ yields the optimal power adaptation

$$
\frac{P(\gamma)}{\bar{P}} = \begin{cases} -1/c_3 (\ln 2) \lambda \bar{P} - 1/\gamma K & P(\gamma) \geq 0, \ k(\gamma) \geq 0, \\ 0 & \text{else,} \end{cases} \tag{9.47}
$$

where

$$
K = -\frac{c_2}{c_4 \ln(P_b/c_1)}. \tag{9.48}
$$

The power adaptation (9.47) can be written in the more simplified form

$$
\frac{P(\gamma)}{\bar{P}} = \begin{cases} \mu - 1/\gamma K & P(\gamma) \geq 0, \ k(\gamma) \geq 0, \\ 0 & \text{else.} \end{cases} \tag{9.49}
$$

The constant μ in (9.49) is determined from the average power constraint (9.12).

Although the analytical expression for the optimal power adaptation (9.49) looks simple, its behavior is highly dependent on the c_4 values in the P_b approximation (9.43). For (9.43) given by the MQAM approximations (9.6) or (9.7), the power adaptation is the water-filling formula given by (9.13). However, water-filling is not optimal in all cases, as we now show.

Based on (6.16), with Gray coding the BER for MPSK is tightly approximated as

$$
P_b \approx \frac{2}{\log_2 M} Q\left(\sqrt{2\gamma} \sin\left(\frac{\pi}{M} \right) \right). \tag{9.50}
$$

However, (9.50) is not in the desired form (9.43). In particular, the Q-function is not easily inverted to obtain the optimal rate and power adaptation for a given target BER. Let us therefore consider the following three P_b bounds for MPSK, which are valid for $k(\gamma) \geq 2$.

$$
\text{Bound 1:} \quad P_b(\gamma) \approx .05 \exp\left[\frac{-6\gamma (P(\gamma)/\bar{P})}{2^{1.9k(\gamma)} - 1} \right]. \tag{9.51}
$$

$$
\text{Bound 2:} \quad P_b(\gamma) \approx .2 \exp\left[\frac{-7\gamma (P(\gamma)/\bar{P})}{2^{1.9k(\gamma)} + 1} \right]. \tag{9.52}
$$

$$
\text{Bound 3:} \quad P_b(\gamma) \approx .25 \exp\left[\frac{-8\gamma (P(\gamma)/\bar{P})}{2^{1.94k(\gamma)}} \right]. \tag{9.53}
$$

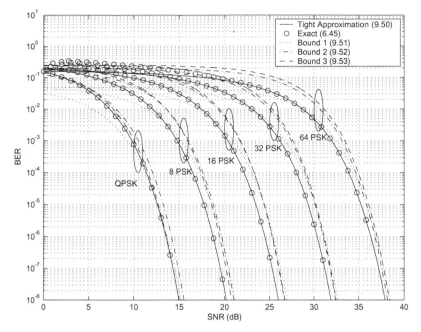

Figure 9.14: BER bounds for MPSK.

The bounds are plotted in Figure 9.14 along with the tight approximation (9.50). We see that all bounds well approximate the exact BER given by (6.45), especially at high SNRs.

In the first bound (9.51), $c_1 = .05$, $c_2 = 6$, $c_3 = 1.9$, and $c_4 = 1$. Thus, in (9.49), $K = -c_2/(c_4 \ln(P_b/c_1))$ is positive as long as the target P_b is less than .05, which we assume. Therefore μ must be positive for the power adaptation $P(\gamma)/\bar{P} = \mu - 1/\gamma K$ to be positive about a cutoff SNR γ_0. Moreover, for K positive, $k(\gamma) \geq 0$ for any $P(\gamma) \geq 0$. Thus, with μ and $k(\gamma)$ positive, (9.49) can be expressed as

$$\frac{P(\gamma)}{\bar{P}} = \begin{cases} 1/\gamma_0 K - 1/\gamma K & P(\gamma) \geq 0, \\ 0 & \text{else,} \end{cases} \tag{9.54}$$

where $\gamma_0 \geq 0$ is a cutoff fade depth below which no signal is transmitted. Like μ, this cutoff value is determined by the average power constraint (9.12). The power adaptation (9.54) is the same water-filling as in adaptive MQAM given by (9.13), which results from the similarity of the MQAM P_b bounds (9.7) and (9.6) to the MPSK bound (9.51). The corresponding optimal rate adaptation, obtained by substituting (9.54) into (9.44), is

$$k(\gamma) = \begin{cases} (1/c_3) \log_2(\gamma/\gamma_0) & \gamma \geq \gamma_0, \\ 0 & \text{else,} \end{cases} \tag{9.55}$$

which is also in the same form as the adaptive MQAM rate adaptation (9.16).

Let us now consider the second bound (9.52). Here $c_1 = .2$, $c_2 = 7$, $c_3 = 1.9$, and $c_4 = -1$. Thus, $K = -c_2/(c_4 \ln(P_b/c_1))$ is negative for a target $P_b < .2$, which we assume. From (9.44), with K negative we must have $\mu \geq 0$ in (9.49) to make $k(\gamma) \geq 0$. Then the optimal power adaptation such that $P(\gamma) \geq 0$ and $k(\gamma) \geq 0$ becomes

$$\frac{P(\gamma)}{\bar{P}} = \begin{cases} \mu - 1/\gamma K & k(\gamma) \geq 0, \\ 0 & \text{else.} \end{cases} \tag{9.56}$$

From (9.44), the optimal rate adaptation then becomes

$$k(\gamma) = \begin{cases} (1/c_3) \log_2(\gamma/\gamma_0) & \gamma \geq \gamma_0, \\ 0 & \text{else,} \end{cases} \tag{9.57}$$

where $\gamma_0 = -1/K\mu$ is a cutoff fade depth below which the channel is not used. Note that for the first bound (9.51) the positivity constraint on power ($P(\gamma) \geq 0$) dictates the cutoff fade depth, whereas for this bound (9.52) the positivity constraint on rate ($k(\gamma) \geq 0$) determines the cutoff. We can rewrite (9.56) in terms of γ_0 as

$$\frac{P(\gamma)}{\bar{P}} = \begin{cases} 1/\gamma_0(-K) + 1/\gamma(-K) & \gamma \geq \gamma_0, \\ 0 & \text{else.} \end{cases} \tag{9.58}$$

This power adaptation is an *inverse water-filling*: since K is negative, less power is used as the channel SNR increases above the optimized cutoff fade depth γ_0. As usual, the value of γ_0 is obtained based on the average power constraint (9.12).

Finally, for the third bound (9.53), $c_1 = .25$, $c_2 = 8$, $c_3 = 1.94$, and $c_4 = 0$. Thus, $K = -c_2/(c_4 \ln(P_b/c_1)) = \infty$ for a target $P_b < .25$, which we assume. From (9.49), the optimal power adaptation becomes

$$\frac{P(\gamma)}{\bar{P}} = \begin{cases} \mu & k(\gamma) \geq 0, \ P(\gamma) \geq 0, \\ 0 & \text{else.} \end{cases} \tag{9.59}$$

This is *on–off* power transmission: power is either zero or a constant nonzero value. By (9.44), the optimal rate adaptation $k(\gamma)$ with this power adaptation is

$$k(\gamma) = \begin{cases} (1/c_3) \log_2(\gamma/\gamma_0) & \gamma \geq \gamma_0, \\ 0 & \text{else,} \end{cases} \tag{9.60}$$

where $\gamma_0 = -\ln(P_b/c_1)/c_2\mu$ is a cutoff fade depth below which the channel is not used. As for the previous bound, it is the rate positivity constraint that determines the cutoff fade depth γ_0. The optimal power adaptation as a function of γ_0 is

$$\frac{P(\gamma)}{\bar{P}} = \begin{cases} K_0/\gamma_0 & \gamma \geq \gamma_0, \\ 0 & \text{else,} \end{cases} \tag{9.61}$$

where $K_0 = -\ln(P_b/c_1)/c_2$. The value of γ_0 is determined from the average power constraint to satisfy

$$\frac{K_0}{\gamma_0} \int_{\gamma_0}^{\infty} p(\gamma) \, d\gamma = 1. \tag{9.62}$$

Thus, for all three P_b approximations in MPSK, the optimal adaptive rate schemes (9.55), (9.57), and (9.60) have the same form whereas the optimal adaptive power schemes (9.54), (9.58), and (9.61) have different forms. The optimal power adaptations (9.54), (9.58), and (9.61) are plotted in Figure 9.15 for Rayleigh fading with a target BER of 10^{-3} and $\bar{\gamma} = 30$ dB. This figure clearly shows the water-filling, inverse water-filling, and on–off behavior

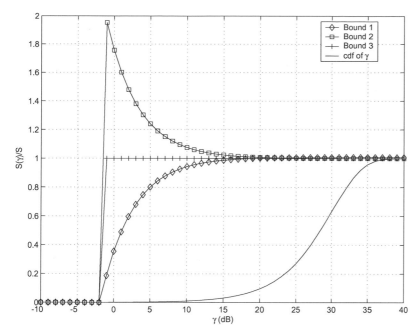

Figure 9.15: Power adaptation for MPSK BER bounds (Rayleigh fading, $P_b = 10^{-3}$, $\bar{\gamma} = 30$ dB).

of the different schemes. Note that the cutoff γ_0 for all these schemes is roughly the same. We also see from this figure that even though the power adaptation schemes are different at low SNRs, they are almost the same at high SNRs. Specifically, we see that for $\gamma < 10$ dB, the optimal transmit power adaptations are dramatically different, whereas for $\gamma \geq 10$ dB they rapidly converge to the same constant value. From the cumulative distribution function of γ, also shown in Figure 9.15, the probability that γ is less than 10 is 0.01. Thus, although the optimal power adaptation corresponding to low SNRs is very different for the different techniques, this behavior has little impact on spectral efficiency because the probability of being at those low SNRs is quite small.

9.4.2 Discrete-Rate Adaptation

We now assume a given discrete set of constellations $\mathcal{M} = \{M_0 = 0, \ldots, M_{N-1}\}$, where M_0 corresponds to no data transmission. The rate corresponding to each of these constellations is $k_j = \log_2 M_j$ ($j = 0, \ldots, N-1$), where $k_0 = 0$. Each rate k_j ($j > 0$) is assigned to a fading region of γ-values $R_j = [\gamma_{j-1}, \gamma_j)$, $j = 0, \ldots, N-1$, for $\gamma_{-1} = 0$ and $\gamma_{N-1} = \infty$. The boundaries γ_j ($j = 0, \ldots, N-2$) are optimized as part of the adaptive policy. The channel is not used for $\gamma < \gamma_0$. We again assume that P_b is approximated using the general formula (9.43). Then the power adaptation that maintains the target BER above the cutoff γ_0 is

$$\frac{P(\gamma)}{\bar{P}} = \frac{h(k_j)}{\gamma}, \quad \gamma_{j-1} \leq \gamma \leq \gamma_j, \tag{9.63}$$

where

$$h(k_j) = -\frac{\ln(P_b/c_1)}{c_2}(2^{c_3 k_j} - c_4). \tag{9.64}$$

The region boundaries $\gamma_0, \ldots, \gamma_{N-2}$ that maximize spectral efficiency are found using the Lagrange equation

$$J(\gamma_0, \gamma_1, \ldots, \gamma_{N-2}) = \sum_{j=1}^{N-1} k_j \int_{\gamma_{j-1}}^{\gamma_j} p(\gamma)\, d\gamma + \lambda \left[\sum_{j=1}^{N-1} \int_{\gamma_{j-1}}^{\gamma_j} \frac{h(k_j)}{\gamma} p(\gamma)\, d\gamma - 1 \right]. \quad (9.65)$$

The optimal rate region boundaries are obtained by solving the following equation for γ_j:

$$\frac{\partial J}{\partial \gamma_j} = 0, \quad 0 \le j \le N - 2. \quad (9.66)$$

This yields

$$\gamma_0 = \frac{h(k_1)}{k_1} \rho \quad (9.67)$$

and

$$\gamma_j = \frac{h(k_{j+1}) - h(k_j)}{k_{j+1} - k_j} \rho, \quad 1 \le i \le N - 2, \quad (9.68)$$

where ρ is determined by the average power constraint

$$\sum_{j=1}^{N-1} \int_{\gamma_{j-1}}^{\gamma_j} \frac{h(k_j)}{\gamma} p(\gamma)\, d\gamma = 1. \quad (9.69)$$

9.4.3 Average BER Target

Suppose now that we relax our assumption that the P_b target must be met on every symbol transmission, requiring instead that just the average P_b be below some target average $\bar{P}_b$. In this case, in addition to adapting rate and power, we can also adapt the instantaneous $P_b(\gamma)$ subject to the average constraint $\bar{P}_b$. This gives an additional degree of freedom in adaptation that may lead to higher spectral efficiencies. We define the average probability of error for adaptive modulation as

$$\bar{P}_b = \frac{\mathbf{E}[\text{number of bits in error per transmission}]}{\mathbf{E}[\text{number of bits per transmission}]}. \quad (9.70)$$

When the bit rate $k(\gamma)$ is continuously adapted this becomes

$$\bar{P}_b = \frac{\int_0^\infty P_b(\gamma) k(\gamma) p(\gamma)\, d\gamma}{\int_0^\infty k(\gamma) p(\gamma)\, d\gamma}, \quad (9.71)$$

and when $k(\gamma)$ takes values in a discrete set this becomes

$$\bar{P}_b = \frac{\sum_{j=1}^{N-1} k_j \int_{\gamma_{j-1}}^{\gamma_j} P_b(\gamma) p(\gamma)\, d\gamma}{\sum_{j=1}^{N-1} k_j \int_{\gamma_{j-1}}^{\gamma_j} p(\gamma)\, d\gamma}. \quad (9.72)$$

We now derive the optimal continuous rate, power, and BER adaptation to maximize spectral efficiency $\mathbf{E}[k(\gamma)]$ subject to an average power constraint $\bar{P}$ and the average BER constraint (9.71). As with the instantaneous BER constraint, this is a standard constrained optimization problem, which we solve using the Lagrange method. We now require two Lagrangians for the two constraints: average power and average BER. Specifically, the Lagrange equation is

$$J(k(\gamma), P(\gamma)) = \int_0^\infty k(\gamma)p(\gamma)\,d\gamma$$

$$+ \lambda_1\left[\int_0^\infty P_b(\gamma)k(\gamma)p(\gamma)\,d\gamma - \bar{P}_b\int_0^\infty k(\gamma)p(\gamma)\,d\gamma\right]$$

$$+ \lambda_2\left[\int_0^\infty P(\gamma)p(\gamma)\,d\gamma - \bar{P}\right]. \tag{9.73}$$

The optimal rate and power adaptation must satisfy

$$\frac{\partial J}{\partial k(\gamma)} = 0 \quad \text{and} \quad \frac{\partial J}{\partial P(\gamma)} = 0, \tag{9.74}$$

with the additional constraint that $k(\gamma)$ and $P(\gamma)$ be nonnegative for all γ.

Assume that P_b is approximated using the general formula (9.43). Define

$$f(k(\gamma)) = 2^{c_3 k(\gamma)} - c_4. \tag{9.75}$$

Then, using (9.43) in (9.73) and solving (9.74), we obtain that the power and BER adaptation that maximize spectral efficiency satisfy

$$\frac{P(\gamma)}{\bar{P}} = \max\left[\frac{f(k(\gamma))}{\frac{\partial f(k(\gamma))}{\partial k(\gamma)}}\lambda_2\bar{P}(\lambda_1\bar{P}_b - 1) - \frac{f(k(\gamma))^2}{c_2\gamma\frac{\partial f(k(\gamma))}{\partial k(\gamma)}k(\gamma)}, 0\right] \tag{9.76}$$

for nonnegative $k(\gamma)$ and

$$P_b(\gamma) = \frac{\lambda_2\bar{P}f(k(\gamma))}{\lambda_1 c_2\gamma k(\gamma)}. \tag{9.77}$$

Moreover, from (9.43), (9.76), and (9.77) we get that the optimal rate adaptation $k(\gamma)$ is either zero or the nonnegative solution of

$$\frac{\lambda_1\bar{P}_b - 1}{\frac{\partial f(k(\gamma))}{\partial k(\gamma)}\lambda_2\bar{P}} - \frac{f(k(\gamma))}{c_2\gamma\frac{\partial f(k(\gamma))}{\partial k(\gamma)}k(\gamma)} = \frac{1}{\gamma c_2}\ln\left[\frac{\lambda_1 c_1 c_2\gamma k(\gamma)}{\lambda_2\bar{P}f(k(\gamma))}\right]. \tag{9.78}$$

The values of $k(\gamma)$ and the Lagrangians λ_1 and λ_2 must be found through a numerical search whereby the average power constraint $\bar{P}$ and average BER constraint (9.71) are satisfied.

In the discrete-rate case, the rate is varied within a fixed set $k_0, \ldots, k_{N-1}$, where k_0 corresponds to no data transmission. We must determine region boundaries $\gamma_0, \ldots, \gamma_{N-2}$ such that we assign rate k_j to the rate region $[\gamma_{j-1}, \gamma_j)$, where $\gamma_{-1} = 0$ and $\gamma_{N-1} = \infty$. Under this rate assignment we wish to maximize spectral efficiency through optimal rate, power, and BER adaptation subject to an average power and BER constraint. Since the set of possible rates and their corresponding rate region assignments are fixed, the optimal rate adaptation corresponds to finding the optimal rate region boundaries γ_j, $j = 0, \ldots, N - 2$. The Lagrangian for this constrained optimization problem is

$$J(\gamma_0, \gamma_1, \ldots, \gamma_{N-2}, P(\gamma)) = \sum_{j=1}^{N-1} k_j \int_{\gamma_{j-1}}^{\gamma_j} p(\gamma)\, d\gamma$$

$$+ \lambda_1 \left[\sum_{j=1}^{N-1} k_j \int_{\gamma_{j-1}}^{\gamma_j} (P_b(\gamma) - \bar{P}_b) p(\gamma)\, d\gamma \right]$$

$$+ \lambda_2 \left[\int_{\gamma_0}^{\infty} P(\gamma) p(\gamma)\, d\gamma - \bar{P} \right]. \tag{9.79}$$

The optimal power adaptation is obtained by solving the following equation for $P(\gamma)$:

$$\frac{\partial J}{\partial P(\gamma)} = 0. \tag{9.80}$$

Similarly, the optimal rate region boundaries are obtained by solving the following set of equations for γ_j:

$$\frac{\partial J}{\partial \gamma_j} = 0, \quad 0 \le j \le N - 2. \tag{9.81}$$

From (9.80) we see that the optimal power and BER adaptation must satisfy

$$\frac{\partial P_b(\gamma)}{\partial P(\gamma)} = \frac{-\lambda_2}{k_j \lambda_1}, \quad \gamma_{j-1} \le \gamma \le \gamma_j. \tag{9.82}$$

Substituting (9.43) into (9.82), we get that

$$P_b(\gamma) = \lambda \frac{f(k_j)}{\gamma k_j}, \quad \gamma_{j-1} \le \gamma \le \gamma_j, \tag{9.83}$$

where $\lambda = \bar{P}\lambda_2/c_2\lambda_1$. This form of BER adaptation is similar to the water-filling power adaptation: the instantaneous BER decreases as the channel quality improves. Now, setting the BER in (9.43) equal to (9.83) and solving for $P(\gamma)$ yields

$$P(\gamma) = P_j(\gamma), \quad \gamma_{j-1} \le \gamma \le \gamma_j, \tag{9.84}$$

where

$$\frac{P_j(\gamma)}{\bar{P}} = \ln \left[\frac{\lambda f(k_j)}{c_1 \gamma k_j} \right] \frac{f(k_j)}{-\gamma c_2}, \quad 1 \le j \le N - 1, \tag{9.85}$$

and $P(\gamma) = 0$ for $\gamma < \gamma_0$. We see by (9.85) that $P(\gamma)$ is discontinuous at the γ_j boundaries.

Let us now consider the optimal region boundaries $\gamma_0, \ldots, \gamma_{N-2}$. Solving (9.81) for $P_b(\gamma_j)$ yields

$$P_b(\gamma_j) = \bar{P}_b - \frac{1}{\lambda_1} - \frac{\lambda_2}{\lambda_1} \frac{P_{j+1}(\gamma_j) - P_j(\gamma_j)}{k_{j+1} - k_j}, \quad 0 \le j \le N - 2, \tag{9.86}$$

where $k_0 = 0$ and $P_0(\gamma) = 0$. Unfortunately, this set of equations can be difficult to solve for the optimal boundary points $\{\gamma_j\}$. However, if we assume that $P(\gamma)$ is continuous at each boundary, then (9.86) becomes

$$P_b(\gamma_j) = \bar{P}_b - \frac{1}{\lambda_1}, \quad 0 \le j \le N - 2, \tag{9.87}$$

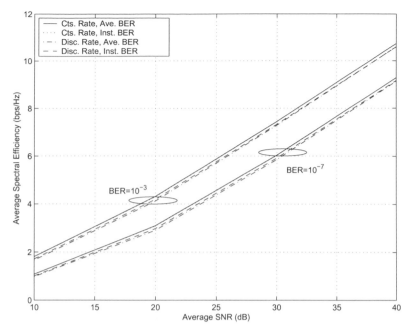

Figure 9.16: Spectral efficiency for different adaptation constraints.

for the Lagrangian λ_1. Under this assumption we can solve for the suboptimal rate region boundaries as

$$\gamma_{j-1} = \frac{f(k_j)}{k_j}\rho, \quad 1 \leq j \leq N - 1, \tag{9.88}$$

for some constant ρ. The constants λ_1 and ρ are found numerically such that the average power constraint

$$\sum_{j=1}^{N-1} \int_{\gamma_{j-1}}^{\gamma_j} \frac{P_j(\gamma)}{\bar{P}} p(\gamma)\, d\gamma = 1 \tag{9.89}$$

and the BER constraint (9.72) are satisfied. Note that the region boundaries (9.88) are suboptimal because $P(\gamma)$ is not necessarily continuous at the boundary regions, so these boundaries yield a suboptimal spectral efficiency.

In Figure 9.16 we plot average spectral efficiency for adaptive MQAM under both continuous and discrete rate adaptation, showing both average and instantaneous BER targets for a Rayleigh fading channel. The adaptive policies are based on the BER approximation (9.7) with a target BER of either 10^{-3} or 10^{-7}. For the discrete-rate cases we assume that six different MQAM signal constellations are available (seven fading regions), given by $\mathcal{M} = \{0, 4, 16, 64, 256, 1024, 4096\}$. We see in this figure that the spectral efficiencies of all four policies under the same instantaneous or average BER target are very close to each other. For discrete-rate adaptation, the spectral efficiency with an instantaneous BER target is slightly higher than with an average BER target even though the latter case is more constrained: that is because the efficiency with an average BER target is calculated using suboptimal rate region boundaries, which leads to a slight efficiency degradation.

9.5 Adaptive Techniques in Combined Fast and Slow Fading

In this section we examine adaptive techniques for composite fading channels consisting of both fast and slow fading (shadowing). We assume the fast fading changes too quickly to accurately measure and feed back to the transmitter, so the transmitter adapts only to the slow fading. The instantaneous SNR γ has distribution $p(\gamma \mid \bar{\gamma})$, where $\bar{\gamma}$ is a short-term average over the fast fading. This short-term average varies slowly because of shadowing and has a distribution $p(\bar{\gamma})$, where the average SNR relative to this distribution is $\bar{\bar{\gamma}}$. The transmitter adapts only to the slow fading $\bar{\gamma}$, hence its rate $k(\bar{\gamma})$ and power $P(\bar{\gamma})$ are functions of $\bar{\gamma}$. The power adaptation is subject to a long-term average power constraint over both the fast and slow fading:

$$\int_0^\infty P(\bar{\gamma})p(\bar{\gamma})\,d\bar{\gamma} = \bar{P}. \tag{9.90}$$

As before, we approximate the instantaneous probability of bit error by the general form (9.43). Since the power and rate are functions of $\bar{\gamma}$, the conditional BER, conditioned on $\bar{\gamma}$, is

$$P_b(\gamma \mid \bar{\gamma}) \approx c_1 \exp\left[\frac{-c_2\gamma P(\bar{\gamma})/\bar{P}}{2^{c_3 k(\bar{\gamma})} - c_4}\right]. \tag{9.91}$$

Since the transmitter does not adapt to the fast fading γ, we cannot require a given instantaneous BER. However, since the transmitter adapts to the shadowing, we can require a target average probability of bit error averaged over the fast fading for a fixed value of the shadowing. This short-term average for a given $\bar{\gamma}$ is obtained by averaging $P_b(\gamma \mid \bar{\gamma})$ over the fast fading distribution $p(\gamma \mid \bar{\gamma})$:

$$\bar{P}_b(\bar{\gamma}) = \int_0^\infty P_b(\gamma \mid \bar{\gamma})p(\gamma \mid \bar{\gamma})\,d\gamma. \tag{9.92}$$

Using (9.91) in (9.92) and assuming Rayleigh fading for the fast fading, this becomes

$$\bar{P}_b(\bar{\gamma}) = \frac{1}{\bar{\gamma}}\int_0^\infty c_1 \exp\left[\frac{-c_2\gamma P(\bar{\gamma})/\bar{P}}{2^{c_3 k(\bar{\gamma})} - c_4} - \frac{\gamma}{\bar{\gamma}}\right]d\gamma = \frac{c_1}{\dfrac{1.5\bar{\gamma}P(\bar{\gamma})/\bar{P}}{2^{c_3 k(\bar{\gamma})} - c_4} + 1}. \tag{9.93}$$

For example, with MQAM modulation with the tight BER bound (9.7), equation (9.93) becomes

$$\bar{P}_b(\bar{\gamma}) = \frac{.2}{\dfrac{1.5\bar{\gamma}P(\bar{\gamma})/\bar{P}}{2^{k(\bar{\gamma})} - 1} + 1}. \tag{9.94}$$

We can now invert (9.93) to obtain the adaptive rate $k(\bar{\gamma})$ as a function of the target average BER $\bar{P}_b$ and the power adaptation $P(\bar{\gamma})$:

$$k(\bar{\gamma}) = \frac{1}{c_3}\log_2\left(c_4 + \frac{K\bar{\gamma}P(\bar{\gamma})}{\bar{P}}\right), \tag{9.95}$$

where

$$K = \frac{c_2}{c_1/\bar{P}_b - 1} \tag{9.96}$$

depends only on the target average BER and decreases as this target decreases. We maximize spectral efficiency by maximizing

$$\mathbf{E}[k(\bar{\gamma})] = \int_0^\infty \frac{1}{c_3} \log_2\left(c_4 + \frac{K\bar{\gamma}P(\bar{\gamma})}{\bar{P}}\right) p(\bar{\gamma}) \, d\bar{\gamma} \qquad (9.97)$$

subject to the average power constraint (9.90).

Let us assume that $c_4 > 0$. Then this maximization and the power constraint are in the exact same form as (9.11) with the fading γ replaced by the slow fading $\bar{\gamma}$. Thus, the optimal power adaptation also has the same water-filling form as (9.13) and is given by

$$\frac{P(\bar{\gamma})}{\bar{P}} = \begin{cases} 1/\bar{\gamma}_0 - c_4/\bar{\gamma}K & \bar{\gamma} \geq c_4\bar{\gamma}_0/K, \\ 0 & \bar{\gamma} < c_4\bar{\gamma}_0/K, \end{cases} \qquad (9.98)$$

where the channel is not used when $\bar{\gamma} < c_4\bar{\gamma}_0/K$. The value of $\bar{\gamma}_0$ is determined by the average power constraint. Substituting (9.98) into (9.95) yields the rate adaptation

$$k(\bar{\gamma}) = \frac{1}{c_3} \log_2\left(\frac{K\bar{\gamma}}{\bar{\gamma}_0}\right), \qquad (9.99)$$

and the corresponding average spectral efficiency is given by

$$\frac{R}{B} = \int_{c_4\gamma_0/K}^\infty \log_2\left(\frac{K\bar{\gamma}}{\bar{\gamma}_0}\right) p(\bar{\gamma}) \, d\bar{\gamma}. \qquad (9.100)$$

Thus, we see that in a composite fading channel where rate and power are adapted only to the slow fading and for $c_4 > 0$ in (9.43), water-filling *relative to the slow fading* is the optimal power adaptation for maximizing spectral efficiency subject to an average BER constraint.

Our derivation has assumed that the fast fading is Rayleigh; however, it can be shown [34] that, with $c_4 > 0$ in (9.43), the optimal power and rate adaptation for any fast fading distribution have the same form. Since we have assumed $c_4 > 0$ in (9.43), the positivity constraint on power dictates the cutoff value below which the channel is not used. As we saw in Section 9.4.1, when $c_4 \leq 0$ the positivity constraint on rate dictates this cutoff, and the optimal power adaptation becomes inverse water-filling for $c_4 < 0$ and on–off power adaptation for $c_4 = 0$.

PROBLEMS

9-1. Find the average SNR required to achieve an average BER of $\bar{P}_b = 10^{-3}$ for 8-PSK modulation in Rayleigh fading. What is the spectral efficiency of this scheme, assuming a symbol time of $T_s = 1/B$?

9-2. Consider a truncated channel inversion variable-power technique for Rayleigh fading with average SNR of 20 dB. What value of σ corresponds to an outage probability of .1? Find the maximum size MQAM constellation that can be transmitted under this policy so that, in non-outage, $P_b \approx 10^{-3}$.

9-3. Find the power adaptation for QPSK modulation that maintains a fixed $P_b = 10^{-3}$ in non-outage for a Rayleigh fading channel with $\bar{\gamma} = 20$ dB. What is the outage probability of this system?

9-4. Consider a variable-rate MQAM modulation scheme with just two constellations, $M = 4$ and $M = 16$. Assume a target P_b of approximately 10^{-3}. If the target cannot be met then no data is transmitted.

    (a) Using the BER bound (9.7), find the range of γ-values associated with the three possible transmission schemes (no transmission, 4-QAM, and 16-QAM) where the BER target is met. What is the cutoff γ_0 below which the channel is not used?

    (b) Assuming Rayleigh fading with $\bar{\gamma} = 20$ dB, find the average data rate of the variable-rate scheme.

    (c) Suppose that, instead of suspending transmission below γ_0, BPSK is transmitted for $0 \le \gamma \le \gamma_0$. Using the loose bound (9.6), find the average probability of error for this BPSK transmission.

9-5. Consider an adaptive modulation and coding scheme consisting of three modulations: BPSK, QPSK, and 8-PSK, along with three block codes of rate $1/2, 1/3$, and $1/4$. Assume that the first code provides roughly 3 dB of coding gain for each modulation type, the second code 4 dB, and the third code 5 dB. For each possible value of SNR $0 \le \gamma \le \infty$, find the combined coding and modulation with the maximum data rate for a target BER of 10^{-3} (you can use any reasonable approximation for modulation BER in this calculation, with SNR increased by the coding gain). Find the average data rate of the system for a Rayleigh fading channel with average SNR of 20 dB, assuming no transmission if the target BER cannot be met with any combination of modulation and coding.

9-6. Show that the spectral efficiency given by (9.11) with power constraint (9.12) is maximized by the water-filling power adaptation (9.13). Do this by setting up the Lagrangian equation, differentiating it, and solving for the maximizing power adaptation. Also show that, with this power adaptation, the rate adaptation is as given in (9.16).

9-7. In this problem we compare the spectral efficiency of nonadaptive techniques with that of adaptive techniques.

    (a) Using the tight BER bound for MQAM modulation given by (9.7), find an expression for the average probability of bit error in Rayleigh fading as a function of M and $\bar{\gamma}$.

    (b) Based on the expression found in part (a), find the maximum constellation size that can be transmitted over a Rayleigh fading channel with a target average BER of 10^{-3}, assuming $\bar{\gamma} = 20$ dB.

    (c) Compare the spectral efficiency of part (b) with that of adaptive modulation shown in Figure 9.3 for the same parameters. What is the spectral efficiency difference between the adaptive and nonadaptive techniques?

9-8. Consider a Rayleigh fading channel with an average SNR of 20 dB. Assume a target BER of 10^{-4}.

    (a) Find the optimal rate and power adaptation for variable-rate variable-power MQAM as well as the cutoff value γ_0/K below which the channel is not used.

    (b) Find the average spectral efficiency for the adaptive scheme derived in part (a).

(c) Compare your answer in part (b) to the spectral efficiency of truncated channel inversion, where γ_0 is chosen to maximize this efficiency.

9-9. Consider a discrete time-varying AWGN channel with four channel states. Assuming a fixed transmit power $\bar{P}$, the received SNR associated with each channel state is $\gamma_1 = 5$ dB, $\gamma_2 = 10$ dB, $\gamma_3 = 15$ dB, and $\gamma_4 = 20$ dB, respectively. The probabilities associated with the channel states are $p(\gamma_1) = .4$ and $p(\gamma_2) = p(\gamma_3) = p(\gamma_4) = .2$. Assume a target BER of 10^{-3}.

(a) Find the optimal power and rate adaptation for continous-rate adaptive MQAM on this channel.

(b) Find the average spectral efficiency with this optimal adaptation.

(c) Find the truncated channel inversion power control policy for this channel and the maximum data rate that can be supported with this policy.

9-10. Consider a Rayleigh fading channel with an average received SNR of 20 dB and a required BER of 10^{-3}. Find the spectral efficiency of this channel using truncated channel inversion, assuming the constellation is restricted to size 0, 2, 4, 16, 64, or 256.

9-11. Consider a Rayleigh fading channel with an average received SNR of 20 dB, a Doppler frequency of 80 Hz, and a required BER of 10^{-3}.

(a) Suppose you use adaptive MQAM modulation on this channel with constellations restricted to size 0, 2, 4, 16, and 64. Using $\gamma_K^* = .1$, find the fading regions R_j associated with each of these constellations. Also find the average spectral efficiency of this restricted adaptive modulation scheme and the average time spent in each region R_j. If the symbol rate is $T_s = B^{-1}$, over approximately how many symbols is each constellation transmitted before a change in constellation size is needed?

(b) Find the exact BER of your adaptive scheme using (9.33). How does it differ from the target BER?

9-12. Consider a Rayleigh fading channel with an average received SNR of 20 dB, a signal bandwidth of 30 kHz, a Doppler frequency of 80 Hz, and a required BER of 10^{-3}.

(a) Suppose the estimation error $\varepsilon = \hat{\gamma}/\gamma$ in a variable-rate variable-power MQAM system with a target BER of 10^{-3} is uniformly distributed between .5 and 1.5. Find the resulting average probability of bit error for this system.

(b) Find an expression for the average probability of error in a variable-rate variable-power MQAM system in which the SNR estimate $\hat{\gamma}$ available at the transmitter is both a delayed and noisy estimate of γ: $\hat{\gamma}(t) = \gamma(t - \tau) + \gamma_\varepsilon(t)$. What joint distribution is needed to compute this average?

9-13. Consider an adaptive trellis-coded MQAM system with a coding gain of 3 dB. Assume a Rayleigh fading channel with an average received SNR of 20 dB. Find the optimal adaptive power and rate policy for this system and the corresponding average spectral efficiency. Assume a target BER of 10^{-3}.

9-14. In Chapter 6, a bound on P_b for nonrectangular MQAM was given as

$$P_b \approx \frac{4}{\log_2 M} Q\left(\sqrt{\frac{3\gamma}{(M-1)}}\right).$$

Find values for c_1, c_2, c_3, and c_4 for the general BER form (9.43) to approximate this bound with $M = 8$. Any curve approximation technique is acceptable. Plot both BER formulas for $0 \leq \gamma \leq 30$ dB.

9-15. Show that the average spectral efficiency $\mathbf{E}[k(\gamma)]$ for $k(\gamma)$ given by (9.44) with power constraint $\bar{P}$ is maximized by the power adaptation (9.47).

9-16. In this problem we investigate the optimal adaptive modulation for MPSK modulation based on the three BER bounds (9.51), (9.52), and (9.53). We assume a Rayleigh fading channel (so that γ is exponentially distributed with $\bar{\gamma} = 30$ dB) and a target BER of $P_b = 10^{-7}$.

    (a) The cutoff fade depth γ_0 must satisfy

$$\int_{\gamma_0/K}^{\infty} \left(\frac{1}{\gamma_0} - \frac{1}{\gamma K} \right) p(\gamma)\, d\gamma \leq 1$$

    for K given by (9.48). Find the cutoff value γ_0 corresponding to the power adaptation for each of the three bounds.

    (b) Plot $P(\gamma)/\bar{P}$ and $k(\gamma)$ as a function of γ for Bounds 1, 2, and 3 for γ ranging from 0 dB to 30 dB. Also state whether the cutoff value below which the channel is not used is based on the power or rate positivity constraint.

    (c) How does the power adaptation associated with the different bounds differ at low SNRs? What about at high SNRs?

9-17. Show that, under discrete rate adaptation for general M-ary modulation, the power adaptation that maintains a target instantaneous BER is given by (9.63). Also show that the region boundaries that maximize spectral efficiency – obtained using the Lagrangin given in (9.65) – are given by (9.67) and (9.68).

9-18. Show that, for general M-ary modulation with an average target BER, the Lagrangian (9.80) implies that the optimal power and BER adaptation must satisfy (9.82). Then show how (9.82) leads to BER adaptation given by (9.83), which in turn leads to the power adaptation given by (9.84) and (9.85). Finally, use (9.81) to show that the optimal rate region boundaries must satisfy (9.86).

9-19. Consider adaptive MPSK where the constellation is restricted to either no transmission or $M = 2, 4, 8, 16$. Assume the probability of error is approximated using (9.51). Find and plot the optimal discrete-rate continuous-power adaptation for $0 \leq \gamma \leq 30$ dB assuming a Rayleigh fading channel with $\bar{\gamma} = 20$ dB and a target P_b of 10^{-4}. What is the resulting average spectral efficiency?

9-20. We assume the same discrete-rate adaptive MPSK as in the previous problem, except now there is an average target P_b of 10^{-4} instead of an instantaneous target. Find the optimal discrete-rate continuous-power adaptation for a Rayleigh fading channel with $\bar{\gamma} = 20$ dB and the corresponding average spectral efficiency.

9-21. Consider a composite fading channel with fast Rayleigh fading and slow log-normal shadowing with an average dB SNR $\mu_{\psi_{dB}} = 20$ dB (averaged over both fast and slow fading) and $\sigma_{\psi_{dB}} = 8$ dB. Assume an adaptive MPSK modulation that adapts only to the shadowing, with a target average BER of 10^{-3}. Using the BER approximation (9.51), find the optimal power and rate adaptation policies as a function of the slow fading $\bar{\gamma}$ that maximize average spectral efficiency while meeting the average BER target. Also determine the average spectral efficiency that results from these policies.

9-22. In Section 9.5 we determined the optimal adaptive rate and power policies to maximize average spectral efficiency while meeting a target average BER in combined Rayleigh fading and shadowing. The derivation assumed the general bound (9.43) with $c_4 > 0$. For the same composite channel, find the optimal adaptive rate and power policies to maximize average spectral efficiency while meeting a target average BER assuming $c_4 < 0$. *Hint:* The derivation is similar to the case of continuous-rate adaptation using the second MPSK bound and results in the same channel inversion power control.

9-23. As in the previous problem, we again examine the adaptative rate and power policies to maximize average spectral efficiency while meeting a target average BER in combined Rayleigh fading and shadowing. In this problem we assume the general bound (9.43) with $c_4 = 0$. For the composite channel, find the optimal adaptive rate and power policies to maximize average spectral efficiency while meeting a target average BER assuming $c_4 = 0$. *Hint:* the derivation is similar to that of Section 9.4.1 for the third MPSK bound and results in the same on–off power control.

REFERENCES

[1]   J. F. Hayes, "Adaptive feedback communications," *IEEE Trans. Commun. Tech.*, pp. 29–34, February 1968.

[2]   J. K. Cavers, "Variable-rate transmission for Rayleigh fading channels," *IEEE Trans. Commun.*, pp. 15–22, February 1972.

[3]   S. Otsuki, S. Sampei, and N. Morinaga, "Square-QAM adaptive modulation/TDMA/TDD systems using modulation level estimation with Walsh function," *Elec. Lett.*, pp. 169–71, February 1995.

[4]   W. T. Webb and R. Steele, "Variable rate QAM for mobile radio," *IEEE Trans. Commun.*, pp. 2223–30, July 1995.

[5]   Y. Kamio, S. Sampei, H. Sasaoka, and N. Morinaga, "Performance of modulation-level-controlled adaptive-modulation under limited transmission delay time for land mobile communications," *Proc. IEEE Veh. Tech. Conf.*, pp. 221–5, July 1995.

[6]   B. Vucetic, "An adaptive coding scheme for time-varying channels," *IEEE Trans. Commun.*, pp. 653–63, May 1991.

[7]   M. Rice and S. B. Wicker, "Adaptive error control for slowly varying channels," *IEEE Trans. Commun.*, pp. 917–26, February–April 1994.

[8]   S. M. Alamouti and S. Kallel, "Adaptive trellis-coded multiple-phased-shift keying for Rayleigh fading channels," *IEEE Trans. Commun.*, pp. 2305–14, June 1994.

[9]   T. Ue, S. Sampei, and N. Morinaga, "Symbol rate and modulation level controlled adaptive modulation/TDMA/TDD for personal communication systems," *Proc. IEEE Veh. Tech. Conf.*, pp. 306–10, July 1995.

[10] H. Matsuoka, S. Sampei, N. Morinaga, and Y. Kamio, "Symbol rate and modulation level controlled adaptive modulation/TDMA/TDD for personal communication systems," *Proc. IEEE Veh. Tech. Conf.*, pp. 487–91, April 1996.

[11] S. Sampei, N. Morinaga, and Y. Kamio, "Adaptive modulation/TDMA with a BDDFE for 2 Mbit/s multi-media wireless communication systems," *Proc. IEEE Veh. Tech. Conf.*, pp. 311–15, July 1995.

[12] S. T. Chung and A. J. Goldsmith, "Degrees of freedom in adaptive modulation: A unified view," *IEEE Trans. Commun.,* pp. 1561–71, September 2001.

[13] A. Furuskar, S. Mazur, F. Muller, and H. Olofsson, "EDGE: Enhanced data rates for GSM and TDMA/136 evolution," *IEEE Wireless Commun. Mag.,* pp. 56–66, June 1999.

[14] A. Ghosh, L. Jalloul, B. Love, M. Cudak, and B. Classon, "Air-interface for 1XTREME/1xEV-DV," *Proc. IEEE Veh. Tech. Conf.,* pp. 2474–8, May 2001.

[15] S. Nanda, K. Balachandran, and S. Kumar, "Adaptation techniques in wireless packet data services," *IEEE Commun. Mag.,* pp. 54–64, January 2000.

[16] H. Sari, "Trends and challenges in broadband wireless access," *Proc. Sympos. Commun. Veh. Tech.,* pp. 210–14, October 2000.

[17] K. M. Kamath and D. L. Goeckel, "Adaptive-modulation schemes for minimum outage probability in wireless systems," *IEEE Trans. Commun.,* pp. 1632–5, October 2004.

[18] J. Hagenauer, "Rate-compatible punctured convolutional codes (RCPC codes) and their applications," *IEEE Trans. Commun.,* pp. 389–400, April 1988.

[19] G. J. Foschini and J. Salz, "Digital communications over fading radio channels," *Bell System Tech. J.,* pp. 429–56, February 1983.

[20] G. D. Forney, Jr., R. G. Gallager, G. R. Lang, F. M. Longstaff, and S. U. Quereshi, "Efficient modulation for band-limited channels," *IEEE J. Sel. Areas Commun.,* pp. 632–47, September 1984.

[21] J. G. Proakis, *Digital Communications,* 4th ed., McGraw-Hill, New York, 2001.

[22] M. Filip and E. Vilar, "Implementation of adaptive modulation as a fade countermeasure," *Internat. J. Sat. Commun.,* pp. 181–91, 1994.

[23] C. C. Tan and N. C. Beaulieu, "On first-order Markov modeling for the Rayleigh fading channel," *IEEE Trans. Commun.,* pp. 2032–40, December 2000.

[24] L. Kleinrock, *Queueing Systems,* vol. I: *Theory,* Wiley, New York, 1975.

[25] W. C. Jakes, Jr., *Microwave Mobile Communications,* Wiley, New York, 1974.

[26] H. S. Wang and P.-C. Chang, "On verifying the first-order Markov assumption for a Rayleigh fading channel model," *IEEE Trans. Veh. Tech.,* pp. 353–7, May 1996.

[27] X. Tang, M.-S. Alouini, and A. Goldsmith, "Effect of channel estimation error on M-QAM BER performance in Rayleigh fading," *IEEE Trans. Commun.,* pp. 1856–64, December 1999.

[28] J. K. Cavers, "An analysis of pilot symbol assisted modulation for Rayleigh fading channels," *IEEE Trans. Veh. Tech.,* pp. 686–93, November 1991.

[29] A. J. Goldsmith and L. J. Greenstein, "Effect of average power estimation error on adaptive MQAM modulation," *Proc. IEEE Internat. Conf. Commun.,* pp. 1105–9, June 1997.

[30] D. L. Goeckel, "Adaptive coding for time-varying channels using outdated fading estimates," *IEEE Trans. Commun.,* pp. 844–55, June 1999.

[31] A. Duel-Hallen, S. Hu, and H. Hallen, "Long-range prediction of fading signals," *IEEE Signal Proc. Mag.,* pp. 62–75, May 2000.

[32] M.-S. Alouini and A. J. Goldsmith, "Adaptive modulation over Nakagami fading channels," *Kluwer J. Wireless Pers. Commun.,* pp. 119–43, May 2000.

[33] S.-G. Chua and A. J. Goldsmith, "Adaptive coded modulation for fading channels," *IEEE Trans. Commun.,* pp. 595–602, May 1998.

[34] S. Vishwanath, S. A. Jafar, and A. J. Goldsmith, "Adaptive resource allocation in composite fading environments," *Proc. IEEE Globecom Conf.,* pp. 1312–16, November 2001.

Multiple Antennas and Space-Time Communications

In this chapter we consider systems with multiple antennas at the transmitter and receiver, which are commonly referred to as multiple-input multiple-output (MIMO) systems. The multiple antennas can be used to increase data rates through multiplexing or to improve performance through diversity. We have already seen diversity in Chapter 7. In MIMO systems, the transmit and receive antennas can both be used for diversity gain. Multiplexing exploits the structure of the channel gain matrix to obtain independent signaling paths that can be used to send independent data. Indeed, the initial excitement about MIMO was sparked by the pioneering work of Winters [1], Foschini [2], Foschini and Gans [3], and Telatar [4; 5] predicting remarkable spectral efficiencies for wireless systems with multiple transmit and receive antennas. These spectral efficiency gains often require accurate knowledge of the channel at the receiver – and sometimes at the transmitter as well. In addition to spectral efficiency gains, ISI and interference from other users can be reduced using smart antenna techniques. The cost of the performance enhancements obtained through MIMO techniques is the added cost of deploying multiple antennas, the space and circuit power requirements of these extra antennas (especially on small handheld units), and the added complexity required for multidimensional signal processing. In this chapter we examine the different uses for multiple antennas and find their performance advantages. This chapter uses several key results from matrix theory: Appendix C provides a brief overview of these results.

10.1 Narrowband MIMO Model

In this section we consider a narrowband MIMO channel. A narrowband point-to-point communication system of M_t transmit and M_r receive antennas is shown in Figure 10.1. This system can be represented by the following discrete-time model:

$$\begin{bmatrix} y_1 \\ \vdots \\ y_{M_r} \end{bmatrix} = \begin{bmatrix} h_{11} & \cdots & h_{1M_t} \\ \vdots & \ddots & \vdots \\ h_{M_r 1} & \cdots & h_{M_r M_t} \end{bmatrix} \begin{bmatrix} x_1 \\ \vdots \\ x_{M_t} \end{bmatrix} + \begin{bmatrix} n_1 \\ \vdots \\ n_{M_r} \end{bmatrix}$$

or simply as $\mathbf{y} = \mathbf{Hx} + \mathbf{n}$. Here $\mathbf{x}$ represents the M_t-dimensional transmitted symbol, $\mathbf{n}$ is the M_r-dimensional noise vector, and $\mathbf{H}$ is the $M_r \times M_t$ matrix of channel gains h_{ij} representing the gain from transmit antenna j to receive antenna i. We assume a channel bandwidth

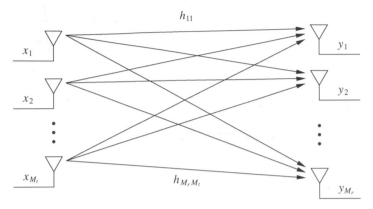

Figure 10.1: MIMO systems.

of B and complex Gaussian noise with zero mean and covariance matrix $\sigma^2 \mathbf{I}_{M_r}$, where typically $\sigma^2 \triangleq \mathbf{E}[n_i^2] = N_0/2$, the power spectral density of the channel noise. For simplicity, given a transmit power constraint P we will assume an equivalent model with a noise power σ^2 of unity and transmit power $P/\sigma^2 = \rho$, where ρ can be interpreted as the average SNR per receive antenna under unity channel gain. This power constraint implies that the input symbols satisfy

$$\sum_{i=1}^{M_t} \mathbf{E}[x_i x_i^*] = \rho, \tag{10.1}$$

or (equivalently) that $\mathrm{Tr}(\mathbf{R_x}) = \rho$, where $\mathrm{Tr}(\mathbf{R_x})$ is the trace of the input covariance matrix $\mathbf{R_x} = \mathbf{E}[\mathbf{xx}^H]$.

Different assumptions can be made about the knowledge of the channel gain matrix $\mathbf{H}$ at the transmitter and receiver, referred to as channel side information at the transmitter (CSIT) and channel side information at the receiver (CSIR), respectively. For a static channel CSIR is typically assumed, since the channel gains can be obtained fairly easily by sending a pilot sequence for channel estimation. More details on estimation techniques for MIMO channels can be found in [6, Chap. 3.9]. If a feedback path is available then CSIR from the receiver can be sent back to the transmitter to provide CSIT: CSIT may also be available in bi-directional systems without a feedback path when the reciprocal properties of propagation are exploited. When the channel is not known to either the transmitter or receiver then some distribution on the channel gain matrix must be assumed. The most common model for this distribution is a zero-mean spatially white (ZMSW) model, where the entries of $\mathbf{H}$ are assumed to be independent and identically distributed (i.i.d.) zero-mean, unit-variance, complex circularly symmetric Gaussian random variables.[1] We adopt this model unless stated otherwise. Alternatively, these entries may be complex circularly symmetric Gaussian random variables

---

[1] A complex random vector $\mathbf{x}$ is circularly symmetric if, for any $\theta \in [0, 2\pi]$, the distribution of $\mathbf{x}$ is the same as the distribution of $e^{j\theta}\mathbf{x}$. Thus, the real and imaginary parts of $\mathbf{x}$ are i.i.d. For $\mathbf{x}$ circularly symmetric, taking $\theta = \pi$ implies that $\mathbf{E}[\mathbf{x}] = 0$. Similarly, taking $\theta = \pi/2$ implies that $\mathbf{E}[\mathbf{xx}^T] = 0$, which by definition means that $\mathbf{x}$ is a *proper* random vector. For a complex random vector $\mathbf{x}$ the converse is also true: if $\mathbf{x}$ is zero-mean and proper then it is circularly symmetric.

with a nonzero mean or with a covariance matrix not equal to the identity matrix. In general, different assumptions about CSI and about the distribution of the $\mathbf{H}$ entries lead to different channel capacities and different approaches to space-time signaling.

Optimal decoding of the received signal requires maximum likelihood demodulation. If the symbols modulated onto each of the M_t transmit antennas are chosen from an alphabet of size $|\mathcal{X}|$, then – because of the cross-coupling between transmitted symbols at the receiver antennas – ML demodulation requires an exhaustive search over all $|\mathcal{X}|^{M_t}$ possible input vectors of M_t symbols. For general channel matrices, when the transmitter does not know $\mathbf{H}$, the complexity cannot be reduced further. This decoding complexity is typically prohibitive for even a small number of transmit antennas. However, decoding complexity is significantly reduced if the channel is known to the transmitter, as shown in the next section.

10.2 Parallel Decomposition of the MIMO Channel

We have seen in Chapter 7 that multiple antennas at the transmitter or receiver can be used for diversity gain. When both the transmitter and receiver have multiple antennas, there is another mechanism for performance gain called *multiplexing gain*. The multiplexing gain of a MIMO system results from the fact that a MIMO channel can be decomposed into a number R of parallel independent channels. By multiplexing independent data onto these independent channels, we get an R-fold increase in data rate in comparison to a system with just one antenna at the transmitter and receiver. This increased data rate is called the multiplexing gain. In this section we describe how to obtain independent channels from a MIMO system.

Consider a MIMO channel with $M_r \times M_t$ channel gain matrix $\mathbf{H}$ that is known to both the transmitter and the receiver. Let $R_{\mathbf{H}}$ denote the rank of $\mathbf{H}$. From Appendix C, for any matrix $\mathbf{H}$ we can obtain its singular value decomposition (SVD) as

$$\mathbf{H} = \mathbf{U}\mathbf{\Sigma}\mathbf{V}^H, \tag{10.2}$$

where the $M_r \times M_r$ matrix $\mathbf{U}$ and the $M_t \times M_t$ matrix $\mathbf{V}$ are unitary matrices[2] and where $\mathbf{\Sigma}$ is an $M_r \times M_t$ diagonal matrix of singular values $\{\sigma_i\}$ of $\mathbf{H}$. These singular values have the property that $\sigma_i = \sqrt{\lambda_i}$ for λ_i the ith largest eigenvalue of $\mathbf{H}\mathbf{H}^H$, and $R_{\mathbf{H}}$ of these singular values are nonzero. Because $R_{\mathbf{H}}$, the rank of matrix $\mathbf{H}$, cannot exceed the number of columns or rows of $\mathbf{H}$, it follows that $R_{\mathbf{H}} \leq \min(M_t, M_r)$. If $\mathbf{H}$ is full rank, which is referred to as a *rich scattering environment*, then $R_{\mathbf{H}} = \min(M_t, M_r)$. Other environments may lead to a low-rank $\mathbf{H}$: a channel with high correlation among the gains in $\mathbf{H}$ may have rank 1.

The parallel decomposition of the channel is obtained by defining a transformation on the channel input and output $\mathbf{x}$ and $\mathbf{y}$ via *transmit precoding* and *receiver shaping*. In transmit precoding the input $\mathbf{x}$ to the antennas is generated by a linear transformation on input vector $\tilde{\mathbf{x}}$ as $\mathbf{x} = \mathbf{V}\tilde{\mathbf{x}}$. Receiver shaping performs a similar operation at the receiver by multiplying the channel output $\mathbf{y}$ by $\mathbf{U}^H$, as shown in Figure 10.2.

The transmit precoding and receiver shaping transform the MIMO channel into $R_{\mathbf{H}}$ parallel single-input single-output (SISO) channels with input $\tilde{\mathbf{x}}$ and output $\tilde{\mathbf{y}}$, since from the SVD we have that

---

[2] $\mathbf{U}$ and $\mathbf{V}$ unitary imply that $\mathbf{U}^H\mathbf{U} = \mathbf{I}_{M_r}$ and $\mathbf{V}^H\mathbf{V} = \mathbf{I}_{M_t}$.

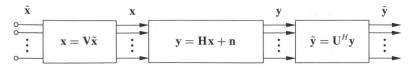

Figure 10.2: Transmit precoding and receiver shaping.

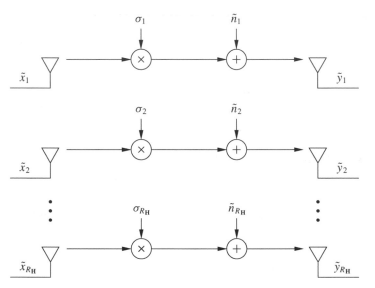

Figure 10.3: Parallel decomposition of the MIMO channel.

$$\tilde{\mathbf{y}} = \mathbf{U}^H(\mathbf{Hx} + \mathbf{n})$$
$$= \mathbf{U}^H(\mathbf{U\Sigma V}^H\mathbf{x} + \mathbf{n})$$
$$= \mathbf{U}^H(\mathbf{U\Sigma V}^H\mathbf{V\tilde{x}} + \mathbf{n})$$
$$= \mathbf{U}^H\mathbf{U\Sigma V}^H\mathbf{V\tilde{x}} + \mathbf{U}^H\mathbf{n}$$
$$= \mathbf{\Sigma\tilde{x}} + \tilde{\mathbf{n}},$$

where $\tilde{\mathbf{n}} = \mathbf{U}^H\mathbf{n}$ and where $\mathbf{\Sigma}$ is the matrix of singular values of $\mathbf{H}$ with σ_i on the ith diagonal and zeros everywhere else. Note that multiplication by a unitary matrix does not change the distribution of the noise – that is, $\mathbf{n}$ and $\tilde{\mathbf{n}}$ are identically distributed. Thus, the transmit precoding and receiver shaping transform the MIMO channel into $R_{\mathbf{H}}$ parallel independent channels, where the ith channel has input $\tilde{x}_i$, output $\tilde{y}_i$, noise $\tilde{n}_i$, and channel gain σ_i. Note that the σ_i are related because they are all functions of $\mathbf{H}$, but since the resulting parallel channels do not interfere with each other we say that the channels with these gains are independent – linked only through the total power constraint. This parallel decomposition is shown in Figure 10.3. Since the parallel channels do not interfere with each other, the optimal ML demodulation complexity is linear in $R_{\mathbf{H}}$, the number of independent paths that need to be demodulated. Moreover, by sending independent data across each of the parallel channels, the MIMO channel can support $R_{\mathbf{H}}$ times the data rate of a system with just one transmit and receive antenna, leading to a multiplexing gain of $R_{\mathbf{H}}$. Note, however, that the performance on each one of the channels will depend on its gain σ_i. The next section will more precisely characterize the multiplexing gain associated with the Shannon capacity of the MIMO channel.

EXAMPLE 10.1: Find the equivalent parallel channel model for a MIMO channel with channel gain matrix

$$\mathbf{H} = \begin{bmatrix} .1 & .3 & .7 \\ .5 & .4 & .1 \\ .2 & .6 & .8 \end{bmatrix}. \tag{10.3}$$

Solution: The SVD of $\mathbf{H} = \mathbf{U\Sigma V}^H$ is given by

$$\mathbf{H} = \begin{bmatrix} -.555 & .3764 & -.7418 \\ -.3338 & -.9176 & -.2158 \\ -.7619 & .1278 & .6349 \end{bmatrix} \begin{bmatrix} 1.3333 & 0 & 0 \\ 0 & .5129 & 0 \\ 0 & 0 & .0965 \end{bmatrix}$$
$$\times \begin{bmatrix} -.2811 & -.7713 & -.5710 \\ -.5679 & -.3459 & .7469 \\ -.7736 & .5342 & -.3408 \end{bmatrix}. \tag{10.4}$$

There are three nonzero singular values and so $R_\mathbf{H} = 3$, leading to three parallel channels with respective channel gains $\sigma_1 = 1.3333$, $\sigma_2 = .5129$, and $\sigma_3 = .0965$. Note that the channels have diminishing gain, with a very small gain on the third channel. Hence, this last channel will either have a high error probability or a low capacity.

10.3 MIMO Channel Capacity

This section focuses on the Shannon capacity of a MIMO channel, which equals the maximum data rate that can be transmitted over the channel with arbitrarily small error probability. Capacity versus outage defines the maximum rate that can be transmitted over the channel with some nonzero outage probability. Channel capacity depends on what is known about the channel gain matrix or its distribution at the transmitter and/or receiver. First the static channel capacity under different assumptions about this channel knowledge will be given, which forms the basis for the subsequent section on capacity of fading channels.

10.3.1 Static Channels

The capacity of a MIMO channel is an extension of the mutual information formula for a SISO channel given by equation (4.3) to a matrix channel. For static channels a good estimate of $\mathbf{H}$ can be obtained fairly easily at the receiver, so we assume CSIR throughout this section. Under this assumption, the capacity is given in terms of the mutual information between the channel input vector $\mathbf{x}$ and output vector $\mathbf{y}$ as

$$C = \max_{p(\mathbf{x})} I(\mathbf{X}; \mathbf{Y}) = \max_{p(\mathbf{x})} [H(\mathbf{Y}) - H(\mathbf{Y} \mid \mathbf{X})] \tag{10.5}$$

for $H(\mathbf{Y})$ and $H(\mathbf{Y} \mid \mathbf{X})$ the entropy in $\mathbf{y}$ and $\mathbf{y} \mid \mathbf{x}$, as defined in Section 4.1.[3] The definition of entropy yields that $H(\mathbf{Y} \mid \mathbf{X}) = H(\mathbf{n})$, the entropy in the noise. Since this noise $\mathbf{n}$ has

---

[3] Entropy was defined in Section 4.1 for scalar random variables, and the definition is identical for random vectors.

fixed entropy independent of the channel input, maximizing mutual information is equivalent to maximizing the entropy in $\mathbf{y}$.

Given covariance matrix $\mathbf{R_x}$ on the input vector $\mathbf{x}$, the output covariance matrix $\mathbf{R_y}$ associated with MIMO channel output $\mathbf{y}$ is given by

$$\mathbf{R_y} = \mathbf{E}[\mathbf{yy}^H] = \mathbf{HR_xH}^H + \mathbf{I}_{M_r}. \tag{10.6}$$

It turns out that, for all random vectors with a given covariance matrix $\mathbf{R_y}$, the entropy of $\mathbf{y}$ is maximized when $\mathbf{y}$ is a zero-mean, circularly symmetric complex Gaussian (ZMCSCG) random vector [5]. But $\mathbf{y}$ is ZMCSCG only if the input $\mathbf{x}$ is ZMCSCG, and hence this is the optimal distribution on $\mathbf{x}$ in (10.5), subject to the power constraint $\mathrm{Tr}(\mathbf{R_x}) = \rho$. Thus we have $H(\mathbf{Y}) = B \log_2 \det[\pi e \mathbf{R_y}]$ and $H(\mathbf{n}) = B \log_2 \det[\pi e \mathbf{I}_{M_r}]$, resulting in the mutual information

$$I(\mathbf{X}; \mathbf{Y}) = B \log_2 \det[\mathbf{I}_{M_r} + \mathbf{HR_xH}^H]. \tag{10.7}$$

This formula was derived in [3; 5] for the mutual information of a multiantenna system, and it also appeared in earlier works on MIMO systems [7; 8] and matrix models for ISI channels [9; 10].

The MIMO capacity is achieved by maximizing the mutual information (10.7) over all input covariance matrices $\mathbf{R_x}$ satisfying the power constraint:

$$C = \max_{\mathbf{R_x}:\mathrm{Tr}(\mathbf{R_x})=\rho} B \log_2 \det[\mathbf{I}_{M_r} + \mathbf{HR_xH}^H], \tag{10.8}$$

where $\det[\mathbf{A}]$ denotes the determinant of the matrix $\mathbf{A}$. Clearly the optimization relative to $\mathbf{R_x}$ will depend on whether or not $\mathbf{H}$ is known at the transmitter. We now consider this optimization under different assumptions about transmitter CSI.

CHANNEL KNOWN AT TRANSMITTER: WATER-FILLING

The MIMO decomposition described in Section 10.2 allows a simple characterization of the MIMO channel capacity for a fixed channel matrix $\mathbf{H}$ known at the transmitter and receiver. Specifically, the capacity equals the sum of capacities on each of the independent parallel channels with the transmit power optimally allocated between these channels. This optimization of transmit power across the independent channels results from optimizing the input covariance matrix to maximize the capacity formula (10.8). Substituting the matrix SVD (10.2) into (10.8) and using properties of unitary matrices, we get the MIMO capacity with CSIT and CSIR as

$$C = \max_{\rho_i:\sum_i \rho_i \leq \rho} \sum_{i=1}^{R_{\mathbf{H}}} B \log_2(1 + \sigma_i^2 \rho_i), \tag{10.9}$$

where $R_{\mathbf{H}}$ is the number of nonzero singular values σ_i^2 of $\mathbf{H}$. Since the MIMO channel decomposes into $R_{\mathbf{H}}$ parallel channels, we say that it has $R_{\mathbf{H}}$ *degrees of freedom*. Since $\rho = P/\sigma^2$, the capacity (10.9) can also be expressed in terms of the power allocation P_i to the ith parallel channel as

$$C = \max_{P_i:\sum_i P_i \leq P} \sum_{i=1}^{R_{\mathbf{H}}} B \log_2\left(1 + \frac{\sigma_i^2 P_i}{\sigma^2}\right) = \max_{P_i:\sum_i P_i \leq P} \sum_{i=1}^{R_{\mathbf{H}}} B \log_2\left(1 + \frac{P_i \gamma_i}{P}\right), \qquad (10.10)$$

where $\gamma_i = \sigma_i^2 P/\sigma^2$ is the SNR associated with the ith channel at full power. This expression indicates that, at high SNRs, channel capacity increases linearly with the number of degrees of freedom in the channel. Conversely, at low SNRs, all power will be allocated to the parallel channel with the largest SNR (or, equivalently, the largest σ_i^2). The capacity formula (10.10) is similar to the case of flat fading (4.9) or frequency-selective fading (4.23). Solving the optimization leads to a water-filling power allocation for the MIMO channel:

$$\frac{P_i}{P} = \begin{cases} 1/\gamma_0 - 1/\gamma_i & \gamma_i \geq \gamma_0, \\ 0 & \gamma_i < \gamma_0, \end{cases} \qquad (10.11)$$

for some cutoff value γ_0. The resulting capacity is then

$$C = \sum_{i:\gamma_i \geq \gamma_0} B \log\left(\frac{\gamma_i}{\gamma_0}\right). \qquad (10.12)$$

EXAMPLE 10.2: Find the capacity and optimal power allocation for the MIMO channel given in Example 10.1, assuming $\rho = P/\sigma^2 = 10$ dB and $B = 1$ Hz.

Solution: From Example 10.1, the singular values of the channel are $\sigma_1 = 1.3333$, $\sigma_2 = 0.5129$, and $\sigma_3 = 0.0965$. Since $\gamma_i = 10\sigma_i^2$, this means we have $\gamma_1 = 17.7769$, $\gamma_2 = 2.6307$, and $\gamma_3 = .0931$. Assuming that power is allocated to all three parallel channels, the power constraint yields

$$\sum_{i=1}^{3}\left(\frac{1}{\gamma_0} - \frac{1}{\gamma_i}\right) = 1 \implies \frac{3}{\gamma_0} = 1 + \sum_{i=1}^{3} \frac{1}{\gamma_i} = 12.1749.$$

Solving for γ_0 yields $\gamma_0 = .2685$, which is inconsistent because $\gamma_3 = .0931 < \gamma_0 = .2685$. Thus, the third channel is not allocated any power. Then the power constraint yields

$$\sum_{i=1}^{2}\left(\frac{1}{\gamma_0} - \frac{1}{\gamma_i}\right) = 1 \implies \frac{2}{\gamma_0} = 1 + \sum_{i=1}^{2} \frac{1}{\gamma_i} = 1.4364.$$

Solving for γ_0 in this case yields $\gamma_0 = 1.392 < \gamma_2$, so this is the correct cutoff value. Then $P_i/P = 1/1.392 - 1/\gamma_i$, so $P_1/P = .662$ and $P_2/P = .338$. The capacity is given by $C = \log_2(\gamma_1/\gamma_0) + \log_2(\gamma_2/\gamma_0) = 4.59$.

Capacity under perfect CSIT and CSIR can also be defined on channels for which there is a single antenna at the transmitter and multiple receive antennas (single-input multiple-output, SIMO) or multiple transmit antennas and a single receive antenna (multiple-input single-output, MISO). These channels can obtain diversity and array gain from the multiple antennas, but no multiplexing gain. If both transmitter and receiver know the channel then the capacity equals that of an SISO channel with the signal transmitted or received over the multiple antennas coherently combined to maximize the channel SNR, as in MRC. This results in capacity $C = B \log_2(1 + \rho\|\mathbf{h}\|^2)$ where the channel matrix $\mathbf{H}$ is reduced to a vector $\mathbf{h}$ of channel gains, the optimal weight vector $\mathbf{c} = \mathbf{h}^H/\|\mathbf{h}\|$, and $\rho = P/\sigma^2$.

CHANNEL UNKNOWN AT TRANSMITTER: UNIFORM POWER ALLOCATION

Suppose now that the receiver knows the channel but the transmitter does not. Without channel information, the transmitter cannot optimize its power allocation or input covariance structure across antennas. If the distribution of $\mathbf{H}$ follows the ZMSW channel gain model, then there is no bias in terms of the mean or covariance of $\mathbf{H}$. Thus, it seems intuitive that the best strategy should be to allocate equal power to each transmit antenna, resulting in an input covariance matrix equal to the scaled identity matrix: $\mathbf{R_x} = (\rho/M_t)\mathbf{I}_{M_t}$. It is shown in [4] that, under these assumptions, this input covariance matrix indeed maximizes the mutual information of the channel. For an M_t-transmit M_r-receive antenna system, this yields mutual information given by

$$I(\mathbf{x}; \mathbf{y}) = B \log_2 \det\left[\mathbf{I}_{M_r} + \frac{\rho}{M_t}\mathbf{H}\mathbf{H}^H \right].$$

Using the SVD of $\mathbf{H}$, we can express this as

$$I(\mathbf{x}; \mathbf{y}) = \sum_{i=1}^{R_{\mathbf{H}}} B \log_2\left(1 + \frac{\gamma_i}{M_t} \right), \tag{10.13}$$

where $\gamma_i = \sigma_i^2 \rho = \sigma_i^2 P/\sigma^2$.

The mutual information of the MIMO channel (10.13) depends on the specific realization of the matrix $\mathbf{H}$, in particular its singular values $\{\sigma_i\}$. The average mutual information of a random matrix $\mathbf{H}$, averaged over the matrix distribution, depends on the probability distribution of the singular values of $\mathbf{H}$ [5; 11; 12]. In fading channels the transmitter can transmit at a rate equal to this average mutual information and ensure correct reception of the data, as discussed in the next section. But for a static channel, if the transmitter does not know the channel realization (or, more precisely, the channel's average mutual information) then it does not know at what rate to transmit such that the data will be received correctly. In this case the appropriate capacity definition is capacity with outage. In capacity with outage the transmitter fixes a transmission rate R, and the outage probability associated with R is the probability that the transmitted data will not be received correctly or, equivalently, the probability that the channel $\mathbf{H}$ has mutual information less than R. This probability is given by

$$P_{\text{out}} = p\left(\mathbf{H} : B \log_2 \det\left[\mathbf{I}_{M_r} + \frac{\rho}{M_t}\mathbf{H}\mathbf{H}^H \right] < R \right). \tag{10.14}$$

This probability is determined by the distribution of the eigenvalues of $\mathbf{H}\mathbf{H}^H$: these eigenvalues are the squares of the singular values of $\mathbf{H}$. The distribution of singular values for matrices is a well-studied problem and is known for common matrices associated with the MIMO channel [12, Sec. 2.1]

As the number of transmit and receive antennas grows large, random matrix theory provides a central limit theorem for the distribution of the singular values of $\mathbf{H}$ [13], resulting in a constant mutual information for all channel realizations. These results were applied to obtain MIMO channel capacity with uncorrelated fading in [14; 15; 16; 17] and with correlated fading in [18; 19; 20]. As an example of this limiting distribution, note that, for fixed M_r, under the ZMSW model the law of large numbers implies that

$$\lim_{M_t \to \infty} \frac{1}{M_t} \mathbf{H}\mathbf{H}^H = \mathbf{I}_{M_r}. \tag{10.15}$$

Substituting this into (10.13) yields that the mutual information in the asymptotic limit of large M_t becomes a constant equal to $C = M_r B \log_2(1 + \rho)$. Defining $M = \min(M_t, M_r)$, this implies that as M grows large, the MIMO channel capacity in the absence of CSIT approaches $C = MB \log_2(1 + \rho)$ and hence grows linearly in M. Moreover, this linear growth of capacity with M in the asymptotic limit of large M is observed even for a small number of antennas [20]. Similarly, as SNR grows large, capacity also grows linearly with $M = \min(M_t, M_r)$ for any M_t and M_r [2]. Since the ZMSW MIMO channel has $R_{\mathbf{H}} = M = \min(M_t, M_r)$ we see that, for this channel, capacity in the absence of CSIT at high SNRs or with a large number of antennas increases linearly with the number of degrees of freedom in the channel. These results are the main reason for the widespread appeal of MIMO techniques: even if the channel realization is not known at the transmitter, the capacity of ZMSW MIMO channels still grows linearly with the minimum number of transmit and receiver antennas – as long as the channel can be accurately estimated at the receiver. Thus, MIMO channels can provide high data rates without requiring increased signal power or bandwidth. Note, however, that transmit antennas are not beneficial at very low SNRs: in this regime capacity scales only with the number of receive antennas, independent of the number of transmit antennas. The reason is that, at these low SNRs, the MIMO system is just trying to collect energy rather than trying to exploit all available degrees of freedom. Hence, for the ZMSW MIMO channel, energy can be spread over all transmit antennas or concentrated in just one (or a few) of these antennas in order to achieve capacity [4]. As SNR increases, power ceases to be a limiting factor and capacity becomes limited by the degrees of freedom in the channel.

Although lack of CSIT does not affect the growth rate of capacity relative to M, at least for a large number of antennas, it does complicate demodulation. Specifically, without CSIT the transmission scheme cannot convert the MIMO channel into noninterfering SISO channels. Recall that decoding complexity is exponential in the number of independent symbols transmitted over the multiple transmit antennas, and this number equals the rank of the input covariance matrix.

Our analysis here under perfect CSIR and no CSIT assumes that the channel gain matrix has a ZMSW distribution; that is, it has mean zero and covariance matrix equal to the identity matrix. When the channel has nonzero mean or a nonidentity covariance matrix, there is a spatial bias in the channel that should be exploited by the optimal transmission strategy, so equal power allocation across antennas is no longer optimal [21; 22; 23]. Results in [21; 24] indicate that, when the channel has a dominant mean or covariance direction, *beamforming* (described in Section 10.4) can be used to achieve channel capacity. This is a fortuitous situation, given the simplicity of beamforming.

10.3.2 Fading Channels

Suppose now that the channel gain matrix experiences flat fading, so the gains h_{ij} vary with time. As in the case of the static channel, the capacity depends on what is known about the channel matrix at the transmitter and receiver. With perfect CSIR and CSIT the transmitter can adapt to the channel fading, and its capacity equals the average over all channel matrix realizations with optimal power allocation. With CSIR and no CSIT, ergodic capacity and capacity with outage are used to characterize channel capacity. These different characterizations are described in more detail in the following sections.

CHANNEL KNOWN AT TRANSMITTER: WATER-FILLING

With CSIT and CSIR, the transmitter optimizes its transmission strategy for each fading channel realization, as in the case of a static channel. The capacity is then just the average of capacities associated with each channel realization, given by (10.8), with power optimally allocated. This average capacity is called the *ergodic* capacity of the channel. There are two possibilities for allocating power under ergodic capacity. A short-term power constraint assumes that the power associated with each channel realization must equal the average power constraint $\bar{P}$. In this case the ergodic capacity becomes

$$
\begin{aligned}
C &= \mathbf{E}_{\mathbf{H}} \left[\max_{\mathbf{R}_{\mathbf{x}} : \mathrm{Tr}(\mathbf{R}_{\mathbf{x}}) = \rho} B \log_2 \det[\mathbf{I}_{M_r} + \mathbf{H} \mathbf{R}_{\mathbf{x}} \mathbf{H}^H] \right] \\
&= \mathbf{E}_{\mathbf{H}} \left[\max_{P_i : \sum_i P_i \leq \bar{P}} \sum_i B \log_2 \left(1 + \frac{P_i \gamma_i}{\bar{P}} \right) \right],
\end{aligned}
\tag{10.16}
$$

where $\gamma_i = \sigma_i^2 \bar{P} / \sigma^2$. A less restrictive constraint is a long-term power constraint, where we can use different powers $P_{\mathbf{H}}$ for different channel realizations $\mathbf{H}$ subject to the average power constraint over all channel realizations $\mathbf{E}_{\mathbf{H}}[P_{\mathbf{H}}] \leq \bar{P}$. The ergodic capacity under this assumption is given by

$$
\begin{aligned}
C &= \max_{\rho_{\mathbf{H}} : \mathbf{E}_{\mathbf{H}}[\rho_{\mathbf{H}}] = \rho} \mathbf{E}_{\mathbf{H}} \left[\max_{\mathbf{R}_{\mathbf{x}} : \mathrm{Tr}(\mathbf{R}_{\mathbf{x}}) = \rho_{\mathbf{H}}} B \log_2 \det[\mathbf{I}_{M_r} + \mathbf{H} \mathbf{R}_{\mathbf{x}} \mathbf{H}^H] \right] \\
&= \max_{P_{\mathbf{H}} : \mathbf{E}_{\mathbf{H}}[P_{\mathbf{H}}] \leq \bar{P}} \mathbf{E}_{\mathbf{H}} \left[\max_{P_i : \sum_i P_i \leq P_{\mathbf{H}}} \sum_i B \log_2 \left(1 + \frac{P_i \gamma_i}{P_{\mathbf{H}}} \right) \right]
\end{aligned}
\tag{10.17}
$$

for $\gamma_i = \sigma_i^2 P_{\mathbf{H}} / \sigma^2$. The short-term power constraint gives rise to a water-filling in space across the antennas, whereas the long-term power constraint allows for a two-dimensional water-filling across both space and time; this is similar to the frequency–time water-filling associated with the capacity of a time-varying frequency-selective fading channel. The bracketed terms in the second lines of (10.16) and (10.17) are functions of the singular values $\{\sigma_i\}$ of the matrix $\mathbf{H}$. Thus the expectation with respect to $\mathbf{H}$ in these expressions will be based on the distribution of these singular values. This distribution for $\mathbf{H}$ a ZMSW matrix, as well as for other types of matrices, can be found in [12, Sec. 2.1].

CHANNEL UNKNOWN AT TRANSMITTER: ERGODIC CAPACITY
AND CAPACITY WITH OUTAGE

Consider now a time-varying channel with random matrix $\mathbf{H}$ that is known at the receiver but not the transmitter. The transmitter assumes a ZMSW distribution for $\mathbf{H}$. The two relevant capacity definitions in this case are ergodic capacity and capacity with outage. Ergodic capacity defines the maximum rate, averaged over all channel realizations, that can be transmitted over the channel for a transmission strategy based only on the distribution of $\mathbf{H}$. This leads to the transmitter optimization problem – that is, finding the optimum input covariance

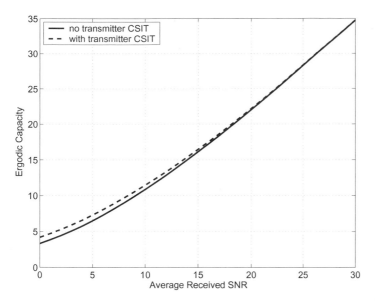

Figure 10.4: Ergodic capacity of 4×4 MIMO channel.

matrix to maximize ergodic capacity subject to the transmit power constraint. Mathematically, the problem is one of characterizing the optimum $\mathbf{R_x}$ to maximize

$$C = \max_{\mathbf{R_x}:\mathrm{Tr}(\mathbf{R_x})=\rho} \mathbf{E_H}[B \log_2 \det[\mathbf{I}_{M_r} + \mathbf{HR_xH}^H]]; \qquad (10.18)$$

here the expectation is with respect to the distribution on the channel matrix $\mathbf{H}$, which for the ZMSW model is i.i.d. zero-mean circularly symmetric and of unit variance.

As in the case of static channels, the optimum input covariance matrix that maximizes ergodic capacity for the ZMSW model is the scaled identity matrix $\mathbf{R_x} = (\rho/M_t)\mathbf{I}_{M_t}$; that is, the transmit power is divided equally among all the transmit antennas, and independent symbols are sent over the different antennas. Thus the ergodic capacity is given by

$$C = \mathbf{E_H}\left[B \log_2 \det\left[\mathbf{I}_{M_r} + \frac{\rho}{M_t}\mathbf{HH}^H\right]\right], \qquad (10.19)$$

where, as when CSIT is available, the expectation with respect to $\mathbf{H}$ is based on the distribution of its singular values. The capacity of the static channel grows as $M = \min(M_t, M_r)$ for M large, so this will also be true of the ergodic capacity because it simply averages the static channel capacity. Expressions for the growth rate constant can be found in [4; 25]. When the channel is not ZMSW, capacity depends on the distribution of the singular values for the random channel matrix; these distributions and the resulting ergodic capacity in this more general setting are studied in [11].

The ergodic capacity of a 4×4 MIMO system with i.i.d. complex Gaussian channel gains is shown in Figure 10.4. This figure shows capacity with both transmitter and receiver CSI and with receiver CSI only. There is little difference between the two plots and the difference decreases with SNR, which is also the case for an SISO channel. Comparing the capacity of this channel to that of the SISO fading channel shown in Figure 4.7, we see that the MIMO

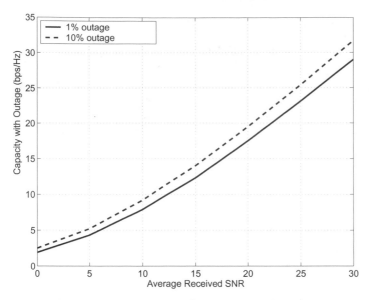

Figure 10.5: Capacity with outage of a 4×4 MIMO channel.

ergodic capacity is four times larger than the SISO ergodic capacity, which is just as expected since $\min(M_t, M_r) = 4$.

If the channel gain matrix is unknown at the transmitter and the entries are complex Gaussian but not i.i.d., then the channel mean or covariance matrix can be used at the transmitter to increase capacity. The basic idea is to allocate power according to the mean or covariance. This channel model is sometimes referred to as mean or covariance feedback. This model assumes perfect receiver CSI, and the impact of correlated fading depends on what is known at the transmitter: if the transmitter knows the channel realization or if it knows *neither* the channel realization nor the correlation structure, then antenna correlation decreases capacity relative to i.i.d. fading. However, if the transmitter knows the correlation structure then, for $M_r = 1$, capacity is increased relative to i.i.d. fading. Details on capacity under these different conditions can be found in [21; 24; 26].

Capacity with outage is defined similarly to the definition for static channels described in Section 10.3.1, although now capacity with outage applies to a slowly varying channel, where the channel matrix **H** is constant over a relatively long transmission time and then changes to a new value. As in the static channel case, the channel realization and corresponding channel capacity are not known at the transmitter, yet the transmitter must still fix a transmission rate in order to send data over the channel. For any choice of this rate R, there will be an outage probability associated with R that equals the probability that the transmitted data will not be received correctly. The outage probability is the same as in the static case, given by (10.14). The capacity with outage can sometimes be improved by not allocating power to one or more of the transmit antennas, especially when the outage probability is high [4]. This is because capacity with outage depends on the tail of the probability distribution. With fewer antennas, less averaging takes place and the spread of the tail increases.

The capacity with outage of a 4×4 MIMO system with i.i.d. complex Gaussian channel gains is shown in Figure 10.5 for outage of 1% and 10%. We see that the difference in

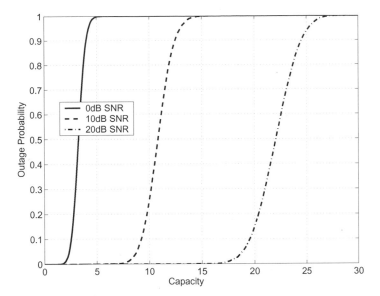

Figure 10.6: Outage probability distribution of a 4 × 4 MIMO channel.

capacity with outage for these two outage probabilities increases with SNR, which can be explained by the distribution curves for capacity shown in Figure 10.6. These curves show that at low SNRs the distribution is very steep, so that the capacity with outage at 1% is very close to that at 10% outage. At higher SNRs the curves become less steep, leading to more of a capacity difference at different outage probabilities.

NO CSI AT THE TRANSMITTER OR RECEIVER
When there is no CSI at either the transmitter or receiver, the linear growth in capacity as a function of the number of transmit and receive antennas disappears, and in some cases adding additional antennas provides negligible capacity gain. Moreover, channel capacity becomes heavily dependent on the underlying channel model, which makes it difficult to obtain generalizations about capacity growth. For an i.i.d. ZMSW block fading channel it is shown in [27] that increasing the number of transmit antennas by more than the duration of the block does not increase capacity. Thus, there is no data rate increase beyond a certain number of transmit antennas. However, when fading is correlated, additional transmit antennas do increase capacity [28]. The results of [27] were extended in [29] to explicitly characterize capacity and the capacity-achieving transmission strategy for the i.i.d. block fading model in the high-SNR regime. It was shown in [29] that, for i.i.d. block fading channels, capacity grows linearly with the number of channel degrees of freedom.

The capacity of more general fading channel models without transmitter or receiver CSI was investigated in [30; 31; 32]. These works indicate that there are three distinct regions associated with the capacity of slowly varying flat fading channels with no CSIT or CSIR. At low SNRs capacity is limited by noise and grows linearly with the number of channel degrees of freedom. At moderate to high SNRs capacity is limited by estimation error, and its growth is also linear in the number of channel degrees of freedom. At very high SNRs the degrees of freedom cease to matter and capacity grows double logarithmically with SNR. In

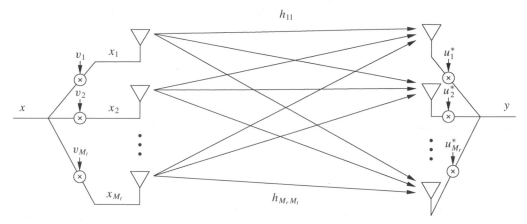

Figure 10.7: MIMO channel with beamforming.

other words, at very high SNRs there is no multiplexing gain associated with multiple antennas for slowly varying channels without transmitter or receiver CSI.

10.4 MIMO Diversity Gain: Beamforming

The multiple antennas at the transmitter and receiver can be used to obtain array and diversity gain (defined in Section 7.1) instead of capacity gain. In this setting the same symbol – weighted by a complex scale factor – is sent over each transmit antenna, so that the input covariance matrix has unit rank. This scheme is also referred to as *MIMO beamforming*.[4] A beamforming strategy corresponds to the precoding and shaping matrices (described in Section 10.2) being just column vectors: $\mathbf{V} = \mathbf{v}$ and $\mathbf{U} = \mathbf{u}$, as shown in Figure 10.7. As indicated in the figure, the transmit symbol x is sent over the ith antenna with weight v_i. On the receive side, the signal received on the ith antenna is weighted by u_i^*. Both transmit and receive weight vectors are normalized so that $\|\mathbf{u}\| = \|\mathbf{v}\| = 1$. The resulting received signal is given by

$$y = \mathbf{u}^H \mathbf{H} \mathbf{v} x + \mathbf{u}^H \mathbf{n}, \tag{10.20}$$

where if $\mathbf{n} = (n_1, \ldots, n_{M_r})$ has i.i.d. elements then the statistics of $\mathbf{u}^H \mathbf{n}$ are the same as the statistics for each of these elements.

Beamforming provides diversity and array gain via coherent combining of the multiple signal paths. Channel knowledge at the receiver is assumed, since this is required for coherent combining. The performance gain then depends on whether or not the channel is known at the transmitter. When the channel matrix $\mathbf{H}$ is known, the received SNR is optimized by choosing $\mathbf{u}$ and $\mathbf{v}$ as the principal left and right singular vectors of the channel matrix $\mathbf{H}$. That is, for $\sigma_1 = \sigma_{\max}$ the maximum singular value of $\mathbf{H}$, $\mathbf{u}$ and $\mathbf{v}$ are (respectively) the first columns of $\mathbf{U}$ and $\mathbf{V}$. The corresponding received SNR can be shown to equal $\gamma = \sigma_{\max}^2 \rho$, where $\sigma_{\max}$ is the largest singular value of $\mathbf{H}$ [6; 33]. The resulting capacity is $C = B \log_2(1 + \sigma_{\max}^2 \rho)$,

---

[4] Unfortunately, beamforming is also used in the smart antenna context of Section 10.8 to describe adjustment of the transmit or receive antenna directivity in a given direction.

corresponding to the capacity of a SISO channel with channel power gain σ_{max}^2. For **H** a ZMSW matrix, it can be shown [6, Chap. 5.4.4] that the array gain of beamforming diversity is between $\max(M_t, M_r)$ and $M_t M_r$ and that the diversity gain is $M_t M_r$.

When the channel is not known to the transmitter, for $M_t = 2$ the Alamouti scheme described in Section 7.3.2 can be used to extract an array gain of M_r and the maximum diversity gain of $2M_r$ [6, Chap. 5.4.3]. For $M_t > 2$, full diversity gain can also be obtained using space-time block codes, as described in Section 10.6.3. Although beamforming has a reduced capacity relative to the capacity-achieving transmit precoding and receiver shaping matrices, the demodulation complexity with beamforming is on the order of $|\mathcal{X}|$ instead of $|\mathcal{X}|^{R_H}$. An even simpler strategy is to use MRC at either the transmitter or receiver and antenna selection on the other end: this was analyzed in [34].

EXAMPLE 10.3: Consider a MIMO channel with gain matrix

$$\mathbf{H} = \begin{bmatrix} .7 & .9 & .8 \\ .3 & .8 & .2 \\ .1 & .3 & .9 \end{bmatrix}.$$

Find the capacity of this channel under beamforming, given channel knowledge at the transmitter and receiver, $B = 100$ kHz, and $\rho = 10$ dB.

Solution: The largest singular value of **H** is $\sigma_{max} = \sqrt{\lambda_{max}}$, where λ_{max} is the maximum eigenvalue of

$$\mathbf{HH}^H = \begin{bmatrix} 1.94 & 1.09 & 1.06 \\ 1.09 & .77 & .45 \\ 1.06 & .45 & .91 \end{bmatrix}.$$

The largest eigenvalue of this matrix is $\lambda_{max} = 3.17$. Thus, $C = B\log_2(1 + \lambda_{max}\rho) = 10^5 \log_2(1 + 31.7) = 503$ kbps.

10.5 Diversity–Multiplexing Trade-offs

The previous sections suggest two mechanisms for utilizing multiple antennas to improve wireless system performance. One option is to obtain capacity gain by decomposing the MIMO channel into parallel channels and multiplexing different data streams onto these channels. This capacity gain is also referred to as a *multiplexing gain*. However, the SNR associated with each of these channels depends on the singular values of the channel matrix. In capacity analysis this is taken into account by assigning a relatively low rate to these channels. However, practical signaling strategies for these channels will typically have poor performance unless powerful channel coding techniques are employed. Alternatively, beamforming can be used, where the channel gains are coherently combined to obtain a very robust channel with high diversity gain. It is not necessary to use the antennas purely for multiplexing or diversity. Some of the space-time dimensions can be used for diversity gain and the remaining dimensions used for multiplexing gain. This gives rise to a fundamental design question in MIMO systems: Should the antennas be used for diversity gain, multiplexing gain, or both?

The diversity–multiplexing trade-off or, more generally, the trade-off between data rate, probability of error, and complexity for MIMO systems has been extensively studied in the

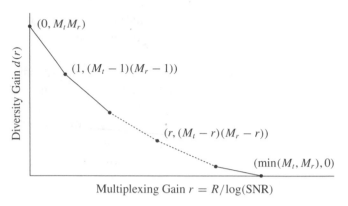

Figure 10.8: Diversity–multiplexing trade-off for high-SNR block fading.

literature – both from a theoretical perspective and in terms of practical space-time code designs [35; 36; 37; 38]. This work has primarily focused on block fading channels with receiver CSI only, since when both transmitter and receiver know the channel the trade-off is relatively straightforward: antenna subsets can first be grouped for diversity gain, and then the multiplexing gain corresponds to the new channel with reduced dimension due to the grouping. For the block fading model with receiver CSI only, as the blocklength grows asymptotically large, full diversity gain and full multiplexing gain (in terms of capacity with outage) can be obtained simultaneously with reasonable complexity by encoding diagonally across antennas [2; 39; 40]. An example of this type of encoding is D-BLAST, described in Section 10.6.4. For finite blocklengths it is not possible to achieve full diversity and full multiplexing gain simultaneously, in which case there is a trade-off between these gains. A simple characterization of this trade-off is given in [38] for block fading channels in the limit of asymptotically high SNR. In this analysis a transmission scheme is said to achieve multiplexing gain r and diversity gain d if the data rate (bps) per unit hertz, $R(\text{SNR})$, and probability of error, $P_e(\text{SNR})$, as functions of SNR satisfy

$$\lim_{\text{SNR}\to\infty} \frac{R(\text{SNR})}{\log_2 \text{SNR}} = r \tag{10.21}$$

and

$$\lim_{\text{SNR}\to\infty} \frac{\log P_e(\text{SNR})}{\log \text{SNR}} = -d, \tag{10.22}$$

where the log in (10.22) can be in any base.[5] For each r, the optimal diversity gain $d_{\text{opt}}(r)$ is the maximum diversity gain that can be achieved by any scheme. It is shown in [38] that if the fading blocklength $T \geq M_t + M_r - 1$ then

$$d_{\text{opt}}(r) = (M_t - r)(M_r - r), \quad 0 \leq r \leq \min(M_t, M_r). \tag{10.23}$$

The function (10.23) is plotted in Figure 10.8. This figure implies that (i) if we use all transmit *and* receive antennas for diversity then we get full diversity gain $M_t M_r$ and (ii) we can use some of these antennas to increase data rate at the expense of diversity gain.

---

[5] The base of the log cancels out of the expression because (10.22) is the ratio of two logs with the same base.

It is also possible to adapt the diversity and multiplexing gains relative to channel conditions. Specifically, in poor channel states more antennas can be used for diversity gain, whereas in good states more antennas can be used for multiplexing. Adaptive techniques that change antenna use to trade off diversity and multiplexing based on channel conditions have been investigated in [41; 42; 43].

EXAMPLE 10.4: Let the multiplexing and diversity parameters r and d be as defined in (10.21) and (10.22). Suppose that r and d approximately satisfy the diversity–multiplexing trade-off $d_{opt}(r) = (M_t - r)(M_r - r)$ at any large finite SNR. For an $M_t = M_r = 8$ MIMO system with an SNR of 15 dB, if we require a data rate per unit hertz of $R = 15$ bps then what is the maximum diversity gain the system can provide?

Solution: With SNR $= 15$ dB, to get $R = 15$ bps we require $r \log_2(10^{1.5}) = 15$, which implies that $r = 3.01$. Therefore, three of the antennas are used for multiplexing and the remaining five are used for diversity. The maximum diversity gain is then $d_{opt}(r) = (M_t - r)(M_r - r) = (8 - 3)(8 - 3) = 25$.

10.6 Space-Time Modulation and Coding

Because a MIMO channel has input–output relationship $\mathbf{y} = \mathbf{H}\mathbf{x} + \mathbf{n}$, the symbol transmitted over the channel each symbol time is a vector rather than a scalar, as in traditional modulation for the SISO channel. Moreover, when the signal design extends over both space (via the multiple antennas) and time (via multiple symbol times), it is typically referred to as a *space-time code*.

Most space-time codes – including all codes discussed in this section – are designed for quasi-static channels, where the channel is constant over a block of T symbol times and the channel is assumed unknown at the transmitter. Under this model, the channel input and output become matrices with dimensions corresponding to space (antennas) and time. Let $\mathbf{X} = [\mathbf{x}_1, \ldots, \mathbf{x}_T]$ denote the $M_t \times T$ channel input matrix with ith column $\mathbf{x}_i$ equal to the vector channel input over the ith transmission time. Let $\mathbf{Y} = [\mathbf{y}_1, \ldots, \mathbf{y}_T]$ denote the $M_r \times T$ channel output matrix with ith column $\mathbf{y}_i$ equal to the vector channel output over the ith transmission time, and let $\mathbf{N} = [\mathbf{n}_1, \ldots, \mathbf{n}_T]$ denote the $M_r \times T$ noise matrix with ith column $\mathbf{n}_i$ equal to the receiver noise vector over the ith transmission time. With this matrix representation, the input–output relationship over all T blocks becomes

$$\mathbf{Y} = \mathbf{H}\mathbf{X} + \mathbf{N}. \tag{10.24}$$

10.6.1 ML Detection and Pairwise Error Probability

Assume a space-time code where the receiver has knowledge of the channel matrix $\mathbf{H}$. Under maximum likelihood detection it can be shown using similar techniques as in the scalar (Chapter 5) or vector (Chapter 8) case that, given received matrix $\mathbf{Y}$, the ML transmit matrix $\hat{\mathbf{X}}$ satisfies

$$\hat{\mathbf{X}} = \arg \min_{\mathbf{X} \in \mathcal{X}^{M_t \times T}} \|\mathbf{Y} - \mathbf{H}\mathbf{X}\|_F^2 = \arg \min_{\mathbf{X} \in \mathcal{X}^{M_t \times T}} \sum_{i=1}^{T} \|\mathbf{y}_i - \mathbf{H}\mathbf{x}_i\|^2, \tag{10.25}$$

where $\|\mathbf{A}\|_F$ denotes the Frobenius norm[6] of the matrix $\mathbf{A}$ and the minimization is taken over all possible space-time input matrices $\mathcal{X}^{M_t \times T}$. The pairwise error probability for mistaking a transmit matrix $\mathbf{X}$ for another matrix $\hat{\mathbf{X}}$, denoted as $p(\hat{\mathbf{X}} \to \mathbf{X})$, depends only on the distance between the two matrices after transmission through the channel and the noise power σ^2; that is,

$$p(\hat{\mathbf{X}} \to \mathbf{X}) = Q\left(\sqrt{\frac{\|\mathbf{H}(\mathbf{X} - \hat{\mathbf{X}})\|_F^2}{2\sigma^2}}\right). \tag{10.26}$$

Let $\mathbf{D_X} = \mathbf{X} - \hat{\mathbf{X}}$ denote the difference matrix between $\mathbf{X}$ and $\hat{\mathbf{X}}$. Applying the Chernoff bound to (10.26), we have

$$p(\hat{\mathbf{X}} \to \mathbf{X}) \le \exp\left[-\frac{\|\mathbf{H}\mathbf{D_X}\|_F^2}{4\sigma^2}\right]. \tag{10.27}$$

Let $\mathbf{h}_i$ denote the ith row of $\mathbf{H}$, $i = 1, \ldots, M_r$. Then

$$\|\mathbf{H}\mathbf{D_X}\|_F^2 = \sum_{i=1}^{M_r} \mathbf{h}_i \mathbf{D_X}\mathbf{D_X}^H \mathbf{h}_i^H. \tag{10.28}$$

Let $\mathcal{H} = \text{vec}(\mathbf{H}^T)^T$, where $\text{vec}(\mathbf{A})$ is defined as the vector that results from stacking the columns of matrix $\mathbf{A}$ on top of each other to form a vector.[7] Hence $\mathcal{H}^T$ is a vector of length $M_r M_t$. Also define $\mathcal{D}_{\mathbf{X}} = \mathbf{I}_{M_r} \otimes \mathbf{D_X}$, where $\otimes$ denotes the Kronecker product. With these definitions,

$$\|\mathbf{H}\mathbf{D_X}\|_F^2 = \|\mathcal{H}\mathcal{D}_{\mathbf{X}}\|_F^2. \tag{10.29}$$

Substituting (10.29) into (10.27) and taking the expectation relative to all possible channel realizations yields

$$p(\mathbf{X} \to \hat{\mathbf{X}}) \le \left(\det\left[\mathbf{I}_{M_t M_r} + \frac{1}{4\sigma^2}\mathbf{E}[\mathcal{D}_{\mathbf{X}}^H \mathcal{H}^H \mathcal{H} \mathcal{D}_{\mathbf{X}}]\right]\right)^{-1}. \tag{10.30}$$

Suppose that the channel matrix $\mathbf{H}$ is random and spatially white, so that its entries are i.i.d. zero-mean, unit-variance, complex Gaussian random variables. Then taking the expectation in (10.30) yields

$$p(\mathbf{X} \to \hat{\mathbf{X}}) \le \left(\frac{1}{\det[\mathbf{I}_{M_t} + .25\rho\boldsymbol{\Delta}]}\right)^{M_r}, \tag{10.31}$$

where $\boldsymbol{\Delta} = (1/P)\mathbf{D_X}\mathbf{D_X}^H$. This simplifies to

$$p(\mathbf{X} \to \hat{\mathbf{X}}) \le \prod_{k=1}^{R_{\boldsymbol{\Delta}}} \left(\frac{1}{1 + .25\gamma\lambda_k(\boldsymbol{\Delta})}\right)^{M_r}; \tag{10.32}$$

here $\gamma = P/\sigma^2 = \rho$ is the SNR per input symbol and $\lambda_k(\boldsymbol{\Delta})$ is the kth nonzero eigenvalue of $\boldsymbol{\Delta}$, $k = 1, \ldots, R_{\boldsymbol{\Delta}}$, where $R_{\boldsymbol{\Delta}}$ is the rank of $\boldsymbol{\Delta}$. In the high-SNR regime (i.e., for $\gamma \gg 1$) this simplifies to

---

[6] The Frobenius norm of a matrix is the square root of the sum of the square of its elements.

[7] So for the $M \times N$ matrix $\mathbf{A} = [\mathbf{a}_1, \ldots, \mathbf{a}_N]$, where $\mathbf{a}_i$ is a vector of length M, $\text{vec}(\mathbf{A}) = [\mathbf{a}_1^T, \ldots, \mathbf{a}_N^T]^T$ is a vector of length MN.

$$p(\mathbf{X} \rightarrow \hat{\mathbf{X}}) \leq \left(\prod_{k=1}^{R_\Delta} \lambda_k(\mathbf{\Delta}) \right)^{-M_r} (.25\gamma)^{-R_\Delta M_r}. \tag{10.33}$$

This equation gives rise to the main criteria for design of space-time codes, described in the next section.

10.6.2 Rank and Determinant Criteria

The pairwise error probability in (10.33) indicates that the probability of error decreases as γ^{-d} for $d = R_\Delta M_r$. Thus, $R_\Delta M_r$ is the diversity gain of the space-time code. The maximum diversity gain possible through coherent combining of M_t transmit and M_r receive antennas is $M_t M_r$. Thus, to obtain this maximum diversity gain, the space-time code must be designed such that the $M_t \times M_t$ difference matrix $\mathbf{\Delta}$ between any two codewords has full rank equal to M_t. This design criterion is referred to as the *rank criterion*.

The coding gain associated with the pairwise error probability in (10.33) depends on the term $\left(\prod_{k=1}^{R_\Delta} \lambda_k(\mathbf{\Delta}) \right)^{M_r}$. Thus, a high coding gain is achieved by maximizing the minimum of the determinant of $\mathbf{\Delta}$ over all input matrix pairs $\mathbf{X}$ and $\hat{\mathbf{X}}$. This criterion is referred to as the *determinant criterion*.

The rank and determinant criteria were first developed in [36; 44; 45]. These criteria are based on the pairwise error probability associated with different transmit signal matrices – rather than the binary domain of traditional codes – and hence often require computer searches to find good codes [46; 47]. A general binary rank criterion was developed in [48] to provide a better construction method for space-time codes.

10.6.3 Space-Time Trellis and Block Codes

The rank and determinant criteria have been primarily applied to the design of space-time trellis codes (STTCs), which are an extension of conventional trellis codes to MIMO systems [6; 45]. They are described using a trellis and are decoded using ML sequence estimation via the Viterbi algorithm. STTCs can extract excellent diversity and coding gain, but the complexity of decoding increases exponentially with the diversity level and transmission rate [49]. Space-time block codes (STBCs) are an alternative space-time code that can also extract excellent diversity gain with linear receiver complexity. Interest in STBCs was initiated by the Alamouti code described in Section 7.3.2, which obtains full diversity order with linear receiver processing for a two-antenna transmit system. This scheme was generalized in [50] to STBCs that achieve full diversity order with an arbitrary number of transmit antennas. However, while these codes achieve full diversity order, they do not provide coding gain and thus have inferior performance to STTCs, which achieve both full diversity gain as well as coding gain. Added coding gain for STTCs as well as STBCs can be achieved by concatenating these codes either in serial or in parallel with an outer channel code to form a turbo code [51; 52; 53]. The linear complexity of the STBC designs in [50] result from making the codes orthogonal along each dimension of the code matrix. A similar design premise is used in [54] to design *unitary space-time modulation* schemes for block fading channels when neither the transmitter nor the receiver has CSI. More comprehensive treatments of space-time coding can be found in [6; 49; 55; 56] and the references therein.

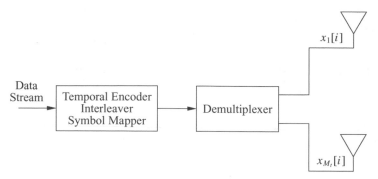

Figure 10.9: Spatial multiplexing with serial encoding.

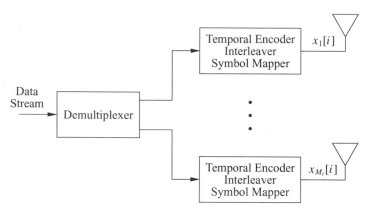

Figure 10.10: Spatial multiplexing with parallel encoding: V-BLAST.

10.6.4 Spatial Multiplexing and BLAST Architectures

The basic premise of spatial multiplexing is to send M_t independent symbols per symbol period using the dimensions of space and time. To obtain full diversity order, an encoded bit stream must be transmitted over all M_t transmit antennas. This can be done through a serial encoding, illustrated in Figure 10.9. With serial encoding the bit stream is temporally encoded over the channel blocklength T to form the codeword $[x_1, \ldots, x_T]$. The codeword is interleaved and mapped to a constellation point, then demultiplexed onto the different antennas. The first M_t symbols are transmitted from the M_t antennas over the first symbol time, the next M_t symbols are transmitted from the antennas over the next symbol time, and this process continues until the entire codeword has been transmitted. We denote the symbol sent over the kth antenna at time i as $x_k[i]$. If a codeword is sufficiently long, it is transmitted over all M_t transmit antennas and received by all M_r receive antennas, resulting in full diversity gain. However, the codeword length T required to achieve this full diversity is $M_t M_r$, and decoding complexity grows exponentially with this codeword length. This high level of complexity makes serial encoding impractical.

A simpler method to achieve spatial multiplexing, pioneered at Bell Laboratories as one of the Bell Labs Layered Space Time (BLAST) architectures for MIMO channels [2], is parallel encoding, illustrated in Figure 10.10. With parallel encoding the data stream is demultiplexed into M_t independent streams. Each of the resulting substreams is passed through

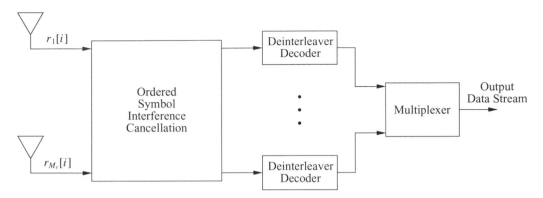

Figure 10.11: V-BLAST receiver with linear complexity.

an SISO temporal encoder with blocklength T, interleaved, mapped to a signal constellation point, and transmitted over its corresponding transmit antenna. Specifically, the kth SISO encoder generates the codeword $\{x_k[i], i = 1, \ldots, T\}$, which is transmitted sequentially over the kth antenna. This process can be considered to be the encoding of the serial data into a vertical vector and hence is also referred to as *vertical encoding* or V-BLAST [57]. Vertical encoding can achieve at most a diversity order of M_r, since each coded symbol is transmitted from one antenna and received by M_r antennas. This system has a simple encoding complexity that is linear in the number of antennas. However, optimal decoding still requires joint detection of the codewords from each of the transmit antennas, since all transmitted symbols are received by all the receive antennas. It was shown in [58] that the receiver complexity can be significantly reduced through the use of *symbol interference cancellation,* as shown in Figure 10.11. This cancellation, which exploits the synchronicity of the symbols transmitted from each antenna, works as follows. First the M_t transmitted symbols are ordered in terms of their received SNR. An estimate of the received symbol with the highest SNR is made while treating all other symbols as noise. This estimated symbol is subtracted out, and the symbol with the next highest SNR estimated while treating the remaining symbols as noise. This process repeats until all M_t transmitted symbols have been estimated. After cancelling out interfering symbols, the coded substream associated with each transmit antenna can be individually decoded, resulting in a receiver complexity that is linear in the number of transmit antennas. In fact, coding is not even needed with this architecture, and data rates of 20–40 bps/Hz with reasonable error rates were reported in [57] using uncoded V-BLAST.

The simplicity of parallel encoding and the diversity benefits of serial encoding can be obtained by using a creative combination of the two techniques called *diagonal encoding* or D-BLAST [2], illustrated in Figure 10.12. In D-BLAST, the data stream is first parallel encoded. However, rather than transmitting each codeword with one antenna, the codeword symbols are rotated across antennas, so that a codeword is transmitted by all M_t antennas. The operation of the stream rotation is shown in Figure 10.13. Suppose the ith encoder generates the codeword $\mathbf{x}_i = [x_i[1], \ldots, x_i[T]]$. The stream rotator transmits each symbol on a different antenna, so $x_i[1]$ is sent on antenna 1 over symbol time i, $x_i[2]$ is sent on antenna 2 over symbol time $i + 1$, and so forth. If the code blocklength T exceeds M_t then the rotation

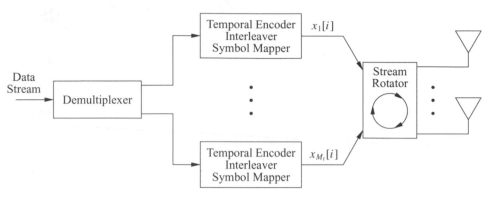

Figure 10.12: Diagonal encoding with stream rotation.

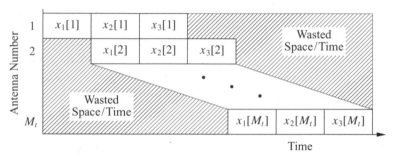

Figure 10.13: Stream rotation.

begins again on antenna 1. As a result, the codeword is spread across all spatial dimensions. Transmission schemes based on D-BLAST can achieve full $M_t M_r$ diversity gain if the temporal coding with stream rotation is capacity-achieving (Gaussian codewords of infinite blocklength T) [6, Chap. 6.3.5]. Moreover, the D-BLAST system can achieve maximum capacity with outage if the wasted space-time dimensions along the diagonals are neglected [6, Chap. 12.4.1]. Receiver complexity is also linear in the number of transmit antennas, since the receiver decodes each diagonal code independently. However, this simplicity comes at a price, as the efficiency loss of the wasted space-time dimensions (illustrated in Figure 10.13) can be large if the frame size is not appropriately chosen.

10.7 Frequency-Selective MIMO Channels

When the MIMO channel bandwidth is large relative to the channel's multipath delay spread, the channel suffers from intersymbol interference; this is similar to the case of SISO channels. There are two approaches to dealing with ISI in MIMO channels. A channel equalizer can be used to mitigate the effects of ISI. However, the equalizer is much more complex in MIMO channels because the channel must be equalized over both space and time. Moreover, when the equalizer is used in conjuction with a space-time code, the nonlinear and noncausal nature of the code further complicates the equalizer design. In some cases the structure of the code can be used to convert the MIMO equalization problem to a SISO problem for which well-established SISO equalizer designs can be used [59; 60; 61].

An alternative to equalization in frequency-selective fading is multicarrier modulation or orthogonal frequency division multiplexing (OFDM). OFDM techniques for SISO channels are described in Chapter 12: the main premise is to convert the wideband channel into a set of narrowband subchannels that exhibit only flat fading. Applying OFDM to MIMO channels results in a parallel set of narrowband MIMO channels, and the space-time modulation and coding techniques just described for a single MIMO channel are applied to this parallel set. MIMO frequency-selective fading channels exhibit diversity across space, time, and frequency, so ideally all three dimensions should be fully exploited in the signaling scheme.

10.8 Smart Antennas

We have seen that multiple antennas at the transmitter and/or receiver can provide diversity gain as well as increased data rates through space-time signal processing. Alternatively, sectorization or phased array techniques can be used to provide directional antenna gain at the transmit or receive antenna array. This directionality can increase the signaling range, reduce delay spread (ISI) and flat fading, and suppress interference between users. In particular, interference typically arrives at the receiver from different directions. Thus, directional antennas can exploit these differences to null or attenuate interference arriving from given directions, thereby increasing system capacity. The reflected multipath components of the transmitted signal also arrive at the receiver from different directions and can also be attenuated, thereby reducing ISI and flat fading. The benefits of directionality that can be obtained with multiple antennas must be weighed against their potential diversity or multiplexing benefits, giving rise to a multiplexing–diversity–directionality trade-off analysis. Whether it is best to use the multiple antennas to increase data rates through multiplexing, increase robustness to fading through diversity, or reduce ISI and interference through directionality is a complex decision that depends on the overall system design.

The most common directive antennas are *sectorized* or *phased* (directional) antenna arrays, and the gain patterns for these antennas – along with an omnidirectional antenna gain pattern – are shown in Figure 10.14. Sectorized antennas are designed to provide high gain across a range of signal arrival angles. Sectorization is commonly used at cellular system base stations to cut down on interference: assuming different sectors are assigned the same frequency band or timeslot, then with perfect sectorization only users within a sector interfere with each other, thereby reducing the average interference by a factor equal to the number of sectors. For example, Figure 10.14 shows a sectorized antenna with a 120° beamwidth. A base station could divide its 360° angular range into three sectors to be covered by three 120° sectorized antennas, in which case the interference in each sector is reduced by a factor of 3 relative to an omnidirectional base station antenna. The price paid for reduced interference in cellular systems via sectorization is the need for handoff between sectors.

Directional antennas typically use antenna arrays coupled with phased array techniques to provide directional gain, which can be tightly contolled with sufficiently many antenna elements. Phased array techniques work by adapting the phase of each antenna element in the array, which changes the angular locations of the antenna beams (angles with large gain) and nulls (angles with small gain). For an antenna array with N antennas, N nulls can be formed to significantly reduce the received power of N separate interferers. If there are $N_I < N$

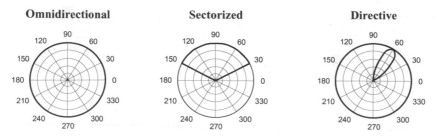

Figure 10.14: Antenna gains for omnidirectional, sectorized, and directive antennas.

interferers, then the N_I interferers can be nulled out using N_I antennas in a phased array, and the remaining $N - N_I$ antennas can be used for diversity gain. Note that directional antennas must know the angular location of the desired and interfering signals in order to provide high or low gains in the appropriate directions. Tracking of user locations can be a significant impediment in highly mobile systems, which is why cellular base stations use sectorized instead of directional antennas.

The complexity of antenna array processing – along with the real estate required for an antenna array – make the use of smart antennas in small, lightweight, low-power, handheld devices unlikely in the near future. However, base stations and access points already use antenna arrays in many cases. More details on the technology behind smart antennas and their use in wireless systems can be found in [62].

PROBLEMS

10-1. Matrix identities are commonly used in the analysis of MIMO channels. Prove the following matrix identities.

    (a) Given an $M \times N$ matrix $\mathbf{A}$, show that the matrix $\mathbf{AA}^H$ is Hermitian. What does this reveal about the eigendecomposition of $\mathbf{AA}^H$?

    (b) Show that $\mathbf{AA}^H$ is positive semidefinite.

    (c) Show that $\mathbf{I}_M + \mathbf{AA}^H$ is Hermitian positive definite.

    (d) Show that $\det[\mathbf{I}_M + \mathbf{AA}^H] = \det[\mathbf{I}_N + \mathbf{A}^H\mathbf{A}]$.

10-2. Find the SVD of the following matrix:

$$\mathbf{H} = \begin{bmatrix} .7 & .6 & .2 & .4 \\ .1 & .5 & .9 & .2 \\ .3 & .6 & .9 & .1 \end{bmatrix}.$$

10-3. Find a 3×3 channel matrix $\mathbf{H}$ with two nonzero singular values.

10-4. Consider the 4×4 MIMO channels given below. What is the maximum multiplexing gain of each – that is, how many independent scalar data streams can be reliably supported?

$$\mathbf{H}_1 = \begin{bmatrix} 1 & 1 & -1 & 1 \\ 1 & 1 & -1 & -1 \\ 1 & 1 & 1 & 1 \\ 1 & 1 & 1 & -1 \end{bmatrix}, \qquad \mathbf{H}_2 = \begin{bmatrix} 1 & 1 & 1 & -1 \\ 1 & 1 & -1 & 1 \\ 1 & -1 & 1 & 1 \\ 1 & -1 & -1 & -1 \end{bmatrix}.$$

10-5. The capacity of a static MIMO channel with only receiver CSI is given by $C = \sum_{i=1}^{R_{\mathbf{H}}} B \log_2(1 + \sigma_i^2 \rho / M_t)$. Show that, if the sum of singular values is bounded, then this expression is maximized when all $R_{\mathbf{H}}$ singular values are equal.

10-6. Consider a MIMO system with the following channel matrix:

$$\mathbf{H} = \begin{bmatrix} .1 & .3 & .4 \\ .3 & .2 & .2 \\ .1 & .3 & .7 \end{bmatrix} = \begin{bmatrix} -.5196 & -.0252 & -.8541 \\ -.3460 & -.9077 & .2372 \\ -.7812 & .4188 & .4629 \end{bmatrix} \begin{bmatrix} .9719 & 0 & 0 \\ 0 & .2619 & 0 \\ 0 & 0 & .0825 \end{bmatrix}$$

$$\times \begin{bmatrix} -.2406 & -.4727 & -.8477 \\ -.8894 & -.2423 & .3876 \\ .3886 & -.8472 & .3621 \end{bmatrix}^H .$$

Note that $\mathbf{H}$ is written in terms of its singular value decomposition (SVD) $\mathbf{H} = \mathbf{U}\mathbf{\Sigma}\mathbf{V}^H$.

   (a) Check if $\mathbf{H} = \mathbf{U}\mathbf{\Sigma}\mathbf{V}^H$. You will see that the matrices $\mathbf{U}$, $\mathbf{\Sigma}$, and $\mathbf{V}^H$ do not have sufficiently large precision, so that $\mathbf{U}\mathbf{\Sigma}\mathbf{V}^H$ is only approximately equal to $\mathbf{H}$. This indicates the sensitivity of the SVD – in particular, the matrix $\mathbf{\Sigma}$ – to small errors in the estimate of the channel matrix $\mathbf{H}$.
   (b) Based on the singular value decomposition $\mathbf{H} = \mathbf{U}\mathbf{\Sigma}\mathbf{V}^H$, find an equivalent MIMO system consisting of three independent channels. Find the transmit precoding and the receiver shaping matrices necessary to transform the original system into the equivalent system.
   (c) Find the optimal power allocation P_i ($i = 1, 2, 3$) across the three channels found in part (b), and find the corresponding total capacity of the equivalent system assuming $P/\sigma^2 = 20$ dB and a system bandwidth of $B = 100$ kHz.
   (d) Compare the capacity in part (c) to that when the channel is unknown at the transmitter and so equal power is allocated to each antenna.

10-7. Use properties of the SVD to show that, for a MIMO channel that is known to the transmitter and receiver both, the general capacity expression

$$C = \max_{\mathbf{R_x}:\mathrm{Tr}(\mathbf{R_x})=\rho} B \log_2 \det[\mathbf{I}_{M_r} + \mathbf{H}\mathbf{R_x}\mathbf{H}^H]$$

reduces to

$$C = \max_{\rho_i:\sum_i \rho_i \leq \rho} \sum_i B \log_2(1 + \sigma_i^2 \rho_i)$$

for singular values $\{\sigma_i\}$ and SNR ρ.

10-8. For the 4×4 MIMO channels given below, find their capacity assuming both transmitter and receiver know the channel, whose SNR $\rho = 10$ dB and bandwidth $B = 10$ MHz:

$$\mathbf{H}_1 = \begin{bmatrix} 1 & 1 & -1 & 1 \\ 1 & 1 & -1 & -1 \\ 1 & 1 & 1 & 1 \\ 1 & 1 & 1 & -1 \end{bmatrix}, \quad \mathbf{H}_2 = \begin{bmatrix} 1 & 1 & 1 & -1 \\ 1 & 1 & -1 & 1 \\ 1 & -1 & 1 & 1 \\ 1 & -1 & -1 & -1 \end{bmatrix}.$$

10-9. Assume a ZMCSCG MIMO system with channel matrix $\mathbf{H}$ corresponding to $M_t = M_r = M$ transmit and receive antennas. Using the law of large numbers, show that

$$\lim_{M \to \infty} \frac{1}{M}\mathbf{H}\mathbf{H}^H = \mathbf{I}_M.$$

Then use this to show that

$$\lim_{M \to \infty} B \log_2 \det\left[\mathbf{I}_M + \frac{\rho}{M} \mathbf{H} \mathbf{H}^H \right] = MB \log_2(1 + \rho).$$

10-10. Plot the ergodic capacities for a ZMCSCG MIMO channel with SNR $0 \leq \rho \leq 30$ dB and $B = 1$ MHz for the following MIMO dimensions: (a) $M_t = M_r = 1$; (b) $M_t = 2$, $M_r = 1$; (c) $M_t = M_r = 2$; (d) $M_t = 2$, $M_r = 3$; (e) $M_t = M_r = 3$. Verify that, at high SNRs, capacity grows linearly as $M = \min(M_t, M_r)$.

10-11. Plot the outage capacities for $B = 1$ MHz and an outage probability $P_{\text{out}} = .01$ for a ZMCSCG MIMO channel with SNR $0 \leq \rho \leq 30$ dB for the following MIMO dimensions: (a) $M_t = M_r = 1$; (b) $M_t = 2$, $M_r = 1$; (c) $M_t = M_r = 2$; (d) $M_t = 2$, $M_r = 3$; (e) $M_t = M_r = 3$. Verify that, at high SNRs, capacity grows linearly as $M = \min(M_t, M_r)$.

10-12. Show that if the noise vector $\mathbf{n} = (n_1, \ldots, n_{M_r})$ has i.i.d. elements then, for $\|\mathbf{u}\| = 1$, the statistics of $\mathbf{u}^H \mathbf{n}$ are the same as the statistics for each of these elements.

10-13. Consider a MIMO system where the channel gain matrix $\mathbf{H}$ is known at the transmitter and receiver. Show that if transmit and receive antennas are used for diversity then the optimal weights at the transmitter and receiver lead to an SNR of $\gamma = \lambda_{\max} \rho$, where $\lambda_{\max}$ is the largest eigenvalue of $\mathbf{H} \mathbf{H}^H$.

10-14. Consider a channel with channel gain matrix

$$\mathbf{H} = \begin{bmatrix} .1 & .5 & .9 \\ .3 & .2 & .6 \\ .1 & .3 & .7 \end{bmatrix}.$$

Assuming $\rho = 10$ dB, find the output SNR when beamforming is used on the channel with equal weights on each transmit antenna and optimal weighting at the receiver. Compare with the SNR under beamforming with optimal weights at both the transmitter and receiver.

10-15. Consider an 8×4 MIMO system, and assume a coding scheme that can achieve the rate–diversity trade-off $d(r) = (M_t - r)(M_r - r)$.

(a) Find the maximum multiplexing rate for this channel, given a required $P_e = \rho^{-d} \leq 10^{-3}$ and assuming that $\rho = 10$ dB.

(b) Given the r derived in part (a), what is the resulting P_e?

10-16. Find the capacity of a SIMO channel with channel gain vector $\mathbf{h} = [.1 \ .4 \ .75 \ .9]$, optimal receiver weighting, $\rho = 10$ dB, and $B = 10$ MHz.

10-17. Consider a 2×2 MIMO system with channel gain matrix $\mathbf{H}$ given by

$$\mathbf{H} = \begin{bmatrix} .3 & .5 \\ .7 & .2 \end{bmatrix}.$$

Assume that $\mathbf{H}$ is known at both transmitter and receiver and that there is a total transmit power of $P = 10$ mW across the two transmit antennas, AWGN with $N_0 = 10^{-9}$ W/Hz at each receive antenna, and bandwidth $B = 100$ kHz.

(a) Find the SVD for $\mathbf{H}$.

(b) Find the capacity of this channel.

(c) Assume that transmit precoding and receiver shaping have been used to transform this channel into two parallel independent channels with a total power constraint P. Find the maximum data rate that can be transmitted over this parallel set assuming MQAM modulation on each channel, with optimal power adaptation across the channels subject to power constraint P. Assume a target BER of 10^{-3} on each channel and that the BER is bounded: $P_b \leq .2e^{-1.5\gamma/(M-1)}$; assume also that the constellation size of the MQAM is unrestricted.

(d) Suppose now that the antennas at the transmitter and receiver are all used for diversity (with optimal weighting at the transmitter and receiver) to maximize the SNR of the combiner output. Find the SNR of the combiner output as well as the BER of a BPSK modulated signal transmitted over this diversity system. Compare the data rate and BER of this BPSK signaling with diversity (assuming $B = 1/T_b$) to the rate and BER from part (c).

(e) Comment on the diversity–multiplexing trade-offs between the systems in parts (c) and (d).

10-18. Consider an $M \times M$ MIMO channel with ZMCSCG channel gains.

(a) Plot the ergodic capacity of this channel for $M = 1$ and $M = 4$ with $0 \leq \rho \leq 20$ dB and $B = 1$ MHz, assuming that both transmitter and receiver have CSI.

(b) Repeat part (a) assuming that only the receiver has CSI.

10-19. Find the outage capacity for a 4×4 MIMO channel with ZMCSCG elements at 10% outage for $\rho = 10$ dB and $B = 1$ MHz.

10-20. Plot the cdf of capacity for an $M \times M$ MIMO channel with $\rho = 10$ dB and $B = 1$ MHz, assuming no transmitter knowledge for $M = 4, 6, 8$. What happens as M increases? What are the implications of this behavior for a practical system design?

REFERENCES

[1] J. Winters, "On the capacity of radio communication systems with diversity in a Rayleigh fading environment," *IEEE J. Sel. Areas Commun.*, pp. 871–8, June 1987.

[2] G. J. Foschini, "Layered space-time architecture for wireless communication in fading environments when using multi-element antennas," *Bell System Tech. J.*, pp. 41–59, Autumn 1996.

[3] G. J. Foschini and M. Gans, "On limits of wireless communications in a fading environment when using multiple antennas," *Wireless Pers. Commun.*, pp. 311–35, March 1998.

[4] E. Telatar, "Capacity of multi-antenna Gaussian channels," AT&T Bell Labs Internal Tech. Memo, June 1995.

[5] E. Telatar, "Capacity of multi-antenna Gaussian channels," *Euro. Trans. Telecommun.*, pp. 585–96, November 1999.

[6] A. Paulraj, R. Nabar, and D. Gore, *Introduction to Space-Time Wireless Communications*, Cambridge University Press, 2003.

[7] L. H. Brandenburg and A. D. Wyner, "Capacity of the Gaussian channel with memory: The multivariate case," *Bell System Tech. J.*, pp. 745–78, May/June 1974.

[8] J. Salz and A. D. Wyner, "On data transmission over cross coupled multi-input, multi-output linear channels with applications to mobile radio," AT&T Bell Labs Internal Tech. Memo, 1990.

[9] B. Tsybakov, "The capacity of a memoryless Gaussian vector channel," *Prob. Inform. Trans.*, 1(1), pp. 18–29, 1965.

[10] J. L. Holsinger, "Digital communication over fixed time-continuous channels with memory, with special application to telephone channels," MIT Res. Lab Elec. Tech. Rep. 430, 1964.

[11] H. Shin and J. H. Lee, "Capacity of multiple-antenna fading channels: Spatial fading correlation, double scattering, and keyhole," *IEEE Trans. Inform. Theory,* pp. 2636–47, October 2003.

[12] A. M. Tulino and S. Verdú, "Random matrix theory and wireless communications," *Found. Trends Commun. Inform. Theory,* 1(1), pp. 1–182, 2004.

[13] V. L. Girko, "A refinement of the central limit theorem for random determinants," *Theory Probab. Appl.,* 42(1), pp. 121–9, 1998.

[14] A. Grant, "Rayleigh fading multiple-antenna channels," *J. Appl. Signal Proc.,* Special Issue on Space-Time Coding (Part I), pp. 316–29, March 2002.

[15] P. J. Smith and M. Shafi, "On a Gaussian approximation to the capacity of wireless MIMO systems," *Proc. IEEE Internat. Conf. Commun.,* pp. 406–10, April 2002.

[16] S. Verdú and S. Shamai (Shitz), "Spectral efficiency of CDMA with random spreading," *IEEE Trans. Inform. Theory,* pp. 622–40, March 1999.

[17] Z. Wang and G. B. Giannakis, "Outage mutual information of space-time MIMO channels," *Proc. Allerton Conf. Commun., Control, Comput.,* pp. 885–94, October 2002.

[18] C.-N. Chuah, D. N. C. Tse, J. M. Kahn, and R. A. Valenzuela, "Capacity scaling in MIMO wireless systems under correlated fading," *IEEE Trans. Inform. Theory,* pp. 637–50, March 2002.

[19] A. Lozano, A. M. Tulino, and S. Verdú, "Multiple-antenna capacity in the low-power regime," *IEEE Trans. Inform. Theory,* pp. 2527–44, October 2003.

[20] A. L. Moustakas, S. H. Simon, and A. M. Sengupta, "MIMO capacity through correlated channels in the presence of correlated interferers and noise: A (not so) large N analysis," *IEEE Trans. Inform. Theory,* pp. 2545–61, October 2003.

[21] S. A. Jafar and A. J. Goldsmith, "Transmitter optimization and optimality of beamforming for multiple antenna systems," *IEEE Trans. Wireless Commun.,* pp. 1165–75, July 2004.

[22] A. Narula, M. Lopez, M. Trott, and G. Wornell, "Efficient use of side information in multiple-antenna data transmission over fading channels," *IEEE J. Sel. Areas Commun.,* pp. 1423–36, October 1998.

[23] E. Visotsky and U. Madhow, "Space-time transmit precoding with imperfect feedback," *Proc. IEEE Internat. Sympos. Inform. Theory,* pp. 357–66, June 2000.

[24] E. Jorswieck and H. Boche, "Channel capacity and capacity-range of beamforming in MIMO wireless systems under correlated fading with covariance feedback," *IEEE Trans. Wireless Commun.,* pp. 1543–53, September 2004.

[25] B. Hochwald and V. Tarokh, "Multiple-antenna channel hardening and its implications for rate feedback and scheduling," *IEEE Trans. Inform. Theory,* pp. 1893–1909, September 2004.

[26] A. J. Goldsmith, S. A. Jafar, N. Jindal, and S. Vishwanath, "Capacity limits of MIMO channels," *IEEE J. Sel. Areas Commun.,* pp. 684–701, June 2003.

[27] T. Marzetta and B. Hochwald, "Capacity of a mobile multiple-antenna communication link in Rayleigh flat fading," *IEEE Trans. Inform. Theory,* pp. 139–57, January 1999.

[28] S. A. Jafar and A. J. Goldsmith, "Multiple-antenna capacity in correlated Rayleigh fading with channel covariance information," *IEEE Trans. Wireless Commun.,* pp. 990–7, May 2005.

[29] L. Zheng and D. N. Tse, "Communication on the Grassmann manifold: A geometric approach to the non-coherent multi-antenna channel," *IEEE Trans. Inform. Theory,* pp. 359–83, February 2002.

[30] R. Etkin and D. Tse, "Degrees of freedom in underspread MIMO fading channels," *Proc. IEEE Internat. Sympos. Inform. Theory,* p. 323, July 2003.

[31] A. Lapidoth and S. Moser, "On the fading number of multi-antenna systems over flat fading channels with memory and incomplete side information," *Proc. IEEE Internat. Sympos. Inform. Theory,* p. 478, July 2002.

[32] T. Koch and A. Lapidoth, "The fading number and degrees of freedom in non-coherent MIMO fading channels: A peace pipe," *Proc. IEEE Internat. Sympos. Inform. Theory,* September 2005.

[33] G. B. Giannakis, Y. Hua, P. Stoica, and L. Tong, *Signal Processing Advances in Wireless and Mobile Communications: Trends in Single- and Multi-user Systems*, Prentice-Hall, New York, 2001.

[34] A. Molisch, M. Win, and J. H. Winters, "Reduced-complexity transmit/receive-diversity systems," *IEEE Trans. Signal Proc.*, pp. 2729–38, November 2003.

[35] H. Gamal, G. Caire, and M. Damon, "Lattice coding and decoding achieve the optimal diversity-multiplexing trade-off of MIMO channels," *IEEE Trans. Inform. Theory*, pp. 968–85, June 2004.

[36] V. Tarokh, N. Seshadri, and A. Calderbank, "Space-time codes for high data rate wireless communication: Performance criterion and code construction," *IEEE Trans. Inform. Theory*, pp. 744–65, March 1998.

[37] H. Yao and G. Wornell, "Structured space-time block codes with optimal diversity-multiplexing tradeoff and minimum delay," *Proc. IEEE Globecom Conf.*, pp. 1941–5, December 2003.

[38] L. Zheng and D. N. Tse, "Diversity and multiplexing: A fundamental trade-off in multiple antenna channels," *IEEE Trans. Inform. Theory*, pp. 1073–96, May 2003.

[39] M. O. Damen, H. El Gamal, and N. C. Beaulieu, "Linear threaded algebraic space-time constellations," *IEEE Trans. Inform. Theory*, pp. 2372–88, October 2003.

[40] H. El Gamal and M. O. Damen, "Universal space-time coding," *IEEE Trans. Inform. Theory*, pp. 1097–1119, May 2003.

[41] R. W. Heath, Jr., and A. J. Paulraj, "Switching between multiplexing and diversity based on constellation distance," *Proc. Allerton Conf. Commun., Control, Comput.*, pp. 212–21, October 2000.

[42] R. W. Heath, Jr., and D. J. Love, "Multi-mode antenna selection for spatial multiplexing with linear receivers," *IEEE Trans. Signal Proc.* (to appear).

[43] V. Jungnickel, T. Haustein, V. Pohl, and C. Von Helmolt, "Link adaptation in a multi-antenna system," *Proc. IEEE Veh. Tech. Conf.*, pp. 862–6, April 2003.

[44] J.-C. Guey, M. P. Fitz, M. Bell, and W.-Y. Kuo, "Signal design for transmitter diversity wireless communication systems over Rayleigh fading channels," *IEEE Trans. Commun.*, pp. 527–37, April 1999.

[45] V. Tarokh, A. Naguib, N. Seshadri, and A. Calderbank, "Space-time codes for high data rate wireless communication: Performance criteria in the presence of channel estimation errors, mobility, and multiple paths," *IEEE Trans. Commun.*, pp. 199–207, February 1999.

[46] S. Baro, G. Bauch, and A. Hansman, "Improved codes for space-time trellis coded modulation," *IEEE Commun. Lett.*, pp. 20–2, January 2000.

[47] J. Grimm, M. Fitz, and J. Korgmeier, "Further results in space-time coding for Rayleigh fading," *Proc. Allerton Conf. Commun., Control, Comput.*, pp. 391–400, September 1998.

[48] H. Gamal and A. Hammons, "On the design of algebraic space-time codes for MIMO block-fading channels," *IEEE Trans. Inform. Theory*, pp. 151–63, January 2003.

[49] A. Naguib, N. Seshadri, and A. Calderbank, "Increasing data rate over wireless channels," *IEEE Signal Proc. Mag.*, pp. 76–92, May 2000.

[50] V. Tarokh, H. Jafarkhani, and A. Calderbank, "Space-time block codes from orthogonal designs," *IEEE Trans. Inform. Theory*, pp. 1456–67, July 1999.

[51] V. Gulati and K. R. Narayanan, "Concatenated codes for fading channels based on recursive space-time trellis codes," *IEEE Trans. Wireless Commun.*, pp. 118–28, January 2003.

[52] Y. Liu, M. P. Fitz, and O. Y. Takeshita, "Full-rate space-time codes," *IEEE J. Sel. Areas Commun.*, pp. 969–80, May 2001.

[53] K. R. Narayanan, "Turbo decoding of concatenated space-time codes," *Proc. Allerton Conf. Commun., Control, Comput.*, pp. 217–26, September 1999.

[54] B. Hochwald and T. Marzetta, "Unitary space-time modulation for multiple-antenna communications in Rayleigh flat fading," *IEEE Trans. Inform. Theory*, pp. 543–64, March 2000.

[55] D. Gesbert, M. Shafi, D.-S. Shiu, P. Smith, and A. Naguib, "From theory to practice: An overview of MIMO space-time coded wireless systems," *IEEE J. Sel. Areas Commun.*, pp. 281–302, April 2003.

[56] E. G. Larsson and P. Stoica, *Space-Time Block Coding for Wireless Communications,* Cambridge University Press, 2003.

[57] P. Wolniansky, G. Foschini, G. Golden, and R. Valenzuela, "V-blast: An architecture for realizing very high data rates over the rich-scattering wireless channel," *Proc. URSI Internat. Sympos. Signal Syst. Elec.,* pp. 295–300, October 1998.

[58] G. Foschini, G. Golden, R. Valenzuela, and P. Wolniansky, "Simplified processing for high spectral efficiency wireless communication employing multi-element arrays," *IEEE J. Sel. Areas Commun.,* pp. 1841–52, November 1999.

[59] G. Bauch and A. Naguib, "Map equalization of space-time coded signals over frequency selective channels," *Proc. IEEE Wireless Commun. Network Conf.,* pp. 261–5, September 1999.

[60] C. Fragouli, N. Al-Dhahir, and S. Diggavi, "Pre-filtered space-time M-BCJR equalizer for frequency selective channels," *IEEE Trans. Commun.,* pp. 742–53, May 2002.

[61] A. Naguib, "Equalization of transmit diversity space-time coded signals," *Proc. IEEE Globecom Conf.,* pp. 1077–82, December 2000.

[62] J. Winters, "Smart antennas for wireless systems," *IEEE Pers. Commun. Mag.,* pp. 23–7, February 1998.

Equalization

We have seen in Chapter 6 that delay spread causes intersymbol interference (ISI), which can cause an irreducible error floor when the modulation symbol time is on the same order as the channel delay spread. Signal processing provides a powerful mechanism to counteract ISI. In a broad sense, equalization defines any signal processing technique used at the receiver to alleviate the ISI problem caused by delay spread. Signal processing can also be used at the transmitter to make the signal less susceptible to delay spread: spread-spectrum and multicarrier modulation fall in this category of transmitter signal processing techniques. In this chapter we focus on equalization; multicarrier modulation and spread spectrum are the topics of Chapters 12 and 13, respectively.

Mitigation of ISI is required when the modulation symbol time T_s is on the order of the channel's rms delay spread σ_{T_m}. For example, cordless phones typically operate indoors, where the delay spread is small. Since voice is also a relatively low–data-rate application, equalization is generally not needed in cordless phones. However, the IS-136 digital cellular standard is designed for outdoor use, where $\sigma_{T_m} \approx T_s$, so equalization is part of this standard. Higher–data-rate applications are more sensitive to delay spread and generally require high-performance equalizers or other ISI mitigation techniques. In fact, mitigating the impact of delay spread is one of the most challenging hurdles for high-speed wireless data systems.

Equalizer design must typically balance ISI mitigation with noise enhancement, since both the signal and the noise pass through the equalizer, which can increase the noise power. Nonlinear equalizers suffer less from noise enhancement than linear equalizers but typically entail higher complexity, as discussed in more detail below. Moreover, equalizers require an estimate of the channel impulse or frequency response to mitigate the resulting ISI. Since the wireless channel varies over time, the equalizer must learn the frequency or impulse response of the channel (training) and then update its estimate of the frequency response as the channel changes (tracking). The process of equalizer training and tracking is often referred to as adaptive equalization, since the equalizer adapts to the changing channel. Equalizer training and tracking can be quite difficult if the channel is changing rapidly. In this chapter we will discuss the various issues associated with equalizer design, to include balancing ISI mitigation with noise enhancement, linear and nonlinear equalizer design and properties, and the process of equalizer training and tracking.

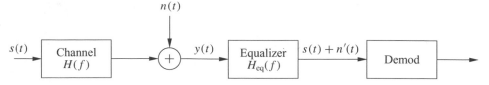

Figure 11.1: Analog equalizer illustrating noise enhancement.

An equalizer can be implemented at baseband, the carrier frequency, or an intermediate frequency. Most equalizers are implemented digitally after A/D conversion because such filters are small, cheap, easily tunable, and very power efficient. This chapter mainly focuses on digital equalizer implementations, although in the next section we will illustrate noise enhancement using an analog equalizer for simplicity.

11.1 Equalizer Noise Enhancement

The goal of equalization is to mitigate the effects of ISI. However, this goal must be balanced so that, in the process of removing ISI, the noise power in the received signal is not enhanced. A simple analog equalizer, shown in Figure 11.1, illustrates the pitfalls of removing ISI without considering this effect on noise. Consider a signal $s(t)$ that is passed through a channel with frequency response $H(f)$. At the receiver front end, white Gaussian noise $n(t)$ is added to the signal and so the signal input to the receiver is $Y(f) = S(f)H(f) + N(f)$, where $N(f)$ is white noise with power spectral density (PSD) $N_0/2$. If the bandwidth of $s(t)$ is B, then the noise power within the signal bandwidth of interest is $N_0 B$. Suppose we wish to equalize the received signal so as to completely remove the ISI introduced by the channel. This is easily done by introducing an analog equalizer in the receiver that is defined by

$$H_{\text{eq}}(f) = 1/H(f). \tag{11.1}$$

The received signal $Y(f)$ after passing through this equalizer becomes

$$[S(f)H(f) + N(f)]H_{\text{eq}}(f) = S(f) + N'(f),$$

where $N'(f)$ is colored Gaussian noise with power spectral density $.5N_0/|H(f)|^2$. Thus, all ISI has been removed from the transmitted signal $S(f)$.

However, if $H(f)$ has a spectral null ($H(f_0) = 0$ for some f_0) at any frequency within the bandwidth of $s(t)$, then the power of the noise $N'(f)$ is infinite. Even without a spectral null, if some frequencies in $H(f)$ are greatly attenuated then the equalizer $H_{\text{eq}}(f) = 1/H(f)$ will greatly enhance the noise power at those frequencies. In this case, even though the ISI effects are removed, the equalized system will perform poorly because of its greatly reduced SNR. Thus, the goal of equalization is to balance mitigation of ISI with maximizing the SNR of the postequalization signal. In general, linear digital equalizers work by approximately inverting the channel frequency response and thus have the most noise enhancement. Nonlinear equalizers do not invert the channel frequency response, so they tend to suffer much less from noise enhancement. In the next section we give an overview of the different types of linear and nonlinear equalizers, their structures, and the algorithms used for updating their tap coefficients in equalizer training and tracking.

EXAMPLE 11.1: Consider a channel with impulse response $H(f) = 1/\sqrt{|f|}$ for $|f| < B$, where B is the channel bandwidth. Given noise PSD $N_0/2$, what is the noise power for channel bandwidth $B = 30$ kHz with and without an equalizer that inverts the channel?

Solution: Without equalization, the noise power is just $N_0 B = 3N_0 \cdot 10^4$. With equalization, the noise PSD is $.5N_0|H_{\text{eq}}(f)|^2 = .5N_0/|H(f)|^2 = .5|f|N_0$ for $|f| < B$. So the noise power is

$$.5N_0 \int_{-B}^{B} |f| \, df = .5N_0 B^2 = 4.5N_0 \cdot 10^8,$$

an increase in noise power of more than four orders of magnitude!

11.2 Equalizer Types

Equalization techniques fall into two broad categories: linear and nonlinear. The linear techniques are generally the simplest to implement and to understand conceptually. However, linear equalization techniques typically suffer from more noise enhancement than nonlinear equalizers and hence are not used in most wireless applications. Among nonlinear equalization techniques, decision-feedback equalization (DFE) is the most common because it is fairly simple to implement and usually performs well. However, on channels with low SNR, the DFE suffers from error propagation when bits are decoded in error, leading to poor performance. The optimal equalization technique is maximum likelihood sequence estimation (MLSE). Unfortunately, the complexity of this technique grows exponentially with the length of the delay spread, so it is impractical on most channels of interest. However, the performance of the MLSE is often used as an upper bound on performance for other equalization techniques. Figure 11.2 summarizes the different equalizer types along with their corresponding structures and tap updating algorithms.

Equalizers can also be categorized as symbol-by-symbol (SBS) or sequence estimators (SEs). SBS equalizers remove ISI from each symbol and then detect each symbol individually. All linear equalizers in Figure 11.2 (as well as the DFE) are SBS equalizers. Sequence estimators detect sequences of symbols, so the effect of ISI is part of the estimation process. Maximum likelihood sequence estimation is the optimal form of sequence detection, but it is highly complex.

Linear and nonlinear equalizers are typically implemented using a transversal or lattice structure. The transversal structure is a filter with $N - 1$ delay elements and N taps featuring tunable complex weights. The lattice filter uses a more complex recursive structure [1]. In exchange for this increased complexity relative to transversal structures, lattice structures often have better numerical stability and convergence properties and greater flexibility in changing their length [2]. This chapter will focus on transversal structures; details on lattice structures and their performance relative to transversal structures can be found in [1; 2; 3; 4].

In addition to the equalizer type and structure, adaptive equalizers require algorithms for updating the filter tap coefficients during training and tracking. Many algorithms have been developed over the years for this purpose. These algorithms generally incorporate trade-offs between complexity, convergence rate, and numerical stability.

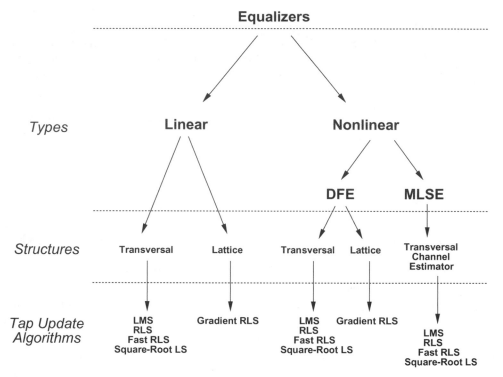

Figure 11.2: Equalizer types, structures, and algorithms.

In the remainder of this chapter – after discussing conditions for ISI-free transmission – we will examine the different equalizer types, their structures, and their update algorithms in more detail. Our analysis of equalization is based on the equivalent lowpass representation of bandpass systems, described in Appendix A.

11.3 Folded Spectrum and ISI-Free Transmission

Equalizers are typically implemented digitally. Figure 11.3 is a block diagram of an equivalent lowpass end-to-end system with a digital equalizer. The input symbol d_k is passed through a pulse-shaping filter $g(t)$ and then transmitted over the ISI channel with equivalent lowpass impulse response $c(t)$. We define the combined channel impulse response $h(t) \triangleq g(t) * c(t)$, and the equivalent lowpass transmitted signal is thus given by $d(t) * g(t) * c(t)$ for the train $d(t) = \sum_k d_k \delta(t - kT_s)$ of information symbols. The pulse shape $g(t)$ improves the spectral properties of the transmitted signal, as described in Section 5.5. This pulse shape is under the control of the system designer, whereas the channel $c(t)$ is introduced by nature and thus outside the designer's control.

At the receiver front end, equivalent lowpass white Gaussian noise $n(t)$ with PSD N_0 is added to the received signal for a resulting signal $w(t)$. This signal is passed through an analog matched filter $g_m^*(-t)$ to obtain the equivalent lowpass output $y(t)$, which is then sampled via an A/D converter. The purpose of the matched filter is to maximize the SNR of the signal before sampling and subsequent processing.[1] Recall from Section 5.1 that in AWGN the

---

[1] Although the matched filter could be more efficiently implemented digitally, the analog implementation before the sampler allows for a smaller dynamic range in the sampler, which significantly reduces cost.

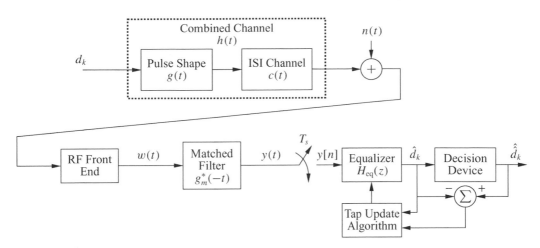

Figure 11.3: End-to-end system (equivalent lowpass representation).

SNR of the received signal is maximized prior to sampling by using a filter that is matched to the pulse shape. This result indicates that, for the system shown in Figure 11.3, SNR prior to sampling is maximized by passing $w(t)$ through a filter matched to $h(t)$, so ideally we would have $g_m(t) = h(t)$. However, since the channel impulse response $c(t)$ is time varying and since analog filters are not easily tunable, it is generally not possible to have $g_m(t) = h(t)$. Thus, part of the art of equalizer design is to chose $g_m(t)$ to get good performance. Often $g_m(t)$ is matched to the pulse shape $g(t)$, which is the optimal pulse shape when $c(t) = \delta(t)$, but this design is clearly suboptimal when $c(t) \neq \delta(t)$. Not matching $g_m(t)$ to $h(t)$ can result in significant performance degradation and also makes the receiver extremely sensitive to timing error. These problems are somewhat mitigated by sampling $y(t)$ at a rate much faster than the symbol rate and then designing the equalizer for this oversampled signal. This process is called *fractionally spaced equalization* [4, Chap. 10.2.4].

The equalizer output provides an estimate of the transmitted symbol. This estimate is then passed through a decision device that rounds the equalizer output to a symbol in the alphabet of possible transmitted symbols. During training, the equalizer output is passed to the tap update algorithm to update the tap values, so that the equalizer output closely matches the known training sequence. During tracking, the round-off error associated with the symbol decision is used to adjust the equalizer coefficients.

Let $f(t)$ be defined as the composite equivalent lowpass impulse response consisting of the transmitter pulse shape, channel, and matched filter impulse responses:

$$f(t) \triangleq g(t) * c(t) * g_m^*(-t). \tag{11.2}$$

Then the matched filter output is given by

$$y(t) = d(t) * f(t) + n_g(t) = \sum d_k f(t - kT_s) + n_g(t), \tag{11.3}$$

where $n_g(t) = n(t) * g_m^*(-t)$ is the equivalent lowpass noise at the equalizer input and T_s is the symbol time. If we let $f[n] = f(nT_s)$ denote samples of $f(t)$ every T_s seconds, then sampling $y(t)$ every T_s seconds yields the discrete-time signal $y[n] = y(nT_s)$ given by

$$y[n] = \sum_{k=-\infty}^{\infty} d_k f(nT_s - kT_s) + n_g(nT_s)$$

$$= \sum_{k=-\infty}^{\infty} d_k f[n-k] + v[n]$$

$$= d_n f[0] + \sum_{k \neq n} d_k f[n-k] + v[n], \tag{11.4}$$

where the first term in (11.4) is the desired data bit, the second term is the ISI, and the third term is the sampled noise. By (11.4) there is no ISI if $f[n-k] = 0$ for $k \neq n$ – that is, $f[k] = \delta[k] f[0]$. In this case (11.4) reduces to $y[n] = d_n f[0] + v[n]$.

We now show that the condition for ISI-free transmission, $f[k] = \delta[k] f[0]$, is satisfied if and only if

$$F_\Sigma(f) \triangleq \frac{1}{T_s} \sum_{n=-\infty}^{\infty} F\left(f + \frac{n}{T_s}\right) = f[0]. \tag{11.5}$$

The function $F_\Sigma(f)$, which is periodic with period $1/T_s$, is often called the *folded spectrum*. If $F_\Sigma(f) = f[0]$ we say that the folded spectrum is *flat*.

To show the equivalence between ISI-free transmission and a flat folded spectrum, begin by observing that

$$f[k] = f(kT_s) = \int_{-\infty}^{\infty} F(f) e^{j2\pi fkT_s} \, df$$

$$= \sum_{n=-\infty}^{\infty} \int_{.5(2n-1)/T_s}^{.5(2n+1)/T_s} F(f) e^{j2\pi fkT_s} \, df$$

$$= \sum_{n=-\infty}^{\infty} \int_{-.5/T_s}^{.5/T_s} F\left(f' + \frac{n}{T_s}\right) e^{j2\pi(f' + n/T_s)kT_s} \, df'$$

$$= \int_{-.5/T_s}^{.5/T_s} e^{j2\pi fkT_s} \left[\sum_{n=-\infty}^{\infty} F\left(f + \frac{n}{T_s}\right) \right] df. \tag{11.6}$$

Equation (11.6) implies that the Fourier series representation of $F_\Sigma(f)$ is

$$F_\Sigma(f) = \frac{1}{T_s} \sum_k f[k] e^{-j2\pi fkT_s}.$$

We first demonstrate that a flat folded spectrum implies that $f[k] = \delta[k] f[0]$. Suppose (11.5) is true. Then substituting (11.5) into (11.6) yields

$$f[k] = T_s \int_{-.5/T_s}^{.5/T_s} e^{j2\pi fkT_s} f[0] \, df = \frac{\sin \pi k}{\pi k} f[0] = \delta[k] f[0], \tag{11.7}$$

which is the desired result. We now show that $f[k] = \delta[k] f[0]$ implies a flat folded spectrum. By (11.6) and the definition of $F_\Sigma(f)$,

$$f[k] = T_s \int_{-.5/T_s}^{.5/T_s} F_\Sigma(f) e^{j2\pi fkT_s} \, df. \tag{11.8}$$

Hence $f[k]$ is the inverse Fourier transform of $F_\Sigma(f)$.[2] Therefore, if $f[k] = \delta[k] f[0]$ then $F_\Sigma(f) = f[0]$.

EXAMPLE 11.2: Consider a channel with combined equivalent lowpass impulse response $f(t) = \mathrm{sinc}(t/T_s)$. Find the folded spectrum and determine if this channel exhibits ISI.

Solution: The Fourier transform of $f(t)$ is

$$F(f) = T_s \, \mathrm{rect}(fT_s) = \begin{cases} T_s & |f| < .5/T_s, \\ .5T_s & |f| = .5/T_s, \\ 0 & |f| > .5/T_s. \end{cases}$$

Therefore,

$$F_\Sigma(f) = \frac{1}{T_s} \sum_{n=-\infty}^{\infty} F\left(f + \frac{n}{T_s}\right) = 1,$$

so the folded spectrum is flat and there is no ISI. We can also see this from the fact that

$$f(nT_s) = \mathrm{sinc}\left(\frac{nT_s}{T_s}\right) = \mathrm{sinc}(n) = \begin{cases} 1 & n = 0, \\ 0 & n \neq 0. \end{cases}$$

Thus, $f[k] = \delta[k]$, our equivalent condition for a flat folded spectrum and zero ISI.

11.4 Linear Equalizers

If $F_\Sigma(f)$ is not flat, we can use the equalizer $H_{\mathrm{eq}}(z)$ in Figure 11.3 to reduce ISI. In this section we assume a linear equalizer implemented via a $2L + 1 = N$-tap transversal filter:

$$H_{\mathrm{eq}}(z) = \sum_{i=-L}^{L} w_i z^{-i}. \tag{11.9}$$

The length of the equalizer N is typically dictated by implementation considerations, since a large N entails more complexity and delay. Causal linear equalizers have $w_i = 0$ when $i < 0$. For a given equalizer size N, the equalizer design must specify (i) the tap weights $\{w_i\}_{i=-L}^{L}$ for a given channel frequency response and (ii) the algorithm for updating these tap weights as the channel varies. Recall that our performance metric in wireless systems is probability of error (or outage probability), so for a given channel the optimal choice of equalizer coefficients would be the coefficients that minimize probability of error. Unfortunately, it is extremely difficult to optimize the $\{w_i\}$ with respect to this criterion. Since we cannot directly optimize for our desired performance metric, we must instead use an indirect optimization that balances ISI mitigation with the prevention of noise enhancement, as discussed with regard to the preceding simple analog example. We now describe two linear equalizers: the zero-forcing (ZF) equalizer and the minimum mean-square error (MMSE) equalizer. The

---

[2] This also follows from the Fourier series representation of $F_\Sigma(f)$.

former equalizer cancels all ISI but can lead to considerable noise enhancement. The latter technique minimizes the expected mean-squared error between the transmitted symbol and the symbol detected at the equalizer output, thereby providing a better balance between ISI mitigation and noise enhancement. Because of this more favorable balance, MMSE equalizers tend to have better BER performance than equalizers using the ZF algorithm.

11.4.1 Zero-Forcing (ZF) Equalizers

From (11.4), the samples $\{y_n\}$ input to the equalizer can be represented based on the discretized combined equivalent lowpass impulse response $f(t) = h(t) * g^*(-t)$ as

$$Y(z) = D(z)F(z) + N_g(z), \tag{11.10}$$

where $N_g(z)$ is the z-transform of the noise samples at the output of the matched filter $G_m^*(1/z^*)$ and

$$F(z) = H(z)G_m^*\left(\frac{1}{z^*}\right) = \sum_n f(nT_s)z^{-n}. \tag{11.11}$$

The zero-forcing equalizer removes all ISI introduced in the composite response $f(t)$. From (11.10) we see that the equalizer to accomplish this is given by

$$H_{ZF}(z) = \frac{1}{F(z)}. \tag{11.12}$$

This is the discrete-time equivalent lowpass equalizer of the analog equalizer (11.1) described before, and it suffers from the same noise enhancement properties. Specifically, the power spectrum $N(z)$ of the noise samples at the equalizer output is given by

$$\begin{aligned}
N(z) &= N_0|G_m^*(1/z^*)|^2|H_{ZF}(z)|^2 = \frac{N_0|G_m^*(1/z^*)|^2}{|F(z)|^2} \\
&= \frac{N_0|G_m^*(1/z^*)|^2}{|H(z)|^2|G_m^*(1/z^*)|^2} = \frac{N_0}{|H(z)|^2}.
\end{aligned} \tag{11.13}$$

We see from (11.13) that if the channel $H(z)$ is sharply attenuated at any frequency within the signal bandwidth of interest – as is common on frequency-selective fading channels – the noise power will be significantly increased. This motivates an equalizer design that better optimizes between ISI mitigation and noise enhancement. One such equalizer is the MMSE equalizer, described in the next section.

The ZF equalizer defined by $H_{ZF}(z) = 1/F(z)$ may not be implementable as a finite–impulse response (FIR) filter. Specifically, it may not be possible to find a finite set of coefficients $w_{-L}, \ldots, w_L$ such that

$$w_{-L}z^L + \cdots + w_L z^{-L} = \frac{1}{F(z)}. \tag{11.14}$$

In this case we find the set of coefficients $\{w_i\}$ that best approximates the zero-forcing equalizer. Note that this is not straightforward because the approximation must be valid for all values of z. There are many ways we can make this approximation. One technique is to represent $H_{ZF}(z)$ as an infinite–impulse response (IIR) filter, $1/F(z) = \sum_{i=-\infty}^{\infty} c_i z^{-i}$, and then set $w_i = c_i$. It can be shown that this minimizes

$$\left| \frac{1}{F(z)} - (w_{-L} z^L + \cdots + w_L z^{-L}) \right|^2$$

at $z = e^{j\omega}$. Alternatively, the tap weights can be set to minimize the peak distortion (worst-case ISI). Finding the tap weights to minimize peak distortion is a convex optimization problem that can be solved by standard techniques – for example, the method of steepest descent [4].

EXAMPLE 11.3: Consider a channel with impulse response
$$h(t) = \begin{cases} e^{-t/\tau} & t \geq 0, \\ 0 & \text{else.} \end{cases}$$
Find a two-tap ZF equalizer for this channel.

Solution: We have
$$h[n] = 1 + e^{-T_s/\tau} \delta[n-1] + e^{-2T_s/\tau} \delta[n-2] + \cdots .$$
Thus,
$$H(z) = 1 + e^{-T_s/\tau} z^{-1} + e^{-2T_s/\tau} z^{-2} + e^{-3T_s/\tau} z^{-3} + \cdots$$
$$= \sum_{n=0}^{\infty} (e^{-T_s/\tau} z^{-1})^n = \frac{z}{z - e^{-T_s/\tau}}.$$
So $H_{eq}(z) = 1/H(z) = 1 - e^{-T_s/\tau} z^{-1}$. The two-tap ZF equalizer therefore has tap weight coefficients $w_0 = 1$ and $w_1 = e^{-T_s/\tau}$.

11.4.2 Minimum Mean-Square Error (MMSE) Equalizers

In MMSE equalization, the goal of the equalizer design is to minimize the average mean-square error (MSE) between the transmitted symbol d_k and its estimate $\hat{d}_k$ at the output of the equalizer. In other words, the $\{w_i\}$ are chosen to minimize $\mathbf{E}[d_k - \hat{d}_k]^2$. Since the MMSE equalizer is linear, its output $\hat{d}_k$ is a linear combination of the input samples $y[k]$:

$$\hat{d}_k = \sum_{i=-L}^{L} w_i y[k-i]. \tag{11.15}$$

As such, finding the optimal filter coefficients $\{w_i\}$ becomes a standard problem in linear estimation. In fact, if the noise input to the equalizer is white then we have a standard Weiner filtering problem. However, because of the matched filter $g_m^*(-t)$ at the receiver front end, the noise input to the equalizer is not white; rather, it is colored with power spectrum $N_0 |G_m^*(1/z^*)|^2$. Therefore, in order to apply known techniques for optimal linear estimation, we expand the filter $H_{eq}(z)$ into two components – a noise-whitening component $1/G_m^*(1/z^*)$ and an ISI-removal component $\hat{H}_{eq}(z)$ – as shown in Figure 11.4.

The purpose of the noise-whitening filter (as the name indicates) is to whiten the noise so that the noise component output from this filter has a constant power spectrum. Since the

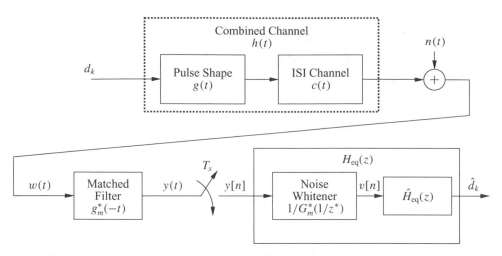

Figure 11.4: MMSE equalizer with noise-whitening filter.

noise input to this filter has power spectrum $N_0|G_m^*(1/z^*)|^2$, the appropriate noise-whitening filter is $1/G_m^*(1/z^*)$. The noise power spectrum at the output of the noise-whitening filter is then $N_0|G_m^*(1/z^*)|^2/|G_m^*(1/z^*)|^2 = N_0$. Note that the filter $1/G_m^*(1/z^*)$ is not the only filter that will whiten the noise, and another noise-whitening filter with more desirable properties (like stability) may be chosen. It might seem odd at first to introduce the matched filter $g_m^*(-t)$ at the receiver front end only to cancel its effect in the equalizer. Recall, however, that the matched filter is meant to maximize the SNR prior to sampling. By removing the effect of this matched filter via noise whitening after sampling, we merely simplify the design of $\hat{H}_{eq}(z)$ to minimize MSE. In fact, if the noise-whitening filter does not yield optimal performance then its effect would be cancelled by the $\hat{H}_{eq}(z)$ filter design, as we shall see in the case of IIR MMSE equalizers.

We assume that the filter $\hat{H}_{eq}(z)$, with input v_n, is a linear filter with $N = 2L + 1$ taps:

$$\hat{H}_{eq}(z) = \sum_{i=-L}^{L} w_i z^{-i}. \tag{11.16}$$

Our goal is to design the filter coefficients $\{w_i\}$ so as to minimize $\mathbf{E}[d_k - \hat{d}_k]^2$. This is the same goal as for the total filter $H_{eq}(z)$ – we just added the noise-whitening filter to make solving for these coefficients simpler. Define $\mathbf{v}^T = (v[k+L], v[k+L-1], \ldots, v[k-L]) = (v_{k+L}, v_{k+L-1}, \ldots, v_{k-L})$ as the row vector of inputs to the filter $\hat{H}_{eq}(z)$ used to obtain the filter output $\hat{d}_k$ and define $\mathbf{w}^T = (w_{-L}, \ldots, w_L)$ as the row vector of filter coefficients. Then

$$\hat{d}_k = \mathbf{w}^T \mathbf{v} = \mathbf{v}^T \mathbf{w}. \tag{11.17}$$

Thus, we want to minimize the mean-square error

$$J = \mathbf{E}[d_k - \hat{d}_k]^2 = \mathbf{E}[\mathbf{w}^T \mathbf{v} \mathbf{v}^H \mathbf{w}^* - 2\,\mathrm{Re}\{\mathbf{v}^H \mathbf{w}^* d_k\} + |d_k|^2]. \tag{11.18}$$

Define $\mathbf{M_v} = \mathbf{E}[\mathbf{v}\mathbf{v}^H]$ and $\mathbf{v_d} = \mathbf{E}[\mathbf{v}^H d_k]$. The matrix $\mathbf{M_v}$ is an $N \times N$ Hermitian matrix, and $\mathbf{v_d}$ is a length-N row vector. Assume $\mathbf{E}|d_k|^2 = 1$. Then the MSE J is

$$J = \mathbf{w}^T \mathbf{M_v} \mathbf{w}^* - 2\,\mathrm{Re}\{\mathbf{v_d}\mathbf{w}^*\} + 1. \tag{11.19}$$

We obtain the optimal tap vector $\mathbf{w}$ by setting the gradient $\nabla_\mathbf{w} J = 0$ and solving for $\mathbf{w}$. By (11.19), the gradient is given by

$$\nabla_\mathbf{w} J = \left(\frac{\partial J}{\partial w_{-L}}, \ldots, \frac{\partial J}{\partial w_L} \right) = 2\mathbf{w}^T \mathbf{M_v} - 2\mathbf{v_d}. \qquad (11.20)$$

Setting this to zero yields $\mathbf{w}^T \mathbf{M_v} = \mathbf{v_d}$ or, equivalently, that the optimal tap weights are given by

$$\mathbf{w}_{\mathrm{opt}} = (\mathbf{M}_\mathbf{v}^T)^{-1} \mathbf{v_d}^T. \qquad (11.21)$$

Note that solving for $\mathbf{w}_{\mathrm{opt}}$ requires a matrix inversion with respect to the filter inputs. Thus, the complexity of this computation is quite high, typically on the order of N^2 to N^3 operations. Substituting in these optimal tap weights, we obtain the minimum mean-square error as

$$J_{\min} = 1 - \mathbf{v_d} \mathbf{M}_\mathbf{v}^{-1} \mathbf{v_d}^H. \qquad (11.22)$$

For an equalizer of infinite length, $\mathbf{v}^T = (v_{n+\infty}, \ldots, v_n, \ldots, v_{n-\infty})$ and $\mathbf{w}^T = (w_{-\infty}, \ldots, w_0, \ldots, w_\infty)$. Then $\mathbf{w}^T \mathbf{M_v} = \mathbf{v_d}$ can be written as

$$\sum_{i=-\infty}^{\infty} w_i(f[j-i] + N_0 \delta[j-i]) = g_m^*[-j], \quad -\infty \le j \le \infty \qquad (11.23)$$

[5, Chap. 7.4]. Taking z-transforms and noting that $\hat{H}_{\mathrm{eq}}(z)$ is the z-transform of the filter coefficients $\mathbf{w}$ yields

$$\hat{H}_{\mathrm{eq}}(z)(F(z) + N_0) = G_m^*(1/z^*). \qquad (11.24)$$

Solving for $\hat{H}_{\mathrm{eq}}(z)$, we obtain

$$\hat{H}_{\mathrm{eq}}(z) = \frac{G_m^*(1/z^*)}{F(z) + N_0}. \qquad (11.25)$$

Since the MMSE equalizer consists of the noise-whitening filter $1/G_m^*(1/z^*)$ plus the ISI-removal component $\hat{H}_{\mathrm{eq}}(z)$, it follows that the full MMSE equalizer (when it is not restricted to be finite length) becomes

$$H_{\mathrm{eq}}(z) = \frac{\hat{H}_{\mathrm{eq}}(z)}{G_m^*(1/z^*)} = \frac{1}{F(z) + N_0}. \qquad (11.26)$$

There are three interesting things to note about this result. First of all, the ideal infinite-length MMSE equalizer cancels out the noise-whitening filter. Second, this infinite-length equalizer is identical to the ZF filter except for the noise term N_0, so in the absence of noise the two equalizers are equivalent. Finally, this ideal equalizer design clearly shows a balance between inverting the channel and noise enhancement: if $F(z)$ is highly attenuated at some frequency, then the noise term N_0 in the denominator prevents the noise from being significantly enhanced by the equalizer. Yet at frequencies where the noise power spectral density N_0 is small compared to the composite channel $F(z)$, the equalizer effectively inverts $F(z)$.

For the equalizer (11.26) it can be shown [5, Chap. 7.4.1] that the minimum MSE (11.22) can be expressed in terms of the folded spectrum $F_\Sigma(f)$ as

$$J_{\min} = T_s \int_{-.5/T_s}^{.5/T_s} \frac{N_0}{F_\Sigma(f) + N_0} \, df. \tag{11.27}$$

This expression for MMSE has several interesting properties. First it can be shown that, as expected, $0 \leq J_{\min} = \mathbf{E}[d_k - \hat{d}_k]^2 \leq 1$. In addition, $J_{\min} = 0$ in the absence of noise ($N_0 = 0$) as long as $F_\Sigma(f) \neq 0$ within the signal bandwidth of interest. Also, as expected, $J_{\min} = 1$ if $N_0 = \infty$.

EXAMPLE 11.4: Find $J_{\min}$ when the folded spectrum $F_\Sigma(f)$ is flat, $F_\Sigma(f) = f[0]$, in the asymptotic limit of high and low SNR.

Solution: If $F_\Sigma(f) = f[0] \triangleq f_0$ then

$$J_{\min} = T_s \int_{-.5/T_s}^{.5/T_s} \frac{N_0}{f_0 + N_0} \, df = \frac{N_0}{f_0 + N_0}.$$

For high SNR, $f_0 \gg N_0$ and so $J_{\min} \approx N_0/f_0 = N_0/E_s$, where E_s/N_0 is the SNR per symbol. For low SNR, $N_0 \gg f_0$ and so $J_{\min} = N_0/(N_0 + f_0) \approx N_0/N_0 = 1$.

11.5 Maximum Likelihood Sequence Estimation

Maximum-likelihood sequence estimation avoids the problem of noise enhancement because it doesn't use an equalizing filter: instead it estimates the sequence of transmitted symbols. The structure of the MLSE is the same as in Figure 11.3, except that the equalizer $H_{eq}(z)$ and decision device are replaced by the MLSE algorithm. Given the combined pulse-shaping filter and channel response $h(t)$, the MLSE algorithm chooses the input sequence $\{d_k\}$ that maximizes the likelihood of the received signal $w(t)$. We now investigate this algorithm in more detail.

Using a Gram–Schmidt orthonormalization procedure, we can express $w(t)$ on a time interval $[0, LT_s]$ as

$$w(t) = \sum_{n=1}^{N} w_n \phi_n(t), \tag{11.28}$$

where $\{\phi_n(t)\}$ form a complete set of orthonormal basis functions. The number N of functions in this set is a function of the channel memory, since $w(t)$ on $[0, LT_s]$ depends on $d_0, \ldots, d_L$. With this expansion we have

$$w_n = \sum_{k=-\infty}^{\infty} d_k h_{nk} + v_n = \sum_{k=0}^{L} d_k h_{nk} + v_n, \tag{11.29}$$

where

$$h_{nk} = \int_0^{LT_s} h(t - kT_s)\phi_n^*(t) \, dt \tag{11.30}$$

and

$$v_n = \int_0^{LT_s} n(t)\phi_n^*(t) \, dt. \tag{11.31}$$

The v_n are complex Gaussian random variables with mean zero and covariance $.5 \, \mathbf{E}[v_n^* v_m] = N_0 \delta[n-m]$. Thus, $\mathbf{w}^N = (w_1, \ldots, w_N)$ has a multivariate Gaussian distribution:

$$p(\mathbf{w}^N \mid d^L, h(t)) = \prod_{n=1}^{N} \left[\frac{1}{\pi N_0} \exp\left[-\frac{1}{N_0} \left| w_n - \sum_{k=0}^{L} d_k h_{nk} \right|^2 \right] \right]. \tag{11.32}$$

Given a received signal $w(t)$ or, equivalently, $\mathbf{w}^N$, the MLSE decodes this as the symbol sequence d^L that maximizes the likelihood function $p(\mathbf{w}^N \mid d^L, h(t))$ (or the log of this function). That is, the MLSE outputs the sequence

$$\hat{d}^L = \arg\max[\log p(\mathbf{w}^N \mid d^L, h(t))]$$

$$= \arg\max\left[-\sum_{n=1}^{N} \left| w_n - \sum_k d_k h_{nk} \right|^2 \right]$$

$$= \arg\max\left[-\sum_{n=1}^{N} |w_n|^2 + \sum_{n=1}^{N} \left(w_n^* \sum_k d_k h_{nk} + w_n \sum_k d_k^* h_{nk}^* \right) \right.$$

$$\left. - \sum_{n=1}^{N} \left(\sum_k d_k h_{nk} \right) \left(\sum_m d_m^* h_{nm}^* \right) \right]$$

$$= \arg\max\left[2\,\mathrm{Re}\left\{ \sum_k d_k^* \sum_{n=1}^{N} w_n h_{nk}^* \right\} - \sum_k \sum_m d_k d_m^* \sum_{n=1}^{N} h_{nk} h_{nm}^* \right]. \tag{11.33}$$

Note that

$$\sum_{n=1}^{N} w_n h_{nk}^* = \int_{-\infty}^{\infty} w(\tau) h^*(\tau - kT_s) \, d\tau = y[k] \tag{11.34}$$

and

$$\sum_{n=1}^{N} h_{nk} h_{nm}^* = \int_{-\infty}^{\infty} h(\tau - kT_s) h^*(\tau - mT_s) \, d\tau = f[k-m]. \tag{11.35}$$

Combining (11.33), (11.34), and (11.35), we have that

$$\hat{d}^L = \arg\max\left[2\,\mathrm{Re}\left\{ \sum_k d_k^* y[k] \right\} - \sum_k \sum_m d_k d_m^* u[k-m] \right]. \tag{11.36}$$

We see from this equation that the MLSE output depends only on the sampler output $\{y[k]\}$ and the channel parameters $u[n-k] = u(nT_s - kT_s)$, where $u(t) = h(t) * h^*(-t)$. Because the derivation of the MLSE is based on the channel output $w(t)$ only (prior to matched filtering), our derivation implies that the receiver matched filter in Figure 11.3 with $g_m(t) = h(t)$ is optimal for MLSE detection (typically the matched filter is optimal for detecting signals in AWGN, but this derivation shows that it is also optimal for detecting signals in the presence of ISI if MLSE is used).

The Viterbi algorithm can be used for MLSE to reduce complexity [4; 5; 6; 7]. However, the complexity of this equalization technique still grows exponentially with the channel delay

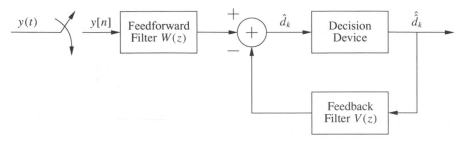

Figure 11.5: Decision-feedback equalizer structure.

spread. A nonlinear equalization technique with significantly less complexity is decision-feedback equalization.

11.6 Decision-Feedback Equalization

The DFE consists of a feedforward filter $W(z)$ with the received sequence as input (similar to the linear equalizer) followed by a feedback filter $V(z)$ with the previously detected sequence as input. The DFE structure is shown in Figure 11.5. In effect, the DFE determines the ISI contribution from the detected symbols $\{\hat{\hat{d}}_k\}$ by passing them through a feedback filter that approximates the composite channel $F(z)$ convolved with the feedforward filter $W(z)$. The resulting ISI is then subtracted from the incoming symbols. Since the feedback filter $V(z)$ in Figure 11.5 sits in a feedback loop, it must be strictly causal or else the system is unstable. The feedback filter of the DFE does not suffer from noise enhancement because it estimates the channel frequency response rather than its inverse. For channels with deep spectral nulls, DFEs generally perform much better than linear equalizers.

Assuming that $W(z)$ has $N_1 + 1$ taps and $V(z)$ has N_2 taps, we can write the DFE output as

$$\hat{d}_k = \sum_{i=-N_1}^{0} w_i y[k-i] - \sum_{i=1}^{N_2} v_i \hat{\hat{d}}_{k-i}.$$

The typical criteria for selecting the coefficients for $W(z)$ and $V(z)$ are either zero-forcing (remove all ISI) or MMSE (minimize expected MSE between the DFE output and the original symbol). We will assume that the matched filter in the receiver is perfectly matched to the channel: $g_m(t) = h(t)$. With this assumption, in order to obtain a causal filter for $V(z)$, we require a spectral factorization of $H(z)G_m^*(1/z^*) = H(z)H^*(1/z^*)$ [4, Chap. 10.1.2]. Specifically, $H(z)H^*(1/z^*)$ is factorized by finding a causal filter $B(z)$ such that $H(z)H^*(1/z^*) = B(z)B^*(1/z^*) = \lambda^2 B_1(z)B_1^*(1/z^*)$, where $B_1(z)$ is normalized to have a leading coefficient of 1. With this factorization, when both $W(z)$ and $V(z)$ have infinite duration, it was shown by Price [8] that the optimal feedforward and feedback filters for a zero-forcing DFE are (respectively) $W(z) = 1/\lambda^2 B_1^*(1/z^*)$ and $V(z) = 1 - B_1(z)$.

For the MMSE criterion, we wish to minimize $\mathbf{E}[d_k - \hat{d}_k]^2$. Let $b_n = b[n]$ denote the inverse z-transform of $B(z)$ given the spectral factorization $H(z)H^*(1/z^*) = B(z)B^*(1/z^*)$ just described. Then the MMSE minimization yields that the coefficients of the feedforward filter must satisfy the following set of linear equations:

$$\sum_{i=-N_1}^{0} q_{li} \hat{w}_i = b^*_{-l}$$

for $q_{li} = \sum_{j=-l}^{0} b_j^* b_{j+l-i} + N_0 \delta[l - i]$ with $i = -N_1, \ldots, 0$ [4, Chap. 10.3.1]. The coefficients of the feedback filter are then determined from the feedforward coefficients by

$$v_k = -\sum_{i=-N_1}^{0} \hat{w}_i b_{k-i}.$$

These coefficients completely eliminate ISI when there are no decision errors – that is, when $\hat{d}_k = d_k$. It was shown by Salz [9] that the resulting minimum MSE is

$$J_{\min} = \exp\left[T_s \int_{-.5/T_s}^{.5/T_s} \ln\left[\frac{N_0}{F_\Sigma(f) + N_0} \right] df \right].$$

In general, the MMSE associated with a DFE is much lower than that of a linear equalizer, assuming the impact of feedback errors is ignored.

Decision-feedback equalizers exhibit feedback errors if $\hat{d}_k \neq d_k$, since the ISI subtracted from the feedback path is not the true ISI corresponding to d_k. Such errors therefore propagate to later bit decisions. Moreover, this error propagation cannot be improved through channel coding, since the feedback path operates on coded channel symbols before decoding. This is because the ISI must be subtracted immediately, which doesn't allow for any decoding delay. Hence the error propagation seriously degrades performance on channels with low SNR. We can address this problem by introducing some delay in the feedback path to allow for channel decoding [10] or by turbo equalization, described in the next section. A systematic treatment of the DFE with coding can be found in [11; 12]. Moreover, the DFE structure can be generalized to encompass MIMO channels [13].

11.7 Other Equalization Methods

Although MLSE is the optimal form of equalization, its complexity precludes widespread use. There has been much work on reducing the complexity of the MLSE [4, Chap. 10.4]. Most techniques either reduce the number of surviving sequences in the Viterbi algorithm or reduce the number of symbols spanned by the ISI through preprocessing or decision-feedback in the Viterbi detector. These reduced-complexity MLSE equalizers have better performance–complexity trade-offs than symbol-by-symbol equalization techniques, and they achieve performance close to that of the optimal MLSE with significantly less complexity.

The turbo decoding principle introduced in Section 8.5 can also be used in equalizer design [14; 15]. The resulting design is called a *turbo equalizer*. A turbo equalizer iterates between a maximum a posteriori (MAP) equalizer and a decoder to determine the transmitted symbol. The MAP equalizer computes the a posteriori probability (APP) of the transmitted symbol given the past channel outputs. The decoder computes the log likelihood ratio (LLR) associated with the transmitted symbol given past channel outputs. The APP and LLR comprise the soft information exchanged between the equalizer and decoder in the turbo iteration.

After some number of iterations, the turbo equalizer converges on its estimate of the transmitted symbol.

If the channel is known at the transmitter, then the transmitter can *pre-equalize* the transmitted signal by passing it through a filter that effectively inverts the channel frequency response. Since the channel inversion occurs in the transmitter rather than the receiver, there is no noise enhancement. It is difficult to pre-equalize in a time-varying channel because the transmitter must have an accurate estimate of the channel, but this approach is practical to implement in relatively static wireline channels. One problem is that the channel inversion often increases the dynamic range of the transmitted signal, which can result in distortion or inefficiency from the amplifier. This problem has been addressed through a precoding technique called *Tomlinson–Harashima precoding* [16; 17].

11.8 Adaptive Equalizers: Training and Tracking

All of the equalizers described so far are designed based on a known value of the combined impulse response $h(t) = g(t) * c(t)$. Since the channel $c(t)$ in generally not known when the receiver is designed, the equalizer must be tunable so it can adjust to different values of $c(t)$. Moreover, since in wireless channels $c(t) = c(\tau, t)$ will change over time, the system must periodically estimate the channel $c(t)$ and update the equalizer coefficients accordingly. This process is called *equalizer training* or *adaptive equalization* [18; 19]. The equalizer can also use the detected data to adjust the equalizer coefficients, a process known as *equalizer tracking*. However, *blind equalizers* do not use training: they learn the channel response via the detected data only [20; 21; 22; 23].

During training, the coefficients of the equalizer are updated at time k based on a known training sequence $[d_{k-M}, \ldots, d_k]$ that has been sent over the channel. The length $M + 1$ of the training sequence depends on the number of equalizer coefficients that must be determined and the convergence speed of the training algorithm. Note that the equalizer must be retrained when the channel decorrelates – that is, at least every T_c seconds, where T_c is the channel coherence time. Thus, if the training algorithm is slow relative to the channel coherence time then the channel may change before the equalizer can learn the channel. Specifically, if $(M + 1)T_s > T_c$ then the channel will decorrelate before the equalizer has finished training. In this case equalization is not an effective countermeasure for ISI, and some other technique (e.g., multicarrier modulation or CDMA) is needed.

Let $\{\hat{d}_k\}$ denote the bit decisions output from the equalizer given a transmitted training sequence $\{d_k\}$. Our goal is to update the N equalizer coefficients at time $k + 1$ based on the training sequence we have received up to time k. We denote these updated coefficients as $\{w_{-L}(k + 1), \ldots, w_L(k + 1)\}$. We will use the MMSE as our criterion to update these coefficients; that is, we will choose $\{w_{-L}(k + 1), \ldots, w_L(k + 1)\}$ as the coefficients that minimize the MSE between d_k and $\hat{d}_k$. Recall that $\hat{d}_k = \sum_{i=-L}^{L} w_i(k + 1)y_{k-i}$, where $y_k = y[k]$ is the output of the sampler in Figure 11.3 at time k with the known training sequence as input. The $\{w_{-L}(k + 1), \ldots, w_L(k + 1)\}$ that minimize MSE are obtained via a Weiner filter [4; 5]. Specifically,

$$\mathbf{w}(k + 1) = \{w_{-L}(k + 1), \ldots, w_L(k + 1)\} = \mathbf{R}^{-1}\mathbf{p}, \tag{11.37}$$

where $\mathbf{p} = d_k[y_{k+L} \cdots y_{k-L}]^T$ and

Table 11.1: Equalizer training and tracking characteristics

Algorithm	No. of multiply operations per symbol time T_s	Complexity	Training convergence time	Tracking
LMS	$2N+1$	Low	Slow ($>10NT_s$)	Poor
MMSE	N^2 to N^3	Very high	Fast ($\sim NT_s$)	Good
RLS	$2.5N^2 + 4.5N$	High	Fast ($\sim NT_s$)	Good
Fast Kalman DFE	$20N+5$	Fairly low	Fast ($\sim NT_s$)	Good
Square-root LS DFE	$1.5N^2 + 6.5N$	High	Fast ($\sim NT_s$)	Good

$$\mathbf{R} = \begin{bmatrix} |y_{k+L}|^2 & y_{k+L}y^*_{k+L-1} & \cdots & y_{k+L}y^*_{k-L} \\ y_{k+L-1}y^*_{k+L} & |y_{k+L-1}|^2 & \cdots & y_{k+L-1}y^*_{k-L} \\ \vdots & \ddots & \ddots & \vdots \\ y_{k-L}y^*_{k+L} & \cdots & \cdots & |y_{k-L}|^2 \end{bmatrix}. \tag{11.38}$$

Note that finding the optimal tap updates in this case requires a matrix inversion, which requires N^2 to N^3 multiply operations on each iteration (each symbol time T_s). In exchange for its complexity, the convergence of this algorithm is very fast; it typically converges in around N symbol times for N the number of equalizer tap weights.

If complexity is an issue then the large number of multiply operations needed for MMSE training can be prohibitive. A simpler technique is the least mean square (LMS) algorithm [4, Chap. 11.1.2]. In this algorithm the tap weight vector $\mathbf{w}(k+1)$ is updated linearly as

$$\mathbf{w}(k+1) = \mathbf{w}(k) + \Delta\varepsilon_k[y^*_{k+L} \cdots y^*_{k-L}], \tag{11.39}$$

where $\varepsilon_k = d_k - \hat{d}_k$ is the error between the bit decisions and the training sequence and Δ is the step size of the algorithm. The choice of Δ dictates the convergence speed and stability of the algorithm. For small values of Δ the convergence is very slow; it takes many more than N bits for the algorithm to converge to the proper equalizer coefficients. However, if Δ is chosen to be large then the algorithm can become unstable, basically skipping over the desired tap weights at every iteration. Thus, for good performance of the LMS algorithm, Δ is typically small and convergence is typically slow. However, the LMS algorithm exhibits significantly reduced complexity compared to the MMSE algorithm, since the tap updates require only about $2N+1$ multiply operations per iteration. Thus, the complexity is linear in the number of tap weights. Other algorithms – such as the root least squares (RLS), square-root least squares, and fast Kalman – provide various trade-offs in terms of complexity and performance that lie between the two extremes of the LMS algorithm (slow convergence but low complexity) and the MMSE algorithm (fast convergence but high complexity). A description of these other algorithms is given in [4, Chap. 11]. Table 11.1 summarizes the specific number of multiply operations and the relative convergence rate of all these algorithms.

Note that the symbol decisions $\hat{d}_k$ output from the equalizer are typically passed through a threshold detector to round the decision to the nearest constellation point. The resulting roundoff error can be used to adjust the equalizer coefficients during data transmission, a process called equalizer tracking. Tracking is based two premises: (i) that if the roundoff error is nonzero then the equalizer is not perfectly trained; and (ii) that the roundoff error can be used to adjust the channel estimate inherent in the equalizer. The procedure works as follows. The equalizer output bits $\hat{d}_k$ and threshold detector output bits $\hat{\tilde{d}}_k$ are used to adjust an estimate of the equivalent lowpass composite channel $F(z)$. In particular, the coefficients of $F(z)$ are adjusted to minimize the MSE between $\hat{d}_k$ and $\hat{\tilde{d}}_k$, using the same MMSE procedures described previously. The updated version of $F(z)$ is then taken to equal the composite channel and used to update the equalizer coefficients accordingly. More details can be found in [4; 5]. Table 11.1 also includes a summary of the training and tracking characteristics for the different algorithms as a function of the number of taps N. Note that the fast Kalman and square-root LS may be unstable in their convergence and tracking, which is the price to be paid for their fast convergence with relatively low complexity.

EXAMPLE 11.5: Consider a five-tap equalizer that must retrain every $.5T_c$, where T_c is the coherence time of the channel. Assume the transmitted signal is BPSK ($T_s = T_b$) with a rate of 1 Mbps for both data and training sequence transmission. Compare the length of training sequence required for the LMS equalizer versus the fast Kalman DFE. For an 80-Hz Doppler, by how much is the data rate reduced in order to enable periodic training for each of these equalizers? How many operations do these two equalizers require for this training?

Solution: The equalizers must retrain every $.5T_c = .5/B_d = .5/80 = 6.25$ ms. For a data rate of $R_b = 1/T_b = 1$ Mbps, Table 11.1 shows that the LMS algorithm requires $10NT_b = 50 \cdot 10^{-6}$ seconds for training to converge and the fast Kalman DFE requires $NT_b = 50 \cdot 10^{-5}$ seconds. If training occurs every 6.25 ms then the fraction of time the LMS algorithm uses for training is $50 \cdot 10^{-6}/6.25 \cdot 10^{-3} = .008$. Thus, the effective data rate becomes $(1 - .008)R_b = .992$ Mbps. The fraction of time used by the fast Kalman DFE for training is $50 \cdot 10^{-5}/6.25 \cdot 10^{-3} = .0008$, resulting in an effective data rate of $(1 - .0008)R_b = .9992$ Mbps. The LMS algorithm requires approximately $(2N+1) \cdot 10N = 550$ operations for training per training period, whereas the fast Kalman DFE requires $(20N + 5) \cdot N = 525$ operations. With processor technology today, this is not a significant difference in terms of CPU requirements.

PROBLEMS

11-1. Design a continuous-time equivalent lowpass equalizer $H_{eq}(f)$ to completely remove the ISI introduced by a channel with equivalent lowpass impulse response $H(f) = 1/f$. Assume your transmitted signal has a bandwidth of 100 kHz. Assuming a channel with equivalent lowpass AWGN of PSD N_0, find the noise power at the output of your equalizer within the 100-kHz bandwidth of interest. Will this equalizer improve system performance?

11-2. This problem investigates the interference generated by ISI and the noise enhancement that occurs in zero-forcing equalization. Consider two multipath channels: the first channel has impulse response

$$h_1(t) = \begin{cases} 1 & 0 \le t < T_m, \\ 0 & \text{else}; \end{cases}$$

the second channel has impulse response

$$h_2(t) = e^{-t/T_m}, \quad 0 \le t < \infty.$$

(a) Assume that the transmitted signal $s(t)$ is an infinite sequence of impulses with amplitude A and time separation $T_b = T_m/2$; that is, $s(t) = \sum_{n=-\infty}^{\infty} A\delta(t - nT_b)$. Calculate the average ISI power over a bit time T_b.

(b) Let $T_m = 10\ \mu s$. Suppose a BPSK signal is transmitted over a channel with impulse response $h_1(t)$. Find the maximum data rate that can be sent over the channel with zero ISI under BPSK modulation with rectangular pulse shaping of pulse width $T = 1\ \mu s$. How would this answer change if the signal bandwidth were restricted to 100 kHz?

11-3. Consider a channel with equivalent lowpass impulse response

$$h(t) = \begin{cases} e^{-t/\tau} & t \geq 0, \\ 0 & \text{else,} \end{cases}$$

where $\tau = 6\ \mu s$. The channel also has equivalent lowpass AWGN with power spectral density N_0.

(a) What is the frequency response of the equivalent lowpass continuous-time ZF linear equalizer for this channel? Assume no matched filter or pulse shaping.

(b) Suppose we transmit a 30-kHz signal over this channel, and assume that the frequency response of the signal is a rectangular pulse shape. What is the ratio of SNR with equalization to SNR without equalization in the bandwidth of our transmitted signal? *Hint:* Recall that a stationary random process with power spectral density $P(f)$ has total power $\int P(f)\, df$ and that, if this process is passed through a filter $G(f)$, then the output process has power spectral density $P(f)|G(f)|^2$.

(c) Approximate the equivalent lowpass MMSE equalizer for this channel using a discrete-time transversal filter with three taps. Use any method you like as long as it reasonably approximates the time-domain response of the MMSE equalizer.

11-4. Consider an FIR ZF equalizer with tap weights $w_i = c_i$, where $\{c_i\}$ is the inverse z-transform of $1/F(z)$. Show that this choice of tap weights minimizes

$$\left| \frac{1}{F(z)} - (w_0 + w_1 z^{-1} + \cdots + w_N z^{-N}) \right|^2$$

at $z = e^{j\omega}$.

11-5. Consider a communication system where the modulated signal $s(t)$ has power 10 mW, carrier frequency f_c, and passband bandwidth $B_s = 40$ MHz. The signal $s(t)$ passes through a frequency-selective fading channel with symmetric frequency response $H(f) = H(-f)$, where

$$H(f) = \begin{cases} 1 & f_c - 20\ \text{MHz} \leq f < f_c - 10\ \text{MHz}, \\ .5 & f_c - 10\ \text{MHz} \leq f < f_c, \\ 2 & f_c \leq f < f_c + 10\ \text{MHz}, \\ .25 & f_c + 10\ \text{MHz} \leq f < f_c + 20\ \text{MHz}, \\ 0 & \text{else.} \end{cases}$$

The received signal is $y(t) = s(t) * h(t) + n(t)$, where $n(t)$ is AWGN with PSD $N_0/2 = 10^{-12}$ W/Hz.

(a) Suppose $y(t)$ is passed through a continuous-time (passband) ZF equalizer. Find the frequency response $H_{eq}(f)$ for this equalizer within the bandwidth of interest ($f_c \pm 20$ MHz).

(b) For the equalizer of part (a), find the SNR at the equalizer output.

(c) Suppose the symbol time for $s(t)$ is $T_s = .5/B_s$ and assume no restrictions on the constellation size. Find the maximum data rate that can be sent over the channel with the ZF equalizer of part (a) such that $P_b < 10^{-3}$.

11-6. Consider an ISI channel with received signal after transmission through the channel given by

$$y(t) = \sum_{i=-\infty}^{\infty} x_i f(t - iT),$$

where $x_i = \pm 1$ and $f(t)$ is the composite impulse response consisting of the pulse-shaping filter, the equivalent lowpass channel, and the receiver matched filter. Assume that $f(t) = \sin(\pi t/T)/(\pi t/T)$, which satisfies the Nyquist criterion for zero ISI. There are two difficulties with this pulse shape: first, it has a rectangular spectrum, which is difficult to implement in practice. Second, the tails of the pulse decay as $1/t$, so timing error leads to a sequence of ISI samples that do not converge. For parts (b) and (c) of this problem we make the assumption that $f(t) = 0$ for $|t| > NT$, where N is a positive integer. This is not strictly correct, since it would imply that $f(t)$ is both time limited and bandlimited. However, it is a reasonable approximation in practice.

(a) Show that the folded spectrum of $f(t)$ is flat.
(b) Suppose that (because of timing error) the signal is sampled at $t = kT + t_0$, where $t_0 < T$. Calculate the response $y_k = y(kT + t_0)$ and separate your answer into the desired term and the ISI terms.
(c) Assume that the polarities of the x_i are such that every term in the ISI is positive (worst-case ISI). Under this assumption, show that the ISI term from part (b) is

$$\text{ISI} \approx \frac{2}{\pi} \sin\left(\frac{\pi t_0}{T}\right) \sum_{n=1}^{N} \frac{n}{n^2 - t_0^2/T^2}$$

and therefore ISI $\to \infty$ as $N \to \infty$.

11-7. Let $g(t) = \text{sinc}(t/T_s)$, $|t| < T_s$. Find the matched filter $g_m(t)$ for $g(t)$. Find the noise-whitening filter $1/G_m^*(1/z^*)$ for this system that must be used in an MMSE equalizer to whiten the noise.

11-8. Show that the minimum MSE (11.22) for an IIR MMSE equalizer can be expressed in terms of the folded spectrum $F_\Sigma(f)$ as

$$J_{\min} = T \int_{-.5/T}^{.5/T} \frac{N_0}{F_\Sigma(f) + N_0} \, df.$$

11-9. Show that the gradient of tap weights associated with the MMSE equalizer is given by

$$\nabla_{\mathbf{w}} J = \left(\frac{\partial J}{\partial w_0}, \dots, \frac{\partial J}{\partial w_N}\right) = 2\mathbf{w}^T \mathbf{M_v} - 2\mathbf{v_d}.$$

Set this equal to zero and solve for the optimal tap weights to obtain

$$\mathbf{w}_{\text{opt}} = (\mathbf{M_v^T})^{-1} \mathbf{v_d^H}.$$

11-10. Show that the MMSE $J_{\min}$ for an IIR MMSE equalizer, given by (11.27), satisfies $0 \le J_{\min} \le 1$.

11-11. Compare the value of the minimum MSE, $J_{\min}$, under both MMSE equalization and DF equalization for a channel with three-tap discrete-time equivalent model $C(z) = 1 + .5z^{-1} + .3z^{-2}$.

11-12. This problem illustrates (i) the noise enhancement of zero-forcing equalizers and (ii) how this enhancement can be mitigated using an MMSE approach. Consider a frequency-selective fading channel with equivalent lowpass frequency response

$$H(f) = \begin{cases} 1 & 0 \leq |f| < 10 \text{ kHz}, \\ 1/2 & 10 \text{ kHz} \leq |f| < 20 \text{ kHz}, \\ 1/3 & 20 \text{ kHz} \leq |f| < 30 \text{ kHz}, \\ 1/4 & 30 \text{ kHz} \leq |f| < 40 \text{ kHz}, \\ 1/5 & 40 \text{ kHz} \leq |f| < 50 \text{ kHz}, \\ 0 & \text{else.} \end{cases}$$

The frequency response is symmetric in positive and negative frequencies. Assume an AWGN channel with equivalent lowpass noise PSD $N_0 = 10^{-9}$.

(a) Find a ZF analog equalizer that completely removes the ISI introduced by $H(f)$.

(b) Find the total noise power at the output of the equalizer from part (a).

(c) Assume an MMSE analog equalizer of the form $H_{eq}(f) = 1/(H(f) + \alpha)$. Find the total noise power at the output of this equalizer for an AWGN input with PSD N_0 for $\alpha = .5$ and for $\alpha = 1$.

(d) Describe qualitatively two effects on a signal that is transmitted over channel $H(f)$ and then passed through the MMSE equalizer $H_{eq}(f) = 1/(H(f) + \alpha)$ with $\alpha > 0$. What design considerations should go into the choice of α?

(e) What happens to the total noise power for the MMSE equalizer in part (c) as $\alpha \rightarrow \infty$? What is the disadvantage of letting $\alpha \rightarrow \infty$ in this equalizer design?

(f) For the equalizer design of part (c), suppose the system has a data rate of 100 kbps and that your equalizer requires a training sequence of 1000 bits to train. What is the maximum channel Doppler such that the equalizer coefficients converge before the channel decorrelates?

11-13. Why does an equalizer that tracks the channel during data transmission still need to train periodically? Specify two benefits of tracking.

11-14. Assume a four-tap equalizer that must retrain every $.5T_c$, where T_c is the channel coherence time. If a digital signal processing chip can perform 10 million multiplications per second and if the convergence rates for the LMS DFE algorithm and the RLS algorithm are, respectively, 1000 iterations (symbol times) and 50 iterations, then what is the maximum data rate for both equalizers assuming BPSK modulation and Doppler spread $B_D = 100$ Hz? For $B_D = 1000$ Hz? Assume that the transmitting speed is equal for training and information sequences.

11-15. In this problem we seek the procedure for updating the channel estimate during tracking. Find the formula for updating the channel coefficients corresponding to the channel $H(z)$ based on minimizing the MSE between $\hat{d}_k$ and $\hat{\hat{d}}_k$.

11-16. Ultrawideband (UWB) systems spread a data signal and its corresponding power over a very wide bandwidth so that the power per hertz of the signal is small (typically below the noise floor). Hence such systems coexist with other systems without causing them much interference. Consider a baseband UWB system with BPSK modulation. The data bits are modulated with a rectangular pulse $g(t)$, where $g(t)$ has a very narrow time duration T as compared to the bit time T_b. For this problem we assume $T = 10^{-9}$. Thus, a UWB signal

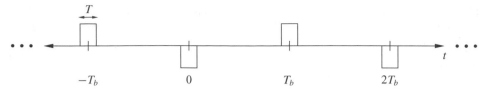

Figure 11.6: UWB signal form for Problem 11-16.

with BPSK modulation would have the form $s(t) = \sum_n d_n g(t - nT_b)$, where d_n takes the value ± 1 and $T_b \gg T$ is the bit time. A sketch of $s(t)$ with a data sequence of alternating 1-bits and 0-bits is shown in Figure 11.6.

(a) What is the approximate bandwidth of $s(t)$ if $T_b = 10^{-5}$?

(b) One of the selling points of UWB is that its signals do not experience flat fading in typical channels. Consider a single-bit transmission $s(t) = d_0 g(t)$. Suppose $s(t)$ is transmitted through a channel that follows a two-ray model $h(t) = \alpha_0 \delta(t) + \alpha_1 \delta(t - \tau)$. Sketch the channel output for $\tau \ll T$ and $\tau \gg T$. Which of your two sketches is more likely to depict the output of a real wireless channel? Why does this imply that UWB signals don't typically experience flat fading?

(c) Consider a channel with a multipath delay spread of $T_m = 20\ \mu s$. For Figure 11.6, what is the *exact* maximum data rate that can be sent over this channel with no ISI? Is the bandwidth of $s(t)$ in the figure less than the channel coherence bandwidth at this data rate?

(d) Let $F(z) = \alpha_0 + \alpha_1 z^{-1} + \alpha_2 z^{-2}$ denote the composite impulse response of the transmitter, channel, and matched filter in this UWB system. Find a two-tap digital equalizer $H_{eq}(z) = w_0 + w_1 z^{-1}$ that approximates an IIR zero-forcing equalizer for $F(z)$. Any reasonable approximation is fine as long as you can justify it.

(e) For the equalizer designed in part (d), suppose the system has a data rate of 100 kbps and that your equalizer requires a training sequence of 1000 bits to train. What is the maximum channel Doppler such that the equalizer coefficients converge before the channel decorrelates?

REFERENCES

[1]   E. H. Satorius and S. T. Alexander, "Channel equalization using adaptive lattice algorithms," *IEEE Trans. Commun.,* pp. 899–905, June 1979.

[2]   F. Ling and J. Proakis, "Adaptive lattice decision feedback equalizers – Their performance and application to time-variant multipath channels," *IEEE Trans. Commun.,* pp. 348–56, April 1985.

[3]   J. Cioffi and T. Kailath, "Fast, recursive-least-squares transversal filters for adaptive filtering," *IEEE Trans. Signal Proc.,* pp. 304–37, April 1984.

[4]   J. G. Proakis, *Digital Communications,* 4th ed., McGraw-Hill, New York, 2001.

[5]   G. L. Stuber, *Principles of Mobile Communications,* 2nd ed., Kluwer, Dordrecht, 2001.

[6]   G. D. Forney, Jr., "Maximum-likelihood sequence estimation of digital sequences in the presence of intersymbol interference," *IEEE Trans. Inform. Theory,* pp. 363–78, May 1972.

[7]   B. Sklar, "How I learned to love the trellis," *IEEE Signal Proc. Mag.,* pp. 87–102, May 2003.

[8]   R. Price, "Nonlinearly feedback-equalized PAM vs. capacity," *Proc. IEEE Internat. Conf. Commun.,* pp. 22.12–22.17, June 1972.

[9] J. Salz, "Optimum mean-square decision feedback equalization," *Bell System Tech. J.,* pp. 1341–73, October 1973.

[10] M. V. Eyuboglu, "Detection of coded modulation signals on linear, severely distorted channels using decision-feedback noise prediction with interleaving," *IEEE Trans. Commun.,* pp. 401–9, April 1988.

[11] J. M. Cioffi, G. P. Dudevoir, V. Eyuboglu, and G. D. Forney, Jr., "MMSE decision-feedback equalizers and coding. Part I: Equalization results," *IEEE Trans. Commun.,* pp. 2582–94, October 1995.

[12] J. M. Cioffi, G. P. Dudevoir, V. Eyuboglu, and G. D. Forney, Jr., "MMSE decision-feedback equalizers and coding. Part II: Coding results," *IEEE Trans. Commun.,* pp. 2595–2604, October 1995.

[13] J. M. Cioffi and G. D. Forney, Jr., "Generalized decision-feedback equalization for packet transmission with ISI and Gaussian noise," in A. Paulraj, V. Roychowdhury, and C. Schaper (Eds.), *Communication, Computation, Control, and Signal Processing,* pp. 79–127, Kluwer, Boston, 1997.

[14] C. Douillard, M. Jezequel, C. Berrou, A. Picart, P. Didier, and A. Glavieux, "Iterative correction of intersymbol interference: Turbo equalization," *Euro. Trans. Telecommun.,* pp. 507–11, September/October 1995.

[15] M. Tüchler, R. Koetter, and A. C. Singer, "Turbo equalization: Principles and new results," *IEEE Trans. Commun.,* pp. 754–67, May 2002.

[16] M. Tomlinson, "A new automatic equalizer emplying modulo arithmetic," *Elec. Lett.,* 7, pp. 138–9, 1971.

[17] H. Harashima and H. Miyakawa, "Matched-transmission techniques for channels with intersymbol interference," *IEEE Trans. Commun.,* pp. 774–80, August 1972.

[18] J. G. Proakis, "Adaptive equalization for TDMA digital mobile radio," *IEEE Trans. Veh. Tech.,* pp. 333–41, May 1991.

[19] S. U. Qureshi, "Adaptive equalization," *Proc. IEEE,* pp. 1349–87, September 1985.

[20] A. Benveniste and M. Goursat, "Blind equalizers," *IEEE Trans. Commun.,* pp. 871–83, August 1984.

[21] C. R. Johnson Jr., "Admissibility in blind adaptive channel equalization," *IEEE Control Syst. Mag.,* pp. 3–15, January 1991.

[22] R. Johnson, P. Schniter, T. J. Endres, J. D. Behm, D. R. Brown, and R. A. Casas, "Blind equalization using the constant modulus criterion: A review," *Proc. IEEE,* pp. 1927–50, October 1998.

[23] L. Tong, G. Zu, and T. Kailath, "Blind identification and equalization based on second-order statistics: A time domain approach," *IEEE Trans. Inform. Theory,* pp. 340–9, March 1994.

Multicarrier Modulation

The basic idea of multicarrier modulation is to divide the transmitted bitstream into many different substreams and send these over many different subchannels. Typically the subchannels are orthogonal under ideal propagation conditions. The data rate on each of the subchannels is much less than the total data rate, and the corresponding subchannel bandwidth is much less than the total system bandwidth. The number of substreams is chosen to ensure that each subchannel has a bandwidth less than the coherence bandwidth of the channel, so the subchannels experience relatively flat fading. Thus, the intersymbol interference on each subchannel is small. The subchannels in multicarrier modulation need not be contiguous, so a large continuous block of spectrum is not needed for high-rate multicarrier communications. Moreover, multicarrier modulation is efficiently implemented digitally. In this discrete implementation, called orthogonal frequency division multiplexing (OFDM), the ISI can be completely eliminated through the use of a cyclic prefix.

Multicarrier modulation is currently used in many wireless systems. However, it is not a new technique: it was first used for military HF radios in the late 1950s and early 1960s. Starting around 1990 [1], multicarrier modulation has been used in many diverse wired and wireless applications, including digital audio and video broadcasting in Europe [2], digital subscriber lines (DSL) using discrete multitone [3; 4; 5], and the most recent generation of wireless LANs [5; 6; 7]. There are also a number of newly emerging uses for multicarrier techniques, including fixed wireless broadband services [8; 9], mobile wireless broadband known as FLASH-OFDM [10], and even for ultrawideband radios, where multiband OFDM is one of two competing proposals for the IEEE 802.15 ultrawideband standard. Multicarrier modulation is also a candidate for the air interface in next-generation cellular systems [11; 12].

The multicarrier technique can be implemented in multiple ways, including vector coding [13; 14] and OFDM [13; 15], all of which are discussed in this chapter. These techniques have subtle differences, but all are based on the same premise of breaking a wideband channel into multiple parallel narrowband channels by means of an orthogonal channel partition.

There is some debate as to whether multicarrier or single-carrier modulation is better for ISI channels with delay spreads on the order of the symbol time. It is claimed in [2] that, for some mobile radio applications, single-carrier modulation with equalization has roughly the same performance as multicarrier modulation with channel coding, frequency-domain interleaving, and weighted maximum likelihood decoding. Adaptive loading was not taken into account in [2], though it has the potential to significantly improve multicarrier performance [16]. But there are other problems with multicarrier modulation that impair its performance, most significantly frequency offset and timing jitter, which degrade the orthogonality of the subchannels. In addition, the peak-to-average power ratio of multicarrier systems is significantly higher than that of single-carrier systems, which is a serious problem when nonlinear amplifiers are used. Trade-offs between multicarrier and single-carrier block transmission systems with respect to these impairments are discussed in [17].

Despite these challenges, multicarrier techniques are common in high–data-rate wireless systems with moderate to large delay spread, as they have significant advantages over time-domain equalization. In particular, the number of taps required for an equalizer with good performance in a high–data-rate system is typically large. Thus, these equalizers are quite complex. Moreover, it is difficult to maintain accurate weights for a large number of equalizer taps in a rapidly varying channel. For these reasons, most emerging high-rate wireless systems use either multicarrier modulation or spread spectrum instead of equalization to compensate for ISI.

12.1 Data Transmission Using Multiple Carriers

The simplest form of multicarrier modulation divides the data stream into multiple substreams to be transmitted over different orthogonal subchannels centered at different subcarrier frequencies. The number of substreams is chosen to make the symbol time on each substream much greater than the delay spread of the channel or, equivalently, to make the substream bandwidth less than the channel coherence bandwidth. This ensures that the substreams will not experience significant ISI.

Consider a linearly modulated system with data rate R and bandwidth B. The coherence bandwidth for the channel is assumed to be $B_c < B$, so the signal experiences frequency-selective fading. The basic premise of multicarrier modulation is to break this wideband system into N linearly modulated subsystems in parallel, each with subchannel bandwidth $B_N = B/N$ and data rate $R_N \approx R/N$. For N sufficiently large, the subchannel bandwidth $B_N = B/N \ll B_c$, which ensures relatively flat fading on each subchannel. This can also be seen in the time domain: the symbol time T_N of the modulated signal in each subchannel is proportional to the subchannel bandwidth $1/B_N$. So $B_N \ll B_c$ implies that $T_N \approx 1/B_N \gg 1/B_c \approx T_m$, where T_m denotes the delay spread of the channel. Thus, if N is sufficiently large, the symbol time is much greater than the delay spread, so each subchannel experiences little ISI degradation.

Figure 12.1 illustrates a multicarrier transmitter.[1] The bit stream is divided into N substreams via a serial-to-parallel converter. The nth substream is linearly modulated (typically via QAM or PSK) relative to the subcarrier frequency f_n and occupies bandwidth B_N. We assume coherent demodulation of the subcarriers so the subcarrier phase is neglected in our analysis. If we assume raised cosine pulses for $g(t)$ we get a symbol time $T_N = (1 + \beta)/B_N$

---

[1] In practice, the complex symbol s_i would have its real part transmitted over the in-phase signaling branch and its imaginary part transmitted over the quadrature signaling branch. For simplicity we illustrate multicarrier transmission based on sending a complex symbol along the in-phase signaling branch.

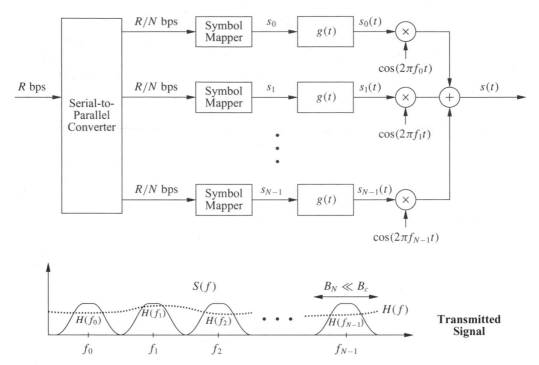

Figure 12.1: Multicarrier transmitter.

for each substream, where β is the rolloff factor of the pulse shape. The modulated signals associated with all the subchannels are summed together to form the transmitted signal, given as

$$s(t) = \sum_{i=0}^{N-1} s_i g(t) \cos(2\pi f_i t + \phi_i), \qquad (12.1)$$

where s_i is the complex symbol associated with the ith subcarrier and ϕ_i is the phase offset of the ith carrier. For nonoverlapping subchannels we set $f_i = f_0 + i(B_N), i = 0, \ldots, N-1$. The substreams then occupy orthogonal subchannels with bandwidth B_N, yielding a total bandwidth $NB_N = B$ and data rate $NR_N \approx R$. Thus, this form of multicarrier modulation does not change the data rate or signal bandwidth relative to the original system, but it almost completely eliminates ISI for $B_N \ll B_c$.

The receiver for this multicarrier modulation is shown in Figure 12.2. Each substream is passed through a narrowband filter (to remove the other substreams), demodulated, and combined via a parallel-to-serial converter to form the original data stream. Note that the ith subchannel will be affected by flat fading corresponding to a channel gain $\alpha_i = H(f_i)$.

Although this simple type of multicarrier modulation is easy to understand, it has several significant shortcomings. First, in a realistic implementation, subchannels will occupy a larger bandwidth than under ideal raised cosine pulse shaping because the pulse shape must be time limited. Let ε/T_N denote the additional bandwidth required due to time limiting of these pulse shapes. The subchannels must then be separated by $(1 + \beta + \varepsilon)/T_N$, and since the multicarrier system has N subchannels, the bandwidth penalty for time limiting is $\varepsilon N/T_N$. In particular, the total required bandwidth for nonoverlapping subchannels is

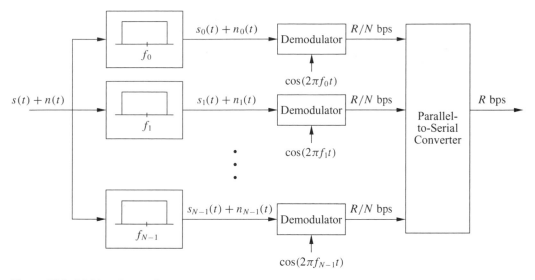

Figure 12.2: Multicarrier receiver.

$$B = \frac{N(1 + \beta + \varepsilon)}{T_N}.\tag{12.2}$$

Thus, this form of multicarrier modulation can be spectrally inefficient. Additionally, near-ideal (and hence expensive) lowpass filters will be required to maintain the orthogonality of the subcarriers at the receiver. Perhaps most importantly, this scheme requires N independent modulators and demodulators, which entails significant expense, size, and power consumption. The next section presents a modulation method that allows subcarriers to overlap and removes the need for tight filtering. Section 12.4 presents the discrete implementation of multicarrier modulation, which eliminates the need for multiple modulators and demodulators.

EXAMPLE 12.1: Consider a multicarrier system with a total passband bandwidth of 1 MHz. Suppose the system operates in a city with channel delay spread $T_m = 20\,\mu s$. How many subchannels are needed to obtain approximately flat fading in each subchannel?

Solution: The channel coherence bandwidth is $B_c = 1/T_m = 1/.00002 = 50$ kHz. To ensure flat fading on each subchannel, we take $B_N = B/N = .1B_c \ll B_c$. Thus, $N = B/.1B_c = 1000000/5000 = 200$ subchannels are needed to ensure flat fading on each subchannel. In discrete implementations of multicarrier modulation, N must be a power of 2 for the DFT (discrete Fourier transform) and IDFT (inverse DFT) operations, in which case $N = 256$ for this set of parameters.

EXAMPLE 12.2: Consider a multicarrier system with $T_N = .2$ ms: $T_N \gg T_m$ for T_m the channel delay spread, so each subchannel experiences minimal ISI. Assume the system has $N = 128$ subchannels. If time-limited raised cosine pulses with $\beta = 1$ are used and if the additional bandwidth required (because of time limiting) to ensure minimal power outside the signal bandwidth is $\varepsilon = .1$, then what is the total bandwidth of the system?

Solution: From (12.2),

$$B = \frac{N(1 + \beta + \varepsilon)}{T_N} = \frac{128(1 + 1 + .1)}{.0002} = 1.344 \text{ MHz.}$$

We will see in the next section that the bandwidth requirements for this system can be substantially reduced by overlapping subchannels.

12.2 Multicarrier Modulation with Overlapping Subchannels

We can improve on the spectral efficiency of multicarrier modulation by overlapping the subchannels. The subcarriers must still be orthogonal so that they can be separated out by the demodulator in the receiver. The subcarriers $\{\cos(2\pi(f_0 + i/T_N)t + \phi_i), i = 0, 1, 2, \ldots\}$ form a set of (approximately) orthogonal basis functions on the interval $[0, T_N]$ for any set of subcarrier phase offsets $\{\phi_i\}$, since

$$\frac{1}{T_N} \int_0^{T_N} \cos\left(2\pi\left(f_0 + \frac{i}{T_N}\right)t + \phi_i\right) \cos\left(2\pi\left(f_0 + \frac{j}{T_N}\right)t + \phi_j\right) dt$$

$$= \frac{1}{T_N} \int_0^{T_N} .5 \cos\left(2\pi \frac{(i-j)t}{T_N} + \phi_i - \phi_j\right) dt$$

$$+ \frac{1}{T_N} \int_0^{T_N} .5 \cos\left(2\pi\left(2f_0 + \frac{i+j}{T_N}\right)t + \phi_i + \phi_j\right) dt$$

$$\approx \frac{1}{T_N} \int_0^{T_N} .5 \cos\left(2\pi \frac{(i-j)t}{T_N} + \phi_i - \phi_j\right) dt$$

$$= .5\delta(i - j), \tag{12.3}$$

where the approximation follows because the integral in the third line of (12.3) is approximately zero for $f_0 T_N \gg 1$. Moreover, it is easily shown that no set of subcarriers with a smaller frequency separation forms an orthogonal set on $[0, T_N]$ for arbitrary subcarrier phase offsets. This implies that the minimum frequency separation required for subcarriers to remain orthogonal over the symbol interval $[0, T_N]$ is $1/T_N$. Since the carriers are orthogonal, by Section 5.1 the set of functions $\{g(t) \cos(2\pi(f_0 + i/T_N)t + \phi_i), i = 0, 1, \ldots, N-1\}$ also form a set of (approximately) orthonormal basis functions for appropriately chosen baseband pulse shapes $g(t)$: the family of raised cosine pulses are a common choice for this pulse shape. Given this orthonormal basis set, even if the subchannels overlap, the modulated signals transmitted in each subchannel can be separated out in the receiver, as we now show.

Consider a multicarrier system where each subchannel is modulated using raised cosine pulse shapes with rolloff factor β. The bandwidth of each subchannel is then $B_N = (1 + \beta)/T_N$. The ith subcarrier frequency is set to $(f_0 + i/T_N), i = 0, 1, \ldots, N-1$, for some f_0, so the subcarriers are separated by $1/T_N$. However, the bandwidth of each subchannel is $B_N = (1 + \beta)/T_N > 1/T_N$ for $\beta > 0$, so the subchannels overlap. Excess bandwidth due to time windowing will increase the subcarrier bandwidth by an additional ε/T_N. However, β and ε do not affect the total system bandwidth resulting from the subchannel overlap except in the first and last subchannels, as illustrated in Figure 12.3. The total system bandwidth with overlapping subchannels is given by

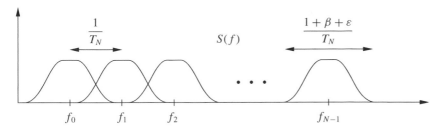

Figure 12.3: Multicarrier with overlapping subcarriers.

$$B = \frac{N + \beta + \varepsilon}{T_N} \approx \frac{N}{T_N}, \tag{12.4}$$

where the approximation holds for N large. Thus, with N large, the impact of β and ε on the total system bandwidth is negligible, in contrast to the required bandwidth of $B = N(1 + \beta + \varepsilon)/T_N$ when the subchannels do not overlap.

EXAMPLE 12.3: Compare the required bandwidth of a multicarrier system with overlapping subchannels versus nonoverlapping subchannels using the same parameters as in Example 12.2.

Solution: In the prior example $T_N = .2$ ms, $N = 128$, $\beta = 1$, and $\varepsilon = .1$. With overlapping subchannels, from (12.4) we have

$$B = \frac{N + \beta + \varepsilon}{T_N} = \frac{128 + 1 + .1}{.0002} = 645.5 \text{ kHz} \approx N/T_N = 640 \text{ kHz}.$$

By comparison, in the prior example the required bandwidth with nonoverlapping subchannels was shown to be 1.344 MHz, more than double the required bandwidth when the subchannels overlap.

Clearly, in order to separate out overlapping subcarriers, a different receiver structure is needed than the one shown in Figure 12.2. In particular, overlapping subchannels are demodulated with the receiver structure shown in Figure 12.4, which demodulates the appropriate symbol without interference from overlapping subchannels. Specifically, if the effect of the channel $h(t)$ and noise $n(t)$ are neglected then, for received signal $s(t)$ given by (12.1), the input to each symbol demapper in Figure 12.4 is

$$\hat{s}_i = \int_0^{T_N} \left(\sum_{j=0}^{N-1} s_j g(t) \cos(2\pi f_j t + \phi_j) \right) g(t) \cos(2\pi f_i t + \phi_i) \, dt$$

$$= \sum_{j=0}^{N-1} s_j \int_0^{T_N} g^2(t) \cos\left(2\pi \left(f_0 + \frac{j}{T_N} \right) t + \phi_j \right) \cos\left(2\pi \left(f_0 + \frac{i}{T_N} \right) t + \phi_i \right) dt$$

$$= \sum_{j=0}^{N-1} s_j \delta(j - i) \tag{12.5}$$

$$= s_i, \tag{12.6}$$

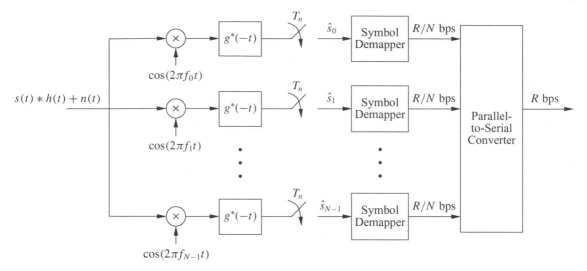

Figure 12.4: Multicarrier receiver for overlapping subcarriers.

where (12.5) follows from the fact that the functions $\{g(t)\cos(2\pi f_j t + \phi_j)\}$ form a set of orthonormal basis functions on $[0, T_N]$. If the channel and noise effects are included, then the symbol in the ith subchannel is scaled by the channel gain $\alpha_i = H(f_i)$ and corrupted by the noise sample, so $\hat{s}_i = \alpha_i s_i + n_i$, where n_i is additive white Gaussian noise with power $N_0 B_N$. This multicarrier system makes much more efficient use of bandwidth than in systems with nonoverlapping subcarriers. However, since the subcarriers overlap, their orthogonality is compromised by timing and frequency offset. Even when relatively small, these effects can significantly degrade performance, as they cause subchannels to interfere with each other. These effects are discussed in more detail in Section 12.5.2.

12.3 Mitigation of Subcarrier Fading

The advantage of multicarrier modulation is that each subchannel is relatively narrowband, which mitigates the effect of delay spread. However, each subchannel experiences flat fading, which can cause large bit error rates on some of the subchannels. In particular, if the transmit power on subcarrier i is P_i and if the fading on that subcarrier is α_i, then the received signal-to-noise power ratio is $\gamma_i = \alpha_i^2 P_i / N_0 B_N$, where B_N is the bandwidth of each subchannel. If α_i is small then the received SNR on the ith subchannel is low, which can lead to a high BER on that subchannel. Moreover, in wireless channels α_i will vary over time according to a given fading distribution, resulting in the same performance degradation as is associated with flat fading for single-carrier systems discussed in Chapter 6. Because flat fading can seriously degrade performance in each subchannel, it is important to compensate for flat fading in the subchannels. There are several techniques for doing this, including coding with interleaving over time and frequency, frequency equalization, precoding, and adaptive loading, all described in subsequent sections. Coding with interleaving is the most common, and it has been adopted as part of the European standards for digital audio and video broadcasting [2; 18]. Moreover, in rapidly changing channels it is difficult to estimate the channel at the receiver and feed this information back to the transmitter. Without channel information at the transmitter, precoding and adaptive loading cannot be done, so only coding with interleaving is effective at fading mitigation.

12.3.1 Coding with Interleaving over Time and Frequency

The basic idea in coding with interleaving over time and frequency is to encode data bits into codewords, interleave the resulting coded bits over both time and frequency, and then transmit the coded bits over different subchannels such that the coded bits within a given codeword all experience independent fading [19]. If most of the subchannels have a high SNR, the codeword will have most coded bits received correctly, and the errors associated with the few bad subchannels can be corrected. Coding across subchannels basically exploits the frequency diversity inherent in a multicarrier system to correct for errors. This technique works well only if there is sufficient frequency diversity across the total system bandwidth. If the coherence bandwidth of the channel is large, then the fading across subchannels will be highly correlated, which will significantly reduce the benefits of coding. Most coding schemes assume channel information in the decoder. Channel estimates are typically obtained by a two-dimensional pilot symbol transmission over both time and frequency [20].

Note that coding with frequency/time interleaving takes advantage of the fact that the data on all the subcarriers are associated with the same user and can therefore be jointly processed. The other techniques for fading mitigation discussed in subsequent sections are all basically flat fading compensation techniques, which apply equally to multicarrier systems as well as to narrowband flat fading single-carrier systems.

12.3.2 Frequency Equalization

In frequency equalization the flat fading α_i on the ith subchannel is basically inverted in the receiver [2]. Specifically, the received signal is multiplied by $1/\alpha_i$, which gives a resultant signal power $\alpha_i^2 P_i/\alpha_i^2 = P_i$. While this removes the impact of flat fading on the signal, it enhances the noise power. Hence the incoming noise signal is also multiplied by $1/\alpha_i$, so the noise power becomes $N_0 B_N/\alpha_i^2$ and the resultant SNR on the ith subchannel after frequency equalization is the same as before equalization. Therefore, frequency equalization does not really change the performance degradation associated with subcarrier flat fading.

12.3.3 Precoding

Precoding uses the same idea as frequency equalization, except that the fading is inverted at the transmitter instead of at the receiver [21]. This technique requires the transmitter to have knowledge of the subchannel flat fading gains α_i ($i = 0, \ldots, N-1$), which must be obtained through estimation [22]. In this case, if the desired received signal power in the ith subchannel is P_i and if the channel introduces a flat fading gain α_i in the ith subchannel, then under precoding the power transmitted in the ith subchannel is P_i/α_i^2. The subchannel signal is corrupted by flat fading with gain α_i, so the received signal power is $P_i\alpha_i^2/\alpha_i^2 = P_i$, as desired. Note that the channel inversion takes place at the transmitter instead of the receiver, so the noise power remains $N_0 B_N$. Precoding is quite common on wireline multicarrier systems like high–bit-rate digital subscriber lines (HDSL). There are two main problems with precoding in a wireless setting. First, precoding is basically channel inversion, and we know from Section 6.3.5 that inversion is not power efficient in fading channels. In particular, an infinite amount of power is needed for channel inversion on a Rayleigh fading channel. The other problem with precoding is the need for accurate channel estimates at the transmitter, which are difficult to obtain in a rapidly fading channel.

12.3.4 Adaptive Loading

Adaptive loading is based on the adaptive modulation techniques discussed in Chapter 9. It is commonly used on slowly changing channels like digital subscriber lines [16], where channel estimates at the transmitter can be obtained fairly easily. The basic idea is to vary the data rate and power assigned to each subchannel relative to that subchannel gain. As in the case of precoding, this requires knowledge of the subchannel fading $\{\alpha_i, i = 0, \ldots, N - 1\}$ at the transmitter. In adaptive loading, power and rate on each subchannel are adapted to maximize the total rate of the system using adaptive modulation such as variable-rate variable-power MQAM.

Before investigating adaptive modulation, let us consider the capacity of the multicarrier system with N independent subchannels of bandwidth B_N and subchannel gain $\{\alpha_i, i = 0, \ldots, N - 1\}$. Assuming a total power constraint P, this capacity is given by[2]

$$C = \max_{P_i : \sum P_i = P} \sum_{i=0}^{N-1} B_N \log_2\left(1 + \frac{\alpha_i^2 P_i}{N_0 B_N}\right). \tag{12.7}$$

The power allocation P_i that maximizes this expression is a water-filling over frequency given by equation (4.24):

$$\frac{P_i}{P} = \begin{cases} 1/\gamma_c - 1/\gamma_i & \gamma_i \geq \gamma_c, \\ 0 & \gamma_i < \gamma_c, \end{cases} \tag{12.8}$$

for some cutoff value γ_c, where $\gamma_i = \alpha_i^2 P/N_0 B_N$. The cutoff value is obtained by substituting the power adaptation formula into the power constraint. The capacity then becomes

$$C = \sum_{i : \gamma_i \geq \gamma_c} B_N \log_2\left(\frac{\gamma_i}{\gamma_c}\right). \tag{12.9}$$

If we now apply the variable-rate variable-power MQAM modulation scheme described in Section 9.3.2 to the subchannels, then the total data rate is given by

$$R = \sum_{i=0}^{N-1} B_N \log_2\left(1 + \frac{K\gamma_i P_i}{P}\right), \tag{12.10}$$

where $K = -1.5/\ln(5P_b)$ for P_b the desired target BER in each subchannel. Optimizing this expression relative to the P_i yields the optimal power allocation,

---

[2] As discussed in Section 4.3.1, this summation is the exact capacity when the α_i are independent. However, in order for the α_i to be independent, the subchannels must be separated by the coherence bandwidth of the channel, which would imply that the subchannels are no longer flat fading. Since the subchannels are designed to be flat fading, the subchannel gains $\{\alpha_i, i = 1, \ldots, N\}$ will be correlated, in which case the capacity obtained by summing over the capacity in each subchannel is an upper bound on the true capacity. We will take this bound to be the actual capacity, since in practice the bound is quite tight.

$$\frac{KP_i}{P} = \begin{cases} 1/\gamma_K - 1/\gamma_i & \gamma_i \geq \gamma_K, \\ 0 & \gamma_i < \gamma_K, \end{cases} \tag{12.11}$$

and corresponding data rate,

$$R = \sum_{i:\gamma_i \geq \gamma_K} B_N \log\left(\frac{\gamma_i}{\gamma_K}\right), \tag{12.12}$$

where γ_K is a cutoff fade depth dictated by the power constraint P and K.

12.4 Discrete Implementation of Multicarrier Modulation

Although multicarrier modulation was invented in the 1950s, its requirement for separate modulators and demodulators on each subchannel was far too complex for most system implementations at the time. However, the development of simple and cheap implementations of the discrete Fourier transform and the inverse DFT twenty years later – combined with the realization that multicarrier modulation could be implemented with these algorithms – ignited its widespread use. In this section, after first reviewing the basic properties of the DFT, we illustrate OFDM, which implements multicarrier modulation using the DFT and IDFT.

12.4.1 The DFT and Its Properties

Let $x[n]$, $0 \leq n \leq N - 1$, denote a discrete time sequence. The N-point DFT of $x[n]$ is defined [23] as

$$\text{DFT}\{x[n]\} = X[i] \triangleq \frac{1}{\sqrt{N}} \sum_{n=0}^{N-1} x[n] e^{-j2\pi ni/N}, \quad 0 \leq i \leq N - 1. \tag{12.13}$$

The DFT is the discrete-time equivalent to the continuous-time Fourier transform, because $X[i]$ characterizes the frequency content of the time samples $x[n]$ associated with the original signal $x(t)$. Both the continuous-time Fourier transform and the DFT are based on the fact that complex exponentials are eigenfunctions for any linear system. The sequence $x[n]$ can be recovered from its DFT using the IDFT:

$$\text{IDFT}\{X[i]\} = x[n] \triangleq \frac{1}{\sqrt{N}} \sum_{i=0}^{N-1} X[i] e^{j2\pi ni/N}, \quad 0 \leq n \leq N - 1. \tag{12.14}$$

The DFT and its inverse are typically performed via hardware using the fast Fourier transform (FFT) and inverse FFT (IFFT).

When an input data stream $x[n]$ is sent through a linear time-invariant discrete-time channel $h[n]$, the output $y[n]$ is the discrete-time convolution of the input and the channel impulse response:

$$y[n] = h[n] * x[n] = x[n] * h[n] = \sum_k h[k] x[n - k]. \tag{12.15}$$

The N-point *circular convolution* of $x[n]$ and $h[n]$ is defined as

$$y[n] = x[n] \circledast h[n] = h[n] \circledast x[n] = \sum_k h[k] x[n - k]_N, \tag{12.16}$$

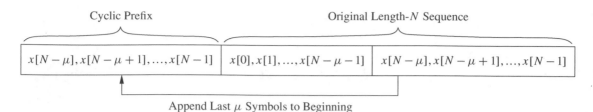

Figure 12.5: Cyclic prefix of length μ.

where $[n-k]_N$ denotes $[n-k]$ modulo N. In other words, $x[n-k]_N$ is a periodic version of $x[n-k]$ with period N. It is easily verified that $y[n]$ given by (12.16) is also periodic with period N. From the definition of the DFT, circular convolution in time leads to multiplication in frequency:

$$\text{DFT}\{y[n] = x[n]\circledast h[n]\} = X[i]H[i], \quad 0 \le i \le N-1, \tag{12.17}$$

where $H[i]$ is the N-point of $\{h[n]\}$. Note that if the sequence $\{h[n]\}$ is of length $k < N$ then it is padded with $N - k$ zeros to obtain a length N for the N-point DFT. By (12.17), if the channel and input are circularly convoluted then, as long as $h[n]$ is known at the receiver, the original data sequence $x[n]$ can be recovered by taking the IDFT of $Y[i]/H[i]$, $0 \le i \le N-1$. Unfortunately, the channel output is not a circular convolution but a linear convolution. However, the linear convolution between the channel input and impulse response can be turned into a circular convolution by adding a special prefix to the input called a *cyclic prefix*, described in the next section.

12.4.2 The Cyclic Prefix

Consider a channel input sequence $x[n] = x[0],\dots,x[N-1]$ of length N and a discrete-time channel with finite impulse response (FIR) $h[n] = h[0],\dots,h[\mu]$ of length $\mu+1 = T_m/T_s$, where T_m is the channel delay spread and T_s the sampling time associated with the discrete time sequence. The cyclic prefix for $x[n]$ is defined as $\{x[N-\mu],\dots,x[N-1]\}$: it consists of the last μ values of the $x[n]$ sequence. For each input sequence of length N, these last μ samples are appended to the beginning of the sequence. This yields a new sequence $\tilde{x}[n]$, $-\mu \le n \le N-1$, of length $N+\mu$, where $\tilde{x}[-\mu],\dots,\tilde{x}[N-1] = x[N-\mu],\dots,x[N-1],x[0],\dots,x[N-1]$, as shown in Figure 12.5. Note that with this definition, $\tilde{x}[n] = x[n]_N$ for $-\mu \le n \le N-1$, which implies that $\tilde{x}[n-k] = x[n-k]_N$ for $-\mu \le n-k \le N-1$.

Suppose $\tilde{x}[n]$ is input to a discrete-time channel with impulse response $h[n]$. The channel output $y[n]$, $0 \le n \le N-1$, is then

$$y[n] = \tilde{x}[n] * h[n]$$

$$= \sum_{k=0}^{\mu} h[k]\tilde{x}[n-k]$$

$$= \sum_{k=0}^{\mu} h[k]x[n-k]_N$$

$$= x[n]\circledast h[n], \tag{12.18}$$

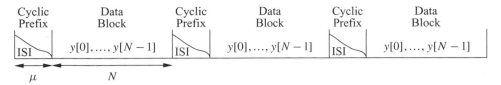

Figure 12.6: ISI between data blocks in channel output.

where the third equality follows from the fact that, for $0 \le k \le \mu$, $\tilde{x}[n-k] = x[n-k]_N$ for $0 \le n \le N-1$. Thus, by appending a cyclic prefix to the channel input, the linear convolution associated with the channel impulse response $y[n]$ for $0 \le n \le N-1$ becomes a circular convolution. Taking the DFT of the channel output in the absense of noise then yields

$$Y[i] = \text{DFT}\{y[n] = x[n] \circledast h[n]\} = X[i]H[i], \quad 0 \le i \le N-1, \tag{12.19}$$

and the input sequence $x[n]$, $0 \le n \le N-1$, can be recovered from the channel output $y[n]$, $0 \le n \le N-1$, for known $h[n]$ by

$$x[n] = \text{IDFT}\left\{ \frac{Y[i]}{H[i]} \right\} = \text{IDFT}\left\{ \frac{\text{DFT}\{y[n]\}}{\text{DFT}\{h[n]\}} \right\}. \tag{12.20}$$

Note that $y[n]$, $-\mu \le n \le N-1$, has length $N+\mu$, yet from (12.20) the first μ samples $y[-\mu], \ldots, y[-1]$ are not needed to recover $x[n]$, $0 \le n \le N-1$, owing to the redundancy associated with the cyclic prefix. Moreover, if we assume that the input $x[n]$ is divided into data blocks of size N with a cyclic prefix appended to each block to form $\tilde{x}[n]$, then the first μ samples of $y[n] = h[n] * \tilde{x}[n]$ in a given block are corrupted by ISI associated with the last μ samples of $x[n]$ in the prior block, as illustrated in Figure 12.6. The cyclic prefix serves to eliminate ISI between the data blocks, because the first μ samples of the channel output affected by this ISI can be discarded without any loss relative to the original information sequence. In continuous time this is equivalent to using a guard band of duration T_m (the channel delay spread) after every block of N symbols of duration NT_s in order to eliminate the ISI between these data blocks.

The benefits of adding a cyclic prefix come at a cost. Since μ symbols are added to the input data blocks, there is an overhead of μ/N and a resulting data-rate reduction of $N/(\mu + N)$. The transmit power associated with sending the cyclic prefix is also wasted because this prefix consists of redundant data. It is clear from Figure 12.6 that any prefix of length μ appended to input blocks of size N eliminates ISI between data blocks if the first μ samples of the block are discarded. In particular, the prefix can consist of all zero symbols, in which case – although the data rate is still reduced by $N/(N+\mu)$ – no power is used in transmitting the prefix. Trade-offs associated with the cyclic prefix versus this all-zero prefix, which is a form of vector coding, are discussed in Section 12.4.5.

The foregoing analysis motivates the design of OFDM. In OFDM the input data is divided into blocks of size N, where each block is referred to as an *OFDM symbol*. A cyclic prefix is added to each OFDM symbol to induce circular convolution of the input and channel impulse response. At the receiver, the output samples affected by ISI between OFDM symbols are removed. The DFT of the remaining samples are used to recover the original input sequence. The details of this OFDM system design are given in the next section.

EXAMPLE 12.4: Consider an OFDM system with total bandwidth $B = 1$ MHz assuming $\beta = \varepsilon = 0$. A single-carrier system would have symbol time $T_s = 1/B = 1\,\mu$s. The channel has a maximum delay spread of $T_m = 5\,\mu$s, so with $T_s = 1\,\mu$s and $T_m = 5\,\mu$s there would clearly be severe ISI. Assume an OFDM system with MQAM modulation applied to each subchannel. To keep the overhead small, the OFDM system uses $N = 128$ subcarriers to mitigate ISI. So $T_N = NT_s = 128\,\mu$s. The length of the cyclic prefix is set to $\mu = 8 > T_m/T_s$ to ensure no ISI between OFDM symbols. For these parameters, find the subchannel bandwidth, the total transmission time associated with each OFDM symbol, the overhead of the cyclic prefix, and the data rate of the system assuming $M = 16$.

Solution: The subchannel bandwidth $B_N = 1/T_N = 7.812$ kHz, so $B_N \ll B_c = 1/T_m = 200$ kHz, ensuring negligible ISI. The total transmission time for each OFDM symbol is $T = T_N + \mu T_s = 128 + 8 = 136\,\mu$s. The overhead associated with the cyclic prefix is $8/128$ which is 6.25%. The system transmits $\log_2 16 = 4$ bits per subcarrier every T seconds, so the data rate is $128 \cdot 4/136 \cdot 10^{-6} = 3.76$ Mbps, which is slightly less than $4B$ as a result of the cyclic prefix overhead.

12.4.3 Orthogonal Frequency-Division Multiplexing (OFDM)

The OFDM implementation of multicarrier modulation is shown in Figure 12.7. The input data stream is modulated by a QAM modulator, resulting in a complex symbol stream $X[0], X[1], \ldots, X[N-1]$. This symbol stream is passed through a serial-to-parallel converter, whose output is a set of N parallel QAM symbols $X[0], \ldots, X[N-1]$ corresponding to the symbols transmitted over each of the subcarriers. Thus, the N symbols output from the serial-to-parallel converter are the discrete frequency components of the OFDM modulator output $s(t)$. In order to generate $s(t)$, the frequency components are converted into time samples by performing an inverse DFT on these N symbols, which is efficiently implemented using the IFFT algorithm. The IFFT yields the OFDM symbol consisting of the sequence $x[n] = x[0], \ldots, x[N-1]$ of length N, where

$$x[n] = \frac{1}{\sqrt{N}} \sum_{i=0}^{N-1} X[i] e^{j2\pi ni/N}, \quad 0 \le n \le N-1. \tag{12.21}$$

This sequence corresponds to samples of the multicarrier signal: the multicarrier signal consists of linearly modulated subchannels, and the right-hand side of (12.21) corresponds to samples of a sum of QAM symbols $X[i]$ each modulated by the carrier $e^{j2\pi it/T_N}, i = 0, \ldots, N-1$. The cyclic prefix is then added to the OFDM symbol, and the resulting time samples $\tilde{x}[n] = \tilde{x}[-\mu], \ldots, \tilde{x}[N-1] = x[N-\mu], \ldots, x[0], \ldots, x[N-1]$ are ordered by the parallel-to-serial converter and passed through a D/A converter, resulting in the baseband OFDM signal $\tilde{x}(t)$, which is then upconverted to frequency f_0.

The transmitted signal is filtered by the channel impulse response and corrupted by additive noise, resulting in the received signal $r(t)$. This signal is downconverted to baseband and filtered to remove the high-frequency components. The A/D converter samples the resulting signal to obtain $y[n] = \tilde{x}[n] * h[n] + v[n]$, $-\mu \le n \le N-1$, where $h[n]$ is the discrete-time equivalent lowpass impulse response of the channel. The prefix of $y[n]$ consisting of the first μ samples is then removed. This results in N time samples whose DFT in the absence of noise is

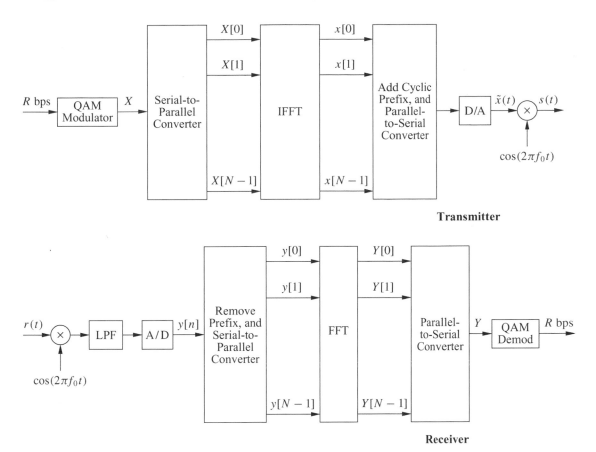

Figure 12.7: OFDM with IFFT/FFT implementation.

$Y[i] = H[i]X[i]$. These time samples are serial-to-parallel converted and passed through an FFT. This results in scaled versions of the original symbols $H[i]X[i]$, where $H[i] = H(f_i)$ is the flat fading channel gain associated with the ith subchannel. The FFT output is parallel-to-serial converted and passed through a QAM demodulator to recover the original data.

The OFDM system effectively decomposes the wideband channel into a set of narrowband orthogonal subchannels with a different QAM symbol sent over each subchannel. Knowledge of the channel gains $H[i]$, $i = 0, \ldots, N-1$, is not needed for this decomposition, in the same way that a continuous-time channel with frequency response $H(f)$ can be divided into orthogonal subchannels without knowledge of $H(f)$ by splitting the total signal bandwidth into nonoverlapping subbands. The demodulator can use the channel gains to recover the original QAM symbols by dividing out these gains: $X[i] = Y[i]/H[i]$. This process is called *frequency equalization*. However, as discussed in Section 12.3.2 for continuous-time OFDM, frequency equalization leads to noise enhancement because the noise in the ith subchannel is also scaled by $1/H[i]$. Hence, while the effect of flat fading on $X[i]$ is removed by this equalization, its received SNR is unchanged. Precoding, adaptive loading, and coding across subchannels (as discussed in Section 12.3) are better approaches to mitigate the effects of flat fading across subcarriers. An alternative to using the cyclic prefix is to use a prefix consisting of all zero symbols. In this case the OFDM symbol consisting of $x[n]$, $0 \le n \le N-1$, is preceded by μ null samples, as illustrated in Figure 12.8. At the receiver the "tail"

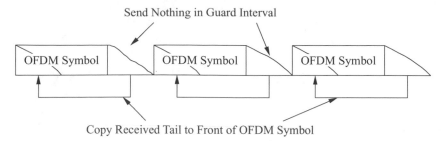

Figure 12.8: Creating a circular channel with an all-zero prefix.

of the ISI associated with the end of a given OFDM symbol is added back in to the beginning of the symbol, which re-creates the effect of a cyclic prefix, so the rest of the OFDM system functions as usual. This zero prefix reduces the transmit power relative to a cyclic prefix by $N/(\mu + N)$, since the prefix does not require any transmit power. However, the noise from the received tail is added back into the beginning of the symbol, which increases the noise power by $(N + \mu)/N$. Thus, the difference in SNR is not significant for the two prefixes.

12.4.4 Matrix Representation of OFDM

An alternate analysis for OFDM is based on a matrix representation of the system. Consider a discrete-time equivalent lowpass channel with FIR $h[n]$, $0 \le n \le \mu$, input $\tilde{x}[n]$, noise $v[n]$, and output $y[n] = \tilde{x}[n] * h[n] + v[n]$. Denote the nth element of these sequences as $h_n = h[n]$, $\tilde{x}_n = \tilde{x}[n]$, $v_n = v[n]$, and $y_n = y[n]$. With this notation the channel output sequence can be written in matrix form as

$$
\begin{bmatrix} y_{N-1} \\ y_{N-2} \\ \vdots \\ y_0 \end{bmatrix} =
\begin{bmatrix}
h_0 & h_1 & \cdots & h_\mu & 0 & \cdots & 0 \\
0 & h_0 & \cdots & h_{\mu-1} & h_\mu & \cdots & 0 \\
\vdots & \vdots & \ddots & \ddots & \ddots & \ddots & \vdots \\
0 & \cdots & 0 & h_0 & \cdots & h_{\mu-1} & h_\mu
\end{bmatrix}
\begin{bmatrix} x_{N-1} \\ \vdots \\ x_0 \\ x_{-1} \\ \vdots \\ x_{-\mu} \end{bmatrix} +
\begin{bmatrix} v_{N-1} \\ v_{N-2} \\ \vdots \\ v_0 \end{bmatrix}, \quad (12.22)
$$

which can be written more compactly as

$$
\mathbf{y} = \mathbf{H}\mathbf{x} + \mathbf{v}. \quad (12.23)
$$

The received symbols $y_{-1}, \ldots, y_{-\mu}$ are discarded because they are affected by ISI in the prior data block and are not needed to recover the input. The last μ symbols of $x[n]$ correspond to the cyclic prefix: $x_{-1} = x_{N-1}, x_{-2} = x_{N-2}, \ldots, x_{-\mu} = x_{N-\mu}$. From this it can be shown that the matrix representation (12.22) is equivalent to

$$
\begin{bmatrix} y_{N-1} \\ y_{N-2} \\ \vdots \\ \\ \vdots \\ \\ y_0 \end{bmatrix} =
\begin{bmatrix}
h_0 & h_1 & \cdots & h_\mu & 0 & \cdots & 0 \\
0 & h_0 & \cdots & h_{\mu-1} & h_\mu & \cdots & 0 \\
\vdots & \vdots & \ddots & \ddots & \ddots & \ddots & \vdots \\
0 & \cdots & 0 & h_0 & \cdots & h_{\mu-1} & h_\mu \\
\vdots & \vdots & \ddots & \ddots & \ddots & \ddots & \vdots \\
h_2 & h_3 & \cdots & h_{\mu-2} & \cdots & h_0 & h_1 \\
h_1 & h_2 & \cdots & h_{\mu-1} & \cdots & 0 & h_0
\end{bmatrix}
\begin{bmatrix} x_{N-1} \\ x_{N-2} \\ \vdots \\ \vdots \\ x_0 \end{bmatrix} +
\begin{bmatrix} v_{N-1} \\ v_{N-2} \\ \vdots \\ \vdots \\ v_0 \end{bmatrix}, \quad (12.24)
$$

which can be written more compactly as

$$\mathbf{y} = \tilde{\mathbf{H}}\mathbf{x} + \boldsymbol{v}. \tag{12.25}$$

This equivalent model shows that the inserted cyclic prefix allows the channel to be modeled as a circulant convolution matrix $\tilde{\mathbf{H}}$ over the N samples of interest. The matrix $\tilde{\mathbf{H}}$ is *normal* ($\tilde{\mathbf{H}}^H\tilde{\mathbf{H}} = \tilde{\mathbf{H}}\tilde{\mathbf{H}}^H$), so it has an eigenvalue decomposition

$$\tilde{\mathbf{H}} = \mathbf{M}\boldsymbol{\Lambda}\mathbf{M}^H, \tag{12.26}$$

where $\boldsymbol{\Lambda}$ is a diagonal matrix of eigenvalues of $\tilde{\mathbf{H}}$ and $\mathbf{M}$ is a unitary matrix whose columns constitute the eigenvectors of $\tilde{\mathbf{H}}$.

It is straightforward to show that the DFT operation on $x[n]$ can be represented by the matrix multiplication

$$\mathbf{X} = \mathbf{Q}\mathbf{x},$$

where $\mathbf{X} = (X[0], \ldots, X[N-1])^T$, $\mathbf{x} = (x[0], \ldots, x[N-1])^T$, and $\mathbf{Q}$ is an $N \times N$ matrix given by

$$\mathbf{Q} = \frac{1}{\sqrt{N}} \begin{bmatrix} 1 & 1 & 1 & \cdots & 1 \\ 1 & W_N & W_N^2 & \cdots & W_N^{N-1} \\ \vdots & \vdots & \vdots & \ddots & \vdots \\ 1 & W_N^{N-1} & W_N^{2(N-1)} & \cdots & W_N^{(N-1)^2} \end{bmatrix} \tag{12.27}$$

for $W_N = e^{-j2\pi/N}$. Since

$$\mathbf{Q}^{-1} = \mathbf{Q}^H, \tag{12.28}$$

the IDFT can be similarly represented as

$$\mathbf{x} = \mathbf{Q}^{-1}\mathbf{X} = \mathbf{Q}^H\mathbf{X}. \tag{12.29}$$

Let $\mathbf{v}$ be an eigenvector of $\tilde{\mathbf{H}}$ with eigenvalue λ. Then

$$\lambda\mathbf{v} = \tilde{\mathbf{H}}\mathbf{v}.$$

The unitary matrix $\mathbf{M}$ has columns that are the eigenvectors of $\tilde{\mathbf{H}}$; that is, $\lambda_i\mathbf{m}_i = \tilde{\mathbf{H}}\mathbf{m}_i$ for $i = 0, 1, \ldots, N-1$, where $\mathbf{m}_i$ denotes the ith column of $\mathbf{M}$. It can also be shown by induction that the columns of the DFT matrix $\mathbf{Q}^H$ are eigenvectors of $\tilde{\mathbf{H}}$, which implies that $\mathbf{Q} = \mathbf{M}^H$ and $\mathbf{Q}^H = \mathbf{M}$. Thus we have that

$$\begin{aligned} \mathbf{Y} &= \mathbf{Q}\mathbf{y} \\ &= \mathbf{Q}[\tilde{\mathbf{H}}\mathbf{x} + \boldsymbol{v}] \\ &= \mathbf{Q}[\tilde{\mathbf{H}}\mathbf{Q}^H\mathbf{X} + \boldsymbol{v}] \\ &= \mathbf{Q}[\mathbf{M}\boldsymbol{\Lambda}\mathbf{M}^H\mathbf{Q}^H\mathbf{X} + \boldsymbol{v}] \\ &= \mathbf{Q}\mathbf{M}\boldsymbol{\Lambda}\mathbf{M}^H\mathbf{Q}^H\mathbf{X} + \mathbf{Q}\boldsymbol{v} \\ &= \mathbf{M}^H\mathbf{M}\boldsymbol{\Lambda}\mathbf{M}^H\mathbf{M}\mathbf{X} + \mathbf{Q}\boldsymbol{v} \\ &= \boldsymbol{\Lambda}\mathbf{X} + \boldsymbol{v}_Q; \end{aligned} \tag{12.30} \tag{12.31}$$

here, since $\mathbf{Q}$ is unitary, it follows that $\mathbf{v}_Q = \mathbf{Q}\mathbf{v}$ has the same noise autocorrelation matrix as $\mathbf{v}$ and hence is generally white and Gaussian, with unchanged noise power. Thus, this matrix analysis also shows that, by adding a cyclic prefix and using the IDFT/DFT, OFDM decomposes an ISI channel into N orthogonal subchannels, and knowledge of the channel matrix $\mathbf{H}$ is not needed for this decomposition.

The matrix representation is also useful in analyzing OFDM systems with multiple antennas [24]. As discussed in Chapter 10, a MIMO channel is typically represented by an $M_r \times M_t$ matrix, where M_t is the number of transmit antennas and M_r the number of receive antennas. Thus, an MIMO-OFDM channel with N subchannels, M_t transmit antennas, M_r receive antennas, and a channel FIR of duration μ can be represented as

$$\mathbf{y} = \mathbf{H}\mathbf{x} + \mathbf{v}, \tag{12.32}$$

where $\mathbf{y}$ is a vector of dimension $M_r N \times 1$ corresponding to N output time samples at each of the M_r antennas, H is a $NM_r \times (N+\mu)M_t$ matrix corresponding to the N flat fading subchannel gains on each transmit–receive antenna pair, $\mathbf{x}$ is a vector of dimension $M_t(N+\mu) \times 1$ corresponding to N input time samples with appended cyclic prefix of length μ at each of the M_t transmit antennas, and $\mathbf{v}$ is the channel noise vector of dimension $M_r N \times 1$. The matrix is in the same form as in the case of OFDM without multiple antennas, so the same design and analysis applies: with MIMO-OFDM the ISI is removed by breaking the wideband channel into many narrowband subchannels. Each subchannel experiences flat fading and so can be treated as a flat fading MIMO channel. The capacity of this channel is obtained by applying the same matrix analysis as for standard MIMO to the augmented channel with MIMO and OFDM [25]. In discrete implementations the input associated with each transmit antenna is broken into blocks of size N, with a cyclic prefix appended for converting linear convolution to circular and eliminating ISI between input blocks. More details can be found in [26].

12.4.5 Vector Coding

In OFDM the $N \times N$ circulant channel matrix $\tilde{\mathbf{H}}$ is decomposed using its eigenvalues and eigenvectors. Vector coding (VC) is a similar technique whereby the original $N \times (N + \mu)$ channel matrix $\mathbf{H}$ from (12.23) is decomposed using a singular value decomposition, which can be applied to a matrix of any dimension. The SVD does not require a cyclic prefix to make the subchannels orthogonal, so it is more efficient than OFDM in terms of energy. However, it is more complex and requires knowledge of the channel impulse response for the decomposition – in contrast to OFDM, which does not require channel knowledge for its decomposition.

From Appendix C, the singular value decomposition of $\mathbf{H}$ can be written as

$$\mathbf{H} = \mathbf{U}\mathbf{\Sigma}\mathbf{V}^H, \tag{12.33}$$

where $\mathbf{U}$ is $N \times N$ unitary, $\mathbf{V}$ is $(N + \mu) \times (N + \mu)$ unitary, and $\mathbf{\Sigma}$ is a matrix with ith diagonal element σ_i equal to the ith singular value of $\mathbf{H}$ and with zeros everywhere else. The singular values of $\mathbf{H}$ are related to the eigenvalues of $\mathbf{H}\mathbf{H}^H$ by $\sigma_i = \sqrt{\lambda_i}$ for λ_i the ith eigenvalue of the matrix $\mathbf{H}\mathbf{H}^H$. Because $\mathbf{H}$ is a block-diagonal convolutional matrix, the rank of $\mathbf{H}$ is N and so $\sigma_i \neq 0$ for all $i = 1, \ldots, N$.

In vector coding, as in OFDM, input data symbols are grouped into vectors of N symbols. Let $X_i, i = 0, \ldots, N-1$, denote the symbol to be transmitted over the ith subchannel

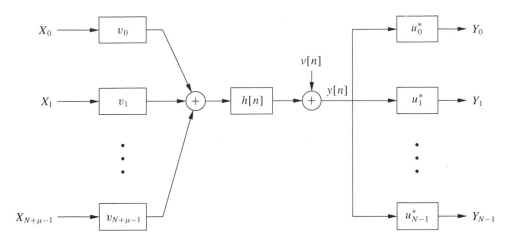

Figure 12.9: Vector coding.

and let $\mathbf{X} = (X_0, \ldots, X_{N-1}, X_n, \ldots, X_{N+\mu-1})$ denote a vector of these symbols with μ symbols $X_N, \ldots, X_{N+\mu-1}$ (typically zeros) appended at the end. The data symbol X_i is multiplied by the ith column of $\mathbf{V}$ to form a vector, and then these vectors are added together. At the receiver, the received vector $\mathbf{y}$ is multiplied by each row of $\mathbf{U}^H$ to yield N output symbols Y_i, $i = 0, 1, \ldots, N - 1$. This process is illustrated in Figure 12.9, where the multiplication by $\mathbf{V}$ and $\mathbf{U}^H$ performs a similar function as the transmit precoding and receiver shaping in MIMO systems.

Mathematically, it can be seen that the filtered transmit and received vectors are

$$\mathbf{x} = \mathbf{V}\mathbf{X}$$

and

$$\mathbf{Y} = \mathbf{U}^H\mathbf{y}. \tag{12.34}$$

As a result, it can be shown through simple linear algebra that the filtered received vector $\mathbf{Y}$ is ISI-free, since

$$\begin{aligned}
\mathbf{Y} &= \mathbf{U}^H\mathbf{y} \\
&= \mathbf{U}^H(\mathbf{H}\mathbf{x} + \boldsymbol{v}) \\
&= \mathbf{U}^H(\mathbf{U}\boldsymbol{\Sigma}\mathbf{V}^H)\mathbf{V}\mathbf{X} + \mathbf{U}^H\boldsymbol{v} \\
&= \boldsymbol{\Sigma}\mathbf{X} + \mathbf{U}^H\boldsymbol{v}.
\end{aligned} \tag{12.35}$$

Hence, each element of $\mathbf{X}$ is effectively passed through a scalar channel without ISI, where the scalar gain of subchannel i is the ith singular value of $\mathbf{H}$. Additionally, the new noise vector $\tilde{\boldsymbol{v}} = \mathbf{U}^H\boldsymbol{v}$ has unchanged noise variance, since $\mathbf{U}$ is unitary. The resulting received vector is thus

$$\begin{bmatrix} Y_{N-1} \\ Y_{N-2} \\ \vdots \\ Y_0 \end{bmatrix} = \begin{bmatrix} \sigma_1 X_{N-1} \\ \sigma_2 X_{N-2} \\ \vdots \\ \sigma_N X_0 \end{bmatrix} + \begin{bmatrix} \tilde{v}_{N-1} \\ \tilde{v}_{N-2} \\ \vdots \\ \tilde{v}_0 \end{bmatrix}. \tag{12.36}$$

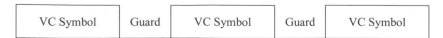

Figure 12.10: Guard interval (null prefix) in vector coding.

We call $\mathbf{X}$, the vector obtained by appending μ extra symbols to each block of N data symbols, a *vector codeword*. In contrast to OFDM, we see from (12.35) and the structure of Σ that the SVD does not require any particular value for these extra symbols; they need not be a cyclic prefix, nor must the "tail" be added back in if these symbols are all zeros. In practice the extra symbols are set to zero to save transmit power, thereby forming a guard band or null prefix between the vector codeword (VC) symbols, as shown in Figure 12.10.

Information and estimation theory have proven that vector coding is the optimal partition of the N-dimensional channel $\mathbf{H}$ [13, Chap. 4]. Thus, the capacity of any other channel partitioning scheme will be upper bounded by vector coding. Despite its theoretical optimality and ability to create ISI-free channels with relatively small overhead and no wasted transmit power, there are a number of important practical problems with vector coding. The two most important problems are as follows.

1. *Complexity.* With vector coding, as in simple multichannel modulation, the complexity scales quickly with N, the number of subcarriers. As seen from Figure 12.9, N transmit precoding and N receive shaping filters are required to implement vector coding. Furthermore, the complexity of finding the singular value decomposition of the $N \times (N + \mu)$ matrix $\mathbf{H}$ increases rapidly with N.

2. *SVD and channel knowledge.* In order to orthogonally partition the channel, the SVD of the channel matrix $\mathbf{H}$ must be computed. In particular, the precoding filter matrix must be known to the transmitter. This means that, every time the channel changes, a new SVD must be computed and the results conveyed to the transmitter. Generally, the computational complexity of the SVD and the delay incurred in getting the channel information back to the transmitter is prohibitive in wireless systems. Since OFDM can perform this decomposition without channel knowledge, OFDM is the method of choice for discrete multicarrier modulation in wireless applications.

EXAMPLE 12.5: Consider a simple two-tap discrete-time channel (i.e. $\mu = 1$) described as

$$H(z) = 1 + .9z^{-1}.$$

Since $\mu + 1 = T_m / T_s = 2$, with $N = 8$ we ensure $B_N \approx 1/(NT_s) \ll B_c \approx 1/T_m$. Find the system matrix representation (12.23) and the singular values of the associated channel matrix $\mathbf{H}$.

Solution: The representation (12.23) for $H(z) = 1 + .9z^{-1}$ and $N = 8$ is given by

$$\begin{bmatrix} y_7 \\ y_6 \\ \vdots \\ y_0 \end{bmatrix} = \begin{bmatrix} 1 & .9 & 0 & \cdots & 0 & \cdots & 0 \\ 0 & 1 & .9 & 0 & 0 & \cdots & 0 \\ \vdots & \vdots & & \ddots & \ddots & \ddots & \vdots \\ 0 & \cdots & 0 & 0 & 0 & 1 & .9 \end{bmatrix} \begin{bmatrix} x_7 \\ x_6 \\ \vdots \\ x_{-1} \end{bmatrix} + \begin{bmatrix} v_7 \\ v_6 \\ \vdots \\ v_0 \end{bmatrix}. \quad (12.37)$$

The singular values of the matrix $\mathbf{H}$ in (12.37) can be found via a standard computer package (e.g. Matlab) as

$$\mathbf{\Sigma} = \text{diag}[1.87, 1.78, 1.65, 1.46, 1.22, .95, .66, .34].$$

The precoding and shaping matrices $\mathbf{U}$ and $\mathbf{V}$ are also easily found. Given $\mathbf{U}$, $\mathbf{V}$, and $\mathbf{\Sigma}$, this communication is ISI-free, with the symbols $X_0, X_1, \ldots, X_{N-1}$ being multiplied by the corresponding singular values as in (12.36).

12.5 Challenges in Multicarrier Systems

12.5.1 Peak-to-Average Power Ratio

The peak-to-average power ratio (PAR) is an important attribute of a communication system. A low PAR allows the transmit power amplifier to operate efficiently, whereas a high PAR forces the transmit power amplifier to have a large *backoff* in order to ensure linear amplification of the signal. This is demonstrated in Figure 12.11, which shows a typical power amplifier response. Operation in the linear region of this response is generally required to avoid signal distortion, so the peak value is constrained to be in this region. Clearly it would be desirable to have the average and peak values be as close together as possible in order for the power amplifier to operate at maximum efficiency. Additionally, a high PAR requires high resolution for the receiver A/D converter, since the dynamic range of the signal is much larger for high-PAR signals. High-resolution A/D conversion places a complexity and power burden on the receiver front end.

The PAR of a continuous-time signal is given by

$$\text{PAR} \triangleq \frac{\max_t |x(t)|^2}{\mathbf{E}_t[|x(t)|^2]} \tag{12.38}$$

and for a discrete-time signal is given by

$$\text{PAR} \triangleq \frac{\max_n |x[n]|^2}{\mathbf{E}_n[|x[n]|^2]}. \tag{12.39}$$

Any constant amplitude signal (e.g., a square wave) has PAR $= 0$ dB. A sine wave has PAR $= 3$ dB because $\max[\sin^2(2\pi t/T)] = 1$ and

$$\mathbf{E}\left[\sin^2\left(\frac{2\pi t}{T}\right)\right] = \frac{1}{T}\int_0^T \sin^2\left(\frac{2\pi t}{T}\right) dt = .5,$$

so PAR $= 1/.5 = 2$.

In general, PAR should be measured with respect to the continuous-time signal using (12.38), since the input to the amplifier is an analog signal. The PAR given by (12.38) is sensitive to the pulse shape $g(t)$ used in the modulation, and it does not generally lead to simple analytical formulas [27]. For illustration we will focus on the PAR associated with the discrete-time signal, since it lends itself to a simple characterization. However, care must be taken when interpreting these results, since they can be quite inaccurate if the pulse shape $g(t)$ is not taken into account.

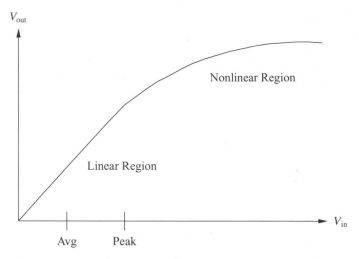

V_{out}

Nonlinear Region

Linear Region

V_{in}

Avg   Peak

Figure 12.11: Typical power amplifier response.

Consider the time-domain samples that are output from the IFFT:

$$x[n] = \frac{1}{\sqrt{N}} \sum_{i=0}^{N-1} X[i] e^{j2\pi in/N}, \quad 0 \le n \le N-1. \tag{12.40}$$

If N is large then the central limit theorem is applicable, and $x[n]$ are zero-mean complex Gaussian random variables because the real and imaginary parts are summed. The Gaussian approximation for IFFT outputs is generally quite accurate for a reasonably large number of subcarriers ($N \ge 64$). For $x[n]$ complex Gaussian, the envelope of the OFDM signal is Rayleigh distributed with variance σ^2, and the phase of the signal is uniform. Since the Rayleigh distribution has infinite support, the peak value of the signal will exceed any given value with nonzero probability. It can then be shown [28] that the probability that the PAR given by (12.39) exceeds a threshold $P_0 = \sigma_0^2/\sigma^2$ is given by

$$p(\text{PAR} \ge P_0) = 1 - (1 - e^{-P_0})^N. \tag{12.41}$$

Let us now investigate how PAR grows with the number of subcarriers. Consider N Gaussian independent and identically distributed random variables x_n ($0 \le n \le N-1$) with zero mean and unit power. The average signal power $\mathbf{E}_n[|x[n]|^2]$ is then

$$\mathbf{E}\left[\frac{1}{\sqrt{N}}|x_0 + x_1 + \cdots + x_{N-1}|^2\right] = \frac{1}{N}\mathbf{E}|x_0 + x_1 + \cdots + x_{N-1}|^2$$
$$= \frac{\mathbf{E}|x_0^2| + \mathbf{E}|x_1^2| + \cdots + \mathbf{E}|x_{N-1}^2|}{N}$$
$$= 1. \tag{12.42}$$

The maximum value occurs when all the x_i add coherently, in which case

$$\max\left[\frac{1}{\sqrt{N}}|x_0 + x_1 + \cdots + x_{N-1}|\right]^2 = \left|\frac{N}{\sqrt{N}}\right|^2 = N. \tag{12.43}$$

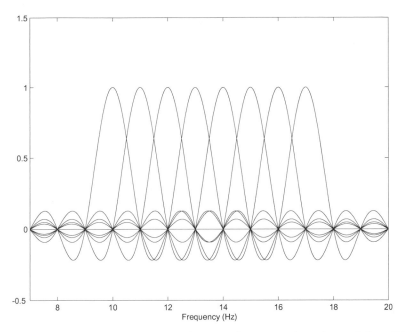

Figure 12.12: OFDM overlapping subcarriers for rectangular pulses with $f_0 = 10$ Hz and $\Delta f = 1$ Hz.

Hence the maximum PAR is N for N subcarriers. In practice, full coherent addition of all N symbols is highly improbable and so the observed PAR is typically less than N – usually by many decibels. Nevertheless, PAR increases approximately linearly with the number of subcarriers. So even though it is desirable to have N as large as possible in order to keep the overhead associated with the cyclic prefix down, a large PAR is an important penalty that must be paid for large N.

There are a number of ways to reduce or tolerate the PAR of OFDM signals, including clipping the OFDM signal above some threshold, peak cancellation with a complementary signal, allowing nonlinear distortion from the power amplifier (and correction for it), and special coding techniques [29]. A good summary of some of these methods can be found in [30].

12.5.2 Frequency and Timing Offset

We have seen that OFDM modulation encodes the data symbols X_i onto orthogonal subchannels, where orthogonality is assured by the subcarrier separation $\Delta f = 1/T_N$. The subchannels may overlap in the frequency domain, as shown in Figure 12.12 for a rectangular pulse shape in time (sinc function in frequency). In practice, the frequency separation of the subcarriers is imperfect and so Δf is not exactly equal to $1/T_N$. This is generally caused by mismatched oscillators, Doppler frequency shifts, or timing synchronization errors. For example, if the carrier frequency oscillator is accurate to 1 part per million then the frequency offset $\Delta f_\varepsilon \approx f_0 (.1 \cdot 10^{-6})$. If $f_0 = 5$ GHz, the carrier frequency for 802.11a WLANs, then $\Delta f_\varepsilon = 500$ Hz, which will degrade the orthogonality of the subchannels because now the received samples of the FFT will contain interference from adjacent subchannels. Next we'll analyze this intercarrier interference (ICI) more closely.

The signal corresponding to subcarrier i can be simply expressed for the case of rectangular pulse shapes (suppressing the data symbol and the carrier frequency) as

$$x_i(t) = e^{j2\pi i t/T_N}. \tag{12.44}$$

An interfering subchannel signal can be written as

$$x_{i+m}(t) = e^{j2\pi(i+m)t/T_N}. \tag{12.45}$$

If the signal is demodulated with a frequency offset of δ/T_N, then this interference becomes

$$x_{i+m}(t) = e^{j2\pi(i+m+\delta)t/T_N}. \tag{12.46}$$

The ICI between subchannel signals x_i and x_{i+m} is simply the inner product between them:

$$I_m = \int_0^{T_N} x_i(t)x_{i+m}^*(t)\, dt = \frac{T_N(1 - e^{-j2\pi(\delta+m)})}{j2\pi(m+\delta)}. \tag{12.47}$$

It can be seen that, in (12.47), $\delta = 0 \Rightarrow I_m = 0$ as expected. The total ICI power on subcarrier i is then

$$\text{ICI}_i = \sum_{m\neq i}|I_m|^2 \approx C_0(T_N\delta)^2, \tag{12.48}$$

where C_0 is some constant. Several important trends can be observed from this simple approximation. First, as T_N increases, the subcarriers grow narrower and hence more closely spaced, which then results in more ICI. Second, the ICI predictably grows with the frequency offset δ, and the growth is about quadratic. Another interesting observation is that (12.48) does not appear to be directly affected by N. But picking N large generally forces T_N to be large also, which then causes the subcarriers to be closer together. Along with the larger PAR that comes with large N, the increased ICI is another reason to pick N as low as possible, assuming the overhead budget can be met. In order to further reduce the ICI for a given choice of N, nonrectangular windows can also be used [31; 32].

The effects from timing offset are generally less than those from the frequency offset [5, Chap. 5], as long as a full N-sample OFDM symbol is used at the receiver without interference from the previous or subsequent OFDM symbols (this is ensured by taking the cyclic prefix length $\mu \gg \sigma_{T_m}/T_s$, where σ_{T_m} is the channel's rms delay spread). It can be shown that the ICI power on subcarrier i due to a receiver timing offset τ can be approximated as $2(\tau/T_N)^2$. Since usually $\tau \ll T_N$, this effect is typically negligible.

12.6 Case Study: The IEEE 802.11a Wireless LAN Standard

The IEEE 802.11a wireless LAN standard, which occupies 300 MHz of bandwidth in the 5-GHz unlicensed band, is based on OFDM [6]. The IEEE 802.11g standard is virtually identical to the 802.11a standard in its link layer design, but it operates in the smaller and more crowded 2.4-GHz unlicensed ISM band [7]. In this section we study the properties of this OFDM design and discuss some of the design choices.

In 802.11a, $N = 64$ subcarriers are generated, although only 48 are actually used for data transmission: the outer 12 are zeroed in order to reduce adjacent channel interference, and 4 are used as pilot symbols for channel estimation. The cyclic prefix consists of $\mu = 16$ samples, so the total number of samples associated with each OFDM symbol, including both data samples and the cyclic prefix, is 80. Coded bits are packetized, and the transmitter gets periodic feedback from the receiver about the packet error rate, which it uses to pick an appropriate error correction code and modulation technique. The same code and modulation must be used for *all* the subcarriers at any given time. The error correction code is a convolutional code with one of three possible code rates: $r = 1/2, 2/3$, or $3/4$. The modulation types that can be used on the subchannels are BPSK, QPSK, 16-QAM, or 64-QAM.

The total system bandwidth of 300 MHz is divided into 20-MHz channels that can be assigned to different users. Since the channel bandwidth B (and sampling rate $1/T_s$) is 20 MHz and since there are 64 subcarriers evenly spaced over that bandwidth, the subcarrier bandwidth is:

$$B_N = \frac{20 \text{ MHz}}{64} = 312.5 \text{ kHz}.$$

Since $\mu = 16$ and $1/T_s = 20$ MHz, the maximum delay spread for which ISI is removed is

$$T_m < \mu T_s = \frac{16}{20 \text{ MHz}} = 0.8 \,\mu\text{s},$$

which corresponds to delay spread in an indoor environment. Including both the OFDM symbol and cyclic prefix, there are $80 = 64 + 16$ samples per OFDM symbol time, so the symbol time per subchannel is

$$T_N = 80 T_s = \frac{80}{20 \cdot 10^6} = 4 \,\mu\text{s}.$$

The data rate per subchannel is $\log_2 M / T_N$. Thus, the minimum data rate for this system – corresponding to BPSK (1 bit/symbol), an $r = 1/2$ code, and taking into account that only 48 subcarriers actually carry usable data – is given by

$$R_{\min} = 48 \text{ subcarriers} \cdot \frac{1/2 \text{ bit}}{1 \text{ coded bit}} \cdot \frac{1 \text{ coded bit}}{\text{subcarrier symbol}} \cdot \frac{1 \text{ subcarrier symbol}}{4 \cdot 10^{-6} \text{ seconds}}$$
$$= 6 \text{ Mbps.} \tag{12.49}$$

The maximum data rate that can be transmitted is

$$R_{\max} = 48 \text{ subcarriers} \cdot \frac{3/4 \text{ bit}}{1 \text{ coded bit}} \cdot \frac{6 \text{ coded bits}}{\text{subcarrier symbol}} \cdot \frac{1 \text{ subcarrier symbol}}{4 \cdot 10^{-6} \text{ seconds}}$$
$$= 54 \text{ Mbps.} \tag{12.50}$$

Naturally, a wide range of data rates between these two extremes is possible.

EXAMPLE 12.6: Find the data rate of an 802.11a system assuming 16-QAM modulation and rate-2/3 coding.

Solution: With 16-QAM modulation, each subcarrier transmits $\log_2(16) = 4$ coded bits per subcarrier symbol and there are a total of 48 subcarriers used for data transmission. With a rate-2/3 code, each coded bit relays 2/3 of an information bit per T_N seconds. Thus, the data rate is given by

$$R_{max} = 48 \text{ subcarriers} \cdot \frac{2/3 \text{ bit}}{1 \text{ coded bit}} \cdot \frac{4 \text{ coded bits}}{\text{subcarrier symbol}} \cdot \frac{1 \text{ subcarrier symbol}}{4 \cdot 10^{-6} \text{ seconds}}$$

$$= 32 \text{ Mbps.} \tag{12.51}$$

PROBLEMS

12-1. Show that the minimum separation Δf for subcarriers $\{\cos(2\pi j \Delta f t + \phi_j), j = 1, 2, \dots\}$ to form a set of orthonormal basis functions on the interval $[0, T_N]$ is $\Delta f = 1/T_N$ for any initial phase ϕ_j. Show that if $\phi_j = 0$ for all j then this carrier separation can be reduced by half.

12-2. Consider an OFDM system operating in a channel with coherence bandwidth $B_c = 10$ kHz.

(a) Find a subchannel symbol time $T_N = 1/B_N = 10 T_m$, assuming $T_m = 1/B_c$. This should ensure flat fading on the subchannels.

(b) Assume the system has $N = 128$ subchannels. If raised cosine pulses with $\beta = 1.5$ are used and if the required additional bandwidth (from time limiting) to ensure minimal power outside the signal bandwidth is $\varepsilon = .1$, then what is the total bandwidth of the system?

(c) Find the total required bandwidth of the system using overlapping carriers separated by $1/T_N$, and compare with your answer in part (b).

12-3. Show from the definition of the DFT that circular convolution of discrete time sequences leads to multiplication of their DFTs.

12-4. Consider a data signal with a bandwidth of .5 MHz and a data rate of .5 Mbps. The signal is transmitted over a wireless channel with a delay spread of $10 \, \mu s$.

(a) If multicarrier modulation with nonoverlapping subchannels is used to mitigate the effects of ISI, approximately how many subcarriers are needed? What is the data rate and symbol time on each subcarrier? (We do not need to eliminate the ISI completely, so $T_s = T_m$ is sufficient for ISI mitigation.)

Assume for the remainder of the problem that the average received SNR (γ_s) on the nth subcarrier is $1000/n$ (linear units) and that each subcarrier experiences flat Rayleigh fading (so ISI is completely eliminated).

(b) Suppose BPSK modulation is used for each subcarrier. If a repetition code is used across all subcarriers (i.e., if a copy of each bit is sent over each subcarrier), then what is the BER after majority decoding? What is the data rate of the system?

(c) Suppose you use adaptive loading (i.e., different constellations on each subcarrier) such that the average BER on each subcarrier does not exceed 10^{-3} (this is averaged over the fading distribution; do not assume that the transmitter and receiver adapt power or rate to the instantaneous fade values). Find the MQAM constellation that can be transmitted over each subcarrier while meeting this average BER target. What is the total data rate of the system with adaptive loading?

12-5. Consider a multicarrier modulation transmission scheme with three nonoverlapping subchannels spaced 200 kHz apart (from carrier to carrier) and with subchannel baseband bandwidth of 100 kHz.

(a) For what values of the channel coherence bandwidth will the subchannels of your multicarrier scheme exhibit flat fading (approximately no ISI)? For which such values will the subcarriers exhibit independent fading? If the subcarriers exhibit correlated fading, what impact will this have on coding across subchannels?

(b) Suppose that you have a total transmit power $P = 300$ mW and that the noise power in each subchannel is 1 mW. With equal power of 100 mW transmitted on each subchannel, the received SNR on each subchannel is $\gamma_1 = 11$ dB, $\gamma_2 = 14$ dB, and $\gamma_3 = 18$ dB. Assume the subchannels do not experience fading, so these SNRs are constant. For these received SNRs, find the maximum signal constellation size for MQAM that can be transmitted over each subchannel for a target BER of 10^{-3}. Assume the MQAM constellation is restricted to be a power of 2 and use the bound BER $\leq .2e^{-1.5\gamma/(M-1)}$ for your calculations. Find the corresponding total data rate of the multicarrier signal assuming a symbol rate on each subchannel of $T_s = 1/B$, where B is the baseband subchannel bandwidth.

(c) For the subchannel SNRs given in part (b), suppose we want to use precoding (to equalize the received SNR in each subchannel) and then send the same signal constellation over each subchannel. What size signal constellation is needed to achieve the same data rate as in part (b)? What transmit power would be needed on each subchannel to achieve the required received SNR for this constellation with a 10^{-3} BER target? How much must the total transmit power be increased over the 300-mW transmit power in part (b)?

12-6. Consider a channel with impulse response

$$h(t) = \alpha_0 \delta(t) + \alpha_1 \delta(t - T_1) + \alpha_2 \delta(t - T_2).$$

Assume that $T_1 = 10 \ \mu s$ and $T_2 = 20 \ \mu s$. You want to design a multicarrier system for the channel with subchannel bandwidth $B_N = B_c/2$. If raised cosine pulses with $\beta = 1$ are used and if the subcarriers are separated by the minimum bandwidth necessary to remain orthogonal, then what is the total bandwidth occupied by a multicarrier system with eight subcarriers? Assuming a constant SNR on each subchannel of 20 dB, find the maximum constellation size for MQAM modulation that can be sent over each subchannel with a target BER of 10^{-3}, assuming also that M is restricted to be a power of 2. Find the corresponding total data rate of the system.

12-7. Show the equivalence of matrix representations (12.22) and (12.24) for OFDM when a cyclic prefix is appended to each OFDM symbol.

12-8. Show that the DFT operation on $x[n]$ can be represented by the matrix multiplication $X[i] = \mathbf{Q}x[n]$, where

$$\mathbf{Q} = \frac{1}{\sqrt{N}} \begin{bmatrix} 1 & 1 & 1 & \cdots & 1 \\ 1 & W_N & W_N^2 & \cdots & W_N^{N-1} \\ \vdots & \vdots & \vdots & \ddots & \vdots \\ 1 & W_N^{N-1} & W_N^{2(N-1)} & \cdots & W_N^{(N-1)^2} \end{bmatrix} \quad (12.52)$$

for $W_N = e^{-j2\pi/N}$.

12-9. This problem shows that the rows of the DFT matrix $\mathbf{Q}$ are eigenvectors of $\mathbf{H}$.

(a) Show that the first row of $\mathbf{Q}$ is an eigenvector of $\mathbf{H}$ with eigenvalue $\lambda_0 = \sum_{i=0}^{\mu} h_i$.

(b) Show that row 2 of $\mathbf{Q}$ is an eigenvector of $\mathbf{H}$ with eigenvalue $\lambda_1 = \sum_{i=0}^{\mu} h_i W_N^i$.

(c) Argue by induction that similar relations hold for all rows of $\mathbf{Q}$.

12-10. Show that appending the all-zero prefix to an OFDM symbol and then adding in the tail of the received sequence, as shown in Figure 12.8, results in the same received sequence as with a cyclic prefix.

12-11. Consider a discrete-time FIR channel with $h[n] = .7 + .5\delta[n - 1] + .3\delta[n - 3]$. Consider an OFDM system with $N = 8$ subchannels.

(a) Find the matrix $\mathbf{H}$ corresponding to the matrix representation of the DMT $\mathbf{y} = \mathbf{H}\mathbf{x} + \mathbf{v}$ given in (12.23).

(b) Find the circulant convolution matrix $\mathbf{H}$ corresponding to the matrix representation in (12.25), as well as its eigenvalue decomposition $\mathbf{H} = \mathbf{M}\Lambda\mathbf{M}^H$.

(c) What are the flat fading channel gains associated with each subchannel in the representation of part (b)?

12-12. Consider a five-tap discrete-time channel

$$H(z) = 1 + .6z^{-1} + .7z^{-2} + .3z^{-3} + .2z^{-4}.$$

Assume this channel model characterizes the maximum delay spread of the channel, and assume a vector coded system is used over the channel with $N = 256$ carriers.

(a) What value of μ is needed for the prefix to eliminate ISI between vector codewords? What is the overhead associated with this μ?

(b) Find the system matrix representation (12.23) and the singular values of the associated channel matrix $\mathbf{H}$.

(c) Find the transmit precoding and receiver shaping matrices, $\mathbf{V}$ and $\mathbf{U}^H$, required to orthogonalize the subchannels.

12-13. Find the PAR of a raised cosine pulse with $\beta = 0, .5, 1$. Which pulse shape has the lowest PAR? Is this pulse shape more or less sensitive to timing errors?

12-14. Find the constant C_0 associated with intercarrier interference in (12.48).

12-15. Suppose the four subchannels in 802.11a used for pilot estimation could be used for data transmission by taking advantage of blind estimation techniques. Assuming the same modulation and coding formats are available, what maximum and minimum data rates could be achieved by including these extra subchannels?

12-16. Find the data rate of an 802.11a system assuming half the available 48 subchannels use BPSK with a rate-1/2 channel code and the others use 64-QAM with a rate-3/4 channel code.

REFERENCES

[1]   J. Bingham, "Multicarrier modulation for data transmission: An idea whose time has come," *IEEE Commun. Mag.,* pp. 5–14, May 1990.

[2]   H. Sari, G. Karam, and I. Jeanclaude, "Transmission techniques for digital terrestrial TV broadcasting," *IEEE Commun. Mag.,* pp. 100–9, February 1995.

[3]   J. S. Chow, J. C. Tu, and J. M. Cioffi, "A discrete multitone transceiver system for HDSL applications," *IEEE J. Sel. Areas Commun.,* pp. 895–908, August 1991.

[4]   J. M. Cioffi, "A multicarrier primer," Stanford University/Amati T1E1 contribution, I1E1.4/91–157, November 1991.

[5]   A. R. S. Bahai, B. R. Saltzberg, and M. Ergen, *Multi-Carrier Digital Communications – Theory and Applications of OFDM,* 2nd ed., Springer-Verlag, New York, 2004.

[6]   IEEE 802.11a-1999: *High-Speed Physical Layer in the 5 GHz Band,* 1999.

[7]   IEEE 802.11g-2003: *Further Higher-Speed Physical Layer Extension in the 2.4 GHz Band,* 2003.

[8]   IEEE 802.16a-2001: *IEEE Recommended Practice for Local and Metropolitan Area Networks,* 2001.

[9]   C. Eklund, R. B. Marks, K. L. Stanwood, and S. Wang, "IEEE Standard 802.16: A technical overview of the WirelessMAN 326 air interface for broadband wireless access," *IEEE Commun. Mag.,* pp. 98–107, June 2002.

[10] M. Corson, R. Laroia, A. O'Neill, V. Park, and G. Tsirtsis, "A new paradigm for IP-based cellular networks," *IT Professional,* pp. 20–9, November/December 2001.

[11] W. Lu, "4G mobile research in Asia," *IEEE Commun. Mag.,* pp. 104–6, March 2003.

[12] T. S. Rappaport, A. Annamalai, R. M. Buehrer, and W. H. Tranter, "Wireless communications: Past events and a future perspective," *IEEE Commun. Mag.,* pp. 148–61, May 2002.

[13] J. M. Cioffi. *Digital Communications,* chap. 4: *Multichannel Modulation,* unpublished course notes, available at ⟨http://www.stanford.edu/class/ee379c/⟩.

[14] S. Kasturia, J. Aslanis, and J. Cioffi, "Vector coding for partial response channels," *IEEE Trans. Inform. Theory,* pp. 741–62, July 1990.

[15] L. J. Cimini, "Analysis and simulation of a digital mobile channel using orthogonal frequency division multiplexing," *IEEE Trans. Inform. Theory,* pp. 665–75, July 1985.

[16] P. S. Chow, J. M. Cioffi, and John A. C. Bingham, "A practical discrete multitone transceiver loading algorithm for data transmission over spectrally shaped channels," *IEEE Trans. Commun.,* pp. 773–5, February–April 1995.

[17] Z. Wang, X. Ma, and G. B. Giannakis, "OFDM or single-carrier block transmissions?" *IEEE Trans. Commun.,* pp. 380–94, March 2004.

[18] R. K. Jurgen, "Broadcasting with digital audio," *IEEE Spectrum,* pp. 52–9, March 1996.

[19] S. Kaider, "Performance of multi-carrier CDM and COFDM in fading channels," *Proc. IEEE Globecom Conf.,* pp. 847–51, December 1999.

[20] P. Hoeher, S. Kaiser, and P. Robertson, "Two-dimensional pilot-symbol-aided channel estimation by Wiener filtering," *Proc. IEEE Internat. Conf. Acous., Speech, Signal Proc.,* pp. 1845–8, April 1997.

[21] A. Scaglione, G. B. Giannakis, and S. Barbarossa, "Redundant filterbank precoders and equalizers. I: Unification and optimal designs," *IEEE Trans. Signal Proc.,* pp. 1988–2006, July 1999.

[22] A. Scaglione, G. B. Giannakis, and S. Barbarossa, "Redundant filterbank precoders and equalizers. II: Blind channel estimation, synchronization, and direct equalization," *IEEE Trans. Signal Proc.,* pp. 2007–22, July 1999.

[23] A. V. Oppenheim, R. W. Schafer, and J. R. Buck, *Discrete-Time Signal Processing,* 2nd ed., Prentice-Hall, Englewood Cliffs, NJ, 1999.

[24] H. Sampath, S. Talwar, J. Tellado, V. Erceg, and A. Paulraj, "A fourth-generation MIMO-OFDM broadband wireless system: Design, performance, and field trial results," *IEEE Commun. Mag.,* pp. 143–9, September 2002.

[25] L. H. Brandenburg and A. D. Wyner, "Capacity of the Gaussian channel with memory: The multivariate case," *Bell System Tech. J.,* pp. 745–78, May/June 1974.

[26] G. L. Stuber, J. R. Barry, S. W. McLaughlin, Y. Li, M. A. Ingram, and T. G. Pratt, "Broadband MIMO-OFDM wireless communications," *Proc. IEEE,* pp. 271–94, February 2004.

[27] H. Ochiai and H. Imai, "On the distribution of the peak-to-average power ratio in OFDM signals," *IEEE Trans. Commun.,* pp. 282–9, February 2001.

[28] D. J. G. Mestdagh and P. M. P. Spruyt, "A method to reduce the probability of clipping in DMT-based transceivers," *IEEE Trans. Commun.,* pp. 1234–8, October 1996.

[29] K. G. Paterson and V. Tarokh, "On the existence and construction of good codes with low peak-to-average power ratios," *IEEE Trans. Inform. Theory,* pp. 1974–87, September 2000.

[30] J. Tellado, *Multicarrier Modulation with Low PAR: Applications to DSL and Wireless,* Kluwer, Boston, 2000.

[31] C. Muschallik, "Improving an OFDM reception using an adaptive Nyquist windowing," *IEEE Trans. Consumer Elec.,* pp. 259–69, August 1996.

[32] A. Redfern, "Receiver window design for multicarrier communication systems," *IEEE J. Sel. Areas Commun.,* pp. 1029–36, June 2002.

13

Spread Spectrum

Although bandwidth is a valuable commodity in wireless systems, *increasing* the transmit signal bandwidth can sometimes improve performance. Spread spectrum is a technique that increases signal bandwidth beyond the minimum necessary for data communication. There are many reasons for doing this. Spread-spectrum techniques can hide a signal below the noise floor, making it difficult to detect. Spread spectrum also mitigates the performance degradation due to intersymbol and narrowband interference. In conjunction with a RAKE receiver, spread spectrum can provide coherent combining of different multipath components. Spread spectrum also allows multiple users to share the same signal bandwidth, since spread signals can be superimposed on top of each other and demodulated with minimal interference between them. Finally, the wide bandwidth of spread-spectrum signals is useful for location and timing acquisition.

Spread spectrum first achieved widespread use in military applications because of its inherent property of hiding the spread signal below the noise floor during transmission, its resistance to narrowband jamming and interference, and its low probability of detection and interception. For commercial applications, the narrowband interference resistance has made spread spectrum common in cordless phones. The ISI rejection and bandwidth-sharing capabilities of spread spectrum are very desirable in cellular systems and wireless LANs. As a result, spread spectrum is the basis for both second- and third-generation cellular systems as well as second-generation wireless LANs.

13.1 Spread-Spectrum Principles

Spread spectrum is a modulation method applied to digitally modulated signals that increases the transmit signal bandwidth to a value much larger than is needed to transmit the underlying information bits. There are many signaling techniques that increase the transmit bandwidth above the minimum required for data transmission – for example, coding and frequency modulation. However, these techniques do not fall in the category of spread spectrum. The following three properties are needed for a signal to be spread-spectrum modulated [1].

■ The signal occupies a bandwidth much larger than is needed for the information signal.
■ The spread-spectrum modulation is done using a *spreading code,* which is independent of the data in the signal.

- Despreading at the receiver is done by correlating the received signal with a synchronized copy of the spreading code.

In order to make these notions precise, we return to the signal space representation of Section 5.1 to investigate embedding an information signal of bandwidth B within a much larger transmit signal bandwidth B_s than is needed. From (5.3), a set of linearly independent signals $s_i(t)$, $i = 1, \ldots, M$, of bandwidth B and time duration T can be written using a basis function representation as

$$s_i(t) = \sum_{j=1}^{N} s_{ij} \phi_j(t), \qquad 0 \leq t < T, \tag{13.1}$$

where the basis functions $\phi_j(t)$ are orthonormal and span an N-dimensional space. One of these signals is transmitted every T seconds to convey $\log_2 M/T$ bits per second. As discussed in Section 5.1.2, the minimum number of basis functions needed to represent these signals is approximately $2BT$. Since the $\{s_i(t)\}_{i=1}^{M}$ are linearly independent, this implies $M \approx 2BT$. To embed these signals into a higher-dimensional space, we chose $N \gg M$. The receiver uses an M-branch structure where the ith branch correlates the received signal with $s_i(t)$. The receiver outputs the signal corresponding to the branch with the maximum correlator output.

Suppose we generate the signals $s_i(t)$ using *random sequences,* so that the sequence of coefficients s_{ij} are chosen based on a random sequence generation, where each coefficient has mean zero and variance E_s/N. Thus, the signals $s_i(t)$ will have their energies uniformly distributed over the signal space of dimension N. Consider an interference or jamming signal within this signal space. The interfering signal can be represented as

$$I(t) = \sum_{j=1}^{N} I_j \phi_j(t), \tag{13.2}$$

with total energy over $[0, T]$ given by

$$\int_0^T I^2(t)\, dt = \sum_{j=1}^{N} I_j^2 = E_J. \tag{13.3}$$

Suppose the signal $s_i(t)$ is transmitted. Neglecting noise, the received signal is the sum of the transmitted signal plus interference:

$$x(t) = s_i(t) + I(t). \tag{13.4}$$

The output of the correlator in the ith branch of the receiver is then

$$x_i = \int_0^T x(t) s_i(t)\, dt = \sum_{j=1}^{N} (s_{ij}^2 + I_j s_{ij}), \tag{13.5}$$

where the first term in this expression represents the signal and the second term represents the interference. It can be shown [1] that the signal-to-interference power ratio (SIR) of this signal is

$$\mathrm{SIR} = \frac{E_s}{E_j} \cdot \frac{N}{M}. \tag{13.6}$$

This result is independent of the distribution of the interferer's energy over the N-dimensional signal space. In other words, by spreading the interference power over a larger dimension N than the required signaling dimension M, the SIR is increased by $G = N/M$, where G is called the *processing gain* or *spreading factor*. In practice, spread-spectrum systems have processing gains on the order of 10–1000. Since $N \approx 2B_sT$ and $M \approx 2BT$, we have $G \approx B_s/B$, the ratio of the spread signal bandwidth to the information signal bandwidth. Processing gain is often defined as this bandwidth ratio or something similar, but its underlying meaning is generally related to the performance improvement of a spread-spectrum system relative to a non-spread system in the presence of interference [2, Chap. 2.1]. Note that block and convolution coding are also techniques that improve performance in the presence of noise or interference by increasing signal bandwidth. An interesting trade-off arises as to whether, given a specific spreading bandwidth, it is more beneficial to use coding or spread spectrum. The answer depends on the specifics of the system design [3].

Spread spectrum is typically implemented in one of two forms: *direct sequence* (DS) or *frequency hopping* (FH). In direct-sequence spread-spectrum (DSSS) modulation, the modulated data signal $s(t)$ is multiplied by a wideband *spreading signal* or *code* $s_c(t)$, where $s_c(t)$ is constant over a time duration T_c and has amplitude equal to 1 or −1. The spreading code bits are usually referred to as *chips*, T_c is called the *chip time*, and $1/T_c$ is called the *chip rate*. The bandwidth $B_c \approx 1/T_c$ of $s_c(t)$ is roughly $B_c/B \approx T_s/T_c$ times larger than the bandwidth B of the modulated signal $s(t)$, and the number of chips per bit, T_s/T_c, is an integer approximately equal to G, the processing gain of the system. Multiplying the modulated signal by the spreading signal results in the convolution of these two signals in the frequency domain. Thus, the transmitted signal $s(t)s_c(t)$ has frequency response $S(f) * S_c(f)$, which has a bandwidth of roughly $B_c + B$. The multiplication of a spreading signal $s_c(t)$ with a modulated data signal $s(t)$ over one symbol time T_s is illustrated in Figure 13.1.

For an AWGN (additive white Gaussian noise) channel, the received spread signal is $s(t)s_c(t) + n(t)$ for $n(t)$ the channel noise. If the receiver multiplies this signal by a synchronized replica of the spreading signal, the result is $s(t)s_c^2(t) + n(t)s_c(t)$. Since $s_c(t) = \pm 1$, we have $s_c^2(t) = 1$. Moreover, $n'(t) = n(t)s_c(t)$ has approximately the same white Gaussian statistics as $n(t)$ if $s_c(t)$ is zero mean and sufficiently wideband (i.e., if its autocorrelation approximates a delta function). Thus, the received signal is $s(t)s_c^2(t) + n(t)s_c(t) = s(t) + n'(t)$, which indicates that spreading and despreading have no impact on signals transmitted over AWGN channels. However, spreading and despreading have tremendous benefits when the channel introduces narrowband interference or ISI.

We now illustrate the narrowband interference and multipath rejection properties of direct-sequence spread spectrum (DSSS) in the frequency domain: more details will be given in later sections. We first consider narrowband interference rejection, as shown in Figure 13.2. Neglecting noise, we see that the receiver input consists of the spread modulated signal $S(f) * S_c(f)$ and the narrowband interference $I(f)$. The despreading in the receiver recovers the data signal $S(f)$. However, the interference signal $I(t)$ is multiplied by the spreading signal $s_c(t)$, resulting in their convolution $I(f) * S_c(f)$ in the frequency domain. Thus, receiver despreading has the effect of distributing the interference power over the bandwidth of the spreading code. The demodulation of the modulated signal $s(t)$ effectively acts as a low-pass filter, removing most of the energy of the spread interference, which reduces its power by the processing gain $G \approx B_c/B$.

Baseband Modulated Signal $x(t)$

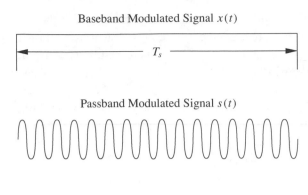

Passband Modulated Signal $s(t)$

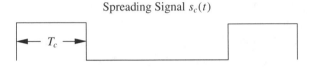

Spreading Signal $s_c(t)$

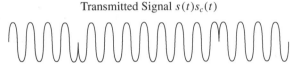

Transmitted Signal $s(t)s_c(t)$

Figure 13.1: Spreading signal multiplication.

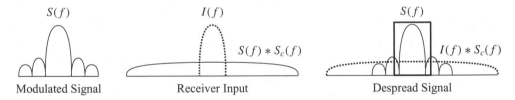

Figure 13.2: Narrowband interference rejection in DSSS.

Figure 13.3 illustrates ISI rejection, which is based on a similar premise. Suppose the spread signal $s(t)s_c(t)$ is transmitted through a two-ray channel with impulse response $h(t) = \alpha\delta(t) + \beta\delta(t - \tau)$. Then $H(f) = \alpha + \beta e^{-j2\pi f\tau}$, resulting in a receiver input in the absence of noise equal to $H(f)[S(f) * S_c(f)]$ in the frequency domain or $[s(t)s_c(t)] * h(t) = \alpha s(t)s_c(t) + \beta s(t - \tau)s_c(t - \tau)$ in the time domain. Suppose that the receiver despreading process multiplies this signal by a copy of $s_c(t)$ synchronized to the first path of this two-ray model. This results in the time-domain signal $\alpha s(t)s_c^2(t) + \beta s(t - \tau)s_c(t - \tau)s_c(t)$. Since the second multipath component $\beta s'(t) = \beta s(t - \tau)s_c(t - \tau)s_c(t)$ includes the product of asynchronized copies of $s_c(t)$, it remains spread out over the spreading code bandwidth, and the demodulation process will remove most of its energy. More precisely, as described in Section 13.2, the demodulation process effectively attenuates the multipath component by the autocorrelation $\rho_c(\tau)$ of the spreading code at delay τ. This autocorrelation can be quite small when $\tau > T_c$, on the order of $1/G \approx T_c/T_s$, resulting in significant mitigation of the

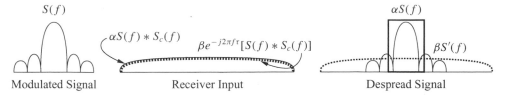

Figure 13.3: ISI rejection in DSSS.

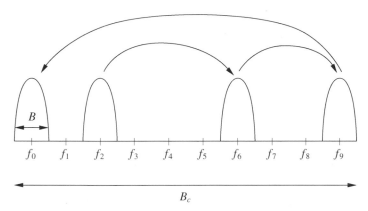

Figure 13.4: Frequency hopping.

ISI when the modulated signal is spread over a wide bandwidth. Since the spreading code autocorrelation determines the ISI rejection of the spread-spectrum system, it is important to use spreading codes with good autocorrelation properties, as discussed in the next section.

The basic premise of frequency-hopping spread spectrum (FHSS) is to hop the modulated data signal over a wide bandwidth by changing its carrier frequency according to the value of a spreading code $s_c(t)$.[1] This process is illustrated in Figure 13.4. The chip time T_c of $s_c(t)$ dictates the time between hops – that is, the time duration over which the modulated data signal is centered at a given carrier frequency f_i before hopping to a new carrier frequency. The hop time can exceed a symbol time, $T_c = kT_s$ for some integer k, which is called slow frequency hopping (SFH); or the carrier can be changed multiple times per symbol, $T_c = T_s/k$ for some integer k, which is called fast frequency hopping (FFH). In FFH there is frequency diversity on every symbol, which protects each symbol against narrowband interference and spectral nulls due to frequency-selective fading. The bandwidth of the FH system is approximately equal to NB, where N is the number of carrier frequencies available for hopping and B is the bandwidth of the data signal. The signal is generated using a frequency synthesizer that determines the modulating carrier frequency from the chip sequence, typically using a form of FM modulation such as CPFSK. In the receiver, the signal is demodulated using a similar frequency synthesizer, synchronized to the chip sequence

---

[1] The concept of frequency hopping was invented during World War II by film star Hedy Lamarr and composer George Antheil. Their patent for a "secret communications system" used a chip sequence generated by a player piano roll to hop between 88 frequencies. The design was intended to make radio-guided torpedos hard to detect or jam.

$s_c(t)$, that generates the sequence of carrier frequencies from this chip sequence for down-conversion. As with DS, FH has no impact on performance in an AWGN channel. However, it does mitigate the effects of narrowband interference and multipath.

Consider a narrowband interferer of bandwidth B at a carrier frequency f_i corresponding to one of the carriers used by the FH system. The interferer and FH signal occupy the same bandwidth only when carrier f_i is generated by the hop sequence. If the hop sequence spends an equal amount of time at each of the carrier frequencies, then interference occurs a fraction $1/N$ of the time, and thus the interference power is reduced by roughly $1/N$. However, the nature of the interference reduction is different in FH versus DS systems. In particular, DS results in a reduced-power interference all the time, whereas FH has a full power interferer a fraction of the time. In FFH systems the interference affects only a fraction of a symbol time, so coding may not be required to compensate for this interference. In SFH systems the interference affects many symbols, so typically coding with interleaving is needed to avoid many simultaneous errors in a single codeword. Frequency hopping is commonly used in military systems, where the interferers are assumed to be malicious jammers attempting to disrupt communications.

We now investigate the impact of multipath on an FH system. For simplicity, we consider a two-ray channel that introduces a multipath component with delay τ. Suppose the receiver synchronizes to the hop sequence associated with the line-of-sight signal path. Then the LOS path is demodulated at the desired carrier frequency. However, the multipath component arrives at the receiver with a delay τ. If $\tau > T_c$ then the receiver will have hopped to a new carrier frequency $f_j \neq f_i$ for downconversion when the multipath component, centered at carrier frequency f_i, arrives at the receiver. Since the multipath occupies a different frequency band than the LOS signal component being demodulated, it causes negligible interference to the demodulated signal. Thus, the demodulated signal does not exhibit either flat or frequency-selective fading for $\tau > T_c$. If $\tau < T_c$ then the impact of multipath depends on the bandwidth B of the modulated data signal as well as on the hop rate. First consider an FFH system where $T_c \ll T_s$. Since we also assume $\tau < T_c$, it follows that $\tau < T_c \ll T_s$. Since all the multipath arrives within a symbol time, the multipath introduces a complex amplitude gain and the signal experiences flat fading. Now consider an SFH system where $T_c \gg T_s$. Since we also assume $\tau < T_c$, all the multipath will arrive while the signal is at the same carrier frequency, so the impact of multipath is the same as if there were no frequency hopping: for $B < 1/\tau$ the signal experiences flat fading, and for $B > 1/\tau$ the signal experiences frequency-selective fading. The fading channel also varies slowly over time, since the equivalent lowpass channel is a function of the carrier frequency and thus changes whenever the carrier hops to a new frequency. In summary, frequency hopping removes the impact of multipath on demodulation of the LOS component whenever $\tau > T_c$. For $\tau < T_c$, an FFH system will exhibit flat fading, whereas an SFH system will exhibit slowly varying flat fading for $B < 1/\tau$ and slowly varying frequency-selective fading for $B > 1/\tau$. The performance analysis under time-varying flat or frequency-selective fading is the same as for systems without hopping, as given in Sections 6.3 and 6.5, respectively.

In addition to their interference and ISI rejection capabilities, both DSSS and FHSS provide a mechanism for multiple access, allowing many users to simultaneously share the spread bandwidth with minimal interference between users. In these multiuser systems, the interference between users is determined by the cross-correlation of their spreading codes. Spreading code designs typically have either good autocorrelation properties to mitigate ISI or good cross-correlation properties to mitigate multiuser interference. However, there is usually a trade-off between optimizing the two features. Thus, the best choice of code design depends on the number of users in the system and the severity of the multipath and interference. Trade-offs between frequency hopping and direct sequence in multiuser systems are discussed in Section 13.5. Frequency hopping is also used in cellular systems to average out interference from other cells.

EXAMPLE 13.1: Consider an SFH system with hop time $T_c = 10\,\mu s$ and symbol time $T_s = 1\,\mu s$. If the FH signal is transmitted over a multipath channel, for approximately what range of multipath delay spreads will the received despread signal exhibit frequency-selective fading?

Solution: Based on the two-ray model analysis, the signal exhibits fading – flat or frequency-selective – only when the delay spread $\tau < T_c = 10\,\mu s$. Moreover, for frequency-selective fading we require $B \approx 1/T_s = 10^6 > 1/\tau$; that is, we require $\tau > 10^{-6} = 1\,\mu s$. So the despread signal will exhibit frequency-selective fading for delay spreads ranging from approximately $1\,\mu s$ to $10\,\mu s$.

13.2 Direct-Sequence Spread Spectrum (DSSS)

13.2.1 DSSS System Model

An end-to-end direct-sequence spread-spectrum system is illustrated in Figure 13.5. The multiplication by $s_c(t)$ and the carrier $\cos(2\pi f_c t)$ could be done in opposite order as well: downconverting prior to despreading allows the code synchronization and despreading to be done digitally, but it complicates carrier phase tracking because this must be done relative to the wideband spread signal.[2] For simplicity we illustrate the receiver only for in-phase signaling; a similar structure is used for the quadrature signal component. The data symbols s_l are first linearly modulated to form the baseband modulated signal $x(t) = \sum_l s_l g(t - lT_s)$, where $g(t)$ is the modulator shaping pulse, T_s the symbol time, and s_l the symbol transmitted over the lth symbol time. Linear modulation is used because DSSS is a form of phase modulation and therefore works best in conjunction with a linearly modulated data signal.

---

[2] A system that performs spreading and despreading on the bandpass modulated signal works as follows. The transmitter consists of a standard narrowband modulator that generates a passband modulated signal, followed by spreading. The receiver consists of despreading, followed by a standard narrowband demodulator. This order of operations makes it straightforward to design a spread-spectrum system using existing narrowband modulators and demodulators, and operations such as carrier phase recovery are not affected by spreading. However, spread-spectrum systems today do as much of the signal processing as possible in the digital domain. Thus, spread-spectrum systems typically modulate the data symbols and multiply by the spreading code at baseband using digital signal processing, followed by A/D conversion and analog upconversion to the carrier frequency. In this case all functions prior to the carrier multiplication in Figure 13.5 are done digitally, with an A/D converter following the multiplication with $s_c(t)$. However, the carrier recovery loop for this system is more difficult to design because it operates on the spread signal. In particular, any nonlinear operation (e.g. squaring) that is used to remove either the data or the spreading sequence in carrier phase recovery can be seriously degraded by the noise associated with the spread signal.

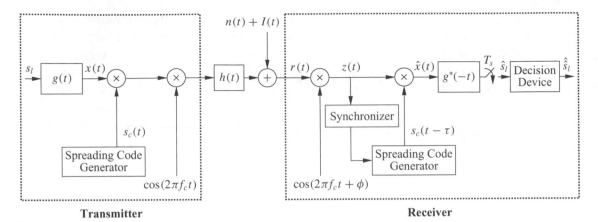

Figure 13.5: DSSS system model.

The modulated signal is then multiplied by the spreading code $s_c(t)$ with chip time T_c, after which it is upconverted through multiplication by the carrier $\cos(2\pi f_c t)$. The spread signal passes through the channel $h(t)$, which also introduces narrowband interference $I(t)$ and zero-mean AWGN $n(t)$ with power spectral density $N_0/2$.

Assume the channel introduces several multipath components: $h(t) = \alpha_0 \delta(t - \tau_0) + \alpha_1 \delta(t - \tau_1) + \cdots$. The received signal is first downconverted to baseband assuming perfect carrier recovery, so that the carrier $\cos(2\pi f_c t + \phi)$ in the receiver has phase ϕ matched to the carrier phase of the incoming signal (e.g., if $h(t) = \delta(t)$ then $\phi = 0$). The carrier recovery loop is typically assumed to lock to the carrier associated with the LOS (or minimum delay) multipath component. The synchronizer then uses the resulting baseband signal $z(t)$ to align the delay τ of the receiver's spreading code generator with one of the multipath component delays τ_i. The spreading code generator then outputs the spreading code $s_c(t - \tau)$, where $\tau = \tau_i$ if the synchronizer is perfectly aligned with the delay associated with the ith multipath component. Ideally the synchronizer would lock to the multipath component with the largest amplitude. However, in practice this requires a complex search procedure, so instead the synchronizer typically locks to the first component it finds with an amplitude above a given threshold. This synchronization procedure can be quite complex, especially for channels with severe ISI or interference, and synchronization circuitry can make up a large part of any spread-spectrum receiver. Synchronization is discussed in more detail in Section 13.2.3.

The multipath component at delay τ is despread via multiplication by the spreading code $s_c(t - \tau)$. The other multipath components are not despread, and most of their energy is removed, as we shortly show. After despreading, the baseband signal $\hat{x}(t)$ passes through a matched filter and decision device. Thus, there are three stages in the receiver demodulation for direct-sequence spread spectrum: downconversion, despreading, and baseband demodulation. This demodulator is also called the single-user *matched filter* detector for DSSS. We now examine the three stages of this detector in more detail.

For simplicity, assume rectangular pulses are used in the modulation ($g(t) = \sqrt{2/T_s}$, $0 \le t \le T_s$). The matched filter $g^*(-t)$ then simply multiplies $\hat{x}(t)$ by $\sqrt{2/T_s}$ and integrates from zero to T_s to obtain the estimate of the transmitted symbol. Since perfect carrier recovery is assumed, the carrier phase offset ϕ in the receiver matches that of the incoming signal.

We also assume perfect synchronization in the receiver. The multipath and interference rejection occurs in the data demodulation process. Specifically, the input to the matched filter is given by

$$\hat{x}(t) = ([x(t)s_c(t)\cos(2\pi f_c t)] * h(t))s_c(t - \tau)\cos(2\pi f_c t + \phi)$$
$$+ n(t)s_c(t - \tau)\cos(2\pi f_c t + \phi) + I(t)s_c(t - \tau)\cos(2\pi f_c t + \phi). \quad (13.7)$$

Without multipath, $h(t) = \delta(t)$, $\phi = 0$, and the receiver ideally synchronizes with $\tau = 0$. Then the spreading/despreading process has no impact on the baseband signal $x(t)$. Specifically, the spreading code has amplitude ± 1, so multiplying $s_c(t)$ by a synchronized copy of itself yields $s_c^2(t) = 1$ for all t. Then, in the absence of multipath and interference (i.e., for $h(t) = \delta(t)$ and $I(t) = 0$),

$$\hat{x}(t) = x(t)s_c^2(t)\cos^2(2\pi f_c t) + n(t)s_c(t)\cos(2\pi f_c t)$$
$$= x(t)\cos^2(2\pi f_c t) + n(t)s_c(t)\cos(2\pi f_c t), \quad (13.8)$$

since $s_c^2(t) = 1$. If $s_c(t)$ is sufficiently wideband then $n(t)s_c(t)$ has approximately the same statistics as $n(t)$: it is a zero-mean AWGN random process with PSD $N_0/2$. The matched filter output over a symbol time for $g(t) = \sqrt{2/T_s}$, $0 \le t \le T_s$, will thus be

$$\hat{s}_l = \int_0^{T_s} \hat{x}(t) * g^*(-t)\, dt$$
$$= \sqrt{\frac{2}{T_s}} \int_0^{T_s} x(t)\cos^2(2\pi f_c t)\, dt + \sqrt{\frac{2}{T_s}} \int_0^{T_s} n(t)s_c(t)\cos(2\pi f_c t)\, dt$$
$$= \frac{2}{T_s} \int_0^{T_s} s_l \cos^2(2\pi f_c t)\, dt + \sqrt{\frac{2}{T_s}} \int_0^{T_s} n(t)s_c(t)\cos(2\pi f_c t)\, dt$$
$$\approx s_l + n_l, \quad (13.9)$$

where s_l and n_l correspond to the data and noise output of a standard demodulator without spreading or despreading and where the approximation assumes $f_c \gg 1/T_s$.

We now consider the interference signal $I(t)$ at the carrier frequency f_c, which can be modeled as $I(t) = I'(t)\cos(2\pi f_c t)$ for some narrowband baseband signal $I'(t)$. We again assume $h(t) = \delta(t)$. Multiplication by the spreading signal perfectly synchronized to the incoming signal yields

$$\hat{x}(t) = x(t)\cos^2(2\pi f_c t) + n(t)s_c(t)\cos(2\pi f_c t) + I'(t)s_c(t)\cos^2(2\pi f_c t), \quad (13.10)$$

where $n(t)s_c(t)$ is assumed to be a zero-mean AWGN process. The demodulator output is then given by

$$\hat{s}_l = \frac{2}{T_s} \int_0^{T_s} s_l s_c^2(t)\cos^2(2\pi f_c t)\, dt + \sqrt{\frac{2}{T_s}} \int_0^{T_s} n(t)s_c(t)\cos(2\pi f_c t)\, dt$$
$$+ \sqrt{\frac{2}{T_s}} \int_0^{T_s} I'(t)s_c(t)\cos^2(2\pi f_c t)\, dt$$
$$\approx s_l + n_l + I_l, \quad (13.11)$$

where s_l and n_l correspond to the data and noise output of a standard demodulator without spreading or despreading and where the approximation assumes $f_c \gg 1/T_s$. The narrowband interference rejection can be seen from the last term of (13.11). In particular, the spread interference $I'(t)s_c(t)$ is a wideband signal with bandwidth of roughly $1/T_c$, and the integration acts as a lowpass filter with bandwidth of roughly $1/T_s \ll 1/T_c$, thereby removing most of the interference power.

Let us now consider ISI rejection. Assume a multipath channel with one delayed component: $h(t) = \alpha_0 \delta(t) + \alpha_1 \delta(t - \tau_1)$. For simplicity, assume $\tau_1 = kT_s$ is an integer multiple of the symbol time. Suppose that the first multipath component is stronger than the second, $\alpha_0 > \alpha_1$, and that the receiver locks to the first component ($\phi = 0$ and $\tau = 0$ in Figure 13.5). Then, in the absence of narrowband interference ($I(t) = 0$), after despreading we have

$$\hat{x}(t) = \alpha_0 x(t) s_c^2(t) \cos^2(2\pi f_c t)$$
$$+ \alpha_1 x(t - \tau_1) s_c(t - \tau_1) \cos(2\pi f_c (t - \tau_1)) s_c(t) \cos(2\pi f_c t)$$
$$+ n(t) s_c(t) \cos(2\pi f_c t). \tag{13.12}$$

Since $\tau_1 = kT_s$, the ISI just corresponds to the signal transmission of the $(l - k)$th symbol; that is, $x(t - \tau_1) = x(t - kT_s) = s_{l-k} g(t - (l - k)T_s)$. The demodulator output over the lth symbol time is then given by

$$\hat{s}_l = \frac{2}{T_s} \int_0^{T_s} \alpha_0 s_l \cos^2(2\pi f_c t)\, dt$$
$$+ \frac{2}{T_s} \int_0^{T_s} \alpha_1 s_{l-k} s_c(t) s_c(t - \tau_1) \cos(2\pi f_c t) \cos(2\pi f_c (t - \tau_1))\, dt$$
$$+ \sqrt{\frac{2}{T_s}} \int_0^{T_s} n(t) s_c(t) \cos(2\pi f_c t)\, dt \tag{13.13}$$
$$\approx \alpha_0 s_l + \alpha_1 s_{l-k} \cos(2\pi f_c \tau_1) \rho_c(\tau_1) + n_l; \tag{13.14}$$

here, as in the case of interference rejection, s_l and n_l correspond to the data symbol and noise output of a standard demodulator without spreading or despreading, and the approximation assumes $f_c \gg 1/T_s$. The middle term $\alpha_1 s_{l-k} \cos(2\pi f_c \tau_1) \rho_c(\tau_1)$ comes from the following integration:

$$\frac{2}{T_s} \int_0^{T_s} s_c(t) s_c(t - \tau_1) \cos(2\pi f_c t) \cos(2\pi f_c (t - \tau_1))\, dt$$
$$= \frac{1}{T_s} \int_0^{T_s} s_c(t) s_c(t - \tau_1)(\cos(2\pi f_c \tau_1) + \cos(4\pi f_c t - 2\pi f_c \tau_1))\, dt$$
$$\approx \cos(2\pi f_c \tau_1) \frac{1}{T_s} \int_0^{T_s} s_c(t) s_c(t - \tau_1)\, dt$$
$$= \cos(2\pi f_c \tau_1) \rho_c(t, t - \tau_1), \tag{13.15}$$

where the approximation is based on $f_c \gg T_s^{-1}$ and where

$$\rho_c(t, t - \tau_1) \triangleq \frac{1}{T_s} \int_0^{T_s} s_c(t) s_c(t - \tau_1)\, dt \tag{13.16}$$

is defined as the *autocorrelation* of the spreading code at time t and delay τ_1 over a symbol time.[3] More generally, the spreading code autocorrelation at time t (assumed to be an integer multiple of T_c) and delay τ, over a period $[0, T]$, is defined as

$$\rho_c(t, t - \tau) \triangleq \frac{1}{T} \int_0^T s_c(t) s_c(t - \tau) \, dt = \frac{1}{N_T} \sum_{n=1}^{N_T} s_c(nT_c) s_c(nT_c - \tau), \qquad (13.17)$$

where $N_T = T/T_c$ is the number of chips over duration T and where the second equality follows because $s_c(t)$ is constant over a chip time T_c. It can be shown that $\rho_c(t, t - \tau)$ is a symmetric function with maximum value at $\tau = 0$. Moreover, if $s_c(t)$ is periodic with period T, then the autocorrelation depends only on the time difference of the spreading codes; that is,

$$\frac{1}{T} \int_0^T s_c(t - \tau_0) s_c(t - \tau_1) \, dt = \rho_c(\tau_1 - \tau_0), \qquad (13.18)$$

so that $\rho_c(t, t - \tau) \triangleq \rho_c(\tau)$ in (13.17) depends only on τ.

By (13.15), if $T = T_s$ and $\rho_c(\tau) = \delta(\tau)$ then the despreading process removes all ISI. Unfortunately, it is not possible to have finite-length spreading codes with autocorrelation equal to a delta function. Hence there has been much work on designing spreading codes with autocorrelation over a symbol time that approximates a delta function. In the next section we discuss spreading codes for ISI rejection, including maximal linear codes, which have excellent autocorrelation properties to minimize ISI effects.

13.2.2 Spreading Codes for ISI Rejection: Random, Pseudorandom, and *m*-Sequences

Spreading codes are generated deterministically, often using a shift register with feedback logic to create a binary code sequence **b** of 1s and 0s. The binary sequence, also called a *chip sequence,* is used to amplitude modulate a square pulse train with pulses of duration T_c – with amplitude 1 for a 1-bit and amplitude -1 for a 0-bit – as shown in Figure 13.6. The resulting spreading code $s_c(t)$ is a sinc function in the frequency domain, corresponding to the Fourier transform of a square pulse. The shift register, consisting of n stages, has a cyclical output with a maximum period of $2^n - 1$. To avoid a spectral spike at DC or biasing the noise in despreading, the spreading code $s_c(t)$ should have no DC component, which requires that the bit sequence **b** have approximately the same number of 1s and 0s. It is also desirable for the number of consecutive 1s or 0s, called a *run,* to be small. Runs are undesirable because, if there is a run of k consecutive 1s or 0s, then the data signal over kT_c is just multiplied by a constant, which reduces the bandwidth spreading (and its advantages) by roughly a factor of k. Ideally

---

[3] Note that, if τ_1 is not an integer multiple of a symbol time, then the middle term in (13.14) becomes more complicated. In particular, assuming $g(t) = \sqrt{2/T_s}$, if $\tau_1 = (k + \kappa)T_s$ for $0 < \kappa < 1$ then $x(t - \tau_1) = \sqrt{2/T_s} s_{l-k-1}$ for $0 \le t \le \kappa T_s$ and $x(t - \tau_1) = \sqrt{2/T_s} s_{l-k}$ for $\kappa T_s \le t \le T_s$. Thus, the middle term of (13.14) becomes

$$\alpha_1 s_{l-k-1} \cos(2\pi f_c \tau_1) \frac{1}{T_s} \int_0^{\kappa T_s} s_c(t) s_c(t - \tau_1) \, dt + \alpha_1 s_{l-k} \cos(2\pi f_c \tau_1) \frac{1}{T_s} \int_{\kappa T_s}^{T_s} s_c(t) s_c(t - \tau_1) \, dt,$$

where each term is a function of the spreading code autocorrelation taken over a fraction of the symbol time.

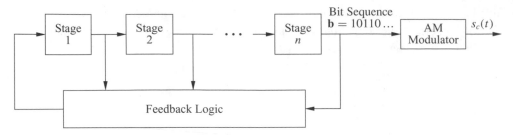

Figure 13.6: Generation of spreading codes.

the chip values change roughly every chip time, which leads to maximal spreading. Based on (13.15), we require spreading codes with $\rho_c(\tau) \approx \delta(\tau)$ in order to minimize ISI effects.

EXAMPLE 13.2: Find the baseband bandwidth of a spreading code $s_c(t)$ with chip time $T_c = 1\,\mu s$.

Solution: The spreading code $s_c(t)$ consists of a sequence of unit-amplitude square pulses of duration T_c modulated with ± 1. The Fourier transform of a unit-amplitude square pulse is $S(f) = T_c \operatorname{sinc}(fT_c)$, with a main lobe of bandwidth $2/T_c$. Thus, the null-to-null baseband bandwidth, defined as the minimum frequency where $S(f) = 0$, is $1/T_c$.

Although DSSS chip sequences must be generated deterministically, properties of random sequences are useful for gaining insight into deterministic sequence design. A random binary chip sequence consists of independent and identically distributed bit values with probability .5 for a 1-bit or a 0-bit. A random sequence of length N can thus be generated, for example, by flipping a fair coin N times and setting the bit to a 1 for heads and a 0 for tails. Random sequences with length N asymptotically large have a number of the properties desired in spreading codes [4]. In particular, such sequences will have an equal number of 1s and 0s, called the *balanced property* of a code. Moreover, the run length in such sequences is generally short. In particular, for asymptotically large sequences, half of all runs are of length 1, a quarter are of length 2, and so forth, so that a fraction $1/2^r$ of all runs are of length r for r finite. This distribution on run length is called the *run-length property* of a code. Random sequences also have the property that, if they are shifted by any nonzero number of elements, the resulting sequence will have half its elements the same as in the original sequence and half its elements different from the original sequence. This is called the *shift property* of a code. Following Golomb [4], a deterministic sequence that has the balanced, run length, and shift properties as it grows asymptotically large is referred to as a *pseudorandom sequence*. Because these three properties are often the most important in system analysis, DSSS analysis is often done using random spreading sequences instead of deterministic spreading sequences owing to the former's analytical tractability [5, Chap. 2.2].

Among all linear codes, spreading codes generated from maximal-length sequences, or m-sequences, have many desirable properties. Maximal-length sequences are a type of cyclic code (see Section 8.2.4). Thus, they are generated and characterized by a generator polynomial, and their properties can be derived using algebraic coding theory [2, Chap. 3.3; 5,

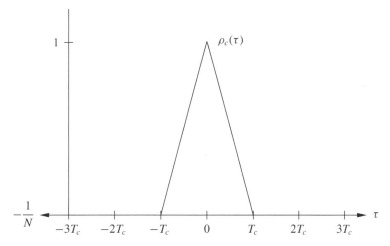

Figure 13.7: Autocorrelation of maximal linear code ($N = T/T_c$).

Chap. 2.2]. These sequences have the maximum period $N = 2^n - 1$ that can be generated by a shift register of length n, so the sequence repeats every NT_c seconds. Moreover, since the sequences are cyclic codes, any time shift of an m-sequence is itself an m-sequence. These sequences also have the property that the modulo-2 addition of an m-sequence and a time shift of itself results in a different m-sequence corresponding to a different time shift of the original sequence. This property is called the *shift-and-add* property of m-sequences. The m-sequences have roughly the same number of 0s and 1s over a period: $2^{n-1} - 1$ 0-bits and 2^{n-1} 1-bits. Thus, spreading codes generated from m-sequences, called *maximal linear codes,* have a very small DC component. Moreover, maximal linear codes have approximately the same run-length property as random binary sequences; that is, the number of runs of length r in an n-length sequence is $1/2^r$ for $r < n$ and $1/2^{r-1}$ for $r = n$. Finally, the balanced and shift-and-add properties of m-sequences can be used to show that m-sequences have the same shift property as random binary sequences. Therefore, since m-sequences have the balanced, run-length, and shift properties of random sequences, they belong to the class of pseudorandom (PN) sequences [5, Chap. 2.2].

The autocorrelation $\rho_c(\tau)$ of a maximal linear spreading code taken over a full period $T = NT_c$ is given by

$$\rho_c(\tau) = \begin{cases} 1 - |\tau|(1 + 1/N)/T_c & |\tau| \leq T_c, \\ -1/N & |\tau| > T_c, \end{cases} \tag{13.19}$$

for $|\tau| < (N-1)T_c$, which is illustrated in Figure 13.7. Moreover, since the spreading code is periodic with period $T = NT_c$, the autocorrelation is also periodic with the same period, as shown in Figure 13.8. Thus, if τ is not within a chip time of kNT_c for any integer k, then $\rho_c(\tau) = -1/N = -1/(2^n - 1)$. By making n sufficiently large, the impact of multipath components associated with delays that are not within a chip time of kNT_c can be mostly removed. For delays τ within a chip time of kNT_c, the attenuation is determined by the autocorrelation $\rho_c(\tau)$, which increases linearly as τ approaches kNT_c. The power spectrum of $s_c(t)$ is obtained by taking the Fourier transform of its autocorrelation $\rho_c(\tau)$, yielding

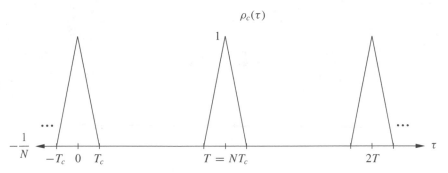

Figure 13.8: Autocorrelation has period $T = NT_c$.

$$P_{s_c}(f) = \sum_{m=-\infty}^{\infty} \frac{N+1}{N^2} \mathrm{sinc}^2\left(\frac{m}{N}\right)\delta\left(f - \frac{m}{T}\right). \qquad (13.20)$$

Since $\rho_c(\tau)$ is periodic, it follows that $P_{s_c}(f)$ is discrete, with samples every $1/T = 1/NT_c$ Hz.

The periodic nature of the autocorrelation $\rho_c(\tau)$ complicates ISI rejection. In particular, from (13.16), the demodulator associated with the data signal in a spread-spectrum system attenuates the ISI by the autocorrelation $\rho_c(\tau)$ taken over a symbol time T_s. Thus, if the code is designed with $N = T_s/T_c$ chips per symbol, then the demodulator computes the autocorrelation over the full period $T_s = NT_c$ and $\rho_c(\tau)$ is as given in (13.19). Setting $N = T_s/T_c$ is sometimes referred to as a *short spreading code* because the autocorrelation repeats every symbol time, as shown in Figure 13.8 for $T = T_s$. However, short codes exhibit significant ISI from multipath components delayed by approximately an integer multiple of a symbol time, in particular the first few symbols after the desired symbol. If the period of the code is extended so that $N \gg T_s/T_c$, then only multipath at very large delays are not fully attenuated, and these multipath components typically have a low power anyway owing to path loss. Setting $N \gg T_s/T_c$ is sometimes referred to as a *long spreading code*. With long spreading codes the autocorrelation (13.17) performed by the demodulator is taken over a partial period $T = T_s \ll NT_c$ instead of the full period NT_c. The autocorrelation of a maximal linear code over a partial period is no longer characterized by (13.19), so multipath delayed by more than a chip time is no longer attenuated by $-1/N$. Moreover, the partial-period autocorrelation is quite difficult to characterize analytically, since it depends on the starting point in the code over which the partial autocorrelation is taken. By averaging over all starting points, it can be shown that the ISI attenuation associated with the partial autocorrelation is roughly equal to $1/G$ for G the processing gain, where $G \approx T_s/T_c$, the number of chips per symbol [6, Chap. 9.2].

Although maximal linear codes have excellent properties for ISI rejection, they have a number of properties that make them suboptimal for exploiting the multiuser capabilities of spread spectrum. In particular, there are fewer maximal linear codes of a given length N than in other codes, which limits the number of users who can share the total system bandwidth for multiuser DSSS based on maximal linear codes. Moreover, maximal linear codes generally have relatively poor cross-correlation properties, at least for some sets of codes. In particular, the normalized code cross-correlation can be as high as .37 [6, Chap. 9.2]. Hence for spread-spectrum systems with multiple users, Gold, Kasami, or Walsh codes are used instead of maximal linear codes because they have superior cross-correlation properties. However, these codes can be less effective at ISI rejection than maximal linear codes. More details on these other types of spreading codes will be given in Section 13.4.1.

EXAMPLE 13.3: Consider a spread-spectrum system using maximal linear codes with period $T = T_s$ and $N = 100$ chips per symbol. Assume the synchronizer has a delay offset of $.5T_c$ relative to the LOS signal component to which it is synchronized. By how much is the power of this signal component reduced by this timing offset?

Solution: For $\tau = .5T_c$ and $N = 100$, the autocorrelation $\rho_c(\tau)$ given by (13.19) is

$$1 - \frac{|\tau|(1 + 1/N)}{T_c} = 1 - \frac{.5T_c(1 + 1/100)}{T_c} = 1 - .5(1.01) = .495.$$

Since the signal component is multiplied by $\rho_c(\tau)$, its power is reduced by $\rho_c^2(\tau) = .495^2 = .245 = -6.11$ dB. This is a significant reduction in power, indicating the importance of accurate synchronization, which is discussed in the next section.

13.2.3 Synchronization

We now examine the operation of the synchronizer in Figure 13.5. This operation is separate from carrier phase recovery, so we will assume that the carrier in the demodulator is coherent in phase with the received carrier. The synchronizer must align the timing of the spreading code generator in the receiver with the spreading code associated with one of the multipath components arriving over the channel. A common method of synchronization uses a feedback control loop, as shown in Figure 13.9. The basic premise of the feedback loop is to adjust the delay τ of the spreading code generator until the function $w(\tau)$ reaches its peak value. At this point, under ideal conditions, the spreading code is synchronized to the input, as we now illustrate.

Consider a channel with impulse response $h(t) = \delta(t - \tau_0)$ that introduces a delay τ_0. Neglecting noise, the signal input to the synchronizer from Figure 13.5 is $z(t) = x(t - \tau_0)s_c(t - \tau_0) \cos(2\pi f_c(t - \tau_0)) \cos(2\pi f_c t + \phi)$, where ϕ is the phase associated with the carrier recovery loop in the receiver. Assuming perfect carrier recovery, $\cos(2\pi f_c(t - \tau_0)) = \cos(2\pi f_c t + \phi)$, so $z(t) = x(t - \tau_0)s_c(t - \tau_0) \cos^2(2\pi f_c t + \phi)$. The feedback loop will achieve synchronization when $\tau = \tau_0$. We will first assume that $x(t)$ is a BPSK signal with rectangular pulse shaping (so $s_l = \pm\sqrt{E_b}$ and $x(t) = \pm\sqrt{2E_b/T_b}$ is constant over a bit time T_b) and that the spreading codes are maximal linear codes. We will then discuss extensions to more general spreading codes and modulated signals. Assume the spreading codes have period $T = NT_c = T_b$, so their autocorrelation over one period is given by (13.19) and shown in Figure 13.7. Then, setting $\lambda = \sqrt{2/T_b}$, we have

$$w(\tau) = \frac{\lambda}{T} \int_0^T s_l s_c(t - \tau_0)s_c(t - \tau) \cos^2(2\pi f_c t + \phi)\, dt$$

$$\approx \frac{.5\lambda s_l}{T} \int_0^T s_c(t - \tau_0)s_c(t - \tau)\, dt = .5\lambda s_l \rho_c(\tau - \tau_0), \qquad (13.21)$$

by (13.18). Since $\rho_c(\tau - \tau_0)$ reaches its maximum at $\tau - \tau_0 = 0$ and since $s_l = \pm\sqrt{E_b}$, the feedback control loop will adjust τ such that $|w(\tau)|$ increases. In particular, suppose

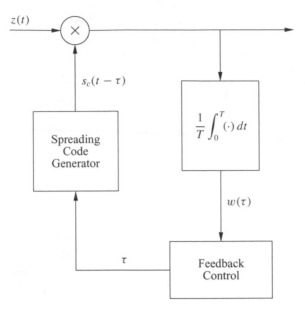

Figure 13.9: Synchronization loop for DSSS.

$|\tau - \tau_0| > T_c$. Then, by (13.19), $\rho_c(\tau - \tau_0) = -1/N$ and the synchronizer is operating outside the triangular region of the autocorrelation function shown in Figure 13.7. The feedback control loop will therefore adjust τ, typically in increments of T_c, until $|w(\tau)|$ increases above $-1/N$. This increase occurs when τ is sufficiently adjusted that $|\tau - \tau_0| < T_c$. At this point the synchronizer is within a chip time of perfect synchronization, which is sometimes referred to as *coarse synchronization* or *acquisition*. In general the channel has many multipath components, in which case coarse synchronization will synchronize to the first multipath component it finds above a given power threshold.

An alternative to the feedback control loop for acquisition is a parallel-search acquisition system. This system has multiple branches that correlate the received signal against a delayed version of the spreading code, where each branch has a different delay equal to an integer multiple of the chip time. The sychronization locks to the branch with the maximum correlator output. A similar structure is used in a RAKE receiver, discussed in the next section, to coherently combine multipath components at different delays. For both synchronization methods, the coarse acquisition often uses short codes with a small period T to reduce acquisition time. If long codes are used, the acquisition time can be shortened by performing the integration in the feedback loop over a fraction of the entire code period. In this case, as long as the partial autocorrelation is small for delays greater than a chip time and is above a given threshold for delays within a chip time, the acquisition loop can compare the partial autocorrelation against the threshold to determine if coarse acquisition has occured. For the fine tuning that follows coarse acquisition, long codes with integration over the full period are typically used to make the synchronization as precise as possible.

Once coarse synchronization is achieved, the feedback control loop makes small adjustments to τ to fine-tune its delay estimate such that $\tau \approx \tau_0$. This is called *fine synchronization* or *tracking*. Suppose through course synchronization we obtain $\tau - \tau_0 = T_c$. Referring to

Figure 13.7, this implies that the synchronizer is operating on the far right edge of the triangular correlation function. As τ is further decreased, $\tau - \tau_0$ decreases toward zero, and the synchronization "walks backwards" toward the peak of the autocorrelation at $\tau - \tau_0 = 0$. Once the peak is attained, the synchronizer locks to the delay τ_0. Because of the time-varying nature of the channel as well as interference, multipath, and noise, τ must be adjusted continuously to optimize synchronization under these dynamic operating conditions. Spread-spectrum tracking often uses the same timing recovery techniques discussed in Section 5.6.3 for narrowband systems.

The acquisition and tracking procedures for more general spreading codes are very similar. Since all periodic spreading codes have an autocorrelation that peaks at zero, the course and fine synchronization will adjust their estimate of the delay to try to maximize the autocorrelation output of the integrator. Synchronization performance is highly dependent on the shape of the autocorrelation function. A sharp autocorrelation facilities accurate fine tuning of the synchronization. Noise, fading, interference, and ISI will complicate both coarse and fine synchronization, since the output of the integrator in Figure 13.9 will be distorted by these factors.

When $s(t)$ is not binary and/or it has a symbol time that is less than the code period, the integrator output will depend on the data symbol(s) over the duration of the integration. This is the same situation as in carrier and timing recovery of narrowband systems with unknown data, discussed in Section 5.6, and similar techniques can be applied in this setting. Note that we have also neglected carrier phase recovery in our analysis, assuming that the receiver has a carrier recovery loop to obtain a coherent phase reference on the received signal. Carrier recovery techniques were discussed in Section 5.6, but these techniques must be modfied for spread-spectrum systems because the spreading codes affect the carrier recovery process [7]. Synchronization is a challenging aspect of spread-spectrum system design, especially in time-varying wireless environments. Much work has been devoted to developing and analyzing spread-spectrum synchronization techniques. Details on the main techniques and their performance can be found in [2, Chaps. 4–5; 8, Chap. 6; 9, Part 4:1–2; 10, Chap. 12.5].

13.2.4 RAKE Receivers

The spread-spectrum receiver shown in Figure 13.5 will synchronize to one of the multipath components in the received signal. The multipath component to which it is synchronized is typically the first one acquired during the coarse synchronization that is above a given threshold. This may not be the strongest multipath component; moreover, the result is that all other multipath components are then treated as interference. A more complicated receiver can have several branches, with each branch synchronized to a different multipath component. This structure is called a *RAKE* receiver,[4] and it typically assumes there is a multipath component at each integer multiple of a chip time. Thus, the time delay of the spreading code between branches is T_c, as shown in Figure 13.10. Note that the carrier phase offset associated with

---

[4] The name RAKE comes from the notion that the multibranch receiver resembles a garden rake and has the effect of raking up the energy associated with the multipath components on each of its branches. The RAKE was invented in the 1950s to deal with the ionospheric multipath on a spread-spectrum high-frequency transcontinental link. The name was coined by the RAKE inventors, Paul Green and Bob Price.

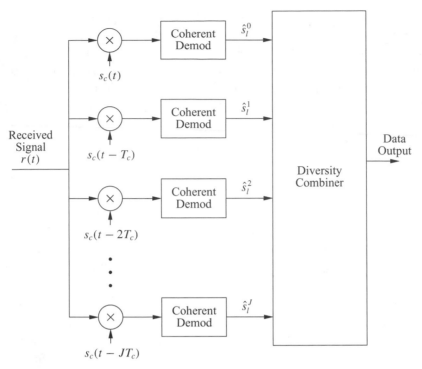

Figure 13.10: RAKE receiver.

each multipath component depends on its delay: this offset is determined by the coherent demodulator in each branch. Alternatively, each branch can be downconverted to baseband prior to multiplication with the spreading code, in which case each branch performs carrier phase recovery prior to downconversion. The RAKE is essentially another form of diversity combining, since the spreading code induces a path diversity on the transmitted signal so that independent multipath components separated by more than a chip time can be resolved. Any of the combining techniques discussed in Chapter 7 may be used, although in practice MRC is the most common.

In order to study the behavior of RAKE receivers, assume a channel model with impulse response $h(t) = \sum_{j=0}^{J} \alpha_j \delta(t - jT_c)$, where α_j is the gain associated with the jth multipath component. This model, described in Section 3.4, can approximate a wide range of multipath environments by matching the statistics of the complex gains to those of the desired environment. The statistics of the α_j have been characterized empirically in [11] for outdoor wireless channels. With this model, each branch of the RAKE receiver in Figure 13.10 synchronizes to a different multipath component and coherently demodulates its associated signal. A larger J implies a higher receiver complexity but also increased diversity. Then, from (13.14) and (13.15), the output of the ith branch demodulator is

$$\hat{s}_l^i = \alpha_i s_l + \sum_{\substack{j=1 \\ i \neq j}}^{J-1} \alpha_j \rho_c(iT_c - jT_c)s_{lj} + n_l^i, \qquad (13.22)$$

where s_l is the symbol transmitted over symbol time $[lT_s, (l+1)T_s]$ (i.e., the symbol associated with the LOS path); s_{lj} is the symbol transmitted over symbol time $[lT_s - jT_c, (l+1)T_s - jT_c]$,

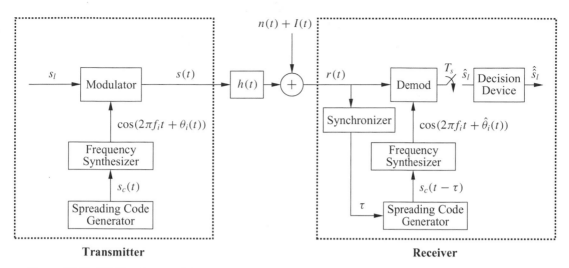

Figure 13.11: FHSS system model.

which we assume to be constant; and n_l^i is the noise sample associated with the demodulator in the ith branch. If s_{lj} is not constant over $[lT_s - jT_c, (l + 1)T_s - jT_c]$ then the ISI term in (13.22) is more complicated and involves partial autocorrelations. However, in all cases the ISI is reduced by roughly the autocorrelation $\rho_c((i - j)T_c)$. The diversity combiner coherently combines the demodulator outputs. In particular, with selection combining the branch output $\hat{s}_l^i$ with the largest path gain a_i is output from the combiner, with equal gain combining all demodulator outputs are combined with equal weighting, and with maximal ratio combining the demodulator outputs are combined with a weight equal to the branch SNR – or to the SINR (signal-to-interference-plus-noise power ratio) if ISI interference is taken into account. If $\rho_c(\tau) \approx 0$ for $|\tau| > T_c$ then we can neglect the ISI terms in each branch, and the performance of the RAKE receiver with J branches is identical to any other J-branch diversity technique. A comprehensive study of RAKE performance for empirically derived channel models was made by Turin in [11].

Spread spectrum is not usually used for diversity alone, since it requires significantly more bandwidth than other diversity techniques. However, if spread-spectrum signaling is chosen for its other benefits – such as its multiuser or interference rejection capabilities – then RAKEs provide a simple mechanism to obtain diversity benefits.

13.3 Frequency-Hopping Spread Spectrum (FHSS)

An end-to-end frequency-hopping spread-spectrum system is illustrated in Figure 13.11. The spreading code is input to the frequency synthesizer in order to generate the hopping carrier signal $c(t) = \cos(2\pi f_i t + \theta_i(t))$, which is then input to the modulator to upconvert the modulated signal to the carrier frequency. The modulator can be coherent, noncoherent, or differentially coherent, although coherent modulation is not as common as noncoherent modulation owing to the difficulties in maintaining a coherent phase reference while hopping the carrier over a wide bandwidth [9, Part 2:2]. At the receiver, a synchronizer is used to synchronize the locally generated spreading code to that of the incoming signal. Once synchronization is achieved, the spreading code is input to the frequency synthesizer to generate the hopping pattern of the carrier, which is then input to the demodulator for downconversion. For a noncoherent or differentially coherent modulator, it is not necessary to synchronize the phase associated with the receive carrier to that of the transmit carrier.

As with DSSS, the synchronization procedure for FH systems is typically performed in two stages. First, a coarse synchronization is done to align the receiver hop sequence to within a fraction of the hop duration T_c associated with the transmitted FH signal. The process is similar to the coarse synchronization of DSSS: the received FH signal plus noise is correlated with the local hopping sequence by multiplying the signals together and computing the energy in their product. If this energy exceeds a given threshold, coarse acquisition is obtained; otherwise, the received FH signal is shifted in time by T_c and the process is repeated. Coarse acquisition can also be done in parallel using multiple hop sequences, each shifted in time by a different integer multiple of T_c. Once coarse acquisition is obtained, fine tuning occurs by continually adjusting the timing of the frequency hopper to maximize the correlation between the receiver hopping sequence and the received signal. More details on FH synchronization and an analysis of system performance under synchronization errors can be found in [9, Part 4].

The impact of multipath on FH systems was discussed in Section 13.1. There we saw that an FH system does not exhibit fading if the multipath components have delay exceeding the hop time, since only one nonfading signal component arrives during each hop. When multipath does cause flat or frequency-selective fading, the performance analysis is the same as for a slowly time-varying non-hopping system. However, the impact of narrowband interference on FH systems, as characterized by the probability of symbol error, is more difficult to determine. In fact, this error probability depends on the exact structure of the interfering signal and how it affects the specific modulation in use, as we now describe.

We will focus on symbol error probability for an SFH system without coding, where the interference, if present, is constant over a symbol time. The analysis for fast frequency hopping is more complicated because interference changes over a symbol time, making it more difficult to characterize its statistics and the resulting impact on the symbol error probability. So consider an SFH system with M out of the N frequency bands occupied by a narrowband interferer. Assuming the signal hops uniformly over the entire frequency band, the probability of any given hop being in the same band as an interferer is then M/N. The probability of symbol error is obtained by conditioning on the presence of an interferer over the given symbol period:

$$
\begin{aligned}
P_s &= p(\text{symbol error} \mid \text{no interference})p(\text{no interference}) \\
&\quad + p(\text{symbol error} \mid \text{interference})p(\text{interference}) \\
&= \frac{N-M}{N}p(\text{symbol error} \mid \text{no interference}) \\
&\quad + \frac{M}{N}p(\text{symbol error} \mid \text{interference}).
\end{aligned}
\tag{13.23}
$$

In the absence of interference, the probability of symbol error just equals that of the modulated data signal transmitted over an AWGN channel with received SNR γ_s, which we will denote as P_s^{AWGN}. Note that γ_s is the received SNR at the input to the demodulator in the absence of interference, so multipath components removed in the despreading process do not affect this SNR. However, γ_s will be affected by the channel gain at the carrier frequency for the multipath components that are not removed by despreading. For most coherent modulations, $P_s^{\text{AWGN}} \approx \alpha_M Q(\sqrt{\beta_M \gamma_s})$ for α_M and β_M dependent on the modulation, as discussed in Section 6.1.6. The P_s^{AWGN} for noncoherent or differentially coherent modulations in AWGN are generally more complex [12, Chap. 1.1]. Given P_s^{AWGN}, it remains only to characterize the probability of error when interference is present, p(symbol error | interference), in order to determine P_s in (13.23). If we denote this probability as P_s^{INT}, then (13.23) becomes

$$P_s = \frac{N - M}{N} P_s^{\text{AWGN}} + \frac{M}{N} P_s^{\text{INT}}. \tag{13.24}$$

Let us now examine P_s^{INT} more closely. This symbol error probability will depend on the exact characteristics of the interference signal. Consider first a narrowband interferer with the same statistics as AWGN within the bandwidth of the modulated signal; an interferer with these characteristics is sometimes referred to as a *partial band noise jammer*. For this type of interferer, P_s^{INT} is obtained by treating the interference as an additional AWGN component with power N_J within the bandwidth of the modulated signal. The total noise power is then $N_0 B + N_J$ and so the SINR in the presence of this interference becomes

$$\gamma_s^{\text{INT}} = \gamma_s \frac{N_0 B}{N_0 B + N_J},$$

which yields

$$P_s^{\text{INT}} = P_s^{\text{AWGN}}(\gamma_s^{\text{INT}}). \tag{13.25}$$

Suppose now that the interference consists of a tone at the hopped carrier frequency with some offset phase. Then the demodulator output $\hat{s}_l$ in Figure 13.11 is given by

$$\hat{s}_l = a_l s_l + n_l + I_l, \tag{13.26}$$

where a_l is the channel gain associated with the received signal after despreading, n_l is the AWGN sample, and $I_l = \sqrt{I} e^{j\phi_l}$ is the interference term with phase offset ϕ_l. Note that, since this is a wideband channel, fading is frequency selective and so (i) the channel gain a_l will depend on the carrier frequency and (ii) some hops may be associated with very poor channel gains. The impact of the additional interference term I_l will depend on the modulation. For example, with coherent MPSK and assuming $\angle s_l = 0$,

$$P_s = 1 - p(|\angle(a_l s_l + n_l + I_l)| \le \pi/M). \tag{13.27}$$

In general, computing P_s for either coherent or noncoherent modulation requires finding the distribution of the random phase $\angle(n_l + I_l)$. This distribution and the resulting P_s are derived in [9, Parts 2–3] for noncoherent, coherent, and differentially coherent modulations for a number of different interference models. Coding or coding with interleaving is often used in FH systems to compensate for frequency-selective fading as well as narrowband interference or jamming. Analysis of coded systems with interference can be found in [9, Part 2:2].

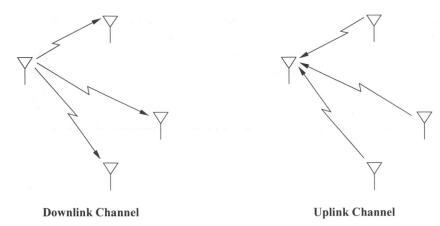

Downlink Channel   **Uplink Channel**

Figure 13.12: Downlink and uplink channels.

13.4 Multiuser DSSS Systems

Spread spectrum can also be used as a mechanism whereby many users can share the same spectrum. Using spreading code properties to support multiple users within the same spread bandwidth is also called spread-spectrum multiple access (SSMA), which is a special case of code-division multiple access (CDMA). In multiuser spread spectrum, each user is assigned a unique spreading code or hopping pattern, which is used to modulate its data signal. The transmitted signals for all users are superimposed in time and in frequency. The spreading codes or hopping patterns can be orthogonal, in which case users do not interfere with each other under ideal propagation conditions, or they can be nonorthogonal, in which case interference between users does exist but is reduced by the spreading code properties. Thus, whereas spread spectrum for single-user systems is spectrally inefficient because it uses more bandwidth than the minimum needed to convey the information signal, spread-spectrum multiuser systems can support an equal or larger number of users in a given bandwidth than other forms of spectral sharing such as time division or frequency division. However, if the spreading mechanisms are nonorthogonal – either by design or through channel distortion – then users interfere with each other. If there is too much interference between users, the performance of all users degrades. Comparison of the spectral efficiency for different spectral-sharing methods in multiuser and cellular systems will be discussed in Chapters 14 and 15.

Performance of multiuser spread spectrum also depends on whether the multiuser system is a *downlink* channel (one transmitter to many receivers) or an *uplink* channel (many transmitters to one receiver). These channel models are illustrated in Figure 13.12: the downlink channel is also called a broadcast channel or forward link, and the uplink channel is also called a multiple access channel or reverse link. The performance differences of DSSS in uplink and downlink channels result from the fact that, in the downlink, all transmitted signals are typically synchronous because they originate from the same transmitter. Moreover, both the desired signal and interference signals pass through the same channel before reaching the desired receiver. In contrast, users in the uplink channel are typically asynchronous, since they originate from transmitters at different locations, and the transmitted signals of the users travel through different channels before reaching the receiver. In this section we will analyze the multiuser properties of DSSS for both downlinks and uplinks. In Section 13.5 we treat multiuser FHSS systems.

13.4.1 Spreading Codes for Multiuser DSSS

Multiuser DSSS is accomplished by assigning each user a unique spreading code sequence $s_{c_i}(t)$. As described in Section 13.2.2, the autocorrelation function of the spreading code determines its multipath rejection properties. The cross-correlation of different spreading codes determines the amount of interference between users whose signals are modulated by these codes. The signals of asynchronous users arrive at the receiver with arbitrary relative delay τ, and the cross-correlation of the codes assigned to user i and user j over one symbol time with this delay is given by

$$\rho_{ij}(t, t - \tau) = \frac{1}{T_s} \int_0^{T_s} s_{c_i}(t) s_{c_j}(t - \tau) \, dt = \frac{1}{N} \sum_{n=1}^{N} s_{c_i}(nT_c) s_{c_j}(nT_c - \tau), \qquad (13.28)$$

where N is the number of chips per symbol time. As with the autocorrelation, if $s_{c_i}(t)$ and $s_{c_j}(t)$ are periodic with period T_s then the cross-correlation depends only on the time difference τ, so $\rho_{ij}(t, t - \tau) \triangleq \rho_{ij}(\tau)$. For synchronous users, their signals arrive at the receiver aligned in time, so $\tau = 0$ and the cross-correlation becomes

$$\rho_{ij}(0) = \frac{1}{T_s} \int_0^{T_s} s_{c_i}(t) s_{c_j}(t) \, dt = \frac{1}{N} \sum_{n=1}^{N} s_{c_i}(nT_c) s_{c_j}(nT_c). \qquad (13.29)$$

Ideally, since interference between users is dictated by the cross-correlation of the spreading code, for $i \neq j$ we would like $\rho_{ij}(\tau) = 0$ for all τ for asynchronous users and $\rho_{ij}(0) = 0$ for synchronous users; this would eliminate interference between users. A set of spreading codes with this *cross-correlation property* is called an *orthogonal* code set, and a set of spreading codes that does not satisfy this property is called a *nonorthogonal* code set. It is not possible to obtain orthogonal codes for asynchronous users, and for synchronous users there is only a finite number of spreading codes that are orthogonal within any given bandwidth. Thus, an orthogonality requirement restricts the number of different spreading codes (and the corresponding number of users) in a synchronous DSSS multiuser system. We now describe the most common chip sequences and their associated spreading codes that are used in multiuser DSSS systems.

GOLD CODES
Gold codes have worse autocorrelation properties than maximal-length codes but better cross-correlation properties if properly designed. The chip sequences associated with a Gold code are produced by the binary addition of two m-sequences each of length $N = 2^n - 1$, and they inherit the balanced, run-length, and shift properties of these component sequences (hence they are pseudorandom sequences). Gold codes take advantage of the fact that, if two distinct m-sequences with time shifts τ_1 and τ_2 are added together modulo 2, then the resulting sequence is unique for every unique value of τ_1 or τ_2. Thus, a very large number of unique Gold codes can be generated, which allows for a large number of users in a multiuser system.

However, if the m-sequences that are modulo-2 added to produce a Gold code are chosen at random, then the cross-correlation of the resulting code may be quite poor. Thus, Gold codes are generated by the chip sequences associated with the modulo-2 addition of *preferred* pairs of m-sequences. These preferred pairs are chosen to obtain good cross-correlation properties in the resulting Gold code. However, the preferred pairs of m-sequences have different autocorrelation properties than general m-sequences. A method for choosing the preferred pairs such that the cross-correlation and autocorrelation functions of the resulting Gold code are bounded was given by Gold in [13] and can also be found in [6, Chap. 9.2; 8, Apx. 7; 14]. The preferred sequences are chosen so that Gold codes have a three-valued cross-correlation over a code period with values

$$\rho_{ij}(\tau) = \begin{cases} -1/N, \\ -t(n)/N, \\ (t(n) - 2)/N, \end{cases} \tag{13.30}$$

where

$$t(n) = \begin{cases} 2^{(n+1)/2} + 1 & n \text{ odd}, \\ 2^{(n+2)/2} + 1 & n \text{ even}. \end{cases} \tag{13.31}$$

The autocorrelation takes on the same three values.

KASAMI CODES

Kasami sequences have properties that are similar to the preferred sequences used to generate Gold codes, and they are also derived from m-sequences. However, the Kasami codes have better cross-correlation properties than Gold codes. There are two different sets of Kasami sequences that are used to generate Kasami codes, the large set and the small set. To generate the small set, we begin with an m-sequence a of length $N = 2^n - 1$ for n even and form a new shorter sequence a' by sampling every $2^{n/2} + 1$ elements of a. The resulting sequence a' will have period $2^{n/2} - 1$. We then generate a small set of Kasami sequences by taking the modulo-2 sum of a with all cyclic shifts of the a' sequence. There are $2^{n/2} - 2$ such cyclic shifts and – by also including the original sequence a as well as $\text{mod}_2(a + a')$ in the set – we obtain a set of $2^{n/2}$ binary sequences of length $2^n - 1$. As with the Gold codes, the autocorrelation and cross-correlation of the Kasami spreading codes obtained from the Kasami sequences over a code period are three-valued, taking on the values

$$\rho_{ij}(\tau) = \begin{cases} -1/N, \\ -s(n)/N, \\ (s(n) - 2)/N, \end{cases} \tag{13.32}$$

where $s(n) = 2^{n/2} + 1$. Since $|s(n)| < |t(n)|$, Kasami codes have better autocorrelation and cross-correlation than Gold codes. In fact, Kasami codes achieve the Welch lower bound for autocorrelation and cross-correlation for any set of $2^{n/2}$ sequences of length $2^n - 1$ and hence are optimal (in terms of minimizing the autocorrelation and cross-correlation) for any such code [9, Part 1:5; 14].

The large set of Kasami sequences is formed in a similar fashion. It has a larger number of sequences than the smaller set and thus can support more users in a multiuser system, but the autocorrelation and cross-correlation properties across the spreading codes generated from this larger set are inferior to those generated from the smaller set. To obtain the large set, we take an m-sequence a of length $N = 2^n - 1$ for n even and form two new sequences a' and a'' by sampling the original sequence every $2^{n/2} + 1$ elements for a' and every $2^{(n+2)/2} + 1$ elements for a''. The set is then constituted by adding a, a', and a'' for all cyclic shifts of a' and a''. The number of such sequences is $2^{3n/2}$ if n is a multiple of 4 and is $2^{3n/2} + 2^{n/2}$ if $\mathrm{mod}_4(n) = 2$. The autocorrelation and cross-correlation of the spreading codes generated from this set over a code period can take on one of five values:

$$\rho(\tau) = \begin{cases} -1/N, \\ (-1 \pm 2^{n/2})/N, \\ (-1 \pm (2^{n/2} + 1))/N. \end{cases} \tag{13.33}$$

Since these values exceed those for codes generated from the small Kasami set, we see that codes generated from the large Kasami set have inferior cross-correlation and autocorrelation properties to those generated from the small Kasami set.

EXAMPLE 13.4: Find the number of sequences and the magnitude of the worst-case cross-correlation over a code period for spreading codes generated from the small and large Kasami sequences with $n = 10$.

Solution: For the small set, there are $2^{n/2} = 2^5 = 32$ sequences. From (13.32), the largest-magnitude cross-correlation of the corresponding Kasami code is

$$\frac{1}{N}[2^{n/2} + 1] = \frac{1}{2^{10} - 1}[2^5 + 1] = .032.$$

For the large set, $\mathrm{mod}_4(10) = 2$, so there are $2^{3n/2} + 2^{n/2} = 2^{15} + 2^{10} = 33792$ sequences, which is three orders of magnitude more codes than in the small set. The largest-magnitude spreading code cross-correlation is thus

$$\frac{1}{N}[2^{n/2} + 2] = \frac{1}{2^{10} - 1}[2^5 + 2] = .033.$$

So there is a slightly larger cross-correlation, the price paid for the significant increase in the number of codes.

WALSH–HADAMARD CODES

Walsh–Hadamard codes of length $N = T_s/T_c$ that are synchronized in time are orthogonal over a symbol time, so that the cross-correlation of any two codes is zero. Thus, synchronous users modulated with Walsh–Hadamard codes can be separated out at the receiver with no interference between them, as long as the channel does not corrupt the orthogonality of the codes. (Delayed multipath components are not synchronous with the LOS paths, and thus the multipath components associated with different users will cause interference between users. The loss of orthogonality can be quantified by the orthogonality factor [15]). Although it is possible to synchronize users on the downlink, where all signals originate from the same transmitter, it is more challenging to synchronize users in the uplink because they are not co-located. Hence, Walsh–Hadamard codes are rarely used for DSSS uplink channels. Walsh–Hadamard sequences of length N are obtained from the rows of an $N \times N$ *Hadamard matrix* $\mathbf{H}_N$. For $N = 2$ the Hadamard matrix is

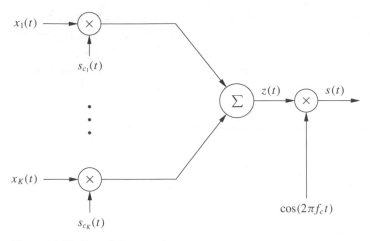

Figure 13.13: Downlink transmitter.

$$\mathbf{H}_2 = \begin{bmatrix} 1 & 1 \\ 1 & -1 \end{bmatrix}.$$

Larger Hadamard matrices are obtained using $\mathbf{H}_2$ and the recursion

$$\mathbf{H}_{2N} = \begin{bmatrix} \mathbf{H}_N & \mathbf{H}_N \\ \mathbf{H}_N & -\mathbf{H}_N \end{bmatrix}.$$

Each row of $\mathbf{H}_N$ specifies a different chip sequence, so the number of spreading codes in a Walsh–Hadamard code is N. Thus, DSSS with Walsh–Hadamard codes can support at most $N = T_s/T_c$ users. Since DSSS uses roughly N times more bandwidth than required for the information signal, approximately the same number of users could be supported by dividing up the total system bandwidth into N nonoverlapping channels (frequency division). Similarly, the same number of users can be supported by dividing time up into N orthogonal timeslots (time division), where each user operates over the entire system bandwidth during his timeslot. Hence, all multiuser techniques that assign orthogonal channels to preclude interference will accommodate approximately the same number of users.

The performance of a DSSS multiuser system depends both on the spreading code properties as well as the channel over which the system operates. In the next section we will study performance of DSSS multiuser systems over downlinks, and performance over uplinks will be treated in Section 13.4.3.

13.4.2 Downlink Channels
The transmitter for a DSSS downlink system is shown in Figure 13.13, with the channel and receiver shown in Figure 13.14. We focus on analyzing the in-phase signal component; a similar analysis applies for the quadrature component. In the downlink the signals of all users are typically sent simultaneously by the transmitter (base station), and each receiver must demodulate its individual signal. Thus we can assume that all signals are synchronous, which allows the use of orthogonal spreading codes such as the Walsh–Hadamard codes. However,

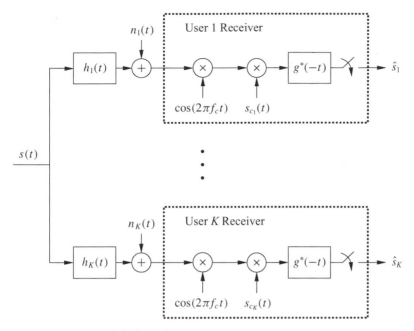

Figure 13.14: Downlink channel and receiver.

the use of orthogonal codes limits the number of users the downlink can support, so such codes are not always used.

Consider a K-user system, where the transmitter sends to K independent users. The baseband modulated signal associated with the kth user is

$$x_k(t) = \sum_l s_{kl} g(t - lT_s), \tag{13.34}$$

where $g(t) = \sqrt{2/T_s}$ is the pulse shape (assumed rectangular), T_s is the symbol time, and s_{kl} is the kth user's symbol over the lth symbol time. The transmitter consists of K branches, where the kth branch multiplies the kth user's signal $x_k(t)$ with the spreading code $s_{c_k}(t)$. The branches are summed together, resulting in the baseband multiuser signal

$$z(t) = \sum_{k=1}^{K} x_k(t) s_{c_k}(t) = \sum_{k=1}^{K} \sqrt{\frac{2}{T_s}} s_{kl} s_{c_k}(t). \tag{13.35}$$

This multiuser signal is multiplied by the carrier to obtain the passband signal $s(t)$, which is transmitted over the channel.

The signal received by user k first passes through its channel, which has impulse response $h_k(t)$ and AWGN. Thus, the received signal at the kth user's receiver is $s(t) * h_k(t) + n_k(t)$. This signal is downconverted, and we assume perfect carrier recovery so that the phase ϕ_k of the carrier in user k's receiver is perfectly matched to that of its incoming signal. Thus we can normalize the delay of the kth user's LOS (or minimum delay) path and its associated phase shift to zero. After downconversion, the signal is then multiplied by the kth user's spreading code $s_{c_k}(t)$, which is assumed to be perfectly synchronized to the kth user's spreading code in the received signal.[5] The signal is then baseband demodulated via a matched filter (i.e., it is multiplied by $\sqrt{2/T_s}$ and integrated over a symbol time). The demodulator output is sampled every T_s to obtain an estimate of the symbol transmitted by the kth user over that symbol time. Comparing Figure 13.5 and Figure 13.14, we see that the kth user's receiver is identical to the matched filter detector in a single-user DSSS system. Thus, in the absence of multiuser interference, the kth user has identical performance as in a single-user DSSS system. However, when multiuser interference is taken into account, the demodulator output includes components associated with the kth user's signal, interference terms from other users' signals, and noise. In particular, the demodulator output associated with the kth user over the lth symbol time is given by

$$
\begin{aligned}
\hat{s}_k &= \sqrt{\frac{2}{T_s}} \int_0^{T_s} [s(t) * h_k(t) + n_k(t)] s_{c_k}(t) \cos(2\pi f_c t)\, dt \\
&= \sqrt{\frac{2}{T_s}} \int_0^{T_s} [z(t) * h_k^{\mathrm{LP}}(t)] s_{c_k}(t) \cos^2(2\pi f_c t)\, dt + \sqrt{\frac{2}{T_s}} \int_0^{T_s} n_k(t) s_{c_k}(t) \cos(2\pi f_c t)\, dt \\
&= \frac{2}{T_s} \int_0^{T_s} \left[\sum_{j=1}^{K} s_{jl} s_{c_j}(t) * h_k^{\mathrm{LP}}(t) \right] s_{c_k}(t) \cos^2(2\pi f_c t)\, dt \\
&\quad + \sqrt{\frac{2}{T_s}} \int_0^{T_s} n_k(t) s_{c_k}(t) \cos(2\pi f_c t)\, dt \\
&= \frac{2}{T_s} \int_0^{T_s} [s_{kl} s_{c_k}(t) * h_k^{\mathrm{LP}}(t)] s_{c_k}(t) \cos^2(2\pi f_c t)\, dt \\
&\quad + \frac{2}{T_s} \int_0^{T_s} \left[\sum_{\substack{j=1 \\ j\neq k}}^{K} s_{jl} s_{c_j}(t) * h_k^{\mathrm{LP}}(t) \right] s_{c_k}(t) \cos^2(2\pi f_c t)\, dt \\
&\quad + \sqrt{\frac{2}{T_s}} \int_0^{T_s} n_k(t) s_{c_k}(t) \cos(2\pi f_c t)\, dt,
\end{aligned}
\tag{13.36}
$$

where $h_k^{\mathrm{LP}}(t)$ is the in-phase component of the equivalent lowpass filter for $h_k(t)$, s_{kl} is the kth user's transmitted symbol over the lth symbol period that is being recovered, and s_{jl} is the transmitted symbol of the jth user over this symbol period, which causes interference. Note that (13.36) consists of three separate terms. The first term corresponds to the received signal of the kth user alone, the second term represents interference from other users in the system, and the last term is the AWGN sample, which we denote as n_k. The first term and the noise sample are characterized by the analysis in Section 13.2 for single-user systems. The second term depends on both the channel $h_k^{\mathrm{LP}}(t)$ and the spreading code properties, as we now show.

---

[5] This synchronization is even more difficult than in the single-user case, since here it must be done in the presence of multiple spread signals. In fact, some spreading code sets are obtained by shifting a single spreading code by some time period. For these systems there must be a control channel to inform the receiver which time shift corresponds to its desired signal. More details on the synchronization for these systems can be found in [8, Chap. 6].

To examine the characteristics of the multiuser interference, let us first assume that the kth user's channel has a real gain of α_k and a normalized delay of zero; that is, $h_k(t) = \alpha_k \delta(t)$. Then (13.36) becomes

$$\hat{s}_k = \frac{2}{T_s} \int_0^{T_s} \alpha_k s_{kl} s_{c_k}^2(t) \cos^2(2\pi f_c t) \, dt$$

$$+ \frac{2}{T_s} \int_0^{T_s} \sum_{\substack{j=1 \\ j \neq k}}^{K} \alpha_k s_{jl} s_{c_j}(t) s_{c_k}(t) \cos^2(2\pi f_c t) \, dt + n_k$$

$$\approx \alpha_k s_{kl} + \alpha_k \sum_{\substack{j=1 \\ j \neq k}}^{K} s_{jl} \rho_{jk}(0) + n_k, \tag{13.37}$$

where $\rho_{jk}(0)$ is the cross-correlation between $s_{c_k}(t)$ and $s_{c_j}(t)$ for a timing offset of zero, since the users are assumed to be synchronous.[6] We define

$$I_{kl} = \alpha_k \sum_{\substack{j=1 \\ j \neq k}}^{K} s_{jl} \rho_{jk}(0) \tag{13.38}$$

as the multiuser interference to the kth user at the demodulator output. We see from (13.37) that the kth user's symbol s_{kl} is attenuated by the channel gain but not affected by the spreading and despreading, exactly as in the single-user case. The noise sample n_k is also the same as in a single-user non-spread system. The interference from other users is attenuated by the kth user's channel gain α_k and the cross-correlation of the codes $\rho_{jk}(0)$. For orthogonal (e.g., Walsh–Hadamard) codes, $\rho_{jk}(0) = 0$ so there is no interference between users. For nonorthogonal codes, $\rho_{jk}(0)$ depends on the specific codes assigned to users j and k; for example, with Gold codes, $\rho_{jk}(0)$ can take on one of three possible values. Note that both the kth user's signal and the interference are attenuated by the same channel gain a_k, since both signal and interference follow the same path from the transmitter to the receiver. As we will see in the next section, this is not the case for DSSS uplink systems.

If the interference in a multiuser system has approximately Gaussian statistics then we can treat it as an additional noise term and determine system performance based on the SINR for each user. However, the Gaussian approximation is often inaccurate, even when the number of interferers is large [16]. Moreover, in fading the interference terms are correlated, since they all experience the same fading α_k. Thus, the interference can only be approximated as conditionally Gaussian, conditioned on the fading. The conditionally Gaussian approximation is most accurate when the number of interferers is large, since the sum of a large number of random variables converges to a Gaussian random variable by the central limit

---

[6] If the users were not synchronous, which is unusual in a downlink channel, then the cross-correlation $\rho_{jk}(0)$ in (13.37) would be replaced by $\rho_{jk}(\tau_{jk})$ for τ_{jk} the relative delay between the received signal from users j and k. This assumes s_{jl} is constant over the integration; if not, the interference term depends on the different symbol values over the integration.

theorem.[7] The SINR for the kth user is defined as the ratio of power associated with the kth user's signal over the average power associated with the multiuser interference and noise at the demodulator output. The kth user's performance is then analyzed based on the bit error rate in AWGN with SNR replaced by the SINR for this user. Moreover, if the interference power is much greater than the system noise power, then we can neglect the noise altogether and determine performance based on an AWGN channel analysis with SNR replaced by the signal-to-interference power ratio (SIR) for each user. The SIR for the kth user is defined as the ratio of power associated with the kth user's signal over the average power associated with the multiuser interference alone. Multiuser spread-spectrum systems where noise can be neglected in the performance analysis are called *interference limited,* since noise is negligible relative to interference in the performance analysis. For both SINR and SIR, obtaining the average interference power depends on the specific spreading sequences and symbol transmissions of the interfering users, which can be highly complex to analyze. As an alternative, average interference power is often computed assuming random spreading sequences. With this assumption it can be shown that the SIR for a synchronous K-user system with N chips per symbol is given by

$$\text{SIR} = \frac{N}{K-1} \approx \frac{G}{K-1},$$ (13.39)

[17, Chap. 2.3], where $G \approx N$ is the processing gain of the system. Note that this matches the SIR expression (13.6) for arbitrary interference with random signal generation. If noise is taken into account, then the SINR is obtained from (13.39) by adding in noise scaled by the energy per symbol E_s:

$$\text{SINR} = \left(\frac{N_0}{E_s} + \frac{K-1}{G}\right)^{-1}.$$ (13.40)

Now consider a more general channel $h_k(t) = \sum_{m=0}^{M-1} \alpha_{km}\delta(t - \tau_{km})$. We assume that $m = 0$ corresponds to the LOS signal path so that, under perfect carrier recovery, we can normalize the delay τ_{k0} and its associated phase shift to zero. The output of the demodulator will again consist of three terms: the first corresponding to the kth user's signal, the second corresponding to the interference from other users, and the last an AWGN noise sample, which is not affected by the channel. The signal component associated with the kth user is analyzed the same way as in Section 13.2 for multipath channels: the delayed signal components are attenuated by the autocorrelation of the kth user's spreading code. The multiuser interference is more complicated than before. In particular, assuming the demodulator is synchronized to the LOS component of the kth user, the demodulator output corresponding to the multiuser interference is given by

$$I_{kl} = \frac{2}{T_s}\int_0^{T_s} \sum_{\substack{j=1 \\ j\neq k}}^{K}\sum_{m=0}^{M-1}\alpha_{km}s_{j(l-l_{km})}s_{c_j}(t - \tau_{km})\cos(2\pi f_c(t - \tau_{km}))s_{c_k}(t)\cos(2\pi f_c t)\,dt$$

$$\approx \sum_{\substack{j=1 \\ j\neq k}}^{K}\sum_{m=0}^{M-1}\alpha_{km}s_{j(l-l_{km})}\cos(2\pi f_c\tau_{km})\rho_{jk}(\tau_{km}),$$ (13.41)

---

[7] This is true even if the random variables are not i.i.d., as long as they decorrelate.

where $s_{j(l-l_{km})}$ is the symbol associated with the jth user over the $(lT_s - \tau_{km})$th symbol time, which we assume to be constant for simplicity. Comparing (13.38) and (13.41), we see that the multipath channel affects the multiuser interference in two ways. First, there are more interference terms: whereas there were $K - 1$ before, we now have $(K - 1)M$; so each interfering user contributes M interference terms, one for each multipath component. In addition, the cross-correlation of the codes is no longer taken at delay $\tau = 0$, even though the users are synchronous. In other words, the multipath destroys the synchronicity of the channel. This is significant because orthogonal codes like the Walsh–Hadamard codes typically have zero cross-correlation only at zero delay. So if a Walsh–Hadamard multiuser system operates in a multipath channel, the users will interfere. Equalization of the spreading codes can be used to mitigate this interference.

EXAMPLE 13.5: Consider a DSSS downlink with bandwidth expansion $N = B_s/B = 100$. Assume the system is interference limited and there is no multipath on any user's channel. Find how many users the system can support under BPSK modulation such that each user has a BER less than 10^{-3}.

Solution: For BPSK we have $P_b = Q(\sqrt{2\gamma_b})$, and $\gamma_b = 6.79$ dB yields $P_b = 10^{-3}$. Since the system is interference limited, we set the SIR equal to the SNR $\gamma_b = 6.79$ dB and solve for K, the number of users:

$$\text{SIR} = \frac{N}{K - 1} = \frac{100}{K - 1} = 10^{.679} = 4.775.$$

Solving for K yields $K \leq 1 + 100/4.77 = 21.96$. Since K must be an integer and we require $P_b \leq 10^{-3}$, it follows that 21.96 must be rounded down to 21 users – although typically a designer would build the system to support 22 users with a slight BER penalty.

13.4.3 Uplink Channels

We now consider DSSS for uplink channels. In multiuser DSSS the spreading code properties are used to separate out the received signals from the different users. The main difference in using DSSS on the uplink versus the downlink is that, in the downlink, both the kth user's signal and the interfering signals from other users pass through the same channel from the transmitter to the kth user's receiver. In an uplink the signals received from each user at the receiver travel through different channels. This gives rise to the *near–far* effect, where users that are close to the uplink receiver can cause a great deal of interference to users farther away, as discussed in more detail below.

The transmitter and channel for each individual user in a K-user uplink are shown in Figure 13.15. As in the downlink, we will focus on in-phase signal analysis. The transmitters are typically not synchronized, since they are not co-located. In general the asynchronous uplink is more complex to analyze than the synchronous uplink and has worse performance. We see from Figure 13.15 that the kth user generates the baseband-modulated signal $x_k(t)$. As in the downlink model, we assume rectangular pulses for $x_k(t)$. The kth user multiplies its baseband signal $x_k(t)$ by its spreading code $s_{c_k}(t)$ and then upconverts to the carrier frequency to form the kth user's transmitted signal $s_k(t)$. (Note that the carrier signals for each user have different phase offsets.) This signal is sent over the kth user's channel, which has

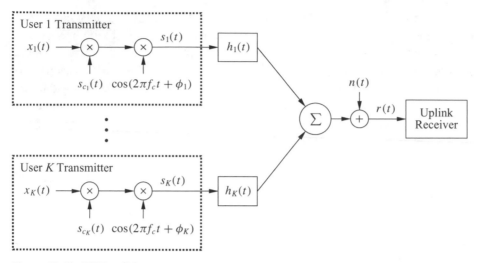

Figure 13.15: DSSS uplink system.

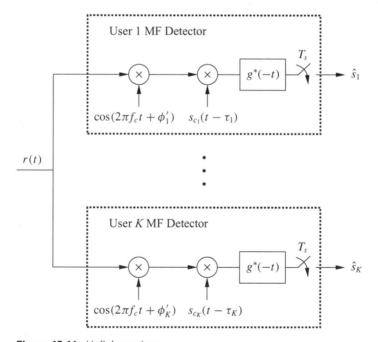

Figure 13.16: Uplink receiver.

impulse response $h_k(t)$. After transmission through their respective channels, all users' signals are summed at the receiver front end together with AWGN $n(t)$.

The uplink received signal is thus given by

$$r(t) = \left[\sum_{k=1}^{K} (x_k(t)s_{c_k}(t)\cos(2\pi f_c t + \phi_k)) * h_k(t) \right] + n(t). \qquad (13.42)$$

The receiver consists of K branches corresponding to the K received signals, as shown in Figure 13.16. We assume (i) that the LOS path of the kth user's channel introduces a delay of τ_k and (ii) that the phase offset resulting from this delay and the transmitter carrier phase offset ϕ_k is matched by the phase offset ϕ_k' obtained by the carrier recovery loop in the kth branch of the receiver. For synchronous users, these delays are all equal. The kth branch downconverts the signal to baseband and then multiplies the received signal by the kth user's spreading code, synchronized to the delay of the kth user's incoming signal. The despread signal is then passed through a matched filter and sampled to obtain an estimate of each user's transmitted symbol over the lth symbol time. Comparing Figures 13.5 and 13.16, we see that the kth branch of the uplink receiver is identical to the matched filter (MF) detector in a single-user DSSS system. Thus, the uplink receiver consists of a bank of K single-user MF detectors, and in the absence of multiuser interference the kth user has identical performance as in a single-user system. With multiuser interference taken into account, the demodulator output of the kth receiver branch over the lth symbol time is given by

$$
\hat{s}_k = \sqrt{\frac{2}{T_s}} \int_0^{T_s} \left[\sum_{j=1}^{K} (x_j(t) s_{c_j}(t)) * h_j^{\mathrm{LP}}(t) \right] s_{c_k}(t - \tau_k) \cos(2\pi f_c t + \phi_k') \cos(2\pi f_c t + \phi_j') \, dt
$$
$$
+ n_k
$$
$$
= \frac{2}{T_s} \int_0^{T_s} [s_{kl} s_{c_k}(t) * h_k^{\mathrm{LP}}(t)] s_{c_k}(t - \tau_k) \cos^2(2\pi f_c t + \phi_k') \, dt
$$
$$
+ \frac{2}{T_s} \int_0^{T_s} \left[\sum_{\substack{j=1 \\ j \neq k}}^{K} s_{ljk} s_{c_j}(t) * h_j^{\mathrm{LP}}(t) \right] s_{c_k}(t - \tau_k) \cos(2\pi f_c t + \phi_k') \cos(2\pi f_c t + \phi_j') \, dt
$$
$$
+ n_k, \tag{13.43}
$$

where n_k is the AWGN sample; $h_j^{\mathrm{LP}}(t)$ is the in-phase component of the equivalent lowpass filter for $h_j(t)$, $j = 1, \ldots, K$, with delay normalized to equal zero for the LOS path; and s_{ljk} is the symbol transmitted over the jth user's channel at time $[lT_s - \tau_j + \tau_k, (l+1)T_s - \tau_j + \tau_k]$, which we assume to be constant. If this symbol takes different values on this interval – that is, if it changes values at lT_s then the ISI term is more complicated and involves partial cross-correlations, but the ISI attenuation is roughly the same. Note that (13.43) consists of three separate terms. The first term corresponds to the received signal of the kth user alone, and the last term is the AWGN sample; these two terms are the same as for a single-user system. The second term represents interference from other users in the system, and the interference of the jth user to the kth user ($j \neq k$) depends on the jth user's equivalent lowpass channel $h_j^{\mathrm{LP}}(t)$ and the spreading code properties, as we now show.

Assume that each user's channel introduces only a gain α_j and a delay τ_j, so $h_j^{\mathrm{LP}}(t) = \alpha_j \delta(t - \tau_j)$. Then the demodulator output for the kth branch over the lth symbol time is

$$
\hat{s}_k = \alpha_k s_{kl} + I_{kl} + n_l, \tag{13.44}
$$

where the first and third terms are the same as for a single-user system with this channel, assuming the spreading code in the receiver is perfectly synchronized to the delay τ_k. Let us now consider the interference term I_{kl}. Substituting $h_j^{\mathrm{LP}}(t) = \alpha_j \delta(t - \tau_j)$ into (13.43) yields

$$I_{kl} = \frac{2}{T_s} \int_0^{T_s} \left[\sum_{\substack{j=1 \\ j \neq k}}^{K} s_{ljk} s_{c_j}(t) * \alpha_j \delta(t - \tau_j) \right] s_{c_k}(t - \tau_k) \cos(2\pi f_c t + \phi_k') \cos(2\pi f_c t + \phi_j') \, dt$$

$$= \frac{1}{T_s} \int_0^{T_s} \left[\sum_{\substack{j=1 \\ j \neq k}}^{K} \alpha_j s_{ljk} s_{c_j}(t - \tau_j) \right] s_{c_k}(t - \tau_k) [\cos(\Delta\phi_{kj}) + \cos(4\pi f_c t + \phi_k' + \phi_j')] \, dt$$

$$\approx \sum_{\substack{j=1 \\ j \neq k}}^{K} \alpha_j \cos(\Delta\phi_{kj}) s_{ljk} \frac{1}{T_s} \int_0^{T_s} s_{c_j}(t - \tau_j) s_{c_k}(t - \tau_k) \, dt$$

$$= \sum_{\substack{j=1 \\ j \neq k}}^{K} \alpha_j \cos(\Delta\phi_{kj}) s_{ljk} \rho_{jk}(\tau_j - \tau_k), \tag{13.45}$$

where $\Delta\phi_{kj} = \phi_k' - \phi_j'$ and the approximation is based on $f_c \gg 1/T_s$. We see from (13.45) that, as with the downlink, multiuser interference in the uplink is attenuated by the cross-correlation of the spreading codes. Since typically $\tau_j \neq \tau_k$, users are asynchronous. It follows that orthogonal codes requiring synchronous reception (e.g., Walsh–Hadamard codes) are not typically used on the uplink. Another important aspect of the uplink is that the kth user's symbol and multiuser interference are attenuated by different channel gains. In particular, the kth user's signal is attenuated by the gain α_k, while the interference from the jth user is attenuated by α_j. If $\alpha_j \gg \alpha_k$ then the interference can significantly degrade performance, even though it is reduced by the spreading code cross-correlation.

We now consider interference-limited uplinks. Suppose initially that all users have the same received power. Then the average SINR for asynchronous users on this channel – assuming random spreading codes with N chips per symbol, random start times, and random carrier phases – is given [18] by

$$\text{SINR} = \left(\frac{K - 1}{3N} + \frac{N_0}{E_s} \right)^{-1}. \tag{13.46}$$

For interference-limited systems we neglect the noise term and obtain the SIR as

$$\text{SIR} = \frac{3N}{K - 1} \approx \frac{3G}{K - 1}, \tag{13.47}$$

where $G \approx N$ is the processing gain of the system. The expressions (13.46) and (13.47) are referred to as the *standard Gaussian approximations* for SINR and SIR. Care must be used in applying these approximations to an arbitrary system, since the SIR and SINR for a given system is heavily dependent on the spreading code properties, timing and carrier phase assumptions, and other characteristics of the system. Modifications to the standard Gaussian approximation have been made to improve its accuracy for practical systems, but these expressions are typically more difficult to work with and don't lead to much greater accuracy than the standard approximations [6, Chap. 9.6]. We can modify (13.47) to approximate the SIR associated with nonrandom spreading codes as

$$\text{SIR} = \frac{3N}{\xi(K-1)} \approx \frac{3G}{\xi(K-1)}, \tag{13.48}$$

where ξ is a constant characterizing the code cross-correlation that depends on the spreading code properties and other system assumptions. Under the standard Gaussian assumption we use $\xi = 1$, whereas for PN sequences $\xi = 2$ [19] or $\xi = 3$ [20], depending on the system assumptions.

Suppose now that all $K-1$ interference terms have channel gain $\alpha \gg \alpha_k$. The SIR for the kth user then becomes

$$\text{SIR}(k) = \frac{\alpha_k^2 3N}{\alpha^2 \xi(K-1)} \approx \frac{\alpha_k^2 3G}{\alpha^2 \xi(K-1)} \ll \frac{3G}{\xi(K-1)}, \tag{13.49}$$

so the kth user in the uplink suffers an SIR penalty of α_k^2/α^2 due to the different channel gains. This phenomenon is called the near–far effect, since users far from the uplink receiver will generally have much smaller channel gains to the receiver than those of the interferers. In fading the α_k are random, which typically reduces the code cross-correlation and hence increases the average SIR. The effect of fading can be captured by adjusting ξ in (13.48) to reflect the average cross-correlation under the fading model. The value of ξ then depends on the spreading code properties, the system assumptions, and the fading characteristics [21].

For multipath channels, $h_j^{\text{LP}}(t) = \sum_{m=1}^{M} \alpha_{jm} \delta(t - \tau_{jm})$. Substituting this into (13.45) yields multiuser interference on the kth branch of

$$I_{kl} \approx \sum_{\substack{j=1 \\ j \neq k}}^{K} \sum_{m=1}^{M} \alpha_{jm} \cos(\Delta\phi_{jkm}) s_{ljm} \rho_{jk}(\tau_{jm} - \tau_k), \tag{13.50}$$

where we assume that the kth branch is synchronized to a channel delay of τ_k, that $\Delta\phi_{jkm}$ is the relative phase offset, and that s_{ljm} is the symbol transmitted by the jth user over time $[lT_s - \tau_k + \tau_{jm}, (l+1)T_s - \tau_k + \tau_{jm}]$, which is assumed constant. This interference also contributes to the near–far effect, since any of the multipath components with a large gain relative to the kth user's signal will degrade SIR.

A solution to the near–far effect in DSSS uplink systems is to use power control based on channel inversion, where the kth user transmits signal power P/α_k^2 so that his received signal power is P, regardless of path loss. This will lead to an SIR given by (13.48) for each user. The disadvantage of this form of power control is that channel inversion can require very large transmit power in some fading channels (e.g., infinite power is required in Rayleigh fading). Moreover, channel inversion can cause significant interference to other systems or users operating on the same frequency. In particular, channel inversion can significantly increase the interference between cells in a cellular system. Despite these problems, channel inversion is used on the mobile-to-base station connection in the IS-95 cellular system standard, although the inversion cannot exceed the maximum transmit power constraint.

EXAMPLE 13.6: Consider a DSSS uplink system with processing gain $G = B_s/B = 100$. Assume the system is interference limited and there is no multipath on any user's channel. Suppose user k has a received power that is 6 dB less than the other users. Find the number of users that the system can support under BPSK modulation such that each user has a BER less than 10^{-3}. Make the computation both for random codes under the standard Gaussian assumption, $\xi = 1$, and for PN codes with $\xi = 3$.

Solution: As in the previous example, we require SIR $= \gamma_b = 6.79$ dB $= 4.775$ for $P_b = 10^{-3}$. We again set the SIR equal to the SNR $\gamma_b = 4.775$ and solve for K to find the maximum number of users that the system can support. Since this is an asynchronous system with $\alpha_k^2/\alpha^2 = .251$ (-6 dB), we have

$$\text{SIR} = \frac{\alpha_k^2 3N}{\alpha^2 \xi(K-1)} = \frac{.251(300)}{\xi(K-1)} = \frac{75.3}{\xi(K-1)} = 4.775.$$

Solving for K yields $K \leq 1 + 75.3/4.77\xi = 16.78$ for $\xi = 1$ and $K \leq 6.26$ for $\xi = 3$, so the system can only support between 6 and 16 users – up to a factor of 3 less than in the prior example for the downlink, which is due to the asynchronicity of the uplink and the near–far effect. This example also illustrates the sensitivity of the system capacity calculation to assumptions about the spreading code properties as captured by ξ. Since the SIR is roughly proportional to $1/(\xi K)$, the number of users the system can support for a given SIR is roughly proportional to $1/\xi$.

13.4.4 Multiuser Detection

Interference signals in SSMA need not be treated as noise. If the spreading code of the interference signal is known, then this knowledge can be used to mitigate the effects of the multiple access interference (MAI). In particular, if all users are detected simultaneously, then interference between users can be subtracted out, which either improves performance or, for a given performance, allows more users to share the channel. Moreover, if all users are detected simultaneously then the near–far effect can aid in detection, since users with strong channel gains are more easily detected (for subsequent cancellation) than if all users had the same channel gains. A DSSS receiver that exploits the structure of multiuser interference in signal detection is called a multiuser detector (MUD). Multiuser detectors are not typically used on downlink channels for several reasons. First, downlink channels are typically synchronous, so they can eliminate all interference by using orthogonal codes as long as the channel doesn't corrupt code orthogonality. Moreover, the kth user's receiver in a downlink is typically limited in terms of power and/or complexity, which makes it difficult to add complex MUD functionality. Finally, the uplink receiver must detect the signals from all users anyway, so any receiver in the uplink is by definition a multiuser detector, albeit not necessarily a good one. By contrast, the kth user's receiver in the downlink need only detect the signal associated with the kth user. For these reasons, work on MUD has primarily focused on DSSS uplink systems, and that is the focus of this section.

Multiuser detection was pioneered by Verdú in [22; 23], where the optimum joint detector for the DSSS asynchronous uplink channel was derived. This derivation assumes an AWGN channel with different channel gains for each user. The optimum detector for this channel chooses the symbol sequences associated with all K users that minimize the MSE between the received signal and the signal that would be generated by these symbol sequences. Because the channel is asynchronous, the entire received waveform must be processed for optimal detection over any one symbol period. The reason is that symbols from other users are not aligned in time, hence all symbols that overlap in the given interval of interest must be considered; by applying to same reasoning to the overlapping symbols, we see that it is not possible to process the signal over any finite interval and still preserve optimality. The optimal MUD for the asynchronous case was shown in [23] to consist of a bank of K single-user matched filter detectors, followed by a Viterbi sequence detection algorithm to jointly detect all users. The Viterbi algorithm has 2^{K-1} states and complexity that grows as 2^K, assuming binary modulation.

For synchronous users, the optimal detection becomes simpler: only one symbol interval needs to be considered in the optimal joint detection, so sequence detection is not needed. In particular, the complexity of this detector grows as $2^K/K$, in contrast to the 2^K growth associated with sequence detection. Consider a two-user synchronous uplink with gain α_k on channel k and binary modulation. The equivalent lowpass received signal over one bit time is

$$r(t) = \alpha_1 b_1 s_{c_1}(t) + \alpha_2 b_2 s_{c_2}(t) + n(t), \qquad (13.51)$$

where b_k is the bit transmitted by the kth user over the given bit time. The optimum (maximum likelihood) detector outputs the pair $\mathbf{b}^* = (b_1^*, b_2^*)$ that satisfies

$$\arg \min_{(b_1, b_2)} \left[-\frac{1}{2\sigma^2} \int_0^{T_s} [r(t) - \alpha_1 b_1 s_{c_1}(t) - \alpha_2 b_2 s_{c_2}(t)]^2 \, dt \right], \qquad (13.52)$$

where σ^2 is the noise power. This is equivalent to finding (b_1^*, b_2^*) to maximize the cost function

$$L(b_1, b_2) = \alpha_1 b_1 r_1 + \alpha_2 b_2 r_2 - \alpha_1 \alpha_2 b_1 b_2 \rho_{12}, \qquad (13.53)$$

where

$$r_k = \int_0^{T_b} r(t) s_{c_k}(t) \, dt \qquad (13.54)$$

and

$$\rho_{jk} = \int_0^{T_b} s_{c_k}(t) s_{c_j}(t) \, dt. \qquad (13.55)$$

This analysis easily extends to K synchronous users. Here we can express $\mathbf{r} = (r_1, \dots, r_K)^T$ in matrix form as

$$\mathbf{r} = \mathbf{RAb} + \mathbf{n} \qquad (13.56)$$

[24], where $\mathbf{b} = (b_1, \dots, b_K)^T$ is the bit vector associated with the K users over the given bit time, $\mathbf{A}$ is a diagonal $K \times K$ matrix of the channel gains α_k, and $\mathbf{R}$ is a $K \times K$ matrix of the cross-correlations between the spreading codes. The optimal choice of bit sequence $\mathbf{b}^*$ is obtained by chosing the sequence to maximize the cost function

$$L(\mathbf{b}) = 2\mathbf{b}^T \mathbf{Ar} - \mathbf{b}^T \mathbf{ARAb}. \qquad (13.57)$$

Unfortunately, maximizing (13.57) for K users also has complexity that grows as 2^K (the same as in the asynchronous case), assuming a search tree is used for the optimization. In addition to the high complexity of the optimal detector, it has the drawback of requiring knowledge of the channel amplitudes α_k.

The complexity of MUD can be decreased at the expense of optimality. Many suboptimal MUDs have been developed with various trade-offs with respect to performance, complexity, and requirements regarding channel knowledge. Suboptimal MUDs fall into two broad categories: linear and nonlinear. Linear MUDs apply a linear operator or filter to the output of the matched filter bank in Figure 13.16. These detectors have complexity that is linear in the number of users, a significant complexity improvement over the optimal detector. The most common linear MUDs are the decorrelating detector [25] and the minimum mean-square error (MMSE) detector. The decorrelating detector simply inverts the matrix $\mathbf{R}$ of cross-correlations, resulting in

$$\hat{\mathbf{b}}^* = \mathbf{R}^{-1}\mathbf{r} = \mathbf{R}^{-1}[\mathbf{R}\mathbf{A}\mathbf{b} + \mathbf{n}] = \mathbf{A}\mathbf{b} + \mathbf{R}^{-1}\mathbf{n}. \qquad (13.58)$$

The inverse exists for most cases of interest. In the absence of noise, the resulting bit sequence equals the original sequence, scaled by the channel gains. In addition to its simplicity, this detector has other appealing features: it completely removes MAI, and it does not require knowledge of the channel gains. However, the decorrelating detector can lead to noise enhancement, since the noise vector is multiplied by the matrix inverse. Thus, decorrelating MUD is somewhat analogous to zero-forcing equalization as described in Section 11.4.1: all MAI can be removed, but at the expense of noise enhancement.

The MMSE detector finds the matrix $\mathbf{D}$ such that multiplication of the filter bank output by $\mathbf{D}$ minimizes the expected MSE between $\mathbf{D}$ and the transmitted bit sequence $\mathbf{b}$. In other words, the matrix $\mathbf{D}$ satisfies

$$\arg \min_{\mathbf{D}} \mathbf{E}[(\mathbf{b} - \mathbf{D}\mathbf{r})^T(\mathbf{b} - \mathbf{D}\mathbf{r})]. \qquad (13.59)$$

The optimizing $\mathbf{D}$ is given [17; 24; 26] by

$$\mathbf{D} = (\mathbf{R} + .5N_0\mathbf{I})^{-1}. \qquad (13.60)$$

Note that, in the absence of noise, the MMSE detector is the same as the decorrelating detector. However, it has better performance at low SNRs because it balances removal of the MAI with noise enhancement. This is analogous to the MMSE equalizer design for ISI channels, described in Section 11.4.2.

Nonlinear MUDs can have much better performance than linear detectors – although not necessarily in all cases, especially with little or no coding [27]. The most common nonlinear MUD techniques are multistage detection, decision-feedback detection, and successive interference cancellation. In a multistage detector, each stage consists of the conventional matched filter bank. The nth stage of the detector uses decisions of the $(n-1)$th stage to cancel the MAI at its input. The multistage detector can be applied to either synchronous [28] or asynchronous [29] systems. The decision-feedback detector is based on the same premise as a decision-feedback equalizer. It consists of a feedforward and feedback filter, where the feedforward filter is the Cholesky factorization of the correlation matrix $\mathbf{R}$. The decision-feedback MUD can be designed for either synchronous [30] or asynchronous [31] systems. These detectors require knowledge of the channel gains and can also suffer from error propagation when decision errors are fed back through the feedback filter. In interference cancellation, an estimate of one or more users is made and then the MAI caused to other users is subtracted out [32]. Interference cancellation can be done in parallel, where all users are detected simultaneously and then cancelled out [33; 34], or sequentially, where users are detected one at a time and then subtracted out from users yet to be detected [35]. Parallel cancellation has a lower latency and is more robust to decision errors. However, its performance suffers owing to the near–far effect, when some users have much weaker received powers than others. Under this unequal power scenario, successive interference cancellation can outperform parallel cancellation [24]. In fact, successive cancellation theoretically achieves Shannon capacity of the uplink channel, as will be discussed in Chapter 14, and in practice approaches Shannon capacity [36]. Successive interference cancellation suffers from error propagation, which can significantly degrade performance, but this degradation can be partially offset through power control [37].

A comprehensive treatment of different MUDs and their performance can be found in [17], and shorter tutorials are provided in [24; 32]. Combined equalization and MUD is treated in [38]. Multiuser detection for multirate CDMA, where different users have different data rates, is analyzed in [39]. Blind, space-time, and turbo multiuser detectors are developed in [40]. Spectral efficiencies of the different detectors have been analyzed in [27].

13.4.5 Multicarrier CDMA

Multicarrier CDMA (MC-CDMA) is a technique that combines the advantages of OFDM and CDMA. It is very effective both at combating ISI and as a mechanism for allowing multiple users to share the same channel. The basic block diagram for a baseband single-user multicarrier CDMA system is shown in Figure 13.17. The data symbol s_l is sent over all N subchannels. On the ith subchannel, s_l is multiplied by the ith chip c_i of a spreading sequence $s_c(t)$, where $c_i = \pm 1$. This is similar to the standard spread-spectrum technique, except that multiplication with the spreading sequence is done in the frequency domain rather than in the time domain. The frequency spread data $(s_l c_1, s_l c_2, \ldots, s_l c_N)$ is then multicarrier modulated in the standard manner: the parallel sequence is passed through an inverse fast Fourier transform (IFFT) and a parallel-to-serial converter and then is D/A converted to produce the modulated signal $s(t)$, where $S(f)$ is as shown in Figure 13.17 for subchannel carrier frequencies $(f_1, \ldots, f_N)$.

Assume the MC-CDMA signal is transmitted through a frequency-selective channel with a constant channel gain of α_i on the ith subchannel and with AWGN $n(t)$. The receiver performs the reverse operations of the transmitter – passing the received signal through an A/D converter, a serial-to-parallel converter, and an FFT – to recover the symbol transmitted over the ith subchannel. The subchannel symbol received on the ith subchannel is multiplied by the ith chip c_i and a weighting factor β_i; these terms are then summed together for the final symbol estimate $\hat{s}_l$.

In a multiuser MC-CDMA system, each user k modulates his signal as in Figure 13.17 but with a unique spreading code $s_{c_k}(t)$. So, for a two-user system, user 1 would use the spreading code $s_{c_1}(t)$ with chips $(c_1^1, \ldots, c_N^1)$ resulting in a transmitted signal $s_1(t)$, and user 2 would use the spreading code $s_{c_2}(t)$ with chips $(c_1^2, \ldots, c_N^2)$ resulting in a transmitted signal $s_2(t)$. If the users transmit simultaneously then their signals are added "in the air" as shown in Figure 13.18, where s_l^1 (resp., s_l^2) is the symbol corresponding to user 1 (resp., user 2) over the lth symbol time. The interference between users in this system is reduced by the

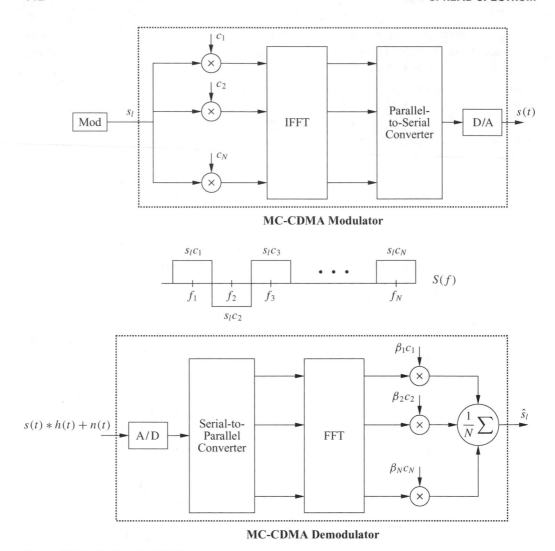

MC-CDMA Modulator

MC-CDMA Demodulator

Figure 13.17: Multicarrier CDMA system.

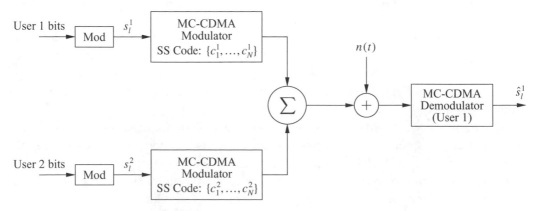

Figure 13.18: Two-user multicarrier CDMA system.

cross-correlation of the spreading codes, as in standard spread spectrum without multicarrier. However, each user benefits from the frequency diversity of spreading its signal over independently fading subchannels. This typically leads to better performance than in standard spread spectrum.

13.5 Multiuser FHSS Systems

Multiuser FHSS is accomplished by assigning each user a unique spreading code sequence $s_{c_i}(t)$ to generate its hop pattern. If the spreading codes are orthogonal and if the users are synchronized in time, then the different users never collide, and performance of each user is the same as in a single-user FH system. However, if the users are asynchronous or if nonorthogonal codes are used, then multiple users will collide by occupying a given channel simultaneously. The symbols transmitted during that time are likely to be in error, so multiuser FHSS typically uses error correction coding to compensate for collisions.

Multiuser FHSS, also referred to as FH-CDMA or FH-SSMA, is mainly applied to uplink channels. This access method is the preferred method for military applications because of the protection from jamming and low probability of interception and detection characteristic of FH systems. FH-CDMA was also proposed as a candidate for second-generation digital cellular systems [2, Chap. 9.4] but was not adopted. Trade-offs between frequency hopping and direct sequence for multiple access are discussed in [8, Chap. 11; 41; 42]. In fact, most analyses indicate that FH-CDMA is inferior to DS-CDMA as a multiple access method in terms of the number of users that can share the channel simultaneously, at least under asynchronous operation. In addition, FH-CDMA systems typically cause interference at much larger distances than DS-CDMA systems, since the interference isn't mitigated by bandwidth spreading [8, Chap. 11]. The main benefit of FH over DS is its robustness to the near–far effect: avoiding interference by hopping – rather than suppressing it by the processing gain – mitigates the performance degradation caused by strong interferers. Hybrid techniques using FH along with another multiple access method have been used to exploit the benefits of both. For example, the GSM digital cellular standard uses a combination of time division and slow frequency hopping, where the frequency hopping is used primarily to average out interference from other cells. FH-CDMA is also used in the Bluetooth system. Bluetooth operates in the unlicensed 2.4-GHz band, and FH was chosen because it can be used with noncoherent FSK modulation, a low-cost and energy-efficient modulation technique.

PROBLEMS

13-1. In this problem we derive the SIR ratio (13.6) for a randomly spread signal with interference. The correlator output of the ith receiver branch in this system is given by (13.5) as

$$x_i = \int_0^T x(t)s_i(t)\,dt = \sum_{j=1}^{N}(s_{ij}^2 + I_j s_{ij}).$$

(a) Show that the conditional expectation of x_i, conditioned on the transmitted signal $s_i(t)$, is $\mathbf{E}[x_i \mid s_i(t)] = E_s$.

(b) Assuming equiprobable signaling ($p(s_i(t)) = 1/M$), show that $\mathbf{E}[x_i] = E_s/M$.

(c) Show that $\text{Var}[x_i \mid s_i(t)] = E_s E_J/N$.

(d) Show that (again with equiprobable signaling) $\text{Var}[x_i] = E_s E_J/NM$.

(e) The SIR is given by

$$\text{SIR} = \frac{\mathbf{E}[x_i]^2}{\text{Var}[x_i]}.$$

Show that

$$\text{SIR} = \frac{E_s}{E_j} \cdot \frac{N}{M}.$$

13-2. Sketch the transmitted DSSS signal $s(t)s_c(t)$ over two bit times $[0, 2T_b]$ assuming that $s(t)$ is BPSK modulated with carrier frequency 100 MHz and $T_s = 1 \, \mu s$. Assume the first data bit is a 1 and the second data bit is a 0. Assume also that there are ten chips per bit and that the chips alternate between ± 1, with the first chip equal to $+1$.

13-3. Consider an FH system transmitted over a two-ray channel, where the reflected path has delay $\tau = 10 \, \mu s$ relative to the LOS path. Assume the receiver is synchronized to the hopping of the LOS path.

(a) For what hopping rates will the system exhibit no fading?

(b) Assume an FFH system with hop time $T_c = 50 \, \mu s$ and symbol time $T_s = .5$ ms. Will this system exhibit no fading, flat fading, or frequency-selective fading?

(c) Assumen a SFH system with hop time $T_c = 50 \, \mu s$ and symbol time $T_s = .5 \, \mu s$. Will this system exhibit no fading, flat fading, or frequency-selective fading?

13-4. In this problem we explore the statistics of the DSSS receiver noise after it is multiplied by the spreading sequence. Let $n(t)$ be a random noise process with autocorrelation function $\rho_n(\tau)$ and let $s_c(t)$ be a zero-mean random spreading code, independent of $n(t)$, with autocorrelation $\rho_c(\tau)$. Let $n'(t) = n(t)s_c(t)$.

(a) Find the autocorrelation and power spectral density of $n'(t)$.

(b) Show that if $\rho_c(\tau) = \delta(\tau)$ then $n'(t)$ is zero mean with autocorrelation function $\rho_n(\tau)$ – that is, it has the same statistics as $n(t)$, so the statistics of $n(t)$ are not affected by its multiplication with $s_c(t)$.

(c) Find the autocorrelation $\rho_{n'}(\tau)$ of $n'(t)$ if $n(t)$ is zero-mean AWGN and $s_c(t)$ is a maximal linear code with autocorrelation given by (13.19). What happens to $\rho_{n'}(\tau)$ as $N \to \infty$ in (13.19)?

13-5. Show that, for any real periodic spreading code $s_c(t)$, its autocorrelation $\rho_c(\tau)$ over one period is symmetric about τ and reaches its maximum value at $\tau = 0$.

13-6. Show that, if $s_c(t)$ is periodic with period T, then the autocorrelation for time-shifted versions of the spreading code depends only on the difference of their time shifts:

$$\frac{1}{T} \int_0^T s_c(t - \tau_0)s_c(t - \tau_1) \, dt = \rho_c(\tau_1 - \tau_0).$$

13-7. Show that, for any periodic spreading code $s_c(t)$ with period T, its autocorrelation $\rho_c(t)$ is periodic with the same period.

13-8. Show that the power spectral density of $s_c(t)$ for a maximal linear spreading code with period $NT_c = T_s$ is given by

$$P_{s_c}(f) = \sum_{m=-\infty}^{\infty} \frac{N+1}{N^2} \operatorname{sinc}^2\left(\frac{m}{N}\right) \delta\left(f - \frac{m}{T_s}\right).$$

Also plot this spectrum for $N = 100$ and $T_s = 1\,\mu$s over the frequency range $-10/T_s \le f \le 10/T_s$.

13-9. Show that m-sequences as well as random binary spreading sequences have the balanced, run-length, and shift properties.

13-10. Suppose that an unmodulated carrier $s(t)$ is spread using a maximal linear code $s_c(t)$ with period T and then is transmitted over a channel with an impulse response of $h(t) = \alpha_0\delta(t - \tau_0) + \alpha_1\delta(t - \tau_1)$. The corresponding received signal $r(t)$ is input to the synchronization loop shown in Figure 13.9. Find the function $w(\tau)$ output from the integrator in this loop as a function of τ. What will determine which of these two multipath components the control loop for coarse acquisition locks to?

13-11. Find the outage probability relative to $P_b = 10^{-6}$ for a three-branch RAKE receiver with DPSK signal modulation, independent Rayleigh fading on each branch, and a branch SNR /bit (prior to despreading) of 10 dB. Assume the code autocorrelation associated with maximal linear codes with $K = N = 2^n - 1 = 15$. Assume also that the code in the first branch is perfectly aligned, but that the code in the second branch is offset by $T_c/4$ and the code in the third branch is offset by $T_c/3$. Assume selection combining diversity and neglect the interference due to other multipath components in your SNR calculations.

13-12. Consider a spread-spectrum signal transmitted over a multipath channel with a LOS component and a single multipath component, where the delay of the multipath relative to the LOS is greater than the chip time T_c. Consider a two-branch RAKE receiver, with one branch corresponding to the LOS component and the other to the multipath component. Assume that, with perfect synchronization in both branches, the incoming signal component at each branch (after despreading) has power that is uniformly distributed between 6 mW and 12 mW. The total noise power in the despread signal bandwidth is 1 mW. Suppose, however, that only the first branch is perfectly synchronized, while the second branch has a timing offset of $T_c/2.366$. The code autocorrelation is that of a maximal linear code with $N \gg 1$. The two branches of the RAKE are combined using maximal-ratio combining with knowledge of the timing offset.

(a) What is the average SNR at the combiner output?
(b) What is the distribution of the combiner output SNR?
(c) What is the outage probability for DPSK modulation with a BER of 10^{-4}?

13-13. This problem illustrates the benefits of RAKE receivers and the optimal choice of multipath components for combining when the receiver complexity is limited. Consider a multipath channel with impulse response

$$h(t) = \alpha_0\delta(t) + \alpha_1\delta(t - \tau_1) + \alpha_2\delta(t - \tau_2).$$

The α_i are Rayleigh fading coefficients, but their expected power varies (because of shadowing) such that $\mathbf{E}[\alpha_0^2] = 5$ with probability .5 and 10 with probability .5, $\mathbf{E}[\alpha_1^2] = 0$ with probability .5 and 20 with probability .5, and $\mathbf{E}[\alpha_2^2] = 5$ with probability .75 and 10 with probability .25 (all units are linear). The transmit power and noise power are such that a spread-spectrum receiver locked to the ith multipath component will have an SNR of α_i^2 in the absence of the other multipath components.

(a) Assuming maximal linear codes, a bit time T_b, and a spread-spectrum receiver locked to the LOS signal component (with zero delay and gain α_0), find the values of τ_1 and τ_2 ($0 \le \tau_1 \le \tau_2 < T_b$) for which their corresponding multipath components will be attenuated by $-1/N$, where N is the number of chips per bit.

For the rest of this problem, assume spreading codes with autocorrelation equal to a delta function.

(b) What is the outage probability of DPSK modulation at an instantaneous $P_b = 10^{-3}$ for a single-branch spread-spectrum receiver locked to the LOS path?

(c) Find the outage probability of DPSK modulation at an instantaneous $P_b = 10^{-3}$ for a three-branch RAKE receiver, where each branch is locked to one of the multipath components and SC is used to combine the paths.

(d) Suppose receiver complexity is limited such that only a two-branch RAKE with SC can be built. Find which two multipath components the RAKE should lock to in order to minimize the outage probability of DPSK modulation at $P_b = 10^{-3}$, and then find this minimum outage probability.

13-14. This problem investigates the performance of a RAKE receiver when the multipath delays are random. Consider a DS spreading code with chip time T_c and autocorrelation function

$$\rho_c(t) = \begin{cases} 1 & -T_c/2 < t < T_c/2, \\ 0 & \text{else.} \end{cases}$$

Suppose this spreading code is used to modulate a DPSK signal with bit time $T_b = 10T_c$. The spread signal is transmitted over a multipath channel, where the channel is modeled using the discrete-time tapped delay model described in Section 3.4, with a tap separation of T_c and a total multipath spread $T_m = 5T_c$. Thus, the model has five multipath "bins", where the ith bin has at most one multipath component of delay $(i - .5)T_c$. The distribution of the multipath component in each bin is independent of the components in all the other bins. The probability of observing a multipath component in bin i is .75 and, conditioned on having a multipath component in bin i, the amplitude of the ith multipath component after despreading is Rayleigh distributed with average SNR/bit of $S_i = 20/i$ for $i = 1, 2, \ldots, 5$ (in linear units). Thus, the average power is decreasing relative to the distance that the multipath component has traveled.

At the receiving end is a five-branch SC-diversity RAKE receiver with each branch synchronized to one of the multipath bins. Assuming a target BER of 10^{-3}, compute the outage probability of the RAKE receiver output. Compare this with the outage probability for the same BER if there is (with probability 1) a multipath component in each bin and if each multipath component after despreading has an average SNR/bit of $S_i = 20$ (linear units).

13-15. Direct-sequence spread-spectrum signals are often used to make channel measurements, since the wideband spread-spectrum signal has good resolution of the individual multipath components in the channel. Channel measurement with spread spectrum, also called *channel sounding*, is performed using a receiver with multiple branches synchronized to the different chip delays. Specifically, an unmodulated spreading sequence $s_c(t)$ is sent through

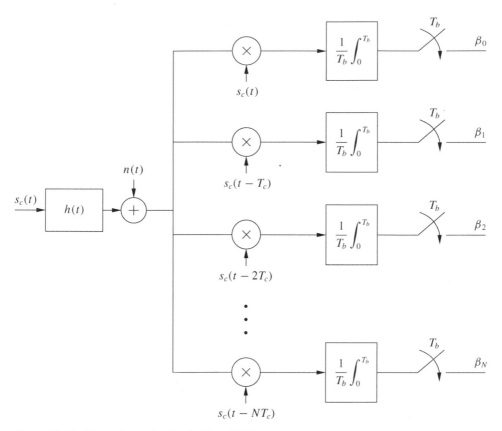

Figure 13.19: Channel sounder for Problem 13-15.

the channel $h(t)$, as shown in Figure 13.19. The receiver has $N+1$ branches synchronized to different chip delays. The β_i output from the ith branch approximates the channel gain associated with that delay, so that the approximate channel model obtained by the channel sounding shown in the figure is $\hat{h}(t) = \sum_{i=0}^{N} \beta_i \delta(t - iT_c)$.

Assume that the autocorrelation function for $s_c(t)$ is

$$\rho_c(\tau) = \frac{1}{T_b} \int_0^{T_b} s_c(t)s_c(t-\tau)\,dt = \begin{cases} 1 - |\tau|/T_c & |\tau| < T_c, \\ 0 & |\tau| \geq T_c. \end{cases}$$

(a) Show that if $h(t) = \sum_{i=0}^{N} \alpha_i \delta(t - iT_c)$ then, in the absence of noise ($n(t) = 0$), the channel sounder of Figure 13.19 will output $\beta_i = \alpha_i$ for all i.

(b) Again neglecting noise, if $h(t) = a\delta(t) + b\delta(t - 1.2T_c) + c\delta(t - 3.5T_c)$, what approximation $\hat{h}(t)$ will be obtained by the channel sounder?

(c) Now assume that the channel sounder yields a perfect estimate of the channel $\hat{h}(t) = h(t) = \beta_0\delta(t) + \beta_1\delta(t - T_c) + \beta_2\delta(t - 2T_c)$, where the β_i are all indepedent Rayleigh fading random variables. Consider a three-branch RAKE receiver with the ith branch perfectly synchronized to the ith multipath component of $h(t)$ and with an average SNR/bit on each branch (after despreading) of 10 dB. Find P_{out} for DPSK modulation with a target BER of 10^{-3} under maximal-ratio combining in the RAKE. Do the same calculation for selection combining in the RAKE.

13-16. Find the values of the autocorrelation and cross-correlation for Gold codes, Kasami codes from the small set, and Kasami codes from the large set for $n = 8$. Also, find the number of such Kasami codes for both the small set and the large set.

13-17. Find the Hadamard matrix for $N = 4$ and show that the spreading codes generated by the rows of this matrix are orthogonal, assuming synchronous users (i.e., show $\rho_{ij}(0) = 0$ $\forall i \neq j$). Also find the cross-correlation between all pairs of users assuming a timing offset of $T_c/2$ between users; that is, find

$$\rho_{ij}\left(\frac{T_c}{2}\right) = \frac{1}{T_s} \int_0^{T_s} s_{c_i}(t) s_{c_j}\left(t - \frac{T_c}{2}\right) dt = \frac{1}{N} \sum_{n=1}^{N} s_{c_i}(nT_c) s_{c_j}\left(nT_c - \frac{T_c}{2}\right)$$

for all pairs of codes.

13-18. Consider an asynchronous DSSS multiple access channel (MAC) system with bandwidth expansion $N = B_s/B = 100$ and $K = 40$ users. Assume the system is interference limited and there is no multipath on any user's channel. Find the probability of error for user k under BPSK modulation, assuming random codes with the standard Gaussian assumption and assuming this user is in a deep fade, with received power that is 6 dB less than the other users. Would this change if the users could be synchronized?

13-19. Show that the vector $\mathbf{r} = (r_1, \ldots, r_K)^T$ for r_k given by (13.54) can be expressed by the matrix equation (13.56). What are the statistics of $\mathbf{n}$ in this expression?

13-20. Show that the maximum likelihood detector for a K-user synchronous MAC receiver choses the vector $\mathbf{b}$ to maximize the cost function given by (13.57).

13-21. Here we illustrate the use of multiple spreading codes in single-user CDMA systems for adaptive modulation or diversity gain. The BER for user k in a K-user DS-CDMA system, where each user transmits his BPSK modulated bit sequence at a rate R bps along his spreading code, is given by:

$$\text{BER}_k = Q\left(\sqrt{\frac{2P_k(\gamma_k)\gamma_k}{\frac{1}{N} \sum_{i=1, i\neq k}^{K} P_i(\gamma_i)\gamma_i + 1}}\right), \tag{13.61}$$

where N is the spreading factor (processing gain), γ_i is the ith user's channel power gain, and $P_i(\gamma_i)$ is i's transmit power when his channel gain is γ_i. Note that noise power has been normalized to unity and that the receiver demodulates each spreading sequence, treating other sequences as noise (matched filter detector). The system has a *single* user who can simultaneously transmit *up to two* spreading sequences, modulating each with an independent BPSK bit stream at rate R bps (on each stream).

(a) Assume that the user's channel fade is γ, and assume that he splits his total transmit power $P(\gamma)$ equally among the transmitted sequences. Note that the user has three options: he can transmit nothing, one BPSK modulated spreading sequence, or both spreading sequences BPSK modulated with independent bits. Based on the BER expression for the multiuser case (13.61), explain why the BER for the single-user multirate DS-CDMA system is given by

$$\text{BER} = Q(\sqrt{2P(\gamma)\gamma}) \tag{13.62}$$

if the user transmits only one spreading sequence and by

$$\text{BER} = Q\left(\sqrt{\frac{P(\gamma)\gamma}{P(\gamma)\gamma/2N + 1}}\right) \tag{13.63}$$

when he transmits both spreading sequences together. What are the rates achieved in both of these cases?

(b) Assume the channel is known perfectly to the transmitter as well as the receiver, and assume that γ is distributed according to the distribution $p(\gamma)$. We want to develop an adaptive rate and power strategy for this channel. Since we do not use error correction coding, the user needs to keep his BER below a threshold P_b^0 for all transmitted bits. Assume an average transmit power constraint of unity: $\int_0^\infty P(\gamma)p(\gamma)\,d\gamma = 1$. We have a finite discrete set of possible rates and so, as with narrowband adaptive modulation, the optimal adaptive rate policy is to send no data when γ is below a cutoff threshold γ_0, one data stream when $\gamma_0 \leq \gamma < \gamma_1$, and both data streams when $\gamma > \gamma_1$. Find the power adaptation strategy that exactly meets the BER target P_b^0 for this adaptive rate strategy as a function of the thresholds γ_0 and γ_1.

(c) Given the adaptive rate and power strategy obtained in part (b), solve the Lagrangian optimization to find γ_0 and γ_1 as a function of the Lagrangian λ.

13-22. You work for a company that wants to design a next-generation cellular system for voice plus high-speed data. The FCC has decided to allocate 100 MHz of spectrum for this system based on whatever standard is agreed to by the various industry players. You have been charged with designing the system and pushing your design through the standards body. You should describe your design in as much detail as possible, paying particular attention to how it will combat the impact of fading and ISI and to its capacity for accommodating both voice and data. Also develop arguments explaining why your design is better than competing strategies.

REFERENCES

[1]   R. Pickholtz, D. Schilling, and L. Milstein, "Theory of spread-spectrum communications – A tutorial," *IEEE Trans. Commun.*, pp. 855–84, May 1982.

[2]   R. L. Peterson, R. E. Ziemer, and D. E. Borth, *Introduction to Spread Spectrum Communications,* Prentice-Hall, Englewood Cliffs, NJ, 1995.

[3]   V. V. Veeravalli and A. Mantravadi, "The coding–spreading trade-off in CDMA systems," *IEEE J. Sel. Areas Commun.*, pp. 396–408, February 2002.

[4]   S. W. Golomb, *Shift Register Sequences,* Holden-Day, San Francisco, 1967.

[5]   A. J. Viterbi, *CDMA Principles of Spread Spectrum Communications,* Addison-Wesley, Reading, MA, 1995.

[6]   G. L. Stuber, *Principles of Mobile Communications,* 2nd ed., Kluwer, Dordrecht, 2001.

[7]   O. C. Mauss, F. Classen, and H. Meyr, "Carrier frequency recovery for a fully digital direct-sequence spread-spectrum receiver: A comparison," *Proc. IEEE Veh. Tech. Conf.*, pp. 392–5, May 1993.

[8]   R. C. Dixon, *Spread Spectrum Systems with Commercial Applications,* 3rd ed., Wiley, New York, 1994.

[9]   M. K. Simon, J. K. Omura, R. A. Scholtz, and B. K. Levitt, *Spread Spectrum Communications Handbook,* McGraw-Hill, New York, 1994.

[10] B. Sklar, *Digital Communications – Fundamentals and Applications,* Prentice-Hall, Englewood Cliffs, NJ, 1988.

[11] G. L. Turin, "Introduction to spread spectrum antimultipath techniques and their application to urban digital radio," *Proc. IEEE,* pp. 328–53, March 1980.

[12] M. K. Simon and M.-S. Alouini, *Digital Communication over Fading Channels: A Unified Approach to Performance Analysis,* Wiley, New York, 2000.

[13] R. Gold, "Optimum binary sequences for spread-spectrum multiplexing," *IEEE Trans. Inform. Theory,* pp. 619–21, October 1967.

[14] E. H. Dinan and B. Jabbari, "Spreading codes for direct sequence CDMA and wideband CDMA cellular networks," *IEEE Commun. Mag.,* pp. 48–54, September 1998.

[15] N. B. Mehta, L. J. Greenstein, T. M. Willis, and Z. Kostic, "Analysis and results for the orthogonality factor in WCDMA downlinks," *IEEE Trans. Wireless Commun.,* pp. 1138–49, November 2003.

[16] S. Verdú, "Demodulation in the presence of multiuser interference: Progress and misconceptions," in D. Docampo, A. Figueiras, and F. Perez-Gonzalez, Eds., *Intelligent Methods in Signal Processing and Communications,* pp. 15–46, Birkhäuser, Boston, 1997.

[17] S. Verdú, *Multiuser Detection,* Cambridge University Press, 1998.

[18] M. Pursley, "Performance evaluation for phase-coded spread-spectrum multiple-access communication – Part I: System analysis," *IEEE Trans. Commun.,* pp. 795–9, August 1977.

[19] R. Pickholtz, L. Milstein, and D. Schilling, "Spread spectrum for mobile communications," *IEEE Trans. Veh. Tech.,* pp. 313–22, May 1991.

[20] K. S. Gilhousen, I. M. Jacobs, R. Padovani, A. J. Viterbi, L. A. Weaver, Jr., and C. E. Wheatley III, "On the capacity of a cellular CDMA system," *IEEE Trans. Veh. Tech.,* pp. 303–12, May 1991.

[21] H. Xiang, "Binary code-division multiple-access systems operating in multipath fading, noisy channels," *IEEE Trans. Commun.,* pp. 775–84, August 1985.

[22] S. Verdú, "Optimum multiuser signal detection," Ph.D. thesis, University of Illinois, Urbana-Champaign, August 1984.

[23] S. Verdú, "Minimum probability of error for asynchronous Gaussian multiple-access channels," *IEEE Trans. Inform. Theory,* pp. 85–96, January 1986.

[24] A. Duel-Hallen, J. Holtzman, and Z. Zvonar, "Multiuser detection for CDMA systems," *IEEE Pers. Commun. Mag.,* pp. 46–58, April 1995.

[25] R. Lupas and S. Verdú, "Linear multiuser detectors for synchronous code-division multiple-access channels," *IEEE Trans. Inform. Theory,* pp. 123–36, January 1989.

[26] M. L. Honig and H. V. Poor, "Adaptive interference suppression," in *Wireless Communications: Signal Processing Perspectives,* ch. 2, Prentice-Hall, Englewood Cliffs, NJ, 1998.

[27] S. Verdú and S. Shamai (Shitz), "Spectral efficiency of CDMA with random spreading," *IEEE Trans. Inform. Theory,* pp. 622–40, March 1999.

[28] M. K. Varanasi and B. Aazhang, "Near-optimum detection in synchronous code division multiple-access communications," *IEEE Trans. Commun.,* pp. 725–36, May 1991.

[29] M. K. Varanasi and B. Aazhang, "Multistage detection in asynchronous code division multiple-access communications," *IEEE Trans. Commun.,* pp. 509–19, April 1990.

[30] A. Duel-Hallen, "Decorrelating decision-feedback multiuser detector for synchronous CDMA," *IEEE Trans. Commun.,* pp. 285–90, February 1993.

[31] A. Duel-Hallen, "A family of multiuser decision-feedback detectors for asynchronous code-division multiple-access channels," *IEEE Trans. Commun.,* pp. 421–34, February–April 1995.

[32] J. G. Andrews, "Interference cancellation for cellular systems: A contemporary overview," *IEEE Wireless Commun. Mag.,* pp. 19–29, April 2005.

[33] D. Divsalar, M. K. Simon, and D. Raphaeli, "Improved parallel interference cancellation for CDMA," *IEEE Trans. Commun.,* pp. 258–68, February 1998.

[34] Y. C. Yoon, R. Kohno, and H. Imai, "A spread-spectrum multiaccess system with cochannel interference cancellation for multipath fading channels," *IEEE J. Sel. Areas Commun.,* pp. 1067–75, September 1993.

[35] P. Patel and J. Holtzman, "Analysis of a simple successive interference cancellation scheme in a DS/CDMA system," *IEEE J. Sel. Areas Commun.,* pp. 796–807, June 1994.

[36] A. J. Viterbi, "Very low rate convolutional codes for maximum theoretical performance of spread-spectrum multiple-access channels," *IEEE J. Sel. Areas Commun.,* pp. 641–9, May 1990.

[37] J. G. Andrews and T. H. Meng, "Optimum power control for successive interference cancellation with imperfect channel estimation," *IEEE Trans. Wireless Commun.,* pp. 375–83, March 2003.

[38] X. Wang and H. V. Poor, "Blind equalization and multiuser detection in dispersive CDMA channels," *IEEE Trans. Commun.,* pp. 91–103, January 1998.

[39] U. Mitra, "Comparison of maximum-likelihood-based detection for two multirate access schemes for CDMA signals," *IEEE Trans. Commun.,* pp. 64–77, January 1999.

[40] X. Wang and H. V. Poor, *Wireless Communication Systems: Advanced Techniques for Signal Reception,* Prentice-Hall, Englewood Cliffs, NJ, 2004.

[41] R. Kohno, R. Meidan, and L. B. Milstein, "Spread spectrum access methods for wireless communications," *IEEE Commun. Mag.,* pp. 58–67, January 1995.

[42] H. El Gamal and E. Geraniotis, "Comparing the capacities of FH/SSMA and DS/CDMA networks," *Proc. Internat. Sympos. Pers., Indoor, Mobile Radio Commun.,* pp. 769–73, September 1998.

14

Multiuser Systems

In multiuser systems the system resources must be divided among multiple users. This chapter develops techniques to allocate resources among multiple users and examines the fundamental capacity limits of multiuser systems. We know from Section 5.1.2 that signals of bandwidth B and time duration T occupy a signal space of dimension $2BT$. In order to support multiple users, the signal space dimensions of a multiuser system must be allocated to the different users.[1] Allocation of signaling dimensions to specific users is called *multiple access*.[2] Multiple access methods perform differently in different multiuser channels, and we will apply these methods to the two basic multiuser channels: downlink channels and uplink channels. Because signaling dimensions can be allocated to different users in an infinite number of different ways, multiuser channel capacity is defined by a *rate region* rather than a single number. This region describes all user rates that can be simultaneously supported by the channel with arbitrarily small error probability. We will discuss multiuser channel capacity regions for both the uplink and the downlink. We also consider random access techniques, whereby signaling dimensions are allocated only to active users, as well as power control, which ensures that users maintain the SINR required for acceptable performance. The performance benefits of multiuser diversity, which exploits the time-varying nature of the users' channels, is also described. We conclude with a discussion of the performance gains and signaling techniques associated with multiple antennas in multiuser systems.

14.1 Multiuser Channels: The Uplink and Downlink

A "multiuser" channel is any channel that must be shared among multiple users. There are two different types of multiuser channels, the *uplink* channel and the *downlink* channel, which are illustrated in Figure 14.1. A downlink, also called a broadcast channel or forward channel, has one transmitter sending to many receivers. Since the signals transmitted to all users originate from the downlink transmitter, the transmitted signal $s(t) = \sum_{k=1}^{K} s_k(t)$, with total power P and bandwidth B, is the sum of signals transmitted to all K users. Thus, the

---

[1] Allocation of signaling dimensions through either multiple access or random access is performed by the "medium access control" layer in the open systems interconnect (OSI) network model [1, Chap. 1.3].

[2] The dimensions allocated to the different users need not be orthogonal, as in multiuser spread spectrum with nonorthogonal spreading codes or the superposition coding technique discussed in Section 14.5.

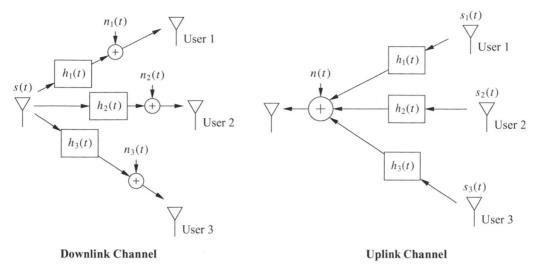

Figure 14.1: Downlink and uplink channels.

total signaling dimensions and power of the transmitted signal must be divided among the different users. Synchronization of the different users is relatively easy in the downlink because all signals originate from the same transmitter, although multipath in the channel can corrupt this synchronization. Another important characteristic of the downlink is that both signal and interference are distorted by the same channel. In particular, user k's signal $s_k(t)$ and all interfering signals $s_j(t)$, $j \neq k$, pass through user k's channel $h_k(t)$ to arrive at k's receiver. This is a fundamental difference between the uplink and the downlink, since in the uplink signals from different users are distorted by different channels. Examples of wireless downlinks include all radio and television broadcasting, the transmission link from a satellite to multiple ground stations, and the transmission link from a base station to the mobile terminals in a cellular system.

An uplink channel, also called a multiple access channel[3] or reverse channel, has many transmitters sending signals to one receiver, where each signal must be within the total system bandwidth B. However, in contrast to the downlink, in the uplink each user has an individual power constraint P_k associated with its transmitted signal $s_k(t)$. In addition, since the signals are sent from different transmitters, these transmitters must coordinate if signal synchronization is required. Figure 14.1 also indicates that the signals of the different users in the uplink travel through different channels, so even if the transmitted powers P_k are the same, the received powers associated with the different users will be different if their channel gains are different. Examples of wireless uplinks include laptop wireless LAN cards transmitting to a wireless LAN access point, transmissions from ground stations to a satellite, and transmissions from mobile terminals to a base station in cellular systems.

Most communication systems are bi-directional and thus consist of both uplinks and downlinks. The radio transceiver that sends to users over a downlink channel and receives from these users over an uplink channel is often referred to as an access point or base station.

---

[3] Note that multiple access techniques must be applied to both multiple access channels – that is, to uplinks as well as to downlinks.

It is generally not possible for radios to receive and transmit on the same frequency band because of the interference that results. Thus, bi-directional systems must separate the uplink and downlink channels into orthogonal signaling dimensions, typically using time or frequency dimensions. This separation is called *duplexing*. In particular, time-division duplexing (TDD) assigns orthogonal timeslots to a given user for receiving from an access point and transmitting to the access point, and frequency-division duplexing (FDD) assigns separate frequency bands for transmitting to and receiving from the access point. An advantage of TDD is that bi-directional channels are typically symmetrical in their channel gains, so channel measurements made in one direction can be used to estimate the channel in the other direction. This is not necessarily the case for FDD in frequency-selective fading: if the frequencies assigned to each direction are separated by more than the coherence bandwidth associated with the channel multipath, then these channels will exhibit independent fading.

14.2 Multiple Access

Efficient allocation of signaling dimensions between users is a key design aspect of both uplink and downlink channels, since bandwidth is usually scarce and/or very expensive. When dedicated channels are allocated to users it is often called multiple access.[4] Applications with continuous transmission and delay constraints, such as voice or video, typically require dedicated channels for good performance to ensure their transmission is not interrupted. Dedicated channels are obtained from the system signal space using a channelization method such as time division, frequency division, code division, or some combination of these techniques. Allocation of signaling dimensions for users with bursty transmissions generally use some form of random channel allocation that does not guarantee channel access. Bandwidth sharing using random channel allocation is called random multiple access or simply *random access,* which will be described in Section 14.3. In general, the choice of whether to use multiple access or random access – and which specific multiple or random access technique to apply – will depend on the system applications, the traffic characteristics of the users in the system, the performance requirements, and the characteristics of the channel and other interfering systems operating in the same bandwidth.

Multiple access techniques divide up the total signaling dimensions into channels and then assign these channels to different users. The most common methods to divide up the signal space are along the time, frequency, and/or code axes. The different user channels are then created by an orthogonal or nonorthogonal division along these axes: time-division multiple access (TDMA) and frequency-division multiple access (FDMA) are orthogonal channelization methods, whereas code-division multiple access (CDMA) can be either orthogonal or nonorthogonal, depending on the code design. Directional antennas, often obtained through antenna array processing, add an additional angular dimension that can also be used to channelize the signal space; this technique is called space-division multiple access (SDMA). The performance of different multiple access methods depends on whether they are applied to

---

[4] An uplink channel is also referred to as a multiple access channel, but multiple access techniques are needed for both uplinks and downlinks.

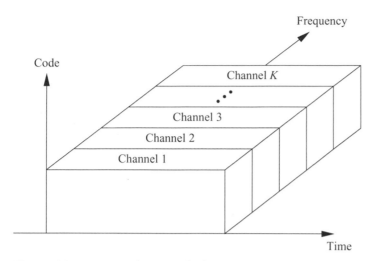

Figure 14.2: Frequency-division multiple access.

an uplink or downlink as well as on their specific characteristics. The TDMA, FDMA, and orthogonal CDMA techniques are all equivalent in the sense that they orthogonally divide up the signaling dimensions and hence create the same number of orthogonal channels. In particular, given a signal space of dimension $2BT$, it follows that N orthogonal channels of dimension $2BT/N$ can be created regardless of the channelization method. As a result, all multiple access techniques that divide the signal space orthogonally have the same capacity when applied to additive white Gaussian noise (AWGN) channels, as will be discussed in Sections 14.5 and 14.6. However, channel impairments such as flat and frequency-selective fading affect these techniques in different ways, which leads to different channel capacities and different performance in typical wireless channels.

14.2.1 Frequency-Division Multiple Access (FDMA)

In FDMA the system signaling dimensions are divided along the frequency axis into nonoverlapping channels, and each user is assigned a different frequency channel; see Figure 14.2. The channels often have guard bands between them to compensate for imperfect filters, adjacent channel interference, and spectral spreading due to Doppler. If the channels are sufficiently narrowband, then the individual channels will not experience frequency-selective fading even if the total system bandwidth is large. Transmission is continuous over time, which can complicate overhead functions such as channel estimation because these functions must be performed simultaneously and in the same bandwidth as data transmission; FDMA also requires frequency-agile radios that can tune to the different carriers associated with the different channels. It is difficult to assign multiple channels to the same user under FDMA, since this requires the radios to simultaneously demodulate signals received over multiple frequency channels. Even so, FDMA is the most common multiple access option for analog communication systems, where transmission is continuous, and serves as the basis for the AMPS and TACS analog cellular phone standards [2, Chap. 11.1]. Multiple access in OFDM systems, called OFDMA, implements FDMA by assigning different subcarriers to different users.

EXAMPLE 14.1: First-generation analog systems were allocated a total bandwidth of $B = 25$ MHz for uplink channels and another $B = 25$ MHz for downlink channels. This bandwidth allocation was split between two operators in every region, so each operator had 12.5 MHz for both their uplink and downlink channels. Each user was assigned $B_c = 30$ kHz of spectrum for its analog voice signal, corresponding to 24 kHz for the FM modulated signal with 3-kHz guard bands on each side. The total uplink and downlink bandwidths also required guard bands of $B_g = 10$ kHz on each side in order to mitigate interference to and from adjacent systems. Find the total number of analog voice users that could be supported in the total 25 MHz of bandwidth allocated to the uplink and the downlink. Also consider a more efficient digital system, with high-level modulation (so that only 10-kHz channels are required for a digital voice signal) and with tighter filtering (so that only 5-kHz guard bands are required on the band edges). How many users can be supported in the same 25 MHz of spectrum for this more efficient digital system?

Solution: For either the uplink or the downlink, guard bands on each side of the users' channels means that a total bandwidth of $NB_c + 2B_g$ is required for N users. Thus, the total number of users that can be supported in the total uplink or downlink bandwidth $B = 25$ kHz is

$$N = \frac{B - 2B_g}{B_c} = \frac{25 \cdot 10^6 - 2 \cdot 10 \cdot 10^3}{30 \cdot 10^3} = 832,$$

or 416 users per operator. Indeed, first-generation analog systems could support 832 users in each cell. The digital system has

$$N = \frac{B - 2B_g}{B_c} = \frac{25 \cdot 10^6 - 2 \cdot 5 \cdot 10^3}{10 \cdot 10^3} = 2499$$

users that can be supported in each cell, a threefold increase over the analog system. The increase is primarily due to the bandwidth savings of the high-level digital modulation, which can accommodate a voice signal in one third the bandwidth of the analog voice signal.

14.2.2 Time-Division Multiple Access (TDMA)

In TDMA, the system dimensions are divided along the time axis into nonoverlapping channels, and each user is assigned a different cyclically repeating timeslot; see Figure 14.3. These TDMA channels occupy the entire system bandwidth, which is typically wideband, so some form of ISI mitigation is required. The cyclically repeating timeslots imply that transmission is not continuous for any user. Therefore, digital transmission techniques that allow for buffering are required. The fact that transmission is not continuous simplifies overhead functions such as channel estimation, since these functions can be performed during the timeslots occupied by other users. Time-division multiple access also has the advantage that it is simple to assign multiple channels to a single user by simply assigning him multiple timeslots.

A major difficulty of TDMA, at least for uplink channels, is the requirement for synchronization among the different users. Specifically, in a downlink channel all signals originate from the same transmitter and pass through the same channel to any given receiver. Thus, for flat-fading channels, if users transmit on orthogonal timeslots then the received signal will maintain this orthogonality. However, in the uplink channel the users transmit over

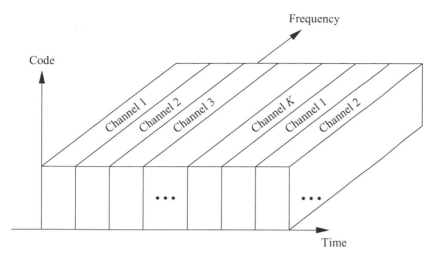

Figure 14.3: Time-division multiple access.

different channels with different respective delays. To maintain orthogonal timeslots in the received signals, the different uplink transmitters must synchronize such that, *after* transmission through their respective channels, the received signals are orthogonal in time. This synchronization is typically coordinated by the base station or access point, and it can entail significant overhead. Multipath can also destroy time-division orthogonality in both uplinks and downlinks if the multipath delays are a significant fraction of a timeslot. Hence TDMA channels often have guard bands between them to compensate for synchronization errors and multipath. Another difficulty of TDMA is that, with cyclically repeating timeslots, the channel characteristics change on each cycle. Thus, receiver functions that require channel estimates, like equalization, must re-estimate the channel on each cycle. When transmission is continuous the channel can be tracked, which is more efficient. Time-division multiple access is used in the GSM, PDC, and IS-136 digital cellular phone standards [2, Chap. 11].

EXAMPLE 14.2: The original GSM design uses 25 MHz of bandwidth for the uplink and for the downlink, the same as AMPs. This bandwidth is divided into 125 TDMA channels of 200 kHz each. Each TDMA channel consists of eight user timeslots; the eight timeslots along with a preamble and trailing bits form a *frame,* which is cyclically repeated in time. Find the total number of users that can be supported in the GSM system and the channel bandwidth of each user. If the root mean square delay spread of the channel is 10 μs, will intersymbol interference mitigation be needed in this system?

Solution: Since there are eight users per channel and 125 channels, the total number of users that can be supported in this system is $125 \cdot 8 = 1000$ users. The bandwidth of each TDMA channel is $25 \cdot 10^6/125 = 200$ kHz. A delay spread of 10 μs corresponds to a channel coherence bandwidth of $B_c \approx 100$ kHz, which is less than the TDMA channel bandwidth of 200 kHz. Thus, ISI mitigation is needed. The GSM specification includes an equalizer to compensate for ISI, but the type of equalizer is at the discretion of the designer.

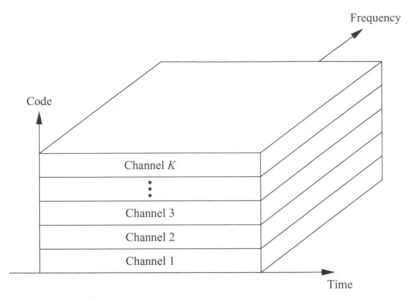

Figure 14.4: Code-division multiple access.

14.2.3 Code-Division Multiple Access (CDMA)

In CDMA the information signals of different users are modulated by orthogonal or nonorthogonal spreading codes. The resulting spread signals simultaneously occupy the same time and bandwidth, as shown in Figure 14.4. The receiver uses the spreading code structure to separate out the different users. The most common form of CDMA is multiuser spread spectrum with either direct sequence or frequency hopping, which are described and analyzed in Sections 13.4 and 13.5.

Downlinks typically use orthogonal spreading codes such as Walsh–Hadamard codes, although the orthogonality can be degraded by multipath. Uplinks generally use nonorthogonal codes owing to the difficulty of user synchronization and the complexity of maintaining code orthogonality in uplinks with multipath [3]. One of the big advantages of nonorthogonal CDMA in uplinks is that little dynamic coordination of users in time or frequency is required, since the users can be separated by the code properties alone. In addition, since TDMA and FDMA carve up the signaling dimensions orthogonally, there is a hard limit on how many orthogonal channels can be obtained. This is also true for CDMA using orthogonal codes, but if nonorthogonal codes are used then there is no hard limit on the number of channels that can be obtained. However, because nonorthogonal codes cause mutual interference between users, the more users that simultaneously share the system bandwidth using nonorthogonal codes, the higher the level of interference, which degrades system performance for all users. A nonorthogonal CDMA scheme also requires power control in the uplink to compensate for the near–far effect. The near–far effect arises in the uplink because the channel gain between a user's transmitter and the receiver is different for different users. Specifically, suppose that one user is very close to his base station or access point while another user is far away. If both users transmit at the same power level, then the interference from the close user will swamp the signal from the far user. Thus, power control is used such that the received signal power of all users is roughly the same. This form of power control, which essentially inverts any attenuation and/or fading on the channel, causes each interferer to contribute an equal amount of power, thereby eliminating the near–far effect. Code-division multiple access systems with nonorthogonal spreading codes can also use MUD to reduce interference between users. Multiuser detection provides considerable performance improvement even under perfect power control, and it works even better when the power control is jointly optimized with the MUD technique [4]. We will see in Sections 14.5 and 14.6 that a form of CDMA with multiuser detection achieves the Shannon capacity of both the uplink and the downlink, although the capacity-achieving transmission and reception strategies for the two channels are quite different. Finally, it is simple to allocate multiple channels to one user with CDMA by assigning that user multiple codes. Code-division multiple access is used for multiple access in the IS-95 digital cellular standards, with orthgonal spreading codes on the downlink and a combination of orthogonal and nonorthogonal codes on the uplink [2, Chap. 11.4]. It is also used in the W-CDMA and CDMA2000 digital cellular standards [5, Chap. 10.5].

EXAMPLE 14.3: The signal-to-interference power ratio (SIR) for a CDMA uplink with nonorthogonal codes under the standard Gaussian assumption was given in (13.47) as

$$\text{SIR} = \frac{3G}{K-1},$$

where K is the number of users and $G \approx 128$ is the ratio of spread bandwidth to signal bandwidth. In IS-95 the uplink channel is assigned 1.25 MHz of spectrum. Thus, the bandwidth of the information signal prior to spreading is $B_s \approx 1.25 \cdot 10^6/128 = 9.765$ kHz. Neglecting noise, if the required SIR on a channel is 10 dB, how many users can the CDMA uplink support? How many could be supported within the same total bandwidth for an FDMA system?

Solution: To determine how many users can be supported, we invert the SIR expression to get

$$K \leq \frac{3G}{\text{SIR}} + 1 = \frac{384}{10} + 1 = 39.4;$$

since K must be an integer, the system can support 39 users. In FDMA we have

$$K = \frac{1.25 \cdot 10^6}{9.765 \cdot 10^3} = 128,$$

so the total system bandwidth of 1.25 MHz can support 128 channels of 9.765 kHz. This calculation implies that FDMA is three times more efficient than nonorthogonal CDMA under the standard Gaussian assumption for code cross-correlation (FDMA is even more efficient under different assumptions about the code cross-correlation). But, in fact, IS-95 typically supports 64 users on the uplink and downlink by allowing variable voice compression rates depending on interference and channel quality and by taking advantage of the fact that interference is not always present (called a voice-activity factor). Although this makes CDMA less efficient than FDMA for a single cell, cellular systems incorporate channel reuse, which can be done more efficiently in CDMA than in FDMA (see Section 15.2).

14.2.4 Space-Division Multiple Access (SDMA)

Space-division multiple access uses direction (angle) as another dimension in signal space, which can be channelized and assigned to different users. This is generally done with directional antennas, as shown in Figure 14.5. Orthogonal channels can be assigned only if the

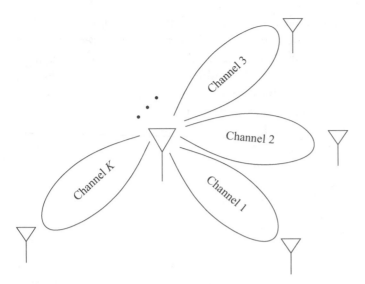

Figure 14.5: Space-division multiple access.

angular separation between users exceeds the angular resolution of the directional antenna. If directionality is obtained by using an antenna array, precise angular resolution requires a large array, which may be impractical for the base station or access point and is certainly unfeasible in small user terminals. In practice, SDMA is often implemented using sectorized antenna arrays, discussed in Section 10.8. In these arrays, the 360° angular range is divided into N sectors. There is high directional gain in each sector and little interference between sectors. Either TDMA or FDMA is used to channelize users within a sector. For mobile users, SDMA must adapt as user angles change; or, if directionality is achieved via sectorized antennas, then a user must be handed off to a new sector when it moves out of its original sector.

14.2.5 Hybrid Techniques

Many systems use a combination of different multiple access schemes to allocate signaling dimensions. OFDMA can be combined with tone hopping to improve frequency diversity [6]. Direct-sequence spread spectrum (DSSS) can be combined with FDMA to break the system bandwidth into subbands; in this hybrid method, different users are assigned to different subbands with their signals spread across the subband bandwidth. Within a subband, the processing gain is smaller than it would be over the entire system bandwidth, so interference and ISI attenuation are reduced. In exchange for this performance loss, this hybrid technique does not require contiguous spectrum between subbands, and it also allows more flexibility in spreading user signals over different-size subbands depending on their requirements. Another hybrid method combines DS-CDMA with FH-CDMA so that the carrier frequency of the spread signal is hopped over the available bandwidth. This reduces the near–far effect because the interfering users change on each hop. Alternatively, TDMA and FH can be combined so that a channel with deep fading or interference is used only on periodic hops, enabling the mitigation of fading and interference effects via error correction coding. This idea is used in the GSM standard, which combines FH with its TDMA scheme in order to reduce the effect of strong interferers in other cells.

There has been much discussion, debate, and analysis about the relative performance of different multiple access techniques for current and future wireless systems (see e.g. [6; 7; 8; 9; 10; 11; 12]). Although analysis and general conclusions can be made for simple system and channel models, it is difficult to come up with a definitive answer as to the best technique for a complex multiuser system under a range of typical operating conditions. Moreover, simplifying assumptions must be made in order to perform a comparative analysis or simulation study, and these assumptions can bias the results in favor of one particular scheme. As with most engineering design questions, the choice of which multiple access technique to use will depend on the system requirements and characteristics along with cost and complexity constraints.

14.3 Random Access

Multiple access techniques are primarily for continuous applications like voice and video, where a dedicated channel facilitates good performance. However, most data applications do not require continuous transmission: data are generated at random time instances, so dedicated channel assignment can be extremely inefficient. Moreover, most systems have many more total users (active plus idle users) than can be accommodated simultaneously, so at any given time channels can only be allocated to users that need them. Random access strategies are used in such systems to efficiently assign channels to the active users.

All random access techniques are based on the premise of packetized data or *packet radio*. In packet radio, user data is collected into packets of N bits, which may include error detection/correction and control bits. Once a packet is formed it is transmitted over the channel. Assuming a fixed channel data rate of R bps, the transmission time of a packet is $\tau = N/R$. The transmission rate R is assumed to require the entire signal bandwidth, and all users transmit their packets over this bandwidth with no additional signaling to separate simultaneously transmitted packets. Thus, if packets from different users overlap in time a *collision* occurs, in which case both packets may be decoded unsuccessfully. Packets may also be decoded in error as a result of noise or other channel impairments. The probability of a packet decoding error is called the *packet error rate*. Analysis of random access techniques typically assumes that, collectively, the users accessing the channel generate packets according to a Poisson process at a rate of λ packets per unit time; that is, λ is the average number of packets that arrive in any time interval $[0, t]$ divided by t. Equivalently, λN is the average number of bits generated in any time inteval $[0, t]$ divided by t. For a Poisson process, the probability that the number of packet arrivals in a time period $[0, t]$, denoted as $X(t)$, is equal to some integer k is given by

$$p(X(t) = k) = \frac{(\lambda t)^k}{k!} e^{-\lambda t}. \tag{14.1}$$

Poisson processes are memoryless, so that the number of packet arrivals during any given time period does not affect the distribution of packet arrivals in any other time period. Note that the Poisson model is not necessarily a good model for all types of user traffic – especially Internet data, where bursty data causes correlated packet arrivals [13].

The traffic *load* on the channel given Poisson packet arrivals at rate λ and packet transmission duration τ is defined as $L = \lambda \tau$. If the channel data rate is R_p packets per second then $\tau = 1/R_p = N/R$ for R the channel data rate in bits per second. Note that L is unit-less: it is the ratio of the packet arrival rate divided by the packet rate that can be transmitted over the channel at the channel's data rate R. We assume that colliding packets are always decoded in error. Thus, if $L > 1$ then on average more packets (or bits) arrive in the system over a given time period than can be transmitted in that period, so systems with $L > 1$ are unstable. If the transmitter is informed by the receiver about packets received in error and retransmits these packets, then the packet arrival rate λ and corresponding load $L = \lambda\tau$ is computed based on arrivals of both new packets and packets that require retransmission. In this case L is referred to as the *total offered load*.

Performance of random access techniques is typically characterized by the *throughput* T of the system. The throughput, which is unitless, is defined as the ratio of the average rate of packets successfully transmitted divided by the channel packet rate R_p. The through-put thus equals the offered load multiplied by the probability of successful packet reception, $T = Lp$(successful packet reception), where this probability is a function of the random ac-cess protocol in use as well as the channel characteristics, which can cause packet errors in the absence of collisions. Thus $T \leq L$. Also, since a system with $L > 1$ is unstable, stable systems have $T \leq L \leq 1$. Observe that the throughput is independent of the channel data rate R, since the load and corresponding throughput are normalized with respect to this rate. This allows analysis of random access protocols to be generic to any underlying link design or channel capacity. For a packet radio with a link data rate of R bps, the *effective data rate* of the system is RT, since T is the fraction of packets or bits successfully transmitted at rate R. The goal of a random access method is to make T as large as possible in order to fully utilize the underlying link rates. Note that in some circumstances overlapping packets do not cause a collision. In particular, short periods of overlap between colliding packets, dif-ferent channel gains on the received packets, and/or error correction coding can allow one or more packets to be successfully received even with a collision. This is called the *capture effect* [14, Chap. 4.3].

Random access techniques were pioneered by Abramson with the ALOHA protocol [15], where data is packetized and users send packets whenever they have data to send. ALOHA is extremely inefficient owing to collisions between users, which leads to very low through-put. The throughput can be doubled by slotting time and synchronizing the users, but even then collisions lead to relatively low throughput values. Modifications of ALOHA protocols to avoid collisions and thereby increase throughput include carrier sensing, collision detec-tion, and collision avoidance. Long bursts of packets can be scheduled to avoid collisions, but this typically takes additional overhead. In this section we will describe these various techniques for random access, their performance, and their design trade-offs.

14.3.1 Pure ALOHA

In pure or unslotted ALOHA, users transmit data packets as soon as they are formed. If we neglect the capture effect, then packets that overlap in time are assumed to be received in error and must be retransmitted. If we also assume packets that do not collide are success-fully received (i.e., if there is no channel distortion or noise), then the throughput equals the offered load multiplied by the probability of no collisions: $T = Lp$(no collisions). Suppose a given user transmits a packet of duration τ during time $[0, \tau]$. Then, if any other user gener-ates a packet during time $[-\tau, \tau]$, that packet (of duration τ) will overlap with the transmitted

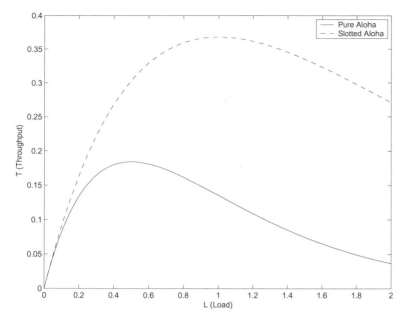

Figure 14.6: Throughput of pure and slotted ALOHA.

packet, causing a collision. The probability of no collisions thus equals the probability that no packets are generated during the time $[-\tau, \tau]$. This probability is given by (14.1) with $t = 2\tau$:

$$p(X(t) = 0) = e^{-2\lambda\tau} = e^{-2L}, \tag{14.2}$$

with corresponding throughput

$$T = Le^{-2L}. \tag{14.3}$$

This throughput is plotted in Figure 14.6, where we see that throughput increases with offered load up to a maximum throughput of approximately .18 for $L = .5$, after which point it decreases. In other words, the data rate is only 18% of what it would be with a single user transmitting continuously on the system. The reason for this maximum is that, for small values of L, there are many idle periods when no user is transmitting, so throughput is small. As L increases, the channel is utilized more but collisions also start to occur. At $L = .5$ there is the optimal balance between users generating enough packets to utilize the channel with reasonable efficiency and these packet generations colliding infrequently. Beyond $L = .5$ the collisions become more frequent, which degrades throughput below its maximum; as L grows very large, most packets experience collisions and throughput approaches zero.

Part of the reason for the inefficiency of pure ALOHA is the fact that users can start their packet transmissions at any time, and any partial overlap of two or more packets destroys the successful reception of all packets. By synchronizing users such that all packet transmissions are aligned in time, the partial overlap of packet transmissions can be avoided. That is the basic premise behind slotted ALOHA.

14.3.2 Slotted ALOHA

In slotted ALOHA, time is assumed to be slotted in timeslots of duration τ, and users can only start their packet transmissions at the beginning of the next timeslot after the packet has formed. Thus, there is no partial overlap of transmitted packets, which increases throughput. Specifically, a packet transmitted over the time period $[0, \tau]$ is successfully received if no other packets are transmitted during this period. This probability is obtained from (14.1) with $t = \tau$: $p(X(t) = 0) = e^{-L}$, with corresponding throughput

$$T = Le^{-L}. \tag{14.4}$$

This throughput is also plotted in Figure 14.6, where we see that throughput increases with offered load up to a maximum of approximately $T = .37$ for $L = 1$, after which point it decreases. Thus, slotted ALOHA has double the maximum throughput as pure ALOHA, and it achieves this maximum at a higher offered load. Although this represents a marked improvement over pure ALOHA, the effective data rate is still less than 40% of the raw transmission rate. This is extremely wasteful of the limited wireless bandwidth, so more sophisticated techniques are needed to increase efficiency.

Note that slotted ALOHA requires synchronization of all nodes in the network, which can entail significant overhead. Even in a slotted system, collisions occur whenever two or more users attempt transmission in the same slot. Error control coding can result in correct detection of a packet even after a collision, but if the error correction is insufficient then the packet must be retransmitted. A study on design optimization between error correction and retransmission is described in [16].

EXAMPLE 14.4: Consider a slotted ALOHA system with a transmission rate of $R = 10$ Mbps. Suppose packets consist of 1000 bits. For what packet arrival rate λ will the system achieve maximum throughput, and what is the effective data rate associated with this throughput?

Solution: The throughput T is maximized for $L = \lambda\tau = 1$, where λ is the packet arrival rate and τ is the packet duration. With a 10-Mbps transmission rate and 1000 bits per packet, $\tau = 1000/10^6 = .1$ ms. Thus, $\lambda = 1/.0001 = 10^4$ packets per second maximizes throughput. The throughput for $L = 1$ is $T = .37$, so the effective data rate is $TR = 3.7$ Mbps. Thus, the data rate is reduced by roughly a factor of 3 (as compared to continuous data transmission) owing to the random nature of the packet arrivals and their corresponding collisions.

14.3.3 Carrier-Sense Multiple Access (CSMA)

Collisions can be reduced by carrier-sense multiple access, where users sense the channel and delay transmission if they detect that another user is currently transmitting. To be effective, detection time and propagation delays in the system must be small [17, Chap. 4.19]. After sensing a busy channel, a user typically waits a random time period before transmitting. This *random backoff* precludes multiple users simultaneously transmitting as soon as the channel is free. Carrier-sense multiple access works only when all users can detect each other's transmissions and the propagation delays are small. Wired LANs have these characteristics, so CSMA is part of the Ethernet protocol. However, the nature of the wireless channel may prevent a given user from detecting the signals transmitted by all other users. This gives rise to the *hidden terminal problem* (illustrated in Figure 14.7), whereby each node

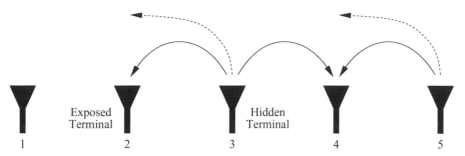

Figure 14.7: Hidden and exposed terminals.

can hear its immediate neighbor but no other nodes in the network. In the figure, node 3 and node 5 each wish to transmit to node 4. Suppose node 5 starts its transmission. Since node 3 is too far away to detect this transmission, it assumes that the channel is idle and begins its transmission, thereby causing a collision with node 5's transmission. Node 3 is said to be "hidden" from node 5 because it cannot detect node 5's transmission.

ALOHA with CSMA also creates inefficiencies in channel utilization from the *exposed terminal problem,* also illustrated in Figure 14.7. Suppose the exposed terminal in this figure – node 2 – wishes to send a packet to node 1 at the same time that node 3 is sending to node 4. When node 2 senses the channel it will detect node 3's transmission and assume the channel is busy, even though node 3 does not interfere with the reception of node 2's transmission by node 1. Thus node 2 will not transmit to node 1 even though no collision would have occurred. Exposed terminals occur only in multihop networks, so we will defer their discussion until Chapter 16.

The collisions introduced by hidden terminals are often avoided in wireless networks by a four-way handshake prior to transmission [18; 19]. This *collision avoidance* is accomplished as follows. A node that wants to send a data packet will first wait for the channel to become available and then transmit a short RTS (request to send) packet. The potential receiver, assuming it perceives an available channel, will immediately respond with a CTS (clear to send) packet that authorizes the initiating node to transmit and also informs neighboring hidden nodes (i.e., nodes that are outside the communication range of the transmitter but within the communication range of the receiver) that they must remain silent for the duration of the transmission. Nodes that overhear the RTS or CTS packet will refrain from transmitting over the expected packet duration. A node can send an RTS packet only if it perceives an idle channel and has not been silenced by another control packet. A node will transmit a CTS packet only if it has not been silenced by another control packet. The RTS/CTS handshake is typically coupled with random backoff to avoid all nodes transmitting as soon as the channel becomes available. In some incarnations [19; 20], including the 802.11 WLAN standard [5, Chap. 14.3], the receiver sends an ACK (acknowledgement) packet back to the transmitter to verify when it has correctly received the packet, after which the channel again becomes available.

Another technique to avoid hidden terminals is busy-tone transmission. In this strategy, users first check to see whether the transmit channel is busy by listening for a "busy tone" on a separate control channel [1, Chap. 4.6]. There is typically not an actual busy tone; instead, a bit is set in a predetermined field on the control channel. This scheme works well in preventing collisions when a centralized controller can be "heard" by users throughout the network. In a flat network without centralized control, more complicated measures are used to ensure that any potential interferer on the first channel can hear the busy tone on the second [21; 22]. Hybrid techniques using handshakes, busy-tone transmission, and power control can also be used [22]. Collisions can also be reduced by combining DSSS with ALOHA. In this scheme, each user modulates his signal with the same spreading code, but if user transmissions are separated by more than a chip time, the interference due to a collision is reduced by the code autocorrelation [23].

14.3.4 Scheduling

Random access protocols work well with bursty traffic, where there are many more users than available channels yet these users rarely transmit. If users have long strings of packets or continuous-stream data, then random access works poorly because most transmissions result in collisions. In this scenario performance can be improved by assigning channels to users in a more systematic fashion through transmission scheduling. In scheduled access, the available bandwidth is channelized into multiple time-, frequency-, or code-division channels. Each node schedules its transmission on different channels in such a way as to avoid conflicts with neighboring nodes while making the most efficient use of the available signaling dimensions.

Even with a scheduling access protocol, some form of ALOHA will still be needed because a predefined mechanism for scheduling is, by definition of random access, unavailable at startup. ALOHA provides a means for initial contact and the establishment of some form of scheduled access for the transmission of relatively large amounts of data. A systematic approach to this initialization that also combines the benefits of random access for bursty data with scheduling for continuous data is *packet-reservation multiple access* (PRMA) [24]. This technique assumes a slotted system with both continuous and bursty users (e.g., voice and data users). Multiple users vie for a given time slot under a random access strategy. A successful transmission by one user in a given timeslot reserves that timeslot for all subsequent transmissions by the same user. If the user has a continuous or long transmission then, after successfully capturing the channel, he has a dedicated channel for the remainder of his transmission (assuming subsequent transmissions are not corrupted by the channel; such corruption causes users to lose their slots and they must then recontend for an unreserved slot, which can entail significant delay and packet dropping [25]). When this user has no more packets to transmit, the slot is returned to the pool of available slots that users attempt to capture via random access. Thus, data users with short transmissions benefit from the random access protocol assigned to unused slots, and users with continuous transmissions get scheduled periodic transmissions after successfully capturing an initial slot. A similar technique using a combined reservation and ALOHA policy is described in [26].

14.4 Power Control

Power control is applied to systems where users interfere with each other. The goal of power control is to adjust the transmit powers of all users such that the SINR of each user meets a given threshold required for acceptable performance. This threshold may be different for different users, depending on their required performance. This problem is straightforward for the downlink, where users and interferers have the same channel gains, but is more complicated in the uplink, where the channel gains may be different. Seminal work on power control for cellular systems and ad hoc networks was done in [27; 28; 29], and power control for the uplink is a special case for which these results can be applied. In the uplink model, the kth transmitter has a fixed channel power gain g_k to the receiver. The quality of each link is determined by the SINR at the intended receiver. In an uplink with K interfering users, we denote the SINR for the kth user as

$$\gamma_k = \frac{g_k P_k}{n + \rho \sum_{j \neq k} g_j P_j}, \quad k = 1, \ldots, K, \tag{14.5}$$

where P_k is the power of the kth transmitter, n is the receiver noise power, and ρ is interference reduction due to signal processing. For example, in a CDMA uplink the interference power is reduced by the processing gain of the code, so $\rho \approx 1/G$ for G the processing gain, whereas in TDMA $\rho = 1$.

Each link is assumed to have a minimum SINR requirement $\gamma_k^* > 0$. This constraint can be represented in matrix form with componentwise inequalities as

$$(\mathbf{I} - \mathbf{F})\mathbf{P} \geq \mathbf{u} \quad \text{with } \mathbf{P} > 0, \tag{14.6}$$

where $\mathbf{P} = (P_1, P_2, \ldots, P_K)^T$ is the vector of transmitter powers,

$$\mathbf{u} = \left(\frac{n\gamma_1^*}{g_1}, \frac{n\gamma_2^*}{g_2}, \ldots, \frac{n\gamma_K^*}{g_K} \right)^T \tag{14.7}$$

is the vector of noise power scaled by the SINR constraints and channel gains, and $\mathbf{F}$ is a matrix with

$$F_{kj} = \begin{cases} 0 & k = j, \\ \gamma_k^* g_j \rho / g_k & k \neq j, \end{cases} \tag{14.8}$$

for $k, j = 1, 2, \ldots, K$.

The matrix $\mathbf{F}$ has nonnegative elements and is irreducible. Let ρ_F be the Perron–Frobenius eigenvalue of $\mathbf{F}$. This is the maximum modulus eigenvalue of $\mathbf{F}$, and for $\mathbf{F}$ irreducible this eigenvalue is simple, real, and positive. Moreover, from the Perron–Frobenius theorem and standard matrix theory [30], the following statements are equivalent.

1. $\rho_F < 1$.
2. There exists a vector $\mathbf{P} > 0$ (i.e., $P_k > 0$ for all k) such that $(\mathbf{I} - \mathbf{F})\mathbf{P} \geq \mathbf{u}$.
3. $(\mathbf{I} - \mathbf{F})^{-1}$ exists and is positive componentwise.

Furthermore, if any of these conditions holds then we also have that $\mathbf{P}^* = (\mathbf{I} - \mathbf{F})^{-1}\mathbf{u}$ is the Pareto optimal solution to (14.6). That is: if $\mathbf{P}$ is any other solution to (14.6), then $\mathbf{P} \geq \mathbf{P}^*$ componentwise. Hence, if the SINR requirements for all users can be met simultaneously then the best power allocation is $\mathbf{P}^*$, which will minimize the transmit power of the users.

In [28] the authors also show that the following iterative power control algorithm converges to $\mathbf{P}^*$ when $\rho_F < 1$ and diverges to infinity otherwise. This iterative Foschini–Miljanic algorithm is given by

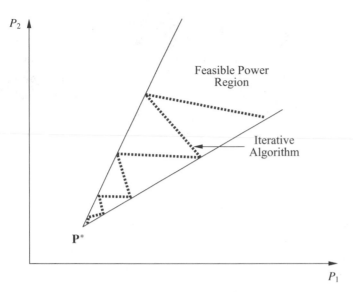

Figure 14.8: Iterative Foschini–Miljanic algorithm.

$$\mathbf{P}(i+1) = \mathbf{F}\mathbf{P}(i) + \mathbf{u} \tag{14.9}$$

for $i = 1, 2, 3, \ldots$. Furthermore, this algorithm can be simplified to a per-user version as follows. Let

$$P_k(i+1) = \frac{\gamma_k^*}{\gamma_k(i)} P_k(i) \tag{14.10}$$

for each link $k \in \{1, 2, \ldots, N\}$. Hence, each transmitter increases power when its SINR is below its target and decreases power when its SINR exceeds its target. Measurements of SINR (or a function of them, such as BER) are typically made at the base station or access points, and a simple "up" or "down" command regarding transmit power can be fed back to each of the transmitters to perform the iterations. It is easy to show that (14.9) and (14.10) are pathwise equivalent and hence the per-user version of the power control algorithm also converges to $\mathbf{P}^*$. The feasible region of power vectors that achieve the SINR targets for a two-user system – along with the iterative algorithm that converges to the minimum power vector in this region – is illustrated in Figure 14.8. We see in this figure that the feasible region consists of all power pairs $\mathbf{P} = (P_1, P_2)$ that achieve a given pair of SINR targets and that the optimal pair $\mathbf{P}^*$ is the minimum power vector in this two-dimensional region.

The Foschini–Miljanic power control algorithm can also be combined with access control [31]. In this combination, access to the system is based on whether or not the new user causes other users to fall below their SINR targets. Specifically, when a new user requests access to the system, the base station or access point determines if a set of transmit powers exists such that he can be admitted without degrading existing users below their desired SINR threshold. If the new user cannot be accommodated in the system without violating the SINR requirements of existing users, then access is denied. If the new user can be accommodated then his power control algorithm and those of existing users are set to a feasible power vector under which all users (new and existing) meet their SINR targets.

A power control strategy for multiple access that takes into account delay constraints is proposed and analyzed in [32]. This strategy optimizes the transmit power relative to channel conditions and delay constraints via dynamic programming. The optimal strategy exhibits three modes: very low transmit power when the channel is poor and the tolerable delay large, higher transmit power when the channel and delay are average, and very high transmit power when the delay constraint is tight. This strategy exhibits significant power savings over constant transmit power while meeting the delay constraints of the traffic.

14.5 Downlink (Broadcast) Channel Capacity

When multiple users share the same channel, the channel capacity can no longer be characterized by a single number. At the extreme, if only one user occupies all signaling dimensions in the channel then the region reduces to the single-user capacity described in Chapter 4. However, since there is an infinite number of ways to divide the channel between many users, the multiuser channel capacity is characterized by a *rate region,* where each point in the region is a vector of achievable rates that can be maintained by all the users simultaneously with arbitrarily small error probability. The union of achievable rate vectors under all multiuser transmission strategies is called the *capacity region* of the multiuser system. The channel capacity is different for uplink channels and downlink channels because of the fundamental differences between these channel models. However, the fact that downlink and uplink channels seem to be mirror images implies that there might be a connection between their capacities. In fact, there is a duality between these channels that allows the capacity region of either channel to be obtained from the capacity region of the other. Note that, in the analysis of channel capacity, the downlink is commonly referred to as the broadcast channel (BC) and the uplink is commonly referred to as the multiple access channel (MAC[5]), and we will use this terminology in our capacity discussions. In this section we describe the capacity region of the BC, Section 14.6 treats the MAC capacity region, and Section 14.7 characterizes the duality between these two channels and how it can be exploited in capacity calculations.

After first describing the AWGN BC model, we will characterize its rate region using superposition code division (CD) with successive interference cancellation, time division (TD), and frequency division (FD). We then obtain the rate regions using DSSS for orthogonal and nonorthogonal codes. The BC and corresponding capacity results under fading are also treated.

We will see that capacity is achieved using superposition CD with interference cancellation. In addition, DSSS with successive interference cancellation has a capacity penalty (relative to superposition coding) that increases with spreading gain. Finally, spread spectrum with orthogonal CD can achieve a subset of the TD and FD capacity regions, but spread spectrum with nonorthogonal coding and no interference cancellation is inferior to all the other spectrum-sharing techniques. The capacity regions in fading depend on what is known about the fading channel at the transmitter and receiver, which is analogous to the case of single-user capacity in fading.

---

[5] MAC is also used as an abbreviation for the "medium access control" layer in networks [1, Chap. 1.2].

14.5.1 Channel Model

We consider a broadcast channel consisting of one transmitter sending different data streams – also called independent information or data – to different receivers. Thus, our model is not applicable to a typical radio or TV broadcast channel, where the same data stream – also called common information or data – is received by all users. However, the capacity results easily extend to include common data, as described in Section 14.5.3. The capacity region of the BC characterizes the rates at which information can be conveyed to the different receivers simultaneously. We mainly focus on capacity regions for the two-user BC, since the general properties and the relative performance of the different spectrum-sharing techniques are the same for any finite number of users [33].

The two-user BC has one transmitter and two distant receivers receiving data at rate R_k, $k = 1, 2$. The channel power gain between the transmitter and kth receiver is g_k, $k = 1, 2$, and each receiver has AWGN of power spectral density (PSD) $N_0/2$. We define the effective noise on the kth channel as $n_k = N_0/g_k$, $k = 1, 2$, and we arbitrarily assume that $n_1 \leq n_2$; that is, we assume the first user has a larger channel gain to its receiver than the second user. Incorporating the channel gains into the noise PSD does not change the SINR for any user, since the signal and interference on each user's channel are attenuated by the same channel gain. Thus, the BC capacity with channel gains $\{g_k\}$ is the same as the BC capacity based on the effective noises $\{n_k\}$ [34]. The fact that the channel gains (or, equivalently, the effective noise of the users) can be ordered makes the channel model a *degraded broadcast channel,* for which a general formula for channel capacity is known [35, Chap. 14.6]. We denote the transmitter's total average power and bandwidth by P and B, respectively.

If the transmitter allocates all the power and bandwidth to one of the users, then clearly the other user will receive a rate of zero. Therefore, the set of simultaneously achievable rates (R_1, R_2) includes the pairs $(C_1, 0)$ and $(0, C_2)$, where

$$C_k = B \log_2\left(1 + \frac{P}{n_k B}\right), \quad k = 1, 2, \tag{14.11}$$

is the single-user capacity in bits per second for an AWGN channel, as given in Section 4.1. These two points bound the BC capacity region. We now consider rate pairs in the interior of the region that are achieved using more equitable methods of dividing the system resources.

14.5.2 Capacity in AWGN

In this section we compute the set of achievable rate vectors of the AWGN BC under TD, FD, and the optimal method of superposition coding, which achieves capacity. In time division, the transmit power P and bandwidth B are allocated to user 1 for a fraction τ of the total transmission time and then to user 2 for the remainder of the transmission. This TD scheme achieves a straight line between the points C_1 and C_2, corresponding to the rate pairs

$$\mathcal{C}_{\text{TD}} = \bigcup_{\{\tau : 0 \leq \tau \leq 1\}} \left(R_1 = \tau B \log_2\left(1 + \frac{P}{n_1 B}\right), R_2 = (1 - \tau) B \log_2\left(1 + \frac{P}{n_2 B}\right)\right). \tag{14.12}$$

This equal-power TD achievable rate region is illustrated in Figures 14.10 and 14.11 (pp. 476–7), where $n_1 B$ and $n_2 B$ differ by 3 dB and 20 dB, respectively. This dB difference, which reflects the difference in the channel gains of the two users, is a crucial parameter in comparing the achievable rates of the different spectrum-sharing techniques, as we discuss in more detail below.

If we also vary the transmit power of each user, subject to an average power constraint P, then we can obtain a larger set of achievable rates. Let P_1 and P_2 denote the power allocated to users 1 and 2, respectively, over their assigned time slots. The average power constraint then becomes $\tau P_1 + (1 - \tau) P_2 = P$. The achievable rate region with TD and variable power (VP) allocation is then

$$C_{\text{TD,VP}} = \bigcup_{\{\tau, P_1, P_2 : 0 \leq \tau \leq 1; \, \tau P_1 + (1-\tau) P_2 = P\}} \left(R_1 = \tau B \log_2 \left(1 + \frac{P_1}{n_1 B} \right), \right.$$
$$\left. R_2 = (1 - \tau) B \log_2 \left(1 + \frac{P_2}{n_2 B} \right) \right). \quad (14.13)$$

In frequency division the transmitter allocates P_k of its total power P and B_k of its total bandwidth B to user k. The power and bandwidth constraints require that $P_1 + P_2 = P$ and $B_1 + B_2 = B$. The set of achievable rates for a fixed frequency division (B_1, B_2) is thus

$$C_{\text{FFD}} = \bigcup_{\{P_1, P_2 : P_1 + P_2 = P\}} \left(R_1 = B_1 \log_2 \left(1 + \frac{P_1}{n_1 B_1} \right), R_2 = B_2 \log_2 \left(1 + \frac{P_2}{n_2 B_2} \right) \right). \quad (14.14)$$

It was shown by Bergmans and Cover [33] that, for n_1 strictly less than n_2 and for any fixed frequency division (B_1, B_2), there exists a range of power allocations $\{P_1, P_2 : P_1 + P_2 = P\}$ whose corresponding rate pairs exceed a segment of the equal-power TD line (14.12). This superiority is also illustrated in Figures 14.10 and 14.11, where the rate regions for fixed FD under two different bandwidth divisions are plotted. The superiority is difficult to distinguish in Figure 14.10, where the users have similar channel gains, but is much more apparent in Figure 14.11, where the users have a 20-dB difference in gain.

The FD achievable rate region is defined as the union of fixed FD rate regions (14.14) over all bandwidth divisions:

$$C_{\text{FD}} = \bigcup_{\{P_1, P_2, B_1, B_2 : P_1 + P_2 = P; \, B_1 + B_2 = B\}} \left(R_1 = B_1 \log_2 \left(1 + \frac{P_1}{n_1 B_1} \right), \right.$$
$$\left. R_2 = B_2 \log_2 \left(1 + \frac{P_2}{n_2 B_2} \right) \right). \quad (14.15)$$

It was shown in [33] that this achievable rate region exceeds the equal-power TD rate region (14.12). This superiority is indicated by the closure of the fixed FD regions in Figures 14.10 and 14.11 – although it is difficult to see in Figure 14.10, where the users have a similar received signal-to-noise ratio. In fact, when $n_1 = n_2$, (14.15) reduces to (14.12) [33]. Thus, optimal power and/or frequency allocation is more beneficial when the users have more disparate channel quality.

Note that the achievable rate region for TD with unequal power allocation given by (14.13) is the same as the FD achievable rate region (14.15). This is seen by letting $B_i = \tau_i B$ and $\pi_i = \tau_i P_i$ in (14.13), where $\tau_1 = \tau$ and $\tau_2 = 1 - \tau$. The power constraint then becomes $\pi_1 + \pi_2 = P$. Making these substitutions in (14.13) yields

$$\mathcal{C}_{\text{TD,VP}} = \bigcup_{\{\pi_1,\pi_2 : \pi_1+\pi_2=P\}} \left(R_1 = B_1 \log_2\left(1 + \frac{\pi_1}{n_1 B_1}\right), R_2 = B_2 \log_2\left(1 + \frac{\pi_2}{n_2 B_2}\right)\right). \quad (14.16)$$

By comparing this with (14.14) we see that, with appropriate choice of P_k and τ_k, any point in the FD achievable rate region can also be achieved through TD with variable power.

Superposition coding with successive interference cancellation is a multiresolution coding technique whereby the user with the higher channel gain can distinguish the fine resolution of the received signal constellation while the user with the lower channel gain can only distinguish the constellation's coarse resolution [33; 35, Chap. 14.6]. An example of a two-level superposition code constellation taken from [36] is 32-QAM with embedded 4-PSK, as shown in Figure 14.9. In this example, the transmitted constellation point is one of the 32-QAM signal points chosen as follows. The data stream intended for the user with the worse SNR (user 2 in our model, since $n_2 > n_1$) provides 2 bits to select one of the 4-PSK superpoints. The data stream intended for the user with the better SNR provides 3 bits to select one of the eight constellation points surrounding the selected superpoint. After transmission through the channel, the user with the better SNR can easily distinguish the quadrant in which the constellation point lies. Thus, the 4-PSK superpoint is effectively subtracted out by this user. However, the user with the worse SNR cannot distinguish between the 32-QAM points around its 4-PSK superpoints. Hence, the 32-QAM modulation superimposed on the 4-PSK modulation appears as noise to this user, and this user can only decode the 4-PSK. These ideas can be easily extended to multiple users using more complex signal constellations. Since superposition coding achieves multiple rates by expanding its signal constellation, it does not require bandwidth expansion.

The two-user capacity region using superposition coding and successive interference cancellation was derived in [33] to be the set of rate pairs

$$\mathcal{C}_{\text{BC}} = \bigcup_{\{P_1,P_2 : P_1+P_2=P\}} \left(R_1 = B \log_2\left(1 + \frac{P_1}{n_1 B}\right), R_2 = B \log_2\left(1 + \frac{P_2}{n_2 B + P_1}\right)\right). \quad (14.17)$$

The intuitive explanation for (14.17) is the same as for the example illustrated in Figure 14.9. Since $n_1 < n_2$, user 1 correctly receives all the data transmitted to user 2. Therefore, user 1 can decode and subtract out user 2's message and then decode its own message. User 2 cannot decode the message intended for user 1, since it has a worse SNR. Thus, user 1's message, with power P_1, contributes an additional noise term to user 2's received message. This message can be treated as an additional AWGN term because the capacity-achieving distributions for the signals associated with each user are Gaussian [33; 35, Chap. 14.1]. This same process is used by the successive interference cancellation method for DSSS described in Section 13.4.4. However, although successive interference cancellation achieves the capacity region (14.17), it is not necessarily the best method to use in practice. The capacity analysis assumes perfect signal decoding and channel estimation, whereas real systems exhibit some decoding and channel estimation error. This error leads to decision-feedback errors in the successive interference cancellation scheme. Thus, multiuser detection methods that do not suffer from this type of error may work better in practice than successive cancellation does.

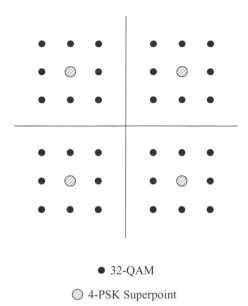

● 32-QAM

⊘ 4-PSK Superpoint

Figure 14.9: 32-QAM with embedded 4-PSK.

The rate region defined by (14.17) was shown in [37] to exceed the regions achievable through either TD or FD when $n_1 < n_2$. Moreover, it was also shown in [37] that this is the maximum achievable set of rate pairs for any type of coding and spectrum sharing and thus (14.17) defines the BC capacity region; hence the notation $\mathcal{C}_{\mathrm{BC}}$. However, if all users have the same SNR, then this capacity region collapses to the equal-power TD line (14.12). Thus, when $n_1 = n_2$, all the spectrum-sharing methods have the same rate region.

The ideas of superposition coding are easily extended to a K-user system for $K > 2$. Assume a BC with K users, each with channel gain g_k. We first order the users relative to their effective noise $n_k = N_0/g_k$. Based on this effective noise ordering, the superposition coding will now have K levels, where the coarsest level can be detected by the user with the largest effective noise (worst SNR), the next level can be detected by the user with the next largest effective noise, and so forth. Each user can remove the effects of the constellation points associated with other users who have noisier channels, but the constellation points transmitted to users with better channels appear as additional noise. Assuming a total power constraint P, the multiuser extension to the two-user region (14.17) is given by

$$\mathcal{C}_{\mathrm{BC}} = \bigcup_{\{P_k : \sum_{k=1}^{K} P_k = P\}} \left\{ (R_1, \ldots, R_K) : R_k = B \log_2 \left(1 + \frac{P_k}{n_k B + \sum_{j=1}^{K} P_j \mathbf{1}[n_k > n_j]} \right) \right\},$$

(14.18)

where $\mathbf{1}[\cdot]$ denotes the indicator function.

We define the *sum-rate* (SR) *capacity* of a BC as the maximum sum of rates taken over all rate vectors in the capacity region:

$$\mathcal{C}_{\mathrm{BCSR}} = \max_{(R_1, \ldots, R_K) \in \mathcal{C}_{\mathrm{BC}}} \sum_{k=1}^{K} R_k.$$

(14.19)

Sum-rate capacity is a single number that defines the maximum throughput of the system (regardless of fairness in the rate allocation between the users). It is therefore much easier to characterize than the K-dimensional capacity region and often leads to important insights. In particular, it can be shown from (14.18) that sum-rate capacity is achieved on the AWGN BC by assigning all power P to the user with the highest channel gain or, equivalently, the lowest effective noise. Defining $n_{\min} \triangleq \min_k n_k$ and $g_{\max} \triangleq \max_k g_k$, this implies that the sum-rate capacity $\mathcal{C}_{\mathrm{BCSR}}$ for the K-user AWGN BC is given by

$$C_{\text{BCSR}} = B \log_2\left(1 + \frac{P}{n_{\min}B}\right) = B \log_2\left(1 + \frac{g_{\max}P}{N_0 B}\right). \qquad (14.20)$$

The sum-rate point is therefore one of the boundary points (14.11) of the capacity region, which is the same for superposition coding, TD, and FD because all resources are assigned to a single user.

EXAMPLE 14.5: Consider an AWGN BC with total transmit power $P = 10$ mW, $n_1 = 10^{-9}$ W/Hz, $n_2 = 10^{-8}$ W/Hz, and $B = 100$ kHz. Suppose user 1 requires a data rate of 300 kbps. Find the rate that can be allocated to user 2 under equal-power TD, equal-bandwidth FD, and superposition coding.

Solution: In equal-power TD, user 1 has a rate of $R_1 = \tau B \log_2(1 + P/n_1 B) = (6.658 \cdot 10^5)\tau$ bps. Setting R_1 to the desired value $R_1 = (6.658 \cdot 10^5)\tau = 3 \cdot 10^5$ bps and solving for τ yields $\tau = 3 \cdot 10^5/6.644 \cdot 10^5 = .451$. Then user 2 gets a rate of $R_2 = (1 - \tau)B \log_2(1 + P/n_2 B) = 1.89 \cdot 10^5$ bps. In equal-bandwidth frequency division we require $R_1 = .5B \log_2(1 + P_1/.5n_1 B) = 3 \cdot 10^5$ bps. Solving for $P_1 = .5n_1 B(2^{R_1/(.5B)} - 1)$ yields $P_1 = 3.15$ mW. Setting $P_2 = P - P_1 = 6.85$ mW, we get $R_2 = .5B \log_2(1 + P_2/.5n_2 B) = 1.94 \cdot 10^5$ bps. Finally, with superposition coding we have $R_1 = B \log_2(1 + P_1/n_1 B) = 3 \cdot 10^5$. Solving for $P_1 = n_1 B(2^{R_1/B} - 1)$ yields $P_1 = .7$ mW. Then

$$R_2 = B \log\left(1 + \frac{P - P_1}{n_2 B + P_1}\right) = 2.69 \cdot 10^5 \text{ bps.}$$

Superposition coding is clearly superior to both TD and FD, as expected, although the performance of these techniques would be closer to that of superposition coding if we optimized the power allocation for TD or the bandwidth allocation for FD.

EXAMPLE 14.6: Find the sum-rate capacity for the system in the prior example.

Solution: We have $P = 10$ mW, $n_1 = 10^{-9}$ W/Hz, $n_2 = 10^{-8}$ W/Hz, and $B = 100$ kHz. The minimum noise is associated with user 1, $n_{\min} = 10^{-9}$. Thus, $C_{\text{BCSR}} = B \log_2(1 + P/n_{\min}B) = 6.644 \cdot 10^5$ bps. This sum-rate is achievable with TD, FD, or superposition coding, which are all equivalent for this sum-rate capacity because all resources are allocated to the first user.

Code division for multiple users can also be implemented using DSSS, as discussed in Section 13.4. In such systems the modulated data signal for each user is modulated by a unique spreading code, which increases the transmit signal bandwidth by approximately G, the processing gain of the spreading code. For orthogonal spreading codes, the cross-correlation between the respective codes is zero, and these codes require a spreading gain of N to produce N orthogonal codes. For a total bandwidth constraint B, the information bandwidth of each user's signal with these spreading codes is thus limited to B/N. The two-user achievable rate region with these orthogonal codes (OC) is then

$$C_{\text{DS,OC}} = \bigcup_{\{P_1, P_2 : P_1 + P_2 = P\}} \left(R_1 = \frac{B}{2} \log_2\left(1 + \frac{P_1}{n_1 B/2}\right), R_2 = \frac{B}{2} \log_2\left(1 + \frac{P_2}{n_2 B/2}\right)\right).$$

$$(14.21)$$

Comparing (14.21) with (14.14), we see that DSSS with orthogonal coding has the same capacity region as fixed FD with the bandwidth equally divided ($B_1 = B_2 = B/2$). From (14.16), TD with unequal power allocation can also achieve all points in this rate region. Thus, DSSS with orthogonal (e.g., Walsh–Hadamard) codes achieves a subset of the TD and FD achievable rate regions. More general orthogonal codes are needed to achieve the same region as these other techniques.

We now consider DSSS with nonorthogonal spreading codes. As discussed in Section 13.4.2, in these systems interference between users is attenuated by the code cross-correlation. Thus, if interference is treated as noise, its power contribution to the SIR is reduced by the square of the code cross-correlation. From (13.6), we will assume that spreading codes with a processing gain of G reduce the interference power by $1/G$. This is a reasonable approximation for random spreading codes, although (as discussed in Section 13.4.1) the exact value of the interference power reduction depends on the nature of the spreading codes and other assumptions [38; 39]. Since the signal bandwidth is increased by G, the two-user BC rate region achievable through DSSS with nonorthogonal codes (NC) and successive interference cancellation (IC) is given by

$$\mathcal{C}_{\text{DS,NC,IC}} \quad \bigcup_{\{P_1, P_2 : P_1 + P_2 = P\}} \left(R_1 = \frac{B}{G} \log_2\left(1 + \frac{P_1}{n_1 B/G}\right), \right.$$
$$\left. R_2 = \frac{B}{G} \log_2\left(1 + \frac{P_2}{n_2 B/G + P_1/G}\right) \right). \quad (14.22)$$

By the convexity of the log function, the rate region defined by (14.22) for $G > 1$ is smaller than the rate region (14.17) obtained using superposition coding, and the degradation increases with increasing values of G. This implies that for DSSS with nonorthogonal coding, the spreading gain should be minimized in order to maximize capacity.

With nonorthogonal coding and no interference cancellation, the receiver treats all signals intended for other users as noise, resulting in the achievable rate region

$$\mathcal{C}_{\text{DS,NC}} = \bigcup_{\{P_1, P_2 : P_1 + P_2 = P\}} \left(R_1 = \frac{B}{G} \log_2\left(1 + \frac{P_1}{n_1 B/G + P_2/G}\right), \right.$$
$$\left. R_2 = \frac{B}{G} \log_2\left(1 + \frac{P_2}{n_2 B/G + P_1/G}\right) \right). \quad (14.23)$$

Again using convexity of the log function, we have that $G = 1$ maximizes this rate region and that the rate region decreases as G increases. Moreover, the radius of curvature for (14.23) is given by

$$\chi = \frac{\dot{R}_1 \ddot{R}_2 - \dot{R}_2 \ddot{R}_1}{(\dot{R}_1^2 + \dot{R}_2^2)^{3/2}}, \quad (14.24)$$

where $\dot{R}_i$ and $\ddot{R}_i$ denote (respectively) the first and second derivatives of R_i with respect to α for $P_1 = \alpha P$ and $P_2 = (1 - \alpha)P$. For $G = 1$, $\chi \geq 0$. Thus, the rate region for nonorthogonal coding without interference cancellation (14.23) is bounded by a convex function with end points C_1 and C_2, as shown in Figures 14.10 and 14.11. Therefore, the achievable rate region

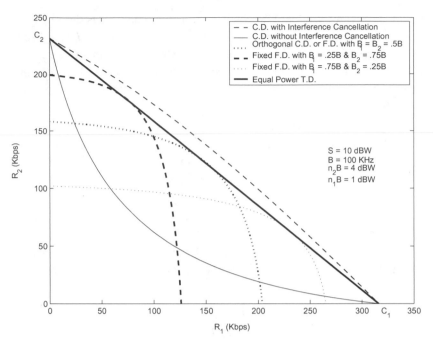

Figure 14.10: Two-user capacity region with 3-dB SNR difference.

for DSSS with nonorthogonal coding and no interference cancellation will lie beneath the regions for TD and FD, which are bounded by concave functions with the same endpoints.

The achievable rate regions for equal-power TD (14.12), FD (14.14), orthogonal DSSS (14.21), and nonorthogonal DSSS with (14.17) and without (14.23) interference cancellation are illustrated in Figures 14.10 and 14.11, where the SNR between the users differs by 3 dB and 20 dB, respectively. For the calculation of (14.23) we assume $G = 1$. Direct-sequence spread spectrum with larger values of the spreading gain will result in a smaller rate region.

14.5.3 Common Data

In many broadcasting applications, common data is sent to all users in the system. For example, television and radio stations broadcast the same data to all users, and in wireless Internet applications many users may want to download the same stock quotes and sports scores. The nature of superposition coding makes it straightforward to develop optimal broadcasting techniques for common data and to incorporate common data into the capacity region for the broadcast channel. In particular, for a two-user BC with superposition coding, the user with the better channel gain always receives the data intended for the user with the worse channel gain, along with his own data. Thus, since common data must be transmitted to both users, we can encode all common data as independent data intended for the user with the worse SNR. Since the user with the better SNR will also receive this data, it will be received by both users.

Under this transmission strategy, if the rate pair (R_1, R_2) is in the capacity region of the two-user BC with independent data defined by (14.17), then for any $R_0 \le R_2$ we can achieve the rate triple $(R_0, R_1, R_2 - R_0)$ for the BC with common and independent data, where R_0

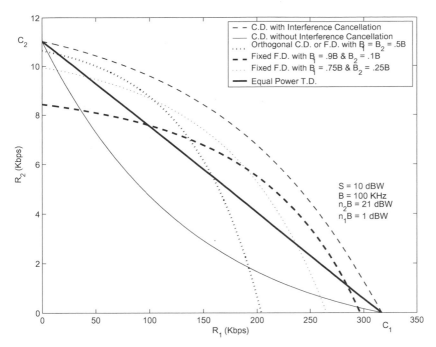

Figure 14.11: Two-user capacity region with 20-dB SNR difference.

is the rate of common data, R_1 is the rate of user 1's independent data, and $R_2 - R_0$ is the rate of user 2's independent data. Mathematically, this gives the three-dimensional capacity region

$$C_{BC} = \bigcup_{\{P_1, P_2 : P_1 + P_2 = P\}} \left(R_0 \leq B \log_2\left(1 + \frac{P_2}{n_2 B + P_1}\right), \; R_1 = B \log_2\left(1 + \frac{P_1}{n_1 B}\right), \right.$$
$$\left. R_2 = B \log_2\left(1 + \frac{P_2}{n_2 B + P_1}\right) - R_0 \right). \quad (14.25)$$

EXAMPLE 14.7: In Example 14.5, we saw that for a broadcast channel with total transmit power $P = 10$ mW, $n_1 = 10^{-9}$ W/Hz, $n_2 = 10^{-8}$ W/Hz, and $B = 100$ kHz, the rate pair $(R_1, R_2) = (3 \cdot 10^5, 2.69 \cdot 10^5)$ is on the boundary of the capacity region. Suppose that user 1 desires an independent data rate of 300 kbps and that a common data rate of 100 kbps is required for the system. At what rate can user 2 get independent data?

Solution: In order for $R_1 = 300$ kbps, we require the same $P_1 = .7$ mW as in Example 14.5.2. The common information rate $R_0 = 10^5 < 2.69 \cdot 10^5$, so from (14.25) it follows that the independent information rate to user 2 is just $R_2 - R_0 = 2.69 \cdot 10^5 - 1 \cdot 10^5 = 1.69 \cdot 10^5$ bps.

14.5.4 Capacity in Fading

We now consider the capacity region of BCs with fading, where the users have independent random channel gains that change over time. As described in Section 4.2 for single-user channels, the capacity of fading broadcast channels depends on what is known about the channel at the transmitter and receiver. However, capacity of a BC is only known for degraded BCs, and this model requires that the channel gains are known to both the transmitter and receiver. Furthermore, superposition coding cannot be used without transmitter knowledge of the channel gains, since if the transmitter does not know the relative channel gains then it does not know which user can receive the coarse constellation point and which can receive the fine one. Thus we will only consider fading BCs where there is perfect channel side information (CSI) about the instantaneous channel gains at the transmitter and at all receivers. We also assume that the channel is slowly fading so that, for a given fading state, the coding strategy that achieves any point in the capacity region for the static BC with this state has sufficient time to drive the error probability close to zero before the channel gains change.[6]

As with the single-user fading channel, there are two notions of capacity for multiuser fading channels with perfect CSI: ergodic (Shannon) capacity and outage capacity. The ergodic capacity region of a multiuser fading channel characterizes the achievable rate vectors averaged over all fading states [34; 40] while the outage capacity region dictates the set of fixed-rate vectors that can be maintained in all fading states subject to a given outage probability [41; 42; 43]. Zero-outage capacity refers to outage capacity with zero outage probability [41] – that is, the set of fixed-rate vectors that can be maintained in all fading states. The ergodic capacity region, which is analogous to ergodic capacity for single-user systems, defines the data-rate vectors that can be maintained over time without any constraints on delay. Hence, in some fading states the data rate may be small or zero, which can be problematic for delay-constrained applications like voice or video. The outage capacity region – analogous to outage capacity in single-user systems – forces a fixed-rate vector in all non-outage fading states, which is perhaps a more appropriate capacity metric for delay-constrained applications. However, the requirement for maintaining a fixed rate even in very deep fades can severely decrease the outage capacity region relative to the ergodic capacity region. In fact, the zero-outage capacity region when all users exhibit Rayleigh fading is zero for all users.

We consider a BC with AWGN and fading where a single transmitter communicates independent information to K users over bandwidth B with average transmit power $\bar{P}$. The transmitter and all receivers have a single antenna. The time-varying power gain of user k's channel at time i is $g_k[i]$. Each receiver has AWGN with PSD $N_0/2$. We define the effective time-varying noise of the kth user as[7] $n_k[i] = N_0/g_k[i]$. The *effective noise vector* at time i is defined as

$$\mathbf{n}[i] = (n_1[i], \ldots, n_K[i]). \tag{14.26}$$

We also call this the *fading state* at time i, since it characterizes the channel gains $g_k[i]$ associated with each user at time i. We will denote the kth element of this vector as $n_k[i]$ or just n_k when the time reference is clear. As with the static channel, the capacity of the fading BC

---

[6] More precisely, the coding strategy that achieves a point in the AWGN BC capacity region uses a block code, and the error probability of the code goes to zero with blocklength. Our slow fading assumption presumes that the channel gains stay constant long enough for the block code associated with these gains to drive the error probability close to zero.

[7] Notice that the noise vector is the instantaneous power of the noise and not the instantaneous noise sample.

can be computed based on its time-varying channel gains or its time-varying effective noise vector. The ergodic BC capacity region is defined as the set of all average rates achievable in a fading channel with arbitrarily small probability of error, where the average is taken with respect to all fading states. In [34], the ergodic capacity region and optimal power allocation scheme for the fading BC is found by decomposing the fading channel into a parallel set of static BCs, one for every possible fading state $\mathbf{n} = (N_0/g_1, \ldots, N_0/g_K)$. In each fading state the channel can be viewed as a static AWGN BC, and time, frequency, or code division can be applied to the channel in each fading state.

Since the transmitter and all receivers know $\mathbf{n}[i]$, superposition coding according to the ordering of the current effective noise vector can be used by the transmitter. Each receiver can perform successive decoding in which the users with larger effective noise are decoded and subtracted out before decoding the desired signal. Furthermore, the power transmitted to each user $P_j(\mathbf{n})$ is a function of the current fading state. Since the transmission scheme is based on superposition coding, it remains only to determine the optimal power allocation across users and over time.

We define a power policy $\mathcal{P}$ over all possible fading states as a function that maps from any fading state $\mathbf{n}$ to the transmitted power $P_k(\mathbf{n})$ for each user. Let $\mathcal{F}_{BC}$ denote the set of all power policies satisfying average power constraint $\bar{P}$:

$$\mathcal{F}_{BC} \triangleq \left\{ \mathcal{P} : \mathbf{E}_{\mathbf{n}}\left[\sum_{k=1}^{K} P_k(\mathbf{n}) \right] \leq \bar{P} \right\}. \tag{14.27}$$

From (14.18), the capacity region assuming a constant fading state $\mathbf{n}$ with power allocation $\mathbf{P}(\mathbf{n}) = \{ P_k(\mathbf{n}) : k = 1, \ldots, K \}$ is given by

$$\mathcal{C}_{BC}(\mathbf{P}(\mathbf{n})) = \left\{ (R_1(\mathbf{P}(\mathbf{n})), \ldots, R_K(\mathbf{P}(\mathbf{n})) : \right.$$
$$\left. R_k(\mathbf{P}(\mathbf{n})) = B \log_2\left(1 + \frac{P_k(\mathbf{n})}{n_k B + \sum_{j=1}^{K} P_j(\mathbf{n}) \mathbf{1}[n_k > n_j]} \right) \right\}. \tag{14.28}$$

Let $\mathcal{C}_{BC}(\mathcal{P})$ denote the set of achievable rates averaged over all fading states for power policy $\mathcal{P}$:

$$\mathcal{C}_{BC}(\mathcal{P}) = \{ (R_1, \ldots, R_K) : R_k \leq \mathbf{E}_{\mathbf{n}}[R_k(\mathbf{P}(\mathbf{n}))] \},$$

where $R_k(\mathbf{P}(\mathbf{n}))$ is as given in (14.28). From [34], the ergodic capacity region of the BC with perfect CSI and power constraint $\bar{P}$ is:

$$\mathcal{C}_{BC}(\bar{P}) = \bigcup_{\mathcal{P} \in \mathcal{F}_{BC}} \mathcal{C}_{BC}(\mathcal{P}). \tag{14.29}$$

It is further shown in [34] that the region $\mathcal{C}_{BC}(\bar{P})$ is convex and that the optimal power allocation scheme is an extension of water-filling with K different water levels for a K-user system.

We can also define achievable rate vectors for TD or FD, although these will clearly lie inside the ergodic capacity region, since superposition coding outperforms both of these techniques in every fading state. The optimal form of TD adapts the power assigned to each user relative to the current fading state. Similarly, the optimal form of FD adapts the bandwidth and power assigned to each user relative to the current fading state. As described in Section 14.5.2, for each fading state, varying the power in TD yields the same rates as varying the power and bandwidth in FD. Thus, the achievable rates for these two techniques averaged over all fading states are the same. Focusing on the FD region, assume a power policy $\mathcal{P} \in \mathcal{F}_{BC}$ that assigns power $P_k(\mathbf{n})$ to the kth user in fading state $\mathbf{n}$. By (14.27), a power policy $\mathcal{P} \in \mathcal{F}_{BC}$ satisfies the average power constraint. Also assume a bandwidth policy $\mathcal{B}$ that assigns bandwidth $B_k(\mathbf{n})$ to user k in state $\mathbf{n}$, and let $\mathcal{G}$ denote the set of all bandwidth policies satisfying the bandwidth constraint of the system:

$$\mathcal{G} \triangleq \left\{ \mathcal{B} : \sum_{k=1}^{K} B_k(\mathbf{n}) = B \ \forall \mathbf{n} \right\}.$$

The set of achievable rates for FD under policies $\mathcal{P}$ and $\mathcal{B}$ is

$$\mathcal{C}_{FD}(\mathcal{P}, \mathcal{B}) = \{(R_1, \ldots, R_K) : R_k \leq \mathbf{E_n}[R_k(\mathbf{P(n)}, \mathcal{B})]\}, \tag{14.30}$$

where

$$R_k(P(\mathbf{n}), \mathcal{B}) = B_k(\mathbf{n}) \log_2 \left(1 + \frac{P_k(\mathbf{n})}{n_k B_k(\mathbf{n})} \right). \tag{14.31}$$

The set of all achievable rates under frequency division with perfect CSI, subject to power constraint $\bar{P}$ and bandwidth constraint B, is then

$$\mathcal{C}_{FD}(\bar{P}, B) = \bigcup_{\mathcal{P} \in \mathcal{F}_{BC}, \, \mathcal{B} \in \mathcal{G}} \mathcal{C}_{FD}(\mathcal{P}, \mathcal{B}). \tag{14.32}$$

The sum-rate capacity for fading BCs is defined as the maximum sum of achievable rates, maximized over all rate vectors in the ergodic BC capacity region. Since sum-rate capacity for the AWGN BC is maximized by transmitting only to the user with the best channel, in fading the sum-rate is maximized by transmitting only to the user with the best channel in each channel state. Clearly superposition CD, TD, and FD are all equivalent in this setting, since all resources are assigned to a single user in each state. We can compute the sum-rate capacity and the optimal power allocation over time from an equivalent single-user fading channel with time-varying effective noise $n[i] = \min_k n_k[i]$ and average power constraint $\bar{P}$. From Section 4.2.4 the optimal power allocation to the user with the best channel at time i is thus a water-filling in time, with cutoff value determined from the distribution of $\min_k n_k[i]$.

The ergodic capacity and achievable rate regions for fading broadcast channels under CD, TD, and FD are computed in [34] for different fading distributions, along with the optimal adaptive resource allocation strategies that achieve the boundaries of these regions. These adaptive transmission policies exploit *multiuser diversity* in that more resources (power, bandwidth, timeslots) are allocated to the users with the best channels in any given fading state. In particular, sum-rate capacity is achieved by allocating all resources in any given state to the user with the best channel. Multiuser diversity will be discussed in more detail in Section 14.8.

The zero-outage BC capacity region defines the set of rates that can be simultaneously achieved for all users in *all* fading states while meeting the average power constraint; it is the multiuser extension of zero-outage capacity defined in Section 4.2.4 for single-user channels. From [42], the power required to support a rate vector $\mathbf{R} = (R_1, R_2, \ldots, R_K)$ in fading state $\mathbf{n}$ is:

$$P^{\min}(\mathbf{R}, \mathbf{n}) = \sum_{k=1}^{K-1} \left[2^{(\sum_{j=k+1}^{K} R_{\pi(j)}/B)} (2^{R_{\pi(k)}/B} - 1) n_{\pi(k)} B \right] + (2^{R_{\pi(K)}/B} - 1) n_{\pi(K)} B, \quad (14.33)$$

where $\pi(\cdot)$ is the permutation such that

$$n_{\pi(1)} < n_{\pi(2)} < \cdots < n_{\pi(K)}.$$

Therefore, the zero-outage capacity region is the union of all rate vectors that meet the average power constraint:

$$\mathcal{C}_{BC}^0(\bar{P}) = \bigcup_{\{\mathbf{R} : \mathbf{E_n}[P^{\min}(\mathbf{R}, \mathbf{n})] \leq \bar{P}\}} \mathbf{R} = (R_1, R_2, \ldots, R_K). \quad (14.34)$$

The boundary of the zero-outage capacity region is the set of all rate vectors $\mathbf{R}$ such that the power constraint is met with equality. For the two-user BC with time-varying AWGN with powers n_1 and n_2, this boundary simplifies to the set of all (R_1, R_2) that satisfy the following equation [42]:

$$\bar{P} = p(n_1 < n_2) \left[\mathbf{E}[n_1 B \mid n_1 < n_2] 2^{R_2/B} (2^{R_1/B} - 1) + \mathbf{E}[n_2 B \mid n_1 < n_2] (2^{R_2/B} - 1) \right]$$
$$+ p(n_1 \geq n_2) \left[\mathbf{E}[n_2 B \mid n_1 \geq n_2] 2^{R_1/B} (2^{R_2/B} - 1) + \mathbf{E}[n_1 B \mid n_1 \geq n_2] (2^{R_1/B} - 1) \right].$$

The boundary is determined solely by $\mathbf{E}[n_1 B \mid n_1 < n_2]$, $\mathbf{E}[n_2 B \mid n_1 < n_2]$, $\mathbf{E}[n_1 B \mid n_1 \geq n_2]$, and $\mathbf{E}[n_2 B \mid n_1 \geq n_2]$. This follows because the power required to achieve a rate vector is a linear function of the noise levels in each state, as seen in (14.33). The zero-outage capacity region depends on the conditional expectations of the noises – as opposed to their unconditional expectations – since every different ordering of noises leads to a different expression for the required power in each state, as can be seen from (14.33).

The outage capacity region of the BC is defined similarly as the zero-outage capacity region, except that users may have some nonzero probability of outage so that they can suspend transmission in some outage states. This provides additional flexibility in the system because, under severe fading conditions, maintaining a fixed rate in *all* fading states can consume a great deal of power. In particular, we saw in Section 4.2.4 that, for a single-user fading channel, maintaining any nonzero fixed rate in Rayleigh fading requires infinite power. By allowing some outage, power can be conserved from outage states to maintain higher rates in non-outage states. The outage capacity region is more difficult to obtain than the zero-outage capacity region, since in any given fading state the transmission strategy must determine which users to put into outage. Once the outage users are determined, the power required to maintain the remaining users is given by (14.33) for the rate vector associated with the $K' \leq K$ users that are not in outage. It is shown in [42] that this decision should be made

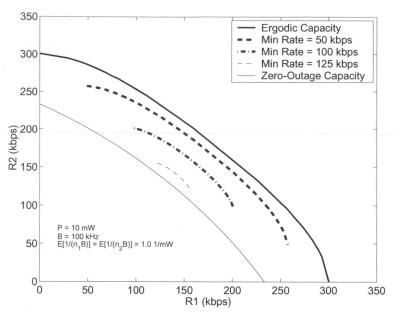

Figure 14.12: Ergodic, zero-outage, and minimum rate BC capacity regions (Rician fading with a *K*-factor of 1, average SNR = 10 dB).

based on a threshold policy, and the resulting outage capacity region is then obtained implicitly based on the threshold policy and the power allocation (14.33) for non-outage users.

The notions of ergodic capacity and outage capacity can also be combined. This combination results in the minimum rate capacity region [44]. A rate vector in this region characterizes the set of all average rate vectors that can be maintained, averaged over all fading states, subject to some minimum rate vector that must be maintained in all states (possible subject to some outage probability). Minimum rate capacity is useful for systems supporting a mix of delay-constrained and delay-unconstrained data. The minimum rates dictate the data rates available for the constrained data that must be maintained in all fading states, while the rates above these minimums are what is available for the unconstrained data, where these additional rates vary depending on the current fading state. The minimum rate capacity region (with zero outage probability) lies between that of the zero-outage capacity region and the ergodic capacity region: for minimum rates of zero it equals the ergodic capacity region; for minimum rates on the boundary of the zero-outage capacity region, it cannot exceed these boundary points. This is illustrated in Figure 14.12, where we plot the ergodic, zero-outage, and minimum rate capacity region for a BC with Rician fading. We see from this figure that the ergodic capacity region is the largest, since it can adapt to the different channel states in order to maximize its average rate (averaged over all fading states). The zero-outage capacity region is the smallest, since it is forced to maintain a fixed rate in all states, which consumes much power when the fading is severe. The minimum rate capacity region lies between the other two and depends on the minimum rate requirements. As the minimum rate vector that must be maintained in all fading states increases, the minimum rate capacity region approaches the zero-outage capacity region, and as this minimum rate vector decreases, the minimum rate capacity region approaches the ergodic capacity region.

14.5.5 Capacity with Multiple Antennas

We now investigate the capacity region for a BC with multiple antennas. We have seen in Section 10.3 that multiple-input multiple-out (MIMO) systems can provide large capacity increases for single-user systems. The same will be true of multiuser systems: in fact, multiple users can exploit multiple spatial dimensions even more effectively than a single user.

Consider a K-user BC in which the transmitter has M_t antennas and each receiver has M_r antennas. The $M_r \times M_t$ channel matrix $\mathbf{H}_k$ characterizes the channel gains between each antenna at the transmitter and each antenna at the kth receiver. The received signal for the kth user is then

$$\mathbf{y}_k = \mathbf{H}_k \mathbf{x} + \mathbf{n}_k, \tag{14.35}$$

where $\mathbf{x}$ is the input to the transmit antennas, and we denote its covariance matrix as $\mathbf{Q_x} = \mathrm{E}[\mathbf{x}\mathbf{x}^H]$. For simplicity, we normalize the bandwidth to unity,[8] $B = 1$ Hz, and assume the noise vector $\mathbf{n}_k$ is circularly symmetric complex Gaussian with $\mathbf{n}_k \sim N(0, \mathbf{I})$.

When the transmitter has more than one antenna, $M_t > 1$, the BC is no longer degraded. In other words, receivers cannot generally be ranked by their channel quality, since receivers have different channel gains associated with the different antennas at the transmitter. The capacity region of the general nondegraded broadcast channels is unknown. However, an achievable region for this channel was proposed in [46; 47] that was later shown [48] to equal the capacity region. The region is based on the notion of *dirty paper coding* (DPC) [49]. The basic premise of DPC is as follows. If the transmitter (but not the receiver) has perfect, noncausal knowledge of interference to a given user, then the capacity of the channel is the same as if there were no interference or, equivalently, as if the receiver had knowledge of the interference and could subtract it out. Dirty paper coding is a technique that allows noncausally known interference to be "presubtracted" at the transmitter but in such a way that the transmit power is not increased. A more practical (and more general) technique to perform this presubtraction is described in [50].

In the MIMO broadcast channel, DPC can be applied at the transmitter when choosing codewords for different users. The transmitter first picks a codeword $\mathbf{x}_1$ for user 1. The transmitter then chooses a codeword $\mathbf{x}_2$ for user 2 with full (noncausal) knowledge of the codeword intended for user 1. Hence the codeword of user 1 can be presubtracted such that user 2 does not see the codeword intended for user 1 as interference. Similarly, the codeword $\mathbf{x}_3$ for user 3 is chosen such that user 3 does not see the signals intended for users 1 and 2 as interference. This process continues for all K users. The ordering of the users clearly matters in such a procedure and needs to be optimized in the capacity calculation. Let $\pi(\cdot)$ denote a permutation of the user indices and let $\mathbf{Q} = [\mathbf{Q}_1, \ldots, \mathbf{Q}_K]$ denote a set of positive semidefinite covariance matrices with $\mathrm{Tr}[\mathbf{Q}_1 + \cdots + \mathbf{Q}_K] \leq P$. Under DPC, if user $\pi(1)$ is encoded first, followed by user $\pi(2)$, and so forth, then the following rate vector is achievable:

$$\mathbf{R}(\pi, \mathbf{Q}) = (R_{\pi(1)}, \ldots, R_{\pi(K)}) :$$

$$R_{\pi(k)} = \log_2 \frac{\det[\mathbf{I} + \mathbf{H}_{\pi(k)}\left(\sum_{j \geq k} \mathbf{Q}_{\pi(j)}\right)\mathbf{H}_{\pi(k)}^H]}{\det[\mathbf{I} + \mathbf{H}_{\pi(k)}\left(\sum_{j > k} \mathbf{Q}_{\pi(j)}\right)\mathbf{H}_{\pi(k)}^H]}, \quad k = 1, \ldots, K. \tag{14.36}$$

---

[8] The capacity of unity-bandwidth MIMO channels has a factor of .5 preceding the log function for real (one-dimensional) channels, with no such factor for complex (two-dimensional) channels [45, Chap. 3.1].

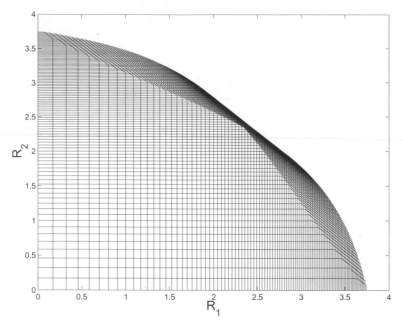

Figure 14.13: MIMO BC capacity region with $\mathbf{H}_1 = [1 \ 0.5]$, $\mathbf{H}_2 = [0.5 \ 1]$, and $P = 10$.

The capacity region $\mathcal{C}$ is then the convex hull of the union of all such rate vectors over all permutations and all positive semidefinite covariance matrices satisfying the average power constraint:

$$\mathcal{C}_{\mathrm{BC}}(P, \mathbf{H}) \triangleq \mathrm{Co}\left(\bigcup_{\pi, \mathbf{Q}} \mathbf{R}(\pi, \mathbf{Q}) \right), \tag{14.37}$$

where $\mathbf{R}(\pi, \mathbf{Q})$ is given by (14.36). The transmitted signal is $\mathbf{x} = \mathbf{x}_1 + \cdots + \mathbf{x}_K$ and the input covariance matrices are of the form $\mathbf{Q}_k = \mathbf{E}[\mathbf{x}_k \mathbf{x}_k^H]$. The DPC implies that $\mathbf{x}_1, \ldots, \mathbf{x}_K$ are uncorrelated, and thus $\mathrm{Tr}[\mathbf{Q}_\mathbf{x}] = \mathrm{Tr}[\mathbf{Q}_1 + \cdots + \mathbf{Q}_K] \leq P$.

One important feature to notice about the rate equations defined by (14.36) is that these equations are neither a concave nor a convex function of the covariance matrices. This makes finding the capacity region difficult, because generally the entire space of covariance matrices that meet the power constraint must be searched [46; 47]. However, as described in Section 14.7, there is a duality between the MIMO BC and the MIMO MAC that can be exploited to greatly simplify this calculation. Figure 14.13 shows the capacity region for a two-user channel computed by exploiting this duality with $M_t = 2$ and $M_r = 1$. The region is defined by the outer boundary, and the lines inside this boundary each correspond to the capacity region of a different dual MIMO MAC channel whose sum power equals the power of the MIMO BC. The union of these dual regions yields the boundary of the MIMO BC region, as will be discussed in Section 14.7.

14.6 Uplink (Multiple Access) Channel Capacity

14.6.1 Capacity in AWGN

The MAC consists of K transmitters, each with power P_k, sending to a receiver over a channel with power gain g_k. We assume that all transmitters and the receiver have a single antenna.

The received signal is corrupted by AWGN with PSD $N_0/2$. The two-user MAC capacity region is the closed convex hull of all vectors (R_1, R_2) satisfying the following constraints [35, Chap. 14.1]:

$$R_k \le B \log_2 \left(1 + \frac{g_k P_k}{N_0 B} \right), \quad k = 1, 2, \tag{14.38}$$

$$R_1 + R_2 \le B \log_2 \left(1 + \frac{g_1 P_1 + g_2 P_2}{N_0 B} \right). \tag{14.39}$$

The first constraint (14.38) is just the capacity associated with each individual channel. The second constraint (14.39) indicates that the sum of rates for all users cannot exceed the capacity of a "superuser" with received power equal to the sum of received powers from all users. For K users, the region becomes

$$\mathcal{C}_{MAC} = \left\{ (R_1, \dots, R_K) : \sum_{k \in S} R_k \le B \log_2 \left(1 + \frac{\sum_{k \in S} g_k P_k}{N_0 B} \right) \forall S \subset \{1, 2, \dots, K\} \right\}. \tag{14.40}$$

Thus, the region described by (14.40) demonstrates that the sum of rates for any subset of the K users cannot exceed the capacity of a superuser with received power equal to the sum of received powers associated with this user subset.

The sum-rate capacity of a MAC is the maximum sum of rates $\sum_{k=1}^{K} R_k$, where the maximum is taken over all rate vectors $(R_1, \dots, R_K)$ in the MAC capacity region. As with the like capacity for the BC, the MAC sum-rate also measures the maximum throughput of the system regardless of fairness and is easier to characterize than the K-dimensional capacity region. It can be shown from (14.40) that sum-rate capacity is achieved on the AWGN MAC by having all users transmit at their maximum power, which yields

$$\mathcal{C}_{MACSR} = B \log_2 \left(1 + \frac{\sum_{k=1}^{K} g_k P_k}{N_0 B} \right). \tag{14.41}$$

The intuition behind this result is that each user in the MAC has an individual power constraint, so not allowing a user to transmit at full power wastes system power. By contrast, the AWGN BC sum-rate capacity (14.20) is achieved by transmitting only to the user with the best channel. However, since all users share the power resource, no power is wasted in this case.

The MAC capacity region for two users is shown in Figure 14.14, where C_k and C_k^* are given by

$$C_k = B \log_2 \left(1 + \frac{g_k P_k}{N_0 B} \right), \quad k = 1, 2; \tag{14.42}$$

$$C_1^* = B \log_2 \left(1 + \frac{g_1 P_1}{N_0 B + g_2 P_2} \right), \tag{14.43}$$

$$C_2^* = B \log_2 \left(1 + \frac{g_2 P_2}{N_0 B + g_1 P_1} \right). \tag{14.44}$$

The point $(C_1, 0)$ is the achievable rate vector when transmitter 1 is sending at its maximum rate and transmitter 2 is silent, and the opposite scenario achieves the rate vector $(0, C_2)$. The corner points (C_1, C_2^*) and (C_1^*, C_2) are achieved using the successive interference cancellation described previously for superposition codes. Specifically, let the first user

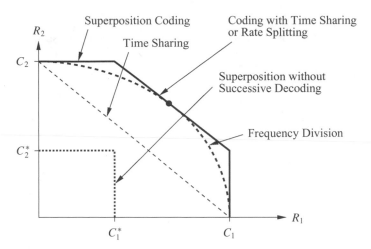

Figure 14.14: Two-user MAC capacity region.

operate at the maximum data rate C_1. Then its signal will appear as noise to user 2; thus, user 2 can send data at rate C_2^* that can be decoded at the receiver with arbitrarily small error probability. If the receiver then subtracts out user 2's message from its received signal, the remaining message component is just user 1's message corrupted by noise, so rate C_1 can be achieved with arbitrarily small error probability. Hence, (C_1, C_2^*) is an achievable rate vector. A similar argument with the user roles reversed yields the rate point (C_1^*, C_2). Time sharing between these two strategies yields any point on the straight line connecting (C_1, C_2^*) and (C_1^*, C_2). This time-sharing line can also be achieved via a technique called *rate splitting*, whereby a given user splits his data stream into multiple substreams and encodes these substreams as if they originated from different virtual users [51]. Note that in the BC the user with the highest SNR must always be decoded last, whereas with MACs the decoding can be done in either order. This is a fundamental difference between the two channels.

A time-division strategy between two transmitters operating at their maximum rates, given by (14.42), yields any rate vector on the straight line connecting C_1 and C_2. With frequency division, the rates depend on the fraction of the total bandwidth that is allocated to each transmitter. Letting B_1 and B_2 denote the bandwidth allocated to each of the two users, we obtain the achievable rate region

$$\mathcal{C}_{\text{FD}} = \bigcup_{\{B_1, B_2 : B_1 + B_2 = B\}} \left(R_1 = B_1 \log_2\left(1 + \frac{g_1 P_1}{N_0 B_1}\right), R_2 = B_2 \log_2\left(1 + \frac{g_2 P_2}{N_0 B_2}\right) \right), \quad (14.45)$$

which is plotted in Figure 14.14. Clearly this region dominates TD, since setting $B_1 = \tau B$ and $B_2 = (1 - \tau)B$ in (14.45) yields $R_1 > \tau C_1$ and $R_2 > (1 - \tau)C_2$. It can be shown [35, Chap. 14.3] that this curve touches the capacity region boundary at one point: this point corresponds to the rate vector that maximizes the sum-rate $R_1 + R_2$. To achieve this rate point, the bandwidths B_1 and B_2 must be proportional to their corresponding received powers $g_1 P_1$ and $g_2 P_2$.

As with the broadcast channel, we can obtain the same achievable rate region with TD as with FD by efficient use of the transmit power. If we take the constraints P_1 and P_2 to be average power constraints then it follows that, since user k uses the channel only a fraction

τ_k of the time, its average power over that time fraction can be increased to P_k/τ_k. The rate region achievable through variable-power TD is then given by

$$\mathcal{C}_{\text{TD,VP}} = \bigcup_{\{\tau_1,\tau_2:\tau_1+\tau_2=1\}} \left(R_1 = \tau_1 B \log_2\left(1 + \frac{g_1 P_1}{N_0 \tau_1 B}\right),\ R_2 = \tau_2 B \log_2\left(1 + \frac{g_2 P_2}{N_0 \tau_2 B}\right) \right),$$

(14.46)

and substituting $B_k \triangleq \tau_k B$ in (14.46) yields the same rate region as in (14.45).

Superposition codes without successive decoding can also be used. With this approach, each transmitter's message acts as noise to the others. Thus, the maximum achievable rate in this case cannot exceed (C_1^*, C_2^*), which is clearly dominated by FD and TD for some bandwidth or time allocations – in particular, the allocation that intersects the capacity region boundary.

EXAMPLE 14.8: Consider a MAC channel in AWGN with transmit power $P_1 = P_2 = 100$ mW for both users and with channel gains of $g_1 = .08$ for user 1 and $g_2 = .001$ for user 2. Assume the receiver noise has $N_0 = 10^{-9}$ W/Hz and the system bandwidth is $B = 100$ kHz. Find the corner points of the MAC capacity region. Also find the rate that user 1 can achieve if user 2 requires a rate of $R_2 = 100$ kbps and of $R_2 = 50$ kbps.

Solution: From (14.42)–(14.44) we have:

$$C_1 = B \log_2\left(1 + \frac{g_1 P_1}{N_0 B}\right) = 6.34 \cdot 10^5,$$

$$C_2 = B \log_2\left(1 + \frac{g_2 P_2}{N_0 B}\right) = 1 \cdot 10^5;$$

$$C_1^* = B \log_2\left(1 + \frac{g_1 P_1}{N_0 B + g_2 P_2}\right) = 5.36 \cdot 10^5,$$

$$C_2^* = B \log_2\left(1 + \frac{g_2 P_2}{N_0 B + g_1 P_1}\right) = 1.77 \cdot 10^3.$$

The maximum rate for user 2 is 100 kbps, so if he requires $R_2 = 100$ kbps then this rate point is associated with the corner point (C_1^*, C_2) of the capacity region; hence user 1 can achieve a rate of $R_1 = C_1^* = 536$ kbps. If user 2 requires only $R_2 = 50$ kbps then the rate point lies on the TD portion of the capacity region. In particular, time sharing between the corner points (C_1, C_2^*) and (C_2, C_1^*) yields, for any τ with $0 \le \tau \le 1$, the rate point $(R_1, R_2) = \tau(C_1, C_2^*) + (1 - \tau)(C_1^*, C_2)$. The time-share value τ that yields $R_2 = 50$ kbps thus satisfies $\tau C_2^* + (1 - \tau)C_2 = R_2$. Solving for τ yields $\tau = (R_2 - C_2)/(C_2^* - C_2) = .51$, about halfway between the two corner points. Then user 1 can achieve rate $R_1 = \tau C_1 + (1 - \tau)C_1^* = 5.86 \cdot 10^5$. This example illustrates the dramatic impact of the near–far effect in MAC channels. Even though both users have the same transmit power, the channel gain of user 2 is much less than the gain of user 1. Hence, user 2 can achieve at most a rate of 100 kbps, whereas user 1 can achieve a rate between 536 and 634 kbps. Moreover, the interference from user 2 has little impact on user 1 owing to the weak channel gain associated with the interference: user 1 sees data rates of $C_1 = 634$ kbps without interference and $C_1^* = 536$ kbps with interference. However, the interference from user 1 severely limits the data rate of user 2, decreasing it by almost two orders of magnitude from $C_2 = 100$ kbps to $C_2^* = 1.77$ kbps.

14.6.2 Capacity in Fading

We now consider the capacity region of a multiple access channel with AWGN and fading, where the channel gains for each user change over time. We assume that all transmitters and the receiver have a single antenna and that the receiver has AWGN with PSD $N_0/2$. Each user has an individual power constraint $\bar{P}_k$, $k = 1, \ldots, K$. The time-varying power gain of user k's channel at time i is $g_k[i]$ and is independent of the fading of other users. We define the fading state at time i as $\mathbf{g}[i] = (g_1[i], \ldots, g_K[i])$, with the time reference dropped when the context is clear. We assume perfect CSI about the fading state at both the transmitter and receiver; the case of receiver CSI only is treated in [52, Chap. 6.3]. Like the broadcast and single-user channels, the fading MAC also has two notions of capacity: the ergodic capacity region, which characterizes the achievable rate vectors averaged over all fading states; and the outage capacity region, which characterizes the maximum rate vector that can be maintained in all states with some nonzero probability of outage.

We first consider the ergodic capacity region as derived in [40]. Define a power policy $\mathcal{P}$ as a function that maps a fading state $\mathbf{g} = (g_1, \ldots, g_K)$ to a set of powers $P_1(\mathbf{g}), \ldots, P_K(\mathbf{g})$, one for each user. Let $\mathcal{F}_{\mathrm{MAC}}$ denote the set of all power policies satisfying the average per-user power constraint $\bar{P}_k$:

$$\mathcal{F}_{\mathrm{MAC}} \triangleq \{\mathcal{P} : \mathbf{E}_{\mathbf{g}}[P_k(\mathbf{g})] \leq \bar{P}_k, \ k = 1, \ldots, K\}.$$

The MAC capacity region assuming a constant fading state $\mathbf{g}$ with power allocation $P_1(\mathbf{g}), \ldots, P_K(\mathbf{g})$ is given by

$$\mathcal{C}_{\mathrm{MAC}}(P_1(\mathbf{g}), \ldots, P_K(\mathbf{g}))$$
$$= \left\{ (R_1, \ldots, R_K) : \sum_{k \in S} R_k \leq B \log_2 \left(1 + \frac{\sum_{k \in S} g_k P_k(\mathbf{g})}{N_0 B} \right) \forall S \subset \{1, 2, \ldots, K\} \right\}.$$
$$(14.47)$$

The set of achievable rates averaged over all fading states under power policy $\mathcal{P}$ is given by

$$\mathcal{C}_{\mathrm{MAC}}(\mathcal{P})$$
$$= \left\{ (R_1, \ldots, R_K) : \sum_{k \in S} R_k \leq \mathbf{E}_{\mathbf{g}} \left[B \log_2 \left(1 + \frac{\sum_{k \in S} g_k P_k(\mathbf{g})}{N_0 B} \right) \right] \forall S \subset \{1, 2, \ldots, K\} \right\}.$$
$$(14.48)$$

The ergodic capacity region is then the union over all power policies that satisfy the individual user power constraints:

$$\mathcal{C}_{\mathrm{MAC}}(\bar{P}_1, \ldots, \bar{P}_K) = \bigcup_{\mathcal{P} \in \mathcal{F}_{\mathrm{MAC}}} \mathcal{C}_{\mathrm{MAC}}(\mathcal{P}). \qquad (14.49)$$

From (14.41), (14.48), and (14.49), the sum-rate capacity of the MAC in fading reduces to

$$\mathcal{C}_{\mathrm{MACSR}} = \max_{\mathcal{P} \in \mathcal{F}_{\mathrm{MAC}}} \mathbf{E}_{\mathbf{g}} \left[B \log_2 \left(1 + \frac{\sum_{k=1}^{K} g_k P_k(\mathbf{g})}{N_0 B} \right) \right]. \qquad (14.50)$$

The maximization in (14.50) is solved using Lagrangian techniques, and the solution reveals that the optimal transmission strategy to achieve sum-rate is to allow only one user to transmit in every fading state [53]. Under this optimal policy, the user that transmits in a given fading state $\mathbf{g}$ is the one with the largest *weighted* channel gain g_k/λ_k, where λ_k is the Lagrange multiplier associated with the average power constraint of the kth user. This Lagrangian is a function of the user's average power constraint and fading distribution. By symmetry, if all users have the same fading distribution and the same average power constraint, then each user has the same λ_k and so the optimal policy is to allow only the user with the best channel g_k to transmit in fading state $\mathbf{g}$. Once it is determined which user should transmit in a given state, the power the user allocates to that state is determined via a water-filling over time. The intuition behind allowing only one user at a time to transmit is as follows. Since users can adapt their powers over time, system resources are best utilized by assigning them to the user with the best channel and allowing that user to transmit at a power commensurate with his channel quality. But when users have unequal average received power this strategy is no longer optimal: users with weak average received SNR would rarely transmit, so their individual power resources would be underutilized.

The MAC zero-outage capacity region, derived in [41], defines the set of rates that can be simultaneously achieved for all users in *all* fading states while meeting the average power constraints of each user. From (14.40), given a power policy $\mathcal{P}$ that maps fading states to user powers, the MAC capacity region in state $\mathbf{g}$ is

$$C_{\text{MAC}}(\mathcal{P}) = \left\{ (R_1, \ldots, R_K) : \sum_{k \in S} R_k \le B \log_2 \left(1 + \frac{\sum_{k \in S} g_k P_k(\mathbf{g})}{N_0 B} \right) \forall S \subset \{1, 2, \ldots, K\} \right\}.$$

(14.51)

Then, under policy $\mathcal{P}$, the set of rates that can be maintained in all fading states $\mathbf{g}$ is

$$C_{\text{MAC}}^0(\mathcal{P}) = \bigcap_{\mathbf{g}} C_{\text{MAC}}(\mathcal{P}).$$

(14.52)

The zero-outage capacity region is then the union of $C_{\text{MAC}}^0(\mathcal{P})$ over all power policies $\mathcal{P} \in \mathcal{F}_{\text{MAC}}$ that satisfy the user power constraints. Thus, the zero-outage MAC capacity region is given by

$$C_{\text{MAC}}^0(\bar{P}_1, \ldots, \bar{P}_K) = \bigcup_{\mathcal{P} \in \mathcal{F}_{\text{MAC}}} \bigcap_{\mathbf{g}} C_{\text{MAC}}(\mathcal{P}).$$

(14.53)

The outage capacity region of the MAC is similar to the zero-outage capacity region, except that users can suspend transmission in some outage states subject to a given nonzero probability of outage. As with the BC, the MAC outage capacity region is more difficult to obtain than the zero-outage capacity region, since in any given fading state the transmission strategy must determine which users to put into outage, the decoding order of the non-outage users, and the power at which these non-outage users should transmit. The MAC outage capacity region is obtained implicitly in [43] by determining whether a given rate vector $\mathbf{R}$ can be maintained in all fading states – subject to a given per-user outage probability – without violating the per-user power constraints. Ergodic and outage capacities can also be combined to obtain the minimum rate capacity region for the MAC. As with the BC, this region characterizes the set of all average rate vectors (averaged over all fading states) that can be maintained subject to some minimum rate vector that must be maintained in all states with some outage probability (possibly zero). The minimum rate capacity region for the fading MAC is derived in [54] using the duality principle that relates capacity regions of the BC and the MAC. This duality principle is described in the next section.

14.6.3 Capacity with Multiple Antennas

We now consider MAC channels with multiple antennas. As in the MIMO BC model, we normalize bandwidth to unity, $B = 1$ Hz, and assume the noise vector $\mathbf{n}$ at the MAC receiver is circularly symmetric complex Gaussian with $\mathbf{n} \sim N(0, \mathbf{I})$. Let the $M_r \times M_t$ matrix $\mathbf{H}_k$ characterize the MIMO MAC channel gains between each antenna at the kth user's transmitter and each antenna at the receiver. Define $\mathbf{H} = [\mathbf{H}_1 \ \dots \ \mathbf{H}_K]$. Then the capacity region of the Gaussian MIMO MAC where user k has channel gain matrix $\mathbf{H}_k$ and power P_k is given [55; 56; 57] by

$$\mathcal{C}_{\text{MAC}}((P_1, \dots, P_K); \mathbf{H})$$

$$= \bigcup_{\{\mathbf{Q}_k \geq 0, \, \text{Tr}(\mathbf{Q}_k) \leq P_k \, \forall k\}} \left\{ (R_1, \dots, R_K) : \right.$$

$$\left. \sum_{k \in S} R_k \leq \log_2 \det \left[\mathbf{I} + \sum_{k \in S} \mathbf{H}_k \mathbf{Q}_k \mathbf{H}_k^H \right] \forall S \subseteq \{1, \dots, K\} \right\}. \quad (14.54)$$

This region is achieved as follows. The kth user transmits a zero-mean Gaussian random vector $\mathbf{x}_k$ with spatial covariance matrix $\mathbf{Q}_k = \mathbf{E}[\mathbf{x}_k \mathbf{x}_k^H]$. Each set of covariance matrices $(\mathbf{Q}_1, \dots, \mathbf{Q}_K)$ corresponds to a K-dimensional polyhedron (i.e., $\{(R_1, \dots, R_K) : \sum_{k \in S} R_k \leq \log_2 \det[\mathbf{I} + \sum_{k \in S} \mathbf{H}_k \mathbf{Q}_k \mathbf{H}_k^H] \forall S \subseteq \{1, \dots, K\}\}$), and the capacity region is equal to the union (over all covariance matrices satisfying the power constraints) of all such polyhedrons. The corner points of this pentagon can be achieved by successive decoding: users' signals are successively decoded and subtracted out of the received signal. Note that the capacity region (14.54) has several similarities with its single-antenna counterpart: it is defined based on the rate sum associated with subsets of users, and the corner points of the region are obtained using successive decoding.

For the two-user case, each set of covariance matrices corresponds to a pentagon, which is similar in form to the capacity region of the single-antenna MAC. For example, the corner point where $R_1 = \log_2 \det[\mathbf{I} + \mathbf{H}_1 \mathbf{Q}_1 \mathbf{H}_1^H]$ and

$$R_2 = \log_2 \det[\mathbf{I} + \mathbf{H}_1 \mathbf{Q}_1 \mathbf{H}_1^H + \mathbf{H}_2 \mathbf{Q}_2 \mathbf{H}_2^H] - R_1$$

$$= \log_2 \det[\mathbf{I} + (\mathbf{I} + \mathbf{H}_1 \mathbf{Q}_1 \mathbf{H}_1^H)^{-1} \mathbf{H}_2 \mathbf{Q}_2 \mathbf{H}_2^H]$$

corresponds to decoding user 2 first (i.e., in the presence of interference from user 1) and decoding user 1 last (without interference from user 2).

14.7 Uplink–Downlink Duality

The downlink and uplink channels shown in Figure 14.1 appear to be quite similar: the downlink is almost the same as the uplink with the direction of the arrows reversed. However, there are three fundamental differences between the two channel models. First, in the downlink there is an additive noise term associated with each receiver, whereas in the uplink there is only one additive noise term (because there is only one receiver). A second fundamental difference is that the downlink has a single power constraint associated with the transmitter, whereas the uplink has different power constraints associated with each user. Finally, on the downlink both the signal and interference associated with each user travel through the same channel, whereas on the uplink these signals travel through different channels, which gives rise to the near–far effect. Despite extensive study of uplink and downlink channels individually, there has been little effort to draw connections between the two models or exploit these connections in analysis and design. In this section we describe a duality relationship between the two channels, and we show how this relationship can be used in capacity analysis and in the design of uplink and downlink transmission strategies.

We say that K-user downlink and uplink, as shown in Figure 14.1 for $K = 3$, are *duals* of each other if all of the following three conditions hold.

1. The channel impulse responses $h_k(t)$, $k = 1, \ldots, K$, in the downlink are the same as in the uplink for all k. For AWGN or flat fading channels, this implies that bi-directional links have the same channel gain in each direction.
2. Each channel in the downlink has the same noise statistics, and these statistics are the same as those of the channel noise in the uplink.
3. The power constraint P on the downlink equals the sum of individual power constraints P_k $(k = 1, \ldots, K)$ on the uplink.

Despite the similarities between the downlink (BC) and uplink (MAC), their capacity regions are quite different. In particular, the two-user AWGN BC capacity region (shown by the largest rate region in Figure 14.10) is markedly different from the two-user AWGN MAC capacity region (shown in Figure 14.14). The capacity regions of dual MACs and BCs are also very different in fading under any of the fading channel capacity definitions: ergodic, outage, or minimum rate capacity. However, despite their different shapes, the capacity regions of the dual channels are both achieved using a superposition coding strategy, and the optimal decoders for the dual channels exploit successive decoding and interference cancellation.

The duality relationship between the two channels is based on exploiting their similar encoding and decoding strategies while bridging their differences by (i) summing the individual MAC power constraints to obtain the BC power constraint and (ii) scaling the BC gains to achieve the near–far effect of the MAC. This relationship was developed in [54], where it was used to show that the capacity region and optimal transmission strategy of either the BC or the MAC can be obtained from the capacity region and optimal transmission strategy of the dual channel. In particular, it was shown in [54] that the capacity region of the AWGN BC with power P and channel gains $\mathbf{g} = (g_1, \ldots, g_K)$ is equal to the capacity region of the dual AWGN MAC with the same channel gains, but where the MAC is subject to a sum power constraint $\sum_{k=1}^{K} P_k \leq P$ instead of individual power constraints $(P_1, \ldots, P_k)$. The sum power constraint in the MAC implies that the MAC transmitters draw power from a single pooled power source with total power P and that power is allocated between the MAC transmitters such that $\sum_{k=1}^{K} P_k \leq P$. Mathematically, the BC capacity region can be expressed as the union of capacity regions for its dual MAC with a pooled power constraint:

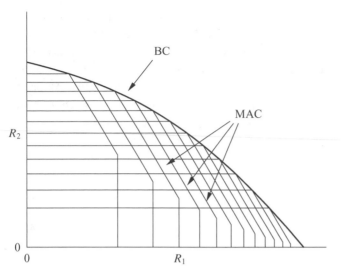

Figure 14.15: AWGN downlink (BC) capacity region as a union of capacity regions for the dual uplink (MAC).

$$\mathcal{C}_{BC}(P, \mathbf{g}) = \bigcup_{\{(P_1, \ldots, P_K): \sum_{i=1}^{K} P_k = P\}} \mathcal{C}_{MAC}(P_1, \ldots, P_K; \mathbf{g}) \qquad (14.55)$$

[54]. Here $\mathcal{C}_{BC}(P, \mathbf{g})$ is the AWGN BC capacity region with total power constraint P and channel gains $\mathbf{g} = (g_1, \ldots, g_K)$, as given by (14.18) with $n_k = N_0/g_k$, and $\mathcal{C}_{MAC}(P_1, \ldots, P_K; \mathbf{g})$ is the AWGN MAC capacity region with individual power constraints $P_1, \ldots, P_K$ and channel gains $\mathbf{g} = (g_1, \ldots, g_K)$, as given by (14.40). This relationship is illustrated for two users in Figure 14.15, where we see the BC capacity region formed from the union of MAC capacity regions with different power allocations between MAC transmitters that sum to the total power P of the dual BC.

In addition to the capacity region relationship of (14.55), it is also shown in [54] that the optimal power allocation for the BC associated with any point on the boundary of its capacity region can be obtained from the allocation of the sum power on the dual MAC that intersects with that point. Moreover, the decoding order of the BC for that intersection point is the reverse decoding order of this dual MAC. Thus, the optimal encoding and decoding strategy for the BC can be obtained from the optimal strategies associated with its dual MAC. This connection between optimal uplink and downlink strategies may have interesting implications for practical designs.

Duality also implies that the MAC capacity region can be obtained from that of its dual BC. This relationship is based on the notion of channel scaling. It is easily seen from (14.40) that the AWGN MAC capacity region is not affected if the kth user's channel gain g_k is scaled by power gain α – as long as its power P_k is also scaled by $1/\alpha$. However, the dual BC is fundamentally changed by channel scaling because the encoding and decoding order of superposition coding on the BC is determined by the order of the channel gains. Thus, the capacity region of the BC with different channel scalings will be different, and it is shown in [54] that the MAC capacity region can be obtained by taking an intersection of the BC with all possible channel scalings α_k on the kth user's channel. Mathematically, we obtain the MAC capacity region from the dual BC as

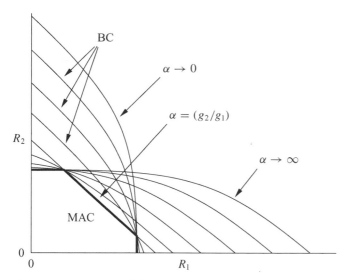

Figure 14.16: AWGN uplink (MAC) capacity region as an intersection of capacity regions for the scaled dual downlink (BC).

$$\mathcal{C}_{\text{MAC}}(P_1, \ldots, P_K; \mathbf{g}) = \bigcap_{(\alpha_1, \ldots, \alpha_K) > 0} \mathcal{C}_{\text{BC}}\left(P = \sum_{k=1}^{K} \frac{P_k}{\alpha_k}; \mathbf{g} = (\alpha_1 g_1, \ldots, \alpha_K g_K)\right), \quad (14.56)$$

where $\mathcal{C}_{\text{MAC}}(P_1, \ldots, P_K; \mathbf{g})$ and $\mathcal{C}_{\text{BC}}(P, \mathbf{g})$ are defined as in (14.55). This relationship is illustrated in Figure 14.16 for two users with channel gain $\mathbf{g} = (g_1, g_2)$. The figure shows that the MAC capacity region is formed from the intersection of BC capacity regions with different channel scalings α applied to the first user.[9] As $\alpha \to 0$, the channel gain αg_1 of the first user goes to zero but the total power $P = P_1/\alpha + P_2$ goes to infinity. Since user 2's channel gain doesn't change, he takes advantage of the increased power and his rate grows asymptotically large with α. The opposite happens as $\alpha \to \infty$: user 1's channel gain grows and the total power $P = P_1/\alpha + P_2 \geq P_2$, so user 1 takes advantage of his increasing channel gain to achieve an asymptotically large rate with any portion of the total power P. All scalings between zero and infinity sketch out different BC capacity regions that intersect to form the MAC region. In particular, when $\alpha = g_2/g_1$, the channel gains of both users in the scaled BC are the same, and this yields the time-sharing segment of the MAC capacity region. The optimal decoding order of the MAC for a given point on its capacity region can also be obtained from the channel scaling associated with the dual scaled BC whose capacity region intersects the MAC capacity region at that point.

These duality relationships are extended in [54] to many other important channel models. In particular, duality applies to flat fading MACs and BCs, so that the ergodic, outage, and minimum rate capacity regions (along with the optimal encoding and decoding strategies) for one channel can be obtained from the regions and strategies for the dual channel. MAC and BC duality also holds for parallel and frequency-selective fading channels, which defines the connection between the capacity regions of frequency-selective fading MACs and BCs [58; 59]. Another important application of duality is to multiple antenna (MIMO) MACs and

---

[9] It is sufficient to take the intersection for scaling over just $K - 1$ users, because scaling by $(\alpha_1, \ldots, \alpha_{K-1}, \alpha_K)$ is equivalent to scaling by $(\alpha_1/\alpha_K, \ldots, \alpha_{K-1}/\alpha_K, 1)$.

BCs [48; 60; 61]. In [60] the notion of duality between the BC and the MAC was extended to MIMO systems such that the MIMO BC capacity region with power constraint P and channel gain matrix $\mathbf{H}$ was shown to equal the union of capacity regions of the dual MAC with channel gain matrix $\mathbf{H}^H$, where the union is taken over all individual power constraints that sum to P. Mathematically, we have

$$\mathcal{C}_{\mathrm{BC}}(P, \mathbf{H}) = \bigcup_{(P_1, \ldots, P_K): \sum_{k=1}^{K} P_k = P} \mathcal{C}_{\mathrm{MAC}}((P_1, \ldots, P_K); \mathbf{H}^H),$$

where $\mathcal{C}_{\mathrm{BC}}(P, \mathbf{H})$ is given by (14.37) and $\mathcal{C}_{\mathrm{MAC}}((P_1, \ldots, P_K); \mathbf{H}^H)$ is given by (14.54). This duality relationship is illustrated in Figure 14.13, where the MIMO BC capacity region is defined by the outer boundary in the figure. The regions inside this boundary are the MIMO MAC capacity region under different individual user power constraints that sum to the total BC power P. Recall that the MIMO BC capacity region is extremely difficult to compute directly, since it is neither concave nor convex over the covariance matrices that must be optimized. In contrast, the optimal MIMO MAC is obtained via a standard convex optimization that is easy to solve [62]. Moreover, duality not only relates the two capacity regions but can also be used to obtain the optimal transmission strategy for a point in the MIMO BC capacity region from a duality transformation of the optimal MIMO MAC strategy that achieves the same point. Thus, for MIMO channels, duality can be exploited to simplify calculation of the capacity region and also to simplify finding the corresponding optimal transmission strategy.

14.8 Multiuser Diversity

Multiuser diversity takes advantage of the fact that, in a system with many users whose channels fade independently, at any given time some users will have better channels than others. By transmitting only to users with the best channels at any given time, system resources are allocated to the users that can best exploit them, which leads to improved system capacity and/or performance. Multiuser diversity was first explored in [53] as a means to increase throughput and reduce error probability in uplink channels, and the same ideas can be applied to downlink channels. The multiuser diversity concept is an extension of the single-user diversity concepts described in Chapter 7. In single-user diversity systems, a point-to-point link consists of multiple independent channels whose signals can be combined to improve performance. In multiuser diversity the multiple channels are associated with different users, and the system typically uses selection diversity to select the user with the best channel in any given fading state. The multiuser diversity gain relies on disparate channels between users and so, the larger the dynamic range of the fading, the higher the multiuser diversity gain. In addition, as with any diversity technique, performance improves with the number of independent channels. Thus, multiuser diversity is most effective in systems with a large number of users.

From Section 14.5, we have seen that the total throughput (sum-rate capacity) of the fading downlink is maximized by allocating the full system bandwidth to the user with the best channel in each fading state. As described in Section 14.6, a similar result holds for the fading uplink if all users have the same fading distribution and average power. If the users have different fading statistics or average powers, then the channel in any given state is allocated to the user with the best weighted channel gain, where the weight depends on the user's channel gain in the given state, his fading statistics, and his average power constraint. The notion of scheduling transmissions to users based on their channel conditions is called *opportunistic scheduling,* and numerical results in [34; 53] show that opportunistic scheduling, when coupled with power control, can significantly increase both uplink and downlink throughput as measured by sum-rate capacity.

Opportunistic scheduling can also improve BER performance [53]. Let $\gamma_k[i]$ for $k = 1, \ldots, K$ denote the SNR for each user's channel at time i. By transmitting only to the user with the largest SNR, the system SNR at time i is $\gamma[i] = \max_k \gamma_k[i]$. It is shown in [63] that, in independent and identically distributed Rayleigh fading, this maximum SNR is roughly $\ln K$ larger than the SNR of any one user as K grows asymptotically large, leading to a multiuser diversity gain in SNR of $\ln K$. Moreover, if $P_s(\gamma)$ denotes the probability of symbol error for the user with the best channel gain at time i, then $P_s(\gamma)$ will exhibit the same diversity gains as selection combining in a single-user system (described in Section 7.2.2) as compared to the probability of error associated with any one user. As the number of users in the system increases, the probability of error approaches that of an AWGN channel without fading, analogous to increasing the number of branches in single-user selection combining diversity.

Scheduling transmission to users with the best channel raises two problems in wireless systems: fairness and delay. If user fade levels change very slowly, then one user will occupy the system for a long period of time. The time between channel uses for any one user could be quite long, and such latency might be unacceptable for a given application. In addition, users with poor average SNRs will rarely have the best channel and thus are rarely allowed to transmit, which leads to unfairness in the allocation of the system resources. A solution to the fairness and delay problems in the downlink, called *proportional fair scheduling,* was proposed in [63]. Suppose that, at time i, each of the K users in the downlink system can support rate $R_k[i]$ if allocated the full power and system bandwidth. Let $T_k[i]$ denote the average throughput of the kth user at time i, averaged using an exponentially weighted lowpass filter parameterized by i_c, which is a parameter of the scheduler design. In the ith time slot, the scheduler then transmits to the user with the largest ratio $R_k[i]/T_k[i]$. If at time i all users have experienced the same average throughput $T_k[i] = T[i]$, then this scheduler transmits to the user with the best channel. Suppose, however, that one user (user j) has experienced poor throughput, so that $T_j[i] \ll T_k[i]$ for $j \neq k$. Then, at time i, user j will likely have a high ratio of $R_j[i]/T_j[i]$ and thus will be favored in the allocation of resources at time i. Assuming that at time i the user k^* has the highest ratio of $R_k[i]/T_k[i]$, the throughput on the next timeslot is updated as

$$T_k[i+1] = \begin{cases} (1 - 1/i_c)T_k[i] + (1/i_c)R_k[i] & k = k^*, \\ (1 - 1/i_c)T_k[i] & k \neq k^*. \end{cases} \tag{14.57}$$

With this scheduling scheme, users with the best channels are still allocated the channel resources when throughput between users is reasonably fair. However, if the throughput of any one user is poor, that user will be favored for resource allocation until his throughput becomes reasonably balanced with that of the other users. Clearly this scheme will have a lower throughput than allocating all resources to the user with the best channel (which maximizes throughput), and the throughput penalty will increase as the users have more disparate average channel qualities. The latency with this scheduling scheme is controlled via the weighting parameter i_c. As this parameter increases the latency also increases, but system throughput increases as well because the scheduler has more flexibility in allocating resources to users. As i_c grows to infinity, the proportional fair scheduler simply reduces to allocating system resources to the user with the best channel, assuming all users have the same average SNR. The proportional fair scheduling algorithm is part of the standard for packet data transmission in CDMA2000 cellular systems [64]; its performance for that system is evaluated in [65]. Alternative methods for incorporating fairness and delay constraints in opportunistic scheduling have been evaluated in [66; 67], along with their performance under practical constraints such as imperfect channel estimates.

14.9 MIMO Multiuser Systems

Multiuser systems with multiple antennas at the transmitter(s) and/or receiver(s) are called MIMO multiuser systems. These multiple antennas can significantly enhance performance in many ways. The antennas can be used to provide diversity gain to improve BER performance. The capacity region of the multiuser channel is increased with multiple transmit and receive antennas, providing multiplexing gain. Finally, multiple antennas can provide directivity gain to spatially separate users, which reduces interference. There typically are trade-offs among these three types of performance gains in MIMO multiuser systems [68].

The multiplexing gain of a MIMO multiuser system characterizes the increase in the uplink or downlink capacity region associated with adding multiple antennas. The capacity regions of MIMO multiuser channels have been extensively studied, motivated by the large capacity gains associated with single-user systems. For AWGN channels the MIMO capacity region for both the uplink and the downlink is described (respectively) in Section 14.6.3 and Section 14.5.5. These results can be extended to find the MIMO capacity region in fading with perfect CSI at all transmitters and receivers. Capacity results and open problems related to MIMO multiuser fading channels under other assumptions about channel CSI are described in [69].

Beamforming was discussed in Section 10.4 as a technique to achieve full diversity in single-user systems at the expense of some capacity loss. In multiuser systems, beamforming has less of a capacity penalty because of the multiuser diversity effect, and in fact beamforming can achieve the sum-rate capacity of the MIMO downlink in the asymptotic limit of a large number of users [70; 71].

Multiuser diversity is based on the idea that in multiuser channels the channel quality varies across users, so performance can be improved by allocating system resources at any given time to the users with the best channels. Design techniques to exploit multiuser diversity were discussed in Section 14.8 for single-antenna multiuser systems. In MIMO multiuser systems, the benefits of multiuser diversity are twofold. First, MIMO multiuser diversity provides improved channel quality, since only users with the best channels are allocated system resources. In addition, MIMO multiuser diversity provides abundant directions where users have good channel gains, so that users chosen for resource allocation in a given state have not only good channel quality but also good spatial separation, thereby limiting interference between them. This twofold diversity benefit allows relatively simple, suboptimal transmitter and receiver techniques to achieve near-optimal performance as the number of users increases [70; 72]. It also eliminates the requirement for multiple receive antennas in downlinks and multiple transmit antennas in uplinks in order to obtain large capacity gains, which reduces complexity and power consumption in the mobile terminal. In particular, the sum-rate capacity gain in MIMO BCs increases roughly linearly with the minimum of the number of transmit antennas and the total number of receive antennas associated with all users [73]. Thus, given a fixed number of transmit antennas at a base station and a large number of users, there is no capacity benefit to having multiple antennas at the user terminals. Similarly, the sum-rate capacity gain in MIMO MACs increases roughly linearly with the minimum of the number of receive antennas and the total number of transmit antennas associated with all users. Hence, in a system with many users and a large number of receive antennas at the base station, adding transmit antennas at the user terminals will not increase capacity. Note that multiuser diversity increases with the dynamic range and rate of the channel fading. By modulating (in a controlled fashion) the amplitude and phase of multiple transmit antennas, the fading rate and dynamic range can be increased, leading to higher multiuser diversity gains. This technique, called *opportunistic beamforming,* is investigated in [63].

Space-time modulation and coding techniques for MIMO multiuser systems have also been developed [74; 75; 76]. The goal of these techniques is to achieve the full range of diversity, multiplexing, and directivity trade-offs inherent in MIMO multiuser systems. Multiuser detection techniques can also be extended to MIMO channels and provide substantial performance gains [77; 78; 79]. In wideband channels, the multiuser MIMO techniques must also cope with frequency-selective fading [76; 80; 81]. Advanced transmission techniques for these wideband channels promise even more significant performance gains than in narrowband channels, since frequency-selective fading provides yet another form of diversity. At low SNRs, however, multiple antennas are mainly used to collect energy. The challenge for MIMO multiuser systems is to develop signaling techniques of reasonable complexity that deliver on the promised performance gains even in practical operating environments.

PROBLEMS

14-1. Consider an FDMA system for multimedia data users. The modulation format requires 10 MHz of spectrum for each user, and guard bands of 1 MHz are required on each side of the allocated spectrum in order to minimize out-of-band interference. What total bandwidth is required to support 100 simultaneous users in this system?

14-2. GSM systems have 25 MHz of bandwidth allocated to their uplink and downlink, divided into 125 TDMA channels with 8 user timeslots per channel. A GSM frame consists of the 8 timeslots, preceded by a set of preamble bits and followed by a set of trail bits. Each timeslot consists of 3 start bits at the beginning, followed by a burst of 58 data bits, then 26 equalizer training bits, another burst of 58 data bits, 3 stop bits, and a guard time corresponding to 8.25 data bits. The transmission rate is 270.833 kbps.

(a) Sketch the structure of a GSM frame and a timeslot within the frame.

(b) Find the fraction of data bits within a timeslot as well as the information data rate for each user.

(c) Find the duration of a frame and the latency between timeslots assigned to a given user in a frame, neglecting the duration of the preamble and trail bits.

(d) What is the maximum delay spread in the channel such that the guard band and stop bits prevent overlap between timeslots?

14-3. Consider a DS CDMA system occupying 10 MHz of spectrum. Assume an interference-limited system with a spreading gain of $G = 100$ and code cross-correlation of $1/G$.

(a) For the MAC, find a formula for the SIR of the received signal as a function of G and the number of users K. Assume that all users transmit at the same power and that there is perfect power control, so all users have the same received power.

(b) Based on your SIR formula in part (a), find the maximum number of users K that can be supported in the system, assuming BPSK modulation with a target BER of 10^{-3}. In your BER calculation you can treat interference as AWGN. How does this compare with the maximum number of users K that an FDMA system with the same total bandwidth and information signal bandwidth could support?

(c) Modify your SIR formula in part (a) to include the effect of voice activity, defined as the percentage of time that users are talking, so interference is multiplied by this percentage. Also find the voice activity factor such that the CDMA system accommodates the same number of users as an FDMA system. Is this a reasonable value for voice activity?

14-4. Consider an FH CDMA system that uses FSK modulation and the same spreading and information bandwidth as the DS CDMA system in Problem 14.3. Thus, there are $G = 100$ frequency slots in the system, each of bandwidth 100 kHz. The hopping codes are random and uniformly distributed, so the probability that a given user occupies a given frequency slot on any hop is .01. As in Problem 14.3, noise is essentially negligible, so the probability of error on a particular hop is zero if only one user occupies that hop. Also assume perfect power control, so the received power from all users is the same.

(a) Find an expression for the probability of bit error when m users occupy the same frequency slot.

(b) Assume there is a total of K users in the system at any time. What is the probability that, on any hop, there is more than one user occupying the same frequency?

(c) Find an expression for the average probability of bit error as a function of K, the total number of users in the system.

14-5. Compute the maximum throughput T for a pure ALOHA and a slotted ALOHA random access system, along with the load L that achieves the maximum in each case.

14-6. Consider a pure ALOHA system with a transmission rate of $R = 10$ Mbps. Compute the load L and throughput T for the system assuming 1000 bit packets and a Poisson arrival rate of $\lambda = 10^3$ packets per second. Also compute the effective data rate (rate of bits successfully received). What other value of load L results in the exact same throughput?

14-7. Consider a three-user uplink channel with channel power gains $g_1 = 1$, $g_2 = 3$, and $g_3 = 5$ from user k to the receiver, $k = 1, 2, 3$. Assume that all three users require a 10-dB SINR. The receiver noise is $n = 1$.

(a) Confirm that the vector equation $(\mathbf{I} - \mathbf{F})\mathbf{P} \geq \mathbf{u}$ given by (14.6) is equivalent to the SINR constraints of each user.

(b) Determine whether a feasible power vector exists for this system such that all users meet the required SINR constraints and, if so, find the optimal power vector $\mathbf{P}^*$ such that the desired SINRs are achieved with minimum transmit power. Assume $\rho = .1$.

14-8. Find the two-user broadcast channel capacity region under superposition coding for transmit power $P = 10$ mW, $B = 100$ kHz, and $N_0 = 10^{-9}$ W/Hz.

14-9. Show that the sum-rate capacity of the AWGN BC is achieved by sending all power to the user with the highest channel gain.

14-10. Derive a formula for the optimal power allocation on a fading broadcast channel to maximize sum-rate.

14-11. Find the sum-rate capacity of a two-user fading BC, where the fading on each user's channel is independent. Assume that each user has (i) a received power of 10 mW and (ii) an effective noise power of 1 mW with probability .5 or 5 mW with probability .5.

14-12. Find the sum-rate capacity for a two-user broadcast fading channel where each user experiences Rayleigh fading. Assume an average received power of $P = 10$ mW for each user, bandwidth $B = 100$ kHz, and $N_0 = 10^{-9}$ W/Hz.

14-13. Consider the set of achievable rates for a broadcast fading channel under frequency division. Given any rate vector in $\mathcal{C}_{\mathrm{FD}}(\mathcal{P}, \mathcal{B})$ for a given power policy $\mathcal{P}$ and bandwidth allocation policy $\mathcal{B}$, as defined in (14.30), find the timeslot and power allocation policy that achieves the same rate vector.

14-14. Consider a time-varying broadcast channel with total bandwidth $B = 100$ kHz. The effective noise for user 1 has $n_1 = 10^{-5}$ W/Hz with probability $3/4$ or $n_1 = 2 \cdot 10^{-5}$ W/Hz with probability $1/4$. The effective noise for user 2 takes the value $n_2 = 10^{-5}$ W/Hz with probability $1/2$ or the value $n_2 = 2 \cdot 10^{-5}$ W/Hz with probability $1/2$. These noise densities are independent of each other over time. The total transmit power is $P = 10$ W.

(a) What is the set of all possible joint noise densities and corresponding probabilities?

(b) Obtain the optimal power allocation between the two users and the corresponding time-varying capacity rate region using time division. Assume that (i) user k is allocated a fixed timeslot τ_k for all time (where $\tau_1 + \tau_2 = 1$) and a fixed average power P over all time and (ii) that each user may change its power within its own timeslot, subject to the average constraint P. Find a rate point that exceeds this region, assuming you don't divide power equally.

(c) Assume now fixed frequency division, where the bandwidth assigned to each user is fixed and evenly divided between the two users: $B_1 = B_2 = B/2$. Assume also that you allocate half the power to each user within his respective bandwidth ($P_1 = P_2 = P/2$) and that you can vary the power over time, subject only to the average power constraint $P/2$. What is the best rate point that can be achieved? Find a rate point that exceeds this region, assuming that you don't share power and/or bandwidth equally.

(d) Is the rate point ($R_1 = 100000$, $R_2 = 100000$) in the zero-outage capacity region of this channel?

14-15. Show that the K-user AWGN MAC capacity region is not affected if the kth user's channel power gain g_k is scaled by α as long as the kth user's transmit power P_k is also scaled by $1/\alpha$.

14-16. Consider a multiple access channel being shared by two users. The total system bandwidth is $B = 100$ kHz. The transmit power of user 1 is $P_1 = 3$ mW, while the transmit power of user 2 is $P_2 = 1$ mW. The receiver noise density is $.001\ \mu$W/Hz. You can neglect any path loss, fading, or shadowing effects.

(a) Suppose user 1 requires a data rate of 300 kbps to see videos. What is the maximum rate that can be assigned to user 2 under time division? Under superposition coding with successive interference cancellation?

(b) Compute the rate pair (R_1, R_2) where the frequency-division rate region intersects the region achieved by code division with successive interference cancellation (assuming $G = 1$).

(c) Compute the rate pair (R_1, R_2) such that $R_1 = R_2$ (i.e., where the two users get the same rate) for time division and for spread-spectrum code division with and without successive interference cancellation for a spreading gain $G = 10$. *Note:* To obtain this region for $G > 1$ you must use the same reasoning on the MAC as was used to obtain the BC capacity region with $G > 1$.

14-17. Show that the sum-rate capacity of the AWGN MAC is achieved by having all users transmit at full power.

14-18. Derive the optimal power adaptation for a two-user fading MAC that achieves the sum-rate point.

14-19. Find the sum-rate capacity of a two-user fading MAC where the fading on each user's channel is independent. Assume each user has (i) a received power of 10 mW and (ii) an effective noise power of 1 mW with probability .5 or 5 mW with probability .5.

14-20. Consider a three-user fading downlink with bandwidth 100 kHz. Suppose the three users all have the same fading statistics, so that their received SNRs when allocated the full power and bandwidth are 5 dB with probability 1/3, 10 dB with probability 1/3, and 20 dB with probability 1/3. Assume a discrete-time system with i.i.d. fading at each time slot.

(a) Find the maximum throughput of this system if (at each time instant) the full power and bandwidth are allocated to the user with the best channel.

(b) Simulate the throughput obtained using the proportional fair scheduling algorithm for i_c values of 1, 5, and 10.

REFERENCES

[1]   D. Bertsekas and R. Gallager, *Data Networks,* 2nd ed., Prentice-Hall, Englewood Cliffs, NJ, 1992.

[2]   T. S. Rappaport, *Wireless Communications – Principles and Practice,* 2nd ed., Prentice-Hall, Englewood Cliffs, NJ, 2001.

[3]   G. Leus, S. Zhou, and G. B. Giannakis, "Orthogonal multiple access over time- and frequency-selective channels," *IEEE Trans. Inform. Theory,* pp. 1942–50, August 2003.

[4] S. Verdú, "Demodulation in the presence of multiuser interference: Progress and misconceptions," in D. Docampo, A. Figueiras, and F. Perez-Gonzalez, Eds., *Intelligent Methods in Signal Processing and Communications,* pp. 15–46, Birkhäuser, Boston, 1997.

[5] W. Stallings, *Wireless Communications and Networks,* 2nd ed., Prentice-Hall, Englewood Cliffs, NJ, 2005.

[6] B. Gundmundson, J. Sköld, and J. K. Ugland, "A comparison of CDMA and TDMA systems," *Proc. IEEE Veh. Tech. Conf.,* pp. 732–5, May 1992.

[7] M. Gudmundson, "Generalized frequency hopping in mobile radio systems," *Proc. IEEE Veh. Tech. Conf.,* pp. 788–91, May 1993.

[8] K. S. Gilhousen, I. M. Jacobs, R. Padovani, A. J. Viterbi, L. A. Weaver, Jr., and C. E. Wheatley III, "On the capacity of a cellular CDMA system," *IEEE Trans. Veh. Tech.,* pp. 303–12, May 1991.

[9] P. Jung, P. W. Baier, and A. Steil, "Advantages of CDMA and spread spectrum techniques over FDMA and TDMA in cellular mobile radio applications," *IEEE Trans. Veh. Tech.,* pp. 357–64, August 1993.

[10] J. Chuang and N. Sollenberger, "Beyond 3G: Wideband wireless data access based on OFDM and dynamic packet assignment," *IEEE Commun. Mag.,* pp. 78–87, July 2000.

[11] K. R. Santhi, V. K. Srivastava, G. SenthilKumaran, and A. Butare, "Goals of true broadband's wireless next wave (4G-5G)," *Proc. IEEE Veh. Tech. Conf.,* pp. 2317–21, October 2003.

[12] M. Frodigh, S. Parkvall, C. Roobol, P. Johansson, and P. Larsson, "Future-generation wireless networks," *IEEE Wireless Commun. Mag.,* pp. 10–17, October 2001.

[13] E. Anderlind and J. Zander, "A traffic model for non-real-time data users in a wireless radio network," *IEEE Commun. Lett.,* pp. 37–9, March 1997.

[14] K. Pahlavan and P. Krishnamurthy, *Principles of Wireless Networks: A Unified Approach,* Prentice-Hall, Englewood Cliffs, NJ, 2002.

[15] N. Abramson, "The ALOHA system – Another alternative for computer communications," *Proc. Amer. Federation Inform. Proc. Soc. Fall Joint Comput. Conf.,* pp. 281–5, November 1970.

[16] A. Chockalingam and M. Zorzi, "Energy consumption performance of a class of access protocols for mobile data networks," *Proc. IEEE Veh. Tech. Conf.,* pp. 820–4, May 1998.

[17] S. Haykin and M. Moher, *Modern Wireless Communications,* Prentice-Hall, Englewood Cliffs, NJ, 2005.

[18] P. Karn, "MACA: A new channel access method for packet radio," *Proc. Comput. Network Conf.,* pp. 134–40, September 1990.

[19] V. Bharghavan, A. Demers, S. Shenkar, and L. Zhang, "MACAW: A media access protocol for wireless LANs," *Proc. ACM SIGCOMM,* vol. 1, pp. 212–25, August 1994.

[20] *IEEE Standard for Wireless LAN Medium Access Control (MAC) and Physical Layer (PHY) Specifications,* IEEE Standard 802.11, 1997.

[21] Z. J. Haas, J. Deng, and S. Tabrizi, "Collision-free medium access control scheme for ad hoc networks, *Proc. Military Commun. Conf.,* pp. 276–80, 1999.

[22] S.-L. Wu, Y.-C. Tseng, and J.-P. Sheu, "Intelligent medium access for mobile ad hoc networks with busy tones and power control," *IEEE J. Sel. Areas Commun.,* pp. 1647–57, September 2000.

[23] N. Abramson, "Wide-band random-access for the last mile," *IEEE Pers. Commun. Mag.,* pp. 29–33, December 1996.

[24] D. J. Goodman, R. A. Valenzuela, K. T. Gayliard, and B. Ramamurthi, "Packet reservation multiple access for local wireless communications," *IEEE Trans. Commun.,* pp. 885–90, August 1989.

[25] N. B. Mehta and A. J. Goldsmith, "Effect of fixed and interference-induced packet error probability on PRMA," *Proc. IEEE Internat. Conf. Commun.,* pp. 362–6, June 2000.

[26] P. Agrawal, "Energy efficient protocols for wireless systems," *Proc. Internat. Sympos. Pers., Indoor, Mobile Radio Commun.,* pp. 564–9, September 1998.

[27] J. Zander, "Performance of optimum transmitter power control in cellular radio systems," *IEEE Trans. Veh. Tech.,* pp. 57–62, February 1992.

[28] G. J. Foschini and Z. Miljanic, "A simple distributed autonomous power control algorithm and its convergence," *IEEE Trans. Veh. Tech.*, pp. 641–6, November 1993.

[29] S. A. Grandhi, R. Vijayan, and D. J. Goodman, "Distributed power control in cellular radio systems," *IEEE Trans. Commun.*, pp. 226–8, February–April 1994.

[30] E. Seneta, *Nonnegative Matrices and Markov Chains*, Springer-Verlag, New York, 1981.

[31] N. Bambos, S. C. Chen, and G. J. Pottie, "Channel access algorithms with active link protection for wireless communication networks with power control," *IEEE/ACM Trans. Network.*, pp. 583–97, October 2000.

[32] S. Kandukuri and N. Bambos, "Power controlled multiple access (PCMA) in wireless communication networks," *Proc. IEEE Infocom Conf.*, pp. 386–95, March 2000.

[33] P. P. Bergmans and T. M. Cover, "Cooperative broadcasting," *IEEE Trans. Inform. Theory*, pp. 317–24, May 1974.

[34] L. Li and A. J. Goldsmith, "Capacity and optimal resource allocation for fading broadcast channels – Part I: Ergodic capacity," *IEEE Trans. Inform. Theory*, pp. 1083–1102, March 2001.

[35] T. Cover and J. Thomas, *Elements of Information Theory*, Wiley, New York, 1991.

[36] L.-F. Wei, "Coded modulation with unequal error protection," *IEEE Trans. Commun.*, pp. 1439–49, October 1993.

[37] P. P. Bergmans, "A simple converse for broadcast channels with additive white Gaussian noise," *IEEE Trans. Inform. Theory*, pp. 279–80, March 1974.

[38] R. Pickholtz, L. Milstein, and D. Schilling, "Spread spectrum for mobile communications," *IEEE Trans. Veh. Tech.*, pp. 313–22, May 1991.

[39] S. Verdú, *Multiuser Detection*, Cambridge University Press, 1998.

[40] D. Tse and S. Hanly, "Multiaccess fading channels – Part I: Polymatroid structure, optimal resource allocation and throughput capacities," *IEEE Trans. Inform. Theory*, pp. 2796–2815, November 1998.

[41] S. Hanly and D. Tse, "Multiaccess fading channels – Part II: Delay-limited capacities," *IEEE Trans. Inform. Theory*, pp. 2816–31, November 1998.

[42] L. Li and A. J. Goldsmith, "Capacity and optimal resource allocation for fading broadcast channels – Part II: Outage capacity," *IEEE Trans. Inform. Theory*, pp. 1103–27, March 2001.

[43] L. Li, N. Jindal, and A. J. Goldsmith, "Outage capacities and optimal power allocation for fading multiple access channels," *IEEE Trans. Inform. Theory*, pp. 1326–47, April 2005.

[44] N. Jindal and A. J. Goldsmith, "Capacity and optimal power allocation for fading broadcast channels with minimum rates," *IEEE Trans. Inform. Theory*, pp. 2895–2909, November 2003.

[45] E. Larsson and P. Stoica, *Space-Time Block Coding for Wireless Communications*, Cambridge University Press, 2003.

[46] G. Caire and S. Shamai, "On the achievable throughput of a multiantenna Gaussian broadcast channel," *IEEE Trans. Inform. Theory*, pp. 1691–1706, July 2003.

[47] W. Yu and J. M. Cioffi, "Trellis precoding for the broadcast channel," *Proc. IEEE Globecom Conf.*, pp. 1344–8, November 2001.

[48] H. Weingarten, Y. Steinberg, and S. Shamai, "The capacity region of the Gaussian MIMO broadcast channel," *Proc. IEEE Internat. Sympos. Inform. Theory*, p. 174, June 2004.

[49] M. Costa, "Writing on dirty paper," *IEEE Trans. Inform. Theory*, pp. 439–41, May 1983.

[50] U. Erez, S. Shamai, and R. Zamir, "Capacity and lattice strategies for cancelling known interference," *Proc. Internat. Sympos. Inform. Theory Appl.*, pp. 681–4, November 2000.

[51] B. Rimoldi and R. Urbanke, "A rate-splitting approach to the Gaussian multiple-access channel," *IEEE Trans. Inform. Theory*, pp. 364–75, March 1996.

[52] D. Tse and P. Viswanath, *Foundations of Wireless Communications*, Cambridge University Press, 2005.

[53] R. Knopp and P. Humblet, "Information capacity and power control in single-cell multiuser communications," *Proc. IEEE Internat. Conf. Commun.*, pp. 331–5, June 1995.

[54] N. Jindal, S. Vishwanath, and A. J. Goldsmith, "On the duality of Gaussian multiple-access and broadcast channels," *IEEE Trans. Inform. Theory*, pp. 768–83, May 2004.

[55] S. Verdú, "Multiple-access channels with memory with and without frame synchronism," *IEEE Trans. Inform. Theory,* pp. 605–19, May 1989.

[56] E. Telatar, "Capacity of multi-antenna Gaussian channels," *Euro. Trans. Telecommun.,* pp. 585–96, November 1999.

[57] W. Yu, W. Rhee, S. Boyd, and J. Cioffi, "Iterative water-filling for vector multiple access channels," *IEEE Trans. Inform. Theory,* pp. 145–52, January 2004.

[58] R. Cheng and S. Verdú, "Gaussian multiaccess channels with ISI: Capacity region and multiuser water-filling," *IEEE Trans. Inform. Theory,* pp. 773–85, May 1993.

[59] A. J. Goldsmith and M. Effros, "The capacity region of broadcast channels with intersymbol interference and colored Gaussian noise," *IEEE Trans. Inform. Theory,* pp. 219–40, January 2001.

[60] S. Vishwanath, N. Jindal, and A. J. Goldsmith, "Duality, achievable rates, and sum-rate capacity of Gaussian MIMO broadcast channels," *IEEE Trans. Inform. Theory,* pp. 2658–68, October 2003.

[61] P. Viswanath and D. N. C. Tse, "Sum capacity of the vector Gaussian broadcast channel and uplink–downlink duality," *IEEE Trans. Inform. Theory,* pp. 1912–21, August 2003.

[62] N. Jindal, W. Rhee, S. Vishwanath, S. A. Jafar, and A. J. Goldsmith, "Sum power iterative water-filling for multi-antenna Gaussian broadcast channels," *IEEE Trans. Inform. Theory,* pp. 1570–9, April 2005.

[63] P. Vishwanath, D. N. C. Tse, and R. Laroia, "Opportunistic beamforming using dumb antennas," *IEEE Trans. Inform. Theory,* pp. 1277–94, June 2002.

[64] TIA/EIA IS-856, "CDMA 2000: High rate packet data air interface specification," November 2000.

[65] A. Jalali, R. Padovani, and R. Pankaj, "Data throughput of CDMA-HDR a high efficiency–high data rate personal communication wireless system," *Proc. IEEE Veh. Tech. Conf.,* pp. 1854–8, May 2000.

[66] M. Andrews, K. Kumaran, K. Ramanan, A. Stolyar, and P. Whiting, "Providing quality of service over a shared wireless link," *IEEE Commun. Mag.,* pp. 150–4, February 2001.

[67] X. Liu, E. K. P. Chong, and N. B. Shroff, "Opportunistic transmission scheduling with resource-sharing constraints in wireless networks," *IEEE J. Sel. Areas Commun.,* pp. 2053–64, October 2001.

[68] D. N. C. Tse, P. Viswanath, and L. Zheng, "Diversity–multiplexing trade-off in multiple-access channels," *IEEE Trans. Inform. Theory,* pp. 1859–74, September 2004.

[69] A. J. Goldsmith, S. A. Jafar, N. Jindal, and S. Vishwanath, "Capacity limits of MIMO channels," *IEEE J. Sel. Areas Commun.,* pp. 684–701, June 2003.

[70] M. Sharif and B. Hassibi, "Scaling laws of sum rate using time-sharing, DPC, and beamforming for MIMO broadcast channels," *Proc. IEEE Internat. Sympos. Inform. Theory,* p. 175, June 2004.

[71] T. Yoo and A. J. Goldsmith, "Optimality of zero-forcing beamforming with multiuser diversity," *Proc. IEEE Internat. Conf. Commun.,* May 2005.

[72] J. Heath, R. W. M. Airy, and A. Paulraj, "Multiuser diversity for MIMO wireless systems with linear receivers," *Proc. Asilomar Conf. Signals, Syst., Comput.,* pp. 1194–9, November 2001.

[73] N. Jindal and A. Goldsmith, "DPC vs. TDMA for MIMO broadcast channels," *IEEE Trans. Inform. Theory,* pp. 1783–94, May 2005.

[74] N. Al-Dhahir, C. Fragouli, A. Stamoulis, W. Younis, and R. Calderbank, "Space-time processing for broadband wireless access," *IEEE Commun. Mag.,* pp. 136–42, September 2002.

[75] M. Brehler and M. K. Varanasi, "Optimum receivers and low-dimensional spreaded modulation for multiuser space-time communications," *IEEE Trans. Inform. Theory,* pp. 901–18, April 2003.

[76] S. N. Diggavi, N. Al-Dhahir, and A. R. Calderbank, "Multiuser joint equalization and decoding of space-time codes," *Proc. IEEE Internat. Conf. Commun.,* pp. 2643–7, May 2003.

[77] H. Dai and H. V. Poor, "Iterative space-time processing for multiuser detection in multipath CDMA channels," *IEEE Trans. Signal Proc.,* pp. 2116–27, September 2002.

[78] S. N. Diggavi, N. Al-Dhahir, and A. R. Calderbank, "On interference cancellation and high-rate space-time codes," *Proc. IEEE Internat. Sympos. Inform. Theory,* p. 238, June 2003.

[79] S. J. Grant and J. K. Cavers, "System-wide capacity increase for narrowband cellular systems through multiuser detection and base station diversity arrays," *IEEE Trans. Wireless Commun.,* pp. 2072–82, November 2004.

[80] Z. Liu and G. B. Giannakis, "Space-time block-coded multiple access through frequency-selective fading channels," *IEEE Trans. Commun.,* pp. 1033–44, June 2001.

[81] K.-K. Wong, R. D. Murch, and K. B. Letaief, "Performance enhancement of multiuser MIMO wireless communication systems," *IEEE Trans. Commun.,* pp. 1960–70, December 2002.

15

Cellular Systems and Infrastructure-Based Wireless Networks

Infrastructure-based wireless networks have base stations, also called access points, deployed throughout a given area. These base stations provide access for mobile terminals to a backbone wired network. Network control functions are performed by the base stations, and often the base stations are connected together to facilitate coordinated control. This infrastructure is in contrast to ad hoc wireless networks, described in Chapter 16, which have no backbone infrastructure. Examples of infrastructure-based wireless networks include cellular phone systems, wireless LANs, and paging systems. Base station coordination in infrastructure-based networks provides a centralized control mechanism for transmission scheduling, dynamic resource allocation, power control, and handoff. As such, it can more efficiently utilize network resources to meet the performance requirements of individual users. Moreover, most networks with infrastructure are designed so that mobile terminals transmit directly to a base station, with no multihop routing through intermediate wireless nodes. In general these single-hop routes have lower delay and loss, higher data rates, and more flexibility than multihop routes. For these reasons, the performance of infrastructure-based wireless networks tends to be much better than in networks without infrastructure. However, it is sometimes more expensive or simply not feasible or practical to deploy infrastructure, in which case ad hoc wireless networks are the best option despite their typically inferior performance.

Cellular systems are a type of infrastructure-based network that make efficient use of spectrum by reusing it at spatially separated locations. The focus of this chapter is on cellular system design and analysis, although many of these principles apply to any infrastructure-based network. We will first describe the basic design principles of cellular systems and channel reuse. System capacity issues are then discussed along with interference reduction methods to increase this capacity. We also explain the performance benefits of dynamic resource allocation. The chapter closes with an analysis of the fundamental rate limits of cellular systems in terms of their Shannon capacity and area spectral efficiency.

15.1 Cellular System Fundamentals

The basic premise behind cellular systems is to exploit the power falloff with distance of signal propagation in order to reuse the same channel at spatially separated locations. Specifically, in cellular systems a given spatial area (such as a city) is divided into nonoverlapping

505

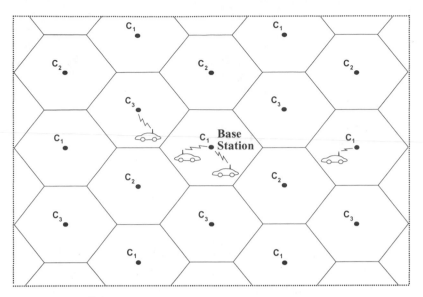

Figure 15.1: Cellular system.

cells, as shown in Figure 15.1. The signaling dimensions of the system are channelized using one of the orthogonal or nonorthogonal techniques discussed in Section 14.2. We will mostly focus on TDMA and FDMA for orthogonal channelization and on CDMA for nonorthogonal channelization. Each channel obtained by channelizing the signaling dimensions is assigned to a unique *channel set C_n*. Every cell is then assigned a channel set, and these sets can be reused at spatially separated locations, as Figure 15.1 illustrates. This reuse of channels is called *frequency reuse* or *channel reuse*. Cells that are assigned the same channel set, called *co-channel cells,* must be spaced far enough apart that interference between users in the cells does not degrade signal quality below tolerable levels. The required spacing depends on the channelization technique, the signal propagation characteristics, and the desired performance for each user.

For the cellular system shown in Figure 15.1, a base station is located near the center of each cell. Under ideal propagation conditions, mobiles within a given cell communicate with the base station in that cell, although in practice the choice is based on the SINR between the mobile and the base station. When a mobile moves between two cells, its call must be *handed off* from the base station in the original cell to the base station in the new cell. The channel from a base station to the mobiles in its cell defines the downlink of the cell, and the channel from the mobiles in a cell to the cell base station defines the uplink of the cell. All base stations in a given region are connected to a mobile telephone switching office, which acts as a central controller. User authentication, allocation of channels, and handoff between base stations are coordinated by the switching office. A handoff is initiated when the signal quality of a mobile to its base station decreases below a given threshold. This occurs when a mobile moves between cells and can also be caused by fading or shadowing within a cell. If no neighboring base station has available channels of acceptable quality then the handoff attempt fails and the call will be dropped.

The cellular system design must include a specific multiple access technique for both the uplink and the downlink. The main multiple access techniques used in cellular systems are time division, frequency division, orthogonal and nonorthogonal code division, and their hybrid combinations. These techniques are sometimes combined with space-division multiple access as well. Uplink and downlink design for a single cell was described in Section 14.2, and many of the same design principles and analyses apply to cellular systems once appropriately modified to include the impact of channel reuse. The trade-offs associated with different multiple access techniques are different in cellular systems than in a single cell, since each technique must cope with interference from outside its cell, known as *intercell* or *co-channel* interference. In addition, systems with nonorthogonal channelization must also deal with interference from within a cell, called *intracell* interference. This intracell interference also arises in systems with orthogonal channelization when multipath, synchronization errors, and other practical impairments compromise the orthogonality.

Although CDMA with nonorthogonal codes has both intracell and intercell interference inherent to its design, all interference is attenuated by the code cross-correlation. In contrast, orthogonal multiple access techniques have no intracell interference under ideal operating conditions. However, their intercell interference has no reduction from processing gain as in spread spectrum systems. The amount of both intercell and intracell interference experienced by a given user is captured by his signal-to-interference-plus-noise power ratio, defined as

$$\text{SINR} = \frac{P_r}{N_0 B + P_I},\qquad(15.1)$$

where P_r is the received signal power and P_I is the received power associated with both intracell and intercell interference. In CDMA systems, P_I is the interference power after despreading. We typically compute the bit error rate of a mobile based on SINR rather than SNR, although this approximation is not precisely accurate if the random interference cannot be characterized by Gaussian statistics.

A larger intercell interference reduces SINR and hence increases user BER. Intercell interference can be kept small by separating cells operating on the same channel by a large distance. However, the number of users that a system can accommodate is maximized by reusing frequencies as often as possible. Thus, the best cellular system design places users that share the same channel at a separation distance where the intercell interference is just below the maximum tolerable level for the required data rate and BER. Good cellular system designs are *interference limited,* meaning that the interference power is much larger than the noise power. Therefore, noise is generally neglected in the study of these systems. In this case SINR reduces to the signal-to-interference power ratio $\text{SIR} = P_r/P_I$. In interference-limited systems, since the BER of users is determined by SIR, the number of users that can be accommodated is limited by the interference they cause to other users. Techniques to reduce interference – such as multiple antenna techniques or multiuser detection – increase the SIR and therefore increase the number of users the system can accommodate for a given BER constraint. Note that the SIR or BER requirement is fairly well-defined for continuous applications such as voice. However, system planning is more complex for data applications owing to the burstiness of the transmissions.

Cell size is another important design choice in cellular systems. We can increase the number of users that can be accommodated within a given system by shrinking the size of a cell, as long as all aspects of the system scale so that the SINR of each user remains the same. Specifically, consider the large and small cells shown in Figure 15.2. Suppose the large cell

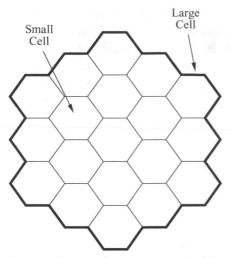

Figure 15.2: Capacity increase by shrinking cell size.

in this figure represents one cell in a cellular system, where each cell accommodates K users. If the cell size is shrunk to the smaller cell size shown in Figure 15.2 (typically by reducing transmit power) and if everything in the system (including propagation loss) scales so that the SINR in the small cells is the same as in the original large cell, then this smaller cell can also accommodate K users. Since there are 19 small cells within the large cell, the new system with smaller cells can accommodate $19K$ users within the area of one large cell – a 19-fold capacity increase. However, propagation characteristics typically change as cell size shrinks, so the system does not scale perfectly. Moreover, a smaller cell size increases the rate at which handoffs occur, which increases the dropping probability if the percentage of failed handoffs stays the same. Smaller cells also increase the load on the backbone network, and more cells per unit area means more base stations, which can increase system cost. Thus, although smaller cells generally increase capacity, they also have their disadvantages. A compromise based on these trade-offs is to embed small "hotspot" cells within large cells that experience high traffic in order to increase their capacity [1, Chap. 3.7].

The cell shape depicted in Figure 15.1 is a hexagon. A hexagon is a *tesselating* cell shape: cells can be laid next to each other with no overlap and so cover the entire geographical region without any gaps. The other tesselating shapes are rectangles, squares, diamonds, and triangles. These regular cell shapes are used to approximate the contours of constant received power around the base station. If propagation follows the free-space or simplified path-loss model where received power is constant along a circle around the base station, then a hexagon provides a reasonable approximation to this circular shape. Hexagons were commonly used to approximate cell shapes for first-generation cellular phone systems, where base stations were placed at the tops of buildings with coverage areas on the order of a few square miles. Systems today have smaller cells, with base stations placed closer to the ground; for these systems, diamonds tend to better approximate the contours of constant power, especially in typical urban street grids [2; 3]. Very small cells and indoor cells are heavily dependent on the propagation environment, making it difficult to accurately approximate contours of constant power using a tesselating shape [4].

15.2 Channel Reuse

Channel reuse is a key element of cellular system design, as it determines how much intercell interference is experienced by different users and therefore establishes the system's capacity and performance. Channel reuse considerations are different for channelization via orthogonal multiple access techniques (e.g., TDMA, FDMA, and orthogonal CDMA) as compared with those of nonorthogonal channelization techniques (nonorthogonal or hybrid orthogonal/nonorthogonal CDMA). In particular, orthogonal techniques have no intracell interference under ideal conditions. However, in TDMA and FDMA, cells using the same channel set are typically spaced several cells away, since co-channel interference from adjacent cells can be very large. In contrast, nonorthogonal channelization exhibits both intercell and intracell interference, but all interference is attenuated by the cross-correlation of the spreading codes, which allows channel sets to be reused in every cell. In CDMA systems with orthogonal codes (typical for a CDMA downlink) the codes, too, are reused in every cell, since the code transmissions from each base station are not synchronized. Thus, the same codes transmitted from different base stations arrive at a mobile with a timing offset, and the resulting intercell interference is attenuated by the code autocorrelation evaluated at that timing offset. This autocorrelation may still be somewhat large. A hybrid technique can also be used where a nonorthogonal code that is unique to each cell is modulated on top of the orthogonal codes used in that cell. The nonorthogonal code then reduces intercell interference by roughly its processing gain. This hybrid approach is used in W-CDMA cellular systems [5]. Throughout this chapter, we will assume that in CDMA systems the same codes are used in every cell, as is typically done in practice. Thus, the reuse distance is unity and we need not address optimizing channel reuse for CDMA systems.

We now discuss the basic premise of channel reuse – cell clustering – and channel assignment. In interference-limited systems, each user's BER is based on his received SIR: the ratio of his received signal power to his intracell and intercell interference power. The received signal powers associated with the desired signal, the intercell interference, and the intracell interference are determined by the characteristics of the channel between the desired or interfering transmitters and the desired receiver. The average SIR is normally computed based on path loss alone, with median shadowing attenuation incorporated into the path-loss models for the signal and interference. Random variations due to shadowing and flat fading are then treated as statistical variations about the path loss.

Since path loss is a function of propagation distance, the *reuse distance D* between cells using the same channel is an important parameter in determining average intercell interference power. Reuse distance is defined as the distance between the centers of cells that use the same channel set. It is a function of cell shape, cell size, and the number of intermediate cells between the two cells sharing the same channel set. Given a required average SINR for a particular performance level, we can find the corresponding minimum reuse distance that meets this performance target. The focus of this section is planning the cellular system layout based on a minimum reuse distance requirement.

Figure 15.3 illustrates the reuse distance associated with a given channel reuse pattern for hexagonal and diamond-shaped cells. Cells that are assigned the same channel set C_n are so indicated in the figure. This pattern of channel reuse for both cell chapes is based on the notion of cell clustering, discussed in more detail below. The reuse distance D between these cells is the minimum distance between the dots at the center of cells using channel set C_n. The radius R of a cell is also shown in the figure. For hexagonal cells, R is defined as the distance from the center of a cell to a vertex of the hexagon. For diamond-shaped cells, R is the distance from the cell center to the middle of a side.

For diamond-shaped cells it is straightforward to compute the reuse distance D based on the number of intermediate cells N_I between co-channel cells and the cell radius R. Specifically, the distance across a cell is $2R$. The distance from a cell center to its boundary

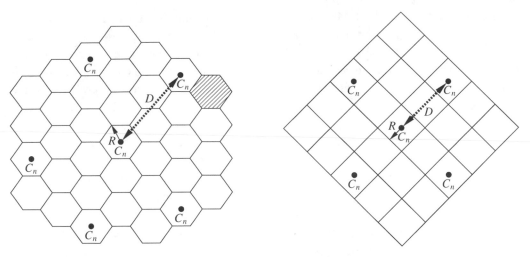

Figure 15.3: Reuse distance D with a given reuse pattern for hexagonal and diamond-shaped cells.

is R and the distance across the N_I intermediate cells between the co-channel cells is $2RN_I$. Thus, $D = R + 2RN_I + R = 2R(N_I + 1)$.

Reuse distance for hexagonally shaped cells is more complicated to determine, since there might not be an integer number of cells between two co-channel cells. In the left side of Figure 15.3, if channel C_n is used in the center cell and again in the shaded cell, then there are two cells between co-channel cells and the reuse distance is easy to find. However, if C_n is reused in the cell adjacent to the shaded cell, as is shown in the figure, then there is not an integer number of cells separating the co-channel cells. The assignment shown in Figure 15.3 is needed to create cell clusters, as discussed in more detail below.

The procedure for channel assignment in hexagonal cells is as follows. Consider the cell diagram in Figure 15.4, where again R is the hexagonal cell radius. Denote the location of each cell by the pair (i, j). Then, assuming cell A to be centered at the origin $(0, 0)$, the location of the cell at (i, j) relative to cell A is obtained by moving i cells along the u-axis, turning 60° counterclockwise, and moving j cells along the v-axis. For example, cell G is located at $(0, 1)$, cell S at $(1, 1)$, cell P at $(-2, 2)$, and cell M at $(-1, -1)$. It is straightforward to show that the distance between cell centers of adjacent cells is $\sqrt{3}R$, and the distance between the centers of the cell located at the point (i, j) and cell A (located at $(0, 0)$) is given by

$$D = \sqrt{3}R\sqrt{i^2 + j^2 + ij}. \tag{15.2}$$

EXAMPLE 15.1: Find the reuse distance D for the channel reuse pattern illustrated in Figure 15.3 for both the diamond- and hexagonally shaped cells as a function of cell radius R.

Solution: For the diamond-shaped cells, there is $N_I = 1$ cell between co-channel cells. Thus, $D = 2R(N_I + 1) = 4R$. For the hexagonal cells shown in Figure 15.3, the reuse pattern moves two cells along the u-axis and then one cell along the v-axis. Thus, $D = \sqrt{3}R\sqrt{2^2 + 1^2 + 2} = 4.58R$.

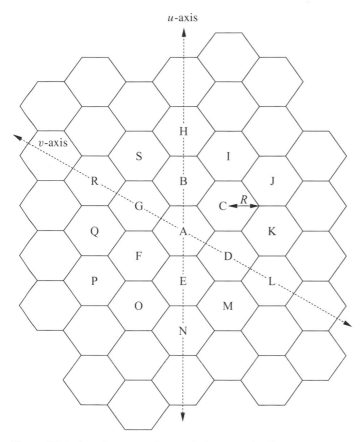

Figure 15.4: Axes for reuse distance in hexagonal cells.

Given a minimum acceptable reuse distance $D_{\min}$, we would like to maintain this distance throughout the cell grid while reusing channels as often as possible. This requires spatially repeating the *cell clusters* for channel assignment, where each cell in the cluster is assigned a unique channel set that is not assigned to any other cell in the cluster. In order to spatially repeat, cell clusters must tesselate. For diamond-shaped cells, a tesselating cell cluster forms another diamond, with K cells on each side, as shown in Figure 15.5 for $K = 4$. The channel set assigned to the nth cell in the cluster is denoted by C_n ($n = 1, \ldots, N$), where N is the number of unique channel sets, and the pattern of channel assignment is repeated in each cluster. This ensures that cells using the same channel are separated by a reuse distance of at least $D = 2KR$. The number of cells per cluster is $N = K^2$, which is also called the *reuse factor*: since $D = 2KR$, we have $N = .25(D/R)^2$. If we let N_c denote the number of channels per cell (i.e., the number of channels in C_n) and N_T the total number of channels, then $N = N_T/N_c$. A small value of N indicates efficient channel reuse (channels reused more often within a given area for a fixed cell size and shape). However, a small N also implies a small reuse distance, since $D = 2KR = 2\sqrt{N}R$, which can lead to substantial intercell interference.

For hexagonal cells, we form cell clusters through the following iterative process. The total bandwidth is first broken into N channel sets $C_1, \ldots, C_N$, where N is the cluster size.

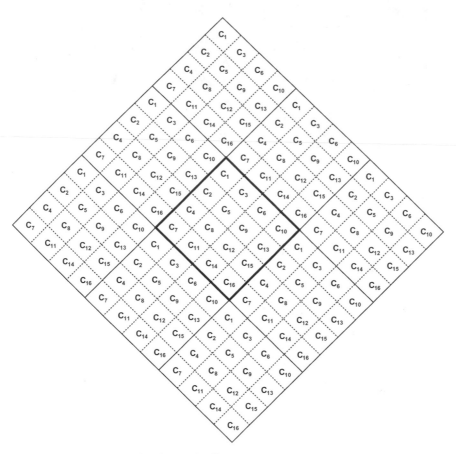

Figure 15.5: Cell clusters for diamond cells.

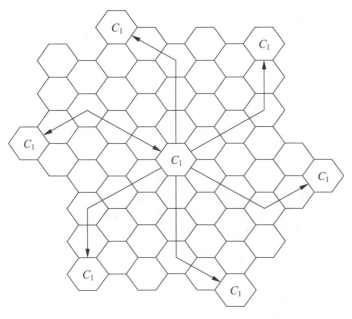

Figure 15.6: Channel assignment in hexagonal cells.

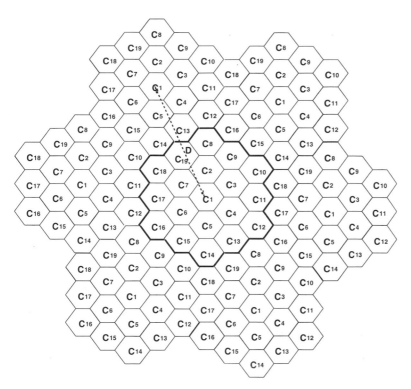

Figure 15.7: Cell clusters for hexagonal cells.

We assign the first channel set C_1 to any arbitrary cell. From this cell, we move i cells along a chain of hexagons in any direction, turn counterclockwise by $60°$, and move j cells along the hexagon chain in this new direction; channel set C_1 is then assigned to this jth cell. Going back to the original cell, we repeat the process in a different direction until we have covered all directions starting from the initial cell. This process is shown in Figure 15.6 for $i = 3$ and $j = 2$. To assign channel set C_1 throughout the region, we repeat the iterative process starting from one of the cells assigned channel set C_1 in a prior iteration until no new assignments can be made starting from any location that has been assigned channel set C_1. Then a new cell that has not been assigned any channel set is selected and assigned channel set C_2, and the iterative channel assignment process for this new channel set is performed. The process is repeated until all cells have been assigned a unique channel set, which results in cell clusters as illustrated in Figure 15.7. The reuse distance between all cells using the same channel set is then $D = \sqrt{3}R\sqrt{i^2 + j^2 + ij}$. We can obtain the approximate cluster size associated with this process by finding the ratio of cluster area to cell area. Specifically, the area of a hexagonal cell is $A_{\text{cell}} = 3\sqrt{3}R^2/2$, and the area of a cluster of hexagonal cells is $A_{\text{cluster}} = \sqrt{3}D^2/2$. Hence, the number of cells per cluster is

$$N = \frac{A_{\text{cluster}}}{A_{\text{cell}}} = \frac{\sqrt{3}D^2/2}{3\sqrt{3}R^2/2} = \frac{1}{3}\left(\frac{D}{R}\right)^2 = \frac{1}{3}\left(\frac{3R^2(i^2 + j^2 + ij)}{R^2}\right) = i^2 + j^2 + ij.$$

As with diamond cells, a small N indicates more efficient channel reuse but also a smaller reuse distance $D = R\sqrt{3N}$, leading to more intercell interference.

15.3 SIR and User Capacity

In this section we compute the SIR of users in a cellular system and the number of users per cell that can be supported for a given SIR target. We neglect the impact of noise on performance under the assumption that the system is interference-limited, although the calculations can easily be extended to include noise. The SIR in a cellular system depends on many factors, including the cell layout, size, reuse distance, and propagation. We will assume the simplified path-loss model (2.39) with reference distance $d_0 = 1$ meter for our path-loss calculations, so $P_r = P_t L d^{-\gamma}$, where L is a constant equal to the average path loss at $d = d_0$ and γ is the path-loss exponent. Since L is typically the same for both the desired signal and interference, its value cancels out in the SIR power ratio; hence we set $L = 1$ in the analysis that follows. The path-loss exponent associated with in-cell signal propagation is denoted by γ_I, while γ_O denotes the path-loss exponent associated with out-of-cell (intercell) signal propagation. These path-loss exponents may be different, depending on the propagation environment and cell size [6]. Using the simplified path-loss model, we will derive expressions for SIR under both orthogonal and nonorthogonal access techniques. We then find user capacity for each model, defined as the maximum number of users per cell that the system can support without violating an SIR target value.

The SIR of a signal is typically used to compute the BER performance associated with that signal. Specifically, the interference is approximated as additive white Gaussian noise (AWGN) and then formulas for the BER versus SNR are applied. For example, from (6.6), performance of uncoded BPSK without fading yields $P_b = Q(\sqrt{2 \cdot \text{SIR}})$, and from (6.58), performance when the desired signal exhibits Rayleigh fading yields $\bar{P}_b \approx .25/\text{SIR}$ for high SIRs. Although we have assumed the simplified path-loss model in the SIR formulas that follow, path-loss models of greater complexity can also be incorporated for a more accurate SIR approximation. However, there are a number of inaccuracies in the model that are not easy to fix. In particular, approximating the interference as Gaussian noise is accurate for a large number of interferers, as is the case for CDMA systems, but not accurate for a small number of interferers, as in TDMA and FDMA systems. Moreover, the performance computation in fading neglects the fact that the interferers also exhibit fading, which results in a received SIR that is the ratio of two random variables. This ratio has a complex distribution that is not well approximated by a Rayleigh distribution or by any other common fading distribution [7]. The complexity of modeling average SIR as well as its distribution under accurate path loss, shadowing, and multipath fading models can be prohibitive. Thus, the SIR distribution is often obtained via simulations [6].

15.3.1 Orthogonal Systems (TDMA/FDMA)

In this section we compute the SIR and user capacity for cellular systems using orthogonal multiple access techniques. In these systems there is no intracell interference, so the SIR is determined from the received signal power and the interference resulting from co-channel cells. Under the simplified path-loss model, the received signal power for a mobile located at distance d from its base station on both the uplink and the downlink is $P_r = P_t d^{-\gamma_I}$. The average intercell interference power is a function of the number of out-of-cell interferers. For simplicity we will neglect interference from outside the first ring of M interfering cells. This approximation is accurate when the path-loss exponent γ_O is relatively large, since then subsequent interfering rings have a much larger path loss than the first ring. We assume that all transmitters send at the same power P_t: the impact of power control will be discussed in Section 15.5.3. Let us assume the user is at distance $d < R$ and there are M interferers at distance d_i, $i = 1, \ldots, M$, from the intended receiver (located at the mobile in the downlink and the base station in the uplink). The resulting SIR is then

$$\text{SIR} = \frac{d^{-\gamma_I}}{\sum_{i=1}^{M} d_i^{-\gamma_O}}. \tag{15.3}$$

This SIR is the ratio of two random variables, whose distribution can be quite complex. However, the statistics of the interference in the denominator has been characterized for several propagation models, and when these interferers are log-normal, their sum is also log-normal [8, Chap. 3]. Note that in general the average SIR for the uplink and downlink may be roughly the same; but the SIR for the uplink, where interferers can all be on the cell boundary closest to the base station they interfere with, generally has a smaller worst-case value than for the downlink, where interference comes from base stations at the cell centers.

The SIR expression can be simplified if we assume that the mobile is on its cell boundary, $d = R$, and all interferers are at the reuse distance D from the intended receiver. Under these assumptions the SIR reduces to

$$\text{SIR} = \frac{R^{-\gamma_I}}{MD^{-\gamma_O}}, \tag{15.4}$$

and if $\gamma_I = \gamma_O = \gamma$ then this simplifies further to

$$\text{SIR} = \frac{1}{M}\left(\frac{D}{R}\right)^{\gamma}. \tag{15.5}$$

Since D/R is a function of the reuse factor N for most cell shapes, this allows us to express the SIR in terms of N. In particular, from Figure 15.5, for diamond-shaped cells we have $M = 8$ and $D/R = \sqrt{4N}$. Substituting these into (15.5) yields SIR $= .125(4N)^{\gamma/2}$. From Figure 15.7, for hexagonal shaped cells we have $M = 6$ and $D/R = \sqrt{3N}$, which yields SIR $= .167(3N)^{\gamma/2}$. Both of these SIR values can be expressed as

$$\text{SIR} = a_1(a_2 N)^{\gamma/2}, \tag{15.6}$$

where $a_1 = .125$ and $a_2 = 4$ for diamond-shaped cells and where $a_1 = .167$ and $a_2 = 3$ for hexagonally shaped cells. This formula provides a simple approximation for the reuse factor required to achieve a given performance. Specifically, given a target SIR value SIR_0 required for a target BER, we can invert (15.6) to obtain the minimum reuse factor that achieves this SIR target as

$$N \geq \frac{1}{a_2}\left(\frac{\text{SIR}_0}{a_1}\right)^{2/\gamma}. \tag{15.7}$$

The corresponding minimum reuse distance D (for a given cell shape) is then calculated based on N and the cell radius R. For path-loss exponent $\gamma = 2$, (15.7) simplifies to $N \geq \text{SIR}_0/a_1a_2$. If the signal has shadow fading then the analysis is more complex, but we can still generally obtain the minimum reuse factor in terms of the SIR requirement subject to some outage probability [9].

The user capacity C_u is defined as the total number of active users per cell that the system can support while meeting a common SIR target for all users. To meet this target in an orthogonal system, the reuse factor N must satisfy (15.7). Then $C_u = N_c$, where N_c is the number of channels assigned to any given cell assuming reuse factor N. The total number of orthogonal channels of bandwidth B_s that can be created from a total system bandwidth of B is $N_T = B/B_s$. Since in orthogonal systems the reuse factor N satisfies $N = N_T/N_c$, this implies that

$$C_u = \frac{N_T}{N} = \frac{B}{NB_s} = \frac{G}{N},\tag{15.8}$$

where $G = B/B_s$ is the ratio of the total system bandwidth to the bandwidth required for an individual user and where N is the minimum integer value satisfying (15.7).

EXAMPLE 15.2: Consider a TDMA cellular system with hexagonally shaped cells and with path-loss exponent $\gamma = 2$ for all signal propagation in the system. Find the minimum reuse factor N needed for a target SIR of 10 dB, and find the corresponding user capacity assuming a total system bandwidth of 20 MHz and a required signal bandwidth of 100 kHz.

Solution: To obtain the reuse factor, we apply (15.7) with $a_1 = .167$ and $a_2 = 3$ to obtain

$$N \geq \frac{\text{SIR}_0}{a_1 a_2} = \frac{10}{.5} = 20.$$

Now setting $G = B/B_s = 20 \cdot 10^6/100 \cdot 10^3 = 200$, we get $C_u = G/N = 10$ users per cell that can be accommodated. Typically $\gamma > 2$, as we consider in the next example.

EXAMPLE 15.3: Consider a TDMA cellular system with diamond-shaped cells, path-loss exponent $\gamma = 4$ for all signal propagation in the system, and BPSK modulation. Assume that the received signal exhibits Rayleigh fading. Suppose the users require $\bar{P}_b = 10^{-3}$. Assuming the system is interference limited, find the minimum reuse factor N that will meet this performance requirement. Also find the user capacity assuming a total system bandwidth of 20 MHz and a per-user signal bandwidth of 100 kHz.

Solution: Treating interference as Gaussian noise, in Rayleigh fading we have $\bar{P}_b \approx .25/\text{SIR}_0$ for SIR_0 the average SIR ratio. The SIR required to meet the $\bar{P}_b$ target is thus $\text{SIR}_0 = .25/10^{-3} = 250$ (approximately 24 dB). Substituting $\text{SIR}_0 = 250$, $a_1 = .125$, $a_2 = 4$, and $\gamma = 4$ into (15.7) yields

$$N \geq \frac{1}{4}\sqrt{\frac{250}{.25}} = 11.18.$$

So a reuse factor of $N = 12$ meets the performance requirement. For the user capacity we have $G = B/B_s = 200$, so $C_u = G/N = 16$ users per cell can be accommodated. Note that the Gaussian assumption for the interference is just an approximation, which (by the central limit theorem) becomes more accurate as the number of interferers grows.

15.3.2 Nonorthogonal Systems (CDMA)

In nonorthogonal systems, codes (i.e. channels) are typically reused in every cell, so the reuse factor is $N = 1$. Since these systems exhibit both intercell and intracell interference, the user capacity is determined by the maximum number of users per cell that can be accommodated for a given target SIR. We will neglect intercell interference from outside the first tier of interfering cells – that is, from cells that are not adjacent to the cell of interest. We will also assume that all signals follow the simplified path-loss model with the same path-loss exponent. This assumption is typically true for interference from adjacent cells, but it ultimately depends on the propagation environment.

Let $N_c = N_T = C_u$ denote the number of channels per cell. In CDMA systems the user capacity is typically limited by the uplink, owing to the near–far problem and the asynchronicity of the codes. Focusing on the uplink, under the simplified path-loss model the received signal power is $P_r = P_t d^{-\gamma}$, where d is the distance between the mobile and its base station. There are $N_c - 1$ asynchronous intracell interfering signals and MN_c asynchronous intercell interfering signals transmitted from mobiles in the M adjacent cells. Let d_i ($i = 1, \ldots, N_c - 1$) denote the distance from the ith intracell interfering mobile to the uplink receiver and let P_i denote this mobile's transmit power. Let d_j ($j = 1, \ldots, MN_c$) denote the distance from the jth intercell interfering mobile to the uplink receiver and let P_j denote this mobile's transmit power. By Section 13.4.3, all interference is reduced by the spreading code cross-correlation $\xi/3G$, where G is the processing gain of the system and ξ is a parameter of the spreading codes with $1 \leq \xi \leq 3$. The total intracell and intercell interference power is thus given by

$$I = \frac{\xi}{3G} \left(\sum_{i=1}^{N_c-1} P_i d_i^{-\gamma} + \sum_{j=1}^{MN_c} P_j d_j^{-\gamma} \right), \tag{15.9}$$

which yields the ratio

$$\text{SIR} = \frac{P_t d^{-\gamma}}{\frac{\xi}{3G} \left(\sum_{i=1}^{N_c-1} P_i d_i^{-\gamma} + \sum_{j=1}^{MN_c} P_j d_j^{-\gamma} \right)}. \tag{15.10}$$

Because all the distances in (15.10) are different, the expression cannot in general be further simplified without additional assumptions. Let us therefore assume perfect power control within a cell, so that the received power of the desired signal and interfering signals within a cell are the same: $P_r = P_t d^{-\gamma} = P_i d_i^{-\gamma}$ for all i. Furthermore, let

$$\lambda = \frac{\sum_{j=1}^{MN_c} P_j d_j^{-\gamma}}{(N_c - 1) P_r} \tag{15.11}$$

denote the ratio of average received power from all intercell interference to that of all intracell interference under this power control algorithm. Using these approximations, we get the following formula for SIR, which is commonly used for the uplink SIR in CDMA systems with power control [10; 11]:

$$\text{SIR} = \frac{1}{\frac{\xi}{3G}(N_c - 1)(1 + \lambda)}. \tag{15.12}$$

Under this approximation, for any SIR target SIR_0 we can determine the user capacity $C_u = N_c$ by setting (15.12) equal to the target SIR and solving for C_u, which yields

$$C_u = 1 + \frac{1}{\frac{\xi}{3G}(1 + \lambda)\text{SIR}_0}. \tag{15.13}$$

Voice signals, because of their statistical nature, need not be continuously active [12]. The fraction of time that a voice user actually occupies the channel is called the *voice activity factor* and is denoted by α, where $0 < \alpha \leq 1$. If the transmitter shuts off during nonactivity then the interference in CDMA – that is, the denominator of (15.13) – is multiplied by α. This increases SIR and therefore user capacity.

EXAMPLE 15.4: Consider a CDMA cellular system with perfect power control within a cell. Assume a target SIR_0 of 10 dB, a processing gain $G = 200$, spreading codes with $\xi = 2$, and equal average power from inside and outside the cell ($\lambda = 1$). Find the user capacity of this system.

Solution: From (15.13) we have

$$C_u = 1 + \frac{1}{\frac{2}{600}(2 \cdot 10)} = 16,$$

so 16 users per cell can be accommodated.

Since (15.13) and (15.8) provide simple expressions for user capacity, it is tempting to compare them for a given SIR target to determine whether TDMA or CDMA can support more users per cell. This was done in Examples 15.2 and 15.4, where for the same SIR target and other system parameters TDMA yielded 10 channels per cell while CDMA yielded 16. However, these capacity expressions are extremely sensitive to the modeling and system assumptions. Increasing λ from 1 to 2 in Example 15.4 reduces C_u for CDMA from 16 to 11, and changing the path-loss exponent in Example 15.2 for TDMA from $\gamma = 2$ to $\gamma = 3$ changes the reuse factor N from 20 to 6, which in turn changes user capacity from 10 to 33. Code-division multiple access systems can trade off spreading and coding, which yields high coding gains and a resulting lower SIR target at the expense of some processing gain [13]; high coding gain is harder to achieve in a TDMA system, since it cannot be traded for spreading gain. Voice activity was not taken into account for the CDMA system, which would lead to a higher capacity. Moreover, the CDMA capacity is derived under an assumption of perfect power control via channel inversion, whereas no power control is assumed for TDMA. The effects of shadowing and fading are also not taken into account, but fading will cause a power penalty in CDMA (owing to the channel inversion power control) and will also affect the intercell interference power for both TDMA and CDMA. All of these factors and trade-offs significantly complicate the analysis, which makes it difficult to draw general conclusions about the superiority of one technique over another in terms of user capacity. Analysis of user capacity for both TDMA and CDMA under various assumptions and models associated with operational systems can be found in [11; 14; 15; 16].

15.4 Interference Reduction Techniques

Since cellular systems are ideally interference limited, any technique that reduces interference increases SIR and user capacity. In this section we describe techniques for interference reduction in cellular systems, including sectorization, smart antennas, interference averaging, multiuser detection, and interference precancellation.

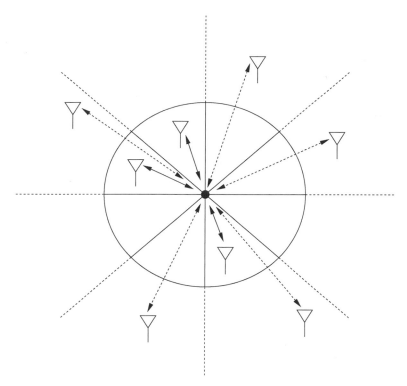

Figure 15.8: Circular cell sectorization for $N = 8$.

A common technique to reduce interference is sectorization. Antenna sectorization uses directional antennas to divide up a base station's 360° omnidirectional antenna into N sectors, as shown in Figure 15.8 for $N = 8$. As the figure indicates, intracell and intercell interference to a given mobile comes primarily from within its sector. Thus, sectorization reduces interference power by roughly a factor of N under heavy loading (interferers in every sector). Handoff is required as mobiles move between sectors, which increases overhead and the probability of call dropping. Sectorization is a common feature in cellular systems, typically with $N = 3$.

Smart antennas generally consist of an antenna array combined with signal processing in both space and time. Smart antennas can form narrow beams that provide high gain to the desired user's signal, and/or can create spatial nulls in the direction of interference signals [17]. However, antenna arrays can also be used for multiplexing gain, which leads to higher data rates, or diversity gain, which improves reliability. These fundamental trade-offs are described in Section 14.9. The use of multiple antennas for interference reduction versus their other performance benefits reflects the trade-offs associated with cellular system design.

Intercell interference in the uplink is often dominated by one or two mobile users located near the cell boundaries closest to the base station serving the desired user. In TDMA or FDMA systems, the impact of these worst-case interferers can be mitigated by superimposing frequency hopping (FH) on top of the TDMA or FDMA channelization. With this FH overlay, all mobiles change their carrier frequency according to a unique hopping pattern. Since the hopping pattern of the worst-case interferers differs from that of the desired user, these interferers would cause intercell interference to the desired signal only when the two hop patterns overlapped in both time and frequency, which seldom occurs. Thus, the FH overlay has the effect of interference averaging, causing intercell interference from any given cell to be averaged relative to all interferer locations. This greatly reduces the effect of interference associated with mobiles on cell boundaries. For this reason, an FH overlay is used in the GSM cellular system.

Another method of mitigating interference is multiuser detection [18]. Multiuser detectors jointly detect the desired signal and some or all interference, so that the detected interference can be mitigated or cancelled out. There are trade-offs between performance and complexity with different multiuser detection methods, especially when there are a large number of interferers. Multiuser detection methods were described in Section 13.4 for the uplink of a single-cell CDMA system. However, these detection methods can also be applied to intercell interference signals – both at the base station [19], where processing complexity is less of a constraint, and in the mobile device to cancel a few dominant interferers [20].

Interference precancellation takes advantage of the fact that, in the downlink, the base station has knowledge of (i) interference between users within its cell and (ii) the interference its transmission causes to mobiles in other cells. This knowledge can be utilized in the base station transmission to presubtract interference between users [21; 22; 23]. Interference presubtraction by the base station has its roots in the capacity-achieving strategy for downlinks, which uses a "dirty paper coding" transmission technique (see Section 14.5.5) for interference presubtraction developed by Costa [24]. Numerical results in [21; 22; 23] indicate that interference presubtraction can lead to an order-of-magnitude capacity increase in cellular systems with a large number of base station antennas. However, the presubtraction requires channel CSI at the transmitter – in contrast to multiuser detection, which does not require transmitter CSI.

15.5 Dynamic Resource Allocation

Cellular systems exhibit significant dynamics in the number of users in any given cell and in their time-varying channel gains. Moreover, as cellular system applications have expanded beyond just voice to include multimedia data, users no longer have uniform data-rate requirements. Thus, resource allocation to users must become more flexible in order to support heterogeneous applications. In dynamic resource allocation the channels, data rates, and power levels in the system are dynamically assigned relative to the current system conditions and user needs. Much work has gone into investigating dynamic resource allocation in cellular systems. In this section we summarize some of the main techniques, including scheduling, dynamic channel allocation, and power control. The references given are but a small sampling of the vast literature on this important topic.

15.5.1 Scheduling

The basic premise of scheduling is to dynamically allocate resources to mobile users according to their required data rates and delay constraints. Meeting these requirements is also called quality-of-service (QoS) support. Schedulers must be efficient in their allocation of resources but also fair. There is generally a trade-off between these two goals since, as discussed in Section 14.8, the most efficient allocation of resources exploits multiuser diversity to allocate resources to the user with the best channel. However, this assumes that the users with the best channels can fully utilize these resources, which may not be the case if their application does not require a high data rate. Moreover, such an allocation is unfair to users with inferior channels.

Scheduling has been investigated for both the uplink and the downlink of both CDMA and TDMA systems. Note that many CDMA cellular systems use a form of TDMA called high data rate (HDR) for downlink data transmission [1, Chap. 2.2], so these CDMA schedulers are based on TDMA channelization. Three different scheduling schemes – round robin, equal latency, and relative fairness – were compared for a TDMA downlink compatible with HDR systems in [25]. TDMA downlink scheduling exploiting multiuser diversity was investigated in [26]. Scheduling issues for the CDMA downlink – including rate adaptation, fairness, and deadline constraints – have been explored in [27; 28; 29]. For CDMA systems, uplink scheduling was investigated in [30; 31] assuming single-user matched filter detection, and improvements using multiuser detection were analyzed in [32]. MIMO provides another degree of freedom in scheduling system resources, as outlined in [33; 34]. Multiple classes of users can also be supported via appropriate scheduling, as described in [35].

15.5.2 Dynamic Channel Assignment

Dynamic channel assignment (DCA) falls into two categories: dynamic assignment of multiple channels within a cell (intracell DCA) and, for orthogonally channelized systems, dynamic assignment of channels between cells (intercell DCA). Intercell DCA is typically not applicable to CDMA systems because channels are reused in every cell. The basic premise of intercell DCA is to make every channel available in every cell, so no fixed channel reuse pattern exists. Each channel can be used in every cell as long as the SIR requirements of each user are met. Thus, channels are assigned to users as needed, and when a call terminates the channel is returned to the pool of available channels for assignment. Intercell DCA has been shown to improve channel reuse efficiency by a factor of 2 or more over fixed reuse patterns, even with relatively simple algorithms [36; 37]. Mathematically, intercell DCA is a combinatorial optimization problem, with channel reuse constraints based on the SIR requirements. Most dynamic channel allocation schemes assume that all system parameters are fixed except for the arrival and departure of calls [38; 39; 40], with channel reuse constraints defined by a connected graph that is constant over time. The channel allocation problem under this formulation is a generalization of the vertex coloring problem and is thus NP-hard [41]. Reduced complexity has been obtained by applying neural networks [42] and simulated annealing [43] to the problem. However, these approaches can suffer from lack of convergence. The superior efficiency of DCA between cells is most pronounced under light loading conditions [38]. As traffic becomes heavier, DCA between cells can suffer from suboptimal allocations that are difficult to reallocate under heavy loading conditions. User mobility also affects performance of intercell DCA, since it causes more frequent channel reassignments and a corresponding increase in dropped calls [44]. Finally, the complexity of DCA between cells, particularly in systems with small cells and rapidly changing propagation conditions and user demands, limits the practicality of such techniques.

Intracell DCA allows dynamic assignment of multiple channels within a cell to a given user. In TDMA systems this is done by assigning a user multiple timeslots; in CDMA, by assigning a user multiple codes and/or spreading factors. Assigning multiple channels to users in TDMA or orthogonal CDMA systems is relatively straightforward and does not change the nature of the channels assigned. Dynamic timeslot assignment was used in [45; 46] to meet voice user requirements while reducing the delay of data services. In contrast, intracell DCA for nonorthogonal CDMA – via either multicode or variable spreading – leads to some performance degradation in the channels used. Specifically, a user who is assigned multiple codes in nonorthogonal CDMA creates self-interference if a single-user matched filter detector is used on each channel. Although this is no worse than if the different codes were assigned to different users, it is clearly suboptimal in a scenario where the same user is assigned multiple codes. For variable-spreading CDMA, a low spreading factor provides a higher data rate, since more of the total system bandwidth is available for data transmission as opposed to spreading. But a reduced spreading factor makes the signal more suscepti-ble to interference from other users. Intracell DCA using variable spreading gain has been analyzed in [47; 48], and the multicode technique was investigated in [49]. A comparison of multicode versus variable-spreading DCA in CDMA systems is given in [50], where it is found that (i) the two techniques are equivalent for the single-user matched filter detector but (ii) multicode is superior for more sophisticated detection techniques.

15.5.3 Power Control

Power control is a critical aspect of wireless system design. As seen in prior chapters, a water-filling power adaptation maximizes capacity of adaptive-rate single-user and multi-user systems in fading. Power control via channel inversion maintains a fixed received SNR in single-user fading channels and also eliminates the near–far effect in CDMA uplinks. However, in cellular systems these power control policies affect intercell interference in dif-ferent ways, as we now describe.

Power control on the downlink has less impact on intercell interference than on the uplink, since the downlink transmissions all originate from the cell center whereas uplink transmis-sions can come from the cell boundaries, which exacerbates interference to neighboring cells. Thus we will focus on the effect of power control on the uplink. Consider the two cells shown in Figure 15.9. Suppose that both mobiles B_1 and B_2 in cell B transmit at the same power. Then the interference caused by the mobile B_1 to the base station in cell A will be relatively large, since it is close to the boundary of cell A, while the interference from B_2 will generally be much weaker owing to the longer propagation distance. If water-filling power adaptation is employed, then B_1 will generally transmit at a lower power than B_2 because it will typi-cally have a worse channel gain than B_2 to the base station in cell B, since it is farther away. This has the positive effect of reducing the intercell interference to cell A. In other words, water-filling power adaptation reduces intercell interference from mobiles near cell bound-aries, the primary source of this interference. A similar phenomenon happens with multiuser diversity, since users transmit only when they have a high channel gain to their base station, which is generally true when they are close to their cell center. Conversely, under channel inversion the boundary mobiles will transmit at a higher power in order to maintain the same received power at the base station as mobiles that are near the cell center. This has the effect of increasing intercell interference from boundary mobiles.

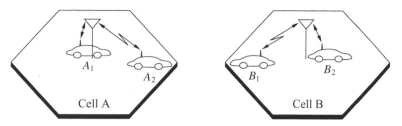

Figure 15.9: Effect of power control on intercell interference.

The power control algorithm discussed in Section 14.4 for maintaining a given SINR target for all users within a single cell can also be extended to multiple cells. Assume there are K users in the system with SINR requirement γ_k^* for the kth user. Focusing on the uplink, the kth user's SIR is given by

$$\gamma_k = \frac{g_k P_k}{n_k + \rho \sum_{k \neq j} g_{kj} P_j}, \quad k, j \in \{1, \ldots, K\}, \qquad (15.14)$$

where g_k is the channel power gain from user k to his base station, $g_{kj} > 0$ is the channel power gain from the jth (intercell or intracell) interfering transmitter to user k's base station, P_k is user k's transmit power, P_j is the jth interferer's transmit power, n_k is the noise power at user k's base station, and ρ is the interference reduction due to signal processing (i.e., $\rho \approx 1/G$ for CDMA and $\rho = 1$ for TDMA). Similar to the case of uplink power control, the SIR constraints can be represented in matrix form as $(\mathbf{I} - \mathbf{F})\mathbf{P} \geq \mathbf{u}$ with $\mathbf{P} > 0$, where $\mathbf{P} = (P_1, P_2, \ldots, P_K)^T$ is the column vector of transmit powers for each user,

$$\mathbf{u} = \left(\frac{\gamma_1^* n_1}{g_1}, \frac{\gamma_2^* n_2}{g_2}, \ldots, \frac{\gamma_K^* n_K}{g_K} \right)^T \qquad (15.15)$$

is the vector of noise powers scaled by the SIR constraints and channel gain, and $\mathbf{F}$ is an irreducible matrix with nonnegative elements given by

$$F_{kj} = \begin{cases} 0 & k = j, \\ \gamma_k^* g_{kj} \rho / g_k & k \neq j, \end{cases} \qquad (15.16)$$

with $k, j \in \{1, 2, \ldots, K\}$. As in uplink power control for a single cell (see Section 14.4), it is shown in [51; 52; 53] that if the Perron–Frobenius (maximum modulus) eigenvalue of $\mathbf{F}$ is less than unity then there exists a vector $\mathbf{P} > 0$ (i.e., $P_k > 0$ for all k) such that the SIR requirements of all users in all cells are satisfied, with $\mathbf{P}^* = (\mathbf{I} - \mathbf{F})^{-1}\mathbf{u}$ the Pareto optimal solution. Thus, $\mathbf{P}^*$ meets the SIR requirements with the minimum transmit power of the users. Moreover, the distributed iterative power control algorithm,

$$P_k(i + 1) = \frac{\gamma_k^*}{\gamma_k(i)} P_k(i), \qquad (15.17)$$

converges to the optimal solution. This is a simple algorithm for power control because it requires only SIR information at each transmitter, where each transmitter increases power when its SIR is below its target and decreases power when its SIR exceeds its target. However, it is important to note that the existence of a feasible power control vector that meets all SIR requirements is less likely in a cellular system than in the single-cell uplink, since there are more interferers contributing to the SIR and since the channel gains range over a larger set of values than in a single-cell uplink. When there is no feasible power allocation, the distributed algorithm will result in all users transmitting at their maximum power and still failing to meet their SIR requirement.

Power control is often combined with scheduling, intercell DCA, or intracell DCA. Intercell DCA and power control for TDMA are analyzed in [54; 55], and these techniques are extended to MIMO systems in [56]. Power control combined with intracell DCA exploiting multiuser diversity is investigated in [57; 58]. Intracell DCA for CDMA via variable spreading is combined with power control in [59; 60]. A comparison of power control combined with either multicode or variable-spreading CDMA is given in [61].

15.6 Fundamental Rate Limits

15.6.1 Shannon Capacity of Cellular Systems

There have been few information-theoretic results on the Shannon capacity of cellular systems, which is due to the difficulty of incorporating channel reuse and the resulting interference into fundamental capacity analysis. Whereas the capacity for the uplink and downlink of an isolated cell is known, as described in Sections 14.5 and 14.6, there is little work on extending these results to multiple cells. The capacity has been characterized in some cases, but the capacity as well as optimal transmission and reception strategies under broad assumptions about channel modeling, base station cooperation, interference characteristics, and transmitter/receiver CSI remain mostly unsolved.

One way to analyze the capacity of a cellular system is to assume that the base stations fully cooperate to jointly encode and decode all signals. In this case the notion of cells does not enter into capacity analysis. Specifically, under the assumption of full base station cooperation, the multiple base stations can be viewed as a single base station with multiple geographically dispersed antennas. Transmission from the multiple-antenna base station to the mobiles can be treated as a MIMO downlink (broadcast channel), and transmission from the mobiles to the multiple-antenna base station can be treated as a MIMO uplink (multiple access channel). The Shannon capacity regions for both of these channels are known, as discussed in Section 14.9, for some channel models and assumptions about channel side information.

The uplink capacity of cellular systems under the assumption of full base station cooperation was first investigated in [62], followed by a more comprehensive treatment in [63]. In both works, propagation between the mobiles and the base stations was characterized using an AWGN channel model, with a channel gain of unity within a cell and a gain of α ($0 \leq \alpha \leq 1$) between cells. The Wyner model of [63] considers one- and two-dimensional arrays of cells and in both cases derives the per-user capacity – defined as the maximum possible rate that all users can maintain simultaneously – as

$$C(\alpha) = \frac{B}{K} \int_0^1 \log_2\left(1 + \frac{KP(1 + 2\alpha\cos(2\pi\theta))^2}{N_0 B}\right) d\theta, \tag{15.18}$$

525 FUNDAMENTAL RATE LIMITS

where B is the total system bandwidth, $N_0/2$ is the noise power spectral density, K is the number of mobiles per cell, and P is the average transmit power of each mobile. It is also shown in both [63] and [62] that uplink capacity is achieved by using orthogonal multiple access techniques (e.g. TDMA) in each cell and then reusing these orthogonal channels in other cells, although this is not necessarily uniquely optimal. The behavior of $C(\alpha)$ as a function of α, the attenuation of the intercell interference, depends on the SNR of the system. The per-user capacity $C(\alpha)$ generally increases with α at high SNRs, since having strong intercell interference aids in decoding and subsequent subtraction of the interference from desired signals. However, at low SNR, $C(\alpha)$ initially decreases with α and then increases. This is because weak intercell interference cannot be reliably decoded and subtracted, so such interference reduces capacity. As the channel gains associated with the intercell interference grows, the joint decoding is better able to decode and subtract out this interference, leading to higher capacity.

An alternate analysis method for capacity of cellular systems assumes no base station cooperation, so that the receivers in each cell treat signals from other cells as interference. This approach mirrors the design of cellular systems in practice. Unfortunately, Shannon capacity of channels with interference is a long-standing open problem in information theory [64; 65], solved only for the special case of strong interference [66]. By treating the interference as Gaussian noise, the capacity of both the uplink and downlink can be determined using the single-cell analysis of Sections 14.5 and 14.6. The Gaussian assumption can be viewed as a worst-case assumption about the interference, since exploiting any known structure of the interference can presumably help in decoding the desired signals and therefore increases capacity. The capacity of a cellular system uplink with fading based on treating interference as Gaussian noise was obtained in [67] for both one- and two-dimensional cellular grids. These capacity results show that, with or without fading, if intercell interference is nonnegligible then an orthogonal multiple access method (e.g. TDMA) within a cell is optimal. This generalization holds also when channel-inversion power control is used within a cell. Moreover, in some cases limited or no reuse of channels in different cells can increase capacity. The effects on capacity for this model when there is partial joint processing between base stations have also been characterized [68].

The results described here provide some insight into the capacity and optimal transmission strategies for the uplink of cellular systems. Unfortunately, no such results are yet available for the downlink under any assumptions about channel modeling or base station cooperation. Although the uplink was the capacity bottleneck for cellular systems providing two-way voice, the downlink is becoming increasingly critical for multimedia downloads. Therefore, a better understanding of the capacity limitations and insights for cellular downlinks would be most beneficial in future cellular system design.

15.6.2 Area Spectral Efficiency

The Shannon capacity regions for cellular systems described in Section 15.6.1 dictate the set of maximum achievable rates on cell uplinks or downlinks. When the capacity region is computed based on the notion of joint processing at the base stations, then there is effectively only one cell with a multiple-antenna base station. However, when capacity is computed based on treating intercell interference as Gaussian noise, the capacity region of both the uplink and downlink become highly dependent on the cellular system structure – in particular, the cell size and channel reuse distance. Area spectral efficiency (ASE) is a capacity measure that allows the cellular structure (and in particular the reuse distance) to be optimized relative to fundamental capacity limits.

Recall that, for both orthogonal and nonorthogonal channelization techniques, the reuse distance D in a cellular system defines the distance between any two cell centers that use the same channel. Since these channels are reused at distance D, the area covered by each channel is roughly the area of a circle with radius $.5D$, which is given by $A = \pi(.5D)^2$. The larger the reuse distance, the less efficient the channel reuse. However, reducing the reuse distance increases intercell interference, thereby reducing the capacity region of each cell if this interference is treated as noise. The ASE captures this trade-off between efficient resource use and the capacity region per cell.

Consider a cellular system with K users per cell, a reuse distance D, and a total bandwidth allocation B. Let $C = (R_1, R_2, \ldots, R_K)$ denote the capacity region, for either the uplink or the downlink, in a given cell when the intercell interference from other cells is treated as Gaussian noise. The corresponding sum-rate, also called the system throughput, is given by

$$C_{\text{SR}} = \max_{(R_1, \ldots, R_K) \in C} \sum_{k=1}^{K} R_k \text{ bps.} \qquad (15.19)$$

The region C and corresponding sum-rate C_{SR} can be obtained for any channelization technique within a cell. Clearly, this capacity region will decrease as intercell interference increases. Moreover, since intercell interference decreases as the reuse distance increases, the size of the capacity region will increase with reuse distance.

The ASE of a cell is defined as the throughput per hertz per unit area that is supported by a cell's resources. Specifically, given the sum-rate capacity just described, the ASE is defined as

$$A_e \triangleq \frac{C_{\text{SR}}/B}{\pi(.5D)^2} \text{ bps/Hz/m}^2. \qquad (15.20)$$

From [67], orthogonal channelization is capacity achieving within a cell, so we will focus on TDMA for computing ASE. If we also assume that the system is interference limited, so that noise can be neglected, then the rate R_k associated with each user in a cell is a function of his received signal-to-interference power $\gamma_k = P_k/I_k$, $k = 1, \ldots, K$. If γ_k is constant then $R_k = \tau_k B \log(1 + \gamma_k)$, where τ_k is the time fraction assigned to user k. Typically γ_k is not constant, since both the interference and signal power of the kth user will vary with propagation conditions and mobile locations. When γ_k varies with time, the capacity region is obtained from optimal resource allocation over time and across users, as described in Sections 14.5.4 and 14.6.2.

As a simple example, consider an AWGN TDMA uplink with hexagonal cells of radius R. Assume all users in the cell are assigned the same fraction of time $\tau_k = 1/K$ and have the same transmit power P. We neglect the impact of intercell interference outside the first tier of interfering cells, so there are six intercell interferers. We take a pessimistic model, where (i) all users in the cell of interest are located at the cell boundary and (ii) all intercell interferers are located at their cell boundaries closest to the cell of interest. Path loss is characterized by the simplified model $P_r = P_t d^{-2}$ within a cell and by $P_r = P_t d^{-\gamma}$ between

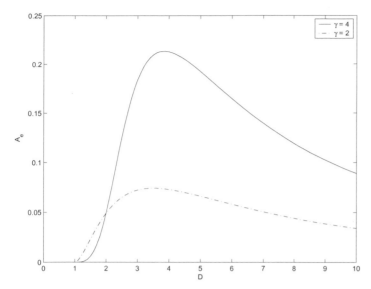

Figure 15.10: Area spectral efficiency for AWGN uplink ($\gamma = 2$ and 4).

cells, where $2 \leq \gamma \leq 4$. The received signal power of the kth user is then $P_k = PR^{-2}$, and the intercell interference power is $I_k = 6P(D-R)^{-\gamma}$. The maximum achievable rate for the kth user in the cell is thus

$$R_k = \frac{B}{K} \log_2 \left(1 + \frac{(D-R)^{\gamma}}{6R^2} \right) \text{ bps,} \tag{15.21}$$

and the ASE is

$$A_e = \frac{\sum_{k=1}^{K} R_k / B}{\pi(.5D)^2} = \frac{\log_2(1 + (D-R)^{\gamma}/6R^2)}{\pi(.5D)^2} \text{ bps/Hz/m}^2. \tag{15.22}$$

Plots of A_e versus D for $\gamma = 4$ and $\gamma = 2$ are shown in Figure 15.10, with the cell radius normalized to $R = 1$. Comparing these plots we see that, as expected, if the intercell interference path loss falls off more slowly then the ASE is decreased. However, it is somewhat surprising that the optimal reuse distance is also decreased.

Suppose now that the interferers are not on the cell boundaries. If all interferers are at a distance $D - R/2$ from the desired user's base station, then the ASE becomes

$$A_e = \frac{\log_2(1 + (D-R/2)^{\gamma}/6R^2)}{\pi(.5D)^2}. \tag{15.23}$$

The ASE in this case is plotted in Figure 15.11 for $\gamma = 4$, along with the ASE for interferers on their cell boundaries. As expected, the ASE is larger for interferers closer to their cell centers than on the boundaries closest to the cell they interfere with, and the optimal reuse distance is smaller.

Area spectral efficiency has been characterized in [69] for a cellular system uplink with orthogonal channelization assuming variable-rate transmission as well as best-case, worst-case, and average intercell interference conditions. The impact of different fading models,

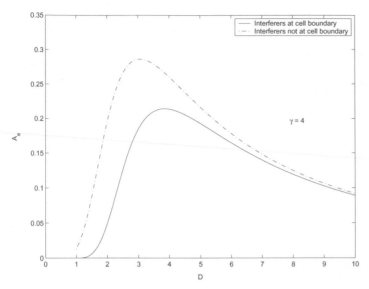

Figure 15.11: ASE for interferers at distance $D - R/2$ and at distance D ($\gamma = 4$).

cell sizes, and system load conditions was also investigated. The results indicate that the optimal reuse factor is unity for both best-case and average interference conditions. That is, channels should be reused in every cell even though there is no interference reduction from spreading. Moreover, the ASE decreases as an exponential of a fourth-order polynomial relative to the cell radius, thus quantifying the capacity gains associated with reducing cell size. A similar framework was used in [70] to characterize the ASE of cellular downlinks.

PROBLEMS

15-1. Consider a city of 10 square kilometers. A macrocellular system design divides the city into square cells of 1 square kilometer, where each cell can accommodate 100 users. Find the total number of users that can be accommodated in the system and the length of time it takes a mobile user to traverse a cell (approximate time needed for a handoff) when moving at 30 km per hour. If the cell size is reduced to 100 square meters and everything in the system scales so that 100 users can be accommodated in these smaller cells, find the total number of users the system can accommodate and the length of time it takes to traverse a cell.

15-2. Show that the reuse distance $D = \sqrt{3}R\sqrt{i^2 + j^2 + ij}$ for the channel assignment algorithm associated with hexagonal cells that is described in Section 15.2.

15-3. Consider a cellular system with diamond-shaped cells of radius $R = 100$ m. Suppose the minimum distance between cell centers using the same frequency must be $D = 600$ m to maintain the required SIR.

    (a) Find the required reuse factor N and the number of cells per cluster.

    (b) If the total number of channels for the system is 450, find the number of channels that can be assigned to each cell.

(c) Sketch two adjacent cell clusters and show a channel assignment for the two clusters with the required reuse distance.

15-4. Consider a cellular system with hexagonal cells of radius $R = 1$ km. Suppose the minimum distance between cell centers using the same frequency must be $D = 6$ km to maintain the required SIR.

(a) Find the required reuse factor N and the number of cells per cluster.
(b) If the total number of channels for the system is 1200, find the number of channels that can be assigned to each cell.
(c) Sketch two adjacent cell clusters and show a channel assignment for the two clusters with the required reuse distance.

15-5. Compute the SIR for a TDMA cellular system with diamond-shaped cells, where the cell radius $R = 10$ m and the reuse distance $D = 60$ m. Assume that the path-loss exponent within the cell is $\gamma_I = 2$ but that the intercell interference has path-loss exponent $\gamma_O = 4$. Compare with the SIR for $\gamma = \gamma_I = \gamma_O = 4$ and for $\gamma = \gamma_I = \gamma_O = 2$. Explain the relative orderings of SIR in each case.

15-6. Find the minimum reuse distance and user capacity for a TDMA cellular system with hexagonally shaped cells, path-loss exponent $\gamma = 2$ for all signal propagation in the system, and BPSK modulation. Assume an AWGN channel model with required $P_b = 10^{-6}$, a total system bandwidth of $B = 50$ MHz, and a required signal bandwidth of 100 kHz for each user.

15-7. Consider a CDMA system with perfect power control, a processing gain of $G = 100$, spreading codes with $\xi = 1$, and the ratio $\lambda = 1.5$. For BPSK modulation with a target P_b of 10^{-6}, under an AWGN channel model find the user capacity of this system with no sectorization and with $N = 3$ sectors.

15-8. In this problem we consider the impact of voice activity factor, which creates a random amount of interference. Consider a CDMA system with

$$ \text{SIR} = \frac{G}{\sum_{i=1}^{N_c-1} \chi_i + N}; $$

here G is the processing gain of the system, the χ_i represent intracell interference and follow a Bernoulli distribution with probability $\alpha = p(\chi_i = 1)$ equal to the voice activity factor, and N characterizes intercell interference and is assumed to be Gaussian, with mean $.247N_c$ and variance $.078N_c$. The probability of outage is defined as the probability that the SIR is below some target SIR_0:

$$ P_{\text{out}} = p(\text{SIR} < \text{SIR}_0). $$

(a) Show that

$$ P_{\text{out}} = p\left(\sum_{i=1}^{N_c-1} \chi_i + N > \frac{G}{\text{SIR}_0} \right). $$

(b) Find an analytical expression for P_{out}.
(c) Using the analytical expression obtained in part (b), compute the outage probability for $N_c = 35$ users, $\alpha = .5$, and a target SIR of $\text{SIR}_0 = 5$ (7 dB). Assume $G = 150$.

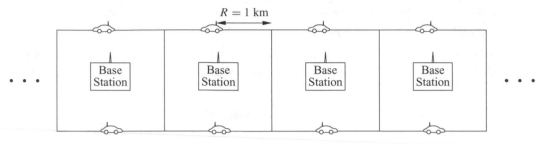

Figure 15.12: One-dimensional cellular system with square cells (Problem 15-11).

(d) Assume now that N_c is sufficiently large that the random variable $\sum_{i=1}^{N_c-1} \chi_i$ can be approximated as a Gaussian random variable. Under this approximation, find the distribution of $\sum_{i=1}^{N_c-1} \chi_i + N$ as a function of N_c as well as an analytical expression for outage probability based on this approximation.

(e) Compute the outage probability using this approximation for $N_c = 35$, $\alpha = .5$, $G = 150$, and a target SIR of $SIR_0 = 5$ (7 dB). Compare with your results in part (c).

15-9. Assume a cellular system with K users, and suppose that the minimum SIR for each user on the downlink is given as $\gamma_1^*, \ldots, \gamma_K^*$. Write down the conditions such that a power control vector exists that will satisfy these constraints.

15-10. Plot the ASE versus reuse distance D ($0 \leq D \leq 10$) of a TDMA uplink with hexagonal cells of radius $R = 1$, assuming that all users in the cell are assigned the same fraction of time and the same transmit power and that all users are located on a cell boundary. Your computation should be based on a path loss of $\gamma = 2$ and interference from only the first ring of interfering cells, where the interferers have probability $1/5$ of being in one of the following five locations: the cell boundary closest to the cell of interest; halfway between the base station and this closest cell boundary; in the cell center; the cell boundary farthest from the cell of interest; or halfway between the base station and this farthest cell boundary. Also plot the ASE and optimal reuse distance for the cases where, with probability 1, the interferers are on the (a) closest or (b) farthest cell boundary. Compare the differences in these plots.

15-11. Consider a one-dimensional cellular system deployed along a highway. The system has square cells of length $2R = 2$ km, as shown in Figure 15.12. This problem focuses on downlink transmission from base stations to mobiles. Assume that each cell has two mobiles located as shown in the figure, so that the mobiles in each cell have the exact same location relative to their respective base stations. Assume a total transmit power at each base station of $P_t = 5$ W, which is evenly divided between the two users in its cell. The total system bandwidth is 100 kHz, and the noise power spectral density at each receiver is 10^{-16} W/Hz. The propagation follows the model $P_r = P_t K (d_0/d)^3$, where $d_0 = 1$ m and $K = 100$. All interference should be treated as AWGN, and interference outside the first ring of interfering cells can be neglected. The system uses a frequency-division strategy, with the bandwidth allocated to each base station evenly divided between the two users in its cell.

(a) For a reuse distance $D = 2$ (frequencies reused every other cell), what bandwidth is allocated to each user in the system?

(b) Compute the minimum reuse distance D required to achieve a 10^{-3} BER for BPSK modulation in fast Rayleigh fading.

(c) Neglecting any fading or shadowing, use the Shannon capacity formula for user rates to compute the area spectral efficiency of each cell under frequency division, where frequencies are reused every other cell ($D = 2$).

15-12. In this problem we investigate the per-user capacity for the uplink of a cellular system for different system parameters. Assume a cellular system uplink where the total system bandwidth is $B = 100$ kHz. Assume in each cell that the noise PSD $N_0/2$ at each base station receiver has $N_0 = 10^{-9}$ W/Hz and that all mobiles have the same transmit power P.

(a) Plot the per-user capacity for the uplink, as given by (15.18), for $K = 10$ users, transmit power $P = 10$ mW per user, and $0 \leq \alpha \leq 1$. Explain the shape of the curve relative to α.

(b) For $\alpha = .5$ and $P = 10$ mW, plot the per-user capacity for $1 \leq K \leq 30$. Explain the shape of the curve relative to K.

(c) For $\alpha = .5$ and $K = 10$ users, plot the per-user capacity for $0 \leq P \leq 100$ mW. Explain the shape of the curve relative to P.

REFERENCES

[1] T. S. Rappaport, *Wireless Communications – Principles and Practice,* 2nd ed., Prentice-Hall, Englewood Cliffs, NJ, 2001.

[2] V. Erceg, A. J. Rustako, and R. S. Roman, "Diffraction around corners and its effects on the microcell coverage area in urban and suburban environments at 900 MHz, 2 GHz, and 4 GHz," *IEEE Trans. Veh. Tech.,* pp. 762–6, August 1994.

[3] A. J. Goldsmith and L. J. Greenstein, "A measurement-based model for predicting coverage areas of urban microcells," *IEEE J. Sel. Areas Commun.,* pp. 1013–23, September 1993.

[4] S. Dehghan and R. Steele, "Small cell city," *IEEE Commun. Mag.,* pp. 52–9, August 1997.

[5] H. Holma and A. Toskala, *WCDMA for UMTS Radio Access for Third Generation Mobile Communications,* 3rd ed., Wiley, New York, 2004.

[6] M. V. Clark, V. Erceg, and L. J. Greenstein, "Reuse efficiency in urban microcellular networks," *IEEE Trans. Veh. Tech.,* pp. 279–88, May 1997.

[7] M. K. Simon, *Probability Distributions Involving Gaussian Random Variables,* Kluwer, Dordrecht, 2002.

[8] G. L. Stuber, *Principles of Mobile Communications,* 2nd ed., Kluwer, Dordrecht, 2001.

[9] R. Prasad and A. Kegel, "Effects of Rician faded and log-normal shadowed signals on spectrum efficiency in microcellular radio," *IEEE Trans. Veh. Tech.,* pp. 274–81, August 1993.

[10] R. Kohno, R. Meidan, and L. B. Milstein, "Spread spectrum access methods for wireless communications," *IEEE Commun. Mag.,* pp. 58–67, January 1995.

[11] K. S. Gilhousen, I. M. Jacobs, R. Padovani, A. J. Viterbi, L. A. Weaver, Jr., and C. E. Wheatley III, "On the capacity of a cellular CDMA system," *IEEE Trans. Veh. Tech.,* pp. 303–12, May 1991.

[12] P. T. Brady, "A statistical analysis of on–off patterns in 16 conversations," *Bell System Tech. J.,* pp. 73–91, January 1968.

[13] V. V. Veeravalli and A. Mantravadi, "The coding–spreading trade-off in CDMA systems," *IEEE J. Sel. Areas Commun.,* pp. 396–408, February 2002.

[14] P. Jung, P. W. Baier, and A. Steil, "Advantages of CDMA and spread spectrum techniques over FDMA and TDMA in cellular mobile radio applications," *IEEE Trans. Veh. Tech.,* pp. 357–64, August 1993.

[15] T. S. Rappaport and L. B. Milstein, "Effects of radio propagation path loss on DS-CDMA cellular frequency reuse efficiency for the reverse channel," *IEEE Trans. Veh. Tech.*, pp. 231–42, August 1992.

[16] B. Gundmundson, J. Sköld, and J. K. Ugland, "A comparison of CDMA and TDMA systems," *Proc. IEEE Veh. Tech. Conf.*, pp. 732–5, May 1992.

[17] J. Winters, "Smart antennas for wireless systems," *IEEE Pers. Commun. Mag.*, pp. 23–7, February 1998.

[18] S. Verdú, *Multiuser Detection*, Cambridge University Press, 1998.

[19] B. M. Zaidel, S. Shamai, and S. Verdú, "Multicell uplink spectral efficiency of coded DS-CDMA with random signatures," *IEEE J. Sel. Areas Commun.*, pp. 1556–69, August 2001.

[20] J. G. Andrews, "Interference cancellation for cellular systems: A contemporary overview," *IEEE Wireless Commun. Mag.*, pp. 19–29, April 2005.

[21] S. Shamai and B. M. Zaidel, "Enhancing the cellular downlink capacity via co-processing at the transmitting end," *Proc. IEEE Veh. Tech. Conf.*, pp. 1745–9, May 2001.

[22] H. Viswanathan, S. Venkatesan, and H. Huang, "Downlink capacity evaluation of cellular networks with known-interference cancellation," *IEEE J. Sel. Areas Commun.*, pp. 802–11, June 2003.

[23] S. A. Jafar, G. J. Foschini, and A. J. Goldsmith, "PhantomNet: Exploring optimal multicellular multiple antenna systems," *J. Appl. Signal Proc.* (EURASIP), pp. 591–605, May 2004.

[24] M. Costa, "Writing on dirty paper," *IEEE Trans. Inform. Theory*, pp. 439–41, May 1983.

[25] E. H. Choi, W. Choi, and J. Andrews, "Throughput of the 1x EV-DO system with various scheduling algorithms," *Proc. Internat. Sympos. Spread Spec. Tech. Appl.*, pp. 359–63, August 2004.

[26] D. Wu and R. Negi, "Downlink scheduling in a cellular network for quality-of-service assurance," *IEEE Trans. Veh. Tech.*, pp. 1547–57, September 2004.

[27] L. Xu, X. Shen, and J. W. Mark, "Dynamic fair scheduling with QoS constraints in multimedia wideband CDMA cellular networks," *IEEE Trans. Wireless Commun.*, pp. 60–73, January 2004.

[28] X. Qiu, L. Chang, Z. Kostic, T. M. Willis, N. Mehta, L. G. Greenstein, K. Chawla, J. F. Whitehead, and J. Chuang, "Some performance results for the downlink shared channel in WCDMA," *Proc. IEEE Internat. Conf. Commun.*, pp. 376–80, April 2002.

[29] A. C. Varsou and H. V. Poor, "HOLPRO: A new rate scheduling algorithm for the downlink of CDMA networks," *Proc. IEEE Veh. Tech. Conf.*, pp. 948–54, September 2000.

[30] E. Villier, P. Legg, and S. Barrett, "Packet data transmissions in a W-CDMA network – Examples of uplink scheduling and performance," *Proc. IEEE Veh. Tech. Conf.*, pp. 2449–53, May 2000.

[31] L. Qian and K. Kumaran, "Uplink scheduling in CDMA packet-data systems," *Proc. IEEE Infocom Conf.*, pp. 292–300, March 2003.

[32] L. Qian and K. Kumaran, "Scheduling on uplink of CDMA packet data network with successive interference cancellation," *Proc. IEEE Wireless Commun. Network Conf.*, pp. 1645–50, March 2003.

[33] H. Boche and M. Wiczanowski, "Queueing theoretic optimal scheduling for multiple input multiple output multiple access channel," *Proc. IEEE Internat. Sympos. Signal Proc. Inform. Tech.*, pp. 576–9, December 2003.

[34] K.-N. Lau, "Analytical framework for multiuser uplink MIMO space-time scheduling design with convex utility functions," *IEEE Trans. Wireless Commun.*, pp. 1832–43, September 2004.

[35] L. F. Chang, X. Qiu, K. Chawla, and C. Jian, "Providing differentiated services in EGPRS through radio resource management," *Proc. IEEE Internat. Conf. Commun.*, pp. 2296–2301, June 2001.

[36] I. Katzela and M. Naghshineh, "Channel assignment schemes for cellular mobile telecommunication systems – A comprehensive survey," *IEEE Pers. Commun. Mag.*, pp. 10–31, June 1996.

[37] D. C. Cox, "Wireless network access for personal communications," *IEEE Commun. Mag.*, pp. 96–115, December 1992.

[38] R. J. McEliece and K. N. Sivarajan, "Performance limits for channelized cellular telephone systems," *IEEE Trans. Inform. Theory,* pp. 21–4, January 1994.

[39] D. Everitt and D. Manfield, "Performance analysis of cellular mobile communication systems with dynamic channel assignment," *IEEE J. Sel. Areas Commun.,* pp. 1172–81, October 1989.

[40] J. Zander and H. Eriksson, "Asymptotic bounds on the performance of a class of dynamic channel assignment algorithms," *IEEE J. Sel. Areas Commun.,* pp. 926–33, August 1993.

[41] M. R. Garey and D. S. Johnson, *Computers and Intractability: A Guide to the Theory of NP-Completeness,* Freeman, New York, 1979.

[42] D. Kunz, "Channel assignment for cellular radio using neural networks," *IEEE Trans. Veh. Tech.,* pp. 188–93, February 1991.

[43] R. Mathar and J. Mattfeldt, "Channel assignment in cellular radio networks," *IEEE Trans. Veh. Tech.,* pp. 647–56, November 1993.

[44] A. Lozano and D. C. Cox, "Distributed dynamic channel assignment in TDMA mobile communication systems," *IEEE Trans. Veh. Tech.,* pp. 1397–1406, November 2002.

[45] L. Chen, U. Yoshida, H. Murata, and S. Hirose, "Dynamic timeslot allocation algorithms suitable for asymmetric traffic in multimedia TDMA/TDD cellular radio," *Proc. IEEE Veh. Tech. Conf.,* pp. 1424–8, May 1998.

[46] Y. Hara, T. Nabetani, and S. Hara, "Performance evaluation of cellular SDMA/TDMA systems with variable bit rate multimedia traffic," *Proc. IEEE Veh. Tech. Conf.,* pp. 1731–4, October 2001.

[47] A. C. Kam, T. Minn, and K.-Y. Siu, "Supporting rate guarantee and fair access for bursty data traffic in W-CDMA," *IEEE J. Sel. Areas Commun.,* pp. 2121–30, November 2001.

[48] U. C. Kozat, I. Koutsopoulos, and L. Tassiulas, "Dynamic code assignment and spreading gain adaptation in synchronous CDMA wireless networks," *Proc. Internat. Sympos. Spread Spec. Tech. Appl.,* pp. 593–7, November 2002.

[49] D. Ayyagari and A. Ephremides, "Cellular multicode CDMA capacity for integrated (voice and data) services," *IEEE J. Sel. Areas Commun.,* pp. 928–38, May 1999.

[50] E. Biglieri, G. Caire, and G. Taricco, "CDMA system design through asymptotic analysis," *IEEE Trans. Commun.,* pp. 1882–96, November 2000.

[51] J. Zander, "Performance of optimum transmitter power control in cellular radio systems," *IEEE Trans. Veh. Tech.,* pp. 57–62, February 1992.

[52] G. J. Foschini and Z. Miljanic, "A simple distributed autonomous power control algorithm and its convergence," *IEEE Trans. Veh. Tech.,* pp. 641–6, November 1993.

[53] S. A. Grandhi, R. Vijayan, and D. J. Goodman, "Distributed power control in cellular radio systems," *IEEE Trans. Commun.,* pp. 226–8, February–April 1994.

[54] S. A. Grandhi, R. D. Yates, and D. J. Goodman, "Resource allocation for cellular radio systems," *IEEE Trans. Veh. Tech.,* pp. 581–7, August 1997.

[55] A. Lozano and D. C. Cox, "Integrated dynamic channel assignment and power control in TDMA mobile wireless communication systems," *IEEE J. Sel. Areas Commun.,* pp. 2031–40, November 1999.

[56] R. Veronesi, V. Tralli, J. Zander, and M. Zorzi, "Distributed dynamic resource allocation with power shaping for multicell SDMA packet access networks," *Proc. IEEE Wireless Commun. Network Conf.,* pp. 2515–20, March 2004.

[57] F. Berggren, S.-L. Kim, R. Jantti, and J. Zander, "Joint power control and intracell scheduling of DS-CDMA nonreal time data," *IEEE J. Sel. Areas Commun.,* pp. 1860–70, October 2001.

[58] H. C. Akin and K. M. Wasserman, "Resource allocation and scheduling in uplink for multimedia CDMA wireless systems," *IEEE/Sarnoff Sympos. Adv. Wired Wireless Commun.,* pp. 185–8, April 2004.

[59] T. H. Hu and M. M. K. Liu, "A new power control function for multirate DS-CDMA systems," *IEEE Trans. Commun.* pp. 896–904, June 1999.

[60] S.-J. Oh, D. Zhang, and K. M. Wasserman, "Optimal resource allocation in multiservice CDMA networks," *IEEE Trans. Wireless Commun.,* pp. 811–21, July 2003.

[61] D. Ayyagari and A. Ephremides, "Optimal admission control in cellular DS-CDMA systems with multimedia traffic," *IEEE Trans. Wireless Commun.*, pp. 195–202, January 2003.

[62] S. V. Hanly and P. Whiting, "Information theory and the design of multi-receiver networks," *Proc. Internat. Sympos. Spread Spec. Tech. Appl.*, pp. 103–6, November 1992.

[63] A. Wyner, "Shannon-theoretic approach to a Gaussian cellular," *IEEE Trans. Inform. Theory*, pp. 1713–27, November 1994.

[64] T. Cover and J. Thomas, *Elements of Information Theory*, Wiley, New York, 1991.

[65] E. C. van der Meulen, "Some reflections on the interference channel," in R. E. Blahut, D. J. Costello, and T. Mittelholzer, Eds., *Communications and Cryptography: Two Sides of One Tapestry*, pp. 409–21, Kluwer, Boston, 1994.

[66] M. H. M. Costa and A. A. El Gamal, "The capacity region of the discrete memoryless interference channel with strong interference," *IEEE Trans. Inform. Theory*, pp. 710–11, September 1987.

[67] S. Shamai and A. D. Wyner, "Information-theoretic considerations for symmetric, cellular, multiple-access fading channels: Part I," *IEEE Trans. Inform. Theory*, pp. 1877–94, November 1997.

[68] S. Shamai and A. D. Wyner, "Information-theoretic considerations for symmetric, cellular, multiple-access fading channels: Part II," *IEEE Trans. Inform. Theory*, pp. 1895–1911, November 1997.

[69] M.-S. Alouini and A. J. Goldsmith, "Area spectral efficiency of cellular mobile radio systems," *IEEE Trans. Veh. Tech.*, pp. 1047–66, July 1999.

[70] M. F. Tariz and A. Nix, "Area spectral efficiency of a channel adaptive cellular mobile radio system in a correlated shadowed environment," *Proc. IEEE Veh. Tech. Conf.*, pp. 1075–9, May 1998.

Ad Hoc Wireless Networks

An ad hoc wireless network is a collection of wireless mobile nodes that self-configure to form a network without the aid of any established infrastructure, as shown in Figure 16.1. Without an inherent infrastructure, the mobiles handle the necessary control and networking tasks by themselves, generally through the use of distributed control algorithms. Multihop routing, whereby intermediate nodes relay packets toward their final destination, can improve the throughput and power efficiency of the network. The Merriam-Webster dictionary lists two relevant definitions for *ad hoc*: "formed or used for specific or immediate problems", and "fashioned from whatever is immediately available". These definitions capture two of the main benefits of ad hoc wireless networks: they can be tailored to specific applications, and they can be formed from whatever network nodes are available. Ad hoc wireless networks have other appealing features as well. They avoid the cost, installation, and maintenance of network infrastructure. They can be rapidly deployed and reconfigured. They also exhibit great robustness owing to their distributed nature, node redundancy, and the lack of single points of failure. These characteristics are especially important for military applications, and much of the groundbreaking research in ad hoc wireless networking was supported by the (Defense) Advanced Research Projects Agency (DARPA) and the U.S. Navy [1; 2; 3; 4; 5; 6]. Many of the fundamental design principles for ad hoc wireless networks were identified and investigated in that early research. However, despite many advances over the last several decades in wireless communications in general – and in ad hoc wireless networks in particular – the optimal design, performance, and fundamental capabilities of these networks remain poorly understood, at least in comparison with other wireless network paradigms.

This chapter begins with an overview of the primary applications for ad hoc wireless networks, as applications drive many of the design requirements. Next, the basic design principles and challenges of these networks are described. The concept of protocol layering is then discussed, along with layer interaction and the benefits of cross-layer design. Fundamental capacity limits and scaling laws for these networks are also outlined. The chapter concludes with a discussion of the unique design challenges inherent to energy-constrained ad hoc wireless networks.

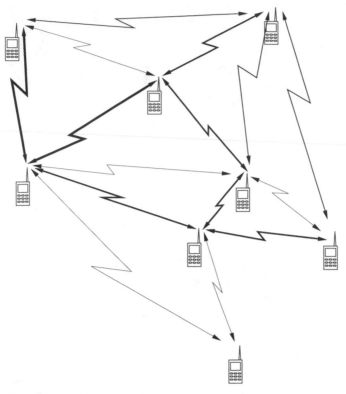

Figure 16.1: Ad hoc network.

16.1 Applications

This section describes some of the most prevalent applications for ad hoc wireless networks. The self-configuring nature and lack of infrastructure inherent to these networks make them highly appealing for many applications, even if it results in a significant performance penalty. The lack of infrastructure is highly desirable for low-cost commercial systems, since it obviates the need for a large investment to get the network up and running, and deployment costs may then scale with network success. Lack of infrastructure is also highly desirable for military systems, where communication networks must be configured quickly as the need arises, often in remote areas. Other advantages of ad hoc wireless networks include ease of network reconfiguration and reduced maintenance costs. However, these advantages must be balanced against any performance penalty resulting from the multihop routing and distributed control inherent to these networks.

We will focus on the following applications: data networks, home networks, device networks, sensor networks, and distributed control systems. This list is by no means comprehensive; in fact, the success of ad hoc wireless networks hinges on making them sufficiently flexible so that there can be accidental successes. Therein lies the design dilemma for these networks. If the network is designed for maximum flexibility to support many applications (a one-size-fits-all network) then it will be difficult to tailor the network to different application requirements. This will likely result in poor performance for some applications, especially those with high rate requirements or stringent delay constraints. On the other hand, if the network is optimized for a few specific applications then designers must predict in advance what these "killer applications" will be – a risky proposition. Ideally an ad hoc wireless network should be sufficiently flexible to support many different applications while adapting its performance to the applications in operation at any given time. An adaptive cross-layer design can provide this flexibility along with the ability to tailor protocol design to different constraints in the nodes.

16.1.1 Data Networks

Ad-hoc wireless data networks primarily support data exchange between laptop computers, palmtops, personal digital assistants (PDAs), and other information devices. These data networks generally fall into three categories based on their coverage area: LANs, MANs, and WANs (for "local", "metropolitan", and "wide" area networks). Infrastructure-based wireless LANs are already quite prevalent, and they deliver good performance at low cost. However, ad hoc wireless data networks have some advantages over these infrastructure-based networks. First, with multihop routing only one access point is needed to connect nodes that are dispersed throughout the area to the backbone wired infrastructure: this reduces cost and installation requirements. In addition, it can sometimes be inefficient for nodes to communicate through an access point or base station. For example, PDAs that are next to each other can exchange information directly rather than routing through an intermediate node.

Wireless MANs typically require multihop routing, often over many hops, because they cover a larger area. The challenge in these networks is to support high data rates in a cost-effective manner and over multiple hops, where the link quality of each hop is different and changes with time. The lack of centralized network control and potential for high-mobility users further complicate this objective. Military programs such as DARPA's GLOMO (global mobile information systems) have invested much time and money in building high-speed ad hoc wireless MANs that support multimedia, with limited success [7; 8]. Ad hoc wireless MANs have also permeated the commercial sector, with Metricom the best example [9]. Although Metricom did deliver fairly high data rates throughout several major metropolitan areas, significant demand never materialized and so the company eventually filed for bankruptcy.

Wireless WANs are needed for applications where network infrastructure to cover a wide area is too costly or impractical to deploy. For example, sensor networks may be dropped into remote areas where network infrastructure cannot be developed. In addition, networks that must be built up and torn down quickly (e.g., for military applications or disaster relief) are unfeasible without an ad hoc approach.

16.1.2 Home Networks

Home networks are envisioned to support communication between PCs, laptops, PDAs, cordless phones, smart appliances, security and monitoring systems, consumer electronics, and entertainment systems anywhere in and around the home. Such networks could enable smart rooms that sense people and movement and adjust light and heating accordingly, as well as "aware homes" that feature network sensors and computers for assisted living of seniors and those with disabilities. Home networks also encompass video or sensor monitoring systems with the intelligence to coordinate and interpret data and alert the home owner and the appropriate police or fire department of unusual patterns; intelligent appliances that coordinate with each other and with the Internet for remote control, software upgrades, and to schedule maintenance; and entertainment systems that allow access to a VCR, set-top box, or PC from any television or stereo system in the home [10; 11; 12; 13].

There are several design challenges for such networks. One of the biggest is the need to support the varied quality-of-service (QoS) requirements for different home networking applications. Quality of service in this context refers to the requirements of a particular application – typically data rates and delay constraints, which can be quite stringent for home entertainment systems. Other major challenges include cost and the need for standardization, since all of the devices being supported on this type of home network must follow the same networking standard. Note that the different devices accessing a home network have very different power constraints: some will have a fixed power source and be effectively unconstrained, while others will have limited battery power and may not be rechargeable. Thus, one of the biggest challenges in home network design is to leverage power in unconstrained devices to take on the heaviest communication and networking burden so that the networking requirements for all nodes in the network, regardless of their power constraints, can be met.

16.1.3 Device Networks

Device networks support short-range wireless connections between devices. Such networks are primarily intended to replace inconvenient cabled connections with wireless connections. Thus, the need for cables and the corresponding connectors between cell phones, modems, headsets, PDAs, computers, printers, projectors, network access points, and other such devices is eliminated. The main technology drivers for such networks are low-cost low-power radios with networking capabilities such as Bluetooth [14; 15], ZigBee [16], and ultrawideband or UWB [17]. The radios are integrated into commercial electronic devices to provide networking capabilities between devices. Some common applications include wireless versions of USB and RS232 connectors, PCMCIA cards, set-top boxes, and cell-phone headsets.

16.1.4 Sensor Networks

Wireless sensor networks consist of small nodes with sensing, computation, and wireless networking capabilities, and as such these networks represent the convergence of these three important technologies. Sensor networks have enormous potential for both consumer and military applications. Military missions require sensors and other intelligence gathering mechanisms that can be placed close to their intended targets. The potential threat to these mechanisms is therefore quite large, so it follows that the technology used must be highly redundant (or robust) and require as little human intervention as possible. An apparent solution to these constraints lies in large arrays of passive electromagnetic, optical, chemical, and biological sensors. These can be used to identify and track targets, and they serve also as a first line of detection for various types of attacks. Such networks can also support the movement of unmanned robotic vehicles. For example, optical sensor networks can provide networked navigation, routing vehicles around obstacles while guiding them into position for defense or attack. The design considerations for some industrial applications are quite similar to those for military applications. In particular, sensor arrays can be deployed and used for remote sensing in nuclear power plants, mines, and other industrial venues.

Examples of sensor networks for the home environment include electricity, gas, and water meters that can be read remotely through wireless connections. The broad use of simple metering devices within the home can help monitor and regulate appliances like air conditioners and hot-water heaters that are significant consumers of power and gas. Simple attachments to power plugs can serve as the metering and communication devices for individual appliances. One can imagine a user tracking various types of information on home energy consumption from a single terminal: the home computer. Television usage and content could be tracked or controlled in similar ways. Another important home application is smoke detectors that could not only monitor different parts of the house but also communicate to trace the spread of the fire. Such information (along with house blueprints) could be conveyed to local firefighters before they arrived on the scene. A similar type of array could be used to detect the presence and spread of gas leaks or other toxic fumes.

Sensor arrays also have great potential for use at the sites of large accidents. Consider, for example, the use of remote sensing in rescue operations following the collapse of a building. Sensor arrays could be rapidly deployed at the site of the accident and be used to track heat, natural gas, and toxic substances. Acoustic sensors and localization techniques could be utilized to detect and locate trapped survivors. It may even be possible to prevent such tragedies altogether through the use of sensor arrays. The collapse of bridges, walkways, and balconies, for example, could be predicted in advance using stress and motion sensors built into the structures from the outset. By inserting a large number of these low-cost, low-power sensors directly into the concrete before it is poured, material fatigue could be detected and tracked over time throughout the structure. Such sensors must be robust and self-configuring and would require a long lifetime, commensurate with the lifetime of the structure.

Most sensors will be deployed with non-rechargeable batteries. The problem of battery lifetime in such sensors may be circumvented by using ultrasmall energy-harvesting radios. Research in this area promises radios smaller than one cubic centimeter, weighing less than 100 grams, and with a power dissipation level below 100 microwatts [18]. This low level of power dissipation enables nodes to extract sufficient power from the environment – energy harvesting – to maintain operation indefinitely. Such radios open up new applications for sensor deployment in buildings, homes, and even the human body.

16.1.5 Distributed Control Systems

Ad hoc wireless networks also enable distributed control applications, with remote plants, sensors and actuators linked together via wireless communication channels. Such networks allow coordination of unmanned mobile units and feature greatly reduced maintenance and reconfiguration costs in comparison to distributed control systems with wired communication links. Ad hoc wireless networks could be used to support coordinated control of multiple vehicles in an automated highway system, remote control of manufacturing and other industrial processes, and coordination of unmanned airborne vehicles for military applications.

Current distributed control designs provide excellent performance as well as robustness to uncertainty in model parameters. However, these designs are based on closed-loop performance that assumes a centralized architecture, synchronous clocked systems, and fixed topology. Consequently, these systems require that the sensor and actuator signals be delivered to the controller with a small, fixed delay. Ad hoc wireless networks cannot provide any performance guarantees in terms of data rate, delay, or loss characteristics: delays are typically random and packets may be lost. Unfortunately, most distributed controllers are not robust to these types of communication errors, and effects of small random delays can be catastrophic [19; 20]. Thus, distributed controllers must be redesigned for robustness to the random delays and packet losses inherent to wireless networks [21]. Ideally, the ad hoc wireless network can be jointly designed with the controller to deliver the best possible end-to-end performance.

16.2 Design Principles and Challenges

The most fundamental aspect of an ad hoc wireless network is its lack of infrastructure, and most design principles and challenges stem from this characteristic. The lack of infrastructure inherent to ad hoc wireless networks is best illustrated by contrast with the most prevalent wireless networks: cellular systems and wireless LANs. Cellular systems divide the geographic area of interest into cells, and mobiles within a cell communicate with a base station in the cell center that is connected to a backbone wired network. Thus, there is no peer-to-peer communication between mobiles. All communication is via the base station through single-hop routing. The base stations and backbone network perform all networking functions, including authentication, call routing, and handoff. Most wireless LANs have a similar, centralized, single-hop architecture: mobile nodes communicate directly with a centralized access point that is connected to the backbone Internet, and the access point performs all networking and control functions for the mobile nodes.[1] In contrast, an ad hoc wireless network has peer-to-peer communication, networking and control functions that are distributed among all nodes, and routing that can exploit intermediate nodes as relays.

Ad hoc wireless networks can form an infrastructure or node hierarchy, either permanently or dynamically. For example, many ad hoc wireless networks form a backbone infrastructure from a subset of nodes in the network to improve network reliability, scalability, and capacity [22]. If a node in this backbone subset leaves the network, the backbone can be reconfigured. Similarly, some nodes may be chosen to perform as base stations for neighboring nodes [14]. Thus, ad hoc wireless networks may create structure to improve network performance, but such structure is not a fundamental design requirement of the network.

A lack of canonical structure is quite common in wired networks. Indeed, most metropolitan area networks (MANs) and wide area networks (WANs), including the Internet, have an ad hoc structure. However, the broadcast nature of the radio channel introduces characteristics in ad hoc wireless networks that are not present in their wired counterparts. In particular, with sufficient transmit power any node can transmit a signal directly to any other node. For a fixed transmit power, the link signal-to-interference-plus-noise power ratio (SINR) between two communicating nodes will typically decrease as the distance between the nodes increases, and it will also depend on the signal propagation and interference environment. Moreover, this link SINR varies randomly over time owing to fading of the signal

---

[1] The 802.11 wireless LAN standard does include ad hoc network capabilities, but this component of the standard is rarely used.

and interference. Link SINR determines the communication performance of the link: the data rate and associated probability of packet error or bit error rate (BER) that can be supported on the link. Links with very low SINRs are not typically used due to their extremely poor performance, leading to partial connectivity among all nodes in the network, as shown in Figure 16.1. However, link connectivity is not a binary decision, as nodes can adapt to the SINR using adaptive modulation or change it using power control. The different SINR values for different links are illustrated by the different line widths in Figure 16.1. Thus, in theory, every node in the network can transmit data directly to any other node. However, this may not be feasible if the nodes are separated by a large distance, and direct transmission even over a relatively short link may have poor performance or cause much interference to other links. Network connectivity also changes as nodes enter and leave the network, and this connectivity can be controlled by adapting the transmit power of existing network nodes to the presence of a new node [23].

The flexibility in link connectivity that results from varying link parameters such as power and data rate has major implications for routing. Nodes can send packets directly to their final destination via single-hop routing as long as the link SINR is above some minimal threshold. However, the SINR can be quite poor under single-hop routing, and this routing method also causes excessive interference to surrounding nodes. In most ad hoc wireless networks, packets are forwarded from source to destination through intermediate relay nodes. Since path loss causes an exponential decrease in received power as a function of distance, intermediate relays can greatly reduce the total transmit power (the sum of transmit power at the source and all relays) needed for end-to-end packet transmission. Multihop routing using intermediate relay nodes is a key feature of ad hoc wireless networks: it allows for communication between geographically dispersed nodes and facilitates the scalability and decentralized control of the network. However, it is much more challenging to support high data rates and low delays over a multihop wireless channel than over the single-hop wireless channels inherent to cellular systems and wireless LANs. This is one of the main difficulties in using an ad hoc wireless network to support applications (e.g. video) that require a high data rate and low delay.

Scalability is required for ad hoc wireless networks with a large number of nodes. The key to scalability lies in the use of distributed network control algorithms, which adjust local performance to account for local conditions. By forgoing the use of centralized information and control resources, protocols can scale as the network grows because they rely on local information only. Work on protocol scalability in ad hoc wireless networks has mainly focused on self-organization [24; 25], distributed routing [26], mobility management [22], and security [27]. Note that distributed protocols often consume a fair amount of energy in local processing and message exchange; this is analyzed in detail for security protocols in [28]. Thus, interesting trade-offs arise concerning how much local processing should be done versus transmitting information to a centralized location for processing. This trade-off is particularly apparent in sensor networks, where nodes close together have correlated data and also coordinate in routing that data through the network. Most experimental work on scalability in ad hoc wireless networks has focused on relatively small networks, fewer than a hundred nodes. Many ad hoc network applications, especially sensor networks, could have hundreds to thousands of nodes or even more. The ability of existing wireless network protocols to scale to such large network sizes remains unclear.

Energy constraints are another big challenge in ad hoc wireless network design [29]. These constraints arise in wireless (e.g. sensor) network nodes powered by batteries that cannot be recharged. Hard energy constraints significantly affect network design considerations. First, there is no longer a notion of data rate, since only a finite number of bits can be transmitted at each node before the battery dies. There is also a trade-off between the duration of a bit and energy consumption: sending bits more slowly conserves transmit energy but may increase circuit energy [30]. Standby operation can consume significant energy, so sleep modes must be employed for energy conservation – but then having nodes go to sleep can complicate network control and routing. In fact, energy constraints affect almost all of the network protocols in some manner, so energy consumption must be optimized over all aspects of the network design.

16.3 Protocol Layers

Protocol layering is a common abstraction in network design. Layering provides design modularity for network protocols, which facilitates standardization and implementation. Unfortunately, the layering paradigm does not work well in ad hoc wireless networks, where many protocol design issues are intertwined. In this section we describe protocol layering as it applies to ad hoc wireless networks as well as the interactions between protocol layers, which motivate the need for cross-layer design.

An international standard called the open systems interconnect (OSI) model was developed as a framework for protocol layering in data networks. The OSI model divides the required functions of the network into seven layers: the application layer, presentation layer, session layer, transport layer, network layer, data link control layer, and physical layer. Each layer is responsible for a separate set of tasks, with a fixed interface between layers to exchange control information and data. The basic premise behind the OSI model is that the protocols developed at any given layer can interoperate with protocols developed at other layers without regard to the details of the protocol implementation. For example, the application layer need not consider how data is routed through the network, or what modulation and coding technique is used on a given link. The set of protocols associated with all layers is referred to as the *protocol stack* of the network. Details of the OSI model and the functionality associated with each of its layers are given in [31, Chap. 1.3].

The Internet has driven the actual implementation of layering, which is built around the transmission control protocol (TCP) for the transport layer and the Internet protocol (IP) for routing at the network layer. Thus, in most networks the OSI layering model has been replaced by a five-layer model, also called the TCP/IP model, that is defined by the main functionality of the TCP and IP protocols. The five layers consist of the application layer, transport layer, network layer, access layer, and physical layer. These layers are illustrated in Figure 16.2, along with their primary functions in ad hoc wireless networks. These functions will be described in more detail below. Note that power control sits at two layers, the physical and access layers, and is part of resource allocation at the network layer as well [32].

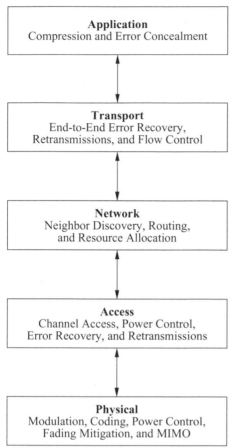

Figure 16.2: Five-layer model for network protocol design.

Thus, power control spans multiple layers of the protocol stack, as discussed in more detail below. Most ad hoc wireless network designs do not use the IP protocol for routing, since routing through a wireless network is much different than for the Internet. Moreover, the addressing and subnetting in the IP protocol is not well suited to ad hoc wireless networks. Transport protocols do not necessarily use TCP either. However, the five-layer model is a common abstraction for the modular design of protocol layers in wireless networks.

The layering principle for protocol design works reasonably well in wired networks like the Internet, where the data rates associated with the physical layer can exceed gigabits per second and packets are rarely lost. However, even in this setting, layering makes it difficult to support high–data-rate applications with hard delay constraints, such as video or even voice. Wireless networks can have very low physical layer data rates and very high packet and bit error probabilities. In this setting, protocol layering can give rise to tremendous inefficiencies and also precludes exploiting interactions between protocol layers for better performance. Cross-layer design considers multiple layers of the protocol stack together, either in terms of a joint design or in information exchange between the layers. Cross-layer design can exhibit tremendous performance advantages over a strictly layered approach. We now describe the layers of the five-layer model and their functionality in wireless networks. We then discuss the basic principles of cross-layer design and the performance advantages of this approach over strict layering.

16.3.1 Physical Layer Design

The physical layer deals primarily with transmitting bits over a point-to-point wireless link, so it is also referred to as the *link layer*. Chapters 5–13 comprehensively covered the design trade-offs associated with the physical layer, including modulation, coding, diversity, adaptive techniques, MIMO, equalization, multicarrier modulation, and spread spectrum. However, the design trade-offs for a link that is part of an ad hoc wireless network have an impact on protocol layers above the physical layer. In fact, few aspects of the physical layer design in an ad hoc wireless network do not affect in some way the protocols associated with the higher layers. We now give some examples of this interaction for physical layer design choices related to packet error rate, multiple antennas, and power control.

In most wireless ad hoc networks, bits are packetized for transmission as described in Section 14.3. The design choices at the physical layer along with the channel and interference conditions determine the link packet error rate (PER). Many access layer protocols retransmit packets received in error, so PER based on the physical layer design affects the retransmission requirements at the access layer. Similarly, as described in Section 10.8, multiple antennas give rise to a multiplexing/diversity/directionality trade-off: the antennas can be used to increase the data rate on the link, to provide diversity in fading so that average BER is reduced, or to provide directionality to reduce fading and the interference a signal causes to other signals. Diversity gain will reduce PER, leading to fewer retransmissions. Multiplexing will increase the link rate, which reduces congestion and delay on the link and benefits all multihop routes using that link. Directionality reduces interference to other links, thereby improving their performance. Thus, the best use of multiple antennas in an ad hoc wireless network clearly transcends just the physical layer; in fact, it simultaneously affects the physical, access, network, and transport layers.

The transmit power of a node at the physical layer also has a broad impact across many layers of the protocol stack. Increasing transmit power at the physical layer reduces PER, thereby affecting the retransmissions required at the access layer. In fact any two nodes in the network can communicate directly with sufficiently high transmit power, so this power drives link connectivity. However, a high transmit power from one node in the network can cause significant interference to other nodes, thereby degrading their performance and possibly breaking their connectivity on some links. In particular, link performance in an ad hoc wireless network is driven by SINR, so the transmit power of all nodes has an impact on the performance of all links in the network. Broadly speaking, the transmit power coupled with adaptive modulation and coding for a given node defines its "local neighborhood" – the collection of nodes that it can reach in a single hop – and thus defines the context in which access, routing, and other higher-layer protocols operate. Therefore, the transmit power of all nodes in the network must be optimized with respect to all layers that it affects. As such, it is a prime motivator for a cross-layer design.

16.3.2 Access Layer Design

The access layer controls how different users share the available spectrum and ensures successful reception of packets transmitted over this shared spectrum. Allocation of signaling dimensions to different users is accomplished through either multiple access or random access, and a detailed discussion of these access techniques can be found in Sections 14.2 and 14.3. Multiple access divides the signaling dimensions into dedicated channels via orthogonal or nonorthogonal channelization methods. The most common of these methods are TDMA, FDMA, and CDMA. The access layer must also provide control functionality in order to assign channels to users and to deny access to users when they cannot be accommodated by the system. In random access, channels are assigned to active users dynamically, and in multihop networks these protocols must contend with hidden and exposed terminals. The most common random access methods are different forms of ALOHA, carrier-sense multiple access (CSMA), and scheduling (see Section 14.3). These random access methods incorporate channel assignment and access denial into their protocols.

As discussed in the prior section, transmit power associated with a single node affects all other nodes. Thus, power control across all nodes in the network is part of the access layer functionality. The main role of power control is to ensure that SINR targets can be met on all links in the network. This is often unfeasible, as discussed in more detail below. The power control algorithms described in Sections 14.4 and 15.5.3 for meeting SINR targets in multiple access and cellular systems, respectively, can be extended to ad hoc networks as follows. Consider an ad hoc wireless network with K nodes and N links between different transmitter–receiver pairs of these nodes.[2] The SINR on link k is given by

$$\gamma_k = \frac{g_{kk}P_k}{n_k + \rho \sum_{j \neq k} g_{kj}P_j}, \quad k, j \in \{1, 2, \ldots, N\}, \tag{16.1}$$

where $g_{kj} > 0$ is the channel power gain from the transmitter of the jth link to the receiver of the kth link, P_k is the power of the transmitter on the kth link, n_k is the noise power of the receiver on the kth link, and ρ is the interference reduction due to signal processing (i.e., $\rho \approx 1/G$ for CDMA with processing gain G and $\rho = 1$ in TDMA). Suppose that the kth link requires an SINR of γ_k^* as determined, for example, by the connectivity and data-rate requirements for that link. Then the SINR constraints for all links can be represented in matrix form as $(\mathbf{I} - \mathbf{F})\mathbf{P} \geq \mathbf{u}$ with $\mathbf{P} > 0$, where $\mathbf{P} = (P_1, P_2, \ldots, P_N)^T$ is the vector of transmit powers associated with the transmitters on the N links,

$$\mathbf{u} = \left(\frac{\gamma_1^* n_1}{g_{11}}, \frac{\gamma_2^* n_2}{g_{22}}, \ldots, \frac{\gamma_N^* n_N}{g_{NN}} \right)^T \tag{16.2}$$

is the vector of noise powers scaled by the SINR constraints and channel gains, and $\mathbf{F}$ is a matrix with

$$F_{kj} = \begin{cases} 0 & k = j, \\ \gamma_k^* g_{kj} \rho / g_{kk} & k \neq j, \end{cases} \tag{16.3}$$

where $k, j \in \{1, 2, \ldots, N\}$. As in the uplink and cellular power control problems described in Section 14.4 and Section 15.5.3, if the Perron–Frobenius (maximum modulus) eigenvalue of $\mathbf{F}$ is less than unity then there exists a vector $\mathbf{P} > 0$ (i.e., $P_k > 0$ for all k) such that the SINR requirements of all links are satisfied, with $\mathbf{P}^* = (\mathbf{I} - \mathbf{F})^{-1}\mathbf{u}$ the Pareto optimal solution [33; 34; 35]. Thus $\mathbf{P}^*$ meets the SINR requirements with the minimum transmit power on all links. Moreover, a distributed iterative power control algorithm, where the transmitter on the kth link updates its transmit power at time $i + 1$ as

$$P_k(i + 1) = \frac{\gamma_k^*}{\gamma_k(i)} P_k(i), \tag{16.4}$$

---

[2] As noted previously, all nodes can communicate with all other nodes, so there are $K(K - 1)$ links for a network with K nodes. However, we assume that only $N \leq K(K - 1)$ of these are used, so we need only consider the SINR on these N links.

can be shown [34; 35] to converge to the optimal solution $\mathbf{P}^*$. This is a simple distributed algorithm for power control in an ad hoc wireless network, since it requires only that the SINR of the kth link be known to the transmitter on that link. Then, if this SINR is below its target the transmitter increases its power, and if it is above this target the transmitter decreases the power. It is quite remarkable that such a simple distributed algorithm converges to a globally optimal power control. However, when the channel gains are not static, SINR constraints can no longer be met with certainty, and it is much more difficult to develop distributed power control algorithms that meet a desired performance target [36]. In particular, the algorithm described by (16.4) can exhibit large fluctuations in link SINR when the channel gains vary over time. More importantly, it is often impossible to meet the SINR constraints of all nodes simultaneously even when the link gains are static, owing to the large number of interferers and the range of channel gains associated with all signals in the network. When the SINR constraints cannot be met, the distributed power control algorithm will diverge such that all nodes transmit at their maximum power and still cannot meet their SINR constraints. This is obviously an undesirable operational state, especially for energy-constrained nodes.

In [37], the distributed power control algorithm assuming static link gains is extended to include distributed admission control. The admission control provides protection for existing link SINR targets when a new user enters the system. In this scheme the active links have a slightly higher SINR target than needed. This buffer is used so that, when a new user who attempts to access the system transmits at a low power level, he will not cause the active links to fall below their minimum SINR targets. The new user gradually ramps up his power and checks if he gets closer to his SINR target as a result. If the new user can be accommodated by the system without violating the SINR constraints of existing links, then the distributed algorithm with this gradual ramp-up will eventually converge to a new $\mathbf{P}^*$ that satisfies the SINR constraints of the new and existing links. However, if the new user cannot be accommodated, his gradual ramp-up will not approach his required SINR and he will eventually be forced to leave the system. Note that the criteria for denial of access is difficult to optimize in time-varying channels with distributed control [36]. These ideas are combined with transmission scheduling in [38; 39] to improve power efficiency and reduce interference. The power control algorithm can also be modified to take into account delay constraints [23; 40]. However, delay constraints are associated with the full multihop route of a packet, so power control should be coordinated with network layer protocols to ensure delay constraints are met on the end-to-end route.

The access layer is also responsible for retransmissions of packets received in error over the wireless link, often referred to as the automatic repeat request (ARQ) protocol. Specifically, data packets typically have an error detection code that is used by the receiver to determine if one or more bits in the packet were corrupted and cannot be corrected. For such packets, the receiver will usually discard the corrupted packet and inform the transmitter via a feedback channel that the packet must be retransmitted. However, rather than discarding the packet, the access layer can save it and use a form of diversity to combine the corrupted packet with the retransmitted packet for a higher probability of correct packet reception. Alternatively, rather than retransmitting the original packet in its entirety, the transmitter can just send some additional coded bits to provide a stronger error correction capability for the receiver to correct for the corrupted bits in the original packet. This technique is called *incremental redundancy,* since the transmitter need only send enough redundant bits to correct for the bits corrupted in the original packet transmission. Diversity and incremental redundancy methods have been shown to substantially improve throughput in comparison with simple retransmissions [41].

16.3.3 Network Layer Design

The network layer is responsible for establishing and maintaining end-to-end connections in the network. This typically requires a network that is *fully connected,* whereby every node in the network can communicate with every other node, although these connections may entail multihop routing through intermediate nodes.[3] The main functions of the network layer in an ad hoc wireless network are neighbor discovery, routing, and dynamic resource allocation. Neighbor discovery is the process by which a node discovers its neighbors when it first enters the network. Routing is another key function of the network layer: the process of determining how packets are routed through the network from their source to their destination. Routing through intermediate nodes is typically done by relaying, although other techniques that better exploit multiuser diversity can also be used. Dynamic resource allocation dictates how network resources such as power and bandwidth are allocated throughout the network, although resource allocation in general occurs at multiple layers of the protocol stack and thus requires cross-layer design.

NEIGHBOR DISCOVERY AND TOPOLOGY CONTROL

Neighbor discovery is one of the first steps in the initialization of a network with randomly distributed nodes. From the perspective of the individual node, this is the process of determining the number and identity of network nodes with which direct communication can be established, given some maximum power level and minimum link performance requirements (typically in terms of data rate and associated BER). Clearly, the higher the allowed transmit power, the greater the number of nodes in a given neighborhood. Note that different addressing schemes at different protocol layers can complicate neighbor discovery, which must resolve these differences.

Neighbor discovery typically begins with a probe of neighboring nodes using some initial transmit power. If this power is not sufficient to establish a connection with $N \geq 1$ neighbors then transmit power is increased and probing repeated. The process continues until N connections are established or the maximum power P_{max} is reached. The parameter N is set based on network requirements for minimal connectivity, while P_{max} is based on the power limitations of each node and the network design. If N and/or P_{max} is small then the network may form in a disconnected manner, with small clusters of nodes that communicate together but cannot reach other clusters. This is illustrated in Figure 16.3, where the dashed circles centered around a node indicate the neighborhood within which it can establish connections with other nodes. If N and P_{max} are large then – although the network is typically fully connected – many nodes are transmitting at higher power than necessary for full network connectivity, which can waste power and increase interference. Once the network is

---

[3] This definition differs from that in graph theory, where every node in a fully connected graph has an edge to every other node.

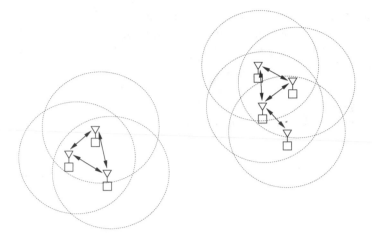

Figure 16.3: Disconnected network.

fully connected, a more sophisticated distributed power control algorithm such as (16.4) can be activated to meet target SINR levels on all links with minimal transmit power. Alternatively, power control can be used to create a desired topology [8].

The exact number of neighbors that each node requires in order to establish a fully connected network depends on the exact network configuration and channel characteristics, but it is generally on the order of six to eight for randomly distributed immobile nodes with channels characterized by path loss alone [2; 31]. The number required for full connectivity under more general assumptions is analyzed in [42; 43; 44]. As node mobility increases, links typically experience large gain variations due to fading. These variations can make it harder for the network to stay fully connected at all times unless the nodes can increase their transmit powers to compensate for instantaneous fading. If the data is tolerant of delay then fading can actually improve network connectivity because it provides network diversity [45]. As network density decreases, network connectivity typically suffers [43; 46; 47; 48]. Connectivity is also heavily influenced by the ability to adapt at the physical layer such parameters as rate, power, and coding, since communication is possible even on links with low SINR if these parameters are adapted.

ROUTING

The routing protocol in an ad hoc wireless network is a significant design challenge – especially under node mobility, where routes must be dynamically reconfigured owing to rapidly changing connectivity. There is broad and extensive work spanning several decades on routing protocols for ad hoc wireless networks, which is difficult to classify in a simple manner. We will focus on three main categories of routing protocols: flooding, proactive (centralized, source-driven, or distributed), and reactive routing [49, Chap. 5].

In *flooding,* a packet is broadcast to all nodes within receiving range. These nodes also broadcast the packet, and the forwarding continues until the packet reaches its ultimate destination. Flooding has the advantage that it is highly robust to changing network topologies and requires little routing overhead. In fact, in highly mobile networks flooding may be the only feasible routing strategy. The obvious disadvantage is that multiple copies of the same packet traverse through the network, wasting bandwidth and battery power of the transmitting nodes. This disadvantage makes flooding impractical for all but the smallest of networks.

The opposite philosophy to flooding is *centralized* route computation. In this approach, information about channel conditions and network topology is determined by each node and forwarded to a centralized location that computes the routing tables for all nodes in the network. These tables are then communicated to the nodes. The criterion used to compute the optimal route depends on the optimization objectives. Common objectives for route optimization include minimum average delay, minimum number of hops, and minimum network congestion [50]. In general these objectives are mapped to the cost associated with each hop along a route. The minimum cost route between the source and destination is then obtained using classic optimization techniques such as the Bellman–Ford or Dijkstra's algorithm (see [51]). Although centralized route computation provides the most efficient routing in terms of the optimality condition, it cannot adapt to fast changes in the channel conditions or network topology, and it also requires considerable overhead for periodically collecting local node information and then disseminating the routing information. Centralized route computation, like flooding, is typically used only in very small networks.

A variation on centralized route computation is *source-driven* routing, where each node obtains connectivity information about the entire network that is then used to calculate the best route from the node to its desired destination. Source-driven routing must also periodically collect network connectivity information, which entails significant overhead. Both centralized and source-driven routing can be combined with hierarchical routing, where nodes are grouped into a hierarchy of clusters and then routing is performed within a cluster at each level of the hierarchy.

Distributed route computation is the most common routing procedure used in ad hoc wireless networks. In this protocol, nodes send their connectivity information to neighboring nodes and then routes are computed from this local information. In particular, nodes determine the next hop in the route of a packet based on this local information. There are several advantages to distributed route computation. First, the overhead of exchanging routing information with local nodes is minimal. In addition, this strategy adapts quickly to link and connectivity changes. The disadvantages of this strategy are that global routes based on local information are typically suboptimal, and *routing loops* – routes where a packet cycles between the same intermediate nodes without ever reaching its final destination – are often common in the distributed route computation. These loops can be avoided by using the destination sequenced distance vector (DSDV) protocol, which incorporates sequence numbers as part of the routing tables [52].

Both centralized and distributed routing require fixed routing tables optimized for a given criterion. Route optimization to minimize hop count is called *distance vector routing,* while optimizing routes with respect to a more general cost function associated with each hop is called *link state routing.* Because of network dynamics, the routing tables obtained by either centralized or distributed route optimization must be updated at regular intervals. An alternate approach is reactive (on-demand) routing, where routes are created only at the initiation of a source node that has traffic to send to a given destination. This eliminates the overhead of maintaining routing tables for routes not currently in use. In this strategy, a source node initiates a route discovery process when it has data to send. This process will determine if one or more routes are available to the destination. The route or routes are maintained until the source has no more data for that particular destination. The advantage of reactive routing is that globally efficient routes can be obtained with relatively little overhead, because these routes need not be maintained at all times. The disadvantage is that reactive routing can entail significant initial delay because the route discovery process is initiated when there is data to send, yet transmission of this data cannot commence until the route discovery process has concluded. The most common protocols for on-demand routing are ad hoc on-demand distance vector routing (AODV) [53] and dynamic source routing (DSR) [54]. Reactive and proactive routing are combined in a hybrid technique called the zone routing protocol (ZRP), which reduces the delay associated with reactive routing as well as the overhead associated with proactive routing [55].

Mobility has a huge impact on routing protocols, since it can cause established routes to no longer exist. High mobility especially degrades the performance of proactive routing, since routing tables quickly become outdated, requiring an enormous amount of overhead to keep them up to date. Flooding is effective in maintaining routes under high mobility but has a huge price in terms of network efficiency. A modification of flooding called multipath routing can be effective without adding significant overhead. In multipath routing, a packet is duplicated on only a few end-to-end paths between its source and destination. Since it is unlikely that the duplicate packets are lost or significantly delayed on all paths simultaneously, the packet has a high probability of reaching its final destination with minimal delay on at least one of the paths [56]. This technique has been shown to perform well under dynamically changing topologies.

The routing protocol is based on an underlying network topology: packets can only be routed over links between two nodes of reasonable quality. However, as described earlier, the definition of connectivity between two nodes is somewhat flexible: it depends on the SINR of the link as well as the physical layer design, which determines the required SINR for data to be reliably transmitted over the link. The access layer also plays a role in connectivity, since it dictates the interference between links. Thus, there is significant interaction between the physical, access, and network layers [57]. This interaction was investigated in [38], where it was found that if CSMA with collision avoidance is coupled with a routing protocol that uses low-SINR links, network throughput is significantly reduced. Another interesting result in [38] is that maintaining a single route between any source–destination pair is suboptimal in terms of total network throughput. Multiplexing between multiple routes associated with any given source–destination pair provides an opportunity to change the interference that pair causes to other end-to-end routes, and this diversity can be exploited to increase network throughput.

Routing algorithms can also be optimized for requirements associated with higher-layer protocols, in particular delay and data-rate requirements of the application layer. Such algorithms are referred to as QoS routing. The goal of QoS routing is to find routes through the network that can satisfy the end-to-end delay and data rate requirements specified by the application. Examples of QoS routing and its performance are given in [58; 59; 60].

Most routing protocols use a decode-and-forward strategy at each relay node, where packets received by the relay are decoded to remove errors through error correction and where retransmissions are requested when errors are detected but cannot be corrected. An alternate strategy is amplify-and-forward, where the relay node simply retransmits the packet it has received without attempting to remove errors or detect corrupted packets. This simplifies the relay design, reduces processing energy at the relay, and reduces delay. However, amplify-and-forward does not work well in a wireless setting, since each wireless link is unreliable and often introduces errors, which are compounded on each hop of a route. An alternative to these two strategies is *cooperative diversity,* where the diversity associated with spatially distributed users is exploited in forwarding packets [61; 62]. This idea was originally proposed in [63; 64], where multiple transmitters cooperate by repeating detected symbols of the others, thereby forming a repetition code with spatial diversity. These ideas have led to more sophisticated cooperative coding techniques [65] along with forms of cooperative diversity other than coding [66]. Finally, *network coding* fuses data received along multiple routes to increase network capacity [67; 68; 69]. Though network coding has been primarily applied to multicasting in wired networks, it can also be used in a wireless setting [70].

RESOURCE ALLOCATION AND FLOW CONTROL

A routing protocol dictates the route a packet should follow from a source node to its destination. When the routing optimization is based on minimum congestion or delay, routing becomes intertwined with flow control, which typically sits at the transport layer. If the routing algorithm sends too much data over a given link, then this link becomes congested and so the routing algorithm should find a different route to avoid this link. Moreover, the delay associated with a given link is a function of the link data rate or capacity: the higher the capacity, the more data that can flow over the link with minimal delay. Because link capacity depends on the resources allocated to the link – in particular, transmit power and bandwidth – we see that routing, resource allocation, and flow control are all interdependent.

The classic metric for delay on a link from node i to node j, neglecting processing and propagation delay, is

$$D_{ij} = \frac{f_{ij}}{C_{ij} - f_{ij}} \tag{16.5}$$

[31, Chap. 5.4], where f_{ij} is the traffic flow assigned to the link and C_{ij} is its capacity (the maximum flow that the link can support). This formula has its roots in queueing theory and provides a good metric in practice, since the closer the flow is to the maximum data rate on a given link, the more likely the link will become congested and incur delay. Another metric on the link between nodes i and j is the link utilization, given by

$$D_{ij} = \frac{f_{ij}}{C_{ij}}. \tag{16.6}$$

As discussed in [31, Chap. 5.4], this metric has properties that are comparable to those of the delay metric (16.5) and is also a quasi-convex function[4] of both flow and capacity, which allows efficient convex optimization methods to be applied in the routing computations. If the data flows across links in the network are fixed, then the routing algorithm can compute the per-hop cost for each link based on the delay metric (16.5) or the utilization metric (16.6) and

---

[4] A function is *quasi-convex* if the set over which its value is below any given threshold is convex.

use these per-hop costs to find the minimum cost route through the network. The difference between these two metrics is that delay grows asymptotically large as the flow approaches link capacity, whereas link utilization approaches unity. Thus, the delay metric (16.5) has a much higher cost than the utilization metric (16.6) when flow is assigned to a link that is operating at close to its capacity. Once the new route is established, it will change the link flows along that route. In most cases this change in flow will not be large, since the contribution of any one node to overall traffic is not large, but in small to moderately sized networks a demanding application such as video can cause significant self-congestion.

The link metrics (16.5) and (16.6) assume a fixed link capacity C_{ij}. However, this capacity is a function of the SINR on the link and also of the bandwidth allocated to that link. By dynamically allocating network resources such as power and bandwidth to congested links, their capacities can be increased and their delay reduced. However, this may take away resources from other links, thereby decreasing their capacity. These changes in link capacity will in turn change the link metrics used to compute optimal routes and thus ultimately affect the overall performance of the network. Hence, the performance of the network depends simultaneously on routing, flow control, and resource allocation.

The joint optimization of flow control, routing, and resource allocation can be formulated as a convex optimization problem over the flow and communications variables, assuming the cost and capacity functions are convex (or quasi-convex). Interior-point convex optimization methods can then be applied to solve for the optimal design. This approach has been investigated in [71; 72; 73] for both TDMA and CDMA wireless networks in order to minimize power, maximize link utilization, or maximize flow utility through joint routing and resource allocation. Similar ideas using iterative optimization were explored in [74]. The maximum throughput in this setting can lead to highly unfair allocation of system resources [75], although the framework can be modified to include fairness constraints [76].

16.3.5 Transport Layer Design

The transport layer provides the end-to-end functions of error recovery, retransmission, reordering, and flow control. Although individual links provide error detection and retransmissions, these mechanisms are not foolproof. The transport layer provides an extra measure of protection by (i) monitoring for corrupt or lost packets on the end-to-end route and (ii) requesting a retransmission from the original source node if a packet is determined to be lost or corrupted. In addition, packets may arrive out of order owing to multipath routing, delays and congestion, or packet loss and retransmission. The transport layer serves to order packets transmitted over an end-to-end route before passing them to the application layer.

The transport layer also provides flow control for the network, allocating flows associated with the application layer to different routes. The transmission control protocol for the transport layer does not work well in wireless networks, since it assumes all lost packets are due to congestion and invokes congestion control as a result. In wired networks, congestion is the primary reason for packet loss and so the TCP works well; this is why it is used for the transport layer of the Internet. However, in wireless networks packets are mostly lost as a result of channel impairments and node mobility. Invoking congestion control in this case can lead to extreme inefficiency [49, Chap. 11.5]. There has been some limited progress on developing mechanisms to improve TCP performance for wireless networks by providing transport-layer feedback about link failures.

In general, flow control in wireless networks is intricately linked to resource allocation and routing, as described in Section 16.3.3. This interdependency is much tighter in wireless networks than in their wired counterparts. In particular, wired networks have links with fixed capacity, whereas the capacity of a wireless link depends on the interference between links. Traffic flows assigned to a given link will cause interference to other links, thereby affecting their capacity and delay. This interdependency makes it difficult to separate out the functions of flow control, resource allocation, and routing into separate network and transport layers, motivating a cross-layer design encompassing both layers.

16.3.5 Application Layer Design

The application layer generates the data to be sent over the network and processes the corresponding data received over the network. As such, this layer provides compression of the application data along with error correction and concealment. The compression must be lossless for data applications but can be lossy for video, voice, or image applications, where some loss can be tolerated in the reconstruction of the original data. The higher the level of compression, the less the data rate burden imposed on the network. However, highly compressed data is more sensitive to errors because most of the redundancy has been removed. Data applications cannot tolerate any loss, so packets that are corrupted or lost in the end-to-end transmission must be retransmitted, which can entail significant delay. Voice, video, and image applications can tolerate some errors, and techniques like error concealment or adaptive playback can mitigate the impact of these errors on the perceived quality at the receiving end [77; 78]. Thus, a trade-off at the application layer is data rate versus robustness: higher rates burden the network but enable data transmissions that are more robust with respect to network performance.

The application layer can also provide a form of diversity through multiple description coding (MDC) [79; 80], which is a form of compression whereby multiple descriptions of the data are generated. The original data can be reconstructed from any of these descriptions with some loss, and the more descriptions that are available, the better the reconstruction. If multiple descriptions of the source data are sent through the network then some of these descriptions can be lost, delayed, or corrupted without significantly degrading overall performance. Thus, MDC provides a form of diversity at the application layer to unreliable network performance. Moreover, MDC can be combined with multipath routing to provide cross-layer diversity in both the application description and the routes over which these descriptions are sent [81]. The trade-off is that, for a given data rate, an MDC entails more distortion than a compression technique that is not geared to providing multiple descriptions. This can be viewed as a performance–diversity trade-off: the application sacrifices some level of performance in order to provide robustness to uncertainty in the network.

Many applications require a guaranteed end-to-end data rate and delay for good performance, collectively referred to as QoS. The Internet today – even with high-speed, high-quality fixed communication links – is unable to deliver guaranteed QoS to applications in terms of guaranteed end-to-end rates or delays. For ad hoc wireless networks – with their low-capacity, error-prone time-varying links, mobile users, and dynamic topologies – the notion of being able to guarantee these forms of QoS is simply unrealistic. Therefore, ad hoc wireless network applications must adapt to the time-varying QoS parameters offered by the network. Although adaptivity in the physical, access, and network layers (as described in previous sections) will provide the best possible QoS to the application, this QoS will vary with time as channel conditions, network topology, and user demands change. Applications should therefore adapt to the QoS that is offered. There can also be some negotiation in which users with a higher priority can obtain a better QoS by lowering the QoS of less-important users.

As a simple example, the network may offer the application a rate–delay trade-off curve that is derived from the capabilities of the lower-layer protocols [82]. The application layer must then decide at which point on this curve to operate. Some applications may be able to tolerate a higher delay but not a lower overall rate. Examples include data applications in which the overall data rate must be high yet some latency is tolerable. Other applications (e.g., a distributed control application) might be extremely sensitive to delay but might be able to tolerate a lower rate (e.g., via a coarser quantization of sensor data). Lossy applications like voice or video might exchange some robustness to errors for a higher data rate. Energy constraints introduce another set of trade-offs related to network performance versus longevity. Hence trade-off curves in network design will typically be multidimensional, incorporating rate, delay, robustness, and longevity. These trade-offs will also change with time as the number of users in the network and the network environment change.

16.4 Cross-Layer Design

The decentralized control, lack of backbone infrastructure, and unique characteristics of wireless links make it difficult to support demanding applications over ad hoc wireless networks, especially applications with high–data-rate requirements and hard delay constraints. The layering approach to wireless (and wired) network design – where each layer of the protocol stack is oblivious to the design and operation of other layers – has not worked well in general, especially under stringent performance requirements. Layering precludes the benefits of joint optimization discussed in prior sections. Moreover, good protocol designs for isolated layers often interact in negative ways across layers, which can significantly degrade end-to-end performance and also make the network extremely fragile to network dynamics and interference. Thus, stringent performance requirements for wireless ad hoc networks can only be met through a cross-layer design. Such a design requires that the interdependencies between layers be characterized, exploited, and jointly optimized. Cross-layer design clearly requires information exchange between layers, adaptivity to this information at each layer, and diversity built into each layer to ensure robustness.

Although cross-layer design can be applied to both wireless and wired networks, wireless ad hoc networks pose unique challenges and opportunities for this design framework owing to the characteristics of wireless channels. The existence of a link between nodes, which can be used to communicate between the nodes or cause them to interfere with each other, can be controlled by adaptive protocols such as adaptive modulation and coding, adaptive signal processing in space, time, or frequency, and adaptive power control. Since higher-layer protocols (access and routing) depend on underlying node connectivity and interference, adaptivity at the physical layer can be exploited by higher-layer protocols to achieve better performance. At the same time, some links exhibit extreme congestion or fading. Higher-layer protocols can bypass such links through adaptive routing, thereby minimizing delays and bottlenecks that arise as a result of weak links. At the highest layer, information about the throughput and delay of end-to-end routes can be used to change the compression rate of the application or send data over multiple routes via MDCs. Thus, higher-layer protocols can adapt to the status of lower layers.

Adaptation at each layer of the protocol stack should compensate for variations at that layer based on the time scale of these variations. Specifically, variations in link SINR are very fast: on the order of microseconds for fast fading. Network topology changes more slowly, on the order of seconds, while variations of user traffic based on their applications may change over tens to hundreds of seconds. The different time scales of the network variations suggest that each layer should attempt to compensate for variation at that layer first. If adapting locally is unsuccessful then information should be exchanged with higher layers for a broader response to the problem. For example, suppose the SINR on a given link in an end-to-end route is low. By the time this connectivity information is relayed to a higher level of the protocol stack (i.e., the network layer for rerouting or the application layer for reduced-rate compression), the link SINR will most likely have changed. Therefore, it makes sense for each protocol layer to adapt to variations that are local to that layer. If this local adaptation is insufficient to compensate for the local performance degradation then the performance metrics at the next layer of the protocol stack will consequently degrade. Adaptation at this next layer may then correct or at least mitigate the problem that could not be fixed through local adaptation. For example, consider again a low SINR link. Link SINR can be measured quite accurately and quickly at the physical layer. The physical layer protocol can therefore respond to the low SINR by increasing transmit power or the level of error correction coding. This will correct for variations in connectivity due to, for example, multipath flat fading. However, if the weak link is caused by something difficult to correct for at the physical layer – for example, the mobile unit is inside a tunnel – then it is better for a higher layer of the network protocol stack to respond by (say) delaying packet transmissions until the mobile leaves the tunnel. Similarly, if nodes in the network are highly mobile then link characteristics and network topology will change rapidly. Informing the network layer of highly mobile nodes might change the routing strategy from unicast to broadcast in the general direction of the intended user. Ultimately, if the network cannot deliver the QoS requested by the application, then the application must adapt to whatever QoS is available or suffer the degradation associated with requiring network performance that cannot be delivered. It is this integrated approach to adaptive networking – how each layer of the protocol stack should respond to local variations given adaptation at higher layers – that constitutes an adaptive cross-layer protocol design.

Diversity is another mechanism that can be exploited in cross-layer design. Diversity is commonly used to provide robustness to fading at the physical layer. However, the basic premise of diversity can be extended across all layers in the network protocol stack. Cooperative diversity provides diversity at the access layer by using multiple spatially distributed nodes to aid in forwarding a given packet. This provides robustness to packet corruption on any one link. Network layer diversity is inherent to multipath routing, where multiple routes through the network are used to send a single packet. This induces a diversity–throughput trade-off at the network layer that is similar to that described for MIMO systems at the physical layer (see Section 10.5). Specifically, a packet transmitted over multiple routes through the network is unlikely to be dropped or significantly delayed simultaneously on all routes. Thus, the packet dropping probability and average delay is decreased by network diversity. However, such a packet utilizes network resources that could be used to send other packets, thereby reducing overall network throughput. Application layer diversity follows from using MDCs to describe the application data: as long as one of the descriptions is received, the source data can be reproduced (albeit with higher distortion than if the reproduction is based on all descriptions). Diversity across all layers of the protocol stack, especially when coupled with adaptive cross-layer design, can ensure reliability and good performance over wireless ad hoc networks despite their inherent challenges.

Cross-layer design across multiple protocol layers below the application layer were discussed in the preceding sections. Cross-layer design that includes the application layer along with lower layers is a difficult challenge requiring interdisciplinary expertise, and there is little work addressing this challenge to date. Moreover, there are potential pitfalls to cross-layer design, including increased complexity and reduced architectural flexibility [83]. Even so, the potential performance gains are significant, as illustrated by the cross-layer designs in [84; 85; 86] for video and image transmission in ad hoc wireless networks.

Cross-layer design is particularly important in energy-constrained networks, where each node has a finite amount of energy that must be optimized across all layers of the protocol stack. Energy constraints pose unique challenges and opportunities for cross-layering. Some of these design issues will be discussed in Section 16.6.4.

16.5 Network Capacity Limits

The fundamental capacity limits of an ad hoc wireless network – the set of maximum data rates possible between all nodes – is a challenging problem in information theory. For a network of K nodes, each node can communicate with $K - 1$ other nodes and so the capacity region has dimension $K(K-1)$ for sending independent information between nodes. The capacity region for sending both independent and common information (multicasting) is much larger. Even for a small number of nodes, the capacity for simple channel configurations (such as the general relay and interference channel) within an ad hoc wireless network remains unsolved [87]. Although rate sums between any two disjoint sets of network nodes can be upper bounded by the corresponding mutual information [87, Thm. 14.10.1], simplifying this formula into a tractable expression for the ad hoc network capacity region is an immensely complex problem.

Given that the entire capacity region appears intractable to find, insights can be obtained by focusing on a less ambitious goal. A landmark result by Gupta and Kumar [88] obtained scaling laws for network throughput as the number of nodes in the network K grows asymptotically large. The authors found that the throughput in terms of bits per second for each node in the network decreases with K at a rate between $1/\sqrt{K \log K}$ and $1/\sqrt{K}$. In other words, the per-node rate of the network goes to zero, although the total network throughput (which is equal to the per-node rate multiplied by K) grows at a rate between $\sqrt{K}/\log K$ and $\sqrt{K}$. This surprising result indicates that, even with optimal routing and scheduling, the per-node rate in a large ad hoc wireless network goes to zero. The reason is that intermediate nodes spend much of their resources forwarding packets for other nodes, so few resources

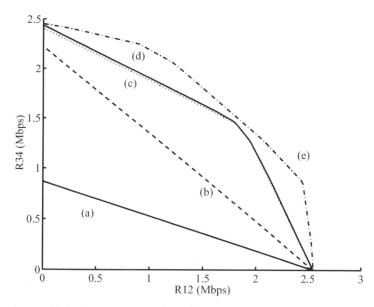

Figure 16.4: Capacity region slice of five-node network along the plane $R_{ij} = 0$, where $\{ij\} \neq \{12\}, \{34\}$ and $i \neq j$. (a) Single-hop routing, no spatial reuse. (b) Multihop routing, no spatial reuse. (c) Multihop routing with spatial reuse. (d) Two-level power control added to (c). (e) Successive interference cancellation added to (c).

are left to send their own data. To some extent this is a pessimistic result: it assumes that nodes choose their destination node at random, whereas in many networks communication between nodes is mostly local. These results were extended in [89] to show that, if mobile nodes can transmit information by physically transporting it close to its desired destination, then node mobility actually increases the per-node rate to a constant; that is, mobility increases network capacity. This increase follows because mobility introduces variation in the network that can be exploited to improve per-user rates. However, in order to exploit the variations due to mobility, significant delay can be incurred. The trade-off between throughput and delay in asymptotically large fixed and mobile networks was characterized in [90; 91]. Similar ideas were applied to finite-size networks and networks with relays in [43; 92].

An alternative approach to scaling laws is to compute achievable rate regions based on suboptimal transmission strategies. This approach was taken in [45] to obtain achievable rate regions based on a time-division strategy associated with all possible rate matrices. The rate matrices describe the set of rates that can be sustained simultaneously by all source–destination pairs at any moment in time. By taking a convex combination of rate matrices at different timeslots, all achievable rates between source–destination pairs under a time-division strategy can be obtained. A rate matrix is a function of the nodes transmitting at that time and the resulting SINR on all links, as well as the transmission strategy. The more capable the transmission strategy, the larger the data rates in a given matrix and the more matrices that are available for use in the time-division scheme. Some of the strategies considered in [45] include variable-rate transmission, single-hop or multihop routing, power control, and successive interference cancellation. The framework can also include the effects of mobility and fading. Figure 16.4 illustrates a two-dimensional slice of a rate region for a network of five nodes randomly distributed in a square area. It is assumed that signal propagation between nodes is governed by the simplified path-loss model with path-loss exponent $\gamma = 4$. This two-dimensional slice of the 20-dimensional rate region indicates the rates achievable between two pairs of nodes – from node 1 to 2 and from node 3 to 4 – when all other nodes in the network may be used to help forward traffic between these nodes but do not generate any independent data of their own. The figure assumes variable-rate transmission based on the link SINRs, and it plots the achievable rate region assuming single-hop or multihop routing, spatial reuse, power control, and successive interference cancellation. We see a substantial capacity increase by adding multihop routing, spatial reuse, and interference cancellation. Power control does not provide a significant increase, because adaptive modulation is already being exploited and so adding power control as well does not make much difference – at least for this particular network configuration.

Network capacity regions under different forms of cooperative diversity have also been explored [93; 94; 95; 96; 97]. Since the capacity region of a general ad hoc network is unknown, capacity under cooperation has mainly been characterized by lower bounds based on achievable rate regions or upper bounds based on the rate-sum mutual information bound. Results show that cooperation can lead to substantial gains in capacity, but the advantages of transmitter and/or receiver cooperation – as well as the most advantageous cooperative techniques to use – are highly dependent on network topology and the availability of channel information.

16.6 Energy-Constrained Networks

Many ad hoc wireless network nodes are powered by batteries with a limited lifetime. Thus, it is important to consider the impact of energy constraints in the design of ad hoc wireless networks. Devices with rechargeable batteries must conserve energy in order to maximize time between recharging. In addition, many interesting applications have devices that cannot be recharged – for example, sensors that are imbedded in walls or dropped into a remote region. Such radios must operate for years solely on battery energy and/or energy that can be harvested from the environment. The μ-AMPs and Picoradio projects are aimed at developing radios for these applications that can operate on less than 100 microwatts and exploit energy harvesting to prolong lifetime [18; 98; 99].

Energy constraints affect the hardware operation, transmit power, and signal processing associated with node operation. The required transmit energy per bit for a given BER target in a noisy channel is minimized by spreading the signal energy over all available time and bandwidth dimensions [100]. However, transmit power is not the only factor in power consumption. The signal processing associated with packet transmission and reception, and even hardware operation in standby mode, consume nonnegligible power as well [24; 101; 102]. This entails interesting energy trade-offs across protocol layers. At the physical layer, many communication techniques that reduce transmit power require a significant amount of signal processing. It is widely assumed that the energy required for this processing is small and continues to decrease with ongoing improvements in hardware technology [24; 103]. However, the results in [101; 102] suggest that these energy costs are still significant. This would indicate that energy-constrained systems must develop energy-efficient processing techniques that minimize power requirements across all levels of the protocol stack and also minimize message passing for network control, since these operations are performed at significant cost to transmitter and receiver energy. Sleep modes for nodes must be similarly optimized, since these modes conserve standby energy but may entail energy costs at other protocol layers due to the associated complications in access and routing. The hardware and operating system design in the node can also be optimized to conserve energy; techniques for this optimization are described in [101; 104]. In fact, energy constraints affect all layers of the protocol stack and hence make cross-layer design even more important if energy-constrained networks are to meet their performance requirements [23; 105; 106; 107]. In this section we describe some of the dominant design considerations for ad hoc wireless networks with energy-constrained nodes.

16.6.1 Modulation and Coding

Modulation and coding choices are typically made based on trade-offs between required transmit power, data rate, BER, and complexity. However, the power consumed within the analog and digital signal processing circuitry can be comparable to the required transmit power for short-range applications. In this case design choices should be based on the total energy consumption, including both the transmit and circuit energy consumption. Modeling circuit energy consumption is quite challenging and depends very much on the exact hardware used [108], which makes it difficult to make broad generalizations regarding trade-offs between circuit and transmit energy. However, the trade-offs certainly exist, especially for short-range applications where transmit energy can be quite low.

Because circuit energy consumption increases with transmission time, minimizing transmission time and then putting nodes to sleep can yield significant energy savings. These ideas were investigated in [98], where it was shown that M-ary modulation may enable energy savings (over binary modulation) for some short-range applications by decreasing the transmission time and shutting down most of the circuitry after transmission. This approach was analyzed for MQAM modulation in [109], which proposed optimal strategies for minimizing the total energy consumption. These ideas were extended in [30] to jointly optimize modulation bandwidth, transmission time, and constellation size for MQAM and MFSK in both AWGN and Rayleigh fading channels. These results indicate that energy consumption is significantly reduced by optimizing transmission time relative to transmission distance: at large distances transmit power dominates, so smaller constellations with larger transmission times are best; but the opposite is true at small transmission distances. As a result, MQAM was slightly more energy efficient than MFSK at short distances because it could transmit over a shorter time duration, but at larger distances MFSK was better owing to the superior energy characteristics of nonlinear amplifiers.

Energy constraints also change the trade-offs inherent to coding. Coding typically reduces the required transmit energy per bit for a given BER target. However, this savings comes at a cost of the processing energy associated with the encoder and decoder. Moreover, some coding schemes (such as block and convolutional codes) encode bits into a codeword that is longer than the original bit sequence; this is sometimes referred to as bandwidth expansion. Although the total transmit energy required for the codeword to achieve a given BER may be less than that required for the uncoded bits, it takes longer for the codeword to be sent, and a longer transmission time consumes more circuit energy. Joint modulation

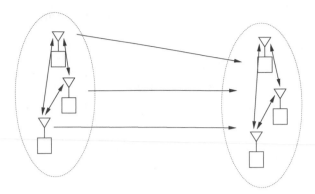

Figure 16.5: Cooperative MIMO.

and coding techniques such as trellis coding do not have bandwidth expansion and hence incur no bandwidth expansion energy penalty. However, their encoder and decoder processing energy must still be taken into account to determine if they yield a net energy savings. The impact of energy constraints on coded MQAM and MFSK was studied in [30]. The results indicate that trellis-coded MQAM provides energy savings at almost all transmission distances of interest (above 1 meter for the hardware parameters considered). However, coding techniques for MFSK are not generally bandwidth efficient, so coding is beneficial for MFSK only at moderate transmissions distances (above 30 meters for the hardware parameters considered).

16.6.2 MIMO and Cooperative MIMO

Multiple-input multiple-output techniques can significantly enhance performance of wireless systems through multiplexing or diversity gain. For a given transmit energy per bit, multiplexing gain provides a higher data rate whereas diversity gain provides a lower BER in fading. However, MIMO systems entail significantly more circuit energy consumption than their single-antenna counterparts, because separate circuitry is required for each antenna signal path and the signal processing associated with MIMO can be highly complex. Thus, it is not clear if MIMO techniques result in performance gain under energy constraints. This question was investigated in [110], where it was found that MIMO does provide energy savings over a single-antenna system for most transmission distances of interest if the constellation size is optimized relative to the distance. The reason is that the MIMO system can support a higher data rate for a given energy per bit, so it transmits the bits more quickly and can then be shut down to save energy.

Many energy-constrained networks consist of very small nodes that cannot support multiple antennas. In this case, nodes that are close together can exchange information to form a multiple-antenna transmitter, and nodes close together on the receive end can cooperate to form a multiple-antenna receiver; see Figure 16.5. As long as the distance between the cooperating nodes is small, the energy associated with their exchange of information is small relative to the energy savings associated with the resulting MIMO system. The energy associated with cooperative MIMO was quantified in [110], where it was shown that this cooperation provides energy savings when the transmit and receive clusters are 10 to 20 times the distance separating the cooperating nodes. When there is less of a difference between the separation of the cooperating nodes and the transmission distance between these clusters, the energy cost required for the local exchange of information exceeds the energy benefits of cooperating. Cooperative MIMO is one form of cooperative diversity. Others were discussed in Section 16.3.3, and these other techniques may provide energy savings comparable to or exceeding those of cooperative MIMO, depending on the network topology.

16.6.3 Access, Routing, and Sleeping

Random access schemes can be made more energy efficient by (i) minimizing collisions and the resulting retransmissions and (ii) optimizing transmit power to the minimum required for successful transmission. One way to reduce collisions is to increase error protection as collisions become more frequent [111]. Alternatively, adaptively minimizing power through probing as part of the random access protocol has been shown to significantly increase energy efficiency [111; 38]. Another method for energy-efficient access is to formulate the distributed access problem using a game-theoretic approach, where energy and delay are costs associated with the game [112]. Several different approaches to energy-efficient access were evaluated in [113]. However, no clear winner emerged because the performance of each protocol is highly dependent on channel characteristics. Delay and fairness constraints can also be incorporated into an energy-efficient access framework, as investigated in [114]. Many of these techniques avoid collisions through a version of TDMA, although setting up channelized access under distributed control can lead to large delays.

If users have long strings of packets or a continuous stream of data, then random access works poorly since most transmissions result in collisions. Hence channels must be assigned to users in a more systematic fashion by transmission scheduling. Energy constraints add a new wrinkle to scheduling optimization. In [100] it was shown that the energy required to send a bit is minimized by transmitting it over all available bandwidth and time dimensions. However, when multiple users wish to access the channel, the system time and bandwidth resources must be shared among all users. More recent work has investigated optimal scheduling algorithms to minimize transmit energy for multiple users sharing a channel [115]. In this work, scheduling was optimized to minimize the transmission energy required by each user subject to a deadline or delay constraint. The energy minimization was based on judiciously varying packet transmission time (and corresponding energy consumption) to meet the delay constraints of the data. This scheme was shown to be significantly more energy efficient than a deterministic schedule with the same deadline constraint.

Energy-constrained networks also require routing protocols that optimize routes relative to energy consumption. If the rate of energy consumption is not evenly distributed across all nodes then some nodes may expire sooner than others, leading to a partitioning of the network. Routing can be optimized to minimize end-to-end energy consumption by applying the standard optimization procedure described in Section 16.3.3, with energy per hop (instead of congestion or delay) as the hop cost [116]. Alternatively, the routes can be computed based on costs associated with the batteries in each node – for example, maximizing the minimum battery lifetime across all nodes in the network [116; 117]. Different cost functions to optimize energy-constrained routing were evaluated via simulation in [116] and were all roughly equivalent. The cost function can also be extended to include the traditional metric of delay along with energy [118]. This method allows the route optimization to trade off between delay and energy consumption through different weighting of their respective contribution to the overall cost function. Note that computation and dissemination of routing tables can entail significant cost: this can be avoided by routing traffic geographically (i.e., in the general direction of its destination), which requires little advance computation [119].

Energy-constrained nodes consume significant power even in standby mode, when they are just passive participants in the network with minimal exchange of data to maintain their network status. The paging industry developed a solution to this problem several decades ago by scheduling "sleep" periods for pagers. The basic idea is that each pager need only listen for transmissions during certain short periods of time. This is a simple solution to implement when a central controller is available, but it is less obvious how to implement such strategies within the framework of distributed network control. Sleep decisions must take into account network connectivity, so it follows that these decisions are local but not autonomous. Mechanisms that support such decisions can be based on neighbor discovery coupled with some means for ordering decisions within the neighborhood. In a given area, the opportunity to sleep should be circulated among the nodes, ensuring that connectivity is not lost through the coincidence of several simultaneous decisions to sleep.

16.6.4 Cross-Layer Design under Energy Constraints

The unique attributes of energy-constrained networks make them prime candidates for cross-layer design. If node batteries cannot be recharged, then each node can transmit only a finite number of bits before it dies, after which time it is no longer available to perform its intended function (e.g. sensing) or to participate in network activities such as routing. Thus, energy must be used judiciously across all layers of the protocol stack in order to prolong network lifetime and meet application requirements.

Energy efficiency at all layers of the protocol stack typically imposes trade-offs between energy consumption, delay, and throughput [120]. However, at any given layer, the optimal operating point on this trade-off curve must be driven by considerations at higher layers. For example, if a node transmits slowly then it conserves transmit energy, but this complicates access for other nodes and increases end-to-end delay. A routing protocol may use a centrally located node for energy-efficient routing, but this will increase congestion and delay on that route and also burn up that node's battery energy quickly, thereby removing it from the network. Ultimately the trade-offs between energy, delay, throughput, and node/network lifetime must be optimized relative to the application requirements. An emergency rescue operation needs on-the-scene information quickly, but typically the network supporting this local information exchange need only last a few hours or days. In contrast, a sensor network embedded into the concrete of a bridge to measure stress and strain must last decades, though the information need only be collected every day or week.

16.6.5 Capacity per Unit Energy

When transmit energy is constrained, it is not possible to transmit any finite number of bits with asymptotically small error probability. This is easy to see intuitively by considering the transmission of a single bit. The only way to ensure that two different values in signal space (representing the two possible bit values) can be decoded with arbitrarily small error is to make their separation arbitrarily large, which requires arbitrarily large energy. Since arbitrarily small error probability is not possible under a hard energy constraint, a different notion of reliable communication is needed for energy-constrained nodes.

A capacity definition for reliable communication under energy constraints was proposed in [121] (see also [100]) as the maximum number of bits per unit energy that can be transmitted so that the probability of error goes to zero with energy. This new notion of capacity per unit energy requires both energy and blocklength to grow asymptotically large for asymptotically small error probability. Thus, a finite energy system transmitting at capacity per unit energy does not have an asymptotically small error probability. Insight into this definition for AWGN channels is obtained by examining the minimum energy per bit required to transmit at the normalized Shannon capacity $C_B = C/B$ bps/Hz [100]. Specifically, the received energy per bit equals the ratio of received power to data rate: $E_b = P/R = P/C$ for transmission at rates approaching the Shannon capacity C. Using this expression in the Shannon capacity formula for an AWGN channel yields

$$C = B \log_2\left(1 + \frac{P}{N_0 B}\right) = B \log_2\left(1 + \frac{E_b C}{N_0 B}\right). \tag{16.7}$$

Inverting (16.7) yields the energy per bit required to transmit at rates approaching the normalized capacity $C_B = C/B$ as

$$\frac{E_b}{N_0}(C_B) = \frac{2^{C_B} - 1}{C_B}. \tag{16.8}$$

As the channel bandwidth B increases, C_B goes to zero, yielding the minimum energy per bit in the wideband limit:

$$\left(\frac{E_b}{N_0}\right)_{min} = \lim_{C_B \to 0} \frac{2^{C_B} - 1}{C_B} = \ln 2 = -1.59 \text{ dB}. \tag{16.9}$$

It was also shown in [100] that a form of on–off signaling such as pulse position modulation achieves this minimum energy per bit in AWGN. Moreover, results in [122; 123] indicate that the minimum energy per bit for reliable communication in flat fading is also given by (16.9), even when the fading is unknown to the receiver. These results indicate that: (i) in the limit of infinite bandwidth, minimum energy per bit is not affected by fading or receiver knowledge; and (ii) on–off signaling is near optimal for minimum energy communication.

Many energy-constrained wireless systems have large but finite bandwidth, and the results obtained for the limiting case of infinite bandwidth can be misleading for designing such systems (see [123; 124]). In particular, the bandwidth required to operate at an energy per bit close to the minimum (16.9) is very sensitive to the fading distribution and what is known about the fading at the receiver. If fading is known at the receiver then coherent QPSK is the optimal signaling scheme with on–off signaling distinctly suboptimal, but an asymptotic form of on–off signaling is optimal without this receiver knowledge.

The capacity per unit energy for single-user channels has been extended to broadcast and multiple access channels in [121; 126; 125]. These results indicate that, in the wideband limit, TDMA is optimal for both channels because it achieves the minimum energy per bit required for reliable communication. However, in the large but finite bandwidth regime, it was shown in [126] that superposition strategies – such as CDMA coupled with multiuser detection – achieve reliable communication with the same minimum energy per bit and less bandwidth than TDMA.

PROBLEMS

16-1. Consider a signal that must be transmitted 1 km. Suppose the path loss follows the simplified model $P_r = P_t d^{-\gamma}$.

(a) Find the required transmit power P_t such that the received power is 10 mW for $\gamma = 2$ and $\gamma = 4$.

(b) Suppose now that a relay node halfway between the transmitter and receiver is used for multihop routing. For $\gamma = 2$ and $\gamma = 4$, if both the transmitter and relay transmit at power P_t, how big must P_t be in order for the relay to receive a signal of 10 mW from the transmitter and the destination to receive a signal of 10 mW from the relay? Find the total power used in the network – that is, the sum of powers at both the transmitter and the relay – for $\gamma = 2$ and $\gamma = 4$.

(c) Derive a general formula for the total power used in the network with N relays, evenly spaced between the transmitter and receiver, such that each relay and the receiver receive 10 mW of power.

16-2. Consider an ad hoc wireless network with three users. Users 1 and 2 require a received SINR of 7 dB while user 3 requires an SINR of 10 dB. Assume that all receivers have the same noise power $n_i = 1$ and that there is no processing gain to reduce interference ($\rho = 1$). Assume a matrix of gain values (indexed by the user numbers) of

$$G = \begin{bmatrix} 1 & .06 & .04 \\ .09 & .9 & .126 \\ .064 & .024 & .8 \end{bmatrix}.$$

(a) Confirm that the equation $(\mathbf{I} - \mathbf{F})\mathbf{P} \geq \mathbf{u}$ is equivalent to the SINR constraints of each user.

(b) Show that a feasible power vector exists for this system such that all users achieve their desired SINRs.

(c) Find the optimal power vector $\mathbf{P}^*$ such that users achieve their desired SINRs with minimum power.

16-3. This problem uses the same setup as in the prior problem. Suppose each user starts out with power 50, so the initial power vector $\mathbf{P}(0) = (P_1(0), P_2(0), P_3(0)) = (50, 50, 50)$. Following the recursive formula (16.4) for each user, plot $P_i(k)$ for each of the users ($i = 1, 2, 3$) over the range $k = 1, \ldots, N$, where N is sufficiently large that the power vector $\mathbf{P}(N)$ is close to its optimal value $\mathbf{P}^*$. Also plot the SINRs of each user for $k = 1, \ldots, N$.

16-4. Assume an infinite grid of network nodes spaced every 10 meters along a straight line. Assume the simplified path-loss model $P_r = P_t d^{-\gamma}$ and that P_r must be at least 10 mW to establish communication between two nodes.

(a) For $\gamma = 2$, find $P_{\max}$ such that every node has a neighborhood of $N = 2$ other nodes. What happens if nodes have a peak power constraint less than this $P_{\max}$?

(b) Find $P_{\max}$ for $\gamma = 2$ and $N = 4$.

(c) Find $P_{\max}$ for $\gamma = 4$ and $N = 4$.

16-5. Consider a geographical region of 100 square meters. Suppose that N nodes are randomly distributed in this region according to a uniform distribution and that each node has transmit power sufficient to communicate with any node within a distance of R meters. Compute the average number of nodes $\mathbf{E}[N]$ as a function of radius R, $1 \le R \le 20$, required for the network to be fully connected. The average $\mathbf{E}[N]$ should be computed based on 100 samples of the random node placement within the region.

16-6. Consider a network with multipath routing, so that every packet is sent over N independent routes from its source to its destination. Suppose delay d along each of the multiple end-to-end routes is exponentially distributed with mean D: $p(d) = e^{-d/D}/D, d > 0$. Find the probability that all N copies of the packet arrive at the destination with a delay exceeding D as a function of N, and evaluate for $N = 1$ and $N = 5$. Also determine how throughput is affected by multipath routing as a function of N. Qualitatively, describe the throughput/delay trade-off associated with multipath routing.

16-7. Show that (16.6) is convex in both f_{ij} and C_{ij}.

16-8. Assume a link with capacity $C = 10$ Mbps. Plot the delay given by (16.5) and by (16.6) for a data flow ranging from 0 to 10 Mbps: $0 \le f_{ij} < 10$ Mbps.

16-9. In this problem we consider the difference in the two delay metrics (16.5) and (16.6). Let λ be the ratio of (16.5) to (16.6).

(a) Find λ as a function of f_{ij}/C_{ij}.

(b) Using the fact that the flow f_{ij} must be less than C_{ij}, find the range of values that λ can take.

(c) Find the value of f_{ij}/C_{ij} such that $\lambda > 10$ – that is, where the delay associated with metric (16.5) exceeds that of (16.6) by an order of magnitude.

(d) Consider a network design where route costs are computed based on either metric (16.5) or (16.6). For which metric will the links be more congested, and why?

16-10. This problem shows the gains from cross-layer design between the network layer and the application layer. For transmission of applications like video, the application layer will try to use a high-rate encoding scheme to improve the quality. Now consider that the application is real-time. Given a capacity assignment made by the network layer, if the rate of transmission is high then there will be congestion on the link that will delay the packets and hence many packets will not reach the decoder in time, leading to poorer quality. Thus we see a trade-off. A simple model of distortion capturing both these effects can be given as

$$\text{Dist}(R) = D_0 + \frac{\theta}{R - R_0} + \kappa e^{-(C-R)T/L}. \tag{16.10}$$

The first two terms correspond to distortion at the application layer due to source encoding, and the last term corresponds to distortion due to delayed packets. Let $D_0 = .38$, $R_0 = 18.3$ kbps, $\theta = 2537$, scaling factor $\kappa = 1$, effective packet length $L = 3040$ bits, and playout deadline $T = 350$ ms. Capacity C and transmission rate R are described below.

(a) If the capacity C takes on the values 45 kbps, 24 kbps, and 60 kbps with probabilities .5, .25, and .25 (respectively), find the optimal rate R that minimizes average distortion Dist(R). Assume full cooperation between the application layer and the network layer: the application layer always knows the instantaneous capacity C associated with the network layer.

(b) Now consider the case when there is no cross-layer optimization and the application layer encodes at a fixed rate of $R = 22$ kbps at all times. Find the average distortion Dist(R) for the same capacity distribution as given in part (a).

(c) Comparing parts (a) and (b), find the percentage increase in distortion when cross-layer optimization is not used.

16-11. Show that the ratio E_b/N_0 as a function of $C_B = C/B$, as given by (16.8), increases with C for B fixed and decreases with B for C fixed. Also show that

$$\lim_{C_B \to 0} \left(\frac{E_b}{N_0} \right)(C_B) = \ln 2.$$

REFERENCES

[1]   R. E. Kahn, S. A. Gronemeyer, J. Burchfiel, and R. C. Kunzelman, "Advances in packet radio technology," *Proc. IEEE,* pp. 1468–96, November 1978.

[2]   L. Kleinrock and J. Silvester, "Optimum transmission radii for packet radio networks or why six is a magic number," *Proc. IEEE Natl. Telecomm. Conf.,* pp. 4.3.1–4.3.5, December 1978.

[3]   A. Ephremides, J. E. Wieselthier, and D. J. Baker, "A design concept for reliable mobile radio networks with frequency hopping signaling," *Proc. IEEE,* pp. 56–73, January 1987.

[4]   J. Jubin and J. D. Tornow, "The DARPA packet radio network protocols," *Proc. IEEE,* pp. 21–32, January 1987.

[5]   M. B. Pursley, "The role of spread spectrum in packet radio networks," *Proc. IEEE,* pp. 116–34, January 1987.

[6]   F. A. Tobagi, "Modeling and performance analysis of multihop packet radio networks," *Proc. IEEE,* pp. 135–55, January 1987.

[7]   B. Leiner, R. Ruther, and A. Sastry, "Goals and challenges of the DARPA Glomo program (global mobile information systems)," *IEEE Pers. Commun. Mag.,* pp. 34–43, December 1996.

[8]   R. Ramanathan and R. Rosales-Hain, "Topology control of multihop wireless networks using transmit power adjustment," *Proc. IEEE Infocom Conf.,* pp. 404–13, March 2000.

[9]   M. Ritter, R. Friday, and M. Cunningham, "The architecture of Metricom's microcellular data network and details of its implementation as the 2nd and 3rd generation ricochet wide-area mobile data service," *Proc. IEEE Emerg. Technol. Sympos. Broadband Commun.,* pp. 143–52, September 2001.

[10] M. N. Huhns, "Networking embedded agents," *IEEE Internet Comput.,* pp. 91–3, January/February 1999.

[11] W. W. Gibbs "As we may live," *Scientific American,* pp. 36, 40, November 2000.

[12] K. J. Negus, A. P. Stephens, and J. Lansford, "HomeRF: Wireless networking for the connected home," *IEEE Pers. Commun. Mag.,* pp. 20–7, February 2000.

[13] A. Schmidt, "How to build smart appliances," *IEEE Pers. Commun. Mag.,* pp. 66–71, August 2001.

[14] J. Haartsen, "The Bluetooth radio system," *IEEE Pers. Commun. Mag.,* pp. 28–36, February 2000.

[15] J. Haartsen and S. Mattisson, "Bluetooth: A new low-power radio interface providing short-range connectivity," *Proc. IEEE,* pp. 1651–61, October 2000.

[16] I. Poole, "What exactly is … ZigBee?" *IEEE Commun. Eng.,* pp. 44–5, August/September 2004.

[17] T. Mitchell, "Broad is the way (ultra-wideband technology)," *IEE Review,* pp. 35–9, January 2001.

[18] J. Rabaey, J. Ammer, J. L. da Silva, Jr., and D. Roundy, "PicoRadio supports ad hoc ultra-low power wireless networking," *IEEE Computer,* pp. 42–8, July 2000.

[19] J. Nilsson, B. Bernhardsson, and B. Wittenmark, "Stochastic analysis and control of real-time systems with random time delays," *Automatica,* pp. 57–64, January 1998.

[20] X. Liu, S. S. Mahal, A. Goldsmith, and J. K. Hedrick, "Effects of communication delay on string stability in vehicle platoons," *Proc. IEEE Internat. Conf. Intell. Transp. Syst.,* pp. 625–30, August 2001.

[21] S. R. Graham, G. Baliga, and P. R. Kumar, "Issues in the convergence of control with communication and computing: Proliferation, architecture, design, services, and middleware," *Proc. IEEE Conf. Decision Control,* pp. 1466–71, December 2004.

[22] S. Basagni, D. Turgut, and S. K. Das, "Mobility-adaptive protocols for managing large ad hoc networks," *Proc. IEEE Internat. Conf. Commun.,* pp. 1539–43, June 2001.

[23] N. Bambos, "Toward power-sensitive network architectures in wireless communications: Concepts, issues, and design aspects," *IEEE Pers. Commun. Mag.,* pp. 50–9, June 1998.

[24] K. Sohrabi, J. Gao, V. Ailawadhi, and G. Pottie, "Protocols for self-organization of a wireless sensor network," *IEEE Pers. Commun. Mag.,* pp. 16–27, October 2000.

[25] L. Subramanian and R. H. Katz, "An architecture for building self-configurable systems," *Proc. Mobile Ad Hoc Network Comput. Workshop,* pp. 63–73, August 2000.

[26] R. Jain, A. Puri, and R. Sengupta, "Geographical routing using partial information for wireless ad hoc networks," *IEEE Pers. Commun. Mag.,* pp. 48–57, February 2001.

[27] L. Zhou and Z. J. Haas, "Securing ad hoc networks," *IEEE Network,* pp. 24–30, November/December 1999.

[28] R. Karri and P. Mishra, "Modeling energy efficient secure wireless networks using network simulation," *Proc. IEEE Internat. Conf. Commun.,* pp. 61–5, May 2003.

[29] *IEEE Wireless Commun. Mag.,* Special Issue on Energy Aware Ad Hoc Wireless Networks (A. J. Goldsmith and S. B. Wicker, Eds.), August 2002.

[30] S. Cui, A. J. Goldsmith, and A. Bahai, "Energy-constrained modulation optimization," *IEEE Trans. Wireless Commun.,* September 2005.

[31] D. Bertsekas and R. Gallager, *Data Networks,* 2nd ed., Prentice-Hall, Englewood Cliffs, NJ, 1992.

[32] V. Kawadia and P. R. Kumar, "Principles and protocols for power control in wireless ad hoc networks," *IEEE J. Sel. Areas Commun.,* pp. 76–88, January 2005.

[33] J. Zander, "Performance of optimum transmitter power control in cellular radio systems," *IEEE Trans. Veh. Tech.,* pp. 57–62, February 1992.

[34] S. A. Grandhi, R. Vijayan, and D. J. Goodman, "Distributed power control in cellular radio systems," *IEEE Trans. Commun.,* pp. 226–8, February–April 1994.

[35] G. J. Foschini and Z. Miljanic, "A simple distributed autonomous power control algorithm and its convergence," *IEEE Trans. Veh. Tech.,* pp. 641–6, November 1993.

[36] T. Holliday, N. Bambos, A. J. Goldsmith, and P. Glynn, "Distributed power control for time varying wireless networks: Optimality and convergence," *Proc. Allerton Conf. Commun., Control, Comput.,* pp. 1024–33, October 2003.

[37] N. Bambos, G. Pottie, and S. Chen, "Channel access algorithms with active link protection for wireless communications networks with power control," *IEEE/ACM Trans. Network.,* pp. 583–97, October 2000.

[38] S. Toumpis and A. J. Goldsmith, "Performance, optimization, and cross-layer design of media access protocols for wireless ad hoc networks," *Proc. IEEE Internat. Conf. Commun.,* pp. 2234–40, May 2003.

[39] T. ElBatt and A. Ephremides, "Joint scheduling and power control for wireless ad hoc networks," *IEEE Trans. Wireless Commun.,* pp. 74–85, January 2004.

[40] S. Kandukuri and N. Bambos, "Power controlled multiple access (PCMA) in wireless communication networks," *Proc. IEEE Infocom Conf.,* pp. 386–95, March 2000.

[41] S. Kallel, "Analysis of memory and incremental redundancy ARQ schemes over a nonstationary channel," *IEEE Trans. Commun.,* pp. 1474–80, September 1992.

[42] V. Rodoplu and T. H. Meng, "Minimum energy mobile wireless networks," *IEEE J. Sel. Areas Commun.,* pp. 1333–44, August 1999.

[43] P. Gupta and P. R. Kumar, "Towards an information theory of large networks: An achievable rate region," *IEEE Trans. Inform. Theory*, pp. 1877–94, August 2003.

[44] F. Xue and P. R. Kumar, "The number of neighbors needed for connectivity of wireless networks," *Wireless Networks*, pp. 169–81, March 2004.

[45] S. Toumpis and A. Goldsmith, "Capacity regions for ad hoc networks," *IEEE Trans. Wireless Commun.*, pp. 736–48, July 2003.

[46] O. Dousse, P. Thiran, and M. Hasler, "Connectivity in ad-hoc and hybrid networks," *Proc. IEEE Infocom Conf.*, pp. 1079–88, June 2002.

[47] B. Krishnamachari, S. B. Wicker, and R. Bejar, "Phase transition phenomena in wireless ad hoc networks," *Proc. IEEE Globecom Conf.*, pp. 2921–5, November 2001.

[48] M. Penrose, *Random Geometric Graphs*, Oxford University Press, 2004.

[49] C.-K. Toh, *Ad Hoc Mobile Wireless Networks: Protocols and Systems*, Prentice-Hall, Englewood Cliffs, NJ, 2002.

[50] S.-J. Lee and M. Gerla, "Dynamic load-aware routing in ad hoc networks," *Proc. IEEE Internat. Conf. Commun.*, pp. 3206–10, June 2001.

[51] L. L. Peterson and B. S. Davie, *Computer Networks – A Systems Approach*, 2nd ed., Morgan-Kaufman, San Mateo, CA, 2000.

[52] C. P. P. Bhagwat, "Highly dynamic destination-sequenced distance vector routing (DSDV) for mobile computers," *Proc. ACM SIGCOMM*, pp. 234–44, September 1994.

[53] C. E. Perkins and E. M. Royer, "Ad hoc on-demand distance vector routing," *Proc. IEEE Workshop Mobile Comput. Syst. Appl.*, pp. 90–100, February 1999.

[54] D. B. Johnson and D. A. Maltz, "Dynamic source routing in ad hoc wireless networks," in T. Imielinsky and H. Korth, Eds., *Mobile Computing*, Kluwer, Dordrecht, 1996.

[55] M. R. Pearlman, Z. J. Haas, and S. I. Mir, "Using routing zones to support route maintenance in ad hoc networks," *Proc. IEEE Wireless Commun. Network Conf.*, pp. 1280–4, September 2000.

[56] A. Tsirigos and Z. J. Haas, "Multipath routing in the presence of frequent topological changes," *IEEE Commun. Mag.*, pp. 132–8, November 2001.

[57] D. Ayyagari, A. Michail, and A. Ephremides, "A unified approach to scheduling, access control and routing for ad-hoc wireless networks," *Proc. IEEE Veh. Tech. Conf.*, pp. 380–4, May 2000.

[58] S. Chen and K. Nahrstedt, "Distributed quality-of-service routing in ad hoc networks," *IEEE J. Sel. Areas Commun.*, pp. 1488–1505, August 1999.

[59] C. R. Lin and J.-S. Liu, "QoS routing in ad hoc wireless networks," *IEEE J. Sel. Areas Commun.*, pp. 1426–38, August 1999.

[60] M. Mirhakkak, N. Schult, and D. Thomson, "Dynamic bandwidth management and adaptive applications for a variable bandwidth wireless environment," *IEEE J. Sel. Areas Commun.*, pp. 1984–97, October 2001.

[61] A. Nosratinia, T. E. Hunter, and A. Hedayat, "Cooperative communication in wireless networks," *IEEE Commun. Mag.*, pp. 74–80, October 2004.

[62] S. Cui and A. J. Goldsmith, "Energy efficient routing using cooperative MIMO techniques," *Proc. IEEE Internat. Conf. Acous., Speech, Signal Proc.*, pp. 805–8, March 2005.

[63] A. Sendonaris, E. Erkip, and B. Aazhang, "User cooperation diversity. Part I: System description," *IEEE Trans. Commun.*, pp. 1927–38, November 2003.

[64] A. Sendonaris, E. Erkip, and B. Aazhang, "User cooperation diversity. Part II: Implementation aspects and performance analysis," *IEEE Trans. Commun.*, pp. 1939–48, November 2003.

[65] M. Janani, A. Hedayat, T. E. Hunter, and A. Nosratinia, "Coded cooperation in wireless communications: Space-time transmission and iterative decoding," *IEEE Trans. Signal Proc.*, pp. 362–71, February 2004.

[66] J. N. Laneman, D. N. C. Tse, and G. W. Wornell, "Cooperative diversity in wireless networks: Efficient protocols and outage behavior," *IEEE Trans. Inform. Theory*, pp. 3062–80, December 2004.

[67] R. Ahlswede, N. Cai, S.-Y. R. Li, and R. W. Yeung, "Network information flow," *IEEE Trans. Inform. Theory*, pp. 1204–16, July 2000.

[68] R. Koetter and M. Medard, "An algebraic approach to network coding," *IEEE/ACM Trans. Network.*, pp. 782–95, October 2003.

[69] S.-Y. R. Li, R. W. Yeung, and N. Cai, "Linear network coding," *IEEE Trans. Inform. Theory*, pp. 371–81, February 2003.

[70] N. Cai and R. W. Yeung, "Network error correction," *Proc. IEEE Inform. Theory Workshop.*, p. 101, June 2003.

[71] R. L. Cruz and A. V. Santhanam, "Optimal routing, link scheduling and power control in multihop wireless networks," *Proc. IEEE Infocom Conf.*, pp. 702–11, April 2003.

[72] M. Johansson, L. Xiao, and S. P. Boyd, "Simultaneous routing and power allocation in CDMA wireless data networks," *Proc. IEEE Internat. Conf. Commun.*, pp. 51–5, May 2003.

[73] L. Xiao, M. Johansson, and S. P. Boyd, "Optimal routing, link scheduling and power control in multihop wireless networks," *IEEE Trans. Commun.*, pp. 1136–44, July 2004.

[74] Y. Wu, P. A. Chou, Q. Zhang, K. Jain, W. Zhu, and S.-Y. Kung, "Network planning in wireless ad hoc networks: A cross-layer approach," *IEEE J. Sel. Areas Commun.*, pp. 136–50, January 2005.

[75] B. Radunović and J. Y. Boudec, "Rate performance objectives of multihop wireless networks," *IEEE Trans. Mobile Comput.*, pp. 334–49, October–December 2004.

[76] D. Julian, M. Chiang, D. O'Neill, and S. Boyd, "QoS and fairness constrained convex optimization of resource allocation for wireless cellular and ad hoc networks," *Proc. IEEE Infocom Conf.*, pp. 477–86, June 2002.

[77] Y. Wang and Q.-F. Zhu, "Error control and concealment for video communication: A review," *Proc. IEEE*, pp. 974–97, May 1998.

[78] E. Steinbach, N. Farber, and B. Girod, "Adaptive playout for low latency video streaming," *Proc. IEEE Internat. Conf. Image Proc.*, pp. 962–5, October 2001.

[79] D. G. Sachs, R. Anand, and K. Ramchandran, "Wireless image transmission using multiple-description based concatenated codes," *Proc. IEEE Data Compress. Conf.*, p. 569, March 2000.

[80] S. D. Servetto, K. Ramchandran, V. A. Vaishampayan, and K. Nahrstedt, "Multiple description wavelet based image coding," *IEEE Trans. Image Proc.*, pp. 813–26, May 2000.

[81] M. Alasti, K. Sayrafian-Pour, A. Ephremides, and N. Farvardin, "Multiple description coding in networks with congestion problem," *IEEE Trans. Inform. Theory*, pp. 891–902, March 2001.

[82] T. Holliday and A. Goldsmith, "Joint source and channel coding for MIMO systems: Is it better to be robust or quick," *Proc. Joint Workshop Commun. Cod.*, p. 24, October 2004.

[83] V. Kawadia and P. R. Kumar, "A cautionary perspective on cross-layer design," *IEEE Wireless Commun. Mag.*, pp. 3–11, February 2005.

[84] W. Kumwilaisak, Y. T. Hou, Q. Zhang, W. Zhu, C.-C. J. Kuo, and Y.-Q. Zhang, "A cross-layer quality-of-service mapping architecture for video delivery in wireless networks," *IEEE J. Sel. Areas Commun.*, pp. 1685–98, December 2003.

[85] T. Yoo, E. Setton, X. Zhu, A. Goldsmith, and B. Girod, "Cross-layer design for video streaming over wireless ad hoc networks," *Proc. IEEE Internat. Workshop Multi. Signal Proc.*, pp. 99–102, September 2004.

[86] W. Yu, K. J. R. Liu, and Z. Safar, "Scalable cross-layer rate allocation for image transmission over heterogeneous wireless networks," *Proc. IEEE Internat. Conf. Acous., Speech, Signal Proc.*, pp. 4.593–4.596, May 2004.

[87] T. Cover and J. Thomas, *Elements of Information Theory*, Wiley, New York, 1991.

[88] P. Gupta and P. R. Kumar, "The capacity of wireless networks," *IEEE Trans. Inform. Theory*, pp. 388–404, March 2000.

[89] M. Grossglauser and D. N. C. Tse, "Mobility increases the capacity of ad-hoc wireless networks," *IEEE/ACM Trans. Network.*, pp. 1877–94, August 2002.

[90] A. El Gamal, J. Mammen, B. Prabhakar, and D. Shah, "Throughput–delay trade-off in wireless networks," *Proc. IEEE Infocom Conf.*, pp. 464–75, March 2004.

[91] S. Toumpis, "Large wireless networks under fading, mobility, and delay constraints," *Proc. IEEE Infocom Conf.*, pp. 609–19, March 2004.

[92]   M. Gastpar and M. Vetterli, "On the capacity of wireless networks: The relay case," *Proc. IEEE Infocom Conf.,* pp. 1577–86, June 2002.

[93]   A. Host-Madsen, "A new achievable rate for cooperative diversity based on generalized writing on dirty paper," *Proc. IEEE Internat. Sympos. Inform. Theory,* p. 317, June 2003.

[94]   A. Host-Madsen, "On the achievable rate for receiver cooperation in ad-hoc networks," *Proc. IEEE Internat. Sympos. Inform. Theory,* p. 272, June 2004.

[95]   M. A. Khojastepour, A. Sabharwal, and B. Aazhang, "Improved achievable rates for user co-operation and relay channels," *Proc. IEEE Internat. Sympos. Inform. Theory,* p. 4, June 2004.

[96]   N. Jindal, U. Mitra, and A. Goldsmith, "Capacity of ad-hoc networks with node cooperation," *Proc. IEEE Internat. Sympos. Inform. Theory,* p. 271, June 2004.

[97]   C. T. K. Ng and A. J. Goldsmith, "Transmitter cooperation in ad-hoc wireless networks: Does dirty-paper coding beat relaying?" *IEEE Inform. Theory Workshop,* pp. 277–82, October 2004.

[98]   A. Chandrakasan, R. Amirtharajah, S. Cho, J. Goodman, G. Konduri, J. Kulik, W. Rabiner, and A. Y. Wang, "Design considerations for distributed microsensor systems," *Proc. IEEE Custom Integrated Circuits Conf.,* pp. 279–86, May 1999.

[99]   J. Rabaey, J. Ammer, J. L. da Silva, Jr., and D. Patel, "PicoRadio: Ad-hoc wireless networking of ubiquitous low-energy sensor/monitor nodes," *IEEE Comput. Soc. Workshop on VLSI,* pp. 9–12, April 2000.

[100] S. Verdú, "On channel capacity per unit cost," *IEEE Trans. Inform. Theory,* pp. 1019–30, September 1990.

[101] P. Agrawal, "Energy efficient protocols for wireless systems," *Proc. Internat. Sympos. Pers., Indoor, Mobile Radio Commun.,* pp. 564–9, September 1998.

[102] W. R. Heinzelman, A. Sinha, and A. P. Chandrakasan, "Energy-scalable algorithms and protocols for wireless microsensor networks," *Proc. IEEE Internat. Conf. Acous., Speech, Signal Proc.,* pp. 3722–5, June 2000.

[103] J. M. Kahn, R. H. Katz, and K. S. Pister, "Emerging challenges: Mobile networking for Smart Dust," *J. Commun. Networks,* pp. 188–96, August 2000.

[104] A. Chandrakasan and R. W. Brodersen, *Low Power Digital CMOS Design,* Kluwer, Norwell, MA, 1995.

[105] A. Ephremides, "Energy concerns in wireless networks," *IEEE Wireless Commun. Mag.,* pp. 48–59, August 2002.

[106] A. J. Goldsmith and S. B. Wicker, "Design challenges for energy-constrained ad hoc wireless networks," *IEEE Wireless Commun. Mag.,* pp. 8–27, August 2002.

[107] W. Stark, H. Wang. A. Worthen, S. Lafortune, and D. Teneketzis, "Low-energy wireless communication network design," *IEEE Wireless Commun. Mag.,* pp. 60–72, August 2002.

[108] E. Shih, P. Bahl, and M. Sinclair, "Wake on wireless: An event driven energy saving strategy for battery operated devices," *Proc. Internat. Conf. Mobile Comput. Network.,* pp. 160–71, September 2002.

[109] C. Schurgers and M. B. Srivastava, "Energy efficient wireless scheduling: Adaptive loading in time," *Proc. IEEE Wireless Commun. Network Conf.,* pp. 706–11, March 2002.

[110] S. Cui, A. J. Goldsmith, and A. Bahai, "Energy-efficiency of MIMO and cooperative MIMO in sensor networks," *IEEE J. Sel. Areas Commun.,* pp. 1089–98, August 2004.

[111] M. Zorzi and R. R. Rao, "Energy-constrained error control for wireless channels," *IEEE Pers. Commun. Mag.,* pp. 27–33, December 1997.

[112] A. B. MacKenzie and S. B. Wicker, "Selfish users in ALOHA: A game-theoretic approach," *Proc. IEEE Veh. Tech. Conf.,* pp. 1354–7, October 2001.

[113] A. Chockalingam and M. Zorzi, "Energy efficiency of media access protocols for mobile data networks," *IEEE Trans. Commun.,* pp. 1418–21, November 1998.

[114] J.-H. Youn and B. Bose, "An energy conserving medium access control protocol for multihop packet radio networks," *IEEE Internat. Conf. Comput. Commun. Network.,* pp. 470–5, October 2001.

[115] E. Uysal-Biyikoglu, B. Prabhakar, and A. El Gamal, "Energy-efficient packet transmission over a wireless link," *IEEE/ACM Trans. Network.*, pp. 487–99, August 2002.

[116] C.-K. Toh, "Maximum battery life routing to support ubiquitous mobile computing in wireless ad hoc networks," *IEEE Commun. Mag.*, pp. 138–47, June 2001.

[117] J. H. Chang and L. Tassiulas, "Energy conserving routing in wireless ad-hoc networks," *Proc. IEEE Infocom Conf.*, pp. 609–19, August 2000.

[118] M. A. Youssef, M. F. Younis, and K. A. Arisha, "A constrained shortest-path energy-aware routing algorithm for wireless sensor networks," *Proc. IEEE Wireless Commun. Network Conf.*, pp. 794–9, March 2002.

[119] M. Zorzi and R. R. Rao, "Geographic random forwarding (GeRaF) for ad hoc and sensor networks: Energy and latency performance," *IEEE Trans. Mobile Comput.*, pp. 349–65, October–December 2003.

[120] S. Cui, R. Madan, A. J. Goldsmith, and S. Lall, "Joint routing, MAC, and link layer optimization in sensor networks with energy constraints," *Proc. IEEE Internat. Conf. Commun.*, May 2005.

[121] R. G. Gallager, "Energy limited channels: Coding, multiaccess and spread spectrum," *Proc. Inform. Syst. Sci. Conf.*, p. 372, March 1988. [See also Technical Report LIDS-P-1714, MIT, November 1987.]

[122] R. S. Kennedy, *Fading Dispersive Communication Channels,* Wiley, New York, 1969.

[123] S. Verdú, "Spectral efficiency in the wideband regime," *IEEE Trans. Inform. Theory,* pp. 1319–43, June 2002.

[124] S. Verdú, "Recent results on the capacity of wideband channels in the low-power regime," *IEEE Wireless Commun. Mag.*, pp. 40–5, August 2002.

[125] A. Lapidoth, I. E. Telatar, and R. Urbanke, "On wide-band broadcast channels," *IEEE Trans. Inform. Theory,* pp. 3250–8, December 2003.

[126] G. Caire, D. Tuninetti, and S. Verdú, "Suboptimality of TDMA in the low-power regime," *IEEE Trans. Inform. Theory,* pp. 608–20, April 2004.

Appendix A: Representation of Bandpass Signals and Channels

Many signals in communication systems are real bandpass signals with a frequency response that occupies a narrow bandwidth $2B$ centered around a carrier frequency f_c with $2B \ll f_c$, as shown in Figure A.1. Since bandpass signals are real, their frequency response has conjugate symmetry: a bandpass signal $s(t)$ has $|S(f)| = |S(-f)|$ and $\angle S(f) = -\angle S(-f)$. However, bandpass signals are not necessarily conjugate symmetric within the signal bandwidth about the carrier frequency f_c; that is, we may have $|S(f_c + f)| \neq |S(f_c - f)|$ or $\angle S(f_c + f) \neq -\angle S(f_c - f)$ for some $f : 0 < f \leq B$. This asymmetry in $|S(f)|$ about f_c (i.e., $|S(f_c + f)| \neq |S(f_c - f)|$ for some $f < B$) is illustrated in the figure. Bandpass signals result from modulation of a baseband signal by a carrier, or from filtering a deterministic or random signal with a bandpass filter. The bandwidth $2B$ of a bandpass signal is roughly equal to the range of frequencies around f_c where the signal has nonnegligible amplitude. Bandpass signals are commonly used to model transmitted and received signals in communication systems. These are real signals because the transmitter circuitry can only generate real sinusoids (not complex exponentials), and the channel simply introduces an amplitude and phase change at each frequency of the real transmitted signal.

We begin by representing a bandpass signal $s(t)$ at carrier frequency f_c in the following form:

$$s(t) = s_I(t) \cos(2\pi f_c t) - s_Q(t) \sin(2\pi f_c t), \tag{A.1}$$

where $s_I(t)$ and $s_Q(t)$ are real lowpass (baseband) signals of bandwidth $B \ll f_c$. This is a common representation for bandpass signals or noise. In fact, modulations such as MPSK and MQAM are commonly described using this representation. We call $s_I(t)$ the *in-phase component* of $s(t)$ and $s_Q(t)$ the *quadrature component* of $s(t)$. Define the complex signal $u(t) = s_I(t) + js_Q(t)$, so $s_I(t) = \mathrm{Re}\{u(t)\}$ and $s_Q(t) = \mathrm{Im}\{u(t)\}$. Then $u(t)$ is a complex lowpass signal of bandwidth B. With this definition we see that

$$s(t) = \mathrm{Re}\{u(t)\} \cos(2\pi f_c t) - \mathrm{Im}\{u(t)\} \sin(2\pi f_c t) = \mathrm{Re}\{u(t)e^{j2\pi f_c t}\}. \tag{A.2}$$

The representation on the right-hand side of this equation is called the *complex lowpass representation* of the bandpass signal $s(t)$, and the baseband signal $u(t)$ is called the *equivalent lowpass signal* for $s(t)$ or its *complex envelope*. Note that $U(f)$ is conjugate symmetric about $f = 0$ only if $u(t)$ is real (i.e., if $s_Q(t) = 0$).

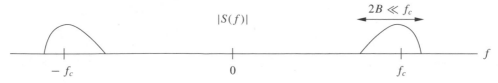

Figure A.1: Bandpass signal $S(f)$.

Using properties of the Fourier transform, we can show that

$$S(f) = .5[U(f - f_c) + U^*(-f - f_c)]. \tag{A.3}$$

Since $s(t)$ is real, it follows that $S(f)$ is conjugate symmetric about $f = 0$. However, the lowpass signals $U(f)$ and $U^*(f)$ are not necessarily conjugate symmetric about $f = 0$, which leads to an asymmetry of $S(f)$ within the bandwidth $2B$ about the carrier frequency f_c, as shown in Figure A.1. In fact, $S(f)$ is conjugate symmetric about the carrier frequency within this bandwidth only if $u(t) = s_I(t)$ – that is, if there is no quadrature component in $u(t)$. We will see shortly that this asymmetry affects the response of bandpass channels to bandpass signals.

An alternate representation of the equivalent lowpass signal is

$$u(t) = a(t)e^{j\phi(t)}, \tag{A.4}$$

with envelope

$$a(t) = \sqrt{s_I^2(t) + s_Q^2(t)} \tag{A.5}$$

and phase

$$\phi(t) = \tan^{-1}\left(\frac{s_Q(t)}{s_I(t)}\right). \tag{A.6}$$

With this representation,

$$s(t) = \text{Re}\{a(t)e^{j\phi(t)}e^{j2\pi f_c t}\} = a(t)\cos(2\pi f_c t + \phi(t)). \tag{A.7}$$

Let us now consider a real channel impulse response $h(t)$ with Fourier transform $H(f)$. If $h(t)$ is real then $H^*(-f) = H(f)$. In communication systems we are mainly interested in the channel frequency response $H(f)$ for $|f - f_c| < B$, since only these frequency components of $H(f)$ affect the received signal within the bandwidth of interest. A *bandpass channel* is similar to a bandpass signal: it has a real impulse response $h(t)$ with frequency response $H(f)$ centered at a carrier frequency f_c with a bandwidth of $2B \ll f_c$. To capture the frequency response of $H(f)$ around f_c, we develop an *equivalent lowpass channel* model similar to the equivalent lowpass signal model, as follows. Because the impulse response $h(t)$ corresponding to $H(f)$ is a bandpass signal, it can be written using an equivalent lowpass representation as

$$h(t) = 2 \text{Re}\{h_l(t)e^{j2\pi f_c t}\}, \tag{A.8}$$

where the extra factor of 2 is to avoid constant factors in the $H(f)$ representation given by (A.9). We call $h_l(t)$ the *equivalent lowpass channel impulse response* for $H(f)$. By (A.2) and (A.3), the representation (A.8) implies that

$$H(f) = H_l(f - f_c) + H_l^*(-f - f_c), \tag{A.9}$$

so $H(f)$ consists of two components: $H_l(f)$ shifted up by f_c, and $H_l^*(f)$ shifted down by f_c. Note that if $H(f)$ is conjugate symmetric about the carrier frequency f_c within the bandwidth $2B$ then $h_l(t)$ will be real and its frequency response $H_l(f)$ will be conjugate symmetric about zero. However, in many wireless channels (e.g., frequency-selective fading channels) $H(f)$ is not conjugate symmetric about f_c, in which case $h_l(t)$ is complex with in-phase component $h_{l,I}(t) = \text{Re}\{h_l(t)\}$ and quadrature component $h_{l,Q}(t) = \text{Im}\{h_l(t)\}$. Note that if $h_l(t)$ is complex then $H_l(f)$ is not conjugate symmetric about zero.

We now use equivalent lowpass signal and channel models to study the output of a bandpass channel with a bandpass signal input. Let $s(t)$ denote the input signal with equivalent lowpass signal $u(t)$. Let $h(t)$ denote the bandpass channel impulse response with equivalent lowpass channel impulse response $h_l(t)$. The transmitted signal $s(t)$ and channel impulse response $h(t)$ are both real, so the channel output $r(t) = s(t) * h(t)$ is also real, with frequency response $R(f) = H(f)S(f)$. Since $S(f)$ is a bandpass signal, $R(f)$ will also be a bandpass signal. Therefore, it has a complex lowpass representation of

$$r(t) = \text{Re}\{v(t)e^{j2\pi f_c t}\}. \tag{A.10}$$

We now consider the relationship between the equivalent lowpass signals corresponding to the channel input $s(t)$, channel impulse response $h(t)$, and channel output $r(t)$. We can express the frequency response of the channel output as

$$
\begin{aligned}
R(f) &= H(f)S(f) \\
&= .5[H_l(f - f_c) + H_l^*(-f - f_c)][U(f - f_c) + U^*(-f - f_c)]. \tag{A.11}
\end{aligned}
$$

For bandpass signals and channels where the bandwidth B of $u(t)$ and $h_l(t)$ is much less than the carrier frequency f_c, we have

$$H_l(f - f_c)U^*(-f - f_c) = 0$$

and

$$H_l^*(-f - f_c)U(f - f_c) = 0.$$

Thus,

$$R(f) = .5[H_l(f - f_c)U(f - f_c) + H_l^*(-f - f_c)U^*(-f - f_c)]. \tag{A.12}$$

Given (A.2) and (A.3), (A.10) implies that

$$R(f) = .5[V(f - f_c) + V^*(-f - f_c)]. \tag{A.13}$$

Equating terms at positive and negative frequencies in (A.12) and (A.13), we get that

$$V(f - f_c) = H_l(f - f_c)U(f - f_c) \tag{A.14}$$

and

$$V^*(-f - f_c) = H_l^*(-f - f_c)U^*(-f - f_c) \tag{A.15}$$

or, equivalently, that

$$V(f) = H_l(f)U(f). \tag{A.16}$$

Taking the inverse Fourier transform then yields

$$v(t) = u(t) * h_l(t). \tag{A.17}$$

Thus, we can obtain the equivalent lowpass signal $v(t)$ for the received signal $r(t)$ by taking the convolution of $h_l(t)$ and $u(t)$. The received signal is therefore given by

$$r(t) = \mathrm{Re}\{(u(t) * h_l(t))e^{j2\pi f_c t}\}. \tag{A.18}$$

Note that $V(f) = H_l(f)U(f)$ is conjugate symmetric about $f = 0$ only if both $U(f)$ and $H_l(f)$ are. In other words, the equivalent lowpass received signal will in general be complex, with nonzero in-phase and quadrature components, if either $u(t)$ or $h_l(t)$ is complex. Moreover, if $u(t) = s_I(t)$ is real (no quadrature component) but the channel impulse response $h_l(t) = h_{l,I}(t) + jh_{l,Q}(t)$ is complex (as with, e.g., frequency-selective fading), then

$$v(t) = s_I(t) * (h_{l,I}(t) + jh_{l,Q}(t)) = s_I(t) * h_{l,I}(t) + js_I(t) * h_{l,Q}(t) \tag{A.19}$$

is complex, so the received signal will have both an in-phase and a quadrature component. More generally, if $u(t) = s_I(t) + js_Q(t)$ and $h_l(t) = h_{l,I}(t) + jh_{l,Q}(t)$ then

$$\begin{aligned} v(t) &= [s_I(t) + js_Q(t)] * [h_{l,I}(t) + jh_{l,Q}(t)] \\ &= [s_I(t) * h_{l,I}(t) - s_Q(t) * h_{l,Q}(t)] + j[s_I(t) * h_{l,Q}(t) + s_Q(t) * h_{l,I}(t)]. \end{aligned} \tag{A.20}$$

Hence the in-phase component of $v(t)$ depends on *both* the in-phase and quadrature components of $u(t)$, and similarly for the quadrature component of $v(t)$. This creates problems in signal detection, since it causes the in-phase and quadrature parts of a modulated signal to interfere with each other in the demodulator.

The main purpose for the equivalent lowpass representation is to analyze bandpass communication systems using the equivalent lowpass models for the transmitted signal, channel impulse response, and received signal. This removes the carrier terms from the analysis – in particular, the dependency of the analysis on the carrier frequency f_c.

Appendix B: Probability Theory, Random Variables, and Random Processes

This appendix provides a brief overview of the main concepts in probability theory, random variables, and random processes that are used throughout the book. More detailed treatments of these broad and deep topics, along with proofs for the properties stated in this appendix, can be found in [1; 2; 3; 4; 5; 6; 7; 8].

B.1 Probability Theory

Probability theory provides a mathematical characterization for random events. Such events are defined on an underlying probability space $(\Omega, \mathcal{E}, p(\cdot))$. The probability space consists of a sample space Ω of possible outcomes for random events; a set of random events $\mathcal{E}$, where each $A \in \mathcal{E}$ is a subset of Ω; and a probability measure $p(\cdot)$ defined on these subsets. Thus, $\mathcal{E}$ is a set of sets, and the probability measure $p(A)$ is defined for every set $A \in \mathcal{E}$. A probability space requires that the set $\mathcal{E}$ be a σ-field. Intuitively, a set $\mathcal{E}$ of sets is a σ-field if it contains all intersections, unions, and complements of its elements.[1] More precisely, $\mathcal{E}$ is a σ-field if: the set of all possible outcomes Ω is one of the sets in $\mathcal{E}$; a set $A \in \mathcal{E}$ implies that $A^c \in \mathcal{E}$; and, for any sets $A_1, A_2, \ldots$ with $A_i \in \mathcal{E}$, we have $\bigcup_{i=1}^{\infty} A_i \in \mathcal{E}$. The set $\mathcal{E}$ must be a σ-field in order for the probability of intersections and unions of random events to be defined. We also require that the probability measure associated with a probability space have the following three fundamental properties.

1. $p(\Omega) = 1$.
2. $0 \leq p(A) \leq 1$ for any event $A \in \mathcal{E}$.
3. If A and B are mutually exclusive (i.e., their intersection is zero), then $p(A \cup B) = p(A) + p(B)$.

Throughout this section, we consider only sets in $\mathcal{E}$, since the probability measure is defined only on these sets.

---

[1] We use the notation $A \cap B$ to denote the intersection of A and B – that is, all elements in both A and B. The union of A and B, denoted $A \cup B$, is the set of all elements in A or B. The complement of a set $A \subset \Omega$, denoted by A^c, is defined as all elements in Ω that are not in the set A.

Several important characteristics of the probability measure $p(\cdot)$ can be derived from its fundamental properties. In particular, $p(A^c) = 1 - p(A)$. Moreover, consider sets $A_1, \ldots, A_n$ where A_i and A_j ($i \neq j$) are disjoint: $A_i \cap A_j = \emptyset$. Then, if $A_1 \cup A_2 \cup \cdots \cup A_n = \Omega$, we have $\sum_{i=1}^n p(A_i) = 1$. We call the set $\{A_1, \ldots, A_n\}$ with these properties a *partition* of Ω. For two sets A_i and A_j that are not disjoint, $p(A_i \cup A_j) = p(A_i) + p(A_j) - p(A_i \cap A_j)$. This leads to the *union bound,* which states that, for any sets $A_1, \ldots, A_n$,

$$p(A_1 \cup A_2 \cup \cdots \cup A_n) \leq \sum_{i=1}^n p(A_i). \tag{B.1}$$

The occurrence of one random event can affect the probability of another random event, since observing one random event indicates which subsets in $\mathcal{E}$ could have contributed to the observed outcome. To capture this effect, we define the probability of event B conditioned on the occurence of event A as $p(B \mid A) = p(A \cap B)/p(A)$, assuming $p(A) \neq 0$. This implies that

$$p(A \cap B) = p(A \mid B)p(B) = p(B \mid A)p(A). \tag{B.2}$$

The conditional probability $p(B \mid A) = p(A \cap B)/p(A)$ essentially normalizes the probability of B with respect to the outcomes associated with A, since it is known that A has occurred. We obtain *Bayes' rule* from (B.2) as

$$p(B \mid A) = \frac{p(A \mid B)p(B)}{p(A)}. \tag{B.3}$$

Independence of events is a function of the probability measure $p(\cdot)$. In particular, events A and B are independent if $p(A \cap B) = p(A)p(B)$. This implies that $p(B \mid A) = p(B)$ and $p(A \mid B) = p(A)$.

B.2 Random Variables

Random variables are defined on an underlying probability space $(\Omega, \mathcal{E}, p(\cdot))$. In particular, a random variable X is a function mapping from the sample space Ω to a subset of the real line. If X takes discrete values on the real line it is called a *discrete* random variable, and if it takes continuous values it is called a *continuous* random variable. The *cumulative distribution function* (cdf) of a random variable X is defined as $P_X(x) \triangleq p(X \leq x)$ for some $x \in \mathbb{R}$. The cdf is derived from the underlying probability space as $p(X \leq x) = p(X^{-1}(-\infty, x))$, where $X^{-1}(\cdot)$ is the inverse mapping from the real line to a subset of Ω: $X^{-1}(-\infty, x) = \{\omega \in \Omega : X(\omega) \leq x\}$. Properties of the cdf are based on properties of the underlying probability measure. In particular, the cdf satisfies $0 \leq P_X(x) = p(X^{-1}(-\infty, x)) \leq 1$. In addition, the cdf is nondecreasing: $P_X(x_1) \leq P_X(x_2)$ for $x_1 \leq x_2$. This is because $P_X(x_2) = p(X^{-1}(-\infty, x_2)) = p(X^{-1}(-\infty, x_1)) + p(X^{-1}(x_1, x_2)) \geq p(X^{-1}(-\infty, x_1)) = P_X(x_1)$.

The *probability density function* (pdf) of a random variable X is defined as the derivative of its cdf: $p_X(x) \triangleq \frac{d}{dx} P_X(x)$. For X a continuous random variable, $p_X(x)$ is a function over the entire real line; for X a discrete random variable, $p_X(x)$ is a set of delta functions at the possible values of X. The pdf, also referred to as the *probability distribution* or *distribution* of X, determines the probability that X lies in a given range of values:

$$p(x_1 < X \leq x_2) = p(X \leq x_2) - p(X \leq x_1) = P_X(x_2) - P_X(x_1) = \int_{x_1}^{x_2} p_X(x)\, dx. \quad \text{(B.4)}$$

Since $P_X(\infty) = 1$ and $P_X(-\infty) = 0$, the pdf integrates to 1:

$$\int_{-\infty}^{\infty} p_X(x)\, dx = 1. \quad \text{(B.5)}$$

Note that the subscript X is often omitted from the pdf and cdf when it is clear from the context that these functions characterize the distribution of X. In such cases the pdf is written as $p(x)$ and the cdf as $P(x)$.

The *mean* or *expected value* of a random variable X is its probabalistic average, defined as

$$\mu_X = \mathbf{E}[X] \triangleq \int_{-\infty}^{\infty} x p_X(x)\, dx. \quad \text{(B.6)}$$

The expectation operator $\mathbf{E}[\cdot]$ is linear and can also be applied to functions of random variables. In particular, the mean of a function of X is given by

$$\mathbf{E}[g(X)] = \int_{-\infty}^{\infty} g(x) p_X(x)\, dx. \quad \text{(B.7)}$$

A function of particular interest is the nth *moment* of X,

$$\mathbf{E}[X^n] = \int_{-\infty}^{\infty} x^n p_X(x)\, dx. \quad \text{(B.8)}$$

The variance of X is defined in terms of its mean and second moment as

$$\mathrm{Var}[X] = \sigma_X^2 \triangleq \mathbf{E}[(X - \mu_X)^2] = \mathbf{E}[X^2] - \mu_X^2. \quad \text{(B.9)}$$

The variance characterizes the average squared difference between X and its mean μ_X. The standard deviation of X, σ_X, is the square root of its variance. From the linearity of the expectation operator it is easily shown that, for any constant c, $\mathbf{E}[cX] = c\mathbf{E}[X]$, $\mathrm{Var}[cX] = c^2\,\mathrm{Var}[X]$, $\mathbf{E}[X + c] = \mathbf{E}[X] + c$, and $\mathrm{Var}[X + c] = \mathrm{Var}[X]$. Thus, scaling a random variable by a constant scales its mean by the same constant and its variance by the constant squared. Adding a constant to a random variable shifts the mean by the same constant but doesn't affect the variance.

The distribution of a random variable X can be determined from its *characteristic function*, defined as

$$\phi_X(\nu) \triangleq \mathbf{E}[e^{j\nu X}] = \int_{-\infty}^{\infty} p_X(x) e^{j\nu x}\, dx. \quad \text{(B.10)}$$

We see from (B.10) that the characteristic function $\phi_X(\nu)$ of X is the inverse Fourier transform of the distribution $p_X(x)$ evaluated at $\nu/(2\pi)$. Thus we can obtain $p_X(x)$ from $\phi_X(\nu)$ as

$$p_X(x) = \frac{1}{2\pi} \int_{-\infty}^{\infty} \phi_X(\nu) e^{-j\nu x}\, dx. \quad \text{(B.11)}$$

This will become significant in finding the distribution for sums of random variables. We can obtain the nth moment of X from $\phi_X(\nu)$ as

$$E[X^n] = (-j)^n \frac{\partial^n \phi_X(\nu)}{\partial \nu^n}\bigg|_{\nu=0}.$$

The *moment generating function* (MGF) of X, defined as $\mathcal{M}_X(\nu) \triangleq E[e^{\nu X}]$, is similar to the characteristic function. However, it may diverge for some values of ν. If the MGF is finite for ν in the neighborhood around zero, then the nth moment of X is obtained as

$$E[X^n] = \frac{\partial^n \mathcal{M}_X(\nu)}{\partial \nu^n}\bigg|_{\nu=0}.$$

Let X be a random variable and $g(x)$ a function on the real line. Let $Y = g(X)$ define another random variable. Then $P_Y(y) = \int_{x:g(x)\leq y} p_X(x)\, dx$. For g monotonically increasing and one-to-one, this becomes $P_Y(y) = \int_{-\infty}^{g^{-1}(y)} p_X(x)\, dx$. For g monotonically decreasing and one-to-one, this becomes $P_Y(y) = \int_{g^{-1}(y)}^{\infty} p_X(x)\, dx$.

We now consider joint random variables. Two random variables must share the same underlying probability space for their joint distribution to be defined. Let X and Y be two random variables defined on the same probability space $(\Omega, \mathcal{E}, p(\cdot))$. Their joint cdf is defined as $P_{XY}(x, y) \triangleq p(X \leq x, Y \leq y)$. Their joint pdf (distribution) is defined as the derivative of the joint cdf:

$$p_{XY}(x, y) \triangleq \frac{\partial^2 P_{XY}(x, y)}{\partial x \partial y}. \tag{B.12}$$

Thus,

$$P_{XY}(x, y) = \int_{-\infty}^{x} \int_{-\infty}^{y} p_{XY}(v, w)\, dv\, dw. \tag{B.13}$$

For joint random variables X and Y, we can obtain the distribution of X by integrating the joint distribution with respect to Y:

$$p_X(x) = \int_{-\infty}^{\infty} p_{XY}(x, y)\, dy. \tag{B.14}$$

Similarly,

$$p_Y(y) = \int_{-\infty}^{\infty} p_{XY}(x, y)\, dx. \tag{B.15}$$

The distributions $p_X(x)$ and $p_Y(y)$ obtained in this manner are sometimes referred to as the *marginal* distributions relative to the joint distribution $p_{XY}(x, y)$. Note that the joint distribution must integrate to unity:

$$\int_{-\infty}^{\infty} \int_{-\infty}^{\infty} p_{XY}(x, y)\, dx\, dy = 1. \tag{B.16}$$

The definitions for joint cdf and joint pdf of two random variables extend in a straightforward manner to any finite number of random variables.

As with random events, observing the value for one random variable can affect the probability of another random variable. We define the conditional distribution of the random variable Y given a realization $X = x$ of random variable X as follows: $p_Y(y \mid X = x) = p_{XY}(x, y)/p_X(x)$. This implies that $p_{XY}(x, y) = p_Y(y \mid X = x)p_X(x)$. Independence between two random variables X and Y is a function of their joint distribution. Specifically,

X and Y are independent random variables if their joint distribution $p_{XY}(x, y)$ factors into separate distributions for X and Y: $p_{XY}(x, y) = p_X(x)p_Y(y)$. For independent random variables, it is easily shown that $\mathbf{E}[f(X)g(Y)] = \mathbf{E}[f(X)]\mathbf{E}[g(Y)]$ for any functions $f(x)$ and $g(y)$.

For X and Y joint random variables with joint pdf $p_{XY}(x, y)$, we define their ijth *joint moment* as

$$\mathbf{E}[X^i Y^j] \triangleq \int_{-\infty}^{\infty} \int_{-\infty}^{\infty} x^i y^j p_{XY}(x, y) \, dx \, dy. \tag{B.17}$$

The *correlation* of X and Y is defined as $\mathbf{E}[XY]$. The *covariance* of X and Y is defined as $\mathrm{Cov}[XY] \triangleq \mathbf{E}[(X - \mu_X)(Y - \mu_Y)] = \mathbf{E}[XY] - \mu_X \mu_Y$. Note that the covariance and correlation of X and Y are equal if either X or Y has mean zero. The *correlation coefficient* of X and Y is defined in terms of their covariance and standard deviations as $\rho \triangleq \mathrm{Cov}[XY]/(\sigma_X \sigma_Y)$. We say that X and Y are *uncorrelated* if their covariance is zero or, equivalently, if their correlation coefficient is zero. Note that uncorrelated random variables (e.g., X and Y with $\mathrm{Cov}[XY] = \mathbf{E}[XY] - \mu_X \mu_Y = 0$) will have a nonzero correlation ($\mathbf{E}[XY] \neq 0$) if their means are not zero. For random variables $X_1, \ldots, X_n$, we define their *covariance matrix* $\boldsymbol{\Sigma}$ as an $n \times n$ matrix with ijth element $\boldsymbol{\Sigma}_{ij} = \mathrm{Cov}[X_i X_j]$. In particular, the ith diagonal element of $\boldsymbol{\Sigma}$ is the variance of X_i: $\boldsymbol{\Sigma}_{ii} = \mathrm{Var}[X_i]$.

Consider two independent random variables X and Y. Let $Z = X + Y$ define a new random variable on the probability space $(\Omega, \mathcal{E}, p(\cdot))$. We can show directly (or by using characteristic functions) that the distribution of Z is the convolution of the distributions of X and Y: $p_Z(z) = p_X(x) * p_Y(y)$. Equivalently, $\phi_Z(v) = \phi_X(v)\phi_Y(v)$. With this distribution it can be shown that $\mathbf{E}[Z] = \mathbf{E}[X] + \mathbf{E}[Y]$ and that $\mathrm{Var}[Z] = \mathrm{Var}[X] + \mathrm{Var}[Y]$. So, for sums of independent random variables, the mean of the sum is the sum of the means and the variance of the sum is the sum of the variances.

A distribution that arises frequently in the study of communication systems is the Gaussian distribution. The Gaussian distribution for a random variable X is defined in terms of its mean μ_X and variance σ_X^2 as

$$p_X(x) = \frac{1}{\sqrt{2\pi}\sigma_X} e^{-[(x - \mu_X)^2 / (2\sigma_X^2)]}. \tag{B.18}$$

The Gaussian distribution, also called the normal distribution, is denoted as $N(\mu_X, \sigma_X^2)$. Note that the tail of the distribution (i.e., the value of $p_X(x)$ as x moves away from μ_X) decreases exponentially. The cdf $P_X(x) = p(X \leq x)$ for this distribution does not exist in closed form. It is defined in terms of the Gaussian Q-function as

$$P_X(x) = p(X \leq x) = 1 - Q\left(\frac{x - \mu_X}{\sigma_X}\right), \tag{B.19}$$

where the Gaussian Q-function, defined by

$$Q(x) \triangleq \int_x^{\infty} \frac{1}{\sqrt{2\pi}} e^{-y^2/2} \, dy, \tag{B.20}$$

is the probability that a Gaussian random variable X with mean 0 and variance 1 is larger than x: $Q(x) = p(X \geq x)$ for $X \sim N(0, 1)$. The Gaussian Q-function is related to the complementary error function as $Q(x) = .5 \operatorname{erfc}(x/\sqrt{2})$. These functions are typically calculated using standard computer math packages.

Let $\mathbf{X} = (X_1, \ldots, X_n)$ denote a vector of jointly Gaussian random variables. Their joint distribution is given by

$$p_{X_1, \ldots, X_n}(x_1, \ldots, x_n) = \frac{1}{\sqrt{(2\pi)^n \det[\boldsymbol{\Sigma}]}} \exp[-.5(\mathbf{x} - \boldsymbol{\mu}_\mathbf{X})^T \boldsymbol{\Sigma}^{-1}(\mathbf{x} - \boldsymbol{\mu}_\mathbf{X})], \qquad (B.21)$$

where $\boldsymbol{\mu}_\mathbf{X} = \mathbf{E}[\mathbf{X}]^T = (\mathbf{E}[X_1], \ldots, \mathbf{E}[X_n])^T$ is the mean of $\mathbf{X}$ and where $\boldsymbol{\Sigma}$ is the $n \times n$ covariance matrix of $\mathbf{X}$ (i.e., $\boldsymbol{\Sigma}_{ij} = \operatorname{Cov}[X_i X_j]$). It can be shown from (B.21) that, for jointly Gaussian random variables X and Y, if $\operatorname{Cov}[XY] = 0$ then $p_{XY}(x, y) = p_X(x)p_Y(y)$. In other words, Gaussian random variables that are uncorrelated are independent.

A complex random variable Z is complex Gaussian if $Z = X + jY$ for X and Y jointly Gaussian real random variables. The distribution of Z is then obtained from the joint distribution of X and Y, as given by (B.21) for the vector (X, Y). Similarly, a complex random vector $\mathbf{Z} = (Z_1, \ldots, Z_N) = (X_1 + jY_1, \ldots, X_N + jY_N)$ is complex Gaussian if the random variables $X_1, \ldots, X_n, Y_1, \ldots, Y_N$ are jointly Gaussian real random variables. The distribution of $\mathbf{Z}$ is obtained from the joint distribution of these random variables, given by (B.21) for the vector $(X_1, \ldots, X_N, Y_1, \ldots, Y_N)$.

The underlying reason why the Gaussian distribution commonly occurs in communication system models is the *central limit theorem* (CLT), which defines the limiting distribution for the sum of a large number of independent random variables with the same distribution. Specifically, let X_i be independent and identically distributed (i.i.d.) joint random variables. Let $Y_n = \sum_{i=1}^n X_i$ and $Z_n = (Y_n - \mu_{Y_n})/\sigma_{Y_n}$. The CLT states that the distribution of Z_n as n goes to infinity converges to a Gaussian distribution with mean 0 and variance 1: $\lim_{n \to \infty} p_{Z_n}(x) = N(0, 1)$. Thus, any random variable equal to the sum of a large number of i.i.d. random components has a distribution that is approximately Gaussian. For example, noise in a radio receiver typically consists of spurious signals generated by the various hardware components, and with a large number of i.i.d. components this noise is accurately modeled as Gauss distributed.

Two other common distributions that arise in communication systems are the uniform distribution and the binomial distribution. A random variable X that is uniformly distributed has pdf $p_X(x) = 1/(b - a)$ for x in the interval $[a, b]$ and zero otherwise. A random phase θ is commonly modeled as uniformly distributed on the interval $[0, 2\pi]$, which we denote as $\theta \sim \mathcal{U}[0, 2\pi]$. The binomial distribution often arises in coding analysis. Let X_i ($i = 1, \ldots, n$) be discrete random variables that take one of two possible values, 0 or 1. Suppose the X_i are i.i.d. with $p(X_i = 1) = p$ and $p(X_i = 0) = 1 - p$. We call X_i a *Bernoulli* random variable. Let $Y = \sum_{i=1}^n X_i$. Then Y is a discrete random variable that takes integer values $k = 0, 1, 2, \ldots$. The distribution of Y is the binomial distribution, given by

$$p(Y = k) = \binom{n}{k} p^k (1 - p)^{n-k}, \qquad (B.22)$$

where

$$\binom{n}{k} \triangleq \frac{n!}{k! \, (n - k)!}. \qquad (B.23)$$

B.3 Random Processes

A random process $X(t)$ is defined on an underlying probability space $(\Omega, \mathcal{E}, p(\cdot))$. In particular, it is a function mapping from the sample space Ω to a set of real functions $\{x_1(t), x_2(t), \ldots\}$, where each $x_i(t)$ is a possible realization of $X(t)$. Samples of $X(t)$ at times $t_0, t_1, \ldots, t_n$ are joint random variables defined on the underlying probability space. Thus, the joint cdf of samples at times $t_0, t_1, \ldots, t_n$ is given by $P_{X(t_0)X(t_1),\ldots,X(t_n)}(x_0, \ldots, x_n) = p\big(X(t_0) \leq x_0, X(t_1) \leq x_1, \ldots, X(t_n) \leq x_n\big)$. The random process $X(t)$ is fully characterized by its joint cdf $P_{X(t_0)X(t_1),\ldots,X(t_n)}(x_0, \ldots, x_n)$ for all possible sets of sample times $\{t_0, t_1, \ldots, t_n\}$.

A random process $X(t)$ is *stationary* if for all T, all n, and all sets of sample times $\{t_0, \ldots, t_n\}$ we have

$$p\big(X(t_0) \leq x_0, X(t_1) \leq x_1, \ldots, X(t_n) \leq x_n\big)$$
$$= p\big(X(t_0 + T) \leq x_0, X(t_1 + T) \leq x_1, \ldots, X(t_n + T) \leq x_n\big).$$

Intuitively, a random process is stationary if time shifts do not affect its probability. Stationarity of a process is often difficult to prove because it requires checking the joint cdf of all possible sets of samples for all possible time shifts. Stationarity of a random process is often inferred from the stationarity of the source generating the process.

The *mean* of a random process is defined as $\mathbf{E}[X(t)]$. Since the mean of a stationary random process is independent of time shifts, it must be constant: $\mathbf{E}[X(t)] = \mathbf{E}[X(t - t)] = \mathbf{E}[X(0)] = \mu_X$. The *autocorrelation* of a random process is defined as $A_X(t, t + \tau) \triangleq \mathbf{E}[X(t)X(t + \tau)]$. The autocorrelation of $X(t)$ is also called its *second moment*. Since the autocorrelation of a stationary process is independent of time shifts, it follows that $A_X(t, t + \tau) = \mathbf{E}[X(t - t)X(t + \tau - t)] = \mathbf{E}[X(0)X(\tau)] \triangleq A_X(\tau)$. Thus, for stationary processes, the autocorrelation depends only on the time difference τ between the samples $X(t)$ and $X(t + \tau)$ and not on the absolute time t. The autocorrelation of a process measures the correlation between samples of the process taken at different times.

Two random processes $X(t)$ and $Y(t)$ defined on the same underlying probability space have a joint cdf characterized by

$$P_{X(t_0),\ldots,X(t_n)Y(t_0'),\ldots,Y(t_m')}(x_0, \ldots, x_n, y_0, \ldots, y_m)$$
$$= p\big(X(t_0) \leq x_0, \ldots, X(t_n) \leq x_n, Y(t_0') \leq y_0, \ldots, Y(t_m') \leq y_m\big) \quad \text{(B.24)}$$

for all possible sets of sample times $\{t_0, \ldots, t_n\}$ and $\{t_0', \ldots, t_m'\}$. Two random processes $X(t)$ and $Y(t)$ are *independent* if for all such sets we have

$$p_{X(t_0),\ldots,X(t_n)Y(t_0'),\ldots,Y(t_m')}\big(X(t_0) \leq x_0, \ldots, X(t_n) \leq x_n, Y(t_0') \leq y_0, \ldots, Y(t_m') \leq y_m\big)$$
$$= p_{X(t_0),\ldots,X(t_n)}\big(X(t_0) \leq x_0, \ldots, X(t_n) \leq x_n\big)$$
$$\cdot p_{Y(t_0'),\ldots,Y(t_m')}\big(Y(t_0') \leq y_0, \ldots, Y(t_m') \leq y_m\big). \quad \text{(B.25)}$$

The cross-correlation between two random processes $X(t)$ and $Y(t)$ is defined as

$$A_{XY}(t, t + \tau) \triangleq \mathbf{E}[X(t)Y(t + \tau)].$$

The two processes are uncorrelated if $\mathbf{E}[X(t)Y(t+\tau)] = \mathbf{E}[X(t)]\mathbf{E}[Y(t+\tau)]$ for all t and τ. As with the autocorrelation, if both $X(t)$ and $Y(t)$ are stationary then the cross-correlation is a function of τ only: $A_{XY}(t, t+\tau) = \mathbf{E}[X(t-t)Y(t+\tau-t)] = \mathbf{E}[X(0)Y(\tau)] \triangleq A_{XY}(\tau)$.

In most analyses of random processes we focus only on the first and second moments. *Wide-sense stationarity* is a notion of stationarity that depends on only the first two moments of a process, and it can be easily verified. Specifically, a process is wide-sense stationary (WSS) if its mean is constant, $\mathbf{E}[X(t)] = \mu_X$, and its autocorrelation depends only on the time difference of the samples, $A_X(t, t+\tau) = \mathbf{E}[X(t)X(t+\tau)] = A_X(\tau)$. Stationary processes are WSS, but WSS processes are not necessarily stationary. For WSS processes, the autocorrelation is a symmetric function of τ, since $A_X(\tau) = \mathbf{E}[X(t)X(t+\tau)] = \mathbf{E}[X(t+\tau)X(t)] = A_X(-\tau)$. Moreover, it can be shown that $A_X(\tau)$ takes its maximum value at $\tau = 0$; that is, $|A_X(\tau)| \le A_X(0) = \mathbf{E}[X^2(t)]$. As with stationary processes, if two processes $X(t)$ and $Y(t)$ are both WSS then their cross-correlation is independent of time shifts and thus depends only on the time difference of the processes: $A_{XY}(t, t+\tau) = \mathbf{E}[X(0)Y(\tau)] = A_{XY}(\tau)$.

The power spectral density (PSD) of a WSS process is defined as the Fourier transform of its autocorrelation function with respect to τ:

$$S_X(f) = \int_{-\infty}^{\infty} A_X(\tau)e^{-j2\pi f\tau}\, d\tau. \tag{B.26}$$

The autocorrelation can be obtained from the PSD through the inverse transform:

$$A_X(\tau) = \int_{-\infty}^{\infty} S_X(f)e^{j2\pi f\tau}\, df. \tag{B.27}$$

The PSD takes its name from the fact that the expected power of a random process $X(t)$ is the integral of its PSD:

$$\mathbf{E}[X^2(t)] = A_X(0) = \int_{-\infty}^{\infty} S_X(f)\, df, \tag{B.28}$$

which follows from (B.27). Similarly, from (B.26) we get that $S_X(0) = \int_{-\infty}^{\infty} A_X(\tau)\, d\tau$. Since $A_X(\tau)$ is real and symmetric, from (B.26) we have that $S_X(f)$ is also symmmetric – that is, $S_X(f) = S_X(-f)$. *White noise* is defined as a zero-mean WSS random process with a PSD that is constant over all frequencies. Thus, a white noise process $X(t)$ has $\mathbf{E}[X(t)] = 0$ and $S_X(f) = N_0/2$ for some constant N_0; this constant is typically referred to as the one-sided white noise PSD. By the inverse Fourier transform, the autocorrelation of white noise is given by $A_X(\tau) = (N_0/2)\delta(\tau)$. In some sense, white noise is the most random of all possible noise processes, since it decorrelates instantaneously.

Random processes are often filtered or modulated, and when the process is WSS the impact of these operations can be characterized in a simple way. In particular, if a WSS process with PSD $S_X(f)$ is passed through a linear time-invariant filter with frequency response $H(f)$, then the filter output is also a WSS process with power spectral density $|H(f)|^2 S_X(f)$. If a WSS process $X(t)$ with PSD $S_X(f)$ is multiplied by a carrier $\cos(2\pi f_c t + \theta)$ with $\theta \sim \mathcal{U}[0, 2\pi]$, the multiplication results in a WSS process $X(t)\cos(2\pi f_c t + \theta)$ with PSD $.25[S_X(f - f_c) + S_X(f + f_c)]$.

A white Gaussian noise process with mean zero and PSD $N_0/2$ can be passed through an ideal bandpass filter centered at frequency f_c with bandwidth $2B$ to create a narrowband noise process $n(t)$. This narrowband process can be represented using the complex lowpass representation as $n(t) = \mathrm{Re}\{n_l(t)e^{j2\pi f_c t}\}$, where $n_l(t) = n_I(t) + jn_Q(t)$ is a complex lowpass Gaussian process with $n_I(t)$ and $n_Q(t)$ real lowpass Gaussian processes. It can be shown that $n_I(t)$ and $n_Q(t)$ are independent, each with mean zero and PSD N_0 (i.e., their PSD is twice that of the original noise process).

Stationarity and WSS are properties of the underlying probability space associated with a random process. We are also often interested in time averages associated with random processes, which can be characterized by different notions of *ergodicity*. A random process $X(t)$ is *ergodic in the mean* if its time-averaged mean, defined as

$$\mu_X^{\text{ta}} = \lim_{T \to \infty} \frac{1}{2T} \int_{-T}^{T} X(t)\,dt, \tag{B.29}$$

is constant for all possible realizations of $X(t)$. In other words, $X(t)$ is ergodic in the mean if $\lim_{T \to \infty} \frac{1}{2T} \int_{-T}^{T} x_i(t)\,dt$ equals the same constant μ_X^{ta} for all possible realizations $x_i(t)$ of $X(t)$. Similarly, a random process $X(t)$ is *ergodic in the nth moment* if its time-averaged nth moment

$$\mu_{X^n}^{\text{ta}} = \lim_{T \to \infty} \frac{1}{2T} \int_{-T}^{T} X^n(t)\,dt \tag{B.30}$$

is constant for all possible realizations of $X(t)$. We can also define the ergodicity of $X(t)$ relative to its time-averaged autocorrelation

$$A_X^{\text{ta}}(\tau) = \lim_{T \to \infty} \frac{1}{2T} \int_{-T}^{T} X(t)X(t+\tau)\,dt. \tag{B.31}$$

Specifically, $X(t)$ is *ergodic in autocorrelation* if $\lim_{T \to \infty} \frac{1}{2T} \int_{-T}^{T} x_i(t)x_i(t+\tau)\,dt$ equals the same value $A_X^{\text{ta}}(\tau)$ for all possible realizations $x_i(t)$ of $X(t)$. Ergodicity of the autocorrelation in higher-order moments requires that the (nm)th-order time-averaged autocorrelation

$$A_X^{\text{ta}}(n, m, \tau) = \lim_{T \to \infty} \frac{1}{2T} \int_{-T}^{T} X^n(t)X^m(t+\tau)\,dt \tag{B.32}$$

is constant for all realizations of $X(t)$. A process that is ergodic in all order moments and autocorrelations is called *ergodic*. Ergodicity of a process requires that its time-averaged nth moment and (ij)th autocorrelation, averaged over all time, be constant for all n, i, and j. This implies that the probability associated with an ergodic process is independent of time shifts and thus the process is stationary. In other words, an ergodic process must be stationary. However, a stationary process can be either ergodic or nonergodic. Since an ergodic process is stationary, it follows that

$$\mu_X^{\text{ta}} = \mathbf{E}[\mu_X^{\text{ta}}]$$

$$= \mathbf{E}\left[\lim_{T \to \infty} \frac{1}{2T} \int_{-T}^{T} X(t)\,dt\right]$$

$$= \lim_{T \to \infty} \frac{1}{2T} \int_{-T}^{T} \mathbf{E}[X(t)]\,dt$$

$$= \lim_{T \to \infty} \frac{1}{2T} \int_{-T}^{T} \mu_X\,dt = \mu_X. \tag{B.33}$$

Thus, the time-averaged mean of $X(t)$ equals its probabilistic mean. Similarly,

$$A_X^{\text{ta}}(\tau) = \mathbf{E}[A_X^{\text{ta}}(\tau)]$$

$$= \mathbf{E}\left[\lim_{T\to\infty} \frac{1}{2T} \int_{-T}^{T} X(t)X(t+\tau)\,dt\right]$$

$$= \lim_{T\to\infty} \frac{1}{2T} \int_{-T}^{T} \mathbf{E}[X(t)(t+\tau)]\,dt$$

$$= \lim_{T\to\infty} \frac{1}{2T} \int_{-T}^{T} A_X(\tau)\,dt = A_X(\tau), \tag{B.34}$$

so the time-averaged autocorrelation of $X(t)$ equals its probabilistic autocorrelation.

B.4 Gaussian Processes

Noise processes in communication systems are commonly modeled as a Gaussian process. A random process $X(t)$ is a Gaussian process if, for all values of T and all functions $g(t)$, the random variable

$$X_g = \int_0^T g(t)X(t)\,dt \tag{B.35}$$

has a Gaussian distribution. A communication receiver typically uses an integrator in signal detection, so this definition implies that, if the channel introduces a Gaussian noise process at the receiver input, then the distribution of the random variable associated with the noise at the output of the integrator will have a Gaussian distribution. The mean of X_g is

$$\mathbf{E}[X_g] = \int_0^T g(t)\mathbf{E}[X(t)]\,dt \tag{B.36}$$

and the variance is

$$\text{Var}[X_g] = \int_0^T \int_0^T g(t)g(s)\mathbf{E}[X(t)X(s)]\,dt\,ds - (\mathbf{E}[X_g])^2. \tag{B.37}$$

If $X(t)$ is WSS then these equations simplify to

$$\mathbf{E}[X_g] = \int_0^T g(t)\mu_X\,dt \tag{B.38}$$

and

$$\text{Var}[X_g] = \int_0^T \int_0^T g(t)g(s)R_X(s-t)\,dt\,ds - (\mathbf{E}[X_g])^2. \tag{B.39}$$

Several important properties of Gaussian random processes can be obtained from this definition. In particular, if a Gaussian random process is input to a linear time-invariant filter, then the filter output is also a Gaussian random process. Moreover, we expect samples $X(t_i)$, $i = 0, 1, \ldots$, of a Gaussian random process to be jointly Gaussian random variables, and indeed this follows from the definition by setting $g(t) = \delta(t - t_i)$ in (B.35). These samples are Gaussian random variables, so if the samples are uncorrelated then they are also independent. In addition, for a WSS Gaussian processes, the distribution of X_g in (B.35) depends only on the mean and autocorrelation of the process $X(t)$. Finally, note that a random process is completely defined by the joint probability of its samples over all sets of sample times. For a Gaussian process, these samples are jointly Gaussian with their joint distribution determined by the mean and autocorrelation of the process. Thus, since the underlying probability of a Gaussian process is completely determined by its mean and autocorrelation, it follows that a WSS Gaussian process is also stationary. Similarly, a Gaussian process that is ergodic in the mean and autocorrelation is an ergodic process.

REFERENCES

[1]   W. Feller, *An Introduction to Probability Theory and Its Applications,* vols. I and II, Wiley, New York, 1968 and 1971.

[2]   R. M. Gray and L. D. Davisson, *Random Processes: A Mathematical Approach for Engineers,* Prentice-Hall, Englewood Cliffs, NJ, 1986.

[3]   W. B. Davenport, Jr., and W. L. Root, *An Introduction to the Theory of Random Signals and Noise,* McGraw-Hill, New York, 1987.

[4]   A. Leon-Garcia, *Probability and Random Processes for Electrical Engineering,* 2nd ed., Addison-Wesley, Reading, MA, 1994.

[5]   P. Billingsley, *Probability and Measure,* 3rd ed., Wiley, New York, 1995.

[6]   R. G. Gallager, *Discrete Stochastic Processes,* Kluwer, Dordrecht, 1996.

[7]   H. Stark and J. W. Woods, *Probability and Random Processes with Applications to Signal Processing,* 3rd ed., Prentice-Hall, Englewood Cliffs, NJ, 2001.

[8]   A. Papoulis and S. U. Pillai, *Probability, Random Variables and Stochastic Processes,* McGraw-Hill, New York, 2002.

Appendix C: Matrix Definitions, Operations, and Properties

This appendix summarizes the definitions, operations, and properties of matrices that are used in the book. More detailed treatments of matrices, along with proofs for the properties stated in this appendix, can be found in [1; 2; 3; 4].

C.1 Matrices and Vectors

An $N \times M$ *matrix* $\mathbf{A}$ is a rectangular array of values with N rows and M columns, written as

$$\mathbf{A} = \begin{bmatrix} a_{11} & \cdots & a_{1M} \\ \vdots & \ddots & \vdots \\ a_{N1} & \cdots & a_{NM} \end{bmatrix}. \tag{C.1}$$

The (ij)th element (or entry) of $\mathbf{A}$ (i.e., the element in the ith row and jth column) is written as $\mathbf{A}_{ij}$. In (C.1) we have $\mathbf{A}_{ij} = a_{ij}$. The matrix elements are also called *scalars* to indicate that they are single numbers. An $N \times M$ matrix is called a *square* matrix if $N = M$, a *skinny* matrix if $N > M$, and a *fat* matrix if $N < M$.

The *diagonal elements* of a matrix are the elements along the diagonal line starting from the top left of the matrix (i.e., the elements $\mathbf{A}_{ij}$ with $i = j$). The *trace* of a square $N \times N$ matrix is the sum of its diagonal elements: $\text{Tr}[\mathbf{A}] = \sum_{i=1}^{N} \mathbf{A}_{ii}$. A square matrix is called a *diagonal matrix* if all elements that are not diagonal elements, referred to as the *off-diagonal* elements, are zero: $\mathbf{A}_{ij} = 0$, $j \neq i$. We denote a diagonal matrix with diagonal elements $a_1, \ldots, a_N$ as $\text{diag}[a_1, \ldots, a_N]$. The $N \times N$ identity matrix $\mathbf{I}_N$ is a diagonal matrix with $\mathbf{I}_{ii} = 1$, $i = 1, \ldots, N$; that is, $\mathbf{I}_N = \text{diag}[1, \ldots, 1]$. The subscript N of $\mathbf{I}_N$ is omitted when the size is clear from the context (e.g., from the size requirements for a given operation like matrix multiplication).

A square matrix $\mathbf{A}$ is called *upper triangular* if all its elements below the diagonal are zero (i.e., $\mathbf{A}_{ij} = 0$, $i > j$). A *lower triangular* matrix is a square matrix where all elements above the diagonal are zero (i.e., $\mathbf{A}_{ij} = 0$, $i < j$). Diagonal matrices are both upper triangular and lower triangular.

Matrices can be formed from entries that are themselves matrices, as long as the dimensions are consistent. In particular, if $\mathbf{B}$ is an $N \times M_1$ matrix and $\mathbf{C}$ is an $N \times M_2$ matrix then we can form the $N \times (M_1 + M_2)$ matrix $\mathbf{A} = [\mathbf{B} \ \mathbf{C}]$. The ith row of this matrix is

$[\mathbf{A}_{i1} \ \ldots \ \mathbf{A}_{i(M_1+M_2)}] = [\mathbf{B}_{i1} \ \ldots \ \mathbf{B}_{iM_1} \ \mathbf{C}_{i1} \ \ldots \ \mathbf{C}_{iM_2}]$. The matrix $\mathbf{A}$ formed in this way is also written as $\mathbf{A} = [\mathbf{B}|\mathbf{C}]$. If we also have a $K \times L_1$ matrix $\mathbf{D}$ and a $K \times L_2$ matrix $\mathbf{E}$, then if $M_1 + M_2 = L_1 + L_2$ we can form the $(N + K) \times (M_1 + M_2)$ matrix

$$\mathbf{A} = \begin{bmatrix} \mathbf{B} & \mathbf{C} \\ \mathbf{D} & \mathbf{E} \end{bmatrix}. \tag{C.2}$$

The matrices $\mathbf{B}, \mathbf{C}, \mathbf{D}$, and $\mathbf{E}$ are called *submatrices* of $\mathbf{A}$. A matrix can be composed of any number of submatrices as long as the sizes are compatible. A submatrix $\mathbf{A}'$ of $\mathbf{A}$ can also be obtained by deleting certain rows and/or columns of $\mathbf{A}$.

A matrix with only one column (i.e., with $M = 1$) is called a *column vector* or just a *vector*. The number of rows of a vector is referred to as its *dimension*. For example, an N-dimensional vector $\mathbf{x}$ is given by

$$\mathbf{x} = \begin{bmatrix} x_1 \\ \vdots \\ x_N \end{bmatrix}. \tag{C.3}$$

The ith element of vector $\mathbf{x}$ is written as $\mathbf{x}_i$. We call an N-dimensional vector with each element equal to 1 a *ones vector* and denote it by $\mathbf{1}_N$. An N-dimensional vector with one element equal to one and the rest equal to zero is called a *unit vector*. In particular, the ith unit vector $\mathbf{e}^i$ has $\mathbf{e}_i^i = 1$ and $\mathbf{e}_j^i = 0$ for $j \neq i$. A matrix with only one row (i.e., with $N = 1$) is called a *row vector*. The number of columns in a row vector is called its dimension, so an M-dimensional row vector $\mathbf{x}$ is given by $\mathbf{x} = [x_1 \ \ldots \ x_M]$ with ith element $\mathbf{x}_i = x_i$. The *Euclidean norm* of an N-dimensional row or column vector, also called its *norm*, is defined as

$$\|\mathbf{x}\| = \sqrt{\sum_{i=1}^{N} |\mathbf{x}_i|^2}. \tag{C.4}$$

C.2 Matrix and Vector Operations

If $\mathbf{A}$ is an $N \times M$ matrix then the *transpose* of $\mathbf{A}$, denoted $\mathbf{A}^T$, is the $M \times N$ matrix defined by $\mathbf{A}_{ij}^T = \mathbf{A}_{ji}$:

$$\mathbf{A}^T = \begin{bmatrix} a_{11} & \cdots & a_{1M} \\ \vdots & \ddots & \vdots \\ a_{N1} & \cdots & a_{NM} \end{bmatrix}^T = \begin{bmatrix} a_{11} & \cdots & a_{N1} \\ \vdots & \ddots & \vdots \\ a_{1M} & \cdots & a_{NM} \end{bmatrix}. \tag{C.5}$$

In other words, $\mathbf{A}^T$ is obtained by transposing the rows and columns of $\mathbf{A}$, so the ith row of $\mathbf{A}$ becomes the ith column of $\mathbf{A}^T$. The transpose of a row vector $\mathbf{x} = [\mathbf{x}_1 \ \ldots \ \mathbf{x}_N]$ yields a vector with the same elements:

$$\mathbf{x}^T = [\mathbf{x}_1 \ \ldots \ \mathbf{x}_N]^T = \begin{bmatrix} x_1 \\ \vdots \\ x_N \end{bmatrix}. \tag{C.6}$$

We therefore often write a column vector $\mathbf{x}$ with elements $\mathbf{x}_i$ as $\mathbf{x} = [\mathbf{x}_1 \ \ldots \ \mathbf{x}_N]^T$. Similarly, the transpose of an N-dimensional vector $\mathbf{x}$ with ith element $\mathbf{x}_i$ is the row vector $[\mathbf{x}_1 \ \ldots \ \mathbf{x}_N]$. Note that for $\mathbf{x}$ a row or column vector, $(\mathbf{x}^T)^T = \mathbf{x}$.

The *complex conjugate* $\mathbf{A}^*$ of a matrix $\mathbf{A}$ is obtained by taking the complex conjugate of each element of $\mathbf{A}$:

$$\mathbf{A}^* = \begin{bmatrix} a_{11} & \cdots & a_{1M} \\ \vdots & \ddots & \vdots \\ a_{N1} & \cdots & a_{NM} \end{bmatrix}^* = \begin{bmatrix} a_{11}^* & \cdots & a_{1M}^* \\ \vdots & \ddots & \vdots \\ a_{N1}^* & \cdots & a_{NM}^* \end{bmatrix}. \tag{C.7}$$

The *Hermitian* of a matrix $\mathbf{A}$, denoted as $\mathbf{A}^H$, is defined as its conjugate transpose: $\mathbf{A}^H = (\mathbf{A}^*)^T$. Note that applying the Hermitian operation twice results in the original matrix: $(\mathbf{A}^H)^H = \mathbf{A}$, so $\mathbf{A}$ is the Hermitian of $\mathbf{A}^H$. A square matrix $\mathbf{A}$ is a *Hermitian matrix* if it equals its Hermitian: $\mathbf{A} = \mathbf{A}^H$. A square matrix $\mathbf{A}$ is a *normal matrix* if $\mathbf{A}^H\mathbf{A} = \mathbf{A}\mathbf{A}^H$. A Hermitian matrix is a normal matrix, since $\mathbf{A}^H = \mathbf{A}$ implies $\mathbf{A}^H\mathbf{A} = \mathbf{A}\mathbf{A}^H$. The complex conjugate and Hermitian operators can also be applied to row or column vectors. In particular, the complex conjugate of a row or column vector $\mathbf{x}$, denoted as $\mathbf{x}^*$, is obtained by taking the complex conjugate of each element of $\mathbf{x}$. The Hermitian of a vector $\mathbf{x}$, denoted as $\mathbf{x}^H$, is its conjugate transpose: $\mathbf{x}^H = (\mathbf{x}^*)^T$.

Two $N \times M$ matrices can be added together to form a new matrix of size $N \times M$. The addition is performed element by element. In other words, if two $N \times M$ matrices $\mathbf{A}$ and $\mathbf{B}$ are added, the resulting $N \times M$ matrix $\mathbf{C} = \mathbf{A} + \mathbf{B}$ has (ij)th element $\mathbf{C}_{ij} = \mathbf{A}_{ij} + \mathbf{B}_{ij}$. Since matrix addition is done element by element, it inherits the commutative and associative properties of addition: $\mathbf{A} + \mathbf{B} = \mathbf{B} + \mathbf{A}$ and $(\mathbf{A} + \mathbf{B}) + \mathbf{C} = \mathbf{A} + (\mathbf{B} + \mathbf{C})$. The transpose of a sum of matrices is the sum of the transposes of the individual matrices: $(\mathbf{A} + \mathbf{B})^T = \mathbf{A}^T + \mathbf{B}^T$. Matrix subtraction is similar: for two $N \times M$ matrices $\mathbf{A}$ and $\mathbf{B}$, $\mathbf{C} = \mathbf{A} - \mathbf{B}$ is an $N \times M$ matrix with (ij)th element $\mathbf{C}_{ij} = \mathbf{A}_{ij} - \mathbf{B}_{ij}$. Two row or column vectors of the same dimension can be added using this definition of matrix addition because such vectors are special cases of matrices. In particular, an N-dimensional column vector $\mathbf{x}$ can be added to another column vector $\mathbf{y}$ of the same dimension to form the new N-dimensional column vector $\mathbf{z} = \mathbf{x} + \mathbf{y}$ with ith element $\mathbf{z}_i = \mathbf{x}_i + \mathbf{y}_i$. Similarly, if $\mathbf{x}$ and $\mathbf{y}$ are row vectors of dimension N then their sum $\mathbf{z} = \mathbf{x} + \mathbf{y}$ is an N-dimensional row vector with ith element $\mathbf{z}_i = \mathbf{x}_i + \mathbf{y}_i$. However, a row vector of dimension $N > 1$ cannot be added to a column vector of dimension N, since these vectors are matrices of different sizes ($1 \times N$ for the row vector, $N \times 1$ for the column vector). The linear combination of column vectors $\mathbf{x}$ and $\mathbf{y}$ of dimension N yields a new N-dimensional column vector $\mathbf{z} = c\mathbf{x} + d\mathbf{y}$ with ith element $\mathbf{z}_i = c\mathbf{x}_i + d\mathbf{y}_i$, where c and d are arbitrary scalars. Similarly, row vectors $\mathbf{x}$ and $\mathbf{y}$ of dimension N can be linearly combined to form the N-dimensional row vector $\mathbf{z} = c\mathbf{x} + d\mathbf{y}$ with ith element $\mathbf{z}_i = c\mathbf{x}_i + d\mathbf{y}_i$ for arbitrary scalars c and d.

A matrix can be multiplied by a scalar, in which case every element of the matrix is multiplied by the scalar. Specifically, multiplication of the matrix $\mathbf{A}$ by a scalar k results in the matrix $k\mathbf{A}$ given by

$$k\mathbf{A} = k\begin{bmatrix} a_{11} & \cdots & a_{1M} \\ \vdots & \ddots & \vdots \\ a_{N1} & \cdots & a_{NM} \end{bmatrix} = \begin{bmatrix} ka_{11} & \cdots & ka_{1M} \\ \vdots & \ddots & \vdots \\ ka_{N1} & \cdots & ka_{NM} \end{bmatrix}. \tag{C.8}$$

A row vector $\mathbf{x}$ multiplied by scalar k yields $k\mathbf{x} = [k\mathbf{x}_1 \ \ldots \ k\mathbf{x}_N]$, and a column vector $\mathbf{x}$ multiplied by scalar k yields $k\mathbf{x} = [k\mathbf{x}_1 \ \ldots \ k\mathbf{x}_N]^T$.

Two matrices can be multiplied together provided they have compatible dimensions. In particular, matrices $\mathbf{A}$ and $\mathbf{B}$ can be multiplied if the number of columns of $\mathbf{A}$ equals the number of rows of $\mathbf{B}$. If $\mathbf{A}$ is an $N \times M$ matrix and $\mathbf{B}$ is a $M \times L$ matrix then their product $\mathbf{C} = \mathbf{A}\mathbf{B}$ is an $N \times L$ matrix with (ij)th element $\mathbf{C}_{ij} = \sum_{k=1}^{M} \mathbf{A}_{ik}\mathbf{B}_{kj}$. Matrix multiplication is not commutative in general (i.e., usually $\mathbf{A}\mathbf{B} \neq \mathbf{B}\mathbf{A}$). In fact, if $\mathbf{A}$ is an $N \times M$ matrix and $\mathbf{B}$ is a $M \times L$ matrix then the product $\mathbf{B}\mathbf{A}$ exists only if $L = N$. In this case $\mathbf{B}\mathbf{A}$ is an $M \times M$ matrix, which may be a different size than the $N \times L$ matrix $\mathbf{A}\mathbf{B}$. Even if $M = L = N$, so that $\mathbf{A}\mathbf{B}$ and $\mathbf{B}\mathbf{A}$ are the same size, they may not be equal. If $\mathbf{A}$ is a square matrix then we can multiply $\mathbf{A}$ by itself. In particular, we define $\mathbf{A}^2 = \mathbf{A}\mathbf{A}$. Similarly $\mathbf{A}^k = \mathbf{A} \cdots \mathbf{A}$ is the product of k copies of $\mathbf{A}$. This implies that $\mathbf{A}^k\mathbf{A}^l = \mathbf{A}^{k+l}$. Multiplication of any matrix by the identity matrix of compatible size results in the same matrix; that is, if $\mathbf{A}$ is an $N \times M$ matrix, then $\mathbf{I}_N\mathbf{A} = \mathbf{A}\mathbf{I}_M = \mathbf{A}$. The transpose of a matrix product is the product of the transpose of the individual matrices in reverse order: $(\mathbf{A}\mathbf{B})^T = \mathbf{B}^T\mathbf{A}^T$. The product of an $N \times M$ matrix $\mathbf{A}$ and its $M \times N$ Hermitian $\mathbf{A}^H$ is a square matrix. In particular, $\mathbf{A}\mathbf{A}^H$ is an $N \times N$ square matrix while $\mathbf{A}^H\mathbf{A}$ is an $M \times M$ square matrix. The *Frobenius norm* of a matrix $\mathbf{A}$, denoted as $\|\mathbf{A}\|_F$, is defined as $\|\mathbf{A}\|_F = \sqrt{\mathrm{Tr}[\mathbf{A}\mathbf{A}^H]} = \sqrt{\mathrm{Tr}[\mathbf{A}^H\mathbf{A}]} = \sqrt{\sum_{i=1}^{N}\sum_{j=1}^{M}|\mathbf{A}_{ij}|^2}$. Matrix multiplication is associative, since $(\mathbf{A}\mathbf{B})\mathbf{C} = \mathbf{A}(\mathbf{B}\mathbf{C})$ as long as the matrix dimensions are compatible for multiplication (hence the parentheses are typically omitted). Matrix multiplication is also distributive: $\mathbf{A}(\mathbf{B} + \mathbf{C}) = \mathbf{A}\mathbf{B} + \mathbf{A}\mathbf{C}$ and $(\mathbf{A} + \mathbf{B})\mathbf{C} = \mathbf{A}\mathbf{C} + \mathbf{B}\mathbf{C}$.

An M-dimensional vector can be multiplied by a matrix with M columns. Specifically, if $\mathbf{A}$ is an $N \times M$ matrix and $\mathbf{x}$ is an M-dimensional vector (i.e., an $M \times 1$ matrix) then their product yields an N-dimensional vector $\mathbf{y} = \mathbf{A}\mathbf{x}$ with ith element $\mathbf{y}_i = \sum_{k=1}^{M}\mathbf{A}_{ik}\mathbf{x}_k$. Note that a matrix must left-multiply a vector, since the dimensions are not compatible for the product $\mathbf{x}\mathbf{A}$. However, if $\mathbf{x}$ is an N-dimensional row vector, then $\mathbf{x}\mathbf{A}$ is a compatible multiplication for $\mathbf{A}$ an $N \times M$ matrix and results in the M-dimensional row vector $\mathbf{y} = \mathbf{x}\mathbf{A}$ with ith element $\mathbf{y}_i = \sum_{k=1}^{N}\mathbf{x}_k\mathbf{A}_{ki}$. An N-dimensional row vector $\mathbf{x}$ can be multiplied by an N-dimensional vector $\mathbf{y}$, which results in a scalar $z = \mathbf{x}\mathbf{y} = \sum_{i=1}^{N}\mathbf{x}_i\mathbf{y}_i$. Note that the transpose of an N-dimensional vector is an N-dimensional row vector. The *inner product* of two N-dimensional vectors $\mathbf{x}$ and $\mathbf{y}$ is defined as $\langle \mathbf{x}, \mathbf{y} \rangle = \mathbf{x}^T\mathbf{y} = \sum_{i=1}^{N}\mathbf{x}_i\mathbf{y}_i$.

Given a matrix $\mathbf{A}$, a subset of rows of $\mathbf{A}$ form a *linearly independent* set if any row in the subset is not equal to a linear combination of the other rows in the subset. Similarly, a subset of columns of $\mathbf{A}$ form a linearly independent set if any column in the subset is not equal to a linear combination of the other columns in the subset. The *rank* $R_\mathbf{A}$ of a matrix $\mathbf{A}$ is equal to the number of rows in the largest subset of linearly independent rows of $\mathbf{A}$, which can be shown to equal the number of columns in the largest subset of linearly independent columns of $\mathbf{A}$. This implies that the rank of an $N \times M$ matrix cannot exceed $\min[N, M]$. An $N \times M$ matrix $\mathbf{A}$ is *full rank* if $R_\mathbf{A} = \min[N, M]$.

The *determinant* of a 2×2 matrix $\mathbf{A}$ is defined as $\det[\mathbf{A}] = \mathbf{A}_{11}\mathbf{A}_{22} - \mathbf{A}_{21}\mathbf{A}_{12}$. For an $N \times N$ matrix $\mathbf{A}$ with $N > 2$, $\det[\mathbf{A}]$ is defined recursively as

$$\det[\mathbf{A}] = \sum_{i=1}^{N} \mathbf{A}_{ij}c_{ij} \tag{C.9}$$

for any j $(1 \leq j \leq N)$. Here c_{ij} is the *cofactor* corresponding to the matrix element $\mathbf{A}_{ij}$, defined as

$$c_{ij} = (-1)^{i+j} \det[\mathbf{A}'], \tag{C.10}$$

where $\mathbf{A}'$ is the submatrix of $\mathbf{A}$ obtained by deleting the ith row and jth column of $\mathbf{A}$.

If $\mathbf{A}$ is an $N \times N$ square matrix and if there is another $N \times N$ matrix $\mathbf{B}$ such that $\mathbf{BA} = \mathbf{I}_N$, then we say that $\mathbf{A}$ is *invertible* or *nonsingular*. We call $\mathbf{B}$ the *inverse* of $\mathbf{A}$, and we denote this inverse as $\mathbf{A}^{-1}$. Thus, $\mathbf{A}^{-1}\mathbf{A} = \mathbf{I}_N$. Moreover, for $\mathbf{A}^{-1}$ defined in this way, we also have that $\mathbf{AA}^{-1} = \mathbf{I}_N$. Only square matrices can be invertible, and the matrix inverse is the same size as the original matrix. A square invertible matrix $\mathbf{U}$ is *unitary* if $\mathbf{UU}^H = \mathbf{I}$, which implies that $\mathbf{U}^H = \mathbf{U}^{-1}$ and thus $\mathbf{U}^H\mathbf{U} = \mathbf{I}$. Not every square matrix is invertible. If a matrix is not invertible, we say it is *singular* or *noninvertible*. The inverse of an inverse matrix is the original matrix: $(\mathbf{A}^{-1})^{-1} = \mathbf{A}$. The inverse of the product of matrices is the product of the inverses in opposite order: $(\mathbf{AB})^{-1} = \mathbf{B}^{-1}\mathbf{A}^{-1}$. The kth power of the inverse is $\mathbf{A}^{-k} = (\mathbf{A}^{-1})^k$.

For a diagonal matrix $\mathbf{D} = \mathrm{diag}[d_1, \ldots, d_N]$ with $d_i \neq 0$ ($i = 1, \ldots, N$), the inverse exists and is given by $\mathbf{D}^{-1} = \mathrm{diag}[1/d_1, \ldots, 1/d_N]$. For a general 2×2 matrix $\mathbf{A}$ with (ij)th element a_{ij}, its inverse exists if $\det[\mathbf{A}] \neq 0$ and is given by

$$\mathbf{A}^{-1} = \begin{bmatrix} a_{11} & a_{12} \\ a_{21} & a_{22} \end{bmatrix}^{-1} = \frac{1}{\det[\mathbf{A}]} \begin{bmatrix} a_{22} & -a_{12} \\ -a_{21} & a_{11} \end{bmatrix}. \tag{C.11}$$

There are more complicated formulas for the inverse of $N \times N$ invertible matrices with $N > 2$. However, matrix inverses are usually obtained using computer math packages.

Matrix inverses are commonly used to solve systems of linear equations. In particular, consider a set of linear equations expressed in matrix form as

$$\mathbf{y} = \mathbf{Ax}. \tag{C.12}$$

If the matrix $\mathbf{A}$ is invertible then, given $\mathbf{y}$, there is a unique vector $\mathbf{x} = \mathbf{A}^{-1}\mathbf{y}$ that satisfies this system of equations.

C.3 Matrix Decompositions

Given a square matrix $\mathbf{A}$, a scalar value λ for which there exists a nonzero vector $\mathbf{x}$ such that $\mathbf{Ax} = \lambda\mathbf{x}$ is called an *eigenvalue* of $\mathbf{A}$. The vector $\mathbf{x}$ is called the *eigenvector* of $\mathbf{A}$ corresponding to λ. The eigenvalues of a matrix $\mathbf{A}$ are all values of λ that satisfy the *characteristic equation* of $\mathbf{A}$, defined as $\det[\mathbf{A} - \lambda\mathbf{I}] = 0$. The polynomial in λ defined by $\det[\mathbf{A} - \lambda\mathbf{I}]$ is called the *characteristic polynomial* of $\mathbf{A}$, so the eigenvalues of $\mathbf{A}$ are the roots of its characteristic polynomial. The characteristic polynomial of an $N \times N$ matrix has N unique roots $r_1, \ldots, r_N$ ($r_i \neq r_j$) if it is of the form $\det[\mathbf{A} - \lambda\mathbf{I}] = (-1)^N(\lambda - r_1) \cdots (\lambda - r_N)$. When the characteristic polynomial includes a term $(\lambda - r_i)^k$, $k > 1$, we say that root r_i has *multiplicity k*. For example, if $\det[\mathbf{A} - \lambda\mathbf{I}] = -(\lambda - r_1)^2(\lambda - r_2)^3$ then root r_1 has multiplicity 2 and root r_2 has multiplicity 3. An $N \times N$ matrix has N eigenvalues $\lambda_1, \ldots, \lambda_N$, although they will not all be unique if any of the roots of the characteristic polynomial have multiplicity greater than 1. It can be shown that the determinant of a matrix equals the product of all its eigenvalues (i.e., an eigenvalue r_i with multiplicity k would contribute r_i^k to the product).

The eigenvalues of a Hermitian matrix are always real, although the eigenvectors can be complex. Moreover, if $\mathbf{A}$ is an $N \times N$ normal matrix then it can be written in the following form:

$$\mathbf{A} = \mathbf{U}\mathbf{\Lambda}\mathbf{U}^H, \tag{C.13}$$

where $\mathbf{U}$ is a unitary matrix whose columns are the eigenvectors of $\mathbf{A}$ and $\mathbf{\Lambda} = \text{diag}[\lambda_1, \ldots, \lambda_K, 0, \ldots, 0]$ is an $N \times N$ diagonal matrix whose first K diagonal elements are the nonzero eigenvalues of $\mathbf{A}$. For $\mathbf{A}$ Hermitian, the $\mathbf{\Lambda}$ in (C.13) has only real elements. We say that a matrix $\mathbf{A}$ is *positive definite* if, for all nonzero vectors $\mathbf{x}$, we have $\mathbf{x}^H\mathbf{A}\mathbf{x} > 0$. A Hermitian matrix is positive definite if and only if all its eigenvalues are positive. Similarly, we say the matrix $\mathbf{A}$ is *positive semidefinite* or *nonnegative definite* if, for all nonzero vectors $\mathbf{x}$, $\mathbf{x}^H\mathbf{A}\mathbf{x} \geq 0$. A Hermitian matrix is nonnegative definite if and only if all of its eigenvalues are nonnegative.

Suppose that $\mathbf{A}$ is an $N \times M$ matrix of rank $R_\mathbf{A}$. Then there exist an $N \times M$ matrix $\mathbf{\Sigma}$ and two unitary matrices $\mathbf{U}$ and $\mathbf{V}$ of size $N \times N$ and $M \times M$ (respectively) such that

$$\mathbf{A} = \mathbf{U}\mathbf{\Sigma}\mathbf{V}^H. \tag{C.14}$$

We call the columns of $\mathbf{V}$ the *right singular vectors* of $\mathbf{A}$ and the columns of $\mathbf{U}$ the *left singular vectors* of $\mathbf{A}$. The matrix $\mathbf{\Sigma}$ has a special form: all elements that are not diagonal elements are zero, so

$$\mathbf{\Sigma}_{N \times M} = \begin{bmatrix} \sigma_1 & \cdots & 0 \\ \vdots & \ddots & \vdots \\ 0 & \cdots & \sigma_M \\ 0 & \cdots & 0 \\ \vdots & \ddots & \vdots \\ 0 & \cdots & 0 \end{bmatrix} \tag{C.15}$$

for $N \geq M$ and

$$\mathbf{\Sigma}_{N \times M} = \begin{bmatrix} \sigma_1 & \cdots & 0 & 0 & \cdots & 0 \\ \vdots & \ddots & \vdots & \vdots & \ddots & \vdots \\ 0 & \cdots & \sigma_N & 0 & \cdots & 0 \end{bmatrix} \tag{C.16}$$

for $N < M$, where $\sigma_i = \sqrt{\lambda_i}$ for λ_i the ith eigenvalue of $\mathbf{A}\mathbf{A}^H$. The values of σ_i are called the *singular values* of $\mathbf{A}$, and $R_\mathbf{A}$ of these singular values are nonzero, where $R_\mathbf{A}$ is the rank of $\mathbf{A}$. The decomposition (C.14) is called the *singular value decomposition* (SVD) of $\mathbf{A}$. The singular values of a matrix are always nonnegative.

Let $\mathbf{A}$ be an $N \times M$ matrix whose ith column we denote as $\mathbf{A}_i$. Treating each column as a submatrix, we can write $\mathbf{A} = [\mathbf{A}_1 \ \mathbf{A}_2 \ \ldots \ \mathbf{A}_M]$. The *vectorization* of the matrix $\mathbf{A}$, denoted as $\text{vec}(\mathbf{A})$, is defined as the (NM)-dimensional vector that results from stacking the columns $\mathbf{A}_i$ $(i = 1, \ldots, M)$ of matrix $\mathbf{A}$ on top of each other to form a vector:

$$\text{vec}(\mathbf{A}) = \begin{bmatrix} \mathbf{A}_1 \\ \vdots \\ \mathbf{A}_M \end{bmatrix} = [\mathbf{A}_{11} \ \mathbf{A}_{21} \ \ldots \ \mathbf{A}_{N1} \ \mathbf{A}_{12} \ \ldots \ \mathbf{A}_{N2} \ \ldots \ \mathbf{A}_{1M} \ \ldots \ \mathbf{A}_{NM}]^T. \tag{C.17}$$

Let $\mathbf{A}$ be an $N \times M$ matrix and $\mathbf{B}$ an $L \times K$ matrix. The *Kronecker product* of $\mathbf{A}$ and $\mathbf{B}$, denoted $\mathbf{A} \otimes \mathbf{B}$, is an $NL \times MK$ matrix defined by

$$\mathbf{A} \otimes \mathbf{B} = \begin{bmatrix} \mathbf{A}_{11}\mathbf{B} & \cdots & \mathbf{A}_{1M}\mathbf{B} \\ \vdots & \ddots & \vdots \\ \mathbf{A}_{N1}\mathbf{B} & \cdots & \mathbf{A}_{NM}\mathbf{B} \end{bmatrix}. \tag{C.18}$$

REFERENCES

[1]   B. Nobel and J. W. Daniel, *Applied Linear Algebra,* Prentice-Hall, Englewood Cliffs, NJ, 1977.

[2]   G. Strang, *Linear Algebra and Its Applications,* 2nd ed., Academic Press, New York, 1980.

[3]   R. A. Horn and C. R. Johnson, *Matrix Analysis,* Cambridge University Press, 1985.

[4]   R. A. Horn and C. R. Johnson, *Topics in Matrix Analysis,* Cambridge University Press, 1991.

Appendix D: Summary of Wireless Standards

This chapter summarizes the technical details associated with the two most prevalent wireless systems in operation today: cellular phones and wireless LANs. It also summarizes the specifications for three short range wireless network standards that have emerged to support a broad range of applications. More details on wireless standards can be found in [1; 2; 3; 4; 5; 6].

D.1 Cellular Phone Standards

D.1.2 First-Generation Analog Systems

In this section we summarize cellular phone standards. We begin with the standards for first-generation (1G) analog cellular phones, whose main characteristics are summarized in Table D.1. Systems based on these standards were widely deployed in the 1980s. While many of these systems have been replaced by digital cellular systems, there are many places throughout the world where these analog systems are still in use. The best known standard is the Advanced Mobile Phone Service (AMPS), developed by Bell Labs in the 1970s and first used commercially in the United States in 1983. After its U.S. deployment, many other countries adopted AMPS as well. This system has a narrowband version, narrowband AMPS (N-AMPS), with voice channels that are one third the bandwidth of regular AMPS. Japan deployed the first commercial cellular phone system in 1979 with the NTT (MCS-L1) standard based on AMPS, but at a higher frequency and with voice channels of slightly lower bandwidth. Europe also developed a similar standard to AMPS called the Total Access Communication System (TACS), which operates at a higher frequency and with smaller bandwidth channels than AMPS. It was deployed in the United Kingdom and in other European countries as well as outside Europe. The frequency range for TACS was extended in the U.K. to obtain more channels, leading to a variation called ETACS. A variation of the TACS system called JTACS was deployed in metropolitan areas of Japan in 1989 to provide higher capacity than the NTT system. JTACS operates at a slightly higher frequency than TACS or ETACS, and it has a bandwidth-efficient version (called NTACS) where voice channels occupy half the bandwidth of the channnels in JTACS. In addition to TACS, countries in Europe had different incompatible standards at different frequencies for analog cellular, including the Nordic Mobile Telephone (NMT) standard in Scandanavia, the Radiocom 2000 (RC2000) standard in

Table D.1: First-generation analog cellular phone standards

Parameter	AMPS	TACS	NMT (450/900)	NTT	C-450	RC2000
Uplink frequencies (MHz)	824–849	890–915	453–458/890–915	925–940[a]	450–455.74	414.8–418[b]
Downlink frequencies (MHz)	869–894	935–960	463–468/935–960	870–885	460–465.74	424.8–428
Modulation	FM	FM	FM	FM	FM	FM
Channel spacing (kHz)	30	25	25/12.5	25	10	12.5
Number of channels	832	1000	180/1999	600	573	256
Multiple access	FDMA	FDMA	FDMA	FDMA	FDMA	FDMA

[a] NTT also operated in several other frequency bands around 900 MHz.
[b] RC2000 also operated in several other frequency bands around 200 MHz.

France, and the C-450 standard in Germany and Portugal. The incompatibilities made it impossible to roam between European countries with a single analog phone, which motivated the need for one unified cellular standard and frequency allocation throughout Europe.

D.1.2 Second-Generation Digital Systems

Next we consider second-generation (2G) digital cellular phone standards, whose main characteristics are summarized in Table D.2. These systems were mostly deployed in the early 1990s. As a result of the incompatibilities in the first-generation analog systems, in 1982 the Groupe Spécial Mobile (GSM) was formed to develop a uniform digital cellular standard for all of Europe. The TACS spectrum in the 900-MHz band was allocated for GSM operation across Europe in order to facilitate roaming between countries. In 1989 the GSM specification was finalized and the system was launched in 1991, although availability was limited until 1992. The GSM standard uses TDMA combined with slow frequency hopping to combat out-of-cell interference. Convolutional coding and parity-check codes along with interleaving are used for error detection and correction. The standard also includes an equalizer to compensate for frequency-selective fading. The GSM standard is used in about 66% of the world's cell phones, with more than 470 GSM operators in 172 countries supporting over a billion users. As the GSM standard became more global, the meaning of the acronym was changed to the Global System for Mobile Communications.

Although Europe got an early jump on developing 2G digital systems, the United States was not far behind. In 1992 the IS-54 digital cellular standard was finalized, with commercial deployment beginning in 1994. This standard uses the same channel spacing (30 kHz) as AMPS to facilitate the analog-to-digital transition for wireless operators, along with a TDMA multiple access scheme to improve handoff and control signaling as compared with analog FDMA. The IS-54 standard (also called the North American Digital Cellular standard) was improved over time, and these improvements evolved into the IS-136 standard, which subsumed the original standard. Similar to the GSM standard, the IS-136 standard uses parity-check codes, convolutional codes, interleaving, and equalization.

A competing standard for 2G systems based on CDMA was proposed by Qualcomm in the early 1990s. The standard, called IS-95 or IS-95a, was finalized in 1993 and deployed

Table D.2: Second-generation digital cellular phone standards

Parameter	GSM	IS-136	IS-95 (cdmaOne)	PDC
Uplink frequencies (MHz)	890–915	824–849	824–849	940–956, 1429–1453
Downlink frequencies (MHz)	935–960	869–894	869–894	810–826, 1477–1501
Carrier separation (kHz)	200	30	1250	25
Number of channels	1000	832[a]	~2500	1600[a]
Modulation	GMSK	$\pi/4$-D-QPSK	BPSK/QPSK	$\pi/4$-D-QPSK
Compressed speech rate (Kbps)	13	7.95	1.2–9.6 (variable)	6.7
Channel data rate (Kbps)	270.833	48.6	(1.2288 Mchips/s)	42
Data code rate	1/2	1/2	1/2 (DL), 1/3 (UL)	1/2
ISI reduction/diversity	Equalizer	Equalizer	RAKE, SHO	Equalizer
Multiple access	TDMA/SFH	TDMA	CDMA	TDMA

[a] Three users per channel.

commercially under the name cdmaOne in 1995. Like IS-136, IS-95 was designed to be compatible with AMPS so that the two systems could coexist in the same frequency band. In CDMA all users are superimposed on top of each other with spreading codes that can separate out the users at the receiver. Thus, channel data rate does not apply to just one user, as in TDMA systems. The channel chip rate is 1.2288 Mchips per second for a total spreading factor of 128 for both the uplink and downlink. The spreading process in IS-95 is different for the downlink (DL) and the uplink (UL), with spreading on both links accomplished through a combination of spread-spectrum modulation and coding. On the downlink, data is first rate-1/2 convolutionally encoded and interleaved, then modulated by one of 64 orthogonal spreading sequences (Walsh functions). Next, a synchronized scrambling sequence unique to each cell is superimposed on top of the Walsh function to reduce interference between cells. The scrambling requires synchronization between base stations. Uplink spreading is accomplished using a combination of a rate-1/3 convolutional code with interleaving, modulation by an orthogonal Walsh function, and modulation by a nonorthogonal user/base station–specific code. The IS-95 standard includes a parity-check code for error detection, as well as power control for the reverse link to avoid the near–far problem. A three-finger RAKE receiver is also specified to provide diversity and to compensate for ISI. A form of base station diversity called soft handoff (SHO), whereby a mobile maintains a connection to both the new and old base stations during handoff and combines their signals, is also included in the standard. CDMA has some advantages over TDMA for cellular systems, including no need for frequency planning, SHO capabilities, the ability to exploit voice activity to increase capacity, and no hard limit on the number of users that can be accommodated in the system. There was much debate about the relative merits of the IS-136 and IS-95 standards throughout the early 1990s, with claims that IS-95 could achieve 20 times the capacity of AMPS whereas IS-136 could only achieve 3 times this capacity. In the end, both systems turned out to achieve approximately the same capacity increase over AMPS.

The 2G digital cellular standard in Japan, called the Personal Digital Cellular (PDC) standard, was established in 1991 and deployed in 1994. It is similar to the IS-136 standard but with 25-kHz voice channels for compatibility with the Japanese analog systems. This system operates in both the 800–900-MHz and 1400-MHz frequency bands. Each channel

Table D.3: 2G enhancements to support 2.5G data capabilities

2G standard	2.5G enhancement	Technique	Data rates (kbps) maximum/realized
GSM	HSCSD	Aggregate timeslots	57.6/14.4–57.6
GSM/IS-136	GPRS	Aggregate timeslots with packet switching	140.8/56
	EDGE	GPRS with variable mod./cod.	384/200
IS-95	IS-95b	Aggregate Walsh functions	115/64

accommodates three users, and six users per channel can be accommodated with a higher compression rate.

D.1.3 Evolution of Second-Generation Systems

In the late 1990s, second-generation systems evolved in two directions: they were ported to higher frequencies as more cellular bandwidth became available in Europe and the United States, and they were modified to support data services in addition to voice. Specifically, in 1994 the FCC began auctioning spectrum in the Personal Communications Service (PCS) band at 1.9 GHz (1900 MHz) for cellular systems. Operators purchasing spectrum in this band could adopt any standard. Different operators chose different standards, so GSM, IS-136, and IS-95 were all deployed at 1.9 GHz in different parts of the country, making nation-wide roaming with a single phone difficult. In fact, many of the initial digital cellphones included an analog AMPS mode in case the digital system was not available. GSM systems operating in the PCS band are sometimes referred to as PCS 1900 systems. The IS-136 and IS-95 (cdmaOne) standards translated to the PCS band go by the same names. Europe allocated additional cellular spectrum in the 1.8-GHz band. The standard for this frequency band, called GSM 1800 or DCS 1800 (for Digital Cellular System), uses GSM as the core standard with some modifications to allow overlays of macrocells and microcells. Note that second-generation cordless phones such as Digital Enhanced Cordless Telecommunications (DECT), the Personal Access Communications System (PACS), and the Personal Handy-phone System (PHS) also operate in the 1.9-GHz frequency band, but these systems are mostly within buildings supporting private branch exchange (PBX) services.

Once digital cellular became available, operators began incorporating data services in addition to voice. The 2G systems with added data capabilities are sometimes referred to as 2.5G systems. The enhancements to 2G systems made to support data services are summarized in Table D.3. GSM systems followed several different upgrade paths to provide data services. The simplest, called High Speed Circuit Switched Data (HSCSD), allows up to four consecutive timeslots to be assigned to a single user, thereby providing a maximum transmission rate of up to 57.6 kbps. Circuit switching is quite inefficient for data, so a more complex enhancement provides for packet-switched data layered on top of the circuit-switched voice. This enhancement is referred to as General Packet Radio Service (GPRS). A maximum data rate of 171.2 kbps is possible with GPRS when all eight timeslots of a GSM frame are allocated to a single user. The data rates of GPRS are further enhanced through variable-rate modulation and coding, referred to as Enhanced Data rates for GSM Evolution (EDGE).

Table D.4: 3G digital cellular phone standards

Parameter	cdma2000				W-CDMA		
	1X	1XEV-DO	1XEV-DV	3X	UMTS	FOMA	J-phone
Channel bandwidth (MHz)	1.25	1.25	1.25	3.75	5	5	5
Chip rate (Mchips/s)	1.2288	1.2288	1.2288	3.6864	3.84	3.84	3.84
Peak data rate (Mbps)	.144	2.4	4.8	5–8	2.4[a]	2.4[a]	2.4[a]
Modulation	QPSK/MPSK (DL), BPSK/QPSK (UL)				QPSK (DL), BPSK (UL)		
Coding	Convolutional (low rate), Turbo (high rate)						
Power control (Hz)	800	800	800	800	1500	1500	1500

[a] 8–10 Mbps with HSDPA.

EDGE provides data rates up to 384 kbps with a bit rate of 48–69.2 kbps per timeslot. GPRS and EDGE are compatible with IS-136 as well as GSM, thus providing a convergent upgrade path for both of these systems.

The IS-95 standard was modified to provide data services by assigning multiple orthogonal Walsh functions to a single user. A maximum of eight functions can be assigned, leading to a maximum data rate of 115.2 kbps, although in practice only about 64 kbps is achieved. This evolution is referred to as the IS-95b standard.

D.1.4 Third-Generation Systems

The fragmentation of standards and frequency bands associated with 2G systems led the International Telecommunications Union (ITU) in the late 1990s to formulate a plan for a single global frequency band and standard for third-generation (3G) digital cellular systems. The standard was named the International Mobile Telephone 2000 (IMT-2000) standard with a desired system rollout in the 2000 timeframe. In addition to voice services, IMT-2000 was to provide Mbps data rates for demanding applications such as broadband Internet access, interactive gaming, and high-quality audio and video entertainment. Agreement on a single standard did not materialize, with most countries supporting one of two competing standards: cdma2000 (backward compatible with cdmaOne), supported by the Third Generation Partnership Project 2 (3GPP2); and wideband CDMA (W-CDMA, backward compatible with GSM and IS-136), supported by the Third Generation Partnership Project 1 (3GPP1). The main characteristics of these two 3G standards are summarized in Table D.4. Both standards use CDMA with power control and RAKE receivers, but the chip rates and other specification details are different. In particular, cdma2000 and W-CDMA are not compatible standards, so a phone must be dual-mode in order to operate with both systems. A third 3G standard, TD-SCDMA, is under consideration in China but is unlikely to be adopted elsewhere. The key difference between TD-SCDMA and the other 3G standards is its use of TDD instead of FDD for uplink/downlink signaling.

The cdma2000 standard builds on cdmaOne to provide an evolutionary path to 3G. The core of the cdma2000 standard is referred to cdma2000 1X or cdma2000 1XRTT, indicating that the radio transmission technology (RTT) operates in one pair of 1.25-MHz radio channels and is thus backward compatible with cdmaOne systems. The cdma2000 1X system doubles the voice capacity of cdmaOne systems and supports high-speed data with projected

Table D.5: 802.11 wireless LAN link layer standards

Parameter	802.11	802.11a	802.11b	802.11g
Bandwidth (MHz)	83.5	300	83.5	83.5
Frequency range (GHz)	2.4–2.4835	5.15–5.25 (lower) 5.25–5.35 (middle) 5.725–5.825 (upper)	2.4–2.4835	2.4–2.4835
Number of channels	3	12 (4 per subband)	3	3
Modulation	BPSK, QPSK DSSS, FHSS	BPSK, QPSK, MQAM OFDM	BPSK, QPSK DSSS	BPSK, QPSK, MQAM OFDM
Coding	Undefined	Convolutional (rate 1/2, 2/3, 3/4)	Barker, CCK	Convolutional (rate 1/2, 2/3, 3/4)
Maximum data rate (Mbps)	2	54	11	54
Range (m)	Undefined	27–30 (lower band)	75–100	30
Random access	CSMA/CA	CSMA/CA	CSMA/CA	CSMA/CA

peak rates of about 300 kbps and actual rates of about 144 kbps. There are two evolutions of this core technology to provide high data rates (referred to as HDR service) above 1 Mbps; these evolutions are referred to as cdma2000 1XEV. The first phase of evolution, cdma2000 1XEV-DO (data only), enhances the cdmaOne system using a separate 1.25-MHz dedicated high-speed data channel that supports downlink data rates up to 3 Mbps and uplink data rates up to 1.8 Mbps for an average combined rate of 2.4 Mbps. The second phase of the evolution, cdma2000 1XEV-DV (data and voice), is projected to support up to 4.8-Mbps data rates as well as legacy 1X voice users, 1XRTT data users, and 1XEV-DO data users – and all within the same radio channel. Another proposed enhancement to cdma2000 is to aggregate three 1.25-MHz channels into one 3.75-MHz channel. This aggregation is referred to as cdma2000 3X, and its exact specifications are still under development.

W-CDMA is the primary competing 3G standard to cdma2000. It has been selected as the 3G successor to GSM and in this context is referred to as the Universal Mobile Telecommunications System (UMTS). W-CDMA is also used in the Japanese FOMA and J-Phone 3G systems. These different systems share the W-CDMA link layer protocol (air interface) but have different protocols for other aspects of the system such as routing and speech compression. W-CDMA supports peak rates of up to 2.4 Mbps, with typical rates anticipated in the 384-kbps range. W-CDMA uses 5-MHz channels, in contrast to the 1.25-MHz channels of cdma2000. An enhancement to W-CDMA called High Speed Data Packet Access (HSDPA) provides data rates of around 9 Mbps, and this may be the precursor to fourth-generation systems.

D.2 Wireless Local Area Networks

Wireless local area networks (WLANs) are built around the family of IEEE 802.11 standards. The main characteristics of this standards family are summarized in Table D.5. The baseline 802.11 standard, released in 1997, occupies 83.5 MHz of bandwidth in the unlicensed 2.4-GHz frequency band. It specifies PSK modulation with FHSS or DSSS. Data rates of up to 2 Mbps are supported, with CSMA/CA used for random access. The baseline standard was expanded in 1999 to create the 802.11b standard, operating in the same 2.4-GHz band using only DSSS. This standard uses variable-rate modulation and coding, with BPSK or QPSK for modulation and channel coding via either Barker sequences or complementary code keying (CCK). This leads to a maximum channel rate of 11 Mbps and a maximum user data rate of about 1.6 Mbps. The transmission range is approximately 100 m. The network architecture in 802.11b is specified as either star or peer-to-peer, although the peer-to-peer feature is not typically used. This standard has been widely deployed, with manufacturers integrating 802.11b wireless LAN cards into many laptop computers.

The 802.11a standard was finalized in 1999 as an extension to 802.11 to improve on the 802.11b data rates. The 802.11a standard occupies 300 MHz of spectrum in the 5-GHz U-NII band. In fact, the 300 MHz of bandwidth is segmented into three 100-MHz subbands: a lower band (5.15–5.25 GHz), a middle band (5.25–5.35 GHz), and an upper band (5.725–5.825 GHz). Channels are spaced 20 MHz apart except on the outer edges of the lower and middle bands, where they are spaced 30 MHz apart. Three maximum transmit power levels are specified: 40 mW for the lower band, 200 mW for the middle band, and 800 mW for the upper band. These restrictions imply that the lower band is mostly suitable for indoor applications only, the middle band for indoor and outdoor, and the high band for outdoor. Variable-rate modulation and coding is used on each channel: the modulation varies over BPSK, QPSK, 16-QAM, and 64-QAM, and the convolutional code rate varies over 1/2, 2/3, and 3/4. This leads to a maximum data rate per channel of 54 Mbps. For indoor systems, the 5-GHz carrier coupled with the power restriction in the lower band reduces the range of 802.11a relative to 802.11b and also makes it more difficult for the signal to penetrate walls and other obstructions. The 802.11a standard uses orthogonal frequency division multiplexing (OFDM) multiple access instead of FHSS or DSSS, and in that sense it diverges from the original 802.11 standard.

The 802.11g standard, finalized in 2003, attempts to combine the best of 802.11a and 802.11b, with data rates of up to 54 Mbps in the 2.4-GHz band for greater range. The standard is backward compatible with 802.11b, so 802.11g access points will work with 802.11b wireless network adapters and vice versa. However, 802.11g uses the OFDM, modulation, and coding schemes of 802.11a. Access points and wireless LAN cards are available with all three standards to avoid incompatibilities. The 802.11a/b/g family of standards are collectively referred to as *Wi-Fi*, for wireless fidelity. Extending these standards to frequency allocations in countries other than the United States falls under the 802.11d standard. There are several other standards in the 802.11 family that are under development; these are summarized in Table D.6.

A potential competitor to the 802.11 standards as well as cellular systems is the emerging IEEE 802.16 standard called WiMAX. This standard promises broadband wireless access with data rates on the order of 40 Mbps for fixed users and 15 Mbps for mobile users, with a range of several kilometers. Details of the specification are still being worked out.

D.3 Wireless Short-Distance Networking Standards

This last section summarizes the main characteristics of ZigBee, Bluetooth, and UWB (ultra-wideband), which have emerged to support a wide range of short-distance wireless network

Table D.6: IEEE 802.11 ongoing standards work

Standard	Scope
802.11e	Provides quality of service (QoS) at the MAC layer
802.11f	Roaming protocol across multivendor access points
802.11h	Adds frequency and power management features to 802.11a to make it more compatible with European operation
802.11i	Enhances security and authentication mechanisms
802.11j	Modifies 802.11a link layer to meet Japanese requirements
802.11k	Provides an interface to higher layers for radio and network measurements that can be used for radio resource management
802.11m	Maintenance of 802.11 standard (technical/editorial corrections)
802.11n	MIMO link enhancements to enable higher throughput

Table D.7: Short-range wireless network standards

Parameter	ZigBee (802.15.4)	Bluetooth (802.15.1)	UWB (802.15.3 proposal)
Frequency range (GHz)	2.4–2.4835	2.4–2.4835	3.1–10.6
Bandwidth (MHz)	83.5	83.5	7500
Modulation	BPSK, OQPSK DSSS	GFSK FHSS	BPSK, QPSK OFDM or DSSS
Maximum data rate (Mbps)	.25	1	100
Range (m)	30	10 (100)	10
Power consumption (mW)	5–20	1 (100)	80–150 mW
Access	CSMA/CA (optional TD)	TD	Undefined
Networking	Mesh/Star/Tree	Subnet clusters (8 nodes)	Undefined

applications. These specifications are designed to be compliant with the IEEE 802.15 standards, a family of IEEE standards for short-distance wireless networking called Wireless Personal Area Networks (WPANs). Bluetooth operates in the 2.4-GHz unlicensed band; ZigBee operates in the same band as well as in the 800-MHz and 900-MHz unlicensed bands; and UWB operates across a broad range of frequencies in an underlay to existing systems. ZigBee and Bluetooth include link, MAC, and higher-layer protocol specifications, whereas UWB specificies just the link layer protocol. Table D.7 summarizes the main characteristics of ZigBee (2.4-GHz band only), Bluetooth, and UWB.

ZigBee consists of link and MAC layer protocols that are compliant with the IEEE 802.15.4 standard, as well as higher-layer protocols for ad hoc networking (mesh, star, or tree topologies), power management, and security. ZigBee supports data rates of up to 250 kbps with PSK modulation and DSSS. ZigBee generally targets applications requiring relatively low data rates, low duty cycles, and large networks. Power efficiency is key, with the goal of nodes operating for months or years on a single battery charge.

In contrast to ZigBee, Bluetooth provides up to 1-Mbps data rate, including three guaranteed low-latency voice channels, using FHSS and FSK modulation with Gaussian pulse shaping (Gaussian FSK or GFSK). Bluetooth normally transmits at a power of 1 mW with a transmission range of about 10 m, although this can be extended to 100 m by increasing the transmit power to 100 mW. Networks are formed in subnet clusters (piconets) of up to eight nodes, with one node acting as a master and the rest as slaves. Time division is used for channel access, with the master node coordinating the frequency-hopping sequence and synchronization with the slave nodes. Extended networks, or *scatternets,* can be formed when one node is part of multiple piconets. However, forming large networks through this approach is difficult owing to the synchronization requirements of FHSS. Portions of the Bluetooth standard were formally adopted by the IEEE as its 802.15.1 standard.

Ultrawideband has significantly higher data rates – up to 100 Mbps – than either ZigBee or Bluetooth. It also occupies significantly more bandwidth and has stringest power restrictions (to prevent it from interfering with primary band users). Thus, it is suitable only for short-range indoor applications. The UWB protocol defines only a link layer technology, so it requires a compatible MAC protocol as well as higher-layer protocols to become part of a wireless network standard. The modulation is BPSK or QPSK, with competing camps recommending either OFDM or DSSS overlayed on the data modulation. Ultrawideband is likely to become the link layer technology for the IEEE 802.15.3 standard, a family of standards for wireless networks supporting imaging and multimedia applications.

REFERENCES

[1] T. S. Rappaport, *Wireless Communications – Principles and Practice,* 2nd ed., Prentice-Hall, Englewood Cliffs, NJ, 2001.
[2] J. D. Vriendt, P. Lainé, C. Lerouge, and X. Xu, "Mobile network evolution: A revolution on the move," *IEEE Commun. Mag.,* pp. 104–11, April 2002.
[3] D. Porcino and W. Hirt, "Ultra-wideband radio technology: Potential and challenges ahead," *IEEE Commun. Mag.,* pp. 66–74, July 2003.
[4] I. Poole, "What exactly is ... ZigBee?" *IEEE Commun. Eng.,* pp. 44–5, August/September 2004.
[5] S. Haykin and M. Moher, *Modern Wireless Communications,* Prentice-Hall, Englewood Cliffs, NJ, 2005.
[6] W. Stallings, *Wireless Communications and Networks,* 2nd ed., Prentice-Hall, Englewood Cliffs, NJ, 2005.

Bibliography

Abou-Faycal, I. C., M. D. Trott, and S. Shamai, "The capacity of discrete-time memoryless Rayleigh fading channels," *IEEE Trans. Inform. Theory,* pp. 1290–1301, May 2001.

Abramson, N., "The ALOHA system – Another alternative for computer communications," *Proc. Amer. Federation Inform. Proc. Soc. Fall Joint Comput. Conf.,* pp. 281–5, November 1970.

Abramson, N., "Wide-band random-access for the last mile," *IEEE Pers. Commun. Mag.,* pp. 29–33, December 1996.

Abrishamkar, F., and Z. Siveski, "PCS global mobile satellites," *IEEE Commun. Mag.,* pp. 132–6, September 1996.

Abu-Dayya, A., and N. Beaulieu, "Analysis of switched diversity systems on generalized-fading channels," *IEEE Trans. Commun.,* pp. 2959–66, November 1994.

Abu-Dayya, A., and N. Beaulieu, "Switched diversity on microcellular Ricean channels," *IEEE Trans. Veh. Tech.,* pp. 970–6, November 1994.

Agrawal, P., "Energy efficient protocols for wireless systems," *Proc. Internat. Sympos. Pers., Indoor, Mobile Radio Commun.,* pp. 564–9, September 1998.

Ahlswede, R., N. Cai, S.-Y. R. Li, and R. W. Yeung, "Network information flow," *IEEE Trans. Inform. Theory,* pp. 1204–16, July 2000.

Akerberg, D., "Properties of a TDMA picocellular office communication system," *Proc. IEEE Globecom Conf.,* pp. 1343–9, December 1988.

Akin, H. C., and K. M. Wasserman, "Resource allocation and scheduling in uplink for multimedia CDMA wireless systems," *IEEE/Sarnoff Sympos. Adv. Wired Wireless Commun.,* pp. 185–8, April 2004.

Alamouti, S., "A simple transmit diversity technique for wireless communications," *IEEE J. Sel. Areas Commun.,* pp. 1451–8, October 1998.

Alamouti, S. M., and S. Kallel, "Adaptive trellis-coded multiple-phased-shift keying for Rayleigh fading channels," *IEEE Trans. Commun.,* pp. 2305–14, June 1994.

Alasti, M., K. Sayrafian-Pour, A. Ephremides, and N. Farvardin, "Multiple description coding in networks with congestion problem," *IEEE Trans. Inform. Theory,* pp. 891–902, March 2001.

Al-Dhahir, N., C. Fragouli, A. Stamoulis, W. Younis, and R. Calderbank, "Space-time processing for broadband wireless access," *IEEE Commun. Mag.,* pp. 136–42, September 2002.

Algans, A., K. I. Pedersen, and P. E. Mogensen, "Experimental analysis of the joint statistical properties of azimuth spread, delay spread, and shadow fading," *IEEE J. Sel. Areas Commun.,* pp. 523–31, April 2002.

Alouini, M.-S., and A. J. Goldsmith, "Area spectral efficiency of cellular mobile radio systems," *IEEE Trans. Veh. Tech.,* pp. 1047–66, July 1999.

Alouini, M.-S., and A. J. Goldsmith, "Capacity of Rayleigh fading channels under different adaptive transmission and diversity combining techniques," *IEEE Trans. Veh. Tech.,* pp. 1165–81, July 1999.

Alouini, M.-S., and A. J. Goldsmith, "Adaptive modulation over Nakagami fading channels," *Kluwer J. Wireless Pers. Commun.*, pp. 119–43, May 2000.

Amitay, N., "Modeling and computer simulation of wave propagation in lineal line-of-sight micro-cells," *IEEE Trans. Veh. Tech.*, pp. 337–42, November 1992.

Ananasso, R., and F. D. Priscoli, "The role of satellites in personal communication services," *IEEE J. Sel. Areas Commun.*, pp. 180–96, February 1995.

Anderlind, E., and J. Zander, "A traffic model for non-real-time data users in a wireless radio network," *IEEE Commun. Lett.*, pp. 37–9, March 1997.

Anderson, C. R., T. S. Rappaport, K. Bae, A. Verstak, N. Tamakrishnan, W. Trantor, C. Shaffer, and L. T. Waton, "In-building wideband multipath characteristics at 2.5 and 60 GHz," *Proc. IEEE Veh. Tech. Conf.*, pp. 24–8, September 2002.

Andrews, J. G., "Interference cancellation for cellular systems: A contemporary overview," *IEEE Wireless Commun. Mag.*, pp. 19–29, April 2005.

Andrews, J. G., and T. H. Meng, "Optimum power control for successive interference cancellation with imperfect channel estimation," *IEEE Trans. Wireless Commun.*, pp. 375–83, March 2003.

Andrews, M., K. Kumaran, K. Ramanan, A. Stolyar, and P. Whiting, "Providing quality of service over a shared wireless link," *IEEE Commun. Mag.*, pp. 150–4, February 2001.

Aulin, T., "A modified model for fading signal at the mobile radio channel," *IEEE Trans. Veh. Tech.*, pp. 182–202, August 1979.

Ayanoglu, E., and R. M. Gray, "The design of joint source and channel trellis waveform coders," *IEEE Trans. Inform. Theory*, pp. 855–65, November 1987.

Ayyagari, D., and A. Ephremides, "Cellular multicode CDMA capacity for integrated (voice and data) services," *IEEE J. Sel. Areas Commun.*, pp. 928–38, May 1999.

Ayyagari, D., and A. Ephremides, "Optimal admission control in cellular DS-CDMA systems with multimedia traffic," *IEEE Trans. Wireless Commun.*, pp. 195–202, January 2003.

Ayyagari, D., A. Michail, and A. Ephremides, "A unified approach to scheduling, access control and routing for ad-hoc wireless networks," *Proc. IEEE Veh. Tech. Conf.*, pp. 380–4, May 2000.

Babich, F., G. Lombardi, and E. Valentinuzzi, "Variable order Markov modeling for LEO mobile satellite channels," *Elec. Lett.*, pp. 621–3, April 1999.

Bahai, A. R. S., B. R. Saltzberg, and M. Ergen, *Multi-Carrier Digital Communications – Theory and Applications of OFDM*, 2nd ed., Springer-Verlag, New York, 2004.

Bambos, N., "Toward power-sensitive network architectures in wireless communications: Concepts, issues, and design aspects," *IEEE Pers. Commun. Mag.*, pp. 50–9, June 1998.

Bambos, N., S. C. Chen, and G. J. Pottie, "Channel access algorithms with active link protection for wireless communication networks with power control," *IEEE/ACM Trans. Network.*, pp. 583–97, October 2000.

Bambos, N., G. Pottie, and S. Chen, "Channel access algorithms with active link protection for wireless communications networks with power control," *IEEE/ACM Trans. Network.*, pp. 583–97, October 2000.

Baro, S., G. Bauch, and A. Hansman, "Improved codes for space-time trellis coded modulation," *IEEE Commun. Lett.*, pp. 20–2, January 2000.

Basagni, S., D. Turgut, and S. K. Das, "Mobility-adaptive protocols for managing large ad hoc networks," *Proc. IEEE Internat. Conf. Commun.*, pp. 1539–43, June 2001.

Bauch, G., and A. Naguib, "Map equalization of space-time coded signals over frequency selective channels," *Proc. IEEE Wireless Commun. Network Conf.*, pp. 261–5, September 1999.

Bello, P. A., "Characterization of randomly time-variant linear channels," *IEEE Trans. Commun. Syst.*, pp. 360–93, December 1963.

Bello, P. A., and B. D. Nelin, "The influence of fading spectrum on the bit error probabilities of incoherent and differentially coherent matched filter receivers," *IEEE Trans. Commun. Syst.*, pp. 160–8, June 1962.

Bello, P. A., and B. D. Nelin, "The effects of frequency selective fading on the binary error probabilities of incoherent and differentially coherent matched filter receivers," *IEEE Trans. Commun. Syst.,* pp. 170–86, June 1963.

Bender, P., P. J. Black, M. S. Grob, R. Padovani, N. T. Sindhushayana, and A. J. Viterbi, "CDMA/HDR: A bandwidth efficient high speed wireless data service for nomadic users," *IEEE Commun. Mag.,* pp. 70–7, July 2000.

Benedetto, S., D. Divsalar, G. Montorsi, and F. Pollara, "Parallel concatenated trellis coded modulation," *Proc. IEEE Internat. Conf. Commun.,* pp. 974–8, June 1996.

Benedetto, S., D. Divsalar, G. Montorsi, and F. Pollara, "Serial concatenation of interleaved codes: Performance analysis, design and iterative decoding," *IEEE Trans. Inform. Theory,* pp. 909–26, May 1998.

Benveniste, A., and M. Goursat, "Blind equalizers," *IEEE Trans. Commun.,* pp. 871–83, August 1984.

Berg, J.-E., R. Bownds, and F. Lotse, "Path loss and fading models for microcells at 900 MHz," *Proc. IEEE Veh. Tech. Conf.,* pp. 666–71, May 1992.

Berggren, F., S.-L. Kim, R. Jantti, and J. Zander, "Joint power control and intracell scheduling of DS-CDMA nonreal time data," *IEEE J. Sel. Areas Commun.,* pp. 1860–70, October 2001.

Bergljung, C., and L. G. Olsson, "Rigorous diffraction theory applied to street microcell propagation," *Proc. IEEE Globecom Conf.,* pp. 1292–6, December 1991.

Bergmans, P. P., "A simple converse for broadcast channels with additive white Gaussian noise," *IEEE Trans. Inform. Theory,* pp. 279–80, March 1974.

Bergmans, P. P., and T. M. Cover, "Cooperative broadcasting," *IEEE Trans. Inform. Theory,* pp. 317–24, May 1974.

Berrou, C., and A. Glavieux, "Near optimum error correcting coding and decoding: Turbo-codes," *IEEE Trans. Commun.,* pp. 1261–71, October 1996.

Berrou, C., A. Glavieux, and P. Thitimajshima, "Near Shannon limit error-correcting coding and decoding: Turbo-codes," *Proc. IEEE Internat. Conf. Commun.,* pp. 54–8, May 1993.

Bertsekas, D., and R. Gallager, *Data Networks,* 2nd ed., Prentice-Hall, Englewood Cliffs, NJ, 1992.

Bhagwat, C. P. P., "Highly dynamic destination-sequenced distance vector routing (DSDV) for mobile computers," *Proc. ACM SIGCOMM,* pp. 234–44, September 1994.

Bharghavan, V., A. Demers, S. Shenkar, and L. Zhang, "MACAW: A media access protocol for wireless LANs," *Proc. ACM SIGCOMM,* vol. 1, pp. 212–25, August 1994.

Biglieri, E., G. Caire, and G. Taricco, "CDMA system design through asymptotic analysis," *IEEE Trans. Commun.,* pp. 1882–96, November 2000.

Billingsley, P., *Probability and Measure,* 3rd ed., Wiley, New York, 1995.

Bingham, J., "Multicarrier modulation for data transmission: An idea whose time has come," *IEEE Commun. Mag.,* pp. 5–14, May 1990.

Blanco, M., and K. Zdunek, "Performance and optimization of switched diversity systems for the detection of signals with Rayleigh fading," *IEEE Trans. Commun.,* pp. 1887–95, December 1979.

Boche, H., and M. Wiczanowski, "Queueing theoretic optimal scheduling for multiple input multiple output multiple access channel," *Proc. IEEE Internat. Sympos. Signal Proc. Inform. Tech.,* pp. 576–9, December 2003.

Börjeson, H., C. Bergljung, and L. G. Olsson, "Outdoor microcell measurements at 1700 MHz," *Proc. IEEE Veh. Tech. Conf.,* pp. 927–31, May 1992.

Brady, P. T., "A statistical analysis of on–off patterns in 16 conversations," *Bell System Tech. J.,* pp. 73–91, January 1968.

Brandenburg, L. H., and A. D. Wyner, "Capacity of the Gaussian channel with memory: The multivariate case," *Bell System Tech. J.,* pp. 745–78, May/June 1974.

Brehler, M., and M. K. Varanasi, "Optimum receivers and low-dimensional spreaded modulation for multiuser space-time communications," *IEEE Trans. Inform. Theory,* pp. 901–18, April 2003.

Buckley, M. E., and S. B. Wicker, "The design and performance of a neural network for predicting decoder error in turbo-coded ARQ protocols," *IEEE Trans. Commun.,* pp. 566–76, April 2000.

Bultitude, R. J. C., and G. K. Bedal, "Propagation characteristics on microcellular urban mobile radio channels at 910 MHz," *IEEE J. Sel. Areas Commun.,* pp. 31–9, January 1989.

Cai, N., and R. W. Yeung, "Network error correction," *Proc. IEEE Inform. Theory Workshop.,* p. 101, June 2003.

Caire, G., and S. Shamai, "On the capacity of some channels with channel state information," *IEEE Trans. Inform. Theory,* pp. 2007–19, September 1999.

Caire, G., and S. Shamai, "On the achievable throughput of a multiantenna Gaussian broadcast channel," *IEEE Trans. Inform. Theory,* pp. 1691–1706, July 2003.

Caire, G., G. Taricco, and E. Biglieri, "Bit-interleaved coded modulation," *IEEE Trans. Inform. Theory,* pp. 927–46, May 1998.

Caire, G., D. Tuninetti, and S. Verdú, "Suboptimality of TDMA in the low-power regime," *IEEE Trans. Inform. Theory,* pp. 608–20, April 2004.

Calderbank, A. R., and N. Seshadri, "Multilevel codes for unequal error protection," *IEEE Trans. Inform. Theory,* pp. 1234–48, July 1993.

Cavers, J. K., "Variable-rate transmission for Rayleigh fading channels," *IEEE Trans. Commun.,* pp. 15–22, February 1972.

Cavers, J. K., "An analysis of pilot symbol assisted modulation for Rayleigh fading channels," *IEEE Trans. Veh. Tech.,* pp. 686–93, November 1991.

Chamberlin, K. C., and R. J. Luebbers, "An evaluation of Longley–Rice and GTD propagation models," *IEEE Trans. Ant. Prop.,* pp. 1093–8, November 1982.

Chan, G. K., "Propagation and coverage prediction for cellular radio systems," *IEEE Trans. Veh. Tech.,* pp. 665–70, November 1991.

Chandrakasan, A., R. Amirtharajah, S. Cho, J. Goodman, G. Konduri, J. Kulik, W. Rabiner, and A. Y. Wang, "Design considerations for distributed microsensor systems," *Proc. IEEE Custom Integrated Circuits Conf.,* pp. 279–86, May 1999.

Chandrakasan, A., and R. W. Brodersen, *Low Power Digital CMOS Design,* Kluwer, Norwell, MA, 1995.

Chang, J. H., and L. Tassiulas, "Energy conserving routing in wireless ad-hoc networks," *Proc. IEEE Infocom Conf.,* pp. 609–19, August 2000.

Chang, L. F., X. Qiu, K. Chawla, and C. Jian, "Providing differentiated services in EGPRS through radio resource management," *Proc. IEEE Internat. Conf. Commun.,* pp. 2296–2301, June 2001.

Chen, A. M., and R. R. Rao, "On tractable wireless channel models," *Proc. Internat. Sympos. Pers., Indoor, Mobile Radio Commun.,* pp. 825–30, September 1998.

Chen, L., U. Yoshida, H. Murata, and S. Hirose, "Dynamic timeslot allocation algorithms suitable for asymmetric traffic in multimedia TDMA/TDD cellular radio," *Proc. IEEE Veh. Tech. Conf.,* pp. 1424–8, May 1998.

Chen, S., and K. Nahrstedt, "Distributed quality-of-service routing in ad hoc networks," *IEEE J. Sel. Areas Commun.,* pp. 1488–1505, August 1999.

Cheng, R., and S. Verdú, "Gaussian multiaccess channels with ISI: Capacity region and multiuser water-filling," *IEEE Trans. Inform. Theory,* pp. 773–85, May 1993.

Chennakeshu, S., and G. J. Saulnier, "Differential detection of $\pi/4$-shifted-DQPSK for digital cellular radio," *IEEE Trans. Veh. Tech.,* pp. 46–57, February 1993.

Cherry, S. M., "WiMax and Wi-Fi: Separate and Unequal," *IEEE Spectrum,* p. 16, March 2004.

Chia, S. T. S., "1700 MHz urban microcells and their coverage into buildings," *Proc. IEEE Ant. Prop. Conf.,* pp. 504–11, York, U.K., April 1991.

Chockalingam, A., and M. Zorzi, "Energy consumption performance of a class of access protocols for mobile data networks," *Proc. IEEE Veh. Tech. Conf.,* pp. 820–4, May 1998.

Chockalingam, A., and M. Zorzi, "Energy efficiency of media access protocols for mobile data networks," *IEEE Trans. Commun.,* pp. 1418–21, November 1998.

Choi, E. H., W. Choi, and J. Andrews, "Throughput of the 1x EV-DO system with various scheduling algorithms," *Proc. Internat. Sympos. Spread Spec. Tech. Appl.,* pp. 359–63, August 2004.

Chow, J. S., J. C. Tu, and J. M. Cioffi, "A discrete multitone transceiver system for HDSL applications," *IEEE J. Sel. Areas Commun.*, pp. 895–908, August 1991.

Chow, P. S., J. M. Cioffi, and John A. C. Bingham, "A practical discrete multitone transceiver loading algorithm for data transmission over spectrally shaped channels," *IEEE Trans. Commun.*, pp. 773–5, February–April 1995.

Chu, M., and W. Stark, "Effect of mobile velocity on communications in fading channels," *IEEE Trans. Veh. Tech.*, pp. 202–10, January 2000.

Chua, S.-G., and A. J. Goldsmith, "Variable-rate variable-power MQAM for fading channels," *IEEE Trans. Commun.*, pp. 1218–30, October 1997.

Chua, S.-G., and A. J. Goldsmith, "Adaptive coded modulation for fading channels," *IEEE Trans. Commun.*, pp. 595–602, May 1998.

Chuah, C.-N., D. N. C. Tse, J. M. Kahn, and R. A. Valenzuela, "Capacity scaling in MIMO wireless systems under correlated fading," *IEEE Trans. Inform. Theory*, pp. 637–50, March 2002.

Chuang, J. C.-I., "The effects of time delay spread on portable radio communications channels with digital modulation," *IEEE J. Sel. Areas Commun.*, pp. 879–89, June 1987.

Chuang, J., and N. Sollenberger, "Beyond 3G: Wideband wireless data access based on OFDM and dynamic packet assignment," *IEEE Commun. Mag.*, pp. 78–87, July 2000.

Chung, S. T., and A. J. Goldsmith, "Degrees of freedom in adaptive modulation: A unified view," *IEEE Trans. Commun.*, pp. 1561–71, September 2001.

Chung, S.-Y., G. D. Forney, T. Richardson, and R. Urbanke, "On the design of low-density parity-check codes within 0.0045 dB of the Shannon limit," *IEEE Commun. Lett.*, pp. 58–60, February 2001.

Cimini, L. J., "Analysis and simulation of a digital mobile channel using orthogonal frequency division multiplexing," *IEEE Trans. Inform. Theory*, pp. 665–75, July 1985.

Cioffi, J. M., "A multicarrier primer," Stanford University/Amati T1E1 contribution, I1E1.4/91–157, November 1991.

Cioffi, J. M., *Digital Communications*, chap. 4: *Multichannel Modulation*, unpublished course notes, available at ⟨http://www.stanford.edu/class/ee379c/⟩.

Cioffi, J. M., G. P. Dudevoir, V. Eyuboglu, and G. D. Forney, Jr., "MMSE decision-feedback equalizers and coding. Part I: Equalization results," *IEEE Trans. Commun.*, pp. 2582–94, October 1995.

Cioffi, J. M., G. P. Dudevoir, V. Eyuboglu, and G. D. Forney, Jr., "MMSE decision-feedback equalizers and coding. Part II: Coding results," *IEEE Trans. Commun.*, pp. 2595–2604, October 1995.

Cioffi, J. M., Jr., and G. D. Forney, "Generalized decision-feedback equalization for packet transmission with ISI and Gaussian noise," in A. Paulraj, V. Roychowdhury, and C. Schaper (Eds.), *Communication, Computation, Control, and Signal Processing*, pp. 79–127, Kluwer, Boston, 1997.

Cioffi, J., and T. Kailath, "Fast, recursive-least-squares transversal filters for adaptive filtering," *IEEE Trans. Signal Proc.*, pp. 304–37, April 1984.

Clark, M. V., V. Erceg, and L. J. Greenstein, "Reuse efficiency in urban microcellular networks," *IEEE Trans. Veh. Tech.*, pp. 279–88, May 1997.

Clarke, R. H., "A statistical theory of mobile radio reception," *Bell System Tech. J.*, pp. 957–1000, July/August 1968.

Corson, M., R. Laroia, A. O'Neill, V. Park, and G. Tsirtsis, "A new paradigm for IP-based cellular networks," *IT Professional*, pp. 20–9, November/December 2001.

Costa, M., "Writing on dirty paper," *IEEE Trans. Inform. Theory*, pp. 439–41, May 1983.

Costa, M. H. M., and A. A. El Gamal, "The capacity region of the discrete memoryless interference channel with strong interference," *IEEE Trans. Inform. Theory*, pp. 710–11, September 1987.

Cover, T., and J. Thomas, *Elements of Information Theory*, Wiley, New York, 1991.

Cowley, W., and L. Sabel, "The performance of two symbol timing recovery algorithms for PSK demodulators," *IEEE Trans. Commun.*, pp. 2345–55, June 1994.

Cox, D. C., "Wireless network access for personal communications," *IEEE Commun. Mag.*, pp. 96–115, December 1992.

Cox, D. C., "Wireless personal communications: What is it?" *IEEE Pers. Commun. Mag.,* pp. 20–35, April 1995.

Cox, R. V., J. Hagenauer, N. Seshadri, and C.-E. W. Sundberg, "Variable rate sub-band speech coding and matched convolutional channel coding for mobile radio channels," *IEEE Trans. Signal Proc.,* pp. 1717–31, August 1991.

Craig, J., "New, simple and exact result for calculating the probability of error for two-dimensional signal constellations," *Proc. Military Commun. Conf.,* pp. 25.5.1–25.5.5, November 1991.

Cruz, R. L., and A. V. Santhanam, "Optimal routing, link scheduling and power control in multihop wireless networks," *Proc. IEEE Infocom Conf.,* pp. 702–11, April 2003.

Csiszár, I., and J. Kórner, *Information Theory: Coding Theorems for Discrete Memoryless Channels,* Academic Press, New York, 1981.

Csiszár, I., and P. Narayan, "The capacity of the arbitrarily varying channel," *IEEE Trans. Inform. Theory,* pp. 18–26, January 1991.

Cui, S., and A. J. Goldsmith, "Energy efficient routing using cooperative MIMO techniques," *Proc. IEEE Internat. Conf. Acous., Speech, Signal Proc.,* pp. 805–8, March 2005.

Cui, S., A. J. Goldsmith, and A. Bahai, "Energy-efficiency of MIMO and cooperative MIMO in sensor networks," *IEEE J. Sel. Areas Commun.,* pp. 1089–98, August 2004.

Cui, S., A. J. Goldsmith, and A. Bahai, "Energy-constrained modulation optimization," *IEEE Trans. Wireless Commun.,* September 2005.

Cui, S., R. Madan, A. J. Goldsmith, and S. Lall, "Joint routing, MAC, and link layer optimization in sensor networks with energy constraints," *Proc. IEEE Internat. Conf. Commun.,* May 2005.

Dai, H., and H. V. Poor, "Iterative space-time processing for multiuser detection in multipath CDMA channels," *IEEE Trans. Signal Proc.,* pp. 2116–27, September 2002.

Damen, M. O., H. El Gamal, and N. C. Beaulieu, "Linear threaded algebraic space-time constellations," *IEEE Trans. Inform. Theory,* pp. 2372–88, October 2003.

Davenport, W. B., Jr., and W. L. Root, *An Introduction to the Theory of Random Signals and Noise,* McGraw-Hill, New York, 1987.

Davey, M. C., and D. MacKay, "Low density parity-check codes over GF(q)," *IEEE Commun. Lett.,* pp. 165–7, June 1998.

Dehghan, S., and R. Steele, "Small cell city," *IEEE Commun. Mag.,* pp. 52–9, August 1997.

Devasirvathan, D. M. J., R. R. Murray, and D. R. Woiter, "Time delay spread measurements in a wireless local loop test bed," *Proc. IEEE Veh. Tech. Conf.,* pp. 241–5, May 1995.

Diggavi, S., "Analysis of multicarrier transmission in time-varying channels," *Proc. IEEE Internat. Conf. Commun.,* pp. 1191–5, June 1997.

Diggavi, S. N., N. Al-Dhahir, and A. R. Calderbank, "Multiuser joint equalization and decoding of space-time codes," *Proc. IEEE Internat. Conf. Commun.,* pp. 2643–7, May 2003.

Diggavi, S. N., N. Al-Dhahir, and A. R. Calderbank, "On interference cancellation and high-rate space-time codes," *Proc. IEEE Internat. Sympos. Inform. Theory,* p. 238, June 2003.

Dinan, E. H., and B. Jabbari, "Spreading codes for direct sequence CDMA and wideband CDMA cellular networks," *IEEE Commun. Mag.,* pp. 48–54, September 1998.

Divsalar, D., M. K. Simon, and D. Raphaeli, "Improved parallel interference cancellation for CDMA," *IEEE Trans. Commun.,* pp. 258–68, February 1998.

Dixon, R. C., *Spread Spectrum Systems with Commercial Applications,* 3rd ed., Wiley, New York, 1994.

Domazetovic, A., L. J. Greenstein, N. Mandayan, and I. Seskar, "A new modeling approach for wireless channels with predictable path geometries," *Proc. IEEE Veh. Tech. Conf.,* September 2002.

Douillard, C., M. Jezequel, C. Berrou, A. Picart, P. Didier, and A. Glavieux, "Iterative correction of intersymbol interference: Turbo equalization," *Euro. Trans. Telecommun.,* pp. 507–11, September/October 1995.

Dousse, O., P. Thiran, and M. Hasler, "Connectivity in ad-hoc and hybrid networks," *Proc. IEEE Infocom Conf.,* pp. 1079–88, June 2002.

Duel-Hallen, A., "Decorrelating decision-feedback multiuser detector for synchronous CDMA," *IEEE Trans. Commun.*, pp. 285–90, February 1993.

Duel-Hallen, A., "A family of multiuser decision-feedback detectors for asynchronous code-division multiple-access channels," *IEEE Trans. Commun.*, pp. 421–34, February–April 1995.

Duel-Hallen, A., J. Holtzman, and Z. Zvonar, "Multiuser detection for CDMA systems," *IEEE Pers. Commun. Mag.*, pp. 46–58, April 1995.

Duel-Hallen, A., S. Hu, and H. Hallen, "Long-range prediction of fading signals," *IEEE Signal Proc. Mag.*, pp. 62–75, May 2000.

Durgin, G., T. S. Rappaport, and H. Xu, "Partition-based path loss analysis for in-home and residential areas at 5.85 GHz," *Proc. IEEE Globecom Conf.*, pp. 904–9, November 1998.

Eklund, C., R. B. Marks, K. L. Stanwood, and S. Wang, "IEEE Standard 802.16: A technical overview of the WirelessMAN 326 air interface for broadband wireless access," *IEEE Commun. Mag.*, pp. 98–107, June 2002.

ElBatt, T., and A. Ephremides, "Joint scheduling and power control for wireless ad hoc networks," *IEEE Trans. Wireless Commun.*, pp. 74–85, January 2004.

El Gamal, A., J. Mammen, B. Prabhakar, and D. Shah, "Throughput–delay trade-off in wireless networks," *Proc. IEEE Infocom Conf.*, pp. 464–75, March 2004.

El Gamal, H., and E. Geraniotis, "Comparing the capacities of FH/SSMA and DS/CDMA networks," *Proc. Internat. Sympos. Pers., Indoor, Mobile Radio Commun.*, pp. 769–73, September 1998.

El Gamal, H., and M. O. Damen, "Universal space-time coding," *IEEE Trans. Inform. Theory,* pp. 1097–1119, May 2003.

Ephremides, A., "Energy concerns in wireless networks," *IEEE Wireless Commun. Mag.*, pp. 48–59, August 2002.

Ephremides, A., J. E. Wieselthier, and D. J. Baker, "A design concept for reliable mobile radio networks with frequency hopping signaling," *Proc. IEEE,* pp. 56–73, January 1987.

Erceg, V., L. J. Greenstein, S. Y. Tjandra, S. R. Parkoff, A. Gupta, B. Kulic, A. A. Julius, and R. Bianchi, "An empirically based path loss model for wireless channels in suburban environments," *IEEE J. Sel. Areas Commun.*, pp. 1205–11, July 1999.

Erceg, V., A. J. Rustako, and R. S. Roman, "Diffraction around corners and its effects on the microcell coverage area in urban and suburban environments at 900 MHz, 2 GHz, and 4 GHz," *IEEE Trans. Veh. Tech.*, pp. 762–6, August 1994.

Erez, U., S. Shamai, and R. Zamir, "Capacity and lattice strategies for cancelling known interference," *Proc. Internat. Sympos. Inform. Theory Appl.*, pp. 681–4, November 2000.

Ertel, R., P. Cardieri, K. W. Sowerby, T. Rappaport, and J. H. Reed, "Overview of spatial channel models for antenna array communication systems," *IEEE Pers. Commun. Mag.*, pp. 10–22, February 1998.

Etkin, R., and D. Tse, "Degrees of freedom in underspread MIMO fading channels," *Proc. IEEE Internat. Sympos. Inform. Theory,* p. 323, July 2003.

European Cooperative in the Field of Science and Technical Research EURO-COST 231, "Urban transmission loss models for mobile radio in the 900 and 1800 MHz bands," rev. 2, The Hague, September 1991.

Everitt, D., and D. Manfield, "Performance analysis of cellular mobile communication systems with dynamic channel assignment," *IEEE J. Sel. Areas Commun.*, pp. 1172–81, October 1989.

Eyuboglu, M. V., "Detection of coded modulation signals on linear, severely distorted channels using decision-feedback noise prediction with interleaving," *IEEE Trans. Commun.*, pp. 401–9, April 1988.

Farvardin, N., and V. Vaishampayan, "Optimal quantizer design for noisy channels: An approach to combined source-channel coding," *IEEE Trans. Inform. Theory,* pp. 827–38, November 1987.

Farvardin, N., and V. Vaishampayan, "On the performance and complexity of channel-optimized vector quantizers," *IEEE Trans. Inform. Theory,* pp. 155–60, January 1991.

Feller, W., *An Introduction to Probability Theory and Its Applications,* vols. I and II, Wiley, New York, 1968 and 1971.

Feuerstein, M., K. Blackard, T. Rappaport, S. Seidel, and H. Xia, "Path loss, delay spread, and outage models as functions of antenna height for microcellular system design," *IEEE Trans. Veh. Tech.,* pp. 487–98, August 1994.

Filip, M., and E. Vilar, "Implementation of adaptive modulation as a fade countermeasure," *Internat. J. Sat. Commun.,* pp. 181–91, 1994.

Fischer, T. R., and M. W. Marcellin, "Joint trellis coded quantization/modulation," *IEEE Trans. Commun.,* pp. 172–6, February 1991.

Fitz, M., "Further results in the unified analysis of digital communication systems," *IEEE Trans. Commun.,* pp. 521–32, March 1992.

Forney, G. D., "Burst error correcting codes for the classic bursty channel," *IEEE Trans. Commun. Tech.,* pp. 772–81, October 1971.

Forney, G. D., Jr., "Maximum-likelihood sequence estimation of digital sequences in the presence of intersymbol interference," *IEEE Trans. Inform. Theory,* pp. 363–78, May 1972.

Forney, G. D., "Coset codes, I: Introduction and geometrical classification, and II: Binary lattices and related codes," *IEEE Trans. Inform. Theory,* pp. 1123–87, September 1988.

Forney, G. D., Jr., R. G. Gallager, G. R. Lang, F. M. Longstaff, and S. U. Quereshi, "Efficient modulation for band-limited channels," *IEEE J. Sel. Areas Commun.,* pp. 632–47, September 1984.

Forney, G. D., Jr., and L.-F. Wei, "Multidimensional constellations – Part I: Introduction, figures of merit, and generalized cross constellations," *IEEE J. Sel. Areas Commun.,* pp. 877–92, August 1989.

Foschini, G. J., "Layered space-time architecture for wireless communication in fading environments when using multi-element antennas," *Bell System Tech. J.,* pp. 41–59, Autumn 1996.

Foschini, G. J., D. Chizhik, M. Gans, C. Papadias, and R. A. Valenzuela, "Analysis and performance of some basic space-time architectures," *IEEE J. Sel. Areas Commun.,* pp. 303–20, April 2003.

Foschini, G. J., and M. Gans, "On limits of wireless communications in a fading environment when using multiple antennas," *Wireless Pers. Commun.,* pp. 311–35, March 1998.

Foschini, G., G. Golden, R. Valenzuela, and P. Wolniansky, "Simplified processing for high spectral efficiency wireless communication employing multi-element arrays," *IEEE J. Sel. Areas Commun.,* pp. 1841–52, November 1999.

Foschini, G. J., and Z. Miljanic, "A simple distributed autonomous power control algorithm and its convergence," *IEEE Trans. Veh. Tech.,* pp. 641–6, November 1993.

Foschini, G. J., and J. Salz, "Digital communications over fading radio channels," *Bell System Tech. J.,* pp. 429–56, February 1983.

Fossorier, M., "Iterative reliability-based decoding of low-density parity check codes," *IEEE J. Sel. Areas Commun.,* pp. 908–17, May 2001.

Fragouli, C., N. Al-Dhahir, and S. Diggavi, "Pre-filtered space-time M-BCJR equalizer for frequency selective channels," *IEEE Trans. Commun.,* pp. 742–53, May 2002.

Fragouli, C., and R. D. Wesel, "Turbo-encoder design for symbol-interleaved parallel concatenated trellis-coded modulation," *IEEE Trans. Commun.,* pp. 425–35, March 2001.

Franks, L. E., "Carrier and bit synchronization in data communication – A tutorial review," *IEEE Trans. Commun.,* pp. 1107–21, August 1980.

Frodigh, M., S. Parkvall, C. Roobol, P. Johansson, and P. Larsson, "Future-generation wireless networks," *IEEE Wireless Commun. Mag.,* pp. 10–17, October 2001.

Fung, V., R. S. Rappaport, and B. Thoma, "Bit error simulation for $\pi/4$ DQPSK mobile radio communication using two-ray and measurement based impulse response models," *IEEE J. Sel. Areas Commun.,* pp. 393–405, April 1993.

Furuskar, A., S. Mazur, F. Muller, and H. Olofsson, "EDGE: Enhanced data rates for GSM and TDMA/136 evolution," *IEEE Wireless Commun. Mag.,* pp. 56–66, June 1999.

Gallager, R. G., "Low-density parity-check codes," *IRE Trans. Inform. Theory*, pp. 21–8, January 1962.

Gallager, R. G., *Information Theory and Reliable Communication*, Wiley, New York, 1968.

Gallager, R. G., "Energy limited channels: Coding, multiaccess and spread spectrum," *Proc. Inform. Syst. Sci. Conf.*, p. 372, March 1988.

Gallager, R. G., *Discrete Stochastic Processes*, Kluwer, Dordrecht, 1996.

Gamal, H., G. Caire, and M. Damon, "Lattice coding and decoding achieve the optimal diversity-multiplexing trade-off of MIMO channels," *IEEE Trans. Inform. Theory*, pp. 968–85, June 2004.

Gamal, H., and A. Hammons, "On the design of algebraic space-time codes for MIMO block-fading channels," *IEEE Trans. Inform. Theory*, pp. 151–63, January 2003.

Garey, M. R., and D. S. Johnson, *Computers and Intractability: A Guide to the Theory of NP-Completeness*, Freeman, New York, 1979.

Gastpar, M., and M. Vetterli, "On the capacity of wireless networks: The relay case," *Proc. IEEE Infocom Conf.*, pp. 1577–86, June 2002.

Gertsman, M. J., and J. H. Lodge, "Symbol-by-symbol MAP demodulation of CPM and PSK signals on Rayleigh flat-fading channels," *IEEE Trans. Commun.*, pp. 788–99, July 1997.

Gesbert, D., M. Shafi, D.-S. Shiu, P. Smith, and A. Naguib, "From theory to practice: An overview of MIMO space-time coded wireless systems," *IEEE J. Sel. Areas Commun.*, pp. 281–302, April 2003.

Ghassemzadeh, S. S., L. J. Greenstein, A. Kavcic, T. Sveinsson, and V. Tarokh, "Indoor path loss model for residential and commercial buildings," *Proc. IEEE Veh. Tech. Conf.*, pp. 3115–19, October 2003.

Ghosh, A., L. Jalloul, B. Love, M. Cudak, and B. Classon, "Air-interface for 1XTREME/1xEV-DV," *Proc. IEEE Veh. Tech. Conf.*, pp. 2474–8, May 2001.

Giannakis, G. B., Y. Hua, P. Stoica, and L. Tong, *Signal Processing Advances in Wireless and Mobile Communications: Trends in Single- and Multi-user Systems*, Prentice-Hall, New York, 2001.

Gibbs, W. W., "As we may live," *Scientific American*, pp. 36, 40, November 2000.

Gilhousen, K. S., I. M. Jacobs, R. Padovani, A. J. Viterbi, L. A. Weaver, Jr., and C. E. Wheatley III, "On the capacity of a cellular CDMA system," *IEEE Trans. Veh. Tech.*, pp. 303–12, May 1991.

Girko, V. L., "A refinement of the central limit theorem for random determinants," *Theory Probab. Appl.*, 42(1), pp. 121–9, 1998.

Goeckel, D. L., "Adaptive coding for time-varying channels using outdated fading estimates," *IEEE Trans. Commun.*, pp. 844–55, June 1999.

Gold, R., "Optimum binary sequences for spread-spectrum multiplexing," *IEEE Trans. Inform. Theory*, pp. 619–21, October 1967.

Goldsmith, A. J., and M. Effros, "Joint design of fixed-rate source codes and multiresolution channel codes," *IEEE Trans. Commun.*, pp. 1301–12, October 1998.

Goldsmith, A. J., and M. Effros, "The capacity region of broadcast channels with intersymbol interference and colored Gaussian noise," *IEEE Trans. Inform. Theory*, pp. 219–40, January 2001.

Goldsmith, A. J., and L. J. Greenstein, "A measurement-based model for predicting coverage areas of urban microcells," *IEEE J. Sel. Areas Commun.*, pp. 1013–23, September 1993.

Goldsmith, A. J., and L. J. Greenstein, "Effect of average power estimation error on adaptive MQAM modulation," *Proc. IEEE Internat. Conf. Commun.*, pp. 1105–9, June 1997.

Goldsmith, A. J., L. J. Greenstein, and G. J. Foschini, "Error statistics of real-time power measurements in cellular channels with multipath and shadowing," *IEEE Trans. Veh. Tech.*, pp. 439–46, August 1994.

Goldsmith, A. J., S. A. Jafar, N. Jindal, and S. Vishwanath, "Capacity limits of MIMO channels," *IEEE J. Sel. Areas Commun.*, pp. 684–701, June 2003.

Goldsmith, A., and M. Medard, "Capacity of time-varying channels with channel side information," *IEEE Trans. Inform. Theory* (to appear).

Goldsmith, A. J., and P. P. Varaiya, "Capacity, mutual information, and coding for finite-state Markov channels," *IEEE Trans. Inform. Theory,* pp. 868–86, May 1996.

Goldsmith, A. J., and P. P. Varaiya, "Capacity of fading channels with channel side information," *IEEE Trans. Inform. Theory,* pp. 1986–92, November 1997.

Goldsmith, A. J., and S. B. Wicker, "Design challenges for energy-constrained ad hoc wireless networks," *IEEE Wireless Commun. Mag.,* pp. 8–27, August 2002.

Golomb, S. W., *Shift Register Sequences,* Holden-Day, San Francisco, 1967.

Goodman, D. J., R. A. Valenzuela, K. T. Gayliard, and B. Ramamurthi, "Packet reservation multiple access for local wireless communications," *IEEE Trans. Commun.,* pp. 885–90, August 1989.

Graham, S. R., G. Baliga, and P. R. Kumar, "Issues in the convergence of control with communication and computing: Proliferation, architecture, design, services, and middleware," *Proc. IEEE Conf. Decision Control,* pp. 1466–71, December 2004.

Grandhi, S. A., R. Vijayan, and D. J. Goodman, "Distributed power control in cellular radio systems," *IEEE Trans. Commun.,* pp. 226–8, February–April 1994.

Grandhi, S. A., R. D. Yates, and D. J. Goodman, "Resource allocation for cellular radio systems," *IEEE Trans. Veh. Tech.,* pp. 581–7, August 1997.

Grant, A., "Rayleigh fading multiple-antenna channels," *J. Appl. Signal Proc.,* Special Issue on Space-Time Coding (Part I), pp. 316–29, March 2002.

Grant, S. J., and J. K. Cavers, "System-wide capacity increase for narrowband cellular systems through multiuser detection and base station diversity arrays," *IEEE Trans. Wireless Commun.,* pp. 2072–82, November 2004.

Gray, R. M., and L. D. Davisson, *Random Processes: A Mathematical Approach for Engineers,* Prentice-Hall, Englewood Cliffs, NJ, 1986.

Grayver, E., and B. Daneshrad, "A low-power all-digital FSK receiver for deep space applications," *IEEE Trans. Commun.,* pp. 911–21, May 2001.

Greenstein, L. G., J. B. Andersen, H. L. Bertoni, S. Kozono, and D. G. Michelson, Eds., *IEEE J. Sel. Areas Commun.,* Special Issue on Channel and Propagation Modeling for Wireless Systems Design, August 2002.

Grimm, J., M. Fitz, and J. Korgmeier, "Further results in space-time coding for Rayleigh fading," *Proc. Allerton Conf. Commun., Control, Comput.,* pp. 391–400, September 1998.

Grossglauser, M., and D. N. C. Tse, "Mobility increases the capacity of ad-hoc wireless networks," *IEEE/ACM Trans. Network.,* pp. 1877–94, August 2002.

Guan, Y. L., and L. F. Turner, "Generalised FSMC model for radio channels with correlated fading," *IEE Proc. Commun.,* pp. 133–7, April 1999.

Gudmundson, M., "Correlation model for shadow fading in mobile radio systems," *Elec. Lett.,* pp. 2145–6, November 7, 1991.

Gudmundson, M., "Generalized frequency hopping in mobile radio systems," *Proc. IEEE Veh. Tech. Conf.,* pp. 788–91, May 1993.

Guey, J.-C., M. P. Fitz, M. Bell, and W.-Y. Kuo, "Signal design for transmitter diversity wireless communication systems over Rayleigh fading channels," *IEEE Trans. Commun.,* pp. 527–37, April 1999.

Gulati, V., and K. R. Narayanan, "Concatenated codes for fading channels based on recursive space-time trellis codes," *IEEE Trans. Wireless Commun.,* pp. 118–28, January 2003.

Gundmundson, B., J. Sköld, and J. K. Ugland, "A comparison of CDMA and TDMA systems," *Proc. IEEE Veh. Tech. Conf.,* pp. 732–5, May 1992.

Gupta, P., and P. R. Kumar, "The capacity of wireless networks," *IEEE Trans. Inform. Theory,* pp. 388–404, March 2000.

Gupta, P., and P. R. Kumar, "Towards an information theory of large networks: An achievable rate region," *IEEE Trans. Inform. Theory,* pp. 1877–94, August 2003.

Gurunathan, S., and K. Feher, "Multipath simulation models for mobile radio channels," *Proc. IEEE Veh. Tech. Conf.,* pp. 131–4, May 1992.

Haartsen, J., "The Bluetooth radio system," *IEEE Pers. Commun. Mag.,* pp. 28–36, February 2000.

Haartsen, J., and S. Mattisson, "Bluetooth: A new low-power radio interface providing short-range connectivity," *Proc. IEEE,* pp. 1651–61, October 2000.

Haas, Z. J., J. Deng, and S. Tabrizi, "Collision-free medium access control scheme for ad hoc networks, *Proc. Military Commun. Conf.,* pp. 276–80, 1999.

Hagenauer, J., "Rate-compatible punctured convolutional codes (RCPC codes) and their applications," *IEEE Trans. Commun.,* pp. 389–400, April 1988.

Hall, E. K., and S. G. Wilson, "Design and analysis of turbo codes on Rayleigh fading channels," *IEEE J. Sel. Areas Commun.,* pp. 160–74, February 1998.

Hanly, S., and D. Tse, "Multiaccess fading channels – Part II: Delay-limited capacities," *IEEE Trans. Inform. Theory,* pp. 2816–31, November 1998.

Hanly, S. V., and P. Whiting, "Information theory and the design of multi-receiver networks," *Proc. Internat. Sympos. Spread Spec. Tech. Appl.,* pp. 103–6, November 1992.

Hara, Y., T. Nabetani, and S. Hara, "Performance evaluation of cellular SDMA/TDMA systems with variable bit rate multimedia traffic," *Proc. IEEE Veh. Tech. Conf.,* pp. 1731–4, October 2001.

Harashima, H., and H. Miyakawa, "Matched-transmission techniques for channels with intersymbol interference," *IEEE Trans. Commun.,* pp. 774–80, August 1972.

Hardy, Q., "Are claims hope or hype?" *Wall Street Journal,* p. A1, September 6, 1996.

Harley, P., "Short distance attenuation measurements at 900 MHz and 1.8 GHz using low antenna heights for microcells," *IEEE J. Sel. Areas Commun.,* pp. 5–11, January 1989.

Hata, M., "Empirical formula for propagation loss in land mobile radio services," *IEEE Trans. Veh. Tech.,* pp. 317–25, August 1980.

Hayes, J. F., "Adaptive feedback communications," *IEEE Trans. Commun. Tech.,* pp. 29–34, February 1968.

Haykin, S., *An Introduction to Analog and Digital Communications,* Wiley, New York, 1989.

Haykin, S., *Communication Systems,* Wiley, New York, 2002.

Haykin, S., and M. Moher, *Modern Wireless Communications,* Prentice-Hall, Englewood Cliffs, NJ, 2005.

Heath, J., R. W. M. Airy, and A. Paulraj, "Multiuser diversity for MIMO wireless systems with linear receivers," *Proc. Asilomar Conf. Signals, Syst., Comput.,* pp. 1194–9, November 2001.

Heath, R. W., Jr., and D. J. Love, "Multi-mode antenna selection for spatial multiplexing with linear receivers," *IEEE Trans. Signal Proc.* (to appear).

Heath, R. W., Jr., and A. J. Paulraj, "Switching between multiplexing and diversity based on constellation distance," *Proc. Allerton Conf. Commun., Control, Comput.,* pp. 212–21, October 2000.

Heegard, C., and S. B. Wicker, *Turbo Coding,* Kluwer, Boston, 1999.

Heinzelman, W. R., A. Sinha, and A. P. Chandrakasan, "Energy-scalable algorithms and protocols for wireless microsensor networks," *Proc. IEEE Internat. Conf. Acous., Speech, Signal Proc.,* pp. 3722–5, June 2000.

Hinedi, S., M. Simon, and D. Raphaeli, "The performance of noncoherent orthogonal M-FSK in the presence of timing and frequency errors," *IEEE Trans. Commun.,* pp. 922–33, February–April 1995.

Hirt, W., and J. L. Massey, "Capacity of the discrete-time Gaussian channel with intersymbol interference," *IEEE Trans. Inform. Theory,* pp. 380–8, May 1988.

Ho, K.-P., and J. M. Kahn, "Combined source-channel coding using channel-optimized quantizer and multicarrier modulation," *Proc. IEEE Internat. Conf. Commun.,* pp. 1323–7, June 1996.

Ho, K.-P., and J. M. Kahn, "Transmission of analog signals using multicarrier modulation: A combined source-channel coding approach," *IEEE Trans. Commun.,* pp. 1432–43, November 1996.

Hochwald, B., and T. Marzetta, "Unitary space-time modulation for multiple-antenna communications in Rayleigh flat fading," *IEEE Trans. Inform. Theory,* pp. 543–64, March 2000.

Hochwald, B., and V. Tarokh, "Multiple-antenna channel hardening and its implications for rate feedback and scheduling," *IEEE Trans. Inform. Theory,* pp. 1893–1909, September 2004.

Hoeher, P., S. Kaiser, and P. Robertson, "Two-dimensional pilot-symbol-aided channel estimation by Wiener filtering," *Proc. IEEE Internat. Conf. Acous., Speech, Signal Proc.*, pp. 1845–8, April 1997.

Holliday, T., N. Bambos, A. J. Goldsmith, and P. Glynn, "Distributed power control for time varying wireless networks: Optimality and convergence," *Proc. Allerton Conf. Commun., Control, Comput.*, pp. 1024–33, October 2003.

Holliday, T., and A. Goldsmith, "Joint source and channel coding for MIMO systems: Is it better to be robust or quick," *Proc. Joint Workshop Commun. Cod.*, p. 24, October 2004.

Holliday, T., A. Goldsmith, and P. Glynn, "Capacity of finite state Markov channels with general inputs," *Proc. IEEE Internat. Sympos. Inform. Theory*, p. 289, July 2003.

Holma, H., and A. Toskala, *WCDMA for UMTS Radio Access for Third Generation Mobile Communications*, 3rd ed., Wiley, New York, 2004.

Holsinger, J. L., "Digital communication over fixed time-continuous channels with memory, with special application to telephone channels," MIT Res. Lab Elec. Tech. Rep. 430, 1964.

Honig, M. L., and H. V. Poor, "Adaptive interference suppression," in *Wireless Communications: Signal Processing Perspectives*, ch. 2, Prentice-Hall, Englewood Cliffs, NJ, 1998.

Hoppe, R., G. Wölfle, and F. M. Landstorfer, "Measurement of building penetration loss and propagation models for radio transmission into buildings," *Proc. IEEE Veh. Tech. Conf.*, pp. 2298–2302, April 1999.

Horn, R. A., and C. R. Johnson, *Matrix Analysis*, Cambridge University Press, 1985.

Horn, R. A., and C. R. Johnson, *Topics in Matrix Analysis*, Cambridge University Press, 1991.

Host-Madsen, A., "A new achievable rate for cooperative diversity based on generalized writing on dirty paper," *Proc. IEEE Internat. Sympos. Inform. Theory*, p. 317, June 2003.

Host-Madsen, A., "On the achievable rate for receiver cooperation in ad-hoc networks," *Proc. IEEE Internat. Sympos. Inform. Theory*, p. 272, June 2004.

Hou, J., P. H. Siegel, and L. B. Milstein, "Performance analysis and code optimization of low-density parity-check codes on Rayleigh fading channels," *IEEE J. Sel. Areas Commun.*, pp. 924–34, May 2001.

Hou, J., P. Siegel, L. Milstein, and H. D. Pfister, "Capacity-approaching bandwidth efficient coded modulation schemes based on low-density parity-check codes," *IEEE Trans. Inform. Theory*, pp. 2141–55, September 2003.

Hu, T. H., and M. M. K. Liu, "A new power control function for multirate DS-CDMA systems," *IEEE Trans. Commun.* pp. 896–904, June 1999.

Huhns, M. N., "Networking embedded agents," *IEEE Internet Comput.*, pp. 91–3, January/February 1999.

IEEE 802.11a-1999: *High-Speed Physical Layer in the 5 GHz Band*, 1999.

IEEE 802.11g-2003: *Further Higher-Speed Physical Layer Extension in the 2.4 GHz Band*, 2003.

IEEE 802.16a-2001: *IEEE Recommended Practice for Local and Metropolitan Area Networks*, 2001.

IEEE J. Sel. Areas Commun., Special Issue on Channel and Propagation Modeling for Wireless Systems Design, April 2002 and August 2002.

IEEE J. Sel. Areas Commun., Special Issue on Ultra-Wideband Radio in Multiaccess Wireless Communications, December 2002.

IEEE Standard for Wireless LAN Medium Access Control (MAC) and Physical Layer (PHY) Specifications, IEEE Standard 802.11, 1997.

IEEE Trans. Inform. Theory, Special Issue on Codes and Graphs and Iterative Algorithms, February 2001.

IEEE Wireless Commun. Mag., Special Issue on Energy Aware Ad Hoc Wireless Networks (A. J. Goldsmith and S. B. Wicker, Eds.), August 2002.

Ikegami, F., S. Takeuchi, and S. Yoshida, "Theoretical prediction of mean field strength for urban mobile radio," *IEEE Trans. Ant. Prop.*, pp. 299–302, March 1991.

Imai, H., and S. Hirakawa, "A new multilevel coding method using error correcting codes," *IEEE Trans. Inform. Theory*, pp. 371–7, May 1977.

Irvine, G. T., and P. J. Mclane, "Symbol-aided plus decision-directed reception for PSK TCM modulation on shadowed mobile satellite fading channels," *IEEE J. Sel. Areas Commun.*, pp. 1289–99, October 1992.

Jafar, S. A., G. J. Foschini, and A. J. Goldsmith, "PhantomNet: Exploring optimal multicellular multiple antenna systems," *J. Appl. Signal Proc.* (EURASIP), pp. 591–605, May 2004.

Jafar, S. A., and A. J. Goldsmith, "Transmitter optimization and optimality of beamforming for multiple antenna systems," *IEEE Trans. Wireless Commun.*, pp. 1165–75, July 2004.

Jafar, S. A., and A. J. Goldsmith, "Multiple-antenna capacity in correlated Rayleigh fading with channel covariance information," *IEEE Trans. Wireless Commun.*, pp. 990–7, May 2005.

Jafarkhani, H., P. Ligdas, and N. Farvardin, "Adaptive rate allocation in a joint source/channel coding framework for wireless channels," *Proc. IEEE Veh. Tech. Conf.*, pp. 492–6, April 1996.

Jain, R., A. Puri, and R. Sengupta, "Geographical routing using partial information for wireless ad hoc networks," *IEEE Pers. Commun. Mag.*, pp. 48–57, February 2001.

Jakes, W. C., Jr., *Microwave Mobile Communications*, Wiley, New York, 1974 [reprinted by IEEE Press].

Jalali, A., R. Padovani, and R. Pankaj, "Data throughput of CDMA-HDR a high efficiency–high data rate personal communication wireless system," *Proc. IEEE Veh. Tech. Conf.*, pp. 1854–8, May 2000.

Jamali, S. H., and T. Le-Ngoc, *Coded-Modulation Techniques for Fading Channels*, Kluwer, New York, 1994.

Janani, M., A. Hedayat, T. E. Hunter, and A. Nosratinia, "Coded cooperation in wireless communications: Space-time transmission and iterative decoding," *IEEE Trans. Signal Proc.*, pp. 362–71, February 2004.

Jin, H., and R. J. McEliece, "Coding theorems for turbo code ensembles," *IEEE Trans. Inform. Theory*, pp. 1451–61, June 2002.

Jindal, N., and A. J. Goldsmith, "Capacity and optimal power allocation for fading broadcast channels with minimum rates," *IEEE Trans. Inform. Theory*, pp. 2895–2909, November 2003.

Jindal, N., and A. Goldsmith, "DPC vs. TDMA for MIMO broadcast channels," *IEEE Trans. Inform. Theory*, pp. 1783–94, May 2005.

Jindal, N., U. Mitra, and A. Goldsmith, "Capacity of ad-hoc networks with node cooperation," *Proc. IEEE Internat. Sympos. Inform. Theory*, p. 271, June 2004.

Jindal, N., W. Rhee, S. Vishwanath, S. A. Jafar, and A. J. Goldsmith, "Sum power iterative waterfilling for multi-antenna Gaussian broadcast channels," *IEEE Trans. Inform. Theory*, pp. 1570–9, April 2005.

Jindal, N., S. Vishwanath, and A. J. Goldsmith, "On the duality of Gaussian multiple-access and broadcast channels," *IEEE Trans. Inform. Theory*, pp. 768–83, May 2004.

Johansson, M., L. Xiao, and S. P. Boyd, "Simultaneous routing and power allocation in CDMA wireless data networks," *Proc. IEEE Internat. Conf. Commun.*, pp. 51–5, May 2003.

Johnson, C. R., Jr., "Admissibility in blind adaptive channel equalization," *IEEE Control Syst. Mag.*, pp. 3–15, January 1991.

Johnson, D. B., and D. A. Maltz, "Dynamic source routing in ad hoc wireless networks," in T. Imielinsky and H. Korth, Eds., *Mobile Computing*, Kluwer, Dordrecht, 1996.

Johnson, R., P. Schniter, T. J. Endres, J. D. Behm, D. R. Brown, and R. A. Casas, "Blind equalization using the constant modulus criterion: A review," *Proc. IEEE*, pp. 1927–50, October 1998.

Jorswieck, E., and H. Boche, "Channel capacity and capacity-range of beamforming in MIMO wireless systems under correlated fading with covariance feedback," *IEEE Trans. Wireless Commun.*, pp. 1543–53, September 2004.

Jubin, J., and J. D. Tornow, "The DARPA packet radio network protocols," *Proc. IEEE*, pp. 21–32, January 1987.

Julian, D.; M. Chiang, D. O'Neill, and S. Boyd, "QoS and fairness constrained convex optimization of resource allocation for wireless cellular and ad hoc networks," *Proc. IEEE Infocom Conf.*, pp. 477–86, June 2002.

Jung, P., P. W. Baier, and A. Steil, "Advantages of CDMA and spread spectrum techniques over FDMA and TDMA in cellular mobile radio applications," *IEEE Trans. Veh. Tech.*, pp. 357–64, August 1993.

Jungnickel, V., T. Haustein, V. Pohl, and C. Von Helmolt, "Link adaptation in a multi-antenna system," *Proc. IEEE Veh. Tech. Conf.*, pp. 862–6, April 2003.

Jurgen, R. K., "Broadcasting with digital audio," *IEEE Spectrum*, pp. 52–9, March 1996.

Kahn, J. M., R. H. Katz, and K. S. Pister, "Emerging challenges: Mobile networking for Smart Dust," *J. Commun. Networks*, pp. 188–96, August 2000.

Kahn, R. E., S. A. Gronemeyer, J. Burchfiel, and R. C. Kunzelman, "Advances in packet radio technology," *Proc. IEEE*, pp. 1468–96, November 1978.

Kaider, S., "Performance of multi-carrier CDM and COFDM in fading channels," *Proc. IEEE Globecom Conf.*, pp. 847–51, December 1999.

Kallel, S., "Analysis of memory and incremental redundancy ARQ schemes over a nonstationary channel," *IEEE Trans. Commun.*, pp. 1474–80, September 1992.

Kam, A. C., T. Minn, and K.-Y. Siu, "Supporting rate guarantee and fair access for bursty data traffic in W-CDMA," *IEEE J. Sel. Areas Commun.*, pp. 2121–30, November 2001.

Kam, P. Y., "Bit error probabilities of MDPSK over the nonselective Rayleigh fading channel with diversity reception," *IEEE Trans. Commun.*, pp. 220–4, February 1991.

Kam, P. Y., "Tight bounds on the bit-error probabilities of 2DPSK and 4DPSK in nonselective Rician fading," *IEEE Trans. Commun.*, pp. 860–2, July 1998.

Kam, P. Y., and H. M. Ching, "Sequence estimation over the slow nonselective Rayleigh fading channel with diversity reception and its application to Viterbi decoding," *IEEE J. Sel. Areas Commun.*, pp. 562–70, April 1992.

Kamath, K. M., and D. L. Goeckel, "Adaptive-modulation schemes for minimum outage probability in wireless systems," *IEEE Trans. Commun.*, pp. 1632–5, October 2004.

Kamio, Y., S. Sampei, H. Sasaoka, and N. Morinaga, "Performance of modulation-level-controlled adaptive-modulation under limited transmission delay time for land mobile communications," *Proc. IEEE Veh. Tech. Conf.*, pp. 221–5, July 1995.

Kandukuri, S., and N. Bambos, "Power controlled multiple access (PCMA) in wireless communication networks," *Proc. IEEE Infocom Conf.*, pp. 386–95, March 2000.

Karn, P., "MACA: A new channel access method for packet radio," *Proc. Comput. Network Conf.*, pp. 134–40, September 1990.

Karri, R., and P. Mishra, "Modeling energy efficient secure wireless networks using network simulation," *Proc. IEEE Internat. Conf. Commun.*, pp. 61–5, May 2003.

Kasturia, S., J. Aslanis, and J. Cioffi, "Vector coding for partial response channels," *IEEE Trans. Inform. Theory*, pp. 741–62, July 1990.

Katzela, I., and M. Naghshineh, "Channel assignment schemes for cellular mobile telecommunication systems – A comprehensive survey," *IEEE Pers. Commun. Mag.*, pp. 10–31, June 1996.

Kawadia, V., and P. R. Kumar, "Principles and protocols for power control in wireless ad hoc networks," *IEEE J. Sel. Areas Commun.*, pp. 76–88, January 2005.

Kawadia, V., and P. R. Kumar, "A cautionary perspective on cross-layer design," *IEEE Wireless Commun. Mag.*, pp. 3–11, February 2005.

Keller, J. B., "Geometrical theory of diffraction," *J. Opt. Soc. Amer.*, 52, pp. 116–30, 1962.

Kennedy, R. S., *Fading Dispersive Communication Channels*, Wiley, New York, 1969.

Khojastepour, M. A., A. Sabharwal, and B. Aazhang, "Improved achievable rates for user cooperation and relay channels," *Proc. IEEE Internat. Sympos. Inform. Theory*, p. 4, June 2004.

Kleinrock, L., *Queueing Systems*, vol. I: *Theory*, Wiley, New York, 1975.

Kleinrock, L., and J. Silvester, "Optimum transmission radii for packet radio networks or why six is a magic number," *Proc. IEEE Natl. Telecomm. Conf.*, pp. 4.3.1–4.3.5, December 1978.

Knopp, R., and P. Humblet, "Information capacity and power control in single-cell multiuser communications," *Proc. IEEE Internat. Conf. Commun.*, pp. 331–5, June 1995.

Koch, T., and A. Lapidoth, "The fading number and degrees of freedom in non-coherent MIMO fading channels: A peace pipe," *Proc. IEEE Internat. Sympos. Inform. Theory,* September 2005.

Koetter, R., and M. Medard, "An algebraic approach to network coding," *IEEE/ACM Trans. Network.,* pp. 782–95, October 2003.

Kohno, R., R. Meidan, and L. B. Milstein, "Spread spectrum access methods for wireless communications," *IEEE Commun. Mag.,* pp. 58–67, January 1995.

Komninakis, C., and R. D. Wesel, "Pilot-aided joint data and channel estimation in flat correlated fading," *Proc. IEEE Globecom Conf.,* pp. 2534–9, November 1999.

Komninakis, C., and R. D. Wesel, "Joint iterative channel estimation and decoding in flat correlated Rayleigh fading," *IEEE J. Sel. Areas Commun.* pp. 1706–17, September 2001.

Kong, H., and E. Shwedyk, "Sequence detection and channel state estimation over finite state Markov channels," *IEEE Trans. Veh. Tech.,* pp. 833–9, May 1999.

Kouyoumjian, R. G., and P. H. Pathak, "A uniform geometrical theory of diffraction for an edge in a perfectly conducting surface," *Proc. IEEE,* pp. 1448–61, November 1974.

Kozat, U. C., I. Koutsopoulos, and L. Tassiulas, "Dynamic code assignment and spreading gain adaptation in synchronous CDMA wireless networks," *Proc. Internat. Sympos. Spread Spec. Tech. Appl.,* pp. 593–7, November 2002.

Krishnamachari, B., S. B. Wicker, and R. Bejar, "Phase transition phenomena in wireless ad hoc networks," *Proc. IEEE Globecom Conf.,* pp. 2921–5, November 2001.

Kschischang, F. R., and D. Frey, "Iterative decoding of compound codes by probability propagation in graphical models," *IEEE J. Sel. Areas Commun.,* pp. 219–30, February 1998.

Kumwilaisak, W., Y. T. Hou, Q. Zhang, W. Zhu, C.-C. J. Kuo, and Y.-Q. Zhang, "A cross-layer quality-of-service mapping architecture for video delivery in wireless networks," *IEEE J. Sel. Areas Commun.,* pp. 1685–98, December 2003.

Kunz, D., "Channel assignment for cellular radio using neural networks," *IEEE Trans. Veh. Tech.,* pp. 188–93, February 1991.

Kurner, T., D. J. Cichon, and W. Wiesbeck, "Concepts and results for 3D digital terrain-based wave propagation models: An overview," *IEEE J. Sel. Areas Commun.,* pp. 1002–12, September 1993.

Laneman, J. N., D. N. C. Tse, and G. W. Wornell, "Cooperative diversity in wireless networks: Efficient protocols and outage behavior," *IEEE Trans. Inform. Theory,* pp. 3062–80, December 2004.

Lapidoth, A., and S. Moser, "On the fading number of multi-antenna systems over flat fading channels with memory and incomplete side information," *Proc. IEEE Internat. Sympos. Inform. Theory,* p. 478, July 2002.

Lapidoth, A., and S. M. Moser, "Capacity bounds via duality with applications to multiple-antenna systems on flat-fading channels," *IEEE Trans. Inform. Theory,* pp. 2426–67, October 2003.

Lapidoth, A., and S. Shamai, "Fading channels: How perfect need 'perfect side information' be?" *IEEE Trans. Inform. Theory,* pp. 1118–34, November 1997.

Lapidoth, A., I. E. Telatar, and R. Urbanke, "On wide-band broadcast channels," *IEEE Trans. Inform. Theory,* pp. 3250–8, December 2003.

Larsson, E. G., and P. Stoica, *Space-Time Block Coding for Wireless Communications,* Cambridge University Press, 2003.

Lau, K.-N., "Analytical framework for multiuser uplink MIMO space-time scheduling design with convex utility functions," *IEEE Trans. Wireless Commun.,* pp. 1832–43, September 2004.

Lawton, M. C., and J. P. McGeehan, "The application of GTD and ray launching techniques to channel modeling for cordless radio systems," *Proc. IEEE Veh. Tech. Conf.,* pp. 125–30, May 1992.

Lee, S.-J., and M. Gerla, "Dynamic load-aware routing in ad hoc networks," *Proc. IEEE Internat. Conf. Commun.,* pp. 3206–10, June 2001.

Lee, W. C. Y., *Mobile Communications Engineering,* McGraw-Hill, New York, 1982.

Lee, W. C. Y., *Mobile Communication Design Fundamentals,* Sams, Indianapolis, IN, 1986.

Lee, W. C. Y., *Mobile Cellular Telecommunications Systems,* McGraw-Hill, New York, 1989.

Leiner, B., R. Ruther, and A. Sastry, "Goals and challenges of the DARPA Glomo program (global mobile information systems)," *IEEE Pers. Commun. Mag.,* pp. 34–43, December 1996.

Leon-Garcia, A., *Probability and Random Processes for Electrical Engineering,* 2nd ed., Addison-Wesley, Reading, MA, 1994.

Leus, G., S. Zhou, and G. B. Giannakis, "Orthogonal multiple access over time- and frequency-selective channels," *IEEE Trans. Inform. Theory,* pp. 1942–50, August 2003.

Li, H.-J., C.-C. Chen, T.-Y. Liu, and H.-C. Lin, "Applicability of ray-tracing techniques for prediction of outdoor channel characteristics," *IEEE Trans. Veh. Tech.,* pp. 2336–49, November 2000.

Li, L., and A. J. Goldsmith, "Capacity and optimal resource allocation for fading broadcast channels – Part I: Ergodic capacity," *IEEE Trans. Inform. Theory,* pp. 1083–1102, March 2001.

Li, L., and A. J. Goldsmith, "Capacity and optimal resource allocation for fading broadcast channels – Part II: Outage capacity," *IEEE Trans. Inform. Theory,* pp. 1103–27, March 2001.

Li, L., and A. J. Goldsmith, "Low-complexity maximum-likelihood detection of coded signals sent over finite-state Markov channels," *IEEE Trans. Commun.,* pp. 524–31, April 2002.

Li, L., N. Jindal, and A. J. Goldsmith, "Outage capacities and optimal power allocation for fading multiple access channels," *IEEE Trans. Inform. Theory,* pp. 1326–47, April 2005.

Li, S.-Y. R., R. W. Yeung, and N. Cai, "Linear network coding," *IEEE Trans. Inform. Theory,* pp. 371–81, February 2003.

Lin, C. R., and J.-S. Liu, "QoS routing in ad hoc wireless networks," *IEEE J. Sel. Areas Commun.,* pp. 1426–38, August 1999.

Lin, S., and J. D. J. Costello, *Error Control Coding,* 2nd ed., Prentice-Hall, Englewood Cliffs, NJ, 2004.

Ling, F., and J. Proakis, "Adaptive lattice decision feedback equalizers – Their performance and application to time-variant multipath channels," *IEEE Trans. Commun.,* pp. 348–56, April 1985.

Liu, C., and K. Feher, "Bit error rate performance of $\pi/4$ DQPSK in a frequency selective fast Rayleigh fading channel," *IEEE Trans. Veh. Tech.,* pp. 558–68, August 1991.

Liu, X., E. K. P. Chong, and N. B. Shroff, "Opportunistic transmission scheduling with resource-sharing constraints in wireless networks," *IEEE J. Sel. Areas Commun.,* pp. 2053–64, October 2001.

Liu, X., S. S. Mahal, A. Goldsmith, and J. K. Hedrick, "Effects of communication delay on string stability in vehicle platoons," *Proc. IEEE Internat. Conf. Intell. Transp. Syst.,* pp. 625–30, August 2001.

Liu, Y., M. P. Fitz, and O. Y. Takeshita, "Full-rate space-time codes," *IEEE J. Sel. Areas Commun.,* pp. 969–80, May 2001.

Liu, Z., and G. B. Giannakis, "Space-time block-coded multiple access through frequency-selective fading channels," *IEEE Trans. Commun.,* pp. 1033–44, June 2001.

Lozano, A., and D. C. Cox, "Integrated dynamic channel assignment and power control in TDMA mobile wireless communication systems," *IEEE J. Sel. Areas Commun.,* pp. 2031–40, November 1999.

Lozano, A., and D. C. Cox, "Distributed dynamic channel assignment in TDMA mobile communication systems," *IEEE Trans. Veh. Tech.,* pp. 1397–1406, November 2002.

Lozano, A., A. M. Tulino, and S. Verdú, "Multiple-antenna capacity in the low-power regime," *IEEE Trans. Inform. Theory,* pp. 2527–44, October 2003.

Lu, W., "4G mobile research in Asia," *IEEE Commun. Mag.,* pp. 104–6, March 2003.

Luebbers, R. J., "Finite conductivity uniform GTD versus knife edge diffraction in prediction of propagation path loss," *IEEE Trans. Ant. Prop.,* pp. 70–6, January 1984.

Lupas, R., and S. Verdú, "Linear multiuser detectors for synchronous code-division multiple-access channels," *IEEE Trans. Inform. Theory,* pp. 123–36, January 1989.

McCune, E., and K. Feher, "Closed-form propagation model combining one or more propagation constant segments," *Proc. IEEE Veh. Tech. Conf.,* pp. 1108–12, May 1997.

McDonald, V. H., "The cellular concept," *Bell System Tech. J.,* pp. 15–49, January 1979.

McEliece, R., D. J. C. MacKay, and J.-F. Cheng, "Turbo decoding as an instance of Pearl's 'belief propagation' algorithm," *IEEE J. Sel. Areas Commun.,* pp. 140–52, February 1998.

McEliece, R. J., and K. N. Sivarajan, "Performance limits for channelized cellular telephone systems," *IEEE Trans. Inform. Theory,* pp. 21–4, January 1994.

McEliece, R. J., and W. E. Stark, "Channels with block interference," *IEEE Trans. Inform. Theory,* pp. 44–53, January 1984.

MacKay, D. J. C., and R. M. Neal, "Near Shannon limit performance of low density parity check codes," *Elec. Lett.,* p. 1645, August 1996.

MacKenzie, A. B., and S. B. Wicker, "Selfish users in ALOHA: A game-theoretic approach," *Proc. IEEE Veh. Tech. Conf.,* pp. 1354–7, October 2001.

McKown, J. W., and R. L. Hamilton, Jr., "Ray tracing as a design tool for radio networks," *IEEE Network,* pp. 27–30, November 1991.

Makrakis, D., P. T. Mathiopoulos, and D. P. Bouras, "Optimal decoding of coded PSK and QAM signals in correlated fast fading channels and AWGN – A combined envelope, multiple differential and coherent detection approach," *IEEE Trans. Commun.,* pp. 63–75, January 1994.

Marsan, M., and G. C. Hess, "Shadow variability in an urban land mobile radio environment," *Elec. Lett.,* pp. 646–8, May 1990.

Marzetta, T., and B. Hochwald, "Capacity of a mobile multiple-antenna communication link in Rayleigh flat fading," *IEEE Trans. Inform. Theory,* pp. 139–57, January 1999.

Mathar, R., and J. Mattfeldt, "Channel assignment in cellular radio networks," *IEEE Trans. Veh. Tech.,* pp. 647–56, November 1993.

Matsuoka, H., S. Sampei, N. Morinaga, and Y. Kamio, "Symbol rate and modulation level controlled adaptive modulation/TDMA/TDD for personal communication systems," *Proc. IEEE Veh. Tech. Conf.,* pp. 487–91, April 1996.

Mauss, O. C., F. Classen, and H. Meyr, "Carrier frequency recovery for a fully digital direct-sequence spread-spectrum receiver: A comparison," *Proc. IEEE Veh. Tech. Conf.,* pp. 392–5, May 1993.

Medard, M., "The effect upon channel capacity in wireless communications of perfect and imperfect knowledge of the channel," *IEEE Trans. Inform. Theory,* pp. 933–46, May 2000.

Mehrotra, A., *Cellular Radio: Analog and Digital Systems,* Artech House, Norwood, MA, 1994.

Mehta, N. B., and A. J. Goldsmith, "Effect of fixed and interference-induced packet error probability on PRMA," *Proc. IEEE Internat. Conf. Commun.,* pp. 362–6, June 2000.

Mehta, N. B., L. J. Greenstein, T. M. Willis, and Z. Kostic, "Analysis and results for the orthogonality factor in WCDMA downlinks," *IEEE Trans. Wireless Commun.,* pp. 1138–49, November 2003.

Mengali, U., and A. N. D'Andrea, *Synchronization Techniques for Digital Receivers,* Plenum, New York, 1997.

Mestdagh, D. J. G., and P. M. P. Spruyt, "A method to reduce the probability of clipping in DMT-based transceivers," *IEEE Trans. Commun.,* pp. 1234–8, October 1996.

Meyr, H., M. Moeneclaey, and S. A. Fechtel, *Digital Communication Receivers,* vol. 2, *Synchronization, Channel Estimation, and Signal Processing,* Wiley, New York, 1997.

Mirhakkak, M., N. Schult, and D. Thomson, "Dynamic bandwidth management and adaptive applications for a variable bandwidth wireless environment," *IEEE J. Sel. Areas Commun.,* pp. 1984–97, October 2001.

Mitchell, T., "Broad is the way (ultra-wideband technology)," *IEE Review,* pp. 35–9, January 2001.

Mitra, U., "Comparison of maximum-likelihood-based detection for two multirate access schemes for CDMA signals," *IEEE Trans. Commun.,* pp. 64–77, January 1999.

Modestino, J. W., and D. G. Daut, "Combined source-channel coding of images," *IEEE Trans. Commun.,* pp. 1644–59, November 1979.

Mohasseb, Y., and M. P. Fitz, "A 3-D spatio-temporal simulation model for wireless channels," *IEEE J. Sel. Areas Commun.,* pp. 1193–1203, August 2002.

Molisch, A., M. Win, and J. H. Winters, "Reduced-complexity transmit/receive-diversity systems," *IEEE Trans. Signal Proc.,* pp. 2729–38, November 2003.

Motley, A. J., and J. M. P. Keenan, "Personal communication radio coverage in buildings at 900 MHz and 1700 MHz," *Elec. Lett.,* pp. 763–4, June 1988.

Moustakas, A. L., S. H. Simon, and A. M. Sengupta, "MIMO capacity through correlated channels in the presence of correlated interferers and noise: A (not so) large N analysis," *IEEE Trans. Inform. Theory,* pp. 2545–61, October 2003.

Muschallik, C., "Improving an OFDM reception using an adaptive Nyquist windowing," *IEEE Trans. Consumer Elec.,* pp. 259–69, August 1996.

Mushkin, M., and I. Bar-David, "Capacity and coding for the Gilbert–Elliot channel," *IEEE Trans. Inform. Theory,* pp. 1277–90, November 1989.

Naguib, A., "Equalization of transmit diversity space-time coded signals," *Proc. IEEE Globecom Conf.,* pp. 1077–82, December 2000.

Naguib, A., N. Seshadri, and A. Calderbank, "Increasing data rate over wireless channels," *IEEE Signal Proc. Mag.,* pp. 76–92, May 2000.

Nanda, S., K. Balachandran, and S. Kumar, "Adaptation techniques in wireless packet data services," *IEEE Commun. Mag.,* pp. 54–64, January 2000.

Narayanan, K. R., "Turbo decoding of concatenated space-time codes," *Proc. Allerton Conf. Commun., Control, Comput.,* pp. 217–26, September 1999.

Narula, A., M. Lopez, M. Trott, and G. Wornell, "Efficient use of side information in multiple-antenna data transmission over fading channels," *IEEE J. Sel. Areas Commun.,* pp. 1423–36, October 1998.

Negus, K. J., A. P. Stephens, and J. Lansford, "HomeRF: Wireless networking for the connected home," *IEEE Pers. Commun. Mag.,* pp. 20–7, February 2000.

Ng, C. T. K., and A. J. Goldsmith, "Transmitter cooperation in ad-hoc wireless networks: Does dirty-paper coding beat relaying?" *IEEE Inform. Theory Workshop,* pp. 277–82, October 2004.

Nilsson, J., B. Bernhardsson, and B. Wittenmark, "Stochastic analysis and control of real-time systems with random time delays," *Automatica,* pp. 57–64, January 1998.

Nobel, B., and J. W. Daniel, *Applied Linear Algebra,* Prentice-Hall, Englewood Cliffs, NJ, 1977.

Nosratinia, A., T. E. Hunter, and A. Hedayat, "Cooperative communication in wireless networks," *IEEE Commun. Mag.,* pp. 74–80, October 2004.

Ochiai, H., and H. Imai, "On the distribution of the peak-to-average power ratio in OFDM signals," *IEEE Trans. Commun.,* pp. 282–9, February 2001.

Oh, S.-J., D. Zhang, and K. M. Wasserman, "Optimal resource allocation in multiservice CDMA networks," *IEEE Trans. Wireless Commun.,* pp. 811–21, July 2003.

Okumura, T., E. Ohmori, and K. Fukuda, "Field strength and its variability in VHF and UHF land mobile service," *Rev. Elec. Commun. Lab.,* pp. 825–73, September/October 1968.

Oppenheim, A. V., R. W. Schafer, and J. R. Buck, *Discrete-Time Signal Processing,* 2nd ed., Prentice-Hall, Englewood Cliffs, NJ, 1999.

Otsuki, S., S. Sampei, and N. Morinaga, "Square-QAM adaptive modulation/TDMA/TDD systems using modulation level estimation with Walsh function," *Elec. Lett.,* pp. 169–71, February 1995.

Owen, F. C., and C. D. Pudney, "Radio propagation for digital cordless telephones at 1700 MHz and 900 MHz," *Elec. Lett.,* pp. 52–3, September 1988.

Padgett, J. E., C. G. Gunther, and T. Hattori, "Overview of wireless personal communications," *IEEE Commun. Mag.,* pp. 28–41, January 1995.

Pahlavan, K., and P. Krishnamurthy, *Principles of Wireless Networks: A Unified Approach,* Prentice-Hall, Englewood Cliffs, NJ, 2002.

Papoulis, A., and S. U. Pillai, *Probability, Random Variables and Stochastic Processes,* McGraw-Hill, New York, 2002.

Parsons, D., *The Mobile Radio Propagation Channel,* Wiley, New York, 1992.

Parsons, J. D., and M. D. Turkmani, "Characterization of mobile radio signals: Model description," *Proc. IEE,* pt. 1, pp. 549–56, December 1991.

Parsons, J. D., and M. D. Turkmani, "Characterization of mobile radio signals: Base station cross-correlation," *Proc. IEE,* pt. 1, pp. 557–65, December 1991.

Patel, P., and J. Holtzman, "Analysis of a simple successive interference cancellation scheme in a DS/CDMA system," *IEEE J. Sel. Areas Commun.,* pp. 796–807, June 1994.

Paterson, K. G., and V. Tarokh, "On the existence and construction of good codes with low peak-to-average power ratios," *IEEE Trans. Inform. Theory,* pp. 1974–87, September 2000.

Pätzold, M., *Mobile Fading Channels,* Wiley, New York, 2002.

Paulraj, A., R. Nabar, and D. Gore, *Introduction to Space-Time Wireless Communications,* Cambridge University Press, 2003.

Pawula, R. F., "A new formula for MDPSK symbol error probability," *IEEE Commun. Lett.,* pp. 271–2, October 1998.

Pawula, R., S. Rice, and J. Roberts, "Distribution of the phase angle between two vectors perturbed by Gaussian noise," *IEEE Trans. Commun.,* pp. 1828–41, August 1982.

Pearlman, M. R., Z. J. Haas, and S. I. Mir, "Using routing zones to support route maintenance in ad hoc networks," *Proc. IEEE Wireless Commun. Network Conf.,* pp. 1280–4, September 2000.

Peleg, M., S. Shamai (Shitz), and S. Galan, "Iterative decoding for coded noncoherent MPSK communications over phase-noisy AWGN channels," *IEE Proc. Commun.,* pp. 87–95, April 2000.

Penrose, M., *Random Geometric Graphs,* Oxford University Press, 2004.

Perkins, C. E., and E. M. Royer, "Ad hoc on-demand distance vector routing," *Proc. IEEE Workshop Mobile Comput. Syst. Appl.,* pp. 90–100, February 1999.

Peterson, L. L., and B. S. Davie, *Computer Networks – A Systems Approach,* 2nd ed., Morgan-Kaufman, San Mateo, CA, 2000.

Peterson, R. L., R. E. Ziemer, and D. E. Borth, *Introduction to Spread Spectrum Communications,* Prentice-Hall, Englewood Cliffs, NJ, 1995.

Pickholtz, R., L. Milstein, and D. Schilling, "Spread spectrum for mobile communications," *IEEE Trans. Veh. Tech.,* pp. 313–22, May 1991.

Pickholtz, R., D. Schilling, and L. Milstein, "Theory of spread-spectrum communications – A tutorial," *IEEE Trans. Commun.,* pp. 855–84, May 1982.

Pimentel, C., and I. F. Blake, "Modeling burst channels using partitioned Fritchman's Markov models," *IEEE Trans. Veh. Tech.,* pp. 885–99, August 1998.

Poole, I., "What exactly is … ZigBee?" *IEEE Commun. Eng.,* pp. 44–5, August / September 2004.

Poon, A. S. Y., and R. W. Brodersen, "The role of multiple-antenna systems in emerging open access environments," *EE Times Commun. Design Conf.,* October 2003.

Poon, L.-S., and H.-S. Wang, "Propagation characteristic measurement and frequency reuse planning in an office building," *Proc. IEEE Veh. Tech. Conf.,* pp. 1807–10, June 1994.

Porcino, D., and W. Hirt, "Ultra-wideband radio technology: Potential and challenges ahead," *IEEE Commun. Mag.,* pp. 66–74, July 2003.

Prasad, R., and A. Kegel, "Effects of Rician faded and log-normal shadowed signals on spectrum efficiency in microcellular radio," *IEEE Trans. Veh. Tech.,* pp. 274–81, August 1993.

Price, R., "Nonlinearly feedback-equalized PAM vs. capacity," *Proc. IEEE Internat. Conf. Commun.,* pp. 22.12–22.17, June 1972.

Proakis, J. G., "Adaptive equalization for TDMA digital mobile radio," *IEEE Trans. Veh. Tech.,* pp. 333–41, May 1991.

Proakis, J. G., *Digital Communications,* 4th ed., McGraw-Hill, New York, 2001.

Proakis, J., and M. Salehi, *Communication Systems Engineering,* Prentice-Hall, Englewood Cliffs, NJ, 2002.

Pursley, M., "Performance evaluation for phase-coded spread-spectrum multiple-access communication – Part I: System analysis," *IEEE Trans. Commun.,* pp. 795–9, August 1977.

Pursley, M. B., "The role of spread spectrum in packet radio networks," *Proc. IEEE,* pp. 116–34, January 1987.

Pursley, M. B., *Introduction to Digital Communications,* Prentice-Hall, Englewood Cliffs, NJ, 2005.

Qian, L., and K. Kumaran, "Scheduling on uplink of CDMA packet data network with successive interference cancellation," *Proc. IEEE Wireless Commun. Network Conf.,* pp. 1645–50, March 2003.

Qian, L., and K. Kumaran, "Uplink scheduling in CDMA packet-data systems," *Proc. IEEE Infocom Conf.,* pp. 292–300, March 2003.

Qiu, X., L. Chang, Z. Kostic, T. M. Willis, N. Mehta, L. G. Greenstein, K. Chawla, J. F. Whitehead, and J. Chuang, "Some performance results for the downlink shared channel in WCDMA," *Proc. IEEE Internat. Conf. Commun.*, pp. 376–80, April 2002.

Qureshi, S. U., "Adaptive equalization," *Proc. IEEE,* pp. 1349–87, September 1985.

Rabaey, J., J. Ammer, J. L. da Silva, Jr., and D. Patel, "PicoRadio: Ad-hoc wireless networking of ubiquitous low-energy sensor/monitor nodes," *IEEE Comput. Soc. Workshop on VLSI,* pp. 9–12, April 2000.

Rabaey, J., J. Ammer, J. L. da Silva, Jr., and D. Roundy, "PicoRadio supports ad hoc ultra-low power wireless networking," *IEEE Computer,* pp. 42–8, July 2000.

Radunović, B., and J. Y. Boudec, "Rate performance objectives of multihop wireless networks," *IEEE Trans. Mobile Comput.,* pp. 334–49, October–December 2004.

Ramanathan, R., and R. Rosales-Hain, "Topology control of multihop wireless networks using transmit power adjustment," *Proc. IEEE Infocom Conf.,* pp. 404–13, March 2000.

Ramsey, J. L., "Realization of optimum interleavers," *IEEE Trans. Inform. Theory,* pp. 338–45, 1970.

Rappaport, T. S., *Wireless Communications – Principles and Practice,* 2nd ed., Prentice-Hall, Englewood Cliffs, NJ, 2001.

Rappaport, T. S., A. Annamalai, R. M. Buehrer, and W. H. Tranter, "Wireless communications: Past events and a future perspective," *IEEE Commun. Mag.,* pp. 148–61, May 2002.

Rappaport, T. S., and L. B. Milstein, "Effects of radio propagation path loss on DS-CDMA cellular frequency reuse efficiency for the reverse channel," *IEEE Trans. Veh. Tech.,* pp. 231–42, August 1992.

Rath, K., and J. Uddenfeldt, "Capacity of digital cellular TDMA systems," *IEEE Trans. Veh. Tech.,* pp. 323–32, May 1991.

Redfern, A., "Receiver window design for multicarrier communication systems," *IEEE J. Sel. Areas Commun.,* pp. 1029–36, June 2002.

Remley, K. A., H. R. Anderson, and A. Weisshar, "Improving the accuracy of ray-tracing techniques for indoor propagation modeling," *IEEE Trans. Veh. Tech.,* pp. 2350–8, November 2000.

Rhodes, S., "Effect of noisy phase reference on coherent detection of offset-QPSK signals," *IEEE Trans. Commun.,* pp. 1046–55, August 1974.

Rice, M., and S. B. Wicker, "Adaptive error control for slowly varying channels," *IEEE Trans. Commun.,* pp. 917–26, February–April 1994.

Rice, S. O., "Mathematical analysis of random noise," *Bell System Tech. J.,* pp. 282–333, July 1944, and pp. 46–156, January 1945.

Richardson, T., A. Shokrollahi, and R. Urbanke, "Design of capacity-approaching irregular low-density parity-check codes," *IEEE Trans. Inform. Theory,* pp. 619–37, February 2001.

Richardson, T., and R. Urbanke, "The capacity of low-density parity-check codes under message passing decoding," *IEEE Trans. Inform. Theory,* pp. 599–618, February 2001.

Rimoldi, B., and R. Urbanke, "A rate-splitting approach to the Gaussian multiple-access channel," *IEEE Trans. Inform. Theory,* pp. 364–75, March 1996.

Ritter, M., R. Friday, and M. Cunningham, "The architecture of Metricom's microcellular data network and details of its implementation as the 2nd and 3rd generation ricochet wide-area mobile data service," *Proc. IEEE Emerg. Technol. Sympos. Broadband Commun.,* pp. 143–52, September 2001.

Robertson, P., and T. Worz, "Bandwidth-efficient turbo trellis-coded modulation using punctured component codes," *IEEE J. Sel. Areas Commun.,* pp. 206–18, February 1998.

Rodoplu, V., and T. H. Meng, "Minimum energy mobile wireless networks," *IEEE J. Sel. Areas Commun.,* pp. 1333–44, August 1999.

Root, W. L., and P. P. Varaiya, "Capacity of classes of Gaussian channels," *SIAM J. Appl. Math.,* pp. 1350–93, November 1968.

Rustako, A. J., Jr., N. Amitay, G. J. Owens, and R. S. Roman, "Radio propagation at microwave frequencies for line-of-sight microcellular mobile and personal communications," *IEEE Trans. Veh. Tech.,* pp. 203–10, February 1991.

Sachs, D. G., R. Anand, and K. Ramchandran, "Wireless image transmission using multiple-description based concatenated codes," *Proc. IEEE Data Compress. Conf.,* p. 569, March 2000.

Salz, J., "Optimum mean-square decision feedback equalization," *Bell System Tech. J.,* pp. 1341–73, October 1973.

Salz, J., and A. D. Wyner, "On data transmission over cross coupled multi-input, multi-output linear channels with applications to mobile radio," AT&T Bell Labs Internal Tech. Memo, 1990.

Sampath, H., S. Talwar, J. Tellado, V. Erceg, and A. Paulraj, "A fourth-generation MIMO-OFDM broadband wireless system: Design, performance, and field trial results," *IEEE Commun. Mag.,* pp. 143–9, September 2002.

Sampei, S., N. Morinaga, and Y. Kamio, "Adaptive modulation/TDMA with a BDDFE for 2 Mbit/s multi-media wireless communication systems," *Proc. IEEE Veh. Tech. Conf.,* pp. 311–15, July 1995.

Santhi, K. R., V. K. Srivastava, G. SenthilKumaran, and A. Butare, "Goals of true broadband's wireless next wave (4G-5G)," *Proc. IEEE Veh. Tech. Conf.,* pp. 2317–21, October 2003.

Sari, H., "Trends and challenges in broadband wireless access," *Proc. Sympos. Commun. Veh. Tech.,* pp. 210–14, October 2000.

Sari, H., G. Karam, and I. Jeanclaude, "Transmission techniques for digital terrestrial TV broadcasting," *IEEE Commun. Mag.,* pp. 100–9, February 1995.

Sasan, I., and S. Shamai, "Improved upper bounds on the ML decoding error probability of parallel and serial concatenated turbo codes via their ensemble distance spectrum," *IEEE Trans. Inform. Theory,* pp. 24–47, January 2000.

Satorius, E. H., and S. T. Alexander, "Channel equalization using adaptive lattice algorithms," *IEEE Trans. Commun.,* pp. 899–905, June 1979.

Scaglione, A., G. B. Giannakis, and S. Barbarossa, "Redundant filterbank precoders and equalizers. I: Unification and optimal designs," *IEEE Trans. Signal Proc.,* pp. 1988–2006, July 1999.

Scaglione, A., G. B. Giannakis, and S. Barbarossa, "Redundant filterbank precoders and equalizers. II: Blind channel estimation, synchronization, and direct equalization," *IEEE Trans. Signal Proc.,* pp. 2007–22, July 1999.

Schaubach, K., N. J. Davis IV, and T. S. Rappaport, "A ray tracing method for predicting path loss and delay spread in microcellular environments," *Proc. IEEE Veh. Tech. Conf.,* pp. 932–5, May 1992.

Schiesel, S., "Paging allies focus strategy on the Internet," *New York Times,* April 19, 1999.

Schmidt, A., "How to build smart appliances," *IEEE Pers. Commun. Mag.,* pp. 66–71, August 2001.

Schurgers, C., and M. B. Srivastava, "Energy efficient wireless scheduling: Adaptive loading in time," *Proc. IEEE Wireless Commun. Network Conf.,* pp. 706–11, March 2002.

Schwartz, M., W. R. Bennett, and S. Stein, *Communication Systems and Techniques,* McGraw-Hill, New York, 1966 [reprinted 1995 by Wiley/IEEE Press].

Seidel, S. Y., and T. S. Rappaport, "914 MHz path loss prediction models for indoor wireless communications in multifloored buildings," *IEEE Trans. Ant. Prop.,* pp. 207–17, February 1992.

Seidel, S. Y., T. S. Rappaport, M. J. Feuerstein, K. L. Blackard, and L. Grindstaff, "The impact of surrounding buildings on propagation for wireless in-building personal communications system design," *Proc. IEEE Veh. Tech. Conf.,* pp. 814–18, May 1992.

Seidel, S. Y., T. S. Rappaport, S. Jain, M. L. Lord, and R. Singh, "Path loss, scattering, and multipath delay statistics in four European cities for digital cellular and microcellular radiotelephone," *IEEE Trans. Veh. Tech.,* pp. 721–30, November 1991.

Sendonaris, A., E. Erkip, and B. Aazhang, "User cooperation diversity. Part I: System description," *IEEE Trans. Commun.,* pp. 1927–38, November 2003.

Sendonaris, A., E. Erkip, and B. Aazhang, "User cooperation diversity. Part II: Implementation aspects and performance analysis," *IEEE Trans. Commun.,* pp. 1939–48, November 2003.

Seneta, E., *Nonnegative Matrices and Markov Chains,* Springer-Verlag, New York, 1981.

Servetto, S. D., K. Ramchandran, V. A. Vaishampayan, and K. Nahrstedt, "Multiple description wavelet based image coding," *IEEE Trans. Image Proc.,* pp. 813–26, May 2000.

Seshadri, N., and C.-E. W. Sundberg, "Multilevel trellis coded modulations for the Rayleigh fading channel," *IEEE Trans. Commun.*, pp. 1300–10, September 1993.

Shamai, S., and A. D. Wyner, "Information-theoretic considerations for symmetric, cellular, multiple-access fading channels: Part I," *IEEE Trans. Inform. Theory*, pp. 1877–94, November 1997.

Shamai, S., and A. D. Wyner, "Information-theoretic considerations for symmetric, cellular, multiple-access fading channels: Part II," *IEEE Trans. Inform. Theory*, pp. 1895–1911, November 1997.

Shamai, S., and B. M. Zaidel, "Enhancing the cellular downlink capacity via co-processing at the transmitting end," *Proc. IEEE Veh. Tech. Conf.*, pp. 1745–9, May 2001.

Shannon, C. E., "A mathematical theory of communication," *Bell System Tech. J.*, pp. 379–423, 623–56, 1948.

Shannon, C. E., "Communications in the presence of noise." *Proc. IRE*, pp. 10–21, 1949.

Shannon, C. E., "Coding theorems for a discrete source with a fidelity criterion," *IRE National Convention Record*, part 4, pp. 142–63, 1959.

Shannon, C. E., and W. Weaver, *The Mathematical Theory of Communication*, University of Illinois Press, Urbana, 1949.

Sharif, M., and B. Hassibi, "Scaling laws of sum rate using time-sharing, DPC, and beamforming for MIMO broadcast channels," *Proc. IEEE Internat. Sympos. Inform. Theory*, p. 175, June 2004.

Shih, E., P. Bahl, and M. Sinclair, "Wake on wireless: An event driven energy saving strategy for battery operated devices," *Proc. Internat. Conf. Mobile Comput. Network.*, pp. 160–71, September 2002.

Shin, H., and J. H. Lee, "Capacity of multiple-antenna fading channels: Spatial fading correlation, double scattering, and keyhole," *IEEE Trans. Inform. Theory*, pp. 2636–47, October 2003.

Simon, M. K., *Probability Distributions Involving Gaussian Random Variables*, Kluwer, Dordrecht, 2002.

Simon, M. K., and M.-S. Alouini, "A unified approach to the performance analysis of digital communications over generalized fading channels," *Proc. IEEE*, pp. 1860–77, September 1998.

Simon, M. K., and M.-S. Alouini, "A unified approach for the probability of error for noncoherent and differentially coherent modulations over generalized fading channels," *IEEE Trans. Commun.*, pp. 1625–38, December 1998.

Simon, M. K., and M.-S. Alouini, *Digital Communication over Fading Channels: A Unified Approach to Performance Analysis*, Wiley, New York, 2000.

Simon, M. K., and D. Divsalar, "Some new twists to problems involving the Gaussian probability integral," *IEEE Trans. Commun.*, pp. 200–10, February 1998.

Simon, M. K., S. M. Hinedi, and W. C. Lindsey, *Digital Communication Techniques: Signal Design and Detection*, Prentice-Hall, Englewood Cliffs, NJ, 1995.

Simon, M. K., J. K. Omura, R. A. Scholtz, and B. K. Levitt, *Spread Spectrum Communications Handbook*, McGraw-Hill, New York, 1994.

Sklar, B., *Digital Communications – Fundamentals and Applications*, Prentice-Hall, Englewood Cliffs, NJ, 1988.

Sklar, B., "How I learned to love the trellis," *IEEE Signal Proc. Mag.*, pp. 87–102, May 2003.

Skolnik, M. I., *Introduction to Radar Systems*, 2nd ed., McGraw-Hill, New York, 1980.

Smith, P. J., and M. Shafi, "On a Gaussian approximation to the capacity of wireless MIMO systems," *Proc. IEEE Internat. Conf. Commun.*, pp. 406–10, April 2002.

Sohrabi, K., J. Gao, V. Ailawadhi, and G. Pottie, "Protocols for self-organization of a wireless sensor network," *IEEE Pers. Commun. Mag.*, pp. 16–27, October 2000.

Sollenberger, N. R., and J. C.-I. Chuang, "Low-overhead symbol timing and carrier recovery for portable TDMA radio systems," *IEEE Trans. Commun.*, pp. 1886–92, October 1990.

Stallings, W., *Wireless Communications and Networks*, 2nd ed., Prentice-Hall, Englewood Cliffs, NJ, 2005.

Stark, H., and J. W. Woods, *Probability and Random Processes with Applications to Signal Processing*, 3rd ed., Prentice-Hall, Englewood Cliffs, NJ, 2001.

Stark, W., H. Wang. A. Worthen, S. Lafortune, and D. Teneketzis, "Low-energy wireless communication network design," *IEEE Wireless Commun. Mag.,* pp. 60–72, August 2002.

Steinbach, E., N. Farber, and B. Girod, "Adaptive playout for low latency video streaming," *Proc. IEEE Internat. Conf. Image Proc.,* pp. 962–5, October 2001.

Strang, G., *Linear Algebra and Its Applications,* 2nd ed., Academic Press, New York, 1980.

Stuber, G. L., *Principles of Mobile Communications,* 2nd ed., Kluwer, Dordrecht, 2001.

Stuber, G. L., J. R. Barry, S. W. McLaughlin, Y. Li, M. A. Ingram, and T. G. Pratt, "Broadband MIMO-OFDM wireless communications," *Proc. IEEE,* pp. 271–94, February 2004.

Subasinghe-Dias, D., and K. Feher, "A coded 16-QAM scheme for fast fading mobile radio channels," *IEEE Trans. Commun.,* pp. 1906–16, February–April 1995.

Subramanian, L., and R. H. Katz, "An architecture for building self-configurable systems," *Proc. Mobile Ad Hoc Network Comput. Workshop,* pp. 63–73, August 2000.

Sundberg, C.-E. W., and N. Seshadri, "Coded modulation for fading channels – An overview," *Euro. Trans. Telecommun.,* pp. 309–24, May/June 1993.

Tan, C. C., and N. C. Beaulieu, "On first-order Markov modeling for the Rayleigh fading channel," *IEEE Trans. Commun.,* pp. 2032–40, December 2000.

Tanabe, N., and N. Farvardin, "Subband image coding using entropy-coded quantization over noisy channels," *IEEE J. Sel. Areas Commun.,* pp. 926–43, June 1992.

Tang, X., M.-S. Alouini, and A. Goldsmith, "Effect of channel estimation error on M-QAM BER performance in Rayleigh fading," *IEEE Trans. Commun.,* pp. 1856–64, December 1999.

Tariz, M. F., and A. Nix, "Area spectral efficiency of a channel adaptive cellular mobile radio system in a correlated shadowed environment," *Proc. IEEE Veh. Tech. Conf.,* pp. 1075–9, May 1998.

Tarng, J. H., W.-S. Liu, Y.-F. Huang, and J.-M. Huang, "A novel and efficient hybrid model of radio multipath-fading channels in indoor environments," *IEEE Trans. Ant. Prop.,* pp. 585–94, March 2003.

Tarokh, V., H. Jafarkhani, and A. Calderbank, "Space-time block codes from orthogonal designs," *IEEE Trans. Inform. Theory,* pp. 1456–67, July 1999.

Tarokh, V., A. Naguib, N. Seshadri, and A. Calderbank, "Space-time codes for high data rate wireless communication: Performance criteria in the presence of channel estimation errors, mobility, and multiple paths," *IEEE Trans. Commun.,* pp. 199–207, February 1999.

Tarokh, V., N. Seshadri, and A. Calderbank, "Space-time codes for high data rate wireless communication: Performance criterion and code construction," *IEEE Trans. Inform. Theory,* pp. 744–65, March 1998.

Teletar, E., "Capacity of multi-antenna Gaussian channels," AT&T Bell Labs Internal Tech. Memo, June 1995.

Telatar, E., "Capacity of multi-antenna Gaussian channels," *Euro. Trans. Telecommun.,* pp. 585–96, November 1999.

Tellado, J., *Multicarrier Modulation with Low PAR: Applications to DSL and Wireless,* Kluwer, Boston, 2000.

TIA/EIA IS-856, "CDMA 2000: High rate packet data air interface specification," November 2000.

Tobagi, F. A., "Modeling and performance analysis of multihop packet radio networks," *Proc. IEEE,* pp. 135–55, January 1987.

Toh, C.-K., "Maximum battery life routing to support ubiquitous mobile computing in wireless ad hoc networks," *IEEE Commun. Mag.,* pp. 138–47, June 2001.

Toh, C.-K., *Ad Hoc Mobile Wireless Networks: Protocols and Systems,* Prentice-Hall, Englewood Cliffs, NJ, 2002.

Toledo, A. F., and A. M. D. Turkmani, "Propagation into and within buildings at 900, 1800, and 2300 MHz," *Proc. IEEE Veh. Tech. Conf.,* pp. 633–6, May 1992.

Toledo, A. F., A. M. D. Turkmani, and J. D. Parsons, "Estimating coverage of radio transmission into and within buildings at 900, 1800, and 2300 MHz," *IEEE Pers. Commun. Mag.,* pp. 40–7, April 1998.

Tomlinson, M., "A new automatic equalizer emplying modulo arithmetic," *Elec. Lett.*, 7, pp. 138–9, 1971.

Tong, L., G. Zu, and T. Kailath, "Blind identification and equalization based on second-order statistics: A time domain approach," *IEEE Trans. Inform. Theory*, pp. 340–9, March 1994.

Toumpis, S., "Large wireless networks under fading, mobility, and delay constraints," *Proc. IEEE Infocom Conf.*, pp. 609–19, March 2004.

Toumpis, S., and A. J. Goldsmith, "Performance, optimization, and cross-layer design of media access protocols for wireless ad hoc networks," *Proc. IEEE Internat. Conf. Commun.*, pp. 2234–40, May 2003.

Toumpis, S., and A. Goldsmith, "Capacity regions for ad hoc networks," *IEEE Trans. Wireless Commun.*, pp. 736–48, July 2003.

Tse, D., and S. Hanly, "Multiaccess fading channels – Part I: Polymatroid structure, optimal resource allocation and throughput capacities," *IEEE Trans. Inform. Theory*, pp. 2796–2815, November 1998.

Tse, D., and P. Viswanath, *Foundations of Wireless Communications*, Cambridge University Press, 2005.

Tse, D. N. C., P. Viswanath, and L. Zheng, "Diversity–multiplexing trade-off in multiple-access channels," *IEEE Trans. Inform. Theory*, pp. 1859–74, September 2004.

Tsirigos, A., and Z. J. Haas, "Multipath routing in the presence of frequent topological changes," *IEEE Commun. Mag.*, pp. 132–8, November 2001.

Tsybakov, B., "The capacity of a memoryless Gaussian vector channel," *Prob. Inform. Trans.*, 1(1), pp. 18–29, 1965.

Tüchler, M., R. Koetter, and A. C. Singer, "Turbo equalization: Principles and new results," *IEEE Trans. Commun.*, pp. 754–67, May 2002.

Tulino, A. M., and S. Verdú, "Random matrix theory and wireless communications," *Found. Trends Commun. Inform. Theory*, 1(1), pp. 1–182, 2004.

Turin, G. L., "Introduction to spread spectrum antimultipath techniques and their application to urban digital radio," *Proc. IEEE*, pp. 328–53, March 1980.

Turin, W., R. Jana, S. S. Ghassemzadeh, V. W. Rice, and V. Tarokh, "Autoregressive modeling of an indoor UWB channel," *Proc. IEEE Conf. UWB Syst. Technol.*, pp. 71–4, May 2002.

Ue, T., S. Sampei, and N. Morinaga, "Symbol rate and modulation level controlled adaptive modulation/TDMA/TDD for personal communication systems," *Proc. IEEE Veh. Tech. Conf.*, pp. 306–10, July 1995.

Ungerboeck, G., "Channel coding with multi-level/phase signals," *IEEE Trans. Inform. Theory*, pp. 55–67, January 1982.

Uysal-Biyikoglu, E., B. Prabhakar, and A. El Gamal, "Energy-efficient packet transmission over a wireless link," *IEEE/ACM Trans. Network.*, pp. 487–99, August 2002.

van der Meulen, E. C., "Some reflections on the interference channel," in R. E. Blahut, D. J. Costello, and T. Mittelholzer, Eds., *Communications and Cryptography: Two Sides of One Tapestry*, pp. 409–21, Kluwer, Boston, 1994.

Varanasi, M. K., and B. Aazhang, "Multistage detection in asynchronous code division multiple-access communications," *IEEE Trans. Commun.*, pp. 509–19, April 1990.

Varanasi, M. K., and B. Aazhang, "Near-optimum detection in synchronous code division multiple-access communications," *IEEE Trans. Commun.*, pp. 725–36, May 1991.

Varsou, A. C., and H. V. Poor, "HOLPRO: A new rate scheduling algorithm for the downlink of CDMA networks," *Proc. IEEE Veh. Tech. Conf.*, pp. 948–54, September 2000.

Vaughan-Nichols, S. J., "Achieving wireless broadband with WiMax," *IEEE Computer*, pp. 10–13, June 2004.

Veeravalli, V. V., and A. Mantravadi, "The coding–spreading trade-off in CDMA systems," *IEEE J. Sel. Areas Commun.*, pp. 396–408, February 2002.

Verdú, S., "Optimum multiuser signal detection," Ph.D. thesis, University of Illinois, Urbana-Champaign, August 1984.

Verdú, S., "Minimum probability of error for asynchronous Gaussian multiple-access channels," *IEEE Trans. Inform. Theory,* pp. 85–96, January 1986.

Verdú, S., "Multiple-access channels with memory with and without frame synchronism," *IEEE Trans. Inform. Theory,* pp. 605–19, May 1989.

Verdú, S., "On channel capacity per unit cost," *IEEE Trans. Inform. Theory,* pp. 1019–30, September 1990.

Verdú, S., "Demodulation in the presence of multiuser interference: Progress and misconceptions," in D. Docampo, A. Figueiras, and F. Perez-Gonzalez, Eds., *Intelligent Methods in Signal Processing and Communications,* pp. 15–46, Birkhäuser, Boston, 1997.

Verdú, S., *Multiuser Detection,* Cambridge University Press, 1998.

Verdú, S., "Spectral efficiency in the wideband regime," *IEEE Trans. Inform. Theory,* pp. 1319–43, June 2002.

Verdú, S., "Recent results on the capacity of wideband channels in the low-power regime," *IEEE Wireless Commun. Mag.,* pp. 40–5, August 2002.

Verdú, S., and S. Shamai (Shitz), "Spectral efficiency of CDMA with random spreading," *IEEE Trans. Inform. Theory,* pp. 622–40, March 1999.

Veronesi, R., V. Tralli, J. Zander, and M. Zorzi, "Distributed dynamic resource allocation with power shaping for multicell SDMA packet access networks," *Proc. IEEE Wireless Commun. Network Conf.,* pp. 2515–20, March 2004.

Villier, E., P. Legg, and S. Barrett, "Packet data transmissions in a W-CDMA network – Examples of uplink scheduling and performance," *Proc. IEEE Veh. Tech. Conf.,* pp. 2449–53, May 2000.

Vishwanath, P., D. N. C. Tse, and R. Laroia, "Opportunistic beamforming using dumb antennas," *IEEE Trans. Inform. Theory,* pp. 1277–94, June 2002.

Vishwanath, S., S. A. Jafar, and A. J. Goldsmith, "Adaptive resource allocation in composite fading environments," *Proc. IEEE Globecom Conf.,* pp. 1312–16, November 2001.

Vishwanath, S., N. Jindal, and A. J. Goldsmith, "Duality, achievable rates, and sum-rate capacity of Gaussian MIMO broadcast channels," *IEEE Trans. Inform. Theory,* pp. 2658–68, October 2003.

Visotsky, E., and U. Madhow, "Space-time transmit precoding with imperfect feedback," *Proc. IEEE Internat. Sympos. Inform. Theory,* pp. 357–66, June 2000.

Viswanath, P., and D. N. C. Tse, "Sum capacity of the vector Gaussian broadcast channel and uplink–downlink duality," *IEEE Trans. Inform. Theory,* pp. 1912–21, August 2003.

Viswanathan, H., S. Venkatesan, and H. Huang, "Downlink capacity evaluation of cellular networks with known-interference cancellation," *IEEE J. Sel. Areas Commun.,* pp. 802–11, June 2003.

Viterbi, A. J., "Error bounds for convolutional codes and asymptotically optimum decoding algorithm," *IEEE Trans. Inform. Theory,* pp. 260–9, April 1967.

Viterbi, A. J., "Very low rate convolutional codes for maximum theoretical performance of spread-spectrum multiple-access channels," *IEEE J. Sel. Areas Commun.,* pp. 641–9, May 1990.

Viterbi, A. J., *CDMA Principles of Spread Spectrum Communications,* Addison-Wesley, Reading, MA, 1995.

Vitetta, G. M., and D. P. Taylor, "Maximum-likelihood decoding of uncoded and coded PSK signal sequences transmitted over Rayleigh flat-fading channels," *IEEE Trans. Commun.,* pp. 2750–8, November 1995.

Vriendt, J. D., P. Lainé, C. Lerouge, and X. Xu, "Mobile network evolution: A revolution on the move," *IEEE Commun. Mag.,* pp. 104–11, April 2002.

Vucetic, B., "An adaptive coding scheme for time-varying channels," *IEEE Trans. Commun.,* pp. 653–63, May 1991.

Vucetic, B., and J. Yuan, *Turbo Codes: Principles and Applications,* Kluwer, Dordrecht, 2000.

Wagen, J.-F., "Signal strength measurements at 881 MHz for urban microcells in downtown Tampa," *Proc. IEEE Globecom Conf.,* pp. 1313–17, December 1991.

Walfisch, J., and H. L. Bertoni, "A theoretical model of UHF propagation in urban environments," *IEEE Trans. Ant. Prop.,* pp. 1788–96, October 1988.

Walker, E. H., "Penetration of radio signals into buildings in cellular radio environments," *Bell Systems Tech. J.,* pp. 2719–34, September 1983.

Wang, H. S., and P.-C. Chang, "On verifying the first-order Markov assumption for a Rayleigh fading channel model," *IEEE Trans. Veh. Tech.,* pp. 353–7, May 1996.

Wang, H. S., and N. Moayeri, "Finite-state Markov channel – A useful model for radio communication channels," *IEEE Trans. Veh. Tech.,* pp. 163–71, February 1995.

Wang, X., and H. V. Poor, "Blind equalization and multiuser detection in dispersive CDMA channels," *IEEE Trans. Commun.,* pp. 91–103, January 1998.

Wang, X., and H. V. Poor, *Wireless Communication Systems: Advanced Techniques for Signal Reception,* Prentice-Hall, Englewood Cliffs, NJ, 2004.

Wang, Y., and Q.-F. Zhu, "Error control and concealment for video communication: A review," *Proc. IEEE,* pp. 974–97, May 1998.

Wang, Z., and G. B. Giannakis, "Outage mutual information of space-time MIMO channels," *Proc. Allerton Conf. Commun., Control, Comput.,* pp. 885–94, October 2002.

Wang, Z., X. Ma, and G. B. Giannakis, "OFDM or single-carrier block transmissions?" *IEEE Trans. Commun.,* pp. 380–94, March 2004.

Webb, W. T., and L. Hanzo, *Modern Quadrature Amplitude Modulation,* IEEE/Pentech Press, London, 1994.

Webb, W. T., and R. Steele, "Variable rate QAM for mobile radio," *IEEE Trans. Commun.,* pp. 2223–30, July 1995.

Wei, L.-F., "Coded modulation with unequal error protection," *IEEE Trans. Commun.,* pp. 1439–49, October 1993.

Wei, L.-F., "Coded M-DPSK with built-in time diversity for fading channels," *IEEE Trans. Inform. Theory,* pp. 1820–39, November 1993.

Weingarten, H., Y. Steinberg, and S. Shamai, "The capacity region of the Gaussian MIMO broadcast channel," *Proc. IEEE Internat. Sympos. Inform. Theory,* p. 174, June 2004.

Weinstein, F. S., "Simplified relationships for the probability distribution of the phase of a sine wave in narrow-band normal noise," *IEEE Trans. Inform. Theory,* pp. 658–61, September 1974.

Weitzen, J., and T. Lowe, "Measurement of angular and distance correlation properties of log-normal shadowing at 1900 MHz and its application to design of PCS systems," *IEEE Trans. Veh. Tech.,* pp. 265–73, March 2002.

Whitteker, J. H., "Measurements of path loss at 910 MHz for proposed microcell urban mobile systems," *IEEE Trans. Veh. Tech.,* pp. 125–9, August 1988.

Wiberg, N., N.-A. Loeliger, and R. Kotter, "Codes and iterative decoding on general graphs," *Euro. Trans. Telecommun.,* pp. 513–25, June 1995.

Wicker, S. B., and S. Kim, *Codes, Graphs, and Iterative Decoding,* Kluwer, Boston, 2002.

Wilson, D. G., *Digital Modulation and Coding,* Prentice-Hall, Englewood Cliffs, NJ, 1996.

Winters, J., "On the capacity of radio communication systems with diversity in a Rayleigh fading environment," *IEEE J. Sel. Areas Commun.,* pp. 871–8, June 1987.

Winters, J., "Signal acquisition and tracking with adaptive arrays in the digital mobile radio system IS-54 with flat fading," *IEEE Trans. Veh. Tech.,* pp. 1740–51, November 1993.

Winters, J., "Smart antennas for wireless systems," *IEEE Pers. Commun. Mag.,* pp. 23–7, February 1998.

Wolfowitz, J., *Coding Theorems of Information Theory,* 2nd ed., Springer-Verlag, New York, 1964.

Wolniansky, P., G. Foschini, G. Golden, and R. Valenzuela, "V-blast: An architecture for realizing very high data rates over the rich-scattering wireless channel," *Proc. URSI Internat. Sympos. Signal Syst. Elec.,* pp. 295–300, October 1998.

Wong, D., and D. C. Cox, "Estimating local mean signal power level in a Rayleigh fading environment," *IEEE Trans. Veh. Tech.,* pp. 956–9, May 1999.

Wong, K.-K., R. D. Murch, and K. B. Letaief, "Performance enhancement of multiuser MIMO wireless communication systems," *IEEE Trans. Commun.,* pp. 1960–70, December 2002.

Wong, W. C., R. Steele, and C.-E. W. Sundberg, *Source-Matched Mobile Communications,* Pentech and IEEE Press, London and New York, 1995.

Wozencraft, J. M., and I. M. Jacobs, *Principles of Communication Engineering,* Wiley, New York, 1965.

Wu, D., and R. Negi, "Downlink scheduling in a cellular network for quality-of-service assurance," *IEEE Trans. Veh. Tech.,* pp. 1547–57, September 2004.

Wu, S.-L., Y.-C. Tseng, and J.-P. Sheu, "Intelligent medium access for mobile ad hoc networks with busy tones and power control," *IEEE J. Sel. Areas Commun.,* pp. 1647–57, September 2000.

Wu, Y., P. A. Chou, Q. Zhang, K. Jain, W. Zhu, and S.-Y. Kung, "Network planning in wireless ad hoc networks: A cross-layer approach," *IEEE J. Sel. Areas Commun.,* pp. 136–50, January 2005.

Wyner, A., "Shannon-theoretic approach to a Gaussian cellular," *IEEE Trans. Inform. Theory,* pp. 1713–27, November 1994.

Xiang, H., "Binary code-division multiple-access systems operating in multipath fading, noisy channels," *IEEE Trans. Commun.,* pp. 775–84, August 1985.

Xiao, L., M. Johansson, and S. P. Boyd, "Optimal routing, link scheduling and power control in multihop wireless networks," *IEEE Trans. Commun.,* pp. 1136–44, July 2004.

Xu, L., X. Shen, and J. W. Mark, "Dynamic fair scheduling with QoS constraints in multimedia wideband CDMA cellular networks," *IEEE Trans. Wireless Commun.,* pp. 60–73, January 2004.

Xue, F., and P. R. Kumar, "The number of neighbors needed for connectivity of wireless networks," *Wireless Networks,* pp. 169–81, March 2004.

Yacoub, M., *Principles of Mobile Radio Engineering,* CRC Press, Boca Raton, FL, 1993.

Yang, L., and G. B. Giannakis, "Ultra-wideband communications: An idea whose time has come," *IEEE Signal Proc. Mag.,* pp. 26–54, November 2004.

Yao, H., and G. Wornell, "Structured space-time block codes with optimal diversity-multiplexing trade-off and minimum delay," *Proc. IEEE Globecom Conf.,* pp. 1941–5, December 2003.

Yoo, T., and A. J. Goldsmith, "Optimality of zero-forcing beamforming with multiuser diversity," *Proc. IEEE Internat. Conf. Commun.,* May 2005.

Yoo, T., E. Setton, X. Zhu, A. Goldsmith, and B. Girod, "Cross-layer design for video streaming over wireless ad hoc networks," *Proc. IEEE Internat. Workshop Multi. Signal Proc.,* pp. 99–102, September 2004.

Yoon, Y. C., R. Kohno, and H. Imai, "A spread-spectrum multiaccess system with cochannel interference cancellation for multipath fading channels," *IEEE J. Sel. Areas Commun.,* pp. 1067–75, September 1993.

Youn, J.-H., and B. Bose, "An energy conserving medium access control protocol for multihop packet radio networks," *IEEE Internat. Conf. Comput. Commun. Network.,* pp. 470–5, October 2001.

Youssef, M. A., M. F. Younis, and K. A. Arisha, "A constrained shortest-path energy-aware routing algorithm for wireless sensor networks," *Proc. IEEE Wireless Commun. Network Conf.,* pp. 794–9, March 2002.

Yu, W., and J. M. Cioffi, "Trellis precoding for the broadcast channel," *Proc. IEEE Globecom Conf.,* pp. 1344–8, November 2001.

Yu, W., K. J. R. Liu, and Z. Safar, "Scalable cross-layer rate allocation for image transmission over heterogeneous wireless networks," *Proc. IEEE Internat. Conf. Acous., Speech, Signal Proc.,* pp. 4.593–4.596, May 2004.

Yu, W., W. Rhee, S. Boyd, and J. Cioffi, "Iterative water-filling for vector multiple access channels," *IEEE Trans. Inform. Theory,* pp. 145–52, January 2004.

Zaidel, B. M., S. Shamai, and S. Verdú, "Multicell uplink spectral efficiency of coded DS-CDMA with random signatures," *IEEE J. Sel. Areas Commun.,* pp. 1556–69, August 2001.

Zander, J., "Performance of optimum transmitter power control in cellular radio systems," *IEEE Trans. Veh. Tech.,* pp. 57–62, February 1992.

Zander, J., and H. Eriksson, "Asymptotic bounds on the performance of a class of dynamic channel assignment algorithms," *IEEE J. Sel. Areas Commun.,* pp. 926–33, August 1993.

Zehavi, E., "8-PSK trellis codes for a Rayleigh channel," *IEEE Trans. Commun.*, pp. 873–84, May 1992.

Zheng, L., and D. N. Tse, "Communication on the Grassmann manifold: A geometric approach to the non-coherent multi-antenna channel," *IEEE Trans. Inform. Theory*, pp. 359–83, February 2002.

Zheng, L., and D. N. Tse, "Diversity and multiplexing: A fundamental trade-off in multiple antenna channels," *IEEE Trans. Inform. Theory*, pp. 1073–96, May 2003.

Zhou, L., and Z. J. Haas, "Securing ad hoc networks," *IEEE Network*, pp. 24–30, November/December 1999.

Ziemer, R., "An overview of modulation and coding for wireless communications," *Proc. IEEE Veh. Tech. Conf.*, pp. 26–30, April 1996.

Zorzi, M., and R. R. Rao, "Energy-constrained error control for wireless channels," *IEEE Pers. Commun. Mag.*, pp. 27–33, December 1997.

Zorzi, M., and R. R. Rao, "Geographic random forwarding (GeRaF) for ad hoc and sensor networks: Energy and latency performance," *IEEE Trans. Mobile Comput.*, pp. 349–65, October–December 2003.

Index